MUSHROOMS
of the Redwood Coast

MUSHROOMS
of the Redwood Coast

A Comprehensive Guide to the
Fungi of Coastal Northern California

NOAH SIEGEL and **CHRISTIAN SCHWARZ**

TEN SPEED PRESS
Berkeley

Contents

Introduction

The rugged edge of the continent upon which Coast Redwoods dwell is fantastically beautiful land. The warm, dry summers and rainy winters that bathe this area's varied terrain create a great diversity of habitats, which are home to mushrooms and other fungi: a tremendously diverse but often overlooked community of organisms.

Mushroom identification as a hobby has an obvious end goal: to be able to attach names to organisms and communicate about them reliably and effectively. But pursuing this goal necessarily involves continually honing and developing increased awareness of other fungi, which leads to outcomes more subtle and more profound. You will learn a few mushrooms your first year, and the next year you will learn a few more; then you will have experiences that challenge you to question your own knowledge, as well as some that strengthen it. You will eventually be surprised at how much you know (and how much more you are left wondering about!). More importantly, with every step you take down this road, your world will expand. You will begin to notice slime molds, mosses, lichens, tiny insects, and the various other minute habitats that surround fungi, and you will soon wonder about all of these as well.

Paired with your persistence, curiosity, and time spent in the woods, this guide will help you become familiar with the names and faces of fungi on the Redwood Coast, and give you a glimpse of what we know about their role in California's ecosystems. We hope that this book will encourage beginners to dive in to the world of California's fungi, as well as assist and encourage more experienced identifiers as they push our collective knowledge forward.

Coast Redwoods are a magnificent and irreplaceable part of our heritage, and we should strive to protect the ecosystems that support them. We hope that this book will serve as a reminder that this coast is also home to a great diversity of fungi. They are rarely given attention in conservation efforts (perhaps due to their intermittent visibility), but fungi are always there. Whether busy on roots, in soil, on moss, or in the heartwood of trees, they are essential to the dynamics of forests.

What Are Mushrooms?

The word *mushroom* is an imprecise but useful term for the visible reproductive organs of certain types of fungi. Fungi are their own kingdom of life, similar to, but separate from, plants and animals. The kingdom Fungi comprises many millions of species of living organisms, ubiquitous in our everyday environments and abundant in nearly every natural habitat. Although fungi are exceptionally important in the functioning of global ecosystems, we generally go about our daily business unaware of their presence. As soon as you start to pay attention, however, you'll find that fungi play many roles in human affairs, both as allies and as adversaries. But what exactly are fungi?

For starters, fungi (along with plants and animals) are eukaryotes. Eukaryotes are organisms whose cells have discrete interior structures; the largest and most important of these is the nucleus. The nucleus contains DNA—the blueprints for the structure, chemistry, function, and organization of an organism. When two organisms share a common ancestor in the recent past, we say that they are closely related; that is, they share many of the instructions encoded in their DNA. The genetic material of more distantly related species differs to a greater degree. The portobello mushrooms on your dinner table are more closely related to the yeast organisms in your bread dough than either is to the sunflowers in your yard. Perhaps even more surprisingly, the portobello mushrooms are more closely related to you than they are to plants, since fungi and animals share a more recent common ancestor than they do with plants!

Fungi were relatively recently separated from the plant kingdom, based on many important biological differences. One such difference has to do with their modes of nutrition: plants are autotrophs, meaning that they produce their own "food." By the process of photosynthesis, plants use sunlight and carbon dioxide to make the sugars that fuel their metabolism and growth. Fungi, on the other hand, cannot photosynthesize—they are heterotrophs. Since fungi cannot produce their own fuel, they rely on organic matter produced by other organisms. Fungi have an interesting method of digestion: they exude powerful enzymes that break down the organic matter in their surroundings and then absorb the products of this breakdown.

Morphology Diagram

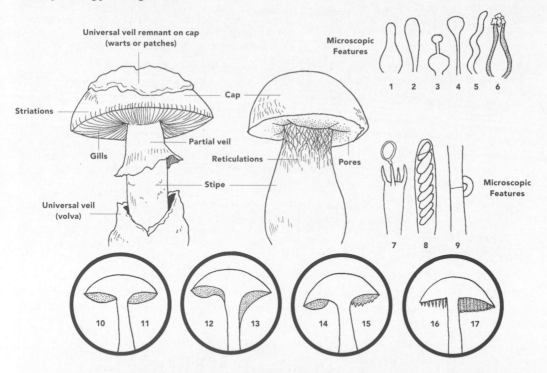

Types of gill attachments and some other kinds of fertile surfaces as seen in cross-sectioned fruitbodies

1. Utriform 2. Clavate 3. Lecythiform/bowling pin shaped 4. Capitates 5. Sinuous 6. Metuloid cystidium with thick walls and crystals at tip 7. Basidium with one spore attached 8. Ascus with eight internal spores 9. Two hyphae joined by a clamp connection 10. Adnate (broadly attached to stipe) 11. Adnexed (finely attached to stipe) 12. Sinuate (notched attachment to stipe) 13. Decurrent (running down stipe) 14. Free (not attached to stipe, separated by a "moat" or gutter) 15. Serrated edge 16. Teeth/spines 17. Tubes and pores

Fungi are also very different from animals. One important difference is found in the composition of the cells: whereas animal cells are surrounded by a layer of lipids (fatty acid molecules), most fungi have cell walls made of a carbohydrate called chitin. This same compound is the main component of the exoskeletons (but not the cells) of shrimp, dragonflies, and all other arthropods. Although there are groups of fungi with very different anatomy, the chitin-bound cells of nearly all the fungi in this book are tubelike structures called hyphae. The vast majority of the fungi in this book have hyphae strung together in long chains and branching networks that form a sort of carpet called the mycelium. Although this can be a very large structure, it is usually invisible to humans because it is buried or embedded in whatever substrate the fungus is growing in and digesting (whether soil, wood, or animal tissue). The thin, elongate cells and highly branched structure of the mycelium have a very high surface-area-to-volume ratio; this anatomy greatly enables efficient diffusion of digestive enzymes and uptake of water and nutrients.

It's important to remember that mushrooms are not representative of the whole fungal organism: they are generally short-lived structures that are produced to achieve sexual reproduction and dispersal, similar to the fruits of a plant. For this reason, mushrooms are referred to as the fruitbodies of the fungi that produce them. A fungus may thrive for many years as a mycelium without ever producing a fruitbody visible to humans!

Fungi are a very diverse group of organisms, but only a small fraction of these species produces structures that we call mushrooms. The majority of fungal spores are microscopic, visible to the human eye only during certain parts of their life cycle when they manifest themselves as discolored splotches or powdery coatings on plant tissues and other organic matter. Examples include the rust disease on your roses or mold on your bread or old leftovers.

The tremendous diversity of the fungal kingdom is paralleled by the many roles that fungi play in natural ecosystems, ranging from harmless decomposers to virulent agents of disease to essential symbiotic partners.

Ecology of Fungi

Most fungi decompose nonliving organic material to fuel their metabolism. Fungi with this nutritional mode are called saprobes and can be found almost everywhere: on old straw bales, in forest humus, and even in Tupperware at the back of your fridge. Saprobic (or saprotrophic) fungi are extremely important in releasing carbon and nitrogen back into local and global nutrient cycles. These and other terms are defined in the Glossary (p. 579).

Most other mushroom-forming fungi are involved in either parasitic or mutualistic symbioses with other organisms. The symbiotic relationships between plants and fungi, called mycorrhizae, are particularly relevant to those seeking to find, identify, and understand the mushrooms in this book.

The term mycorrhiza is a combination of the Greek words *myco*, "a fungus," and *rhiza*, "roots"; it refers to a mutualistic relationship in which the hyphae of a fungus live in close physical contact with the root tips of a plant. In these relationships, the fungal partner exchanges water and nutrients for sugars produced by the plant host. This reciprocity is particularly elegant because it pairs the physiological abilities

and limitations of each partner in a complementary way: The fungal partner, with its cobwebby mycelium, has a high surface area and powerful enzymes to gather water and liberate nutrients from the soil, but cannot produce its own sugars. The plant, on the other hand, can produce sugar, but doesn't have the same capacity to gather resources from the soil.

Scientists have identified and classified many variations of the mycorrhizal relationship, but there are two main categories: arbuscular mycorrhizae and ectomycorrhizae. Arbuscular mycorrhizal fungi penetrate the root cells of their host plants, whereas ectomycorrhizal fungi form sheaths around them. The two groups also have chemical and ecological differences: arbuscular mycorrhizal fungi are often better phosphorus scavengers, while ectomycorrhizal fungi tend to be more adept at gathering nitrogen. Although there are many exceptions, arbuscular mycorrhizae are more common in drier climates and tropical forests, whereas ectomycorrhizal relationships dominate in temperate forests.

All the mycorrhizal mushrooms in this book are of the ectomycorrhizal type. However, not all ectomycorrhizal fungi produce large fruitbodies; many form thin crusts under logs or inconspicuous spore-bearing structures in the soil. Likewise, only a fraction of mushroom-producing fungi are ectomycorrhizal. Although some species are more ecologically flexible, many of these relationships are obligate: the fungi and plants involved cannot complete their life cycles without each other. Mycorrhizal fungi enable many plants to thrive in habitats where they would otherwise not survive due to drought stress or poor nutrition.

These mutualistic relationships are restricted to some extent—there are a limited number of plant species with which any particular fungus can form an ectomycorrhizal symbiosis (and vice versa). For example, the Death Cap (*Amanita phalloides*) can form mycorrhizae with tens of different host plants, while *Suillus caerulescens* only grows with Douglas-fir. Most mycorrhizal fungi are at least moderately specific, each species growing primarily with one or two host tree or shrub species, and at lower frequency with a smaller number of other hosts.

The ecological impacts of these symbioses extend well beyond the individual tree and fungus. In some cases, parasitic plants (such as *Corallorhiza* orchids) exploit the mycelium of a mycorrhizal fungus, absorbing sugars that were originally produced by the tree with which the fungus is partnered. In northern California's old-growth forests, the threatened Spotted Owl predates heavily on Red-backed Voles, which in turn feed primarily on the hypogeous fruitbodies of mycorrhizal *Rhizopogon* fungi. The spores of these truffle-like fungi survive passage through the guts of both animals, and viable spores can be found in both the pellets and the droppings of the owls; thus the owls help disperse the fungi that sustain populations of their own prey. Communication between plants via chemical signaling through the fungal mycelium that connects them has been observed in both forest and experimental settings. This has led some researchers to conceive of mycelial networks as a sort of "Natural Internet" through which forest trees and understory plants sense environmental changes and reallocate resources.

As you might guess by now, knowing how to identify the trees in your area is extremely useful when seeking and identifying mushrooms. Some very similar species of *Suillus* are more easily told apart by their host tree than by their

Life Cycle Diagram

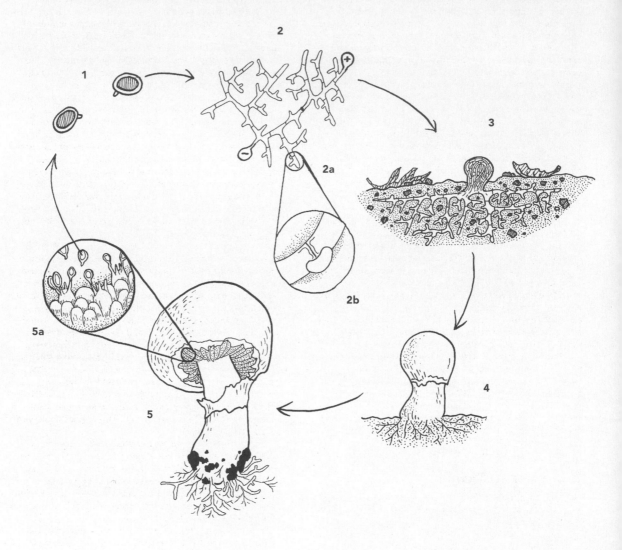

1. Fungal spores are floating in the air.

2. Spores that have landed on a suitable surface germinate into a primary mycelium—a network of branching hyphae. Each hypha (2a) is a microscopic, cylindrical cell with cell walls made of chitin. Clamp connections (2b) are sometimes present, bridging the wall between two adjacent hypha; these structures help fungi regulate the distribution of genetic material throughout their mycelia. If two such mycelia encounter each other and are of compatible mating types, they can fuse into a mycelium with two different sets of genetic information, one from each parent.

3. After a period of growth and given appropriate conditions (including plenty of water), a mature mycelium can produce a tiny primordium, a knot of hyphae that eventually grows into a mushroom.

4. The primordium grows into a "button" or "pin," and then expands into a mature fruitbody.

5. The fertile surface—gills, in this case—produces spores on microscopic cells, in this case, basidia (5a). Other kinds of fungi have different fertile cells (for example, the fertile cells of morels are called asci).

morphology (at least for beginners). In other cases, you can rule out entire groups of fungi based on the habitat of your collection. For example, the ectomycorrhizal *Amanita augusta* will never appear in pure Coast Redwood forest (because redwoods do not form ectomycorrhizal relationships), and the spruce-associated *Cortinarius callimorphus* doesn't occur in Santa Cruz County (where there are no spruce trees).

At the other end of the symbiotic spectrum are the parasitic fungi, whose growth causes negative effects on the health of their host organisms. The most well-known parasitic fungi infect plants and are especially familiar to those who work in agriculture or the timber industry. Honey mushrooms (*Armillaria* species) and conifer root rot (*Heterobasidion* spp.) cause disease on a large scale in natural ecosystems, sometimes killing entire stands of trees.

Fungi also parasitize animals, and sometimes entire groups of mammals are affected at landscape scales. The health of bat populations in northeastern North America is at serious risk due to White Nose Syndrome, a disease caused by the fungus *Geomyces destructans*. Some of the most spectacular animal-parasite fungi specialize on invertebrates: *Cordyceps* species parasitize ants, butterflies, beetles, and other insects. One of these species, *Ophiocordyceps sinensis*, is a prized medicinal mushroom in Asia.

Some fungi even specialize on digesting other fungi! Many of these "mycoparasitic" fungi distort, discolor, or otherwise transform the fruitbodies of their hosts. Some of these include the Lobster Mushroom (*Hypomyces lactifluorum*) and the very rare Body-snatchers (*Squamanita* species).

Finally, we should mention that some fungi (although very few of the species covered in this book) exist in a mutualistic relationship with green algae or cyanobacteria—such species are said to be lichenized. Although the majority of lichens don't produce typical mushroom-shaped fruitbodies, some do, and there is increasing evidence that many members of the family Hygrophoraceae are involved in such relationships (Seitzman et al., 2011).

Humans and Fungi

The fungi most familiar to us are the varieties that we value as food and those that we fear as poisons. The fabled Death Cap and Destroying Angel mushrooms are in the genus *Amanita*, but many other genera also contain toxic species. Toxic *Amanita* species cause a few very serious poisonings every year in California. With appropriate caution, effort, and the help of knowledgeable friends, you can easily keep yourself from joining the ranks of these unfortunate people (see Collecting for the Table, p. 13). There is a broad range of symptoms caused by mushroom poisoning, depending on the species, quantity, and dose involved, as well as the health and biochemistry of the affected individual.

Although for decades the white Button Mushroom (*Agaricus bisporus*) and its variants have been the most common mushrooms in grocery stores in the United States, many other mushrooms have been added to their shelves in recent years. Such species as Shiitake (*Lentinula edodes*), Enoki (*Flammulina velutipes*), Maitake (*Grifola frondosa*), and oysters (*Pleurotus* species) have long histories of cultivation and consumption in other countries, and many Americans now propagate these species in their homes and backyards.

Many excellent references for learning the craft of mushroom cultivation and cookery can be found in print and online.

On the other hand, many species of edible mushrooms are difficult or impossible to cultivate in commercial quantities. Some of the most famous are Chanterelles (*Cantharellus* species), Porcini (*Boletus edulis* and others), Black Trumpets (*Craterellus* species), and Morels (*Morchella* species). These delicacies fetch high prices and are the quarry of fairly large seasonal commercial picking outfits (sometimes causing significant tension with conservation-minded land managers).

In some cultures (particularly in parts of Asia and Mexico), psychotropic fungi play an important role in spiritual and religious practices. Some of the most famous of these fungi are *Amanita muscaria* and various species of *Psilocybe*, better known as "magic mushrooms." There has been a recent resurgence of medical interest in the use of the psilocybin compound produced by species in the latter genus to treat cluster headaches (Sewell, Halpern, and Pope, 2006), to mitigate the psychological trauma associated with terminal illness (Grob et al., 2011), and to disrupt patterns of alcohol and narcotic abuse.

Some other fungi are less visible, but much more important in our daily lives. Many microfungi cause illness or disease in humans, pets, and livestock; these are collectively called mycoses. These include yeast infections, ringworm, Valley Fever (coccidioidomycosis), and aspergillosis. A variety of plant-pathogenic fungi do serious damage to crops around the world. The fungus *Claviceps purpurea* produces toxic "ergot" on cereal grains, and *Phragmidium* species produce bright orange rusts on the roses in your garden. The global economic burden caused by these parasitic fungi is considerable, and much research is devoted to mitigating their effects. Dry rot fungi (*Serpula* species) seriously compromise the structural integrity of wood in buildings around the world. Other fungi are more minor nuisances—an ascomycete called *Rhizopus* produces the blackish fuzz we find on old bread.

But many other fungi are our allies: blue cheeses (and many other cheeses) are made with the help of ascomycete molds, and without *Saccharomyces* yeasts to ferment plant sugars into alcohol, we wouldn't have beer or wine, or leavened bread. The famous antibiotic compound penicillin, used to treat many bacterial infections, was first isolated from a fungus called *Penicillium*.

Fungal spores are abundant in the air we breathe, and products derived from or processed by fungi are ubiquitous in our everyday environments. Fungi are literally everywhere around us—a compelling reason to learn more about them!

The Redwood Coast

We have chosen to cover a geographic area that roughly coincides with the distribution of the Coast Redwood (*Sequoia sempervirens*), a magnificent icon of California's temperate rainforest. This species includes the tallest trees on Earth (an individual called Hyperion measures 115.6 meters tall!). Not only do some individuals reach great age (upward of two thousand years), but the evolutionary lineage as a whole is an ancient one. As if this weren't impressive enough, the most vigorous Coast Redwood forests have some of the greatest biomass productivity of any terrestrial ecosystem.

Although *Sequoia sempervirens* and its close relatives once grew around the northern hemisphere, this species'

Map of the Redwood Coast

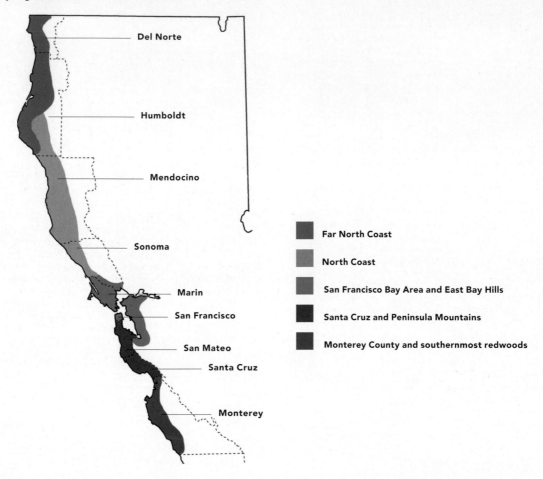

Del Norte

Humboldt

Mendocino

Sonoma

Marin

San Francisco

San Mateo

Santa Cruz

Monterey

- Far North Coast
- North Coast
- San Francisco Bay Area and East Bay Hills
- Santa Cruz and Peninsula Mountains
- Monterey County and southernmost redwoods

modern range consists of a narrow band between extreme southern Monterey County and the southwestern corner of Oregon. Often referred to as the Fog Belt because of its frequently dense maritime moisture layer, this zone experiences a Mediterranean weather regime: warm summers with little precipitation and wet, cool winters with rare prolonged periods of freezing temperatures.

In many parts of their range, Coast Redwoods are dependent on fog to tide them over through the dry summers, and the architecture of their foliage is such that they can effectively harvest this moisture out of the air to meet their prodigious water requirements. The range of the Coast Redwood extends far inland in the wet, mountainous areas of the southern Cascade Range in Del Norte County, and even farther (albeit more scattered) in the drier eastern reaches of the San Francisco Bay. Ronald Lanner's excellent *Conifers of California* identifies Little Redwood Creek (a branch of the Chetco River in Oregon) as the absolute northernmost outpost of the Coast Redwood, while the southernmost trees are found just north of Soda Springs Creek in the Santa Lucia Mountains of Monterey County. The restricted geographic range of the Coast Redwood, as well as their dependence on a heavy summer fog

regime, has led the International Union for Conservation of Nature to list them as a vulnerable species.

Many other habitat types are found on the Redwood Coast, including chaparral, Douglas-fir forest, pine forest, oak woodland, mixed evergreen forest, coastal grassland, and riparian woods. In addition to these native habitats, parks, gardens, and horticultural areas are also important habitat for mushrooms. While some of the fungi included in this book occur primarily in *Sequoia* forests, most are found in other natural habitats on the Redwood Coast, and some grow only in human-modified habitats (parks, lawns, gardens, and other landscaped areas). Since no one is really sure how many thousands of species occur in this area, this book is necessarily incomplete. Our focus is centered on the gilled mushrooms, boletes, and chanterelle-like species, particularly those that are common, conspicuous, and distinctive. But we've also made a great effort to find and photograph uncommon and inconspicuous fungi, some of which require more experience and closer inspection to identify correctly. There are many species we decided not to include due to some combination of being inconspicuous, poorly understood, or rare.

Trees of the Redwood Coast

A big step toward understanding mushrooms is learning to recognize their habitats. As discussed in the section on Ecology of Fungi, many mushrooms are mycorrhizal and only associate with a one or two tree species, while others have a broader (but still limited) range of hosts. Even among those fungi that don't form mycorrhizal associations, some are quite picky: for example, *Crinipellis piceae* only occurs on Sitka Spruce needles, and *Leptonia carnea* is very rarely found outside redwood forest.

Our range covers four major bioregions: the conifer-dominated zone on the Far North Coast (distinguished by the presence of Sitka Spruce, Grand Fir, and Western Hemlock); the vast mixed-evergreen zone (dominated by Douglas-fir, Tanoak, and Madrone forests inland, often flanked by pine forest on the coast); the Coast Live Oak woodlands that occur from southern Mendocino County southward into Mexico; and the majestic Coast Redwood forest, whose range circumscribes the scope of our book (see The Redwood Coast on p. 5). Some of the most important trees and their identifying features are illustrated and described briefly below. Other common trees that we don't profile in detail include cottonwood, willow, and alder—each of these has a small but distinctive suite of associated fungi. On the other hand, Bigleaf Maple (*Acer macrophyllum*) and California Buckeye (*Aesculus californica*) sometimes form inviting groves but almost never produce much in the way of mushroom fruitings.

Ectomycorrhizal Northern Conifers

Western Hemlock (*Tsuga heterophylla*): Widespread tree of the Far North Coast, but some found south to Salt Point on the Sonoma coast. Distinctively drooping growing tip, unlike other conifers on the coast. Needles are short, flat, and of differing lengths on the same twig; each has two whitish lines on the paler underside. Grayish bark with moderately shallow furrows. Small cones with relatively large scales

Sitka Spruce (*Picea sitchensis*): Large conifer with overlapping bark scales, swooping branches, and crowns with upright tips. Can be recognized by the sharp, three-sided needles (roll them between your fingers) with a silvery green top and green underside, and hanging cones. Occurs on north-facing slopes in scattered river valleys on the Central Coast of Mendocino County, becoming more widespread in the northern part of the Lost Coast and then common farther north along the coast.

Grand Fir (*Abies grandis*): Shows a classic Christmas-tree shape, with a pointed growing tip and whorled branches that splay out nearly flat. Smooth bark when young, often covered with sap blisters, then becoming furrowed (similar to Western Hemlock). Needles are also flat, but longer than those of hemlock (up to 3 cm). Upper sides of needles are bright green; undersides are paler and with two whitish lines. Needles are usually splayed out flat on lower branches. Large cones borne upright on the upper branches, but rarely seen whole on the ground, as they tend to break up at maturity.

Ectomycorrhizal Hardwoods

Live oaks (*Quercus* spp.): Those that in our area have green leaves year-round. In addition to the following tree-forming species, there are a few shrubby species that also host many fungi but are rather difficult to survey.

Coast Live Oak (*Quercus agrifolia*): This beautiful evergreen oak is widespread from southern Mendocino County south into Mexico, mostly occurring in drier, warmer habitats. Recognized by its rounded, dark green canopy, frequently stout and multibranched trunk, and often gnarled gray limbs covered in lichens. The leaves are leathery, dark green, round to spoon-shaped (edges often curling inward), and have short, sharp points around their edges. The veins on the underside usually bear a distinct beige fuzz at their branching points ("hairy armpits"). This tree probably hosts the most diverse suite of mycorrhizal fungi of any of our hardwoods, a substantial subset of which are also found in association with other live oaks, as well as Tanoak and Chinquapin to a lesser extent.

Interior Live Oak (*Q. wislizeni*): A common tree away from the immediate coast between Santa Cruz and Mendocino Counties. Tends to be a tree of mixed forest and often has a rather upright (less sprawling) stature than Coast Live Oak. Community of associated fungi with this species seems to be less diverse than with other oaks. **Canyon Live Oak (*Q. chrysolepis*)** occurs in river valleys and higher elevations away from the coast in the northern part of our range. Has oblong, flat leaves with a shiny dark green upper surface and a dull yellowish underside. Acorn cups are distinctly yellowish when fresh. Associates with many of the same fungi as Coast Live Oak. **Deciduous oaks** are less frequent on the coast, mostly occurring in higher

elevations and inland areas. They are abundant in some areas that we have covered less thoroughly in this book. The two species you're most likely to encounter are **Black Oak (*Q. kelloggii*)**, which has lobed leaves with pointed tips, and **Valley Oak (*Q. lobata*)**, which has deeply lobed leaves with rounded tips.

Tanoak (*Notholithocarpus densiflorus*): Forests of this species mixed with Douglas-fir and Madrone dominate much of the interior areas of Sonoma, Mendocino, and Humboldt Counties. Bark starts off smooth and gray, then develops dark vertical furrows as the trees mature. Large, green, oblong leaves with a distinctly toothed margin and pronounced veins and extensive tan fuzz on the underside. The acorn cups are hairy-scaly looking, distinct from most of the true oaks (*Quercus*) in our area. Tanoak and true oaks are fairly close relatives and have many fungal associates in common. Tanoaks are best known for hosting large crops of Black Trumpets, Matsutake, and Queen Boletes.

Chinquapin (*Chrysolepis chrysophylla*): This tree has vertically furrowed grayish bark and smooth, narrowly oblong leaves without teeth and with a distinct golden cast to the undersides. Extremely spiny cases surround the nuts, and these husks litter the forest floor around Chinquapin groves. Hosts a much smaller subset of the fungi found with oaks and Tanoak, as well as a few fungi rarely found in other habitats (for example, *Fistulina*).

Madrone (*Arbutus menziesii*): A gorgeous tree that often occurs in mixed forest with Tanoak and/or Douglas-fir, and occasionally at meadow edges among oaks. Recognized by its dark reddish brown, finely scaled trunk and smooth, shiny, often muscular-looking yellowish or orangish branches. The shiny green leaves are the largest of any hardwood in our area. Bears large clusters of white bell-shaped flowers in spring and round, pebbly-skinned red berries throughout much of the year. The smooth, shiny reddish branches of manzanita species are similar (and the two genera are closely related), but manzanita species have much smaller leaves and are never as large as Madrone.

Douglas-fir (*Pseudotsuga menziesii*): The Douglas-fir is the most widespread tree in our range. Common from Santa Cruz northward (with a few scattered patches to the south); often growing in mixed forests with Tanoak and Madrone. Can be identified by its soft, light green needles that grow in all directions from the twig (not in a single plane like Coast Redwood) and medium-size cones (often littering the ground) with distinctive three-tipped "tongues" of papery tissue poking out between the scales. Bark of young trees is slightly rough and grayish, but trunks of mature trees become deeply and dramatically furrowed.

Manzanita (*Arctostaphylos* spp.): About 100 species of Manzanita are found in California, many of which are extremely range restricted and difficult to identify. In stature, they range from creeping shrubs to small trees. Easy to recognize as a group by their maroon, chestnut red, or orangey trunks and branches that are smooth and shiny, and often covered in peeling and curled strips and flakes of bark skin. Leaves generally small and oval, and often dull green or silvery green. Small, pale, bell-shaped flowers appear in clusters in late winter and spring. The most common species in our area are good mycorrhizal hosts, but can be very tough to explore, both because they form impenetrable thickets and because they often grow in exposed and harsh areas that have short and irregular windows of fungal fruiting.

Pines (*Pinus* spp.): Members of this genus have very diverse mycorrhizal communities that are often quite distinct from other conifers. Pines can be recognized by their long, thin needles in clusters (the number of needles per cluster is an important identifying feature), woody cones, and gray orange-brown or dark reddish brown scaly bark. There are four common coastal species in our area, as well a few others more often found in drier, inland habitat. **Monterey Pine (*P. radiata*)** occurs in coastal forests in Monterey and Santa Cruz Counties (but much more widely as a planted tree). It has long needles in clusters of three and egg-shaped cones 7–17 cm long, that lack sharp spines. **Knobcone Pine (*P. attenuata*)** is similar and also has needles in clusters of three, but has whorls of three to six asymmetrically humped cones covered in stout curved spines. The cones are held very tightly to the branches and trunk, and sometimes engulfed by the wood in age (only to be released by wildfire). Uncommon on the immediate coast; mostly found in drier zones and higher elevations. It is a short-lived, fast-growing tree with a distinctive thin, rangy stature.

Bishop Pine (*P. muricata*) is also similar, but has long needles in clusters of two and spiny, egg-shaped cones 5–10 cm long. Mostly found in scattered populations along the coast north of San Francisco Bay to northern Mendocino County. **Shore Pine (*P. contorta* subsp. *contorta*)** is restricted to the sandy dune habitat of the Far North Coast, with scattered populations in Sonoma and Mendocino Counties. The short, irregular, often stunted trunks, short needles in bundles of two, and small cones are distinctive. **Ponderosa Pine (*P. ponderosa*)** is rare in our area, primarily found in a few areas around the Santa Cruz Sandhills, and drier or higher-elevation inland habitats. Recognized by its tall, straight trunk with large orange-brown "puzzle-piece" bark scales, long needles in clusters of three, and large, spiny cones.

Monterey Cypress (*Hesperocyparis macrocarpa*): The Monterey Cypress has an extremely restricted natural range, comprising two small populations on the coast near Monterey. However, it is widely planted as an ornamental, and a few naturalized, dense groves are found along the coast, especially north of Santa Cruz. Recognized by the horizontal and often flat-topped stature when mature, and by the dense sprays of dark green, overlapping scalelike needles and small, globose cones with tight plates. The fungi that prefer cypress duff contribute significantly to the diversity and unique nature of California's mycoflora; many are endemic to our area and are now very rare. *Lepiota*, *Leucoagaricus*, *Agaricus*, and *Pseudobaeospora* are especially diverse in this habitat. Many redwood-associated fungi can be found here as well, especially *Leptonia* and *Hygrocybe*.

Non-Ectomycorrhizal Trees

Coast Redwood (*Sequoia sempervirens*): A truly majestic tree that is both a keystone ecological species and a dominant part of the California coast landscape. See the section on our area and the accompanying map on p. 6. Easy to recognize by its pointed, stiff green needles arranged in a single flat plane, in combination with the fibrous-hairy, often deeply furrowed reddish brown bark. Does not form ectomycorrhizal relationships with fungi, but the dark, damp understory hosts a number of colorful Waxy Caps and *Leptonia*, some of which appear to be endemic to the redwood forest (although many also occur under Monterey Cypress).

California Bay Laurel (*Umbellularia californica*): Often found in mixed forests, this tree forms especially dense groves in steep drainages and on open hillsides in the southern part of our area. Growing throughout our range, but rare on the conifer-dominated North Coast. Distinctive stature, with an often swollen and bulging lower trunk and many branches diverging near the ground. Leaves lance-shaped with smooth edges, glossy bright to dark green (but often covered in a sooty black mold). The leaves have a very strong, spicy odor when crushed (it can cause headaches!). Generally a very poor mushroom habitat, but many of the *Hygrocybe* and *Leptonia* found under Coast Redwood and Monterey Cypress are occasionally found in this duff. The decaying logs host an interesting assemblage of hardwood-rotting species that deserve closer investigation.

Finding Mushrooms

On the Redwood Coast, most mushrooms fruit during the cool, wet part of the year between early fall and early spring. This fruiting season begins and ends earlier in the northernmost parts of our range, spanning the months from September to early February. This contrasts with the southern part of our range, where mushroom season usually starts in earnest in mid-November and often persists through early March. No two mushroom seasons are alike, and the actual pattern of macrofungal fruiting in a given year can be expected to deviate significantly from anyone's idea of a typical pattern.

There are many challenges inherent in mushroom hunting, not the least of which is that you are nearly guaranteed to get wet and chilly when you search for mushrooms in California: remember to check weather conditions and dress appropriately. Other inherent risks of outdoor endeavors also apply: carry plenty of water (and snacks!), tell someone where you're going and when you expect to be back, and carry maps and/or a GPS unit if you are unfamiliar with the area. Getting lost in the mountains during a cold and wet winter storm can be fatal. Poison Oak is extremely abundant in our area, and the rashes it can cause can be extremely uncomfortable. Ticks are also a major concern (especially in areas where Lyme disease is common). A quick search online will turn up lots of information for avoiding as well as for dealing with ticks and Poison Oak. Mountain Lions are very unlikely to be encountered and even less likely to threaten your safety, but they are certainly worth keeping in mind when you are walking alone in the woods, especially near dusk or dawn.

Perhaps most frustrating for some people is the fact that the mushrooms themselves are unpredictable. An afternoon spent dashing about on a forested hill above the wind-whipped Pacific while uncovering King Boletes from beneath deep carpets of pine needles is primarily thrilling. And there's a very different sensation—one of magic and marvel—when a glittering spot of blue among the thatch of redwood twigs turns out to be *Leptonia carnea*.

The knowledge and intuition required to find specific kinds of mushrooms take time to develop. Different species have their own favored fruiting times within the mushroom season, as well as different habitat preferences. If you intend to search for a particular species of fungus, doing a little research on its seasonality and habitat preferences will greatly increase your chance of success.

The seasonality descriptions found in this book are intended to be useful generalizations. If you look hard enough, macrofungi of some kind or another can be found fruiting *somewhere* during any month. The onset and persistence of the mushroom season vary, depending on many factors, particularly the occurrence of dry spells, heat waves, and cold snaps. The mushroom fruiting season also varies by the shape of the land: valleys and west- or north-facing slopes tend to be cooler and moister (at least in our Coast Ranges), while ridgetops and south- or east-facing slopes tend to be warmer and drier.

Although the seasonal patterns of mushroom fruitings can seem mysterious and even frustrating for beginners, after a few seasons of careful observation and experience, you'll develop a sense for when and where different species appear. Joining a local mycological society or amateur mushroom club (see the list on p. 576) is a great first step. Members of these clubs have years of accumulated knowledge and can provide you with support, companionship, guidance, and validation as you go about building and refining your knowledge of mushrooms. Your learning process will also be accelerated by taking field notes! This is an especially important realm of natural history record-keeping, since there are few existing data on fungal fruiting dates and locations. Websites like mushroomobserver.org (see Resources, p. 576) are very useful for consolidating and exploring this kind of information.

Collecting Mushrooms

Like beachcombing, tree-climbing, rose-smelling, and swimming, appreciating mushrooms is one of those beautiful activities that fundamentally requires no equipment. Go for a walk in the winter woods, looking for mushrooms and inspecting them to your heart's delight. Enjoy the cloaking mists, the dark and drenched duff, and the brilliant company of mushrooms; you'll soon intuitively learn much about their patterns and habits. However, as your interest grows and you start to wonder about their identities, you'll find that certain equipment is indispensable (see the section on tools of the trade, p. 12).

If you are gathering mushrooms to eat (assuming that you are *absolutely certain* of their identity), the pace of your collecting can become rather frenzied, as you dash from one productive area to the next, with your focus centered on a mental image of your target species. Once you've gathered enough to make a meal (or two, or five), the next step will be cleaning and preparing your mushrooms for cooking or storing (see the section on Collecting for the Table, p. 13).

If you are searching for rare and unusual fungi, or surveying to find as many species as possible, the pace of your collecting should be much slower. Many minuscule and inconspicuous fungi will be missed by a hurried collector. Pause every minute to stop and look closely at the area around your feet. Don't neglect the backsides and undersides of rotting logs, or the trunks and branches of trees above you.

Please keep in mind that mushroom collecting is unlawful in many places, and that there aren't consistent regulations even for parks of the same type. For example, some state parks have a limit for the harvest of wild mushrooms, while others ban the activity outright; national parks are similarly uneven. Likewise, private property boundaries can be difficult to determine in the woods, and you may inadvertently end up speaking with an angry landowner. In all cases, keeping the woods clean, making sure the duff layer is relatively undisturbed as you harvest mushrooms, and respecting property boundaries will go a long way toward ensuring that we retain the few rights we have to collect mushrooms on public wildlands.

Identifying Mushrooms

Beginners should stick to large, distinctive-looking species when learning to identify mushrooms. Attempting to classify what David Arora calls LBMs (little brown mushrooms) is very difficult for novices. With a great deal of practice, however, many little brown mushrooms can be identified to genus or even species without a microscope. Inevitably, the littlest, brownest mushrooms (as well as some larger ones) will require a microscope to positively identify.

There is a general set of procedures you should follow to maximize your chances of identifying the mushrooms you find. This set of procedures begins while you are in the field. The following is a simple protocol geared to help you develop your identification skills.

Collect a small number of species to start. Beginners will often come home from the woods with a basket brimming with mushrooms and find themselves overwhelmed, discouraged by the seemingly impenetrable variability and diversity. Keep identification manageable by bringing home fewer than five species on your first outings.

Even though you might only collect five species, this does not mean you should only collect five mushrooms! Try to collect multiple fruitbodies of the same species. Most often, a single species will produce multiple fruitbodies in the same area, allowing you to collect several in good condition and of different developmental stages. Take brief field notes—date, location, surrounding trees, and substrate are the most important. The substrate is the material from which the mushroom was fruiting. Soil, wood, leaf litter, and other mushrooms are examples of different types of substrates. Digging up the very base of the mushroom can sometimes reveal buried wood—investigate carefully. Other important habitat includes the kinds of surrounding trees and plants, the elevation, and the distance from the coast.

Take photos of your mushrooms in the field. This is probably the single most useful step to aid later identification efforts. See the section on mushroom photography (p. 13) for more tips and techniques.

Keep collections of different species separate from each other. Waxed paper bags work well in relatively dry weather; otherwise tackle boxes are ideal, especially for small fruitbodies. When you return home, spread your collections out, keeping different species separate as best you can tell. Dedicate one or two of the more mature fruitbodies in each collection to making a spore print (see Making Spore Prints).

Write a simple description for each of the collections, perhaps including line drawings for any features of the mushroom you find distinctive (see the section on making descriptions, p. 15). It is especially important that you write down notes on odor and taste at this point, since these characteristics often disappear rapidly.

Please note that tasting mushrooms is safe if a small, clean piece of the cap is nibbled and held on the tongue, and then spit out completely (distinctive tastes should be detectable within 20 seconds or so).

After writing a brief description of your mushroom, use the Pictorial Key to the Major Sections (see p. 17). We've decided not to include more extensive keys in order to make space for more photos. After you've identified the general group to which your mushroom belongs, page through the species descriptions looking for photos that closely match your mushrooms. Then read the text descriptions, and pay attention to the similar species mentioned in the comments. Once you feel fairly sure you've identified your mushrooms to species, seek verification from an experienced person (try online mushroom enthusiast groups, but beware the phenomenon of "Wikipedia Experts" who seem authoritative, but lack real robust knowledge of identification and often make subtle but critical mistakes).

Making Spore Prints

A spore print can be a critical first step in identifying practically any mushroom you may find. The goal is to assess the color of the spore deposit, which is simply a thick layer of spores. Follow these steps to make a spore print.

After removing the stipe (if necessary), place the gills, pores, teeth, or other fertile surface face down on a piece of glass, foil, clear plastic, or white paper. Let the mushroom sit undisturbed for six to eight hours. Smaller mushrooms may benefit from being covered by a glass or plastic container to prevent moisture loss and wind disturbance.

Carefully lift the mushroom off the surface to reveal the spore deposit. White or pale-colored spore prints can be difficult to see on white paper (tilting the paper can help) but are obvious if glass or clear plastic is placed on a dark background.

Observe the color of the spore deposit, preferably on a white or black background. Colored backgrounds can distort color perception dramatically.

Be aware that there is some subtlety in interpreting spore print colors—general terms are somewhat useful, but more subtle distinctions can be necessary, especially between different shades of brown.

Tools of the Trade

A number of tools are useful for collecting and identifying macrofungi.

A wicker **basket** is the best way to carry many collections at once. Relatively flat baskets are best, since they maximize the number of mushrooms that can be carried in a single layer (minimizing the risk of crushing). Cloth tote bags can crush mushrooms under their own weight.

Field notebooks are essential for anyone seriously interested in identification. Water-resistant notebooks (Rite in the Rain, for example) are especially useful in wet weather but are relatively expensive.

Tackle boxes are the best system for keeping collections separate and for keeping delicate specimens in good shape. Downsides include the awkward shape of the compartments when attempting to store larger fruitbodies, and poor breathability, which can accelerate bacterial decomposition (so don't keep your collection in plastic longer than absolutely necessary).

Waxed paper bags are the best receptacle for larger fruitbodies, since they can better accommodate irregular shapes. Downsides include the difficulty of transporting multiple bags full of mushrooms (unless you've got a basket) as well as their tendency to disintegrate in wet weather.

Hand lenses are fun in the field, and indispensable back home when examining the smaller features of mushrooms, including details of gill edges and surface textures of many different structures. A 10x lens is usually strong enough for such purposes.

KOH (potassium hydroxide) is available as pellets from biological supply companies. When combined with water (3–10 g KOH per 95 mL water), it can be stored in a glass dropper bottle for about a year (after which it starts to become cloudy and weak). The tissues of some mushroom species change color when KOH is applied; these color changes can be very useful in identification. Be sure to take photographs of and make notes about any such color changes.

The importance of a **camera** for learning macrofungal identification is hard to overstate. A completely serviceable camera can be bought for less than two hundred dollars, and even many smartphones have good enough built-in cameras to get you started. See the section on photography (p. 13) for more information.

Serious collectors will need a **food dehydrator** in order to desiccate specimens to be deposited in personal or institutional herbaria. See the section on Making Collections below.

Making Collections

Making well-documented herbarium collections is one of the most valuable ways in which amateur mycologists can contribute to our understanding of the diversity and distribution of mushrooms. Each collection should be photographed and paired with a basic set of data: 1) species name (if known), 2) precise location, and 3) notes on habitat and morphology. The collection should be dried on relatively low heat (30°–50° C) until completely dry and crisp by using a commercial food dehydrator. The most crucial step is to pair a unique collection number with the photograph, the notes, and the physical collection itself. There are many ways of doing this, and even the simplest can be difficult to handle when a large number of collections are made. Be sure to figure out a system that works for you and practice with a small number of collections first. Once you've got the hang of it, stick to it! Writing a short document explaining your system will help herbarium managers receive your specimens. Although tailored for plants, Tim Ross's *Herbarium Specimens as Documents: Purposes and General Collecting Techniques* is a brief, thorough, and humorous document covering the entire process. Reprints can be ordered at www.socalbot.org/crossosoma.php for very low cost.

Photographing Mushrooms

There are innumerable styles of mushroom photography, but photographers looking to document mushrooms for identification purposes should follow a few basic principles: All parts of the mushroom should be visible, either by taking a series of photos of a single specimen or by placing multiple fruitbodies of the same species at different angles in a single photo. A range of young and old fruitbodies should be shown. A cross-section can also be helpful to show the color of the flesh, and before-and-after photos showing KOH or other chemical reactions are also useful.

While automatic modes often do a fine job of determining camera settings, learning to use your camera's manual mode will give you a lot more control in exposing correctly and capturing color accurately. In general, a low ISO setting (less than 640) and high f-stop (greater than 10) are desirable. A tripod is the single most valuable accessory you can pair with your camera to increase the quality of your pictures. Using a tripod allows for longer exposures (better for depth of field, even light, and accurate colors) that would be blurry if the camera were handheld. This is especially important since many mushrooms fruit in dark habitats beneath dense forest canopies.

Collecting for the Table

When collecting wild mushrooms for food, one rule supersedes all others: *when in doubt, throw it out*. If you are not sure that your mushroom is edible, don't eat it. Assuming you are familiar with the species you are gathering, here are a few pointers on picking mushrooms for the table:

1. Clean your mushrooms in the field. "Field dressing" your mushrooms—cutting off the dirty lower parts of the stipe and brushing off the cap—will make the final job of cleaning them in the kitchen easier.
2. Collect only fresh fruitbodies, but be mindful that picking young, unopened fruitbodies is less sustainable than picking slightly older ones that have had a chance to release some spores (and younger fruitbodies are smaller!).
3. Keep different species separate—a jumble of dirty mushrooms can be a headache to sort later and can lead to dangerous mistakes.
4. Clean your mushrooms carefully. Vigorous brushing will tear many mushrooms apart, and too much water can make them soggy. However, don't hesitate to use enough water to get your mushrooms clean! If they do become a bit waterlogged, cooking them longer can help drive off the excess moisture.
5. Cooking mushrooms is a fine art well beyond the scope of this book, and there are many excellent references available. However, we'd like to point out that the dry sauté is a very simple technique that is the starting point for many mushroom recipes. In this method, the mushrooms, either sliced or diced, are cooked by themselves over relatively high heat in a bare pan until they release their water and it evaporates. Only after this initial step is oil added.

Taxonomy and Cladistics

When we identify mushrooms by using a field guide, consulting websites, or examining microscopic features to compare against technical references, we are engaging in the science of taxonomy. A taxonomic system is a hierarchical arrangement of units called taxa (singular: taxon), which are grouped together according to shared characteristics. Carl Linnaeus (1707–1778) is referred to as the father of biological taxonomy because his system of naming and classifying organisms quickly became the standard in the world of academic biology. The great Swedish mycologist Elias Fries (1794–1878) took this broader system and refined it for use in the classification of macrofungi. Among the most important characters in Fries' system are the shape of the fertile surface and the color of the spore deposit, as well as the colors and texture of the fruitbody.

In modern biological taxonomy, each species of living organism is given a two-part name. The first part is the name of the genus to which the species belongs and is always capitalized. A genus is a collection of species that are similar to each other with regard to a defined set of characters. The second word is the name of the particular species within the genus. What defines a species is a very complicated topic, particularly for fungi (more on that later in this section). In the case of the white Button Mushroom found at your grocery store, the binomial name is *Agaricus bisporus*.

What set of characters define the genus *Agaricus*? With the naked eye, we can recognize *Agaricus* by their free gills, dark brown spores, and presence of a partial veil (see diagram of fruitbody morphology, p. 2). All members of this genus, despite the specific differences between them, share these characters. If we were trying to identify a mushroom that didn't show these features, we would have very good reason to rule out the genus *Agaricus* as a candidate. Listing the characters that distinguish the species *bisporus* from the rest of the members of the genus *Agaricus* is a larger and more complicated task, and the species descriptions in this book are dedicated to just such endeavors.

Every named species also belongs to a named group at each of the taxonomic levels above it, although this is less relevant for most mushroom identifiers. In the case of *Agaricus bisporus*, we can trace the taxonomic hierarchy upward, with each category being more general (and thus more species rich) than the preceding one. The genus *Agaricus* belongs to the family Agaricaceae (which also includes *Lepiota*, *Chlorophyllum*, and *Leucoagaricus*), which belongs to the order Agaricales (nonrussuloid mushrooms), a subset of the class Agaricomycetes (which includes *Russula* and *Lactarius*), itself a member of the division Basidiomycota (all fungi that produce spores on basidia). All are members of kingdom Fungi.

KINGDOM
Fungi
> **PHYLUM**
> Basidiomycota
>> **CLASS**
>> Agaricomycetes
>>> **ORDER**
>>> Agaricales
>>>> **FAMILY**
>>>> Agaricaceae
>>>>> **GENUS**
>>>>> *Agaricus*
>>>>>> **SPECIES**
>>>>>> *bisporus*

This morphological approach—which classifies mushrooms by shared physical characteristics—is powerful and especially useful for amateurs and field workers. However, it has been superseded in academic biology by the cladistic approach, which groups species according to genetic data. The cladistic system groups organisms based on their shared evolutionary histories rather than by shared morphological characters. Although these two criteria are not independent (because a species's evolutionary history strongly influences its morphology), they are far from equivalent. For example, two organisms that are closely related may appear quite different to our eyes. Dolphins barely resemble bats at all, but do appear quite similar in body shape, behavior, and habitat to sharks. But bats and dolphins are both mammals and are therefore more closely related to each other than either is to sharks, which are fish. This phenomenon in which distantly related organisms show similar morphology is called convergent evolution; it occurs when different organisms adapt and evolve in response to similar environmental pressures over very long periods of time.

Convergent evolution is a common phenomenon in the world of fungi and is one of the main reasons that cladistic and morphological approaches to classification don't always align well. For example, although most mushrooms of the genus *Russula* have a cap, a stipe, and spore-producing gills, they are not very closely related to the majority of other gilled mushrooms—their similarity in fruitbody shape is due to convergent evolution.

But how do we know this? What information requires us to reject our intuition about the classification of *Russula*? Most often, it is the use of genetic evidence obtained by sequencing an organism's DNA. The "blueprints" for many aspects of an organism's morphology, biochemistry, and behavior are encoded in the specific order of nucleotides in its DNA molecules. But the order of these nucleotides (and thus the information they contain) differs from species to species, and to a lesser degree between individuals of the same species. For this reason, DNA evidence can be used to illuminate similarities and differences between organisms that would otherwise have been hidden by the effects of divergent or convergent evolution. Generally speaking, the greater the percentage of shared DNA sequence between two species, the more closely related they are.

Determining the order of nucleotides that make up part of the DNA of a particular organism is called sequencing. After we obtain such a sequence, careful interpretation can provide us with clues about an organism's ancestors, giving us another method of building a family tree. When we say that two organisms are each other's closest relatives, we are implying that they share a more recent common ancestor with each other than with other organisms. This kind of evidence is central to our understanding of biological phenomena and is of especially great influence when we go about revising taxonomic arrangements. The complexities of cladistics are beyond the scope of this book, but it is important to keep in mind that cladistic classifications often conflict with our expectations when using morphology-based methods to classify mushrooms. While DNA-based cladistics are very powerful tools for building accurate family trees, the resulting reclassification of familiar organisms can be frustrating for field biologists.

The practical outcome of all this is that you'll soon become familiar with mushroom identifiers saying "that's what we call it here," "it needs a new name," "our local version of the species," or "it used to be called so-and-so." These expressions reflect one of the most frustrating facts of taxonomy and systematics: the scientific names of organisms are frequently changed. Why is this?

Scientists may give mushrooms new names for a variety of reasons: the original name may have been incorrect in some way (for example, a single mushroom species may have been given two names at different times), or the same name may have been mistakenly given to two different species. Much of the time, however, there is no error involved, but rather the new names reflect the reclassification of a group of species based on new data. It is worth noting that the degree of difference that separates two genera in one part of the tree of life might be equivalent to the difference between two families in another group of organisms. Keep in mind that, at their core, these distinctions are arbitrary.

For example, it has been found that within the genus *Boletus* there are many distinctive smaller groups, each of which could easily be considered to constitute a genus on its own. Those pretty red and yellow boletes identified as *Boletus dryophilus* have been given their own genus in older field guides and are now known as *Xerocomellus dryophilus*. Another example is the case of the Gilled Boletes (currently known as *Phylloporus*). Analysis of genetic data has shown that they fit in among the group of boletes in the genus *Xerocomus*. However, until the current species housed in *Phylloporus* are formally transferred into *Xerocomus* in a scientific publication, many people consider it most prudent to continue using the existing names.

Many of the species found in this book occur only in the western United States, but because of early mycologists' habit of borrowing the names of European species to refer to American mushrooms, we are left in an odd position: Californian mushrooms are often quite different in morphology, microscopic features, ecology, and genetics from their European namesakes, but until a new name is published, the most effective way of referring to them may be to use an epithet that we know to be incorrect.

We recommend taking a practical view of things: incorrect names will be used to convey information, and if they succeed in doing so without creating any new confusion, don't worry too much about it. The need for new names and critical study of California's mushrooms is widely acknowledged and is slowly but surely being addressed.

How to Read the Species Descriptions

Our species descriptions are written in a standard format intended to make it easier for readers to find specific pieces of information about the characters of each species, since each subcategory within a description (cap dimensions, for example) will appear in approximately the same place on every page. The descriptive terminology used is for the most part consistent with all groups of mushrooms, but some groups have such distinctively different morphology that they require their own set of categories and terms.

The dimensions we list for the various parts of fungal fruitbodies are fairly general. (For more on mushroom morphology, see the chart on p. 2.) We have tried to list an average range, along with any specific extreme dimensions we've recorded (noted in parentheses). Our use of general measurements rather than highly specific ones reflects the fact that fungal fruitbodies are dramatically affected by the conditions under which they are formed—a single mycelium may produce miniature or oversized fruitbodies depending on whether fruiting was preceded by drought or torrential rain. Species that produce larger fruitbodies generally show a larger range of variation in size, while those that produce smaller fruitbodies are usually more consistent. While we've intentionally kept our descriptions of size fairly general, our descriptions of colors tend to be a bit more subtle and specific. The text descriptions of colors are very important, since a single photo can't hope to depict the full range of variation in a species. Keep in mind that some species never show much deviation from a specific palette of colors, and others seem to be able to produce a rainbow of pigments.

Mushrooms are extremely sensitive to the environmental conditions in which they develop. They are remarkably tolerant of adversity and are often able to produce and release spores even after being subjected to parching air, soaking rain, freezing temperatures, insect infestation, and bright sun.

Under such conditions, you can expect that fruitbodies will look a bit worse for the wear—differing from what might be considered "typical." What this means for you, esteemed mushroom identifier, is that you should always try to take into account the effect that such factors may have had on the shape, size, textures, and colors of a given fruitbody. Rain often washes out pigments, making them lighter or leaching them into other parts of the fruitbody; wind often dries out slime layers and causes caps to crack; sun often causes pigments to become paler; drought often results in miniature fruitbodies. As you become familiar with the mushrooms in your area, you'll develop an intuitive sense for these variations.

General Format of the Species Descriptions

Name and author: After each species name, we list the authors of that name—the people responsible for the original concept of the species and for its current generic placement. If a researcher moves a species from one genus into another, the name is added to the author citation (the rules for how these are formatted are complex; for more information, see iapt-taxon.org/nomen/main.php). The term "group" is used when there are presumed to be a number of different species going by the same name, and for which we have yet to develop criteria to adequately distinguish them. Some researchers propose names for species that have yet to be officially published; these are called provisional names and are denoted "nom. prov."; they usually come with a fairly well-established species concept. In many cases, we have given species "tag names"; these are more or less equivalent to provisional names, but are in plain English (intended to be easier to remember). Lastly, we know that many of California's mushrooms are incorrectly using European names—for such species, we've put "sensu CA" after the name. The authors believe that such species will likely be renamed after proper research.

Cap: Most of the species in this book produce fruitbodies that have some sort of clear cap or "head" that is described in this section. Those mushrooms with radically different structures will have other headings to better reflect their shape.

Gills and other fertile surfaces: This section will have many different names depending on the morphology of the species in question: spines, tubes, pores, or simply fertile surface.

Stipe: If the fruitbody in question has a stipe (whether clearly defined or not), it is described in this section. Some species produce fruitbodies that lack a distinct stipe and won't have an entry for this section.

Veil (partial veil and volva): Veil tissues protect the developing fruitbody—universal veils enclose the entire fruitbody as it grows, while partial veils cover only the fertile surface until it matures enough to release spores. Only a few genera of mushrooms produce fruitbodies with universal veils; however, mushrooms with partial veils are common and found in many genera. The texture of these tissues ranges from thick and membranous (most often) to thin and fibrous, cobwebby, cottony, hairy, or fluffy. In many cases, descriptions of the universal veil tissue will also be included in other sections— as warts or a patch when found on the cap, or as a volva if it encloses the base of the stipe in age. Referring back to the morphology diagram (p. 2) should clarify things in case of confusion.

Flesh: Color and texture characteristics are noted in this section. It is best observed by cutting a fruitbody in half and breaking the stipe or pulling apart the flesh. This is destructive, so be sure to take note of other features before you mangle your mushroom! Note that by cutting and rubbing the tissues of a fruitbody, you expose the cell contents to oxygen in the air; in some species this will produce a distinctive (and sometimes surprising!) color change. In such cases, pay attention—this is an important identification character. Many mushrooms darken to some degree when damaged, but you'll soon learn to distinguish trivial bruising from more pronounced color changes. Some common reactions are bluing, reddening, yellowing, and browning; blackening and purple staining are less common.

Odor and Taste: These are very important characteristics for any mushroom identifier to pay attention to, but are unfortunately rather variably sensed and described. Each person's senses are a bit different, and it's common for two people to describe the smell or taste of the same mushroom very differently. In general, we have found that descriptions of taste seem to be more consistent between people than descriptions of smell, which often vary dramatically. Even if you don't agree with our descriptions of a smell or taste, it's worth "calibrating" your senses to these descriptions, since much of the terminology we use is consistent with commonly established descriptions in prior literature. If you've confirmed beyond doubt that the mushroom you're holding is *Agaricus augustus* (described as smelling like almonds in almost all literature), but you find it to smell more like hair gel, keep this in mind when you encounter "almond" in other species descriptions, as you may again interpret the species as smelling like hair gel. Remember that practically all mushrooms will begin to smell unpleasant in age as they begin to decay. Overall, try to tune your senses and make your descriptions consistent, and make comparisons with common scents (fruits, vegetables, herbs and spices, cleaning products, familiar chemicals). Don't balance your entire identification on your interpretation of the smell. To safely taste a mushroom, bite off a small, clean piece of the cap flesh and nibble it on the tip of your tongue for 5–10 seconds, and then spit it out completely. It is perfectly safe to taste mushrooms (even poisonous ones) in this manner, as long as you aren't ingesting the tissue.

KOH or other chemical reactions: The fruitbodies of some mushrooms change color when exposed to various chemicals (often called reagents). We've included potassium hydroxide (KOH) reactions for many species, since this chemical is relatively easy to obtain and is often useful in identification. Although there's a wide variety of other chemicals that have been used by mushroom taxonomists, we only mention a few in our descriptions—many are hard to find, extremely toxic, and/or less useful for the average identifier. Keep in mind that the intensity of the reaction varies based on a number of factors: age of the fruitbody, degree of rain soaking, and strength of the chemical being used (KOH often becomes weak after a year or two).

Spore deposit: The color of the spore deposit is a very important feature to note when you are first starting to learn mushroom identification—see the section on Making a Spore Print, (p. 12). However, it takes a while to make a spore print, so it won't be something you can determine in the field.

Microscopy: Since most people don't have access to microscopes, we've included just a few of the most useful kinds of microscopic details. Spore shape and size are usually the most crucial features to note. For basic definitions and illustrations of microscopic structures and chemical reactions (amyloid, inamyloid, dextrinoid), see the diagram on p. 2 and the Glossary. For a more detailed look at mushroom microscopy, see Resources (p. 576). There's still much to be learned about the range and significance of variation in microscopic features of mushrooms—our data are representative but not definitive. Please seek out measurements from other sources and tell us what you learn!

Ecology: This section includes a summary of what the authors understand about the range, habitat, and seasonality of the species. Our knowledge is based most heavily on our own experience with the species, readings of primary texts, and discussions with knowledgeable identifiers and people who spend lots of time in the field. These are by no means definitive descriptions, and you are encouraged to report discrepancies between our accounts and your own observations to public forums like mushroomobserver.org. The same goes for our interpretations of commonality—many species are reported as rare until their habitat preferences become well known. Likewise a species may be under-reported if its identification poses difficulties for most people (or if it lacks a name!).

Comments: This is a very important section. It summarizes the most useful identification features for a given species, and gives a rundown of similar species that you should consider and compare. It also includes occasional additional information unrelated to identification (interesting natural history notes, cultural uses, and the like).

Synonym and Misapplied Name: The scientific process of describing and naming species is far from perfect. Sometimes this leads to a single species of mushroom being given multiple names. Ideally, only one name should be used for each species. In this section, we list some of the commonly used or well-known (but incorrect, imprecise, or otherwise suboptimal) names for the species. Unfortunately, many species have undergone name changes recently, and older, "incorrect" names may be much more familiar than the most scientifically accurate name. For some undescribed mushrooms, we've used "tag names" to provide a descriptive but informal way to refer to them. In some descriptions, we use "group" names to refer to complexes of multiple species. In these cases, precise identification is likely to be difficult, complicated, or impossible given our current state of knowledge. We will keep track of nomenclatural updates relevant to the species found in this book on our website (www.redwoodcoastmushrooms.org). Be sure to check in frequently; things change fast in the world of mushroom identification!

How to Use the Pictorial Key to the Major Sections

We've chosen to use a visual, rather than dichotomous key strategy in this book. Use the following pictorial key to help you narrow down the options when looking for a description that matches a mushroom you are trying to identify. Although the small images are intended to give you an idea of the group, no single image can accurately represent a group of species, so don't get too hung up if your mushroom doesn't match it perfectly. Instead, focus on spore print color (see p. 12) and stature. Carefully read the text for each group, paying particular attention to the structure of the fertile surface (gills, pores, teeth, or otherwise), gill attachment (if applicable), texture, and general colors.

Pictorial Key to the Major Sections

Chanterelles and Gomphoids—Spores white to ocher-buff. Blocky-triangular or trumpet-shaped to vase-shaped fruitbodies with low, ridgelike, decurrent "gills." However, one *Craterellus* has a smooth fertile surface. Colors range from bright orange, yellow, beige to cream, and violet to black. Texture from thin and somewhat tough to firm and rubbery with stringy flesh. See p. 23.

Amanita—Spores white. Small to large mushrooms, most often with a ring zone or skirt on the stipe (except in one subgroup), scaly rings, a cup or saclike volva, or other veil remnants around the base of the stipe, and a patch or warts on the cap. Cap colors range from dull to very bright. See p. 32.

Cystoderma and **Cystodermella**—Spores white. Smallish mushrooms with narrowly attached, whitish gills, a partial veil on the stipe (sometimes only an obscure veil zone), and a granular, powdery, or scaly cap texture. Cap colors range from golden tan to brown or pinkish red. See p. 54.

Lepiota and **Allies**—Spores white (grayish green in *Chlorophyllum molybdites*). Rather small to large mushrooms with distinctly free gills, a partial veil present as a collar, skirt, or obscure ring zone, but lacking universal veil tissue. Colors mostly whitish to beige, some with gray, black, or brown scales (one species bright yellow and another small group with pinkish to purple cap colors). See p. 54.

Agaricus and **Melanophyllum**—Spores chocolate brown in *Agaricus*, grayish olive to reddish brown in *Melanophyllum*. *Agaricus* are small to very large mushrooms with distinctly free gills, a skirt or partial veil zone on the stipe, and often with distinct staining reactions when cut and/or distinctive odors. *Melanophyllum* have rather small fruitbodies with powdery gray-brown caps, vivid red to dull reddish gills, and a partial veil (often obscure). See p. 73.

Dark-spored Mushrooms—Spores dark purplish gray, purplish black, or black. Fruitbodies ranging from very small and fragile to medium sized and fleshy-fibrous. Gills often pale to grayish, becoming dark gray or black. Especially common in urban habitats, on wood chips, on dung, or in disturbed areas. See p. 92.

Brown-spored Decomposers—Spores dull brown to rusty brown. Fruitbodies tiny to huge, dull to colorful, and with or without partial veil tissue. If your mushroom has brown or rusty spores and is growing on wood, it should be in this section. Otherwise, start by reading the following short descriptions for *Inocybe*, *Hebeloma*, *Phaeocollybia*, and *Cortinarius* in this pictorial key. If none of those seem promising, your mushroom should be in this section. See p. 125.

Mycorrhizal Brown-spored Mushrooms (*Inocybe*)—Spores dull brown. Small to medium sized, with dry, silky, or scaly to shaggy, often conical caps. Stipes cylindrical, often powdery looking, many with small bulb at base. Odor frequently of green corn or spermatic; others sweet, fishy, or indistinct. Partial veil present and silky to cobwebby, but usually sparse and often visible only on young fruitbodies. Not growing on wood. See p. 148.

Mycorrhizal Brown-spored Mushrooms (*Hebeloma*)—Spores dull brown. Small to fairly large mushrooms with relatively smooth, often viscid caps and attached gills (often beaded with droplets when young). Odors often pungent, radishlike, or chocolatey-musty. Stipes slender or more often club shaped. Not growing on wood. See p. 148.

Mycorrhizal Brown-spored Mushrooms (*Phaeocollybia*)—Spores rusty to ocher-brown. Small to large mushrooms with distinctive long, tapering, rooting, and cartilaginous stipes (many filled with white pith). Caps often conical or bell shaped, often viscid. Fruitbodies dull to very colorful. Veil absent. Found only in forest settings; uncommon. See p. 148.

Cortinarius—Spores rusty (bright orange-brown). A large and varied group of species united by the presence of a cortina—a silky or cobwebby partial veil (membranous in one species). Fruitbodies range from very small to very large. Colors very dull to very colorful. Caps conical to rounded, dry to viscid. Beware that the cortina can be inconspicuous or disappear in age. Found in forest habitats, not growing on wood. See p. 166.

Large *Entoloma*—Spores pinkish tan to salmon. Fruitbodies medium sized to large and stout, caps dry to viscid, gills attached or decurrent, and lacking any veil tissue. Colors white to gray or brown, but one species is sky blue to midnight blue. Veil absent. Found in forest settings, not growing on wood. See p. 203.

Small *Entoloma*—Spores pinkish tan to salmon. Fruitbodies small to medium sized, with attached gills and lacking any veil tissue. Caps dry, smooth, finely scaly, or silky. Colors ranging from dull brown to rosy pink, yellowish tan, blue-gray to iridescent purple, or blue to blackish blue. Veil absent. Found in forest settings, very rarely on wood. See p. 203.

***Pluteus* and Allies—**Spores pinkish to salmon buff. Gills distinctly free, usually growing on wood. One common species and a few rare species in our area have universal veil sheathing the base of the stipe (volva tissue) and grow on organic matter and lignin-rich debris. One small, extremely rare species with a volva grows on decaying mushrooms. See p. 220.

***Russula*—**Spores white to cream, yellowish, or ocher. Often stout, squat mushrooms with convex to uplifted caps. Texture usually chalky, brittle, or crumbly and lacking any "milk" (latex) when broken. Some can have very hard, firm flesh, but will still break like chalk rather than splitting into fibers like most other gilled mushrooms. Colors dull to bright. Veil absent. Found in forest habitats. See p. 227.

***Lactarius*—**Spores white to cream to light yellowish. Often stout and squat with uplifted caps that are centrally depressed. Texture crumbly, brittle, or chalky, usually producing "milk" (latex) when broken or cut. This feature can be hard to see, so check fresh fruitbodies and look closely. The milk is often distinctly colored and sometimes changes color when exposed to air. Veil absent. Not growing on wood. See p. 227.

Waxy Caps I—Spores white. Gills attached, relatively thick and slightly greasy or waxy feeling, and often well spaced. Caps dry to very slimy, dull to bright. Stipes slender to robust, but always solid. Partial veil present or absent. Found in forest settings. See p. 269. This group can be difficult to distinguish from other white-spored mushrooms (also check Waxy Caps II and the large species in White-Spored Multitude).

Waxy Caps II—Spores white. Gills attached, relatively thick, often waxy feeling and well spaced. Caps dry to very slimy, mostly brightly colored (red, orange, yellow, pink, green, or blue), a few duller. Stipes usually slender and hollow (two large, red-capped species have solid stipes). Partial veil always absent. Found in forest settings, commonly under redwood and cypress. See p. 269.

The White-spored Multitude (Small Species)—Spores white. Fruitbodies tiny to small (a few medium sized but with thin stipes); caps usually less than 2.5 cm across. Caps often translucent-striate; colors range from white to brown to dark gray or brightly colored. Mostly litter decomposers, found in both urban settings and forest habitats. See description of white-spored groups on p. 296.

The White-spored Multitude (Large Species)—Spores white. Fruitbodies medium to large. Gill attachment variable—decurrent to broadly or narrowly attached or notched. Partial veil present or absent, colors variable (most dull, but some bright pinkish, orange, or purplish). Growing on ground or wood. Also check Pleurotoid section. See description of white-spored groups on p. 298.

Pleurotoid Mushrooms—Spores white to pink or brown. Caps fan or spatula shaped, and either lacking stipes or with stipes that are lateral, off-center, or short and rudimentary. Most with true gills, but some with poorly defined ridges or wrinkled. Fruitbodies very small to very large. Most grow directly on wood, but some on moss, grass stems, twigs, or woody debris. See p. 397.

Gilled Bolete Relatives—Spores olive brown to dark grayish olive or black. Gills decurrent (and often widely spaced), thick to thin and regular or wavy and branched or forking. Most species with medium-sized fruitbodies; a few species smallish and slender, others quite large. On ground or woody debris. Some species found in urban areas, others only in forest habitats. See p. 410.

Boletes—Spores olive brown (yellowish brown or pinkish tan in a few species). Undersides of caps with a spongelike layer of tubes with round or slightly angular pores; stipes usually well developed and central. Texture of flesh ranges from soft to firm, but usually distinctly fleshy (not tough or leathery). Found in humus in forest settings, rarely on wood. See p. 420.

Polypores and Allies—Spores whitish or brown. Underside of cap with pores (lacking in two species), stipe off-center to central, sometimes short or rudimentary. Flesh very thin to thick and firm, usually tough or leathery (unlike softer-fleshed boletes). Fruitbodies small and dainty to very large and robust. Colors range from whitish gray to creamy, to brown or reddish brown to sky blue or royal blue. On wood or in humus in forest settings. See p. 458.

Shelflike, Conklike, and Rosette-Forming Polypores—Spores whitish or brown. Underside with distinct pores (smooth to wrinkled in one species). Often forming rosettes, lobed clumps, or overlapping shelves; typically lacking a distinct stipe. Flesh tough and leathery to corky or hard and woody. Most fruiting directly from wood (sometimes buried). See p. 458.

Crust Fungi—Spore prints not usually obtainable. A ubiquitous group comprising many unrelated lineages. Fruitbody structures simple but variable; usually flat to slightly shelving and usually irregular in outline. Fertile surfaces smooth, pimply pored, or toothed. Colors dull to brilliant. Often fruiting inconspicuously on the undersides of rotting logs, on standing dead timber, or on other fungi. See p. 485.

Tooth Fungi—Spores pale to brownish. Fertile surface composed of "teeth"—hanging icicle-like structures. Some have distinct or rudimentary stipes, others lack stipes. Some grow on ground (often in rosettes), while others emerge directly from wood and lack any stipe. *Mucronella* are tiny and highly reduced, composed only of teeth hanging from deadwood. See p. 489.

Coral Fungi—Spores pale to ocher-tan, prints often difficult to obtain. Fruitbodies coral-like, with branches and upward-pointing tips (or in one species ruffly, flattened lobes). Degree of branching and shape of tips variable; colors range from white to dull tan to brilliant pinks, yellows, and oranges (one species greenish). Textures range from thin and fragile to tough or firm and fleshy. Found in forest habitats. See p. 503.

Club Fungi—Spores prints usually difficult to obtain. Fruitbodies simple, unbranched, and upright. Some are cylindrical, fruiting singly or in clusters, while others have distinct round, oval, paddle-shaped, or irregular heads. Textures range from thin and fragile to tough and velvety to pimply, powdery, or wiry. Colors variable, from tan to purple, black, yellow orange, green, or white. See p. 503.

Puffballs, Earthballs, and Earthstars—Spores forming a solid or powdery mass inside the fruitbody. Spores yellowish to olive, rusty, brown, purplish, or black. Shape variable. Puffballs and earthballs have round to pear-shaped fruitbodies. Earthstars have a thin-skinned spore sac perched on a star- or saucer-shaped pedestal. *Battarrea* has a long, shaggy stipe and a cap covered in rusty spore powder. See p. 526.

Stinkhorns—Spores held in an olive-brown, foul-smelling slime. Fruitbodies very diverse in structure and color, but all have a terrible odor. Found almost exclusively in urban and disturbed areas in nutrient-rich garden soil or in wood chips. See p. 526.

Bird's Nests—Spore prints not obtainable. Fruitbodies are tiny to small leathery cups with small, egglike spore packets in the middle. Textures range from smooth to hairy, granular, or grooved. Colors always dull, mostly beige, gray, brown, and blackish. On dung, rich soil, or woody debris; found in both disturbed urban areas as well as in forest habitats. See p. 526.

Truffles—Included here are those species that have lost most of their adaptations for fruiting aboveground. Instead of gills, pores, or teeth elevated on a stipe, they form highly reduced "sequestrate" fruitbodies. These are often rounded or knobby and have complex, convoluted inner fertile tissue (sometimes powdery or gelatinous). Many different lineages of mushrooms have trufflelike representatives. See p. 541.

Jelly Fungi—Spore prints often difficult to obtain. Fruitbodies are tongue shaped, amorphous blobby, brainlike clumps, or gelatinous, bumpy, and sheetlike masses. Most share a distinctive watery- or rubbery-gelatinous texture. Colors range from black to yellow, orange, or brown; many are slightly translucent. (A few gelatinous species with barrel- or bowl-shaped fruitbodies are in the Cup Fungi section.) See p. 547.

Morels, False Morels, and Elfin Saddles—Fruitbodies with clearly defined smooth or fluted and grooved stipes and caps with thimblelike, honeycombed, saddlelike, or brainlike structure. Species that produce fruitbodies with stipes and simple cup-shaped "caps" are in Cup Fungi section. See p. 552.

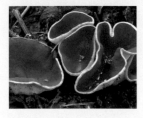

Cup Fungi—Spore print often difficult to obtain; pale to black. Fruitbodies with upward-facing cup- or bowl-shaped fertile surfaces. Some have distinct stipes. Size, color, and texture quite varied. Most are thin fleshed and relatively small; others are robust with thick, rubbery, or gelatinous flesh. *Annulohypoxylon* have hard, round, charcoal-like fruitbodies. Spores produced inside cylindrical cells (asci), usually in groups of eight. See p. 559.

1 • *Chanterelles and Gomphoids*

Cantharellus is a genus of well-known and well-loved mushrooms found in forests around the world. All these true chanterelles share a distinctive fertile surface made of shallow, veinlike "gills" that run deeply down the stipe. The fruitbodies are usually trumpet shaped or triangular in cross section, and the caps are often ruffled and wavy. These mushrooms are extremely popular as edibles because of their bright colors, their fruity odor and flavor, and their abundance (in good years anyway). All are mycorrhizal. Relatively few species occur in California, but there has been a recent surge in the description of new species from the United States, and it's possible that undescribed species remain to be found in our area.

Craterellus species produce tubular fruitbodies, with a hollow running from the center of the cap through the stipe, and have much thinner flesh than the true chanterelles. One species, *C. calicornucopioides*, has a smooth fertile surface, without any specialized structures. All are mycorrhizal.

Gomphus and *Turbinellus* can look very similar to chanterelles, but tend to be more strongly funnel shaped, with extremely decurrent "gills" and lacking a distinct stipe. The lilac-purple fertile surface of *Gomphus* and the woolly-scaly cap textures of *Turbinellus* also help distinguish them from any of the chanterelles.

Important identification characters to note include shape, stature (stocky or slender?), and coloration of all surfaces (especially when young), as well as surrounding trees and habitat.

Cantharellus californicus

Arora & Dunham

CALIFORNIA GOLDEN CHANTERELLE

CAP: 5–30+ cm across, at first convex and knoblike, soon becoming broadly convex to flat with an inrolled, lobed margin, to strongly uplifted overall and with a very wavy or strikingly ruffled margin. Bright yellowish orange or dull yellowish, sometimes paler whitish yellow, especially in dry weather. Developing darker orange or brownish areas as fruitbody ages. Surface smooth, finely velvety at first when dry, but soon smooth, moist to waterlogged in wet weather.
GILLS: Deeply decurrent, shallow, blunt, and veinlike often forking and with many cross veins. Pallid yellow when young or in dry weather; mature specimens usually evenly yellow to creamy yellowish, in dry weather sometimes slightly pinkish.
STIPE: 4–15 cm long, 3–5 cm thick, cylindrical or tapered toward base, continuous with cap flesh. Often short and stout. Pale yellowish white to yellow, smooth to fibrillose.
FLESH: Thick, firm, fibrous; rubbery when young, in age often waterlogged. White, slowly bruising orange to orange-brown where cut. **ODOR:** Indistinct to slightly fruity, most pronounced in dry weather or in very fresh young fruitbodies.
TASTE: Indistinct. **SPORE DEPOSIT:** Creamy to light yellowish.
MICROSCOPY: Spores 8–11 x 4–5.5 μm, ellipsoid, smooth.

ECOLOGY: Solitary, in pairs or clumps forming arcs and rings; prefers deep duff of live oaks in open woodland, but also in dark redwood forest with a mixed hardwood understory. Also occurs with Tanoak and deciduous oaks. Often difficult to spot due to its habit of fruiting, developing, and remaining under the duff. Only a small part of the cap may be visible (but the gold color practically glows against the dark leaf humus). The fruitbodies are produced early in fall, but usually remain small and hidden until after a few more rains and then develop slowly, often halting their growth in dry spells and resuming their expansion when the rains return. Abundance varies widely from year to year.

EDIBILITY: Edible and popular, but the quality varies with the degree of rain soaking and coating with mud and grit.

COMMENTS: In most habitats where it occurs, *C. californicus* is the only chanterelle species. North of Monterey Bay and in some inland areas, forests with Tanoak and live oak mixed with conifers and/or Madrone can host three or four chanterelle species. In areas like these, the evenly yellow coloration, chunky stature, and large size of this species distinguish it from the more slender *C. formosus* and the mostly creamy white *C. subalbidus*. *C. cascadensis* has a bright yellow cap, whitish gills, and a club-shaped stipe. *C. roseocanus* has brighter peachy orange gills and grayish white to pinkish orange caps when young; it grows with conifers. The only seriously toxic species that is sometimes confused with chanterelles in our area is *Omphalotus olivascens*, which is orange with olive-green tones, has true gills and dingy grayish olive flesh, and grows in clusters on wood (sometimes buried and inconspicuous!).

Cantharellus formosus

Corner

PACIFIC GOLDEN CHANTERELLE

CAP: 3–10 (15) cm across, bun shaped with an inrolled margin and often a depressed center at first, becoming broadly convex but margin often strongly downcurved, expanding to flat and then uplifted with a depressed center and strongly wavy margin. Bright yellow to yellow-orange or dull brownish beige, often with ocher-brown scales. Much duller when young and dry, brighter when wet. Surface moist to dry, smooth but with matted-fibrillose scales that sometimes lift up in dry weather. **GILLS:** Deeply decurrent, blunt edged, but relatively thin and fairly broad for a chanterelle, with few cross veins when young (more in age). Dull whitish pink to light yellow when young, pinkish buff to dingy yellowish beige in age. **STIPE:** 4–12 cm long, 1–2.5 cm thick, cylindrical or slightly tapered toward base, usually straight and tall, sometimes slightly curvy but still upright. Whitish cream to pale yellow or orange-yellow. Surface dry, often breaking into chevrons over upper stipe. **FLESH:** Fairly thick, solid, stringy. White, bruising slowly yellow and then brownish when cut. **ODOR:** Indistinct or fruity-fragrant. **TASTE:** Indistinct. **SPORE DEPOSIT:** Creamy or pale yellowish. **MICROSCOPY:** Spores 7–9 x 4.5–6 µm, ellipsoid, smooth. Basidia 4- to 6-spored.

ECOLOGY: Usually in small groups or scattered in large arcs and troops in moss and duff of conifer forests. First fruiting in late summer on the Far North Coast, continuing into winter. Common north of the San Francisco Bay under Sitka Spruce, Douglas-fir, and Western Hemlock.

EDIBILITY: Edible and very good, usually much cleaner than other chanterelles due to its habit of fruiting above the duff, which also makes it easier to find!

COMMENTS: The tall, straight, slender stipe, decurrent gills, yellow to brownish ocher cap, and growth under conifers help identify this chanterelle. *C. cascadensis* has a brighter yellow cap, whitish gills, and a short, chunky, club-shaped stipe. *C. roseocanus* has brighter orange gills, a pinkish orange cap, and a shorter, chunkier stature. *C. californicus* is much larger and chunkier, and grows under oaks. *Craterellus tubaeformis* can look similar but has a thin-fleshed, hollow stipe with a cartilaginous texture. *C. formosus* is occasionally confused with *Chroogomphus tomentosus*, which has a velvety-shaggy cap margin and olive-blackish spores. The false chanterelle, *Hygroporopsis aurantiaca* has much thinner flesh and very crowded narrow gills that fork repeatedly.

NOMENCLATURE NOTE: Genetic data suggest that there are a few different species going by the name *C. formosus*, but little is known about the differences between them.

Cantharellus cascadensis

Dunham, O'Dell & R. Molina

HYBRID CHANTERELLE

CAP: (4) 7–15 cm across, convex and lobed with an inrolled margin when young, expanding to broadly convex or flattened and uplifted with a depressed center, usually quite wavy with a thin, ruffled margin. Bright, clear yellow to egg-yolk yellow, often mottled with watery yellow or creamy patches, occasionally extensively creamy yellow if buried in the duff. Surface moist, smooth to slightly matted-scaly at the center in age. **GILLS:** Deeply decurrent, blunt, variably shallow to somewhat deeper (but never forming "true" bladelike gills), with many cross veins and wrinkles. Whitish to creamy, sometimes with a light pinkish hue in age or when dry. **STIPE:** 6–10 cm long, 2–4 cm thick, chunky, squat, often narrower at apex and distinctly club shaped with a rounded base (up to 5 cm thick). Whitish to creamy, fibrous to smooth. **FLESH:** Fibrous, firm, solid, stringy. White (a bit yellowish in cap), bruising slowly yellow and then ocher-tan after being cut. **ODOR:** Indistinct to fruity-fragrant. **TASTE:** Indistinct. **SPORE DEPOSIT:** Whitish to creamy. **MICROSCOPY:** Spores 7–13 x 5–8 μm, ellipsoid to subglobose, smooth. Basidia 4- to 8-spored.

ECOLOGY: In small groups or scattered in large arcs in deep humus of a variety of conifers. More typical of higher elevation forest, but locally common on the coast from Mendocino County northward; most often with Western Hemlock, Grand Fir, and Douglas-fir. Rare south of Mendocino County; it has been recorded in mixed evergreen forest in the Santa Cruz Mountains. It's the most common chanterelle in the Southern Cascade and Sierra ranges. Fruiting in fall and early winter.

EDIBILITY: Edible and excellent; often with a fruity flavor, more typically associated with eastern North American chanterelles.

COMMENTS: This species has been referred to as the "Hybrid Chanterelle," as it has the bright cap color of *C. formosus*, but a pale underside reminiscent of *C. subalbidus*. The egg-yolk yellow cap with wavy-ruffly edges, whitish, decurrent ridge-like gills, stout, club-shaped stipe, and growth with conifers help identify this species. The more widely distributed *C. subalbidus* can be very similar, but usually has a creamy white to ivory cap (but stains yellowish), and an equal or tapered stipe. *C. formosus* has a yellow-orange cap and yellowish to pinkish buff gills, and is usually smaller with a distinctly tall, slender stipe with a tapered base. *C. roseocanus* often has pinkish tones on the cap when young and has bright peach-orange to orange gills.

Cantharellus roseocanus

(Redhead, Norvell & Danell) Redhead, Norvell & Moncalvo

RAINBOW CHANTERELLE

CAP: 3–10 (15) cm across, convex with an inrolled and lobed margin when young, expanding to plane or slightly uplifted and wavy, usually very ruffled in age. Center of cap often with a mass of lumpy tissue, frequently with small "rosecombs" of gill tissue as well. Orange with an extensively rosy pink margin when young and wet. Pale grayish with a whitish bloom when young and dry, becoming more yellowish orange or dingy ocher, margin frequently fading to grayish and remaining inrolled. Surface moist, smooth, with a few matted scales in age. **GILLS:** Shallow, ridge like, deeply decurrent, with extensive wavy cross veins. Usually bright peachy orange to apricot orange when young, becoming bright orange to yellow-orange. Sometimes paler yellow when young. **STIPE:** 4 10 cm long, 1.5–3 cm thick, cylindrical, tapered, or more often with a club-shaped base. Straight to curved or very irregular. White to yellowish or dingy ocher. Surface moist to dry, smooth when young, wrinkled, bumpy and often grooved in age. Gill tissue often breaking up in chevrons near the apex. **FLESH:** Fibrous, solid, stringy. White, bruising slowly yellowish and then ocher-brown when cut. **ODOR:** Fruity, fragrant. **TASTE:** Indistinct. **SPORE DEPOSIT:** Pale creamy to yellowish. **MICROSCOPY:** Spores (6) 7.5 to 10 (11 5) x 4.5–5.5 μm, ellipsoid, smooth. Basidia 4- to 6-spored.

ECOLOGY: In clumps or scattered in troops, usually buried in duff under conifers. Common under Sitka Spruce on the Far North Coast, and under Bishop Pine farther south. Commonly fruiting in the summer fog drip period, continuing into early fall, with scattered winter and spring fruiting.

EDIBILITY: Edible and excellent, especially if clean young specimens can be found.

COMMENTS: This odd species can be very beautiful in color while being lumpy and mangled in shape. The rich peachy orange to bright orange of the gills is quite different from other chanterelles that grow in the same habitat. *C. formosus* is taller and usually more slender, and has duller, light beige-yellow gills, and a yellow to ocher-brown cap. *C. californicus* can be similar, but is often larger, lacks the rich pinkish tones, and grows under live oak. *C. cascadensis* is similar in shape, but has whitish gills. *C. roseocanus* is also called the summer chanterelle.

Cantharellus subalbidus

A. H. Sm. & Morse

WHITE CHANTERELLE

CAP: 6–15 (20) cm across, convex when young with a lobed and inrolled margin, expanding to broadly convex with a downcurved margin, becoming flat to uplifted and irregular in age with a wavy, ruffled margin. Whitish to creamy, becoming dingy ivory with yellowish blotches and ocher streaks in age. Surface matted-tomentose when young, soon moist and smooth or bumpy. **GILLS:** Deeply decurrent, blunt edged with extensive cross veins. Whitish to light cream or beige with a light pinkish wash. **STIPE:** 5–12 cm long, 2–4 cm thick, cylindrical or slightly tapered downward, usually rather stout, sometimes curvy or sinuous. White to ivory creamy, becoming dingy beige, with yellow to ocher blotches in age. Smooth to fibrillose. **FLESH:** Fibrous, thick, stringy. White, bruising yellowish and then orangish brown. **ODOR:** Mild to pleasantly fruity-fragrant. **TASTE:** Indistinct. **SPORE DEPOSIT:** Whitish. **MICROSCOPY:** Spores 7–9 x 5–5.5 μm, ellipsoid, smooth. Basidia 4- to 6-spored.

ECOLOGY: Solitary or in clusters, often in arcs and large troops in duff. Very common under Douglas-fir and other conifers on the North Coast and in mixed Tanoak-Madrone woods in inland areas and the southern part of our range. Uncommon south of the San Francisco Bay Area. Fruiting in fall and winter.

EDIBILITY: Edible and very good.

COMMENTS: The creamy to pale ivory colors of the entire fruitbody distinguish this species from other chanterelles. Occasional species become extensively yellow bruised in age and can be confused with *C. cascadensis*, which shares the pallid gills and stout stature. Other stout, white mushrooms that grow in forest settings have thinner, blade-like gills, and a few smell distinctly unfruity (*Leucopaxillus albissimus* and *Clitocybe nebularis*).

Craterellus tubaeformis (sensu CA)
YELLOWFOOT

CAP: 2–6 (8) cm across, convex with a small dimple at the center of the cap when young, expanding to broadly convex with a strongly downcurved margin, wavier and wrinkly-ruffled around margin; more deeply depressed at center and uplifted in age. Brown to ocher-brown or yellowish brown, light orange to pale yellowish near the edge (in one form bright orange-yellow when young, becoming pinkish beige). Surface dry, wrinkled, obscurely matted-tomentose to smooth, margin often finely fringed or fuzzy. **GILLS:** Subdecurrent to decurrent, shallow, blunt, forking with cross veins. Color variable, usually light pinkish when young to dingy orange-beige, sometimes light whitish gray, usually strongly contrasting against upper stipe. **STIPE:** 6–10 (12) cm long, 0.5–1.5 cm thick, tall and upright, straight to slightly curved, usually irregularly shaped and strongly grooved, base slightly enlarged. Bright yellow-orange, or at times dull beige-yellow, whitish at extreme base. Smooth. **FLESH:** Very thin and pliable in cap, stipe cartilaginous and entirely hollow. **ODOR:** Indistinct. **TASTE:** Indistinct. **SPORE DEPOSIT:** Whitish. **MICROSCOPY:** Spores 9–11 x 6–10 μm, ellipsoid, smooth.

ECOLOGY: Scattered in clumps, arcs, and large troops in needle duff and moss around stumps and woody debris in conifer forests. Very common on the North Coast, not recorded south of Sonoma County. Fruiting from winter into early spring.

EDIBILITY: Edible and good. It is usually abundant, easy to pick, and very clean.

COMMENTS: The small, brownish yellow, dimpled caps, shallow, decurrent, vein-like gills, tall, hollow, orange-yellow stipes, and growth in troops under conifers distinguish this species. *Cantharellus formosus* is somewhat similar in color and often found in the same areas, but is larger and has solid white flesh. *Chrysomphalina* and *Xeromphalina* species can look similar and also are found in large troops on woody debris, but have thin, solid stipes and deeper, thinner-edged gills. The true *C. tubaeformis* is a European species not known to occur in California; the western North American Yellowfoot represents a distinct species. Although the name *C. neotubaeformis* has been proposed, it hasn't been officially published, and there may be more than one species involved.

Craterellus atrocinereus
D. Arora & J. L. Frank
BLACK CHANTERELLE

CAP: 4–8 cm across, broadly convex but depressed at center, becoming nearly flat or uplifted with a central hollow running through to base of stipe. Ruffled and wavy around margin in age. Dark grayish black to grayish brown, in age paler grayish to beige-brown. Surface dry, innately fibrillose to matted-shaggy with a slightly ragged surface in age, often splitting. **UNDERSIDE:** With shallow, ridge-like decurrent "gills" that are strongly cross veined, often appearing netted. Pale grayish to darker gray, often with a bluish cast when young, more beige-brown in age or when dry. **STIPE:** 5–10 cm long, 1.5–4.5 cm thick (but with a large hollow). Often swollen toward base or highly irregular. Overall pallid grayish with a bluish cast when young, soon dark gray-brown to blackish especially where handled. **FLESH:** Thin, gray, somewhat brittle near margin of cap, tougher in stipe (especially at base where flesh is most solid). **ODOR:** Slightly fragrant, fruity-cheesy. **TASTE:** Mild or fruity to fragrant-cheesy. **SPORE DEPOSIT:** Whitish. **MICROSCOPY:** Spores 8–11 x 4.5–6 μm, ellipsoid, smooth.

ECOLOGY: Solitary or in small clumps under live oak or occasionally Tanoak. Mostly found around the Monterey Bay, with scattered records northward into Mendocino County. Uncommon. Fruiting from late winter through spring (sometimes quite late, into May).

EDIBILITY: An excellent edible, similar in flavor to *C. calicornucopioides*.

COMMENTS: *C. atrocinereus* is very similar to *C. calicornucopioides* in most respects, but the strongly veined fertile surface handily distinguishes this species. The development of the fruitbody is also quite different—this species lacks a tubular-cylindrical phase when young. It is occasionally confused with the Blue Chanterelle (*Polyozellus multiplex*), which has much thicker flesh and is intensely bluish purple overall. The Blue Chanterelle is quite rare in our area, occurring in the northern Coast Ranges.

MISAPPLIED NAME: *Craterellus cinereus* (Pers.) Pers.

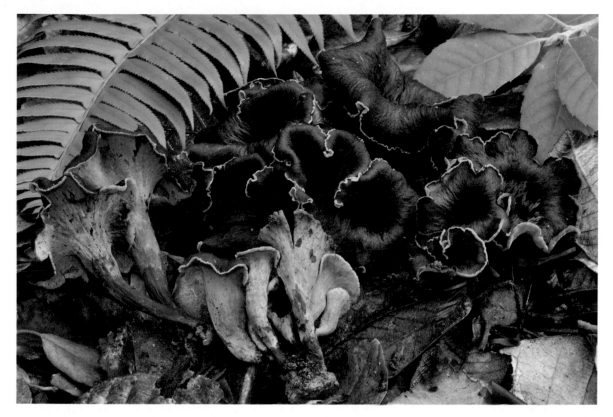

Craterellus calicornucopioides

D. Arora & J. L. Frank

BLACK TRUMPET

CAP: 3–10 (13) cm broad; fruitbody at first tubular, soon expanding to broadly convex to funnel shaped; center of cap deeply depressed, running into the hollow found through the entire trumpet-shaped fruitbody. Margin usually strongly wavy and ruffled, often splitting in age. Dark blackish gray to dark grayish brown to black, at times warm brown, in drier weather gray to beige-gray. Surface dry, innately fibrillose, with flattened matted-tomentose scales, more uplifted and ragged in age. **UNDERSIDE:** Smooth to slightly wrinkled, not distinct from stipe. Dark bluish gray, brownish gray, or light gray, with a pale sheen. **STIPE:** 4–15 cm long, 1–5 cm thick (but with a large central hollow), flesh usually 0.1–0.5 cm thick. At first cylindrical, soon trumpet shaped, straight to strongly sinuous and irregular, often grooved or compressed and irregular in cross section. Almost entirely hollow, flesh thickest and toughest at base. **ODOR:** Strong and fragrant, fruity or floral when fresh; more cheesy-pungent in age or when dried. **TASTE:** Mild. **SPORE DEPOSIT:** Off-white to pale creamy. **MICROSCOPY:** Spores 11–15 x 7–11 µm, ellipsoid, smooth.

ECOLOGY: Solitary or in tufts, tight clusters, or rosettes, often scattered in large troops or arcs in duff of Tanoak and live oak. Especially fond of steep Tanoak-covered hillsides with exposed mineral soil and moss, but also found in dense live oak groves with thick duff. Can be very difficult to spot amid leaf litter until you learn to recognize the ruffly outline surrounding the dark

center. Common from the Santa Cruz Mountains northward, with one southern record from Carmel Valley. Can fruit multiple times in one season and at any point from early winter into spring, but often with distinct peaks: late November into February or early March from Sonoma County north; mid-January through early April in areas south of Sonoma County.

EDIBILITY: Delicious, one of the best mushrooms for the table, with a strong, fragrant, earthy flavor when fresh. The fruitbodies rehydrate beautifully and can be kept dried for a very long time—the flavor becomes richer, cheesier, and almost trufflelike; some people prefer to dry all their collections.

COMMENTS: This well-loved species is one of the best mushrooms for beginning identifiers to learn. It can be very abundant, is exceptionally tasty, and has no toxic look-alikes. The hollow trumpet-shaped fruitbodies and dark colors distinguish it from all but *C. atrocinereus*, which has a strongly veined fertile surface and is not tubular when young; it's an equally good edible, but far less common. *C. atrocinereus* is sometimes found intermixed with *C. calicornucopioides*, but primarily grows under Coast Live Oak in spring. A cream to bright yellow form of *C. calicornucopioides* is occasionally found intermixed with normal-colored fruitbodies. Some patches even produce half-and-half fruitings including both dark and yellow fruitbodies!

MISAPPLIED NAME: *Craterellus cornucopioides* (L.) Pers.

Gomphus clavatus

(Pers.) Gray

PIG'S EAR

CAP: 4–20 cm across, flat when young, soon becoming uplifted, usually with overlapping fanlike lobes forming rosettes. Margin thin, wavy, ruffled. Brownish at first, often with moldy-looking pale lilac patches near center, soon dingy tan, olive, ocher-olive, beige when dry or in age. Finely matted at first, velvety to smooth in age, moist to dry. **UNDERSIDE:** Made of ridges running to base of fruitbody; ridges often poorly defined or crinkly, blunt edged, and with many cross veins. Beautiful purple to lavender at first, fading to light lilac-gray or dingy beige in age and often dusted with ocher-tan spores. **STIPE:** 5–10 (15) cm long, 2–4 cm thick (up to 8 cm thick where fruitbodies are fused), wedge shaped, tapered to base, usually irregularly branched in fused clusters. Mostly covered in gill tissue. Lower part a bit velvety or smooth, deep royal purple to amethyst near base; extreme base white and often with a few, pointed spikes (undeveloped fruitbodies). **FLESH:** Thin to thick, fibrous, solid-stringy, whitish to dingy. **ODOR:** Indistinct. **TASTE:** Indistinct. **SPORE DEPOSIT:** Ocher-tan. **MICROSCOPY:** Spores 10–15 x 4–7.5 µm, elongate-ellipsoid, ornamented with discrete warts.

ECOLOGY: In fused clusters, often in troops or forming arcs and/or large rings in deep duff of conifer forests. Common on the North Coast, occurring south to Sonoma County. Fruiting from fall into winter.

EDIBILITY: Edible. Some people consider it a delicacy, while others find it bland and not worth collecting. Young specimens are more tender.

COMMENTS: This unique species can be recognized by its purple-veined or ridged underside and tan to olive-beige cap, and fruitbodies that fruit in clumps and rosettes. It is practically unmistakable; a few *Thelephora* can be brownish purple, but are much thinner fleshed. *Polyozellus multiplex* is only vaguely similar—even faded specimens are much deeper royal blue to purple-black. Although *G. clavatus* is sometimes marketed under the name Violet Chanterelle, it is more closely related to the coral fungi in the genus *Ramaria*.

Turbinellus floccosus
(Schwein.) Earle ex Giachini & Castellano
WOOLLY FALSE CHANTERELLE

CAP: 5-15 (20) cm across, extremely young fruitbodies shaped like pointed cylinders but usually found after flattening to a truncate cylinder, then becoming deeply depressed at center, tall and vase shaped at maturity. Margin thin and slightly wavy. Dark red to reddish orange when young, soon orange, margin often bright yellow-orange, sometimes duller orange-beige. Surface slightly viscid when wet, otherwise moist to dry, with large matted-tomentose scales, breaking into concentric rings of curved scales around central hollow, more matted near margin. **UNDERSIDE:** With deeply decurrent ridges, crinkly looking, branched, and extensively cross veined. Pale creamy to light beige, occasionally bright yellow when very young. **STIPE:** 8–16 cm long, 3.5–5 cm thick, vase shaped, tapered toward base, usually quite upright, but sometimes with a sinuous lower part. Creamy to beige or with yellow-orange tones, with dingy vinaceous brown areas in age. Smooth to fibrillose. **FLESH:** Thin in cap, stipe thick, solid, fibrous, white. **ODOR:** Indistinct. **TASTE:** Indistinct. **SPORE DEPOSIT:** Pale ocher. **MICROSCOPY:** Spores 11–20.5 x 6–10 µm, elongate-ellipsoid. With broad, fused warts up to 0.5 µm tall.

ECOLOGY: Occasionally solitary, but more often in fused clumps or clusters (often with small pointed fruitbodies at base), in scattered troops, arcs, and rings in duff in mixed evergreen forests. Very common in fall and early winter from Sonoma County northward.

EDIBILITY: Not recommended; it causes gastrointestinal distress in some people.

COMMENTS: This bizarre mushroom is instantly recognizable by its bright red to orange, shaggy-scaly cap, tall vase-shaped fruitbodies, and wrinkled veiny fertile surface. Faded specimens can be confused with *T. kauffmanii*, but that species is usually pale beige to tan, with more dramatic blocks of scales on the cap. True chanterelles have smooth, paler orange to yellow caps and are not vase shaped. *Turbinellus* are related to the Pig's Ear (*Gomphus clavatus*), coral fungi in the genus *Ramaria*, and the hypogeous *Gautieria*; they are not closely related to true chanterelles (*Cantharellus* spp.).

SYNONYM: *Gomphus floccosus* (Schwein.) Singer.

Turbinellus kauffmanii
(A. H. Sm.) Giachini

CAP: 5–20 (30) cm across, fruitbody cylindrical with a flat or slightly depressed cap when young, soon developing a deeply depressed center and becoming broadly convex and then uplifted to vase shaped. Margin thin, often a bit wavy in age. Dull beige to tan, sometimes whiter when young. Surface dry, with concentric rings of blocky to flat, pointed scales, those deeper in sunken center of cap much larger and often curled back with tips touching, those near margin more flattened and turning brownish. Older specimens can lose the pronounced upright-blocky scaly look. **UNDERSIDE:** With crinkly looking deeply decurrent ridges that are branched, and extensively cross veined. Whitish when very young, soon creamy beige to tan. **STIPE:** 7–15 cm long, 2–5 cm thick, cylindrical or with a tapered or club-shaped base. Almost entirely covered in gill tissue, pale beige to tan. **FLESH:** Thin in cap. Stipe thick, firm, fibrous. Whitish, or with orangish tan stains in age. **ODOR:** Indistinct. **TASTE:** Mild. **SPORE DEPOSIT:** Ocher-buff to apricot. **MICROSCOPY:** Spores 12.5–17.5 x 6–8 µm, elongate-ellipsoid, ornamented with short warts and branched ridges.

ECOLOGY: Solitary or more often in clumps in needle duff of conifers. Uncommon in our area, mostly encountered on the Far North Coast, with a few records south to Mendocino County. Fruiting in late fall and early winter.

EDIBILITY: Unknown, likely causing gastric upset like *T. floccosus*.

COMMENTS: The large, vase-shaped fruitbody, beige-tan to light brownish cap with dramatic blocky, upright scales, and pallid veiny "gills" distinguish this species. Specimens that have become more matted-scaly could be confused for faded *T. floccosus*, but that species usually has traces of reddish orange colors somewhere on the cap.

SYNONYM: *Gomphus kauffmanii* (A. H. Sm.) R. H. Petersen.

2 • Amanita

Amanita contains some of the most beautiful, most conspicuous, and best-known mushrooms in the world. With a little practice, it's usually easy to recognize *Amanita* in the field. Although they can be very different from one another, a difficult-to-define *Amanita*-ness unites them. They often have universal veil remnants on the cap (as warts or a skull patch) and/or at the base of the stipe (as a volva, cup, or scaly rings). In most species (except for those in the section *Vaginatae*) a partial veil is present, at least when young. Almost all produce white spore prints, and almost all species are ectomycorrhizal (only one species on the Redwood Coast is not). *Amanita* is equally famous and notorious for containing deadly toxic species, delicious edibles, and psychotropic species.

Important identification characters include overall coloration and stature, presence and distribution of universal and partial veil tissues, presence or absence of grooves (sulcations) around the cap edge, shape of the base of the stipe, and identity of nearby trees. The size, shape, and amyloid reactions of the spores, as well as the presence or absence of clamp connections, are the most important microscopic features.

The genus *Amanita* has seven major sections, six of which are common in our area. Familiarizing yourself with these subgroups will make the job of species-level identification much easier. As a bonus, they represent fairly accurate "natural" groupings of evolutionarily related taxa.

Section *Amanita*—The iconic *Amanita muscaria* is in this section. Most have strongly warted and intensely colored caps. Volva tissue is usually concentrically ringed and scaly or tight and cup- or collarlike around a swollen stipe base. Most species in this group are toxic, often containing ibotenic acid and/or muscarine. Spores not amyloid.

Section *Validae*—Fruitbodies variably staining reddish or brownish. Partial veil prominent; volva usually weakly developed. Spores amyloid.

Section *Lepidella*—Fruitbodies often large, usually entirely white and without any contrasting pigments, but often discoloring yellowish in age. Universal and partial veil tissues poorly developed, often powdery and messy. Spores amyloid. One odd species in our area has small fruitbodies and is found in grass.

Section *Phalloideae*—This section includes the deadliest mushrooms in the genus: those that contain amatoxins, including the Death Cap (*Amanita phalloides*). Cap usually bald and not sulcate, volva usually saclike, base of stipe often strongly swollen bulbed. Spores amyloid.

Section *Caesareae*—Large, robust fruitbodies with striate-sulcate cap margins, hollow or pith-filled stipes, and a saclike volva at base of stipe (which is not strongly swollen or bulbous). Spores not amyloid.

Section *Vaginatae*—Edges of caps sulcate, caps often some shade of gray to brown (although some are ocher-tan, whitish, or pink). Partial veil absent or very rudimentary; volva and universal veil tissue sheathing and sometimes staining rusty orange. Spores not amyloid.

Section *Amidella*—Very rare in our area. Usually showing a saclike volva and hanging tissue around the margin of the cap, and often bruising reddish or pinkish when cut. Spores amyloid.

Amanita muscaria subsp. *flavivolvata*

Singer

FLY AGARIC

CAP: 5–20 cm across, round to tightly convex when young, becoming broadly convex, plane, or occasionally uplifted and/or wavy in age. Typical caps blood red to cherry red, sometimes duller when young. Pigments in the cap are water soluble and light sensitive, so exposed caps often become more orange, or in more extreme cases yellow or even whitish. In general, cap color remains darker and redder near the center. Surface slightly viscid to dry, covered in coarse, discrete off-white to yellowish universal veil warts; these warts are like oatmeal flakes, or sometimes rather solid and angular. Margin distinctly sulcate in age, obscurely so when young, and then often covered by universal veil warts congregated at margin of cap. **GILLS:** Finely attached (less commonly free), close to crowded. White at first, becoming creamy white or yellowish in age. Edges of gills covered in dense, pointed fuzz when very young and tightly packed, and a fine ragged layer of cells in age. **STIPE:** 7–20 (25) cm long, 1.5–4 cm thick at apex, but the swollen base may be substantially wider. Very straight and usually cylindrical over most of its length, gradually enlarged toward bulb. Bulb usually distinctly swollen and egg shaped or tapered downward. Upper stipe (above partial veil) usually white and smooth, lower stipe creamy white to yellowish, surface often bumpy or ragged. **PARTIAL VEIL:** Present as a whitish, membranous hanging skirt, usually with adhering patches of yellow universal veil tissue around

edge. **VOLVA:** Whitish to yellow concentric rings of square or pointed scales on lower stipe. **FLESH:** Firm to soft, whitish; stipe solid. **ODOR:** Indistinct. **TASTE:** Mild or slightly sweet. **SPORE DEPOSIT:** White. **MICROSCOPY:** 9–13 x 6.5–8.5 μm, broadly ellipsoid, occasionally rounder or more elongate, inamyloid. Basidia clamped.

ECOLOGY: Solitary or in clumps, in small to large arcs, rings, or scattered in troops. Generally under pine and spruce, but also found with oaks and even some exotic ornamentals like *Arbutus unedo* and *Eucalyptus globulus*. Very common throughout our range. Usually fruiting in a flush from midfall to midwinter, although scattered fruitings may occur into spring. The highly conspicuous fruitbodies of this species are often used as an indicator of King Bolete (*Boletus edulis*) habitat. There's also an association between this mushroom and the Peppery Boletes (*Chalciporus* spp.), a pattern that has held true in every place it has been looked for (including New Zealand, where neither species is native!). The exact nature of this relationship remains unclear.

EDIBILITY: This species can be considered extremely toxic, mind altering, or a delectable edible. If eaten without special preparation, it is quite toxic—muscimol being the primary compound that causes sweating, nausea and vomiting, diarrhea, and psychotropic alterations ranging from delirium to comalike sleep (often accompanied by vivid dreams). However, if this mushroom is detoxified properly, it can be made into a tasty edible. The procedure involves peeling the cap cuticle and boiling in water twice for seven

minutes (the toxin is water soluble, so discard the water after each round) The mushrooms can then be sautéed and used in normal recipes; most people find them delicious when prepared this way. *Caution is warranted!* (Please read Rubel and Arora's article, available at davidarora.com/uploads /rubel_arora_muscaria_revised.pdf.)

COMMENTS: The bright red cap with white patches and concentric, scaly rings of volval tissue distinguish the most common form of this species from all others. The nominate subspecies of *A. muscaria* is native to Eurasia, but may occur rarely in our area with planted trees; its veil tends to be brighter and purer white, not yellowish. Other variants of *A. muscaria* in our area include the yellow-capped form, often called *A. muscaria* var. *formosa* (lower left), which is similar in all respects except for the lemon yellow to light yellow cap color, and the white variety (*A. muscaria* var. *alba*) with an eggshell white or creamy white cap (lower right). *A. aprica* has a small cuplike volva and a nonsulcate, bright yellow to peachy orange cap (never truly red). Pale members of the *A. gemmata* complex are sometimes confused for the rarer yellow form of *A. muscaria*, but they have a cuplike volva and often are more beige-tan, with less yellow pigment.

Host Specificity

Many mushrooms are involved in reciprocally beneficial symbiotic relationships called mycorrhizae, in which sugars produced by a plant are exuded from its roots and absorbed by a partnered fungus, whose mycelium is tightly wrapped around the plant's root tips. Meanwhile, the plant partner gains access (through its root tips) to some of the water and soil nutrients aggregated by the mycelium of the fungus. There has been much co-evolution between plants and their mycorrhizal fungi, leading to patterns of host specificity. Some fungi are extremely specific to one plant partner, or a few closely related species: *Rhizopogon* and *Suillus* are specialists, being almost entirely restricted to partnerships with members of the pine family. On the contrary, *Amanita phalloides* is a generalist, and can form partnerships with a wide range of trees including both hardwoods and conifers. This pattern has many implications for ecology (how are forest tree communities assembled?) as well as for evolution (differentiation of new species of mycorrhizal fungi can be driven by adaptation to new hosts). Some native mycorrhizal fungi have begun to "jump" to exotic mycorrhizal trees in our area (such as *Eucalyptus*). This is a fascinating time to watch inchoate ecological relationships (and potentially new evolutionary lineages) develop.

Amanita aprica

J. Lindgr. & Tulloss

SUNSHINE AMANITA

CAP: 3–12 cm across, rounded to convex when young, becoming broadly convex to flat, or occasionally uplifted in age. Color varying from bright citrine yellow to apricot orange to nearly peach; usually darkest at center, fading in age. Surface slightly viscid to dry, smooth, covered in universal veil tissue that is often present as a frosty-looking layer. **GILLS:** Narrowly attached to free, often with a fine line of tissue descending the upper stipe at the base of each gill, close to very crowded, edges often with veil tissue. White, becoming creamy in age. **STIPE:** 5–10 cm long, 1.5–3 cm thick at apex, 1.5–6 cm thick at base. Usually relatively short, cylindrical to slightly enlarged or bulbous and swollen toward base. In most cases noticeably stocky (especially for an *Amanita*). White. **PARTIAL VEIL:** A hanging, membranous, white skirt, usually quite thin and ragged, often disappearing entirely in age. **VOLVA:** A small, collared cup held tightly to the base, often with a few concentric rings of tissue above. **FLESH:** Firm to soft, fibrous in stipe, white. **ODOR:** Indistinct. **TASTE:** Indistinct. **SPORE DEPOSIT:** White. **MICROSCOPY:** 9.5–13 × 6.5–8.5 µm ellipsoid or elongated, inamyloid. Basidia rarely clamped.

ECOLOGY: Sporadic fruiter in small numbers on coast, occurring primarily in fall under live oak and Madrone in the southern part of our range and in spruce-fir woods in spring on the Far North Coast.

EDIBILITY: Poisonous, containing muscarine and ibotenic acid.

COMMENTS: The bright yellow to orange cap with frosty universal veil tissue and the stocky stature help distinguish this species. Some forms of *A. muscaria* could cause confusion, but are taller and more gracile, and have more distinct warts and many concentric rings of pointed scales above the stipe base. Members of the *A. gemmata* group usually have duller caps and more discrete warts (not frosty looking). The typical yellow form of this species is more common during spring in the Sierra Nevada. There's a distinct possibility that two species are involved. A much smaller but similar undescribed species with a yellow-orange cap and indistinct partial veil grows under manzanita at scattered locations around the San Francisco Bay Area.

Amanita "pseudobreckonii"

N. Siegel & C. F. Schwarz nom. prov.

CAP: 3–8 cm across, egg shaped at first, becoming convex to flat. Margin finely grooved (sometimes obscure). Pale straw yellow with a pinkish tan tone, sometimes more peachy or cool yellow when young. Surface viscid to dry, irregularly covered in relatively thick, white patches of flocculent-membranous universal veil tissue, often smooth in age. **GILLS:** Narrowly attached to free, close to crowded, edges often finely frosted when young. White in color. **STIPE:** 6–15 cm long, 0.8–2 cm thick at apex, club shaped, gradually enlarged downward to swollen but relatively small basal bulb. White to pale creamy white. Surface dry, smooth above partial veil, finely flocculent below. **PARTIAL VEIL:** Indistinct or absent, however it has a membranous, sheathing collar of universal veil tissue that often resembles a partial veil; becomes an upward-flaring ring of tissue set very low on stipe, breaking up in age. **VOLVA:** Leaving a short rimlike collar of tissue around the top of the bulb when young, and/or loose patches around the stipe base. **FLESH:** Soft, fibrous, and pith filled in stipe, creamy white with yellowish under cap skin. **ODOR:** Indistinct. **TASTE:** Mild. **SPORE DEPOSIT:** White. **MICROSCOPY:** Spores 9.5–12 × 8.5–10 µm, subglobose, inamyloid. Basidia rarely clamped.

ECOLOGY: Solitary or scattered in small groups in duff under conifers, especially Sitka Spruce. Very common on the Far North Coast, uncommon elsewhere. Often fruiting in large flushes in early fall and spring, with scattered fruitings through winter.

EDIBILITY: Likely toxic.

COMMENTS: The color is a distinctive but difficult-to-describe mix of pale straw yellow with pinkish tan or peachy tones. The thick, flocculent-membranous universal veil remnants and the tall, slender stipe are helpful in identification. It has been mistakenly called *A. breckonii* (a supposedly similar species that differs in having long ellipsoid spores and clamped basidia). There's no other way to put it—the gemmatoid *Amanita* are a taxonomic mess in California (see the comments section of *A. gemmata*), with a lot of work needed to sort out and describe the species.

MISAPPLIED NAMES: *Amanita gemmata* var. *exannulata*.

Amanita gemmata group
GEMMED AMANITA

CAP: 5–12 cm across, rounded to convex when young, becoming broadly convex to flat. Margin finely sulcate, especially in age. Color variable, warm yellow to yellowish orange, creamy yellow to cooler yellowish beige. Surface slightly viscid to dry, smooth, covered with variable amounts of small, pale warts or patches of flat, felty, or frosted universal veil tissue. These wipe off easily, especially in wet weather, leaving the cap bare at times. **GILLS:** Narrowly to broadly attached, rarely truly free, close to crowded. White to pale creamy white. **STIPE:** 5–11 cm long, 1–2 cm thick at apex, gradually enlarged downward to swollen or bulbous base. White to pale creamy white. Surface dry, smooth above the partial veil, with fine tufts or scales or smooth below. **PARTIAL VEIL:** A thick, membranous, skirtlike ring on the upper stipe, often falling off. Generally white in color, but often has thicker, creamy beige patches around lower edge. **VOLVA:** A rimmed or short-collared membranous cup around bulbous stipe base. **FLESH:** Firm to soft, fleshy-fibrous, white. **ODOR:** Indistinct. **TASTE:** Indistinct. **SPORE DEPOSIT:** White. **MICROSCOPY:** The typical *gemmata* described above has inamyloid spores 9–13 x 7–8 (10.5) μm that are ellipsoid to elongate, but there is much variation within the group.

ECOLOGY: Solitary or scattered in small troops. Common, especially in late fall and early winter. Growing with various conifers in the northern part of our range, often with pine and live oak southward.

EDIBILITY: Poisonous.

COMMENTS: The creamy yellow to tan cap, whitish warts, skirt-like ring on the stipe, and rimmed bulb help distinguish this group of *Amanita*. This is a very confusing species complex: The true *A. gemmata* does not occur in California. It's unclear how much overlap there is between some of the *gemmata* and *pantherina* taxa in our area, and there appear to be at least four undescribed species in the *gemmata* complex. The most distinct is *A. "pseudobreckonii"*, with a pale straw yellow cap with pinkish tan or peachy tones and a very indistinct or absent partial veil. *A. breckonii* is a poorly understood species with elongate spores (Q > 1.5), described from under Monterey Pine. Yellow forms of *A. muscaria* are generally much warmer colored and have more turnip-shaped bulbs, with multiple scaly rings of tissue on the lower stipe.

Amanita pantherina group
THE PANTHER

CAP: 5–10 cm across, rounded to convex when young, becoming broadly convex to flat. Margin finely sulcate, especially in age. Color variable, often warm brown to beige, but can be anywhere from dark brown to creamy tan or pale creamy beige, often darker on the disc, and fading in age. Surface slightly viscid to dry, smooth, covered with variable amounts of small, pale warts or patches of flat, felty, or frosted universal veil tissue. These wipe off easily, especially in wet weather, leaving the cap bare at times. **GILLS:** Narrowly to broadly attached, often with a fine line descending the stipe at each gill, rarely truly free, close to crowded. White to pale creamy white. **STIPE:** 5–11 cm long, 1–2 cm thick at apex, gradually enlarged downward to a club-shaped base with a collarlike rim. White to pale creamy white. Surface dry, smooth above the partial veil, with fine scales or smooth below. **PARTIAL VEIL:** A membranous, skirtlike ring on the upper stipe. White, but often with thicker, tan patches around the lower edge. **VOLVA:** A short-collared membranous rim of tissue around top of bulb at stipe base. **FLESH:** Firm to soft, fibrous, white. **ODOR:** Indistinct. **TASTE:** Indistinct. **SPORE DEPOSIT:** White. **MICROSCOPY:** Spores 9–14 (16) x 5.5–9 (11) μm, ellipsoid to elongate, inamyloid.

ECOLOGY: Solitary or in small groups under hardwoods and conifers throughout our range. Fruiting through the wet season, but more abundant in late winter and spring.

EDIBILITY: Poisonous; contains similar toxins as *Amanita muscaria* subsp. *flavivolvata* but in higher concentrations.

COMMENTS: The brown to beige cap, whitish warts, skirtlike ring on the stipe, and the rimmed or collared bulb help identify The Panther. *A. gemmata* is similar in many respects, except for its lighter cap color. However, pale specimens of *A. pantherina* and dark specimens of *A. gemmata* can be nearly impossible to distinguish.

MISAPPLIED NAME: *A. pantherina* is a European species that doesn't occur in North America. Rod Tulloss has been using *A. ameripanthera* for at least one western North American Panther, but the name hasn't been formally published yet.

Amanita augusta

Bojantchev & R. M. Davis

WESTERN YELLOW-VEIL

CAP: 5–15 cm across, round when young, convex to flat in age. Cap color quite variable, from very dark brown to bright lemon yellow; older fruitbodies can become dull beige-brown or tan. Typical mature caps show a warm brown center, become golden brown outward, and are yellow at margin. Surface slightly viscid to greasy or dry, partially covered in fluffy or dandruffy universal veil patches (sometimes more solid and upright or pyramidal) that are easily removed by rain or handling. **GILLS:** Finely to broadly attached (rarely free), close to crowded. White or creamy ivory at first, soon with yellow tones. **STIPE:** 8–15 cm long, 2–3 cm thick at apex, swollen or enlarged toward base. Base distinctly enlarged (but not abruptly bulbous) and also tapered downward, resulting in a spindle-shaped appearance. Creamy white, but lower stipe covered with many yellow or gray, pointed, curved scales. Often developing reddish stains at base. **PARTIAL VEIL:** A creamy white membranous skirt, usually with a ring of yellow universal veil scales adhering to edge. **VOLVA:** Sometimes very indistinct, otherwise appearing as concentric rings of flat scales, often as a fused "sheet" that detaches from base of stipe and remains in soil. **FLESH:** Firm to soft, whitish. Bruising slowly dull reddish, especially around base of stipe. **ODOR:** Indistinct. **TASTE:** Indistinct. **SPORE DEPOSIT:** White. **MICROSCOPY:** Spores 8.5–9.5 × 6–7 μm, broadly ellipsoid to ellipsoid, amyloid. Basidia not clamped at base.

ECOLOGY: Solitary or in small groups in a variety of habitats. Very common in early fall under Sitka Spruce in the northern part of our range, then with pine and oak in early fall through midwinter and with occasional spring fruitings farther south.

EDIBILITY: Edible but not recommended. Local experience with this species is growing, but still quite recent. If consumed, it should always be thoroughly cooked. Be especially cautious not to confuse with members of the *A. gemmata* and *A. pantherina* species complexes, which are toxic.

COMMENTS: The cap color ranging from dark brown to bright yellow, creamy gills, skirtlike partial veil, yellowish universal veil remnants, and reddish staining at the stipe base are distinctive. Few other mushrooms approach this species in appearance. *A. gemmata* and its brethren might cause confusion, but they are rarely as richly colored and have white (not yellow) universal veil patches.

MISAPPLIED NAMES: *Amanita franchetii* (Boud.) Fayod and *Amanita aspera.*

Amanita novinupta

Tulloss & J. Lindgr

BLUSHING BRIDE

CAP: 5–15 cm across, rounded or tightly convex when young, becoming broadly convex to flat in age. Usually white or off-white with pinkish blotches, often becoming pinkish or extensively brick red overall in age. Surface dry to slightly viscid, covered in white warts. **GILLS:** Finely to broadly attached (occasionally seeding), often with a fine line of tissue descending upper stipe at base of each gill, close to crowded. White at first, becoming dull beige in age; at all stages may have blotches of pinkish coloration. **STIPE:** 6–12 cm long, 1.5–2 cm thick at apex, swollen base wider (up to 4 cm). Stipe tapered upward, the base distinctly enlarged and usually also tapered downward. Lower stipe often with bands of scales, white to pinkish with redder tones between bands; upper stipe (above partial veil) paler white. **PARTIAL VEIL:** A membranous skirt, usually vertically striate and bright white. **VOLVA:** Usually indistinct, often present only as an obscure pigmented line, sometimes more apparent and then appearing as poorly differentiated rings or even a simple cup. **FLESH:** Pale whitish, bruising pink or dull reddish very slowly, then becoming dark, dingy reddish brown (especially visible in larva tunnels). **ODOR:** Usually indistinct. **TASTE:** Mild. **SPORE DEPOSIT:** White. **MICROSCOPY:** Spores 8–11 x 5.5–7.5 µm, ellipsoid or slightly elongate, amyloid. Basidia usually unclamped.

ECOLOGY: Solitary or in small groups under live oak throughout much of our range. Primarily fruiting in spring, but also found regularly in fall. Often grows in same areas as *A. ocreata* and *A. velosa* in spring. An undescribed species that has been mistakenly called *A. novinupta* occurs with Sitka Spruce on the Far North Coast.

EDIBILITY: Edible. Take caution not to confuse with sunburned specimens of the deadly *A. ocreata*. Local experience gathering this mushroom for the table is limited, but folks who have offer mixed reviews: some people find it to be an excellent edible, while others are unenthused.

COMMENTS: The combination of warts on the cap, membranous partial veil, and the overall whitish color that becomes predominantly pinkish to reddish in age helps identify this species. There is genetic evidence to suggest that a few (undescribed) similar species are going under the name *A. novinupta*, but consistent features to separate them are still completely unresearched. Both *A. ocreata* and *A. velosa* have much more pronounced, sac-like volvas; additionally, *A. ocreata* does not have cap scales or warts, while *A. velosa* usually has a single, contiguous patch of universal veil tissue on the cap. A poorly known *Amanita* in Section *Amidella* occurs very rarely around Santa Cruz County and northern Monterey County; it is also whitish and develops pinkish to reddish stains on the cap in age, but differs in having a large, saclike volva. Also compare with the similar *A. "sponsus"* (p. 40).

Amanita "sponsus"

Tulloss, C. F. Schwarz & N. Siegel nom. prov.

THE GROOM

CAP: 5–15 cm across, rounded and tightly convex when young, becoming broadly convex and then flat in age. Pure white at first, warts soon becoming brown to dark brown or reddish brown, surface between warts becoming eggshell white to beige. Surface dry, covered in low warts that are at first upright but soon flat and very thin, and tightly bound to the cap surface. **GILLS:** Finely to broadly attached (rarely truly free), almost always with a fine line of tissue descending upper stipe at base of each gill, close. White when young, becoming dull cream or pale beige in age. **STIPE:** 5–10 cm long, 1.5–2 cm thick at apex. Base may be substantially wider (up to 4.5 cm). Base moderately swollen, club shaped; extreme base often tapered and thus spindle shaped overall, occasionally with a chiseled-looking cleft. Upper stipe whitish, with distinct bands of chevrons as the surface tissue breaks up; lower stipe white, smooth, with scattered scales or dandruff. **PARTIAL VEIL:** A membranous skirt, usually vertically striate and bright white. **VOLVA:** Indistinct, usually present only as a few rings around the top of the stipe base. **FLESH:** Firm to quite soft, whitish, slowly bruising brownish. **ODOR:** Indistinct or slightly unpleasant. **TASTE:** Indistinct. **SPORE DEPOSIT:** White. **MICROSCOPY:** Spores 8–10.5 x 5–7 μm, ellipsoid to elongate, amyloid. Basidia usually unclamped.

ECOLOGY: Scattered or in small groups under Coast Live Oak, occurring in fall and spring, infrequent in winter. Common in southern California, occurring north to the San Francisco Bay Area, common in Santa Cruz County.

EDIBILITY: Unknown; do not experiment.

COMMENTS: The overall white color with diffuse brown tones in age, chevrons of broken tissue above the grooved partial veil, and extremely obscure volva are distinctive. This species is similar to *A. novinupta*, but never develops a smooth cap (the warts are persistent, actually intergrown with the cap surface), and the cap and stipe stain brownish (not dusky rose red or pinkish) in age. *A. pruittii* is much smaller, grows in open grassy meadows, and differs in microscopic details. *A. silvicola* tends to remain pure white, grows with conifers, and has more flocculent veil tissue. The lack of a saclike volva may also lead to confusion with the rare *Leucoagaricus amanitoides*, which differs by its squatter stature, smoother cap, and different microscopic features.

Amanita magniverrucata

Thiers & Ammirati

PINECONE AMANITA

CAP: 5–15 (20) cm across, rounded to tightly convex when young, becoming broadly convex to flat in age. Pure white when young, becoming off-white and developing creamy yellow or even limited tan or brownish tones in age. Surface dry, covered in very large white warts that are generally pyramidal or at least pointed (less sharply defined if very old or rain beaten), arranged in more or less concentric rings, usually largest and most upright at center of cap. Surface between warts sometimes smooth in age. **GILLS:** Finely attached (rarely free), close to crowded. Eggshell white at first, becoming creamy to pale pinkish white, may develop strong yellowish cream or tan tones in age. **STIPE:** 7.5–15 cm long, 2–3.5 cm thick at apex, much larger (up to 6 cm) near base, straight to slightly curved, distinctly swollen and then tapered into soil (turnip shaped) at extreme base. Eggshell white to slightly creamy. Surface dry, slightly shaggy or with floccose scales. **PARTIAL VEIL:** A ragged, disintegrating ring on upper stipe, often very obscure in fully grown fruitbodies. **VOLVA:** 4–6 concentric rings of strikingly pointed, curved scales around swollen base of stipe. **FLESH:** Thick, firm, whitish. Stipe solid. **ODOR:** Usually indistinct when young, becoming unpleasant in age. **TASTE:** Not sampled. **SPORE DEPOSIT:** White. **MICROSCOPY:** Spores 8–12.5 x 6–8 μm, ellipsoid to elongate, amyloid. Basidia sometimes clamped.

ECOLOGY: Solitary or scattered in small groups, often fruiting in fall and early winter, but scattered fruitings continue into spring. Occurring from Humboldt County south, but uncommon through much of our range; more common in southern California. Occurs with pine and oak.

EDIBILITY: Many *Lepidella* are seriously toxic.

COMMENTS: The huge, upright, often pyramidal warts covering the cap and curved rings of scales around the base of the stipe distinguish this species from other white *Amanita* in our area. Only exceptionally weathered fruitbodies are problematic and could possibly be mistaken for *A. smithiana*, *A. silvicola*, or even the *Amanita* 'Sandhills'. However, those species don't have the persistent curved and pointed scales around the stipe base

Amanita silvicola

Kauffman

WOODLAND AMANITA

CAP: 3.5–10 cm across, rounded when young, becoming broadly convex to nearly plane. Evenly white colored, slightly grayish or creamy in age. Surface dry, cottony when young, soon matted, rarely truly smooth. Sometimes with discernible low warts of matted fibrils. Margin with an often ragged overhanging band of sterile tissue. **GILLS:** Finely to more broadly attached, close, edge finely fringed. White, becoming creamy in age. **STIPE:** 4–12 cm long, 1.5–3.5 cm thick at apex, up to 5 cm at swollen base. Cylindrical with a distinctly enlarged club-shaped or truncate bulbous base, rarely with a noticeable rim. White to creamy overall. Surface dry, floccose, with cottony chevrons as it elongates. **PARTIAL VEIL:** Fragile, powdery-cottony tissue usually clinging to cap margin or as a ragged ring zone on stipe. **VOLVA:** Usually indistinct. **FLESH:** Soft, moderately thick. Whitish to cream. **ODOR:** Indistinct at first, unpleasant in age. **TASTE:** Indistinct. **SPORE DEPOSIT:** White. **MICROSCOPY:** Spores 7–10 (12.5) x 4–6 (8.5) μm, ellipsoid, amyloid. Basidia lacking clamps.

ECOLOGY: Solitary or in small groups, occasionally clustered in coniferous forest, especially with Western Hemlock, but also with Sitka Spruce and Douglas-fir; less often with pine or Tanoak. Common in the northern part of our range, uncommon south of Sonoma County but occurring south to Santa Cruz County. Fruiting from fall into early winter.

EDIBILITY: Likely toxic.

COMMENTS: This species can be told by its smallish size, short and stout stipes with club-shaped or truncate (not rooting) stipe bases, and flocculent to matted cap surfaces (never smooth, even in age). Perhaps most similar is the *Amanita* 'Sandhills', found only in the Santa Cruz Sandhills, which is much larger and has a more tapered stipe base. Also frequently confused with *A. smithiana*, which has a strongly tapered, often rooting stipe base (difficult to see if not carefully removed from soil), smooth areas between cap warts in age, and often more pinkish tones to the cap; it is usually substantially larger as well.

Amanita smithiana

Bas

SMITH'S AMANITA

CAP: 5–15 cm across, rounded to tightly convex when young, becoming broadly convex to flat. Pure white when young, often developing brownish or ocher-brown streaks or fibrils outward in age, especially when wet. Surface dry to moist, with poorly defined, radially arranged fibrillose scales or small, fluffy warts. Often with a very smooth cap with a metallic silvery gray color in dry weather. **GILLS:** Finely to broadly attached. Close to crowded, short gills present and often truncate. Pure white when young, becoming dull beige-yellow in age. **STIPE:** 6–20 (30) cm long, 2–3.5 cm thick at apex, enlarging toward a swollen, bulbous base. Bulb with a swollen upper portion and a long, tapered "root" that is almost always broken off when collected. White to creamy white. Surface scruffy, covered with flocculent scales. **PARTIAL VEIL:** Usually remaining attached to cap margin, with some ragged remnants usually visible on upper stipe, very powdery. **VOLVA:** Indistinct, usually appearing as bands of irregular, ragged, white to creamy scales on lower stipe. **FLESH:** Thick, firm, whitish. **ODOR:** Usually indistinct to fishy or foul at maturity, reportedly to occasionally smell sweet-spicy (like cinnamon or Matsutake) when young. **TASTE:** Not sampled. **SPORE DEPOSIT:** White. **MICROSCOPY:** Spores 8.5–12 x 5.5–8 μm, broadly ellipsoid or elongate, amyloid. Basidia clamped.

ECOLOGY: Solitary or in small groups under conifers and Tanoak. Fairly infrequent in the southern part of our range, but quite common in Douglas-fir, Tanoak, and Western Hemlock forests of the north. Fruiting primarily in fall, but also in winter.

EDIBILITY: Extremely toxic, and potentially deadly! Causes severe gastrointestinal distress, followed by possible kidney failure.

COMMENTS: This *Amanita* can be recognized by its tall stature, bulbous but strongly tapering base, white color, and ragged veil remnants ornamenting most of the fruitbody. The definition between the smooth cap cuticle and the powdery universal veil warts is more distinct than in many other *Amanita* in Section *Lepidella*. It is most similar to *A. silvicola*, but that species is usually smaller overall, and with an abruptly bulbed or truncate (not strongly rooting) stipe base. Additionally, *A. silvicola* does not have clamps at the base of the basidia and has smaller, narrower spores on average.

Amanita "Sandhills"

CAP: 5–15 cm across, hemispherical, rounded to tightly convex when young, becoming broadly convex to flat or slightly wavy in age. White when young, soon eggshell white to creamy white, developing yellowish tones in age. Surface dry, covered in coarse, poorly defined, fluffy white warts that become inconspicuous in age. Edge of cap adorned with a flap of cottony tissue that hangs beyond edge of gills; sometimes also with large, hanging triangles of fluffy veil tissue that disappear in age. **GILLS:** Finely to broadly attached, rarely free, close to crowded, many short gills present. Pale creamy at first, becoming distinctly yellow to dingy ocher in age. Edges of gills lined with a fine layer of ragged white material that is most apparent in young fruitbodies. **STIPE:** 9–18 cm long, 2.5–3 cm thick at apex, 3–5 cm thick at widest point, gradually swollen toward base and then tapered again below swelling, forming an indistinct and usually broken-off rooting base. White to pale cream, sometimes with faint pinkish tones, yellowing somewhat in age. Surface finely to strongly powdery or fluffy-scaly. **PARTIAL VEIL:** A small ring of ragged tissue near middle of stipe, more often hanging on cap margin, or indistinct, especially in age. **VOLVA:** An indistinct flocculent band around top of stipe base when young, disappearing in age. **FLESH:** Thick but soft and often waterlogged, whitish. **ODOR:** Indistinct when young, but as with many *Lepidella* stronger and chlorinelike in age. **TASTE:** Unknown. **SPORE DEPOSIT:** White. **MICROSCOPY:** Spores 9–12.5 (16) × 5–6.5 (8.5) μm, long ellipsoid, elongate or even cylindrical, amyloid. Basidia clamped at base.

ECOLOGY: Solitary or scattered, rarely in large numbers. Locally common in mixed live oak and Ponderosa Pine forests of the sandhills habitat in Santa Cruz County. Fruiting in late fall and early winter.

EDIBILITY: Unknown, likely quite toxic.

COMMENTS: Besides the unique habitat preference, identification may be easiest by excluding other *Lepidella*: *A. magniverrucata* has pyramidal cap warts and curved scales around the stipe base; *A. pruittii* is much smaller and grows in meadows and fields. *A. smithiana* has warts with smooth cap skin between them in age, and develops scattered dingy blushes. It lacks the elongate-cylindrical spores of this species. *A. silvicola* is smaller, with less distinct warts, is usually whiter overall, and has a club-shaped stipe base.

Amanita pruittii

Tulloss, J. Lindgr. & Arora

MEADOW AMANITA

CAP: 2–8 (15) cm across, rounded at first, soon broadly convex to plane, margin slightly wavy in age. Off-white to grayish, then dingy tan-gray with yellow tones in age. Surface dry to moist, at first covered in powdery grayish warts that soon collapse or wash off in rainy conditions, but remain tightly bound to cap tissue in dry weather. Margin not sulcate and often with overlapping veil tissue when young. **GILLS:** Finely to broadly attached (rarely truly free), often with a fine line of tissue descending upper stipe from base of each gill, fairly thick, close to well spaced. Pale cream when young, becoming dull yellowish beige in age. **STIPE:** 2–8 cm long, 0.7–2 cm thick at apex, slightly wider toward base. White to creamy at first, yellowish in age. **PARTIAL VEIL:** Inconspicuous floccose ring zone, soon disappearing. **VOLVA:** Tissue usually very obscure, when present appearing only as ragged scales over a downward-tapering stipe base. **FLESH:** Soft, often waterlogged. Whitish. **ODOR:** Mild when young, unpleasant in age. **TASTE:** Indistinct. **SPORE DEPOSIT:** White. **MICROSCOPY:** Spores (7) 8–11 (14) x 6–9 (11) µm, subglobose to broadly ellipsoid, amyloid.

ECOLOGY: Solitary or in small groups in dead grass thatch in waterlogged meadows or in short lawn grass or cow pastures, or, in one case, under Monterey Cypress in nutrient-poor soil. Not mycorrhizal with woody plants. Uncommon to rare but some patches can be very prolific. Found from Santa Cruz County to Washington state, likely also present farther south. Fruiting from early fall into early winter.

EDIBILITY: Unknown. Other members of the Section *Lepidella* are known to be toxic.

COMMENTS: The messy grayish white cap, yellowing gills in age, and grassland habitat render it quite distinct. Although easy to recognize, these mushrooms aren't much to look at: small, dingy, squat, and quick to decay. However, very fresh young specimens can be sort of charming to the eyes of a *Hygrocybe*-weary taxonomist.

SYNONYM: *Aspidella pruittii* (A.H. Sm. ex Tulloss, J. Lindgr. & Arora) Redhead & Vizzini.

Amanita farinosa (sensu CA)

POWDERY AMANITA

CAP: 4–8 cm across, rounded to convex when young, becoming broadly convex to flat, or uplifted in age. Margin finely sulcate when young, distinctly so in age. Pale gray, gray to grayish tan. Surface dry, completely covered in powdery, grayish to grayish tan universal veil tissue that binds to the surface, but can be sparse in age. **GILLS:** Narrowly attached or notched, close to subdistant, partial gills numerous and abruptly truncate, edges often ragged or finely frosted. White to creamy white, occasional discoloring or staining pale brownish. **STIPE:** 4–9 cm long, 0.8–2 cm thick at apex, club shaped, gradually enlarged downward to swollen or bulbous base. Bulb is round to elongate, sometimes with a distinct rim at top. White to pale creamy white with grayish to grayish tan veil remnants at base. Surface dry, covered with whitish scales or punctations. **PARTIAL VEIL:** Present only when very young, consisting of a powder tissue, disappearing as soon as it expands, leaving no sign of a ring on stipe. **VOLVA:** A powdery rim of tissue around top of bulb and/or as loose patches around stipe base in soil. **FLESH:** Thin, soft, fibrous, and hollow in stipe. White. **ODOR:** Indistinct. **TASTE:** Indistinct. **SPORE DEPOSIT:** White. **MICROSCOPY:** Spores 7.5–9.5 × 5.5–7.5 µm, ellipsoid, inamyloid.

ECOLOGY: Solitary or scattered, rarely abundant, on ground or in duff in mossy coniferous woods on the Far North Coast. Very rare in California, known from a handful of sites from Mendocino County northward. Fruiting in early fall.

EDIBILITY: Unknown. Possibly poisonous; do not experiment.

COMMENTS: This small, rare *Amanita* can easily be identified by the gray to grayish tan cap with powdery veil remnants, white gills, and club-shaped or slightly bulbous, stout stipe. The small grassland species, *A. pruittii*, can superficially resemble *A. farinosa*; besides habitat differences, it has a more flocculent veil tissue on the cap and a stout stipe with an elongated rooting bulb. Some *Lepiota* species look similar, but have finely scaly caps and more distinct free gills and are usually more slender in appearance. The true *A. farinosa* of eastern North American oak forests is much smaller, and has broader spores. Although the western taxon is clearly distinct, a name for it has yet to be published.

Amanita phalloides

(Vaill. ex Fr.) Link.

DEATH CAP

CAP: 3–12 (20) cm across, egg shaped to tightly convex when young, becoming broadly convex to plane. Margin not sulcate, but can rarely appear so along some parts of the cap (especially on old, rain-beaten, or dry fruitbodies) due to collapse of cap tissue onto gills. Color quite variable, ranging from pale greenish yellow to yellowish white to dingy grayish olive or grayish green when young, often with darker fibrils or streaks. Older caps often duller yellowish green, grayish green, brownish yellow to medium brown. Caps often radially mottled or with obscure dark streaks in any combination of the above colors. Margin usually remaining dingy white or at least paler. Surface slightly viscid when wet, quickly becoming dry, smooth, appearing metallic in dry weather. Although not typical, weather conditions can cause a large proportion of fruitbodies in some fruitings to have large, but thin, universal veil patches on cap. **GILLS:** Finely attached to free, close to crowded. White at first, becoming pale creamy white, often with pale yellowish blushes around margin or with an obscure tinge of pink in age. **STIPE:** 5–13 cm long, 1–2 cm thick at apex, tapered upward from a distinctly bulbous, round, swollen base. Upper stipe (above ring) usually white, pale creamy white to pale yellowish. Lower stipe variably white, creamy yellow to yellowish green. Surface dry, mostly smooth, but often with darker yellowish chevrons. **PARTIAL VEIL:** A thin, membranous hanging skirt, sometimes only present as a ragged collar when veil remains attached to cap margin. Underside yellowish green, upper surface white to pale creamy white. **VOLVA:** Spacious, saclike, relatively thin, covering bulbous stipe base. Exterior white, interior usually yellowish. **FLESH:** Firm, relatively thick to thin, soft, whitish or yellow just under cap skin. Stipe solid. **ODOR:** Mild when fresh, distinctive but difficult to describe. Some people call it sickly sweet; others musty metallic or like rotting potatoes, becoming stronger and unpleasant in age. **TASTE:** Not sampled. **KOH:** No reaction. **SPORE DEPOSIT:** White. **MICROSCOPY:** Spores 8–10 × 6–8 μm, nearly round to ellipsoid, amyloid. Basidia not clamped.

ECOLOGY: In scattered troops or in large flushes in fall and early winter. Smaller fruitings can occur almost any time of year, especially in irrigated areas with oaks. This species was accidentally introduced from Europe on the roots of Cork Oaks and has successfully invaded many of California's native habitats. Although it is very common around the San Francisco Bay Area and Monterey Bay area, it is otherwise irregularly distributed (but can be locally abundant). It is very common in association with Coast Live Oak, but has been found with an amazing variety of native trees and shrubs, as well as ornamental hardwoods (*Arbutus*, *Tilia*, *Corylus*, and *Carpinus*). Plantings of these latter trees account for most records in western North America outside California. *A. phalloides* is relatively uncommon but increasing southward (recently arrived in San Diego), and uncommon north of Sonoma County, but spreading quickly. It has been found with pines in Mendocino County, and although recorded north to British Columbia, has yet to be found in California's Humboldt and Del Norte Counties. It is likely that Death Caps will soon colonize the entire Redwood Coast.

EDIBILITY: Deadly poisonous! If you suspect you have eaten this species, don't wait for symptoms to develop further; contact a poison control center and your local hospital right away. This species contains amatoxins, molecules that seriously damage the liver, and prompt treatment and accurate identification of the cause of poisoning are essential. A few deaths occur every year in California due to consumption of this species; dogs are frequent victims as well.

COMMENTS: This common and stately *Amanita* can easily be identified by the combination of a bare, yellowish green cap, whitish gills, and bulbous stipe base encased in a thin white sack of universal veil tissue. Perhaps the most similar species is *Volvopluteus gloiocephalus*, which sometimes has a greenish gray, metallic cap. However it does not have a partial veil at any stage, and its gills are distinctly free and soon become pinkish tan; it is fairly common in wood chips but is rarely found in the woods. *A. phalloides* var. *alba* (pictured above) is identical in all respects except that it is pure white on all parts. It has been reported from Point Reyes in Marin County and the San Francisco Bay Area. The white form or pale, weathered specimens could be confused with *A. ocreata*, but can be distinguished by their fall fruiting habits (keep in mind the possibility of a rare fall appearance of *A. ocreata*) and lack of a bright yellow KOH reaction on the cap. Some novice gatherers worry about confusing atypical forms of this species with the Coccora (*Amanita calyptroderma*, p. 48)—see the comments section under that species for tips on separating these two species.

Amanita porphyria

Alb. & Schwein.

GRAY-VEILED AMANITA

CAP: 2.5–12 cm across, rounded to round-conical when young, becoming convex to nearly flat with a low umbo or slightly uplifted in age. Color variable, but distinctive; ranging from dull grayish lilac with a pale sheen to dusky purple. Washed-out older caps can show more tan or beige. Small percentage of fruitbodies redder or even yellow-brown. Surface slightly greasy to dry, with thin, pallid grayish universal veil tissue. **GILLS:** Finely attached to obscurely free, close to subdistant. White to eggshell white or slightly creamier in age. **STIPE:** 5–15 cm long, 1–2 cm thick at apex, more or less equal, or often slightly enlarged at apex and base (thinnest at middle). Base of stipe with abrupt rimmed or cleft bulb. White with pale, grayish or lilac bands or chevrons (especially lower stipe). **PARTIAL VEIL:** Thin, membranous skirt, with felty, grayish purple undersurface. **VOLVA:** A short collar (but often indistinct) on abruptly bulbous stipe base. **FLESH:** Whitish, fibrous, solid. **ODOR:** Like raw potatoes or radish. **TASTE:** Mildly or strongly like the odor. **KOH:** Sometimes turning cap surface bright lilac. **SPORE DEPOSIT:** White. **MICROSCOPY:** Spores 8–9.5 × 7.5–9 μm, globose to subglobose, amyloid.

ECOLOGY: Solitary or scattered, but rarely in large numbers under conifers, especially Sitka Spruce on the Far North Coast, and pines elsewhere. Locally common on the North Coast, uncommon elsewhere. Rather flexible in fruiting times: November through January in most areas, routinely during spring and summer in fog-bound areas.

EDIBILITY: Unknown.

COMMENTS: The slender stature, grayish lilac to purple colors, grayish partial veil, and blocky, abruptly bulbous stipe base distinguish this species from other local *Amanita*. Browner or faded specimens with wartlike, discrete patches on the cap could be confused with *Amanita* in the *gemmata* and *pantherina* groups, but those tend to be stouter and should have white (not gray) veil tissue and distinctly sulcate margins at maturity.

Amanita ocreata

Peck

WESTERN DESTROYING ANGEL

CAP: 5–15 cm across, round to tightly convex when young, becoming broadly convex to plane, or occasionally slightly uplifted at maturity. Eggshell white to creamy white when young, sometimes with more pronounced creamy yellow tones. Center of the cap often with a pinkish orange blush that sometimes extends over half the distance to margin. In other cases, the entire cap is completely white. Surface slightly viscid to dry, smooth, typically without any universal veil remnants, but can have a thin, very tightly appressed patch of whitish veil tissue. Margin usually not sulcate, but occasionally can be in age as cap tissue collapses against gills. **GILLS:** Finely attached (rarely free), close to crowded. White at first, but often with a creamy or pinkish cast, especially in age. **STIPE:** 6–15 cm long, 1–3 cm thick, but thicker at swollen base; up to 6 cm thick. Usually straight or slightly curved, ending in a distinctly swollen, bulbous base that occasionally elongates slightly. Whitish to creamy white, often with pale pinkish or orangish blushes in age. Surface smooth or slightly scaly below the veil. **PARTIAL VEIL:** A simple hanging skirt, membranous but rather fragile, usually ragged or disintegrating entirely in age, and leaving only an inconspicuous ring zone. **VOLVA:** Thin but large and spacious, totally engulfing the stipe base and bulbous lower stipe. Whitish. **FLESH:** Firm, thick, white. Stipe firm, but often becoming noticeably soft in age, almost always solid, rarely chambered or hollow. **ODOR:** Mild when young, can be pungent or sickly sweet in age. **TASTE:** Not sampled.

KOH: Bright yellow on cap (sometimes slowly), paler yellow on stipe surface. **SPORE DEPOSIT:** White. **MICROSCOPY:** Spores 8.5–12 × 6–8.5 µm, round to broadly ellipsoid, sometimes elongate, amyloid. Basidia not clamped or rarely so.

ECOLOGY: Solitary or in scattered groups or arcs in duff, generally under live oak, but also under deciduous oaks, especially in open woodlands near meadow edges. Common from late winter through spring, especially in the southern part of our area. Quite rare north of Sonoma County and then primarily found inland (following distribution of oaks).

EDIBILITY: Deadly poisonous! See text box on Amatoxins.

COMMENTS: This is an important species for beginners to be familiar with, because it is extremely toxic and can be abundant under oaks in spring. The whitish overall colors in combination with the bulbous stipe base, thin white volva, ring on the stipe, and white gills distinguish this mushroom from most others in our area. Occasionally, the fruitbodies of *A . ocreata* develop a pinkish cap and an obscurely and irregularly sulcate cap margin, and can retain a thin patch of veil tissue. These highly atypical fruitbodies have been confused with *A. velosa*, an edible species that shares the same habitat and fruiting season. However, the cap color of *A. ocreata* is never entirely and evenly coral pink or pinkish tan; nor is the margin of the cap sulcate when young. Additionally, the universal veil on the cap of *A. velosa* tends to be thicker, the stipe never carries a true skirtlike partial veil (ring), and the base of the stipe is never strongly swollen or bulbous as in *A. ocreata*. Additionally, the

stipe of *A. ocreata* is usually distinctly solid, whereas that of *A. velosa* is almost always hollow. *A. phalloides* var. *alba* could also be confused with *A. ocreata*, but it fruits in the fall and early winter, and its cap does not turn yellow in KOH.

Amatoxins

Amatoxins are a group of molecules responsible for the toxicity of *Amanita phalloides* and *A. ocreata*, which cause the most serious cases of mushroom poisoning in our area. However, amatoxins are also found in *Galerina marginata* (which has caused some poisonings due to confusion with *Psilocybe* spp.), some *Lepiota* species, and a few members of other genera as well.

Amatoxins dramatically reduce a cell's ability to create new RNA, which is necessary for protein synthesis. When this happens, metabolism slows to a halt, and the cell breaks down. The cause of death is usually attributed to massive liver and kidney damage.

Victims of amatoxin poisoning often don't realize that anything is amiss until 12–24 hours after eating toxic mushrooms, when gastrointestinal symptoms set in. Unfortunately, these initial symptoms often disappear for an intermission of another 12–24 hours, and many victims don't seek treatment. When this "recovery" period has passed, liver and kidney damage has occurred, and more severe symptoms set in. This can lead to a slow death over the course of a week or more.

The pattern of gastrointestinal distress, a brief "recovery," and then return of severe symptoms highlights the importance of seeking treatment IMMEDIATELY in cases where ingestion of amatoxin-containing mushrooms is suspected; this usually involves a massive IV injection of fluids as soon as possible to help rid the body of amatoxins. Treatment with silibinin (an extract of milk thistle) may be the best hope. This compound has been used widely in Europe to treat amatoxin poisonings, but has yet to be approved by the FDA; experimental trials in Santa Cruz have shown great potential in mitigating organ damage and preventing fatalities.

Amanita calyptratoides
Peck

CANDLESTICK AMANITA

CAP: 3–7 cm across, rounded to convex when young, becoming broadly convex to flat in age, margin sulcate. Dull grayish, brownish to tan, sometimes with a lilac sheen or paler, to nearly whitish gray. Surface slightly viscid to dry, smooth, typically with a single patch of white universal veil tissue **GILLS:** Narrowly attached, soft, moderately close. White to creamy white or slightly pinkish cream at maturity. **STIPE:** 4.5–9 cm long, 1–2.5 cm thick, usually quite uniformly cylindrical or with a slightly enlarged base. White to creamy yellowish, often described as looking waxy or slightly translucent like a candle. Surface above veil often with vertical lines of tissue **PARTIAL VEIL:** A membranous skirt, but disappearing very quickly and leaving only a ragged ring zone, or often no trace at all. **VOLVA:** Saclike, thin, edge ragged and often flaring, often with a thicker-edged inner layer. **FLESH:** Thin, soft, and often fragile, whitish to creamy. Stipe hollow in cross section, usually stuffed with whitish pith, at least when young. **ODOR:** Indistinct to slightly fishy. **TASTE:** Indistinct. **SPORE DEPOSIT:** White. **MICROSCOPY:** 9.5–14 × 6.5–9 μm, ellipsoid, inamyloid. Basidia clamped.

ECOLOGY: Solitary or in small groups under live oak in fall and early winter. Fairly rare over much of the Redwood Coast; quite common in southern California.

EDIBILITY: Edible but uncommon, fragile, and very perishable.

COMMENTS: The waxy-looking stipe, dull grayish brown cap with a white universal veil patch and clearly sulcate margin, thin-edged volva, and smallish stature distinguish this species from others in our area. Its relationships within *Amanita* are somewhat unclear—the weak but clearly present partial veil and slightly enlarged stipe base (not totally elongating) preclude placement in Section *Vaginatae* and are more characteristic of Section *Caesareae*. Rod Tulloss, American amanitologist of note, places it in its own group, Stirps *Calyptratoides*. *A. vaginata* has a cooler gray cap (without extensive tan-beige tones), a more elongated stipe that is not distinctly hollow, and lacks a partial veil.

Amanita calyptroderma

G. F. Atk. & V. G. Ballen

COCCORA

CAP: 6–20 (30) cm across, round when young, becoming broadly convex to flat or uplifted in age. Color somewhat variable but usually warm, radiant coppery brown with golden brown tones outward and yellow around the margin. Fruitbodies with olive or darker brown, or even pigmentless (white) forms are sometimes found. Surface smooth, slightly greasy to dry, center typically covered with a strikingly large and thick patch of universal veil tissue (a "skullcap"), this patch occasionally breaking into smaller warts. Margin with distinct radial grooves (sulcate) even when young (although this may be obscure in the very young buttons). **GILLS:** Finely to broadly attached, close to crowded. Creamy white to pale yellowish cream at first, developing yellower tones in age. **STIPE:** 8–15 (20) cm long, 2.5–4.5 cm thick at apex, up to 6 cm thick at base, cylindrical or weakly tapered upward from base, which is enclosed in the volva. Uniformly creamy white to yellowish cream. **PARTIAL VEIL:** Thin but persistent membranous skirt, often collapsing against the stipe in age, not usually wearing off. **VOLVA:** Membranous, persisting as a large, thick, saclike sock on stipe base. **FLESH:** Fairly firm, whitish. Stipe predominantly hollow or filled with cottony fibers (pith) or a clear gel. **ODOR:** Indistinct when young, quite fishy and unpleasant in age. **TASTE:** Slightly savory or weakly metallic. **SPORE DEPOSIT:** White to creamy white. **MICROSCOPY:** Spores 9–12 x 5.5–7 μm, ellipsoid, inamyloid. Basidia clamped.

ECOLOGY: Solitary, scattered in groups or large arcs and troops, fruiting in synchronous flushes from early fall into early winter. Common throughout our range (less so on the spruce-dominated Far North Coast). Prefers Madrone, live oak, and Tanoak, but can also be found with Douglas-fir and manzanita.

EDIBILITY: Edible and good, but not everyone likes it. The texture is fairly firm, which appeals to most everyone, but the slightly fishy flavor turns some people's stomachs (others love it). Some collections also have a metallic aftertaste. Don't collect it for the table until you are perfectly familiar with *A. phalloides*, the deadly poisonous Death Cap.

COMMENTS: This is one of the loveliest and most impressive mushrooms on the California coast. The spacious, cottony volva, coppery yellow cap with a prominent universal veil patch, sulcate cap margin, and hollow stipe distinguish it from all but the Spring Coccora (*A. vernicoccora*). The latter species was long considered to be a pale spring form of this species, but is now known to be genetically distinct. It differs by its consistently lighter cap (pale chartreuse yellow) and peak fruiting during spring. *A. vernicoccora* sometimes appears in low numbers during fall, whereas spring fruitings of *A. calyptroderma* are rare. Concerns about confusion with *A. phalloides* (the Death Cap) are primarily due to a lack of familiarity with both species. The two can be safely told apart in all but the most exceptional cases by using a *combination* of the following characters: The Coccora has a warmer-colored cap that is darker at the center and brighter at the margin (overall dominated by yellow-brown tones), whereas that of

the Death Cap is typically pale greenish yellow to whitish at the margin (rarely olive, but can be brownish with age). The Coccora has a very thick, large, cleanly peeling universal veil patch on the cap; the Death Cap is usually bald, and when a universal veil patch is present, it tends to be a small, flat, tightly pressed-on patch of tissue. The stipe is almost always hollow in the Coccora and only very rarely partially hollow in the Death Cap. The margin of the cap in the Coccora is distinctly sulcate even when young, whereas the Death Cap has a plain cap margin, though this may be the weakest character (but can develop "pseudostriations" in age as the cap tissue collapses against the gills).

Coccora versus Death Cap Identification Checklist

Coccora (edible)

- Cap coppery or warm, rich yellow-brown
- Cap with a thick, soft patch of white universal veil tissue that is easily peeled from the cap
- Cap margin distinctly grooved (sulcate) all the way around
- Stipe with a large and distinct central hollow, usually filled with cottony pith when young

Death Cap (deadly!)

- Cap greenish-yellow (can be olive or whitish-yellow or rarely nearly brownish)
- Cap lacking a thick patch of universal veil tissue (occasionally with a small, thin, tightly held patch)
- Cap margin not grooved (occasional older specimens very weakly lined)
- Stipe solid (rarely with random small hollows)

Amanita vernicoccora

Bojantchev & R. M. Davis

SPRING COCCORA

CAP: 6–15 (20) cm across, rounded when young, becoming broadly convex to plane, or depressed at the center in age. Margin distinctly sulcate even when young. Pale yellow, corn yellow to clear yellow, fading to straw yellow or dingy yellowish white. Surface slightly viscid to dry, typically with a single, large patch of thick, white universal veil tissue. **GILLS:** Finely to broadly attached (rarely free), close to crowded. Creamy white at first, developing yellowish tones in age. **STIPE:** 5–15 cm long, 2–3 cm thick at apex, cylindrical and straight, slightly larger at base. Off-white to pale cream with yellowish tones. **PARTIAL VEIL:** A thin, membranous hanging skirt, concolorous with the stipe, often collapsed in age or when handled. **VOLVA:** Thick, white, membranous, saclike; spaciously enclosing stipe base. **FLESH:** Thick and firm when young, soft in age; whitish. Stipe hollow and filled with cottony pith or clear gel. **ODOR:** Usually indistinct when young, unpleasant and fishy in age. **TASTE:** Mild to sweet. **SPORE DEPOSIT:** White. **MICROSCOPY:** Spores 9.5–12.5 x 6–8 μm, ellipsoid or slightly elongate, inamyloid. Basidia clamped.

ECOLOGY: Solitary, scattered, or in small groups under hardwoods, primarily live oak and deciduous oaks such as Black Oak. The vast majority of fruiting occurs in spring, but a few are occasionally found in fall and winter.

EDIBILITY: Edible and very similar to *A. calyptroderma*. Extreme care should be taken with your identification, as pale forms can resemble the deadly poisonous *A. ocreata*, near which they sometimes grow.

COMMENTS: For a long time, this taxon was thought to be a pale, spring-fruiting form of *A. calyptroderma*. The range of cap colors is less variable in this species than in *A. calyptroderma*; only the lightest forms of that species approach the most richly colored specimens of *A. vernicoccora*. While *A. vernicoccora* can fruit in low numbers in fall, *A. calyptroderma* very rarely fruits during spring. Faded or weathered specimens of *A. vernicoccora* could possibly be mistaken for the deadly poisonous *A. ocreata*. The latter has a nonsulcate, whitish or yellowish-washed cap lacking a thick, membranous patch of universal veil tissue (although it sometimes has a thin patch of veil) and, typically, has a solid stipe.

Amanita velosa

(Peck) Lloyd

SPRINGTIME AMANITA

CAP: 3–11 cm across, rounded and tightly convex when young, becoming domed to broadly convex or flat in age. Margin distinctly sulcate, even when young. Beautiful coral pink to peachy pink, sometimes warm light brownish tan, often a paler beige-pink in age. A pure white to creamy white form is sometimes encountered as well. Surface slightly greasy or sticky to dry, typically with a single large, more or less central patch of white universal veil tissue. This patch sometimes breaks up into more than one patch or even into small, scattered warts. **GILLS:** Finely attached, sometimes seceding and appearing free, close to crowded. Pale creamy at first, becoming pinkish cream. Edges with an ephemeral, very fine band of ragged tissue. **STIPE:** 5.5–10 cm long, 1.5–3 cm thick at apex, more or less equal or somewhat wider at base. White to off-white, covered in pointed white scales or chevrons over most of its length. **PARTIAL VEIL:** Usually entirely absent; occasionally an evanescent spiky annular zone is present. **VOLVA:** A white, fairly thick, membranous, saclike cup or sheath around the stipe base. White, occasionally with orangish stains, inner surface of volva sometimes a similar color to cap, often slightly darker or duskier. **FLESH:** Firm, relatively thin, whitish. Stipe hollow or stuffed with a pith (at least partially so). **ODOR:** Indistinct. **TASTE:** Sweet to slightly nutty. **SPORE DEPOSIT:** White. **MICROSCOPY:** Spores 9–12 x 7.5–10 μm, subglobose to ellipsoid, inamyloid. Basidia unclamped.

ECOLOGY: Solitary or scattered in duff, most abundant in short-grass meadows bordering live oak groves, often fairly far from edge of canopy. Common in the range of Coast Live Oak, occasionally with Black Oak. Primarily fruiting in spring, with occasional winter fruitings.

EDIBILITY: Outstandingly delicious, one of the best edible mushrooms of any kind, anywhere. The flesh is firm (although not as firm as *A. calyptroderma*), but the flavor is much finer: slightly sweet, nutty, a good choice for delicately flavored dishes. As for all edible *Amanita*, familiarize yourself with the toxic species before eating.

COMMENTS: The coral pink to peachy pink cap with a sulcate margin and thick white "skullcap" of universal veil tissue, the hollow stipe (sometimes stuffed with cottony tissue), and the membranous, saclike cup at the stipe base combine to make this a very distinctive species. Confusion with the deadly *A. ocreata* is a risk for those who are unfamiliar with both species. Once you get to know them, these two species are not very similar, and the real danger lies in the fact that the two species will routinely fruit in the same place, at the same time. Most fruitbodies, when carefully observed, should be readily identifiable in most cases. Pay attention to each and every mushroom you're picking for the table. For more tips on safely separating *A. velosa* from *A. ocreata*, see the comments under *A. ocreata*. The less similar *A. "pseudobreckonii"* has a straw yellow to peachy cap and a swollen stipe base that lacks the sheathing cuplike volva; it grows with conifers north of San Francisco Bay.

Amanita vaginata group
GRISETTE

CAP: 5–15 cm across, egg shaped to rounded-conical when young, soon convex to nearly flat, sometimes with a weak umbo. Dull, medium gray to pale warm gray, sometimes more grayish tan. Surface smooth, dry, usually with a single large patch of white universal veil tissue. Entire margin with distinct radial grooves. **GILLS:** Finely attached to free, close to crowded. White at first, becoming pale cream in age. **STIPE:** 7–15 cm long, 1.5–2.5 (3) cm thick at apex. Lower parts may be slightly wider. Cylindrical and rather slender in most cases, although some variants are quite stout. Whitish to creamy. Surface smooth or evenly covered in obscure white or pale gray fibrils or chevrons. **PARTIAL VEIL:** Absent. **VOLVA:** Relatively thin, forming a saclike cup on stipe base. The form illustrated here doesn't develop ocher stains on the volva (other forms do, and in the northern form, the volva turns grayish). **FLESH:** Soft, whitish. Stipe stuffed with pith when young, often becoming hollow. **ODOR:** Usually mild, fishy in age. **TASTE:** Mild. **SPORE DEPOSIT:** White. **MICROSCOPY:** Spores 9–12 x 9–12 µm, globose to teardrop shaped, inamyloid.

ECOLOGY: Solitary or scattered in groups in duff under both conifers and hardwoods, especially live oak. Fruiting from early fall into winter.

EDIBILITY: Edible but perishable and not generally highly regarded, although some people are enthusiastic about the flavor. This species is much less risky than edible *Amanita* of other sections, but is not recommended as a target for beginners.

COMMENTS: This is a complex group of taxa in California. They can be recognized by their grayish sulcate caps, white gills and stipe, and the white saclike volvas. In northern form, the veil typically goes grayish in age, while most others stain orange. *A. pachycolea* is usually taller, with a darker and more deeply sulcate cap, as well as orange-staining universal vein tissue; it is more frequently found under pine. The volva of *A. constricta* flares upward above a tightly constricted midpoint that "chokes" the stipe base and it also usually has dark gray hairs or other ornamentation on the stipe.

Amanita protecta
Tulloss & G. Wright
PROTECTED GRISETTE

CAP: 5–12 cm across, rounded to convex when young, becoming broadly convex to flat in age. Margin distinctly sulcate. Entirely whitish gray at first (when covered in universal veil tissue), then gray to inky gray or medium grayish tan, often with dark brown areas. Surface dry, at first completely covered with thick, white to pale gray to grayish orange universal veil tissue, which often breaks into smaller patches or warts to reveal a smooth, darker gray surface. **GILLS:** Finely attached, close to subdistant. Whitish at first, then dingy pale grayish white. Edge of gills with a beaded appearance due to the presence of small, dandruffy patches of dark grayish tissue. **STIPE:** 6–12 cm long, 1.5–3 cm thick, cylindrical or enlarged toward base. Entire stipe ornamented with vertical lines of grayish, pointed scales or chevrons of ragged tissue with a whitish base color. **PARTIAL VEIL:** Absent. **VOLVA:** Present as concentric bands of grayish ragged or scaly tissue, or as a whitish membrane that discolors ocher-orange or is not apparent at all. **FLESH:** Soft, pale whitish. Stipe at least partially hollow. **ODOR:** Mild at first, fishy in age. **TASTE:** Indistinct. **SPORE DEPOSIT:** White. **MICROSCOPY:** Spores 9.5–13 x 8.5–11.5 µm, globose to broadly ellipsoid, inamyloid.

ECOLOGY: Solitary or in groups, primarily in areas with willow (*Salix* spp.) and possibly with live oak (although we have never found it without willow present). Appearing throughout mushroom season, it is most abundant in winter. It is widely distributed, known from almost every county on the California coast, from San Diego at least to Humboldt Counties, but rather rare north of the San Francisco Bay Area and more common to the south.

EDIBILITY: Likely nontoxic.

COMMENTS: Fruitbodies of this species can be distinguished from other members of the Section *Vaginatae* by their stockier stature and thicker flesh, more obviously ornamented stipe, and felty or crumbly grayish universal veil that develops strong ocher-orange stains. Other gray grisettes can be very similar (such as *A. constricta*), but usually have less prominent stipe decoration and more membranous saclike universal veil tissue at the base of the stipe, and/or show lighter orange staining that is not as pronounced as in *A. protecta*.

Amanita constricta
Thiers & Ammirati
CONSTRICTED GRISETTE

CAP: 5–15 (25) cm across, buttons egg shaped to conical expanding to broadly convex and soon flat, or even slightly depressed at the center with an uplifted margin. Margin of cap with distinct grooves (sulcate). Dark grayish or brownish gray, sometimes appearing slightly dusted with whitish powder. Surface dry to slightly greasy, smooth, sometimes with a flat, grayish, membranous universal veil patch. **GILLS:** Finely attached and soon seceding, less commonly truly free or broadly attached, close to crowded. Off-white, becoming pale grayish. **STIPE:** 8–15 (25) cm long, 1–3 cm thick, equal or narrowest at apex. Whitish to off-white, often covered in variably dense parallel lines of gray scales, sometimes small and inconspicuous or restricted to base of stipe, or pronounced over entire length of stipe and darker gray. **PARTIAL VEIL:** Absent. **VOLVA:** A membranous but somewhat fragile sac tightly adhering stipe to the base but with a characteristic prominently flaring upper lip. Inner surface grayish, outer surface white, sometimes with rusty stains. **FLESH:** Thin, rather soft, whitish. Stipe usually at least partially hollow or stuffed with cottony fluff. **ODOR:** Mild when young, rather fishy and unpleasant in age. **TASTE:** Mild or slightly sweet. **SPORE DEPOSIT:** White. **MICROSCOPY:** Spores 9.5–13 x 8–11.5 µm, globose to broadly ellipsoid, inamyloid.

ECOLOGY: Solitary or in scattered troops under oaks or conifers. Common throughout most of our area, but status and distribution unreliable due to confusing taxonomy of this section. Primarily fruiting in late fall and winter.

EDIBILITY: Edible and fairly good, but rather perishable.

COMMENTS: The diagnostic feature separating this species (or species complex) from other grisettes is the constricted volva with a flaring upper portion. If this has disappeared (and it often does, being quite fragile), separation from *A. vaginata* can be very difficult. There appear to be two species going by the name *A. constricta*: a smaller variant with oaks and a large, darker conifer associate (shown here).

Amanita pachycolea
D. E. Stuntz in Thiers & Ammirati
WESTERN GRISETTE

CAP: 4–20 cm across, egg shaped to conical when young, expanding to broadly convex with a low, distinct umbo, nearly flat or even uplifted in age. Generally very dark brown to dark gray to grayish brown, but can be evenly gray or brown, often with a paler "halo" midway between center and edge. Surface slightly greasy to dry, smooth, sometimes with one or more patches of thick white to pale grayish universal veil tissue, which often develops rust-colored stains. Margin distinctly and deeply sulcate, grooves approaching half the length from margin to cap center. **GILLS:** Finely attached to distinctly free, close to subdistant. Eggshell white to cream, or whitish gray. **STIPE:** 9–20 cm long, 1.5–2.5 cm thick at apex, gradually wider toward base, slender, straight to slightly curved. Gray with a whitish base color. **PARTIAL VEIL:** Absent. **VOLVA:** Thick, membranous, tall cylindrical saclike cup encasing stipe base, sometimes constricted in middle to form an hourglass shape. White to pale grayish white, often developing blotches and staining orange-brown. **FLESH:** Firm, but becoming soft, thin in cap, pale whitish. Stipe usually stuffed when young with white pith, hollow in age. **ODOR:** Indistinct. **TASTE:** Mild to sweet. **SPORE DEPOSIT:** White. **MICROSCOPY:** Spores 10–13.5 x 8.5–12 µm, globose or broadly ellipsoid, inamyloid.

ECOLOGY: Solitary, scattered, or in small groups in coastal pine forest. Common and widespread; fruiting in flushes throughout the wet season, especially following the first soaking rains in fall.

EDIBILITY: Edible and highly regarded for its flavor. The texture is soft, and these mushrooms are more perishable than most. For best results, only young specimens should be collected, and then cleaned gently and used soon after picking. Be cautious when collecting *Amanita* for the table, especially if you are a beginner!

COMMENTS: The tall stature and large size; cap colors ranging from gray to brown to tawny (typically with a pronounced "halo"); long grooves on the cap margin; growth with pine; and thick, rusty-stained volva tissue characterize this species.

Limacella glioderma (sensu CA)

CAP: 3–8 cm across, egg shaped to rounded at first, becoming broadly convex to plane, often with a low, broad umbo. Orangish brown to reddish brown at first, fading to cinnamon to pinkish brown. Surface viscid at first, becoming moist to dry and often breaking up in an irregular, patchy manner, exposing the pinkish flesh. Very fragile, breaking easily. **GILLS:** Narrowly attached to free, close, partial gills numerous. Creamy, or with pinkish to cinnamon tones. **STIPE:** 4–12 (15) cm long, 0.5–1.5 cm thick, often tall and slender (dry weather specimens often more squat), more or less equal, or tapering at base. Pale creamy or whitish beige at apex, with pinkish cinnamon bands of veil tissue when young, overall dingy pinkish to orangish in age. Surface silky smooth above partial veil, with darker brownish pink matted chevrons below. **PARTIAL VEIL:** Whitish cortina-like fibrils, leaving an annular zone on the stipe or disappearing altogether. **FLESH:** Thin, very fragile. Cinnamon in cap, whitish in stipe. **ODOR:** Strongly farinaceous. **TASTE:** Farinaceous. **KOH:** No reaction. **SPORE DEPOSIT:** White. **MICROSCOPY:** Spores 3.5–5 x 3.5–6.5 μm, globose to subglobose, smooth, inamyloid.

ECOLOGY: Solitary or scattered in duff. Fruiting from early fall into winter in mixed forests. Rare; distribution not well known, but present over much of the southern half of our range.

EDIBILITY: Unknown; do not experiment.

COMMENTS: The pinkish brown cap, creamy gills, obscure partial veil, strong farinaceous odor and taste, and fragile flesh set this oddball apart. A few *Tricholoma* are similar, but have firmer flesh, less pinkish red coloration, and more ellipsoid spores. An undescribed *Limacella* occasionally found under northern conifers is extremely similar, but differs in its more viscid stipe and cap. *L. glischra* (sensu CA) is paler, and both cap and stipe are covered with glutinous slime when wet. *Limacella* are rare in our area and generally fruit in small numbers. They are closely related to *Amanita*, differing primarily in having a slimy rather than membranous veil.

Limacella glischra (sensu CA)

CAP: 3–10.5 cm across, rounded at first, soon becoming convex and then plane, with a low, broad umbo. Orange-brown to brown overall at first, soon darkly pigmented only near center and pale buff to off-white toward margin. Surface smooth, viscid, and covered with glutinous slime when wet, or tacky in dry weather. **GILLS:** Free, close to crowded, partial gills numerous. Cream colored. **STIPE:** 7–15 cm long, 0.7–1.5 cm thick, often tall and slender, cylindrical. Pale buff at apex, pale orangish brown to light brown over lower portion. Surface smooth, covered with glutinous slime below partial veil, dry above. **PARTIAL VEIL:** A thin, gelatinous layer, continuous with the cap slime when young, leaving an annular zone on stipe or disappearing altogether in age. **FLESH:** Thin, fragile. Ivory colored, discoloring slowly brownish in stipe. **ODOR:** Slightly farinaceous or cucumber-like. **TASTE:** Mildly farinaceous. **KOH:** No reaction. **SPORE DEPOSIT:** White. **MICROSCOPY:** Spores 4–6.5 μm, globose, smooth, inamyloid.

ECOLOGY: Solitary or scattered in thick duff or soil. Fruiting under conifers in fall. Very rare, known from scattered locations on the North Coast and in the mountains.

EDIBILITY: Unknown; do not experiment.

COMMENTS: This mushroom can be recognized by its slimy, buff-colored cap with a darker orange-brown center, free gills, and slimy, orange-brown stipe. *Limacella illinita* is almost identical in size, stature, and sliminess, but has a white or pale grayish to yellowish white cap and stipe; it is also quite rare, occurring under conifers on the far North Coast. Partly because they are all fairly rare, the *Limacella* in our range are poorly understood, but Rod Tulloss's excellent work (available at www.amanitaceae.org) is helping to straighten things out.

3 • *Lepiota and Allies*

We include *Cystodermella* and *Cystoderma* in this section because their finely scaly-looking caps, partial veil, white gills and spores, and growth in humus in forest settings make them most likely to be confused with *Lepiota*. However, *Cystodermella* and *Cystoderma* are not closely related to the other mushrooms in this section, and their family-level relationships remain unclear. The rest of the description below pertains specifically to *Lepiota* and similar white-spored genera in the Agaricaceae.

Free gills, a partial veil (usually membranous but occasionally obscure or even cortina-like), and a white spore print help distinguish species in this group. Although especially diverse in cypress and redwood habitats, members of this group can be found in practically all habitats, including urban and cultivated areas. All members of this group are saprobic.

Although the different genera can often be told apart (with some practice), the evolutionary relationships of members of this group are still unclear; extensive taxonomic rearrangements (and, thus, name changes) are likely. Generally speaking, the fruitbodies of *Lepiota* and *Leucocoprinus* are smaller and more slender than those of *Leucoagaricus* (with some exceptions), while those of *Chlorophyllum* are the largest and most robust.

The most important macroscopic characters to note when identifying Lepiotoid fungi include the texture of the cap surface, the structure of the partial veil, stature and size, and any discoloration or staining reactions. Microscopically, note the size and shape of the spores, as well as their reactions in Melzer's reagent (dextrinoid or inamyloid), as well as the presence or absence and shape of cheilocystidia.

Else Vellinga has studied this group extensively, and has contributed a great deal to our knowledge of Californian Lepiotoid fungi; her work has shown that many species remain undescribed (some rare, endemic, and likely endangered). Many of her publications and technical identification resources for the group are available at nature.berkeley.edu/brunslab/people/ev.html.

Cystodermella cinnabarina

(Alb. & Schwein.) Harmaja

CINNABAR POWDERCAP

CAP: 2–6 cm across, rounded at first, becoming broadly convex to flat, sometimes lobed or slightly wavy in age. Color variable, usually beautiful coral pink to bright brick red, sometimes duller reddish brown. Surface dry, densely granular-scaly, sometimes cracking into concentrically arranged zone, often with dingier smooth patches in age or when weathered. Margin sometimes with small, hanging "teeth" or patches of white partial veil tissue. **GILLS:** Narrowly attached, sometimes finely notched. White at first, soon a bit dingier to light beige. **STIPE:** 4–7.5 cm long, 0.7–1 cm thick, base often enlarged and club shaped (to 2 cm thick). Light buff, covered with silvery whitish fibrils, becoming extensively pinkish tan, lower three-quarters covered in rings and fluffy chevrons of brick red to pinkish, granular scales. **FLESH:** Moderately thick, white, fibrous. **KOH:** Indistinct or reddish brown. **ODOR:** Indistinct. **TASTE:** Not sampled. **SPORE DEPOSIT:** Dull whitish or off-white. **MICROSCOPY:** Spores 4–5 x 2.5–3 µm, ellipsoid, inamyloid. Cheilocystidia long, pointed, abundant.

ECOLOGY: Fruiting in moss beds and needle duff under conifers. Common in the fall and early winter on the Far North Coast in Sitka Spruce forest, but absent elsewhere in our range.

EDIBILITY: Unknown.

COMMENTS: The somewhat stocky stature, granular or scaly, pinkish red cap and stipe, and white gills help distinguish this species. *Cystodermella granulosa*, which may also occur in our area, is very similar (perhaps a bit duller pinkish red to orange-brown) and also has inamyloid spores, but lacks the cystidia on its gills. *Cystoderma* species in our area are differently colored (not brick red or pinkish), are more slender, have amyloid spores, and lack cheilocystidia.

Cystoderma fallax

A. H. Sm. & Singer

SHEATHED POWDERCAP

CAP: 1.5–5 cm across, domed at first, becoming broadly convex to nearly plane, often wavy in age. Dark ocher-tan to mustardy olive-tan or slightly darker dingy brown. Surface dry, densely granular-scaly. **GILLS:** Broadly attached or finely notched, close, broad. White, creamy to light tan. **STIPE:** 3–7 cm long, 0.4–0.7 cm thick, cylindrical, sometimes slightly enlarged and/or curved near base. White to creamy above ring but becoming tan to dingy pinkish brown. Ocher-tan to light leather brown below ring. Silky smooth above ring, lower part covered in layer of fine granular chevrons. **PARTIAL VEIL:** Continuation of granular sheath of lower stipe, forming a conspicuous upward-flaring, sheath. Upper surface white, smooth to slightly striate. **FLESH:** Fairly thin, white. **ODOR:** Indistinct. **TASTE:** Not sampled. **KOH:** Dark rusty to reddish brown or nearly black on cap. **SPORE DEPOSIT:** White. **MICROSCOPY:** Spores 3.5–5 x 3–4 µm, ellipsoid, amyloid.

ECOLOGY: Solitary or scattered in needle duff or mossy areas under conifers, especially Douglas-fir and pine. Common throughout our area, occurring at least as far south as Santa Cruz County. Fruiting in late fall and winter.

EDIBILITY: Unknown.

COMMENTS: The granular cap and stipe, brownish to tan colors, and large, sheathing ring help distinguish this species; it is the most widespread *Cystoderma* in California. *C. amianthinum* has a much weaker partial veil and is usually paler and brighter.

Cystoderma amianthinum

(Scop.) Fayod

COMMON POWDERCAP

CAP: 1–4 cm across, domed, expanding to broadly convex, often with a small but distinct umbo. Dingy ocher-tan to bright ocher at center, paler buff to light yellowish outward. Edge often with hanging "teeth" of white veil tissue. Surface dry, distinctly finely granular or powdery looking, often extensively radially wrinkled as well. **GILLS:** Obscurely free or narrowly attached, moderately spaced, broad. White, becoming beige or light tan in age. **STIPE:** 3–7 cm long, 0.3–0.5 cm thick, cylindrical. Light beige to tan, often extensively covered with silvery white fibrils, dingier pinkish brown in age. Smooth above the ring zone, but with many pale, buff, or tan chevrons breaking up over lower half. **PARTIAL VEIL:** Rather weak, forming a small upward-flaring ring or often just an indistinct zone where sheathing layer of scales ends. **FLESH:** Thin, soft, white. **ODOR:** Indistinct, like musty earth or often like green corn. **TASTE:** Mild. **SPORE DEPOSIT:** White. **MICROSCOPY:** Spores 5–6 × 3–3.5 μm, ellipsoid, amyloid.

ECOLOGY: Solitary or scattered in troops in coniferous forests, almost always fruiting in beds of lush moss. Quite common on the Far North Coast from late fall through winter. Occurs south to at least Marin County.

EDIBILITY: Unknown.

COMMENTS: The small size, granular ocher-tan cap, white gills, layer of scales or chevrons over the lower stipe, and growth in moss help distinguish this species. *C. fallax* has a flaring, persistent ring and is usually larger and slightly darker in color. A few *Inocybe* are superficially similar and often share the green-corn odor, but all develop dingy brownish gills and are radially silky or scaly rather than granular.

The *Cystoderma* Body-Snatcher

This species of *Cystoderma* is susceptible to infection by a member of the genus *Squamanita*, which includes some of the rarest and most amazing mushrooms in the world. These mushrooms overtake the fruitbodies of their hosts, and burst up through their stipes, replacing the caps with their own. It appears that there are no confirmed records of *Squamanita* in California, but *S. paradoxa* has been recorded on *C. amianthinum* in the Pacific Northwest. Please keep an eye out for this genus on the Redwood Coast (particularly in the northern reaches). *Squamanita paradoxa* has a purplish or grayish lilac cap with a whitish fringe around the margin, white gills, and a pale upper stipe; however, the lower stipe (which belongs to *C. amianthinum*) is golden and granular.

Cystolepiota seminuda

(Lasch) Bon

PETITE PALE PARASOL

CAP: 0.5–2 cm across, convex, expanding to nearly flat and often with a wavy margin. Bright white or with a pinkish blush at first, becoming slightly yellowish to light, dingy beige-tan in age, especially at center. Surface dry, finely powdery to granular at first but wearing off in age and then nearly smooth or with flattened patches. Margin often with "teeth" of veil tissue hanging over edge. **GILLS:** Free, fairly close, broad, edges smooth. White to creamy. **STIPE:** 2–4 cm long, 0.1–0.3 cm thick, cylindrical, thin, often curvy. Creamy whitish at first, soon developing distinct pinkish tones upward from base, overall pinkish beige to light tan in age. Covered in a fine layer of white powder at first. This soon wears off and then the surface is silky to nearly smooth. **PARTIAL VEIL:** Remaining attached to edge of cap, never forming a ring on stipe. **FLESH:** Extremely thin, whitish, and fragile in cap. Stipe fibrous, creamy to pinkish tan. **ODOR:** Indistinct. **TASTE:** Not sampled. **SPORE DEPOSIT:** White. **MICROSCOPY:** Spores 4–5 x 2.5–3 µm, ellipsoid, smooth, not dextrinoid. Cheilocystidia absent. Cap cells round, with a few scattered, irregularly shaped upright cells.

ECOLOGY: Solitary or in scattered troops in humus and needle duff, most often found under redwood from late fall into early spring. Common throughout our range, but easily overlooked.

EDIBILITY: Unknown.

COMMENTS: The finely powdery cap and stipe and the pinkish color near the base of the stipe make this tiny mushroom easy to recognize (if you have good eyesight). Other *Cystolepiota* in our area are darker in color and somewhat larger; most seem to prefer rich humus in moist microhabitats, especially in redwood forests. *C. fumosifolia* has fawn brown colors on the cap and stipe, and yellow contents in its cheilocystidia; it is known from our area but is rare. *C. petasiformis* is whitish to pale grayish cream and has a fluffy (and often conical) pile of powder at the center of the cap when young. Unlike the other species mentioned here, it lacks clamp connections. A similar but somewhat more frequently encountered species in the southern part of our range is slightly larger, with darker tan to brown powdery granular surfaces; it matches *C. oregonensis* fairly well, but that species is supposed to be reddish brown in age. *C. bucknallii* is a small, slender species with a beautiful lilac bluish to light purple color, very finely granular cap, coal-tar odor, and elongated, dextrinoid spores; it is a very rare species occasionally found on the Far North Coast. *Leucocoprinus cretaceus* and *L. cepistipes* are somewhat similar but larger and have stipes with swollen bases and distinct rings (when young), and large spores with a germ pore.

Leucocoprinus birnbaumii

(Burl.) Boisselet & Guinb

FLOWER POT PARASOL

CAP: 1–4 cm across, conical-convex or egg shaped when young, expanding to broadly convex, often with a low, round umbo. Bright yellow when fresh, whitish to creamy outward in age. Surface dry, finely scaly, becoming radially fibrillose, often pleated near margin. **GILLS:** Distinctly free, close to subdistant, edges smooth. Whitish to light yellow. **STIPE:** 3–5.5 cm long, 0.2–0.4 cm thick, cylindrical to enlarged or weakly club shaped near base. Light yellowish to lemon yellow. Surface dry, silky-fibrillose to smooth. **PARTIAL VEIL:** A thin, yellow, flaring ring. **FLESH:** Thin, fibrous, yellowish, stipe hollow. **ODOR:** Indistinct or like burnt rubber. **TASTE:** Not sampled. **SPORE DEPOSIT:** White. **MICROSCOPY:** Spores 8–11 x 5–7 µm, ellipsoid, smooth, with a germ pore, dextrinoid. Cheilocystidia club shaped, bottle shaped, or swollen and pointed.

ECOLOGY: Solitary or in small clumps in cultivated habitats, most often encountered in watered houseplants (even in rather small containers), often fruiting in warm weather. Also found outdoors in warm microhabitats.

EDIBILITY: Unknown.

COMMENTS: The small size, bright yellow colors, tendency to wither quickly, and growth in houseplant pots distinguish this species. Other *Leucocoprinus* in our area are not yellow, have darker colors at the cap center, and/or grow in different habitats.

Lepiota sequoiarum group

WHITE PARASOL

CAP: 2–5.5 cm across, domed or egg shaped at first, becoming broadly convex and then expanding to nearly flat, usually with a distinct low, round umbo. White to pale creamy, often with a yellowish wash over center. Surface smooth at first, with radially splitting silky fibrils in age, sometimes with a ragged band of white tissue hanging off margin. **GILLS:** Free, fairly close, broad, edges smooth. White. **STIPE:** 3–6 cm long, 0.2–0.5 cm thick, cylindrical, often curvy, wider toward base. White to creamy, dingier in age. Smooth or covered with silky fibrils. **PARTIAL VEIL:** Forming a small but distinct, upward-flaring whitish ring, often with a collarlike lower portion, sometimes disappearing. **FLESH:** Very thin, cap fragile, stipe hollow. White, bruising slowly olive brownish where injured. **ODOR:** Indistinct. **TASTE:** Not sampled. **KOH:** Indistinct. **SPORE DEPOSIT:** White. **MICROSCOPY:** Spores 7–9.5 (10.5) × 3.5–4.5 µm, ellipsoid to almond shaped, without a germ pore, dextrinoid. Cheilocystidia cylindrical or elongated and bottle shaped. Clamps absent.

ECOLOGY: Solitary or scattered in small groups in duff, usually under redwood, fruiting from early fall into winter. Fairly common throughout our range, but seems to fruit much more abundantly in some years than others.

EDIBILITY: Unknown.

COMMENTS: The smooth to silky-fibrillose caps, small size, white colors overall, and ring on the stipe distinguish this species complex. Some species in the group have cheilocystidia with crystals at the tip, others without. Two other similar small, whitish species are found in our area, but grow primarily in urban or disturbed areas (not in forests): *Leucocoprinus cepistipes* has small scales on the cap with distinct ridges and pleats around the edge, and has a fairly smooth stipe. *L. cretaceus* has a distinctly powdery or cottony-shaggy cap and stipe; it is fairly rare in our area, primarily fruiting on old manure and compost piles. *L. sequoiarum* and its close relatives likely belong in the genus *Leucoagaricus*.

Leucocoprinus brebissonii

(Godey) Locq.

CAP: 1–7 cm across, oval to egg shaped at first, expanding to broadly convex, usually with an indistinct, low umbo. Surface dry, dark gray or mousy tan-gray and very finely felty near center, whitish and radially fibrillose outward, covered with many flat, light grayish or tan scales and a large, unbroken, dark, circular patch over umbo. Edge of cap usually distinctly striate or pleated. **GILLS:** Free, close, broad, edges smooth. Whitish to pale creamy. **STIPE:** 3–7 cm long, 0.2–0.4 cm thick, cylindrical, slightly enlarged or club shaped at base. White to light creamy. Surface smooth or slightly fibrillose. **PARTIAL VEIL:** A thin whitish ring with a collarlike lower part, often remaining attached to cap edge or disappearing. **FLESH:** White, very thin, fragile, hollow in stipe. **ODOR:** Indistinct. **TASTE:** Not sampled. **SPORE DEPOSIT:** White. **MICROSCOPY:** Spores 9–12 x 5.5–7 µm, ellipsoid to almond shaped, smooth, with a large germ pore.

ECOLOGY: Scattered in small troops in humus and wet leaf litter in a variety of forested settings, and occasionally in rich soil in urban areas. Fruiting in late summer and early fall. Common (and more frequently in forests) on the Far North Coast, uncommon elsewhere. Found at least as far south as the San Francisco Bay Area.

EDIBILITY: Unknown.

COMMENTS: This species can be recognized by its pleated whitish cap with a distinct, dark grayish circular patch at the center and by its habit of fruiting early in the season. Often confused with members of the *Lepiota atrodisca* group, but those can be distinguished by their finely hairy cap surface (lacking flat scales) and substantially smaller spores without a germ pore. See comments under *L. atrodisca*, as well as under *L. cristata* for other similar species. *Leucoagaricus meleagris* also grows in gardens and urban settings, but has larger, densely clustered fruitbodies that have fine dark grayish brown scales on the cap, stain yellow to pinkish red when bruised, and have spores with an inconspicuous germ pore.

Lepiota atrodisca group
BLACK-EYED PARASOL

CAP: 1.5–5 cm across, egg shaped to domed when young, soon expanding to broadly convex and then becoming flat, occasionally weakly umbonate. Surface dry, at first densely covered with fine, pointed clumps of blackish gray fibrils, soon expanding to show a radially fibrillose white background. In age, dark grayish to nearly black at center, then mousy gray outward, whiter toward margin. Often with a narrow band of shaggy white tufts hanging over margin when young. **GILLS:** Free, close, edges smooth or very finely beaded with a white band of cystidia. White to light creamy. **STIPE:** 3–7.5 cm long, 0.2–0.4 cm thick at apex, cylindrical with a slightly enlarged base, straight to curved, smooth or with a few flattened whitish chevrons of tissue just below ring. White to creamy, becoming dingy beige, occasionally covered with many droplets of golden liquid when young. **PARTIAL VEIL:** A thin white membrane forming an upward-flaring ring with a grayish to black, finely fuzzy edge. **FLESH:** Whitish, very thin, hollow in stipe. **ODOR:** Indistinct or slightly rubbery. **TASTE:** Not sampled. **SPORE DEPOSIT:** White. **MICROSCOPY:** Spores 6–8 x 3–5 µm, ellipsoid, smooth, slightly dextrinoid.

ECOLOGY: Solitary or scattered in troops in humus and duff, typically under redwood. Common and widespread, occurring throughout our range. Fruiting primarily in fall and early winter.

EDIBILITY: Unknown.

COMMENTS: The small size, dark blackish gray fibrillose-scaly cap, and grayish edged ring on the stipe help distinguish this species. Recent genetic data suggest that there are a handful of taxa under this name in our area and they likely belong in the genus *Leucoagaricus*. These are currently only distinguishable by genetic signature, but careful investigation by the community of mushroom enthusiasts should turn up other ways to tell them apart. See comments under *L. cristata* for another group of species that are somewhat similar in stature and color patterning but with brownish rather than black or gray tones.

Lepiota flammeotincta

Kauffman

FLAMING PARASOL

CAP: 1.5–7.5 cm across, domed at first, expanding to broadly convex and then flattening and often splitting in age. Surface dry, innately fibrillose, blackish brown at center, with dark brown, flattened, pointed tufts of fibrils radially arranged outward, often showing paler whitish beige in between when young, but soon more extensively brown (with little pale color remaining in age). Staining brilliant cherry red to blood red when touched, quickly fading to brown. **GILLS:** Free, close to crowded, broad. White, not staining when damaged. **STIPE:** 4.5–12 cm long, 0.4–0.7 cm thick at apex, enlarged and club shaped near base. Often tall, straight to curvy. Whitish to cream at first, very soon developing tan and brown areas mottled with fibrils. Staining brilliant scarlet to deep orange-red immediately when touched, quickly fading to brown. **PARTIAL VEIL:** A thin, flaring ring on stipe, whitish above, tan to brown below, staining red. **FLESH:** Fairly thin, white, fibrous, soft, stipe soon hollow. Whitish cream, staining red when cut. **ODOR:** Indistinct or slightly chemical. **TASTE:** Not sampled. **KOH:** Red becoming green on all parts, often obscured by inherent staining. **SPORE DEPOSIT:** White. **MICROSCOPY:** Spores 6–9 × 3.5–5.5 μm, ellipsoid, elongate-ellipsoid, or slightly tapered at one end, smooth, dextrinoid. Cheilocystidia abundant, cylindrical, walls often distinctly wavy, some narrowly club shaped.

ECOLOGY: Solitary or scattered in small groups in needle duff or mossy humus under conifers. Especially abundant in early fall under redwood, pine, and Sitka Spruce. Common throughout our range.

EDIBILITY: Unknown.

COMMENTS: The dark, fibrillose cap, white gills, and brilliant red staining of the cap and stipe (but not the gills) help identify this species. This is one of the more distinct members of an often difficult-to-identify species complex. *Leucoagaricus flammeotinctoides* is very similar and also common, but has gills that stain orangey red, and has cheilocystidia without wavy walls. *L. pyrrhulus* has a dense, even coating of fine dark fibrils on the cap, gills that do not stain when damaged (but turn coppery when dried), and almond-shaped spores. *L. pyrrhophaeus* has gills that stain orange when bruised (but remain pale when dried) and has ellipsoid spores. Both of the two latter species tend to be smaller and more slender, and have variably shaped cheilocystidia (including bottle- and urn-shaped cells as well as club-shaped ones). *L. erythrophaeus* immediately stains reddish on the cap and stipe and orangey on the gills, but is larger and stouter, with gills that are fused into a "collar" near the stipe. It has long (up to 90 μm), club-shaped cheilocystidia, and a different cap surface with many upright cells (a trichoderm). *Lepiota castanescens* stains less intensely orange-red and has spores with a distinct nipple at the end. Expect to see many of these names reshuffled into the genera *Leucoagaricus* or *Leucocoprinus* in the near future.

Lepiota castanea

Quél.

CHESTNUT PARASOL

CAP: 1–4 cm across, rounded at first, expanding to broadly convex and then flat, usually with a distinct umbo. Surface dry, with a dark reddish brown to ocher-brown center, extensively creamy beige to nearly white toward margin with many small, flat, fibrillose tufts surrounding umbo. Margin often with a band of cottony whitish tissue hanging over edge of gills. **GILLS:** Free, close to subdistant, broad, edges smooth. White to creamy, soon with blotches of apricot to light orange coloration, occasionally evenly pale orange. Bruising orangish when damaged. **STIPE:** 3–6 cm long, 0.2–0.4 cm thick, cylindrical. Whitish to cream at first, but soon dingy pinkish tan with silvery white fibrils overall, developing darker pinkish brown tones near apex, bruising or with orange blotches in age. Base with many scattered dark reddish to orange-brown scales. **FLESH:** Thin, fragile. Stipe hollow with a white pith. White to creamy, with light ocher tones. **ODOR:** Indistinct. **TASTE:** Not sampled. **SPORE DEPOSIT:** White. **MICROSCOPY:** Spores 9–12 x 4–5 μm, elongate-ellipsoid, but with a bulging, asymmetrical knob at one end. Cheilocystidia fat, bottle shaped with a round tip or club shaped. Cap covered with upright brown-walled cells.

ECOLOGY: Solitary or scattered in small troops in duff, moss, and rich humus of many forest types, but especially found of Douglas-fir, Sitka Spruce, and other conifers. Fruiting throughout the season, especially from late fall into early spring.

EDIBILITY: Deadly toxic; contains amatoxins (as in *Amanita phalloides*, *A. ocreata*, and *Galerina marginata*).

COMMENTS: The ocher-brown to reddish brown cap, scattered dark scales over the lower stipe, light ocher to rusty stains, and lack of a ring on the stipe make this *Lepiota* fairly easy to recognize in our area. *L. felina* has darker brown colors (lacking ocher tone) and has a distinct flaring ring on the stipe.

Lepiota felina

(Pers.) P. Karst.

CAP: 1.3–4 cm across, domed to oval when young, expanding to broadly convex, not usually umbonate. Surface dry, slightly tomentose, very dark brown when young, cracking into scales as it expands. Scales paler brown to tan toward margin with extensive whitish to creamy color between scales. Center of cap usually retaining a blackish brown circular patch in age. Edge of cap often with overhanging band of fibrillose white tissue. **GILLS:** Free, close, broad, edges smooth. White to light creamy. **STIPE:** 3–7 cm long, 0.3–0.6 cm thick, cylindrical. White to creamy, becoming tan, silky-smooth above partial veil, with dark brown scales arranged in rings and/or chevrons around base. **PARTIAL VEIL:** Whitish with a dark edge, cottony or slightly membranous when unopened, soon forming a thin, flaring skirt high on the stipe, usually with a light brown underside or sometimes with multiple dark brown scales around edge. **FLESH:** Thin, fibrous, white. **ODOR:** Indistinct. **TASTE:** Not sampled. **SPORE DEPOSIT:** White. **MICROSCOPY:** Spores 6–8.5 x 4.5–5 μm, ellipsoid, one side very slightly bulging, other side straight. Cheilocystidia club shaped or swollen with a slightly pointed tip. Cap surface long with upright dark-walled elements.

ECOLOGY: Solitary, in small clusters, or scattered in groups in moss or conifer duff. Quite rare in our area, with a few records from the Far North Coast. Fruiting in late fall and early winter.

EDIBILITY: Unknown; to be avoided, as close look-alikes are deadly toxic.

COMMENTS: The distinctive pattern of dark scales adorning the cap, ring, and lower stipe makes this lovely species fairly easy to recognize. *Lepiota atrodisca* shares the dark-edged ring on the stipe, but the cap is covered in fine blackish gray hairs and is generally more fragile. *L. phaeoderma* also has a dark cap surface cracking up into scales, but lacks a distinct ring on the stipe, and its cap cuticle is made up of club-shaped cells. Neither of these has dark scales around the base of the stipe. *L. phaeoderma* is an uncommon to rare species, known from San Mateo County under Monterey Cypress, and mixed woods elsewhere. *L. castanea* is very common and has dark scales on the cap and stipe, but they are more reddish brown or ocher in color, and the fruitbodies are often extensively orange stained.

Lepiota cristata

(Bolton) P. Kumm.

CAP: 2–6.5 cm across, oval or rounded-conical when young, expanding to broadly convex, occasionally slightly wavy. Often with a distinct low, round umbo. Surface dry, deep reddish brown or dark brown when young, but soon cracking into small flat scales (often concentrically arranged), revealing a whitish to creamy, radially fibrillose background, often extensively white and without scales near margin. **GILLS:** Free, fairly close, broad, whitish to creamy. **STIPE:** 3–8 cm long, 0.3–0.5 cm thick (up to 0.7 cm thick at base), cylindrical, fairly straight, base only slightly larger. White to light beige, developing tan or pale pinkish tan tones, often overlaid with silvery white fibrils. Smooth to silky. **PARTIAL VEIL:** Membranous ring, usually fairly persistent but not particularly conspicuous (not flaring). Upperside whitish, underside often light tan or with a few brownish scales. Sometimes remaining attached to cap edge instead of stipe or disappearing entirely. **FLESH:** White, very thin, fibrous. **ODOR:** Usually strong like burnt rubber or burning hair. **TASTE:** Not sampled. **SPORE DEPOSIT:** White. **MICROSCOPY:** Spores 5.5–8.5 x 2.5–3.5 µm, with a rounded triangular outline, or bullet shaped with a knob above apiculus, weakly dextrinoid. Cheilocystidia club shaped. Pileipellis a layer of upright club-shaped cells.

ECOLOGY: In clusters or scattered in troops in humus, grassy areas, or wood chips in disturbed areas, especially along trails and roadsides and in urban habitats. Generally uncommon in our area, but can be locally frequent, especially in fall and early winter. Probably can be found throughout the year in watered areas.

EDIBILITY: Unknown. Very similar to some extremely toxic species.

COMMENTS: The somewhat large size (for a *Lepiota*), warm brown, strongly contrasting cap scales, burnt rubber odor, distinct ring on the stipe, growth in disturbed areas, and spurred spores help distinguish this species (but see comments under *L. castaneidisca*). A number of more fragile, slender species share the brown-centered cap but lack the bullet-shaped spores: *Leucoagaricus ophthalmus* has the cells of its cap surface in an irregular upright arrangement (a trichoderm) with scattered clumps of elongated, bottle-shaped cells with brownish walls; it has short, club-shaped cheilocystidia. *Lepiota oculata* and *Leucoagaricus paraplesius* also have upright cells in the cap surface, but both have more elongate-cylindrical cheilocystidia (those of the latter species are slightly more enlarged at the tip and distinctly narrowed toward the base). *Leucoagaricus infuscatus* has sausage-shaped cap cells in chains arranged horizontally (a cutis) and has ellipsoid-rounded spores (the three species mentioned prior have more almond-shaped spores that are often distinctly tapered at both ends). All of these are found in our area, but *Leucoagaricus ophthalmus* and *Lepiota oculata* are the most common and widespread of the bunch.

Lepiota castaneidisca

Murrill

CAP: 1–4.5 cm across, egg shaped to domed, expanding to convex or nearly flat, occasionally weakly uplifted, sometimes with a small, low umbo. Light brick to pinkish brown or pinkish tan when young, soon extensively pinkish beige to creamy outward. Surface dry, nearly smooth when young but almost immediately cracking into fine, flat scales usually very small and not strongly contrasting against pale background. Edge of cap often with overhanging band of shaggy white veil tissue (mostly disappearing in age). **GILLS:** Free, somewhat close, broad. White to creamy. **STIPE:** 5–6 cm long, 0.2–0.5 cm thick, fairly straight, cylindrical with a slightly enlarged base. White to cream with silvery fibrils, becoming beige and often pinkish tan near apex. Surface smooth, silky. Mycelium at base fine and cottony with a few very fine rhizomorphs. **PARTIAL VEIL:** White, fairly persistent ring with a slightly fuzzy or smooth edge. **FLESH:** White, very thin, fibrous. **ODOR:** Indistinct or slightly like burnt rubber. **TASTE:** Not sampled. **SPORE DEPOSIT:** White. **MICROSCOPY:** Spores 5–9 x 3–4 µm, bullet shaped with a rounded knob above apiculus, dextrinoid. Cheilocystidia club shaped, often fairly narrow. Cells of pileipellis club shaped, upright, densely packed.

ECOLOGY: Solitary or more often in small groups or scattered in troops. Fruiting from late fall through winter in rich, moist duff under Monterey Cypress, Coast Redwood, and sometimes other forest types. Quite common throughout our area.

EDIBILITY: Likely toxic.

COMMENTS: Telling this species apart from its relatives is a challenge: The pinkish tan cap that cracks into small, inconspicuous scales, small but distinct white ring on a smooth stipe, and bullet-shaped spores are a good suite of clues. *L. cristata* is slightly larger, with darker, more orange-brown scales, and is more common in disturbed habitats. *L. thiersii* has ellipsoid (not spurred) spores, lacks a distinct ring on the stipe, and lacks cheilocystidia. Although very similar in color and stature, *L. neophana* has a smooth cap (not breaking up into scales), often has spiky mycelium at the base of the stipe, and has ellipsoid spores. *L. scaberula* has pale ocher scales on the cap and lacks a distinct ring on the stipe; it occurs around the San Francisco Bay Area under Monterey Cypress and redwood. Else Vellinga examined specimens identified as *L. castaneidisca* and found two clearly differentiated genetic groups, but could not establish any microscopic or macroscopic means to tell them apart. *L. subincarnata* has darker pinkish cap scales, is usually more stout, and lacks a distinct ring on the stipe (but may have flattened chevrons). *L. luteophylla* is a very rare species known from under Monterey Cypress in San Mateo County; it lacks cheilocystidia and has nondextrinoid spores, but is most easily distinguished by its smooth (not scaly) brown cap and intensely yellow gills.

Lepiota neophana

Morgan

CAP: 1–5 cm across, oval when young, soon expanding to bell shaped and then broadly umbonate-convex to nearly flat, sometimes with a narrow sterile margin exceeding gills. Brick red to pinkish brown, very evenly colored (not breaking into plaques), in age becoming pale creamy beige with a white edge. Very slightly viscid at first, soon dry, smooth to finely wrinkled. **GILLS:** Free, close, broad. White. **STIPE:** 3–7 cm long, 0.2–0.6 cm thick, cylindrical with base very slightly larger. Whitish to light pinkish beige. Surface smooth, silky with vertical fibrils. Base often with white, spiky mycelium. **PARTIAL VEIL:** A thin white membrane, often with a pinkish tan to brown underside, remaining as an upward-flaring collar-like band or a flattened, contrastingly whitish zone on stipe. Sometimes remaining attached to edge of cap, occasionally disappearing. **FLESH:** Thin, white to dingy. **ODOR:** Indistinct or slightly like burnt rubber. **TASTE:** Not sampled. **SPORE DEPOSIT:** White. **MICROSCOPY:** Spores 4–5.5 x 2.5–3.5 μm, average 5 x 3 μm, ellipsoid, smooth, nondextrinoid. Cap cells upright, densely packed cells, pear shaped (swollen with a narrow base).

ECOLOGY: Solitary or in small troops in moist duff of disturbed areas under live oak, Monterey Cypress, and California Bay Laurel. Uncommonly encountered from midfall through spring, mostly known from south of Sonoma County, but likely more broadly distributed.

EDIBILITY: Unknown, to be avoided.

COMMENTS: The small, slender stature, evenly brick-colored to light pinkish beige cap, white gills, and ring on the stipe help distinguish this species. *L. castaneidisca* is quite similar in stature and color, but its cap surface usually breaks into small scales or plaques; it also has very differently shaped spores (spurred or bullet shaped). See comments under *L. castaneidisca* for other similar taxa. The slightly viscid cap of *L. neophana* (when young) and its spiky basal mycelium are unusual for a *Lepiota* and further distinguish it, but these features can be hard to see. This species is most likely to be confused with various *Leucoagaricus* species, most of which are distinctly larger and often show pronounced staining reactions. *Limacella* species can look similar but are larger and/or distinctly viscid overall.

Lepiota subincarnata

J. E. Lange

DEADLY PARASOL

CAP: 1.5–6 cm across, tightly convex when young, becoming broadly convex and then expanding to nearly flat or slightly uplifted, sometimes with a wavy margin. Surface dry, distinctly pinkish or light pinkish brown, often with a narrow, bright white band around edge of cap when young. At first entirely felty or matted-tomentose, soon breaking up into concentric rings of fibrillose scales with paler pinkish beige or creamy background showing through between them, overall fading to dull ocher or beige-tan in age or dry weather. **GILLS:** Free, closely spaced, broad, edges often eroded. White to creamy, becoming light beige. **STIPE:** 2.5–6 cm long, 0.4–0.6 cm thick, cylindrical. Whitish when young, soon light beige-tan to pinkish but covered with silvery whitish fibrils. Silky-smooth above annular zone, with chevrons of brown or pinkish tan tissue near base. **PARTIAL VEIL:** Silky-cobwebby at first, soon breaking and forming an indistinct zone around middle of stipe. **FLESH:** Whitish cream, thin, fibrous, hollow in stipe. **ODOR:** Indistinct, sweetish or a bit like burnt rubber. **TASTE:** Not sampled. **SPORE DEPOSIT:** White. **MICROSCOPY:** Spores 6–7.5 x 3–4 μm, ellipsoid or elongated oval. Cheilocystidia club shaped.

ECOLOGY: In clusters and troops in wood chips and rich soil in disturbed and urban areas. Locally common in the San Francisco Bay Area, but has also been found in mossy duff under Sitka Spruce on the Far North Coast. Rather uncommon through rest of our region. Fruiting throughout the wet season, but most common in winter.

EDIBILITY: Deadly toxic; contains amatoxin (the same compound found in *Amanita phalloides*, *A. ocreata*, and *Galerina marginata*).

COMMENTS: This deadly poisonous species is recognizable by its finely felty pinkish cap extensively covered in concentric rings of scales, somewhat stout stature (for a *Lepiota*), and lack of a distinct ring or skirt on the stipe. See comments under *L. cristata* and *L. castaneidisca* for similar species.

Lepiota "Cypress Scaly"

CAP: 1–5 cm across, egg shaped when young, soon broadly convex, expanding to plane, often with a weakly inrolled margin adorned with whitish teeth of partial veil tissue. Surface dry, densely covered in pointed, scaly tufts when young, soon becoming matted-felty, scales flattening toward margin or falling off, overall smoother at maturity. Warm ocher-tan to light beige, often white toward margin, covered in reddish brown, felty-fibrillose scales. **GILLS:** Distinctly free, very close to crowded. White. **STIPE:** 2.5–5 cm long, 0.6–1.2 cm thick, cylindrical. White to creamy, silky-smooth or slightly cobwebby near apex, lower half shaggy looking, with many tan to brick brown or dark brown pointed scales, often arranged in weak rings. **PARTIAL VEIL:** White felty-cobwebby tissue covering gills when young, not forming a ring on stipe in age. **FLESH:** White, fibrous. **ODOR:** Indistinct. **TASTE:** Mild. **SPORE DEPOSIT:** White. **MICROSCOPY:** Spores 4–4.5 x 2.5–3 μm, ellipsoid, dextrinoid. Cap with many inflated, thin-walled brownish cells in chains. Cheilocystidia present.

ECOLOGY: In small groups in rich needle humus of Monterey Cypress, uncommon to rare around the San Francisco Bay Area. Fruiting primarily in winter and early spring, but can likely fruit at any time during the rainy season.

EDIBILITY: Unknown.

COMMENTS: The small but stocky stature, whitish cream color with many brown scales, rings on the cap and stipe, and lack of a prominent ring on the stipe distinguish this attractive species. It is a close genetic match for *L. carinii*, a species known from Europe. A number of other similar species occur in our area, including a slightly larger species with more extensively pinkish brown to reddish brown colors, which grows under redwood. *Echinoderma asperum* is quite a bit larger (caps 6+ cm wide); it has a cobwebby-membranous partial veil that soon forms a very large, hanging skirt; very crowded gills; a strong smell of burnt rubber; and spores up to 9 μm long. It is rare in our area, but has been found in wood chip beds and other disturbed areas with rich soil in Alameda, Santa Cruz, and San Mateo Counties.

Lepiota spheniscispora

Vellinga

PENGUIN-SPORED PARASOL

CAP: 3–8 cm across, round-conical when young, becoming broadly convex or flat, sometimes with a low, round umbo, occasionally slightly wavy in age. Ocher-tan to golden brown overall, darker warm reddish to brick brown or dark orange-brown at center, more golden outward to nearly white at edge. Surface dry, plush near disc, covered densely in fine scales, often splitting, with bare areas near margin of cap. Extreme margin often with hanging scales or flakes of cap cuticle and veil tissue. **GILLS:** Distinctly free (often with a wide "gutter" around the stipe), fairly close, broad. White to creamy. **STIPE:** 5–10 cm long, 0.5–1.5 cm thick, straight to slightly curved, slightly club shaped. Rhizomorphs and grayish or whitish mycelium often abundant at base. Pale cream to tan, surface cottony, areas near base more shaggy and often bearing brownish scales or chevrons. **PARTIAL VEIL:** Layer of cottony-fluffy tissue between stipe and cap edge when very young, not forming a ring or skirt. **FLESH:** White, soft, fibrous. **ODOR:** Mild or slightly rubbery. **TASTE:** Not sampled. **SPORE DEPOSIT:** Whitish. **MICROSCOPY:** Spores 13–18.5 x 4.5–5.5 μm, dextrinoid, elongate and somewhat spindle shaped, but curvy and with a small pointed apiculus, overall resembling a penguin in profile. Cheilocystidia club shaped or fat and bottle shaped.

ECOLOGY: Solitary or in small groups widely scattered over large areas under live oak in early fall. Seems to prefer warmer microhabitats than *L. magnispora* and thus more common in the southern part of our area.

EDIBILITY: Possibly toxic.

COMMENTS: Very similar to *L. magnispora*, but with practice can be told apart by its more evenly colored cap and (usually) less markedly contrasting umbo, less scaly stipe surface, and slightly different fruiting habits. This species is more frequent in open woodlands, especially under live oak and eucalyptus, early in the season, while *L. magnispora* fruits in darker, denser forests (often under redwoods) later in the season. Most other *Lepiota* are smaller, have a more skirtlike partial veil, and/or lack the silky-shaggy stipe texture.

Lepiota magnispora

Murrill

CAP: 2.5–10 cm across, rounded or egg shaped when young, becoming convex with a distinct round umbo, nearly plane with a low umbo, or occasionally uplifted in age. Dull yellowish beige to pale tan but covered in orange-brown to reddish brown scales. Umbo much darker ocher-brown, reddish brown to mahogany red. Surface dry, nearly smooth when very young but almost immediately cracking into fairly coarse scales concentrated at center, margin of cap often with cottony whitish to yellow veil remnants. **GILLS:** Free, close, soft, and sometimes eroded. Whitish when young, then creamy to pale yellowish. **STIPE:** 5.5–12 cm long, 0.6–1.5 cm thick, cylindrical, straight to slightly curved, base slightly swollen. Pale yellow to yellowish beige overall, in age ocher-tan, cottony-floccose overall or with shaggy bands and chevrons, often with adhering ocher-brown scales bound to soft fibrils especially around base of stipe. **PARTIAL VEIL:** Copious amount of cottony, creamy to whitish tissue between gills and stipe (enclosed in membranous layer of brown scales when very young), soon collapsing or disappearing, rarely leaving a distinct ring zone and never forming a membranous ring. **FLESH:** White, fibrous and stringy, soft, stipe with an indistinct central hollow. **ODOR:** Mild or often like burnt rubber. **TASTE:** Not sampled. **SPORE DEPOSIT:** White. **MICROSCOPY:** Spores 15.5–20.5 x 4.5–6 μm, elongate to spindle shaped, but with a curved side and a potbellied side, as well as a small, pointed apiculus (overall looking like a penguin). Cheilocystidia absent.

ECOLOGY: Solitary, scattered, or in small clusters in deep, moist redwood duff from late fall through winter; also common under Sitka Spruce and Douglas-fir. Common throughout our area.

EDIBILITY: Possibly toxic.

COMMENTS: The scaly shaggy stipe, red-brown umbo, egg-shaped young caps, and yellowish hues to the veil and stipe make this a striking denizen of coastal conifer forests. Although larger and more warmly hued than most other *Lepiota*, confusion is possible with the extremely similar *L. sphenicispora*, which has a more golden tan cap with a less defined umbo (but this feature is variable and takes practice to use reliably). The latter species mostly grows under live oak, has a smoother stipe, and fruits earlier in fall and later in spring. It rarely has such egg-shaped young caps as *L. magnispora*.

Lepiota rubrotinctoides

Murrill

CAP: 3–6.5 cm across, egg shaped to domed at first, becoming broadly convex to plane, occasionally slightly uplifted with a wavy margin in age. Usually with a distinct round umbo. Blackish red, wine red, reddish brown, or pinkish red at center, lighter pinkish red to coral or orangish outward with a distinctive tendency to split radially. White to creamy beige toward margin. Surface dry, smooth. **GILLS:** Free, close, broad, edges smooth. White. **STIPE:** 4–9 cm long, 0.6–1 cm thick at apex, cylindrical with an enlarged, club-shaped base (up to 2 cm thick). White to creamy. Smooth or slightly fibrillose. **PARTIAL VEIL:** A prominent, upward-flaring ring, lower part forming a wide collarlike band. Edge finely fuzzy, sometimes with orange or pinkish colors. **FLESH:** Fibrous, fairly thick. Stipe with a silky-stuffed central hollow. White, bruising occasionally weakly orange around base of stipe in age. **ODOR:** Indistinct. **TASTE:** Mild. **SPORE DEPOSIT:** White. **MICROSCOPY:** Spores 7–8.5 × 4–5 (5.5) μm, ellipsoid to almond shaped, smooth, dextrinoid. Cheilocystidia club shaped, elongate.

ECOLOGY: Solitary or scattered in small troops in humus, leaf litter, or coniferous needle duff in forested areas. Common and widespread. Fruiting very early in the season; soon after the first rains, rarely found in winter.

EDIBILITY: Nontoxic.

COMMENTS: The fairly large fruitbodies, strong contrast between the cap center and the margin, warm pink to orange-brown colors, and big, white funnel-shaped ring on the stipe characterize this species. However, there are a dozen or more similar species in our area, only one or two of which have published names; they are never quite as large as *L. rubrotinctoides* and have pinkish to dull orange or brownish colors. One variant with a slightly more scaly, apricot-colored cap has been provisionally named *L. armeniacus* by Else Vellinga. This is a great example of a species complex that amateurs can help clarify. Remember to make good pictures, take detailed notes, and voucher your dried specimens. By doing so, you can help build robust concepts for these species and even give them names! *Leucoagaricus pardalotus* also has a tendency to split radially toward the margin, but is smaller, with very dark reddish brown to nearly black cap scales and fibrils (lacking coral pinkish tones).

Lepiota roseolivida

Murrill

ROSY PARASOL

CAP: 1.5–4 cm across, broad, rounded to broadly ovoid at first, becoming broadly conical or conical with a low, poorly defined umbo, finally broadly convex or nearly plane with an umbo. Purplish at center, wine colored to rosy pink outward, fibrils pinkish lilac over a whitish or creamy surface. Surface dry, densely velvety at disc, fibrillose overall, with fine hairs (often flattened into chevrons toward margin). **GILLS:** Free, whitish. **STIPE:** 4–8.5 cm long, 0.3–0.9 cm thick, cylindrical, nearly equal overall, sometimes slightly enlarged at base or club shaped. White, smooth. **PARTIAL VEIL:** A thin, membranous ring or band, usually flared upward, sometimes disappearing in age. White with a distinct fuzzy, pinkish purple edge. **FLESH:** Thin, fibrous, white. **ODOR:** Indistinct or slightly rubbery as in many small *Lepiota*. **TASTE:** Not sampled. **KOH:** Indistinct or obscurely brownish or green. **SPORE DEPOSIT:** White. **MICROSCOPY:** Spores 6–9.5 x 4–5.5 μm, almond shaped or elongate-ellipsoid, dextrinoid. Cheilocystidia abundant, cylindrical to club shaped or like elongated bowling pins.

ECOLOGY: Solitary or scattered in small groups in duff of Monterey Cypress and redwood, usually in dark, moist microhabitats. It also has been found in the leaf litter of shrubs, including Toyon (*Heteromeles arbutifolia*) and Laurel Sumac (*Malosma laurina*). Quite uncommon in our area, mostly occurring on the Central Coast in fall and winter.

EDIBILITY: Unknown.

COMMENTS: Although there are a number of pinkish to purplish rose-capped *Lepiota* and *Leucoagaricus*, this beautiful species stands out due to its slender, diminutive stature, and finely hairy cap surface with deep purplish tones at the center. *Lepiota decorata* is larger and stouter and has a less fibrillose texture and a more evenly rosy pink cap.

NOMENCLATURE NOTE: This species likely belongs in the genus *Leucoagaricus*, but hasn't been transferred yet.

Lepiota decorata

Zeller

PINK PARASOL

CAP: 2–5 (6) cm across, domed at first, buttons often with a squared-off lower edge, expanding to broadly convex, occasionally nearly flat, sometimes lobed. Surface dry, dull rosy pink, purplish pink, lilac, or grapefruit pink, breaking up into fine, flat scales with whitish to light cream surface showing between scales. Often bruising orangish red when scratched or rubbed. **GILLS:** Free, close, broad, edges smooth. White to creamy. **STIPE:** 3.5–8.5 cm long, 0.7–1.5 cm thick, cylindrical with a slightly enlarged, club-shaped base. White to creamy beige with many pinkish fibrils below veil, often bruising orangish red when young. Surface silky to smooth. **PARTIAL VEIL:** A whitish membranous ring when young, forming a flaring ring in age with a pinkish underside (nearly same color as cap), especially dark around edge. **FLESH:** Fibrous, white. Stipe stuffed or with a narrow hollow. **ODOR:** Indistinct. **TASTE:** Not sampled. **SPORE DEPOSIT:** White. **MICROSCOPY:** Spores 5.5–7.5 x 3–5 μm, ellipsoid, smooth, dextrinoid. Cheilocystidia club shaped. Pileipellis elements long and cylindrical, mostly horizontal but can be more upright, especially in age.

ECOLOGY: Solitary, in small clusters, or in small troops, rarely abundant, in rich humus in a wide variety of habitats. It has been found under old-growth conifers, under alder and Bigleaf Maple, in live oak duff, and in eucalyptus litter. Occurring throughout our range from fall into spring.

EDIBILITY: Unknown.

COMMENTS: The medium-size fruitbodies with rosy pink colors and the pink-edged flaring ring on the stipe help distinguish this species. A very similar species is regularly found in rich humus under Monterey Cypress in the San Francisco Bay Area; genetic data indicate that it is a distinct, unnamed species.

NOMENCLATURE NOTE: This species likely belongs in the genus *Leucoagaricus*, but hasn't been transferred yet.

Leucoagaricus cupresseus

(Burl.) Boisselet & Guinb

CAP: 4–10 cm across, rounded at first, soon expanding to broadly convex or slightly square (broad and flat at center) with a low, round umbo. Flat in age. Blackish brown at center when young, dark brown outward, with a mottled appearance as it expands, extensively creamy white near margin or overall in age. Surface dry, plush-velvety at first, cracking into fine fibrillose scales outward, nearly smooth in age or when weathered. **GILLS:** Distinctly free, close, broad. White, becoming creamy, often bruising light yellow and then light orange-red. **STIPE:** 4–12 cm long, 1–2 cm thick at apex, cylindrical, distinctly enlarged into a bulb at base. Variable in stature, usually stout and upright, often curved near base. Pale creamy at first but soon extensively dingy tan, developing brown chevrons and streaks, weakly staining orangish to reddish when scraped. **PARTIAL VEIL:** A flaring ring attached to a broad band or collar over middle of stipe. Frequently contrastingly white against the dingy stipe, edge often dark brown. **FLESH:** Thick, fibrous. Stipe with a cylindrical hollow. White, bruising irregularly orange to reddish in flesh of cap and stipe. **ODOR:** Indistinct or pleasantly nutty, musty in age. **TASTE:** Mild. **KOH:** Reddish with a greenish gray ring on stipe, greenish on gills, slowly green on cap. **SPORE DEPOSIT:** White. **MICROSCOPY:** Spores 6.5–8 x 4–5 µm, ellipsoid or oval, sometimes a bit tapered at one end with an obscure nipple, dextrinoid. Cheilocystidia variable, club shaped or bowling pin shaped, or swollen but with a narrow, wavy, cylindrical tip.

ECOLOGY: Solitary or scattered in small groups, almost always in duff under Monterey Cypress, occasionally in rich humus of nonnative trees or in garden soil. Mostly occurs on the coast from Monterey County north into Marin County. Fruiting from later fall into early spring.

EDIBILITY: Unknown.

COMMENTS: This species is recognizable by its large size, white gills, stout stipe with a prominent ring, dark brownish plush-textured cap, and growth under cypress. Similar species that are apparently restricted to Monterey Cypress habitat include two smaller-capped, slender-stemmed species: *L. hesperius*, with a warmer reddish brown cap and gills that stain dark red, and *L. dyscritus*, with a cap surface made of chains of short cells rather than long elements like all the others listed here. *L. adelphicus* is a bit larger than the latter two, but still more slender than *L. cupresseus*, with gills that don't bruise reddish and remain pale when dried; it is more common than the latter two and occurs in other habitats besides cypress. *L. marginatus* is another species known from Monterey Cypress habitats, but our concept of the species remains weak; it seems to be extremely similar to *L. cupresseus* but with a paler cap that is reddish lilac at the center. *Lepiota fuliginescens* has grayish brown patches on the cap and swollen cheilocystidia with long and irregularly constricted, or small, inflated tips; it is not uncommon in our area and grows in a wide range of habitats. *Lepiota castanescens* is a slender species with a brownish red cap surface that stains orangey red, and has almond-shaped spores with a small nipple.

Leucoagaricus barssii

(Zeller) Vellinga

CAP: 3.5–10 cm across, domed at first, becoming broadly convex, occasionally flat. Cool medium gray to grayish tan, paler toward edge or paler overall in age. Surface dry, densely fibrillose to matted at center, splitting up into radially arranged fibrillose scales as cap expand. Margin often with an overhanging band of shaggy white tissue. **GILLS:** Free, close, broad, edge smooth. White. **STIPE:** 3–15 cm long, 1–2 cm thick at apex, usually club shaped at base or spindle shaped (swollen at middle and tapered downward). Dingy white to light grayish or with tan tones. Surface dry, smooth. **FLESH:** Fibrous, firm to soft, hollow in stipe. White, bruising orange to reddish when cut or scraped, especially at base of stipe. **ODOR:** Indistinct or musty in age. **TASTE:** Mild. **SPORE DEPOSIT:** White. **MICROSCOPY:** 7.5–9 × 4.5–5.5 µm, ellipsoid or oval, smooth, without a germ pore, dextrinoid. Cheilocystidia club shaped or cylindrical.

ECOLOGY: Solitary, scattered, or in small clusters, arcs, and rings. Most often in lawns, gardens, and rich humus along trails in disturbed forested areas. Common in the San Francisco Bay Area southward into southern California. Fruiting from early fall into winter.

EDIBILITY: Edible.

COMMENTS: The shaggy grayish cap; free, white gills; and ring on the stipe help identify this species. Often misidentified as the more common *L. leucothites*, but that species usually has a much smoother and often paler cap, and is usually a bit taller in stature. Weathered old specimens can be tough to tell apart. *Agaricus xanthodermus* and *A. californicus* often grow in the same habitat, and the young buttons can look very similar, but the gills soon become pink and then brown in the latter two species.

Leucoagaricus leucothites

(Vittad.) Wasser

CAP: 3.5–10 cm across, marshmallow shaped (rather boxy) to rounded at first, expanding to broadly convex and then flat or occasionally uplifted in age. Typically white or light creamy beige, occasionally cool gray to light grayish. Surface dry, when young very finely velvety and in some places breaking into tiny scales, but more often appearing perfectly smooth and metallic. Edge of cap sometimes with a narrow, ragged fringe of white veil tissue. **GILLS:** Free, white, close to crowded, broad, soft. White, creamy to dingy beige, developing pinkish tones, sometimes extensively stained pinkish gray or salmon in age. **STIPE:** 4–10 (12) cm long, 0.7–1.5 cm thick, cylindrical, with an enlarged, club-shaped base (up to 3 cm thick). Straight and tall or slightly curvy. Whitish to gray, developing dingy beige or tan tones. Base often with felty white mycelium and white rhizomorphs. **FLESH:** Fibrous, white, fairly thick, firm to soft. Stipe with a central hollow. Slowly pinkish to orangey red on outside of stipe base and in stipe flesh when cut or scraped. **ODOR:** Indistinct. **TASTE:** Mild. **SPORE DEPOSIT:** White. **MICROSCOPY:** Spores 6.5–8 x 4.5–5.5 µm, ellipsoid to oval, smooth, with a germ pore, dextrinoid. Cheilocystidia irregular, club shaped.

ECOLOGY: Solitary or more often scattered in troops, arcs, and rings (sometimes in rather large numbers). Most frequently in grass lawns, also in short-grass meadows, rich garden humus, and occasionally even wood chips in disturbed areas. Rarely in forested settings and then usually along edges of trails. Common throughout our area in early fall, sometimes also in spring and summer.

EDIBILITY: Not recommended. Although some people eat (and enjoy) it, many have adverse reactions. If you decide to eat it, sample it cautiously the first time and cook it well. Remember that the lawns it is so fond of are often treated with herbicides and pesticides!

COMMENTS: The whitish to light gray, smooth caps, flaring ring on a club-shaped stipe, and growth in troops in grassy areas help distinguish this species. The similar *L. barsii* has a shaggier and often more grayish cap; it is somewhat less common. Various *Agaricus* can appear macroscopically similar and grow in the same habitat, but have pinkish and then dark brown gills at maturity.

Chlorophyllum olivieri

(Barla) Vellinga

OLIVE SHAGGY-PARASOL

CAP: 6–18 cm across, egg shaped to rounded when young, expanding to broadly convex and then flat, sometimes with low, broad umbo. At first with light brick brown to tan skin, soon breaking up into clumps of fibrils to show light whitish beige background, rapidly developing extensive olive grayish to dingy brownish tan tones overall, but usually retaining a smooth, pinkish tan to light brown area at center of cap. Surface dry, smooth only when very young, then extensively shaggy-fibrillose. Margin often with an overhanging band of shaggy whitish tissue. **GILLS:** Free, close, broad, soft, edges smooth. White to cream, but easily bruised and turning light orange to pinkish. **STIPE:** 8–20 (25) cm long, 0.8–1.5 cm thick at apex, cylindrical, tall, straight, and slender but with a distinctly swollen to abruptly bulbous base up to 4 cm thick. White to light beige, developing tan tones. Smooth, dry. Base often with copious cottony mycelium and adhering substrate. **PARTIAL VEIL:** A whitish membrane splitting to form a flaring ring with a shaggy, double edge, often grayish tan to brown on underside and edge. **FLESH:** Thin to fairly thick, soft. Whitish, rather quickly bruising pinkish red. Stipe with a narrow, cylindrical hollow. Orange to pinkish or reddish in all parts when cut or scraped (especially stipe flesh and gills). **ODOR:** Mild or pleasantly earthy. **TASTE:** Mild. **SPORE DEPOSIT:** White. **MICROSCOPY:** 8–11 x 5.5–7 μm, ellipsoid, with a small indented "thumbprint"-like germ pore. Cheilocystidia balloon shaped with a very narrow base, sometimes with additional inflated cells near base.

ECOLOGY: In troops, arcs, and rings (occasionally solitary), mostly in rich leaf litter and forest humus in disturbed areas, usually producing largest flushes in fall and early winter. Introduced from Europe to the Pacific Northwest, it is becoming more common in our area and has been recorded as far south as Monterey County.

EDIBILITY: Edible and very good. However, *Chlorophyllum* can cause gastrointestinal upsets due to allergic reactions, so be sure to eat only fresh specimens, cook them thoroughly, eat only a small amount the first time, and make sure it's not *C. molybdites*!

COMMENTS: The stature and texture of this mushroom definitely qualify it for the common name "Shaggy Parasol" and help distinguish it from all but the other *Chlorophyllum* species. It is likely to be confused with *C. brunneum*, which is usually thicker stemmed and slightly shorter and has more contrasting dark brown scales on a light whitish beige cap. *C. brunneum* is more common in our area (it is edible but may be slightly more likely to cause stomach upsets). Mushrooms in the genus *Macrolepiota* look very similar to this species, especially in the tall stature, but have a "snakeskin" pattern of brownish scales on the stipe.

Chlorophyllum brunneum

(Farl. & Burt) Vellinga

WESTERN SHAGGY-PARASOL

CAP: 5–20 cm across, rounded-convex or slightly squared when very young, soon expanding to broadly convex and then flattening out. Medium brown to dull brick brown when young, extensively cracking into plaques of flat brownish scales on top of shaggy whitish beige radial fibrils. In age often dingy brown overall. Surface slightly greasy when very young, soon dry. **GILLS:** Free, close, broad, edges smooth. Whitish when young, becoming dull beige, then brownish in age, often with orangey blotches where bruised, sometimes pale blue-green near stipe. **STIPE:** 8–20 (25) cm long, 2–3 cm thick at apex, 4–6 cm thick at base, cylindrical, with an abrupt bulb with an upturned edge, often forming a "gutter" around top of bulb. Creamy to tan, dingy brownish in age, often shiny with vertical brownish streaks. Bulb covered in felty white mycelium. **PARTIAL VEIL:** Thin but very tenacious, persistent ring, upper surface whitish, underside often with adhering brown patch, edge spiky-shaggy. **FLESH:** Pale creamy beige, fibrous, quite firm when fresh. Narrow cylindrical hollow running through stipe. Quickly orangish when scraped or cut, becoming dingy red and then brown. **ODOR:** Often pleasantly savory, then musty in age. **TASTE:** Mild to pleasantly earthy. **SPORE DEPOSIT:** Whitish. **MICROSCOPY:** Spores 10–13 x 7–8.5 μm, ellipsoid to almond shaped with a small germ pore, dextrinoid. Cheilocystidia club shaped.

ECOLOGY: Solitary, in pairs, or in fused clumps, arcs, or large troops. Favors thick duff of Monterey Cypress, eucalyptus, acacia, oak, and pine. Can be locally abundant, especially along coast, but distribution patchy. Largest flushes often in fall and spring, with scattered fruitings throughout the year.

EDIBILITY: Edible and excellent, but can cause serious gastrointestinal upsets (even in people who've eaten them without problems in the past). Use only very fresh caps, cook them thoroughly, and don't eat large quantities. Also, be aware of the habitat in which they were growing—*Chlorophyllum* are known to accumulate heavy metals.

COMMENTS: The large size, free, whitish gills, brown cap scales, prominent ring on the stipe, orange-staining flesh, and abruptly bulbous base help distinguish this species from most other mushrooms. Often misidentified as *C. rachodes*, but that species has a more gradually swollen stipe base (not rimmed or abrupt) and has shorter, rounder cheilocystidia and slightly smaller spores; it is quite rare in our area. *C. olivieri* has a taller, slender stature, and the scales on the cap don't contrast strongly with the fibrillose background. It also has very round, balloonlike cheilocystidia. Occasionally *C. brunneum* will be encountered with light teal-green hues on its gills or near the stipe apex, and this may cause confusion with *C. molybdites*, a toxic, grass-dwelling species with evenly greenish gray gills at maturity and greenish spores. *Leucoagaricus americanus* is somewhat similar, but stains yellow and then red and usually has a spindle-shaped stipe without a bulbous base; the whole fruitbody turns pink-magenta in old age. This species is rare in California, usually found in gardens or wood chip beds in warmer weather.

Chlorophyllum molybdites

(G. Mey.) Massee

GREEN-GILL PARASOL

CAP: 5–20 cm across, tightly convex or egg shaped when young, soon broadly convex, becoming flat, often with a distinct, rounded umbo; sometimes uplifted in age. Surface dry, entirely covered by a light brown to tan skin when young that soon breaks up into scales. Scales of some specimens uplift around edges and "float" on small flaky granules, or peel off almost entirely. Light leather brown to tan at first, soon restricted to scales; surface in between scales whitish to light beige when mature. Developing dingy grayish tones in age. **GILLS:** Free, close. Whitish when young, becoming light bluish green and then dingy olive-gray, browner in old age. **STIPE:** 7–18 cm long, 1–1.5 cm thick at apex, cylindrical, usually quite even and straight, base most often distinctly swollen (up to 3 cm thick). White, becoming dingy tan to pale brownish, especially where handled, quite smooth. **PARTIAL VEIL:** Neat whitish membrane when young, soon forming a flaring ring with a tattered edge, usually whitish with a brown underside in age. **FLESH:** Whitish, fibrous, stipe firm but with a central hollow, cap often quite soft in age. Weakly orange where cut, reacting stronger when fresh and young, then oxidizing to dark reddish brown. **ODOR:** Indistinct or strongly like *Agaricus bisporus* (Button Mushroom), especially in age. **TASTE:** Mild. **SPORE DEPOSIT:** Dull light olive-green to grayish green. **MICROSCOPY:** Spores 8–13 x 6.5–9 μm, broadly ellipsoid to ovoid, smooth, dextrinoid, with a small germ pore. Cheilocystidia club shaped or spindle shaped.

ECOLOGY: Usually in small scattered groups or sometimes in impressive arcs or rings with 20+ fruitbodies, rarely solitary. Mostly fruiting in summer and early fall on lawns or large open grassy areas such as soccer fields. Else Vellinga found an isolated patch of this species in the warm microhabitat formed where a steam vent emerges from a lawn at the University of California, Berkeley! Rare in our area (at least along the coast), but likely to become more common as climate warms.

EDIBILITY: Poisonous to a majority of people, causing severe vomiting and diarrhea. Reported as the most common cause of mushroom poisoning in the United States.

COMMENTS: This species seems to be expanding north from the hot and humid regions of the United States. It apparently needs prolonged periods of warm nighttime temperatures to fruit, which may be why it has so far remained rare in northern California. It can be confused with *C. brunneum*, but that species usually has darker cap scales and gills that remain white in age (and produce a white spore deposit); it is much more common in our area and unlike *C. molybdites* rarely grows in grassy lawns.

4 • *Agaricus and Melanophyllum*

Agaricus is a large and cosmopolitan genus, especially diverse in Mediterranean and subtropical climates around the world. Most are easily recognized by their free gills, their dark brown spores, and the presence of a partial veil (sometimes disappearing in age or obscure even when young). *Agaricus* occur in practically all forest types in our area but are less diverse in cold or high-elevation habitats, and are common on lawns and in urban areas. A few species are commercially cultivated and economically important. *Melanophyllum* species are quite rare in our area, and can be told apart from *Agaricus* by their small, slender fruitbodies with powdery surfaces, reddish gills, and greenish olive spore deposits.

When identifying *Agaricus* species, take note of the overall coloration of the fruitbody, structure of the partial veil (if present), texture of the cap and stipe surfaces, any distinctive odor, any staining reaction when cut or bruised, and habitat details.

The genus *Agaricus* is probably more diverse in California than anywhere else in the United States, with particularly interesting assemblages of rare and endemic species associated with Monterey Cypress and Coast Redwood habitats. Probably somewhere near a hundred species occur in California, but many remain undescribed. To make the job of identifying *Agaricus* easier, you can use these artificial groupings to narrow down your options.

Yellow stainers—Species that discolor yellow to orangey yellow (may be slow or weak) when scratched or bruised. Two subsections of this group are listed below; species in both subsections turn yellow in KOH (potassium hydroxide).

Phenol-scented species—Usually smell unpleasant or chemical, often described as "phenolic" or like pen ink (although not everyone can detect this smell). The staining reaction of these species is usually clear yellow (soon fading away or leaving reddish brown discolorations). Be aware that in older fruitbodies (and even in fresh fruitbodies of *A. californicus*), the yellow staining reaction can be slow or nearly absent.

A. hondensis, A. "deardorffensis", A. xanthodermus, A. californicus

Almond-scented species—Smell sweet or almondlike to most people (although some can develop horsey, musty, or urinelike overtones in age). The staining reactions of the fruitbodies tend to be gold to orangey yellow and often persist or turn amber. Although most are large and robust, a few of the almond-scented species are very small, and the odor can be difficult to detect unless you crush a piece of the

cap; many of these smaller
species have pinkish or dingy purplish colors on the cap and less
pronounced yellow stain.

*A. diminutivus, A. micromegathus, A. semotus, A. smithii, A.
augustus, A. albolutescens, A. fissuratus, A. crocodilinus*

Red stainers—Species that stain strongly red when cut, bruised, or
scratched. Most are generally uncommon in our area; they tend to be found
in Monterey Cypress habitats. Look in the nonstaining section for a few others
that turn only weakly orangey or light pinkish when cut and rubbed(but not
immediately "bleeding").

*A. arorae, A. benesii, A. "brunneofibrillosus", A. pattersoniae,
A. bernardii, A. agrinferus*

Nonstaining species—Species that don't strongly change color when cut
and rubbed or scratched. Some of these blush pinkish or orangey when cut
and rubbed or exposed to air, but don't turn red or yellow as strongly as do the
species listed in the sections above.

*A. lilaceps, A. fuscovelatus, A. subrutilescens, A. bitorquis, A.
bernardii, A. agrinferus, A. bisporus, A. "incultorum", A. campestris*

The latter two species tend to produce smaller, less dense, and more fragile
fruitbodies and have weak partial veils (usually only forming an obscure
fibrillose zone).

Melanophyllum haematospermum

(Bull.) Kreisel

CAP: 1–4 cm across, domed at first, soon becoming broadly convex and then nearly flat. Margin sometimes lobed, occasionally uplifted. Dingy grayish brown to tan or beige, often with olive tones, margin paler. Surface dry, finely matted-felty at first, soon cracked into small, powdery, granular scales. Margin often with "teeth" of pale, powdery tissue hanging over edge of gills, soon wearing off into a narrow band. **GILLS:** Free, close to well spaced, sometimes fairly thick and with eroded edges in age. Blood red to dingy coral red, brightest when young, becoming dingy, mottled pinkish brown and then dark purplish brown. **STIPE:** 2–5.5 cm long, 0.2–0.4 cm thick, cylindrical, thin, and usually fairly straight. Metallic coral pink to magenta-red or blood red at apex, lower part extensively covered with pale grayish tan to dingy brown powdery granules (similar in color and texture to cap surface). **PARTIAL VEIL:** A whitish membrane covered in gray powder, soon rupturing and adhering to edge of cap, not leaving a ring on stipe (at most a weak annular zone). **FLESH:** Very thin, fibrous, light beige in cap, reddish on either side of whitish pith filling hollow stipe. **ODOR:** Indistinct. **TASTE:** Not sampled. **SPORE DEPOSIT:** Olive-green, supposedly more bluish green when taken from very fresh mushrooms, older prints turning reddish brown to black. **MICROSCOPY:** Spores 4.5–7 × 2.5–3 µm, ellipsoid or slightly elongate, smooth or very slightly rough. Light brown, not changing color in Melzer's. Cap surface a layer of thin-walled, round, brownish cells.

ECOLOGY: Solitary or more often scattered in small troops in rich soil, compost, or wood chips, mostly in gardens, landscaped areas, or disturbed areas along roads and trails. Quite rare but with widely scattered records over our entire range. Seasonality uncertain, probably mostly fruiting in fall. Likely introduced.

EDIBILITY: Unknown.

COMMENTS: The small size, powdery grayish tan cap, and brilliant red gills when young help identify this species. Very few other mushrooms produce olive-green spore prints. A very similar but smaller variant grows under redwood; it appears to be genetically distinct but is not named. A few *Cystolepiota* in our area share the powdery cap, small stature, and preference for disturbed, nutrient-rich habitats, but all have pale cream or whitish gills.

Agaricus diminutivus

Peck

DIMINUTIVE AGARICUS

CAP: 1–2.5 cm across, egg or bell shaped to rounded as first, soon broadly convex to nearly plane. Pale white to pale tan, with orange-tan tones at center, but also with pale pinkish and yellowish to yellow-orange tones. Staining yellowish when bruised. Surface dry, smooth to radially fibrillose, interwoven fibrils splitting in age. Margin fringed and hanging past edge of gills. **GILLS:** Free, close. Pale pinkish tan at first, becoming pink and then turning brown in age. **STIPE:** 2.5–5 cm long, 0.3–0.7 cm thick, mostly cylindrical but base sometimes enlarged. Whitish to pale creamy buff with patchy yellow-orange tones. Surface dry, hairy or cottony texture below veil. **PARTIAL VEIL:** Short, thin, hanging skirt, margin often fringed, with yellow-orange tones. **FLESH:** Very thin, fragile, white, bruising faintly yellow. **ODOR:** Usually faintly sweet, like anise. **TASTE:** Mild. **KOH:** Yellow. **SPORE DEPOSIT:** Dark brown. **MICROSCOPY:** Spores (4) 5–5.5 (6.5) x 3.5–4.5 µm, ellipsoid, smooth. Partial veil tissue with yellow to golden orange hyphae in KOH.

ECOLOGY: Solitary or in small scattered groups on ground or in duff in wooded areas, occurring under both oaks and conifers. Widespread but never abundant. Fruiting in fall.

EDIBILITY: Nontoxic, but uncommon and tiny cap.

COMMENTS: There are likely a number of species going under this name, but all are strikingly dainty compared to other *Agaricus*. The pale pinkish tones of the cap, yellow discolorations, slender, upright stature, and faint anise odor (difficult to detect unless crushed) are important identification features. *A. micromegethus* prefers more open, grassy habitat and is not as tall and slender, while *A. semotus* is more robust and has distinct white rhizomorphs. Other small, slender *Agaricus* stain reddish or differ in color and/or odor.

Agaricus micromegethus

Peck

CAP: 2–5 cm across, buttons convex and often somewhat squared off, soon becoming flat. Dull pinkish brown to pale, dingy wine red, or extensively whitish with scattered pinkish purple fibrils. Surface dry, dull, covered in fine hairs, sometimes flattening and forming darker brown-pink scales. **GILLS:** Free, close to crowded. Pale pinkish when young, becoming pinkish brown and then dark brown in age. **STIPE:** 2–5.5 cm long, 0.3–1 cm thick, equal to slightly swollen at base. Whitish to off-white, sometimes with dingy tan areas. Surface dry, smooth. **PARTIAL VEIL:** A thin, whitish, hanging, slightly flared skirt with a delicately fringed, yellowish edge. **FLESH:** Thin, fragile, whitish, not staining. **ODOR:** Weakly of almonds, but generally unnoticed unless crushed. **TASTE:** Mild. **KOH:** Yellow, but sometimes weakly so. **SPORE DEPOSIT:** Dark chocolate brown. **MICROSCOPY:** Spores (4) 4.5–5 (5.5) x (3) 3.5–4 µm, ellipsoid to broadly ellipsoid, smooth.

ECOLOGY: Solitary or scattered in small troops, in grassy areas or disturbed, open settings such as trailsides with short grass at the edges of woodlands. Fruiting in fall and winter. Uncommon through much of our range, more common in the southern part.

EDIBILITY: Presumably edible, but small and uncommonly encountered.

COMMENTS: The small size, pinkish cap, slight odor of almonds, persistent partial veil, and small spores distinguish this species. *A. diminutivus* is usually smaller with a taller, more slender stature, and is often paler. *A. semotus* is larger, with a smoother and often paler cap, and has a more robust, club-shaped stipe with white rhizomorphs at the base. *A. subrutilescens* is much larger, does not stain yellow, and lacks an almond odor.

Agaricus semotus (sensu CA)

CAP: 2–7 cm across, ovoid or slightly marshmallow shaped when young, becoming broadly convex to flat. Dull creamy overall with flattened pinkish purple fibrils near center of cap, staining golden yellow overall in age. Surface dry, covered in flattened-fibrillose scales. **GILLS:** Free, close. Whitish to pink when young, then pale chocolate brown and dark brown in age. **STIPE:** 3–10 cm long, 0.5–1 cm thick, cylindrical with a club-shaped base. Base often with conspicuous white rhizomorphs. Off-whitish, staining gold-yellow. **PARTIAL VEIL:** Relatively persistent as a thin, cream-colored, membranous skirt, sometimes breaking and remaining attached to cap. **FLESH:** Thin, fragile, dingy off-white. **ODOR:** Like almond extract. **TASTE:** Mild to slightly sweet. **KOH:** Yellow. **SPORE DEPOSIT:** Dark chocolate brown. **MICROSCOPY:** Spores (4) 4.5–5.5 (6) x 3–4 (5) µm, ellipsoid, smooth.

ECOLOGY: Solitary or scattered in small groups in duff. Most common under live oak, but occurring under a variety of hardwoods and conifers. Widespread and locally common. Fruiting from early fall into winter, occasionally in spring.

EDIBILITY: Edible, but rarely collected.

COMMENTS: Although *A. semotus* is similar to *A. diminutivus* and *A. micromegethus*, the paler cap (sometimes nearly cream colored overall, with only traces of pinkish purple fibrils), slightly larger size, and conspicuous white rhizomorphs distinguish this species.

Agaricus arorae
Kerrigan
ARORA'S AGARICUS

CAP: 3–10 cm across, convex, often with a broad, flattened center and appearing "square shouldered." Pallid tan to brown with slightly darker, appressed, matted fibrils appearing somewhat fibrillose-scaly as the cap expands. Surface dry. **GILLS:** Free, close to crowded. Pale dingy pinkish tan or grayish tan when young, becoming brown to dark brown in age, often bruising reddish where cut or bruised. **STIPE:** 5–10 cm long, 0.6–1.8 cm thick, cylindrical with a slightly enlarged, club-shaped base. Whitish at first, soon dingier to tan. Surface dry, smooth or with chevrons of whitish tissue above veil. **PARTIAL VEIL:** A thin, white, membranous skirt, often with a slightly fringed or ragged edge. **FLESH:** White, thin, soft, fragile in cap. Fibrous, with a small central hollow in stipe. Bruising distinctly reddish in flesh when cut and scraped. **ODOR:** Mild, fruity, or spicy. Kerrigan notes an odor of almonds when KOH is applied. **TASTE:** Mild. **KOH:** Yellow on cap surface and in flesh at base of stipe. **SPORE DEPOSIT:** Dark brown. **MICROSCOPY:** Spores 4–5.5 x 3–3.5 (4) μm, ellipsoid, smooth.

ECOLOGY: Solitary or in groups, fruiting from early fall into winter in disturbed forest settings or near trails under Douglas-fir and redwood. Rick Kerrigan reports it from under oaks as well. Quite rare; described from Santa Cruz County, and mostly found on the Central Coast.

EDIBILITY: Likely nontoxic.

COMMENTS: This *Agaricus* is quite distinctive due to its slender stature, brownish cap, reddening flesh, yellow KOH reaction, and preference for mixed evergreen forests. *A. arorae* is most likely to be confused with *A. "brunneofibrillosus"*, but that species is stockier, has brown bands around the stipe base, and usually stains red more dramatically. Other red-staining *Agaricus* species in our area are larger and/or stouter, and all lack the yellow KOH reaction.

Agaricus benesii
(Pilát) Pilát
BLEEDING WHITE AGARICUS

CAP: 4–10 cm across, rounded and inrolled at first, becoming broadly convex to nearly plane. White to dull whitish, becoming pale buff to tan-brown. Bruising quickly pinkish to bright reddish on all parts. Surface dry, covered with distinct, uplifted scales that sometimes become flat or disappear in age. **GILLS:** Free, close to crowded. Pale pinkish when young, becoming brown to dark brown. **STIPE:** 4–15 cm long, 1–2.5 cm thick, quite tall and cylindrical, equal or with an enlarged, club-shaped base. Whitish or off-white, quickly staining pinkish to bright reddish. Surface dry, shaggy or scaly below the veil, more or less smooth above. **PARTIAL VEIL:** A white, floccose membrane forming a hanging skirt in age. **FLESH:** Thin to moderately thick, firm when young, quite tough in stipe. White, staining pinkish to bright reddish. **ODOR:** Indistinct. **TASTE:** Mild. **KOH:** Reaction indistinct. **SPORE DEPOSIT:** Dark chocolate brown. **MICROSCOPY:** Spores 4.5–5.5 (6.5) x 3.5–4.5 μm, ellipsoid, smooth.

ECOLOGY: Solitary, scattered, or in loose clusters in duff under pines and especially Monterey Cypress. Rather uncommon, known from the San Francisco Bay Area south to Santa Cruz County. Fruiting in fall and early winter, occasionally in wet springs.

EDIBILITY: Edible, but not commonly eaten. Caution is warranted.

COMMENTS: This rather unusual *Agaricus* is easily distinguished by its tallish, upright stature, white color, shaggy stipe, and distinct red staining. This name is used here tentatively, since it has not yet been shown to be conspecific with the (European) specimens upon which the concept is based. *A. bernardii* is also whitish and stains red, but is much bulkier and more squat, smells briny, and has a smoother cap and stipe.

Agaricus "brunneofibrillosus"

Kerrigan nom. prov.

BLEEDING BROWN AGARICUS

CAP: 3–10 cm across, buttons rounded-convex, often with a squared-off appearance with an inrolled margin, becoming more broadly convex to more or less flat in age. Dull pinkish tan to brownish tan to dark brown, often with patches of white veil tissue, especially near margin. Surface dry to moist, innately fibrillose or with obscure appressed-fibrillose scales. **GILLS:** Free, close to crowded. Pale grayish tan when very young, then medium pinkish tan and finally darker brown, often with reddish blotches where damaged. **STIPE:** 3–12 cm long, 1.5–4 cm thick, equal, often with an enlarged, club-shaped base. Whitish or off-white, often with creamy tan tones. Mostly smooth, but with obscure vertical lines above the annulus and 2–4 bands of short, upward-flaring tan-brown fibrils at base. **PARTIAL VEIL:** A white, membranous, hanging or flaring skirt. Edge of skirt often thick and fringed. **FLESH:** White, fibrillose, rather firm. Turns red strongly and rapidly when cut or rubbed, especially in stipe and in cap flesh above gills. **ODOR:** Like portobello mushrooms, often somewhat fishy or unpleasant in age. **TASTE:** Mild. **KOH:** Reaction indistinct. **SPORE DEPOSIT:** Dark chocolate brown. **MICROSCOPY:** Spores 4.5–6 (7.5) x 3.5–5 μm, ellipsoid, smooth.

ECOLOGY: Solitary, in scattered clusters, or in large troops and arcs, almost always in deep, rich needle duff of Monterey Cypress. Known from San Luis Obispo County to Mendocino County, most common around the San Francisco Bay Area.

Can be quite abundant locally, but distribution is rather patchy. Fruiting from late fall into spring.

EDIBILITY: Edible, with an excellent flavor, but some people find the texture unpleasantly chewy. Should be sampled with caution, since many of the edible cypress *Agaricus* cause allergic reactions in some people.

COMMENTS: Probably endemic to California, this *Agaricus* is distinguished from other members of the genus by the intensity of the red staining; its evenly brown cap; often with white veil patches; and its preference for cypress. *A. pattersoniae* is larger and stouter, with a more distinctly scaly brown cap, pinkish to red-staining flesh, and a flocculent veil. *A. arorae* also has a brown cap, but stains intensely red when cut, is more slender, and stains yellow in KOH. *A. lilaceps* is usually larger, often has yellow or orange tones on the underside of its partial veil, and shows only a modest pinkish staining reaction when cut.

MISAPPLIED NAME: *A. fuscofibrillosus* (a European species).

Agaricus pattersoniae

Peck

PATTERSON'S AGARICUS

CAP: 7–16 cm across, convex at first, becoming broadly convex to nearly plane. Background whitish, but covered evenly with golden brown to brown appressed scales. Surface dry, scales becoming flattened in age, but distinctly differentiated from background at all ages. **GILLS:** Free, close to crowded. Pinkish gray when young, becoming reddish pink and finally brown to dark brown at maturity. **STIPE:** 5–10 cm long, 3–6 cm thick, equal or with an enlarged, club-shaped base. Whitish to creamy off-white, with one or more rings of brown scales over lower part. **PARTIAL VEIL:** Thick, white, floccose-membranous tissue, with a second short hanging collar near stipe, soon forming a hanging skirt, often with grooves on upper surface before collapsing in age. **FLESH:** Thick, firm, white with dingy pale tan areas, staining pinkish red. **ODOR:** Mild or like portobello mushrooms. **TASTE:** Mild. **KOH:** No reaction. **SPORE DEPOSIT:** Dark chocolate brown. **MICROSCOPY:** Spores (4.5) 6.5–8 (9) x (4.5) 5.5–6.5 μm, ellipsoid to broadly ellipsoid, smooth.

ECOLOGY: Solitary, scattered, or in small clusters, often in arcs or groups under Monterey Cypress. Rare and range restricted, only known from a handful of cypress groves from San Francisco to Monterey County. Fruiting from late fall into spring.

EDIBILITY: Edible and very good, but very rare and likely sensitive to overpicking. If you are lucky enough to find it, please exercise restraint.

COMMENTS: Although somewhat similar to other brownish, red-staining *Agaricus* found with cypress, this species can be distinguished by the appressed-fibrillose scales covering the cap (the cap is smoother in *A.* "*brunneofibrillosus*" and *A. lilaceps*), the pinkish red staining (stronger than in *A. fuscovelatus* but weaker than in *A.* "*brunneofibrillosus*"), the robust stature, and the whitish, floccose partial veil (often streaked with orange in *A. lilaceps*, more tenacious and purplish gray to brown in *A. fuscovelatus*).

Agaricus lilaceps

Zeller

GIANT CYPRESS AGARICUS

CAP: 6–30 cm across, buttons strongly rounded and inrolled, becoming broadly convex and then flat. Dull whitish gray to brownish or golden brown, covered with darker, flattened tan-brown fibrils, but sometimes smooth and evenly dark brown. Often developing areas of persistent orange or red tones. Said to show lilac hues in some areas or in cold weather. Surface dry, covered with distinct, appressed fibrils. **GILLS:** Free, close to crowded. Dull pinkish brown to reddish tan when young, becoming chocolate brown and then dark brown to grayish black in age. **STIPE:** 6–30 cm long, 3.5–9 cm thick, usually with an enlarged, club-shaped base. Whitish to off-white, often with dull orange stains. Surface dry, slightly scurfy when young to smooth in age. **PARTIAL VEIL:** Membranous, whitish, but underside often with greasy or dry yellowish to orange patches, becoming brownish in age, often remaining intact even after cap has expanded, becoming a hanging skirt or collapsing in age. **FLESH:** Thick, very firm, whitish, staining absent or irregularly reddish. **ODOR:** Mild, like portobello mushrooms or musty in age. **TASTE:** Mild. **KOH:** No reaction or sometimes pale yellow. **SPORE DEPOSIT:** Dark chocolate brown. **MICROSCOPY:** Spores 5–6 (7) x 4–5 (5.5) μm, broadly ellipsoid to ellipsoid, smooth.

ECOLOGY: Solitary, scattered, or in loose clusters or troops in duff, usually under Monterey Cypress, also occurring under acacia, eucalyptus, and Ngaio (*Myoporum laetum*). Uncommon and local on Central Coast, at least as far north as San Francisco. Fruiting from late fall into spring.

EDIBILITY: Edible and good, meaty and flavorful. However, it can cause severe allergic reactions with some people, as do many otherwise edible cypress-dwelling *Agaricus*).

COMMENTS: This species is distinguished primarily by its often enormous size and cypress habitat, as well as by its smooth, brownish cap, frequently yellow to orange colors on the underside of the partial veil, and the weak pinkish red staining. *A. pattersoniae* is another hefty cypress lover, but usually has a more fibrillose-scaly cap and brownish bands around the base of the stipe, and never shows the orange colors on the partial veil. *A. fuscovelatus* is much smaller, with a straighter, cylindrical stipe, and a tenacious grayish lilac partial veil.

Agaricus fuscovelatus

Kerrigan

PURPLE-VEILED AGARICUS

CAP: 3–10 cm across, young caps irregularly conical to convex, with an inrolled margin, becoming more broadly convex, eventually flat. Covered in pale pinkish tan appressed-fibrillose scales over a dull whitish background, becoming dull brownish overall, sometimes darker brown overall. Surface dry, covered with scales, especially when young and fresh, scales flattening when exposed or in age, cap then appearing nearly smooth. **GILLS:** Free, close to crowded. Pale pinkish brown when young, often with reddish blotches, becoming darker pinkish brown and finally dark brown. **STIPE:** 3–12 cm long, 2–4 cm thick, usually cylindrical, base sometimes slightly enlarged. Dingy whitish, often with dull yellow-brown or reddish brown tones. Slightly shaggy overall, scales smaller and flatter on upper part of stipe. **PARTIAL VEIL:** A very conspicuous, flaring skirt positioned high up on the stipe. Usually with chocolate brown to purple or purplish gray patches on underside, sometimes beaded with golden droplets when very fresh. The veil tends to remain partially attached to edge of cap even after the cap has expanded. **FLESH:** Thin, firm, rather rubbery, whitish, bruising dull reddish brown irregularly throughout stipe, cap, and gills. **ODOR:** Like portobello mushrooms or indistinct. **TASTE:** Mild. **KOH:** No reaction. **SPORE DEPOSIT:** Dark chocolate brown. **MICROSCOPY:** Spores (5.5) 7–8 (9) x 5–7 μm, ellipsoid, smooth.

ECOLOGY: Scattered or in small to medium clusters, usually buried in deep needle duff of Monterey Cypress. Often not showing above ground until mature. Known from San Francisco southward, but rather uncommon. Fruiting from late fall into winter.

EDIBILITY: Edible, but chewy and tough, and often with a bitter metallic aftertaste.

COMMENTS: A California endemic, this unusual *Agaricus* is a delight to find, at least for connoisseurs of the genus. It is distinguished by its tenacious, brownish gray or lilac partial veil, round-conical cap (when young), tough cylindrical stipe, cypress habitat, and weak staining. *A. lilaceps* is much larger, with a smoother cap, firmer flesh, and a whitish to orange-blotched partial veil. *A. pattersoniae* is usually more robust, has pinkish red-staining flesh, and has a paler, floccose partial veil.

Agaricus subrutilescens

(Kauffman) Hotson & D. E. Stuntz

WINE-COLORED AGARICUS

CAP: 4–15 cm across, buttons rounded to hemispherical and sometimes lobed, becoming broadly convex, rarely flattening completely. Dull wine red to pinkish or reddish brown or purplish brown or entirely dark to medium brown, background color pale off-white to creamy white. Surface dry to moist, covered with distinct, appressed-fibrillose scales. **GILLS:** Free, close to crowded. Pale pinkish when young, becoming pinkish tan and finally brown to dark brown. **STIPE:** 6.5–17 cm long, 1–3 cm thick, cylindrical or with a swollen base. Whitish or off-white with creamy tones. Rather shaggy, especially below partial veil. **PARTIAL VEIL:** Fairly thick, membranous, white skirt, but often collapsing in age. **FLESH:** Fleshy, often soft in age. White in cap, bruising mostly absent, gills sometimes becoming red or darker pink where injured, stipe often hollow at maturity, slightly pinkish to brownish pink when cut. **ODOR:** Indistinct to faintly fruity or sweet. **TASTE:** Mild. **KOH:** Pale greenish gray on cap. **SPORE DEPOSIT:** Dark chocolate brown. **MICROSCOPY:** Spores (4) 5–5.5 (6.5) x 3–4 μm, ellipsoid, smooth.

ECOLOGY: Solitary, scattered, or in loose troops, occasionally clustered. Common and widespread in a variety of woodland habitats, but most abundant in pine and redwood forest. Fruiting from late fall through spring, most abundant in midwinter.

EDIBILITY: Edible and very good. More delicately flavored than *A. bitorquis*, but more savory than *A. augustus*. Be careful to eat only a small amount the first time you try it since some people are allergic.

COMMENTS: The tall stature, shaggy stipe, dark reddish brown to purplish brown cap colors, lack of a sweet or phenolic odor, and absence of strong bruising reactions set this species apart from most other *Agaricus* species; the greenish gray KOH reaction, woodland habitat, and winter fruiting habits further serve to support the identification. *A. thiersii* is very similar, but typically has a browner cap with larger scales and a preference for pine and oak duff. *A. hondensis* has a smooth stipe, bruises yellowish, and has a faint phenol odor. *A. augustus* also has a shaggy stipe, but is more golden brown overall, bruises yellow, and smells strongly almondy.

Agaricus augustus

Fr.

THE PRINCE

CAP: 7–30 cm across, tightly convex or squarish when young, becoming broadly convex, often flatter around center of cap, flat or uplifted in age. Warm brown to golden yellow or golden brown with orange blushes, covered with tan to brownish fibrous scales. Bruising yellow to golden yellow to nearly orange and remaining so; orange or reddish brown when rain beaten. Surface dry, covered in flattened fibrous scales. **GILLS:** Free, close to crowded. At first pale, dull tan to pinkish gray, maturing to dark brown to almost black. **STIPE:** 7–25 (35) cm long, 2–5 cm thick, cylindrical or tapered toward the top, base sometimes swollen to club shaped. White to pale creamy yellow, developing distinct yellow or even yellow-orange tones. Smooth above annulus, area below usually distinctly scaly. Scales fibrous and dull yellowish. **PARTIAL VEIL:** A membranous ring, often with small, squarish cottony patches on underside and around margin. **FLESH:** Fairly thick, firm when young, softer in age. Whitish, bruising yellow to golden yellow. **ODOR:** Usually strong and sweet or pleasant like almond extract or marzipan. **TASTE:** Mild to slightly sweet. **KOH:** Yellow. **SPORE DEPOSIT:** Dark brown. **MICROSCOPY:** Spores 7–9.5 (10.5) x 4.5–6 (7) μm, ellipsoid, smooth.

ECOLOGY: Solitary, scattered, or in clusters in duff, in gardens, or along roads and trails; it has a preference for redwood and cypress duff. Common throughout our range, frequently fruiting in spring, summer, and early fall, generally absent in cold winter months.

EDIBILITY: Edible and considered one of the finest mushrooms by many people. Others dislike the strong almondy flavor of young specimens. Eat only a small amount the first time you try it, as some people experience allergic reactions including swollen lips, itching, and vomiting.

COMMENTS: A handsome species as well as a delicious edible, this mushroom can be recognized by its golden yellow to tawny brown colors, scaly lower stipe, and almondy odor. *A. perobscurus* also shows yellowish colors and smells almondy, but its cap is covered with fine, inky gray to blackish (not golden brown) fibrils; it is fairly rare, typically fruiting in the winter under cypress and redwood. *A. smithii* is similar, but has a smoother cap and stipe, and grows primarily under spruce. *A. subrutilescens* also has a wooly-scaly lower stipe, but is usually brown to lilac-gray or light wine reddish (lacking golden tones), and does not smell distinctly almondy. *A. "deardorffensis"* and *A. hondensis* have smoother stipes, smell phenolic, and do not stain amber overall in age.

Agaricus smithii

Kerrigan

GOLDEN SPRUCE AGARICUS

CAP: 5–20 cm across, buttons rather egg shaped (like tapered ovals), becoming convex or umbonate-convex. Usually an attractive orange-brown, dull yellow, gold, or warm tan-brown over a paler base with a darker disc. Base color darkens orangish in age. Surface dry, often covered with completely flattened, dull tan-brown fibrils. Bruising weakly yellowish. **GILLS:** Free, close to crowded. Pallid when young, becoming pinkish tan and finally brown to dark brown. **STIPE:** 8–15 (20) cm long, 1–2 cm thick, base up to 5 cm, equal, quite tall and slender, but with an enlarged, bulbous base. Whitish or creamy with golden tan tones. Finely scaly or scruffy when young to smooth. **PARTIAL VEIL:** Fairly thick, white, with yellow-gold patches, membranous, forming a flaring or hanging skirt set rather high on stipe. **FLESH:** Whitish, firm, fragile, moderately thick. Stipe firm, stuffed, fleshy-fibrous, occasionally hollow in age. **ODOR:** Almondy. **TASTE:** Sweet-nutty to mild. **KOH:** Yellow, especially on cap. **SPORE DEPOSIT:** Dark chocolate brown. **MICROSCOPY:** Spores (6.5) 7.5–8.5 (10) x (4.5) 5–5.5 (6.5) μm, ellipsoid, smooth.

ECOLOGY: Scattered in troops, rings, or arcs in duff in coniferous woodlands, almost always under Sitka Spruce. Common from Mendocino County northward (although a few collections have been made in Sonoma County in mixed evergreen forest). Often produces two or three flushes from fall through winter.

EDIBILITY: A great edible, highly regarded by many. The almond flavor that is sometimes overpowering in *A. augustus* isn't as strong in this species, making it more appealing to some people.

COMMENTS: This interesting *Agaricus* is easily recognized by the egg-shaped cap when young, tawny to gold or orange cap colors, tall stipe (often with yellowish rhizomorphs at the base), almondy odor, and growth under Sitka Spruce. It could be confused with *A. augustus*, but that species usually has a more scaly and sometimes square-shaped cap, as well as a much scalier stipe. A similar-looking species occurs in mixed evergreen forests further south; it may be *A. summensis*. *Agaricus hondensis* has a paler, pinkish fawn cap, phenolic odor, and white partial veil; it is more widespread and is toxic.

Agaricus hondensis

Murrill

FELT-RINGED AGARICUS

CAP: 4–15 cm across, buttons round to convex, becoming broadly convex and eventually flat. Golden brown when very young, soon a dull dingy whitish background overlaid by pale tawny or pale tan to pinkish brown scales. Surface dry to moist, covered with tightly flattened fibrils and scales. **GILLS:** Free, close to crowded. Very pale pinkish tan when young, becoming pinkish brown and finally dark brown. **STIPE:** 4.5–15 (20) cm long, 1.5–4 cm thick, almost always with an enlarged, club-shaped base, extreme base often bulbous and; very much enlarged and irregularly contorted and knobbed. Whitish to grayish white, sometimes shiny, discoloring to dingy brownish at apex in age. Surface smooth or with silky appressed fibrils at base. **PARTIAL VEIL:** A thick, whitish, felty-spongy, flaring to hanging skirt, edge usually thick and flat. **FLESH:** Firm, thick, whitish, bruising irregularly pale yellowish in all parts, most readily seen when stipe base is scratched. **ODOR:** Faintly to strongly phenolic. **TASTE:** Mild. **KOH:** Bright to pale yellow on all parts. **SPORE DEPOSIT:** Dark chocolate brown. **MICROSCOPY:** Spores (3.5) 4.5–6 x 3–4 (5.5) μm, ellipsoid, smooth.

ECOLOGY: Solitary, in small clusters, or scattered in rings or arcs, fruiting in a variety of woodland habitats, especially in oak and mixed-evergreen woods or under cypress along the coast. First appearing in late fall, peaking in abundance during winter, and persisting into early spring.

EDIBILITY: Like most phenol-smelling *Agaricus*, this species is poisonous for most people.

COMMENTS: Usually the most common woodland *Agaricus* in winter months, this lovely species is recognized by its tawny to pinkish tan or light brown cap with appressed fibrils, large felty ring, faint phenolic odor, and weak yellowing staining. *A. "deardorffensis"* and its close relatives have grayish to blackish scales and slightly stronger yellow-staining reactions, especially at the stipe base; they also tend to fruit earlier and later in the season, whereas *A. hondensis* peaks in the midwinter months. *A. xanthodermus* has a smooth whitish cap, stains yellow much more strongly, has a thinner partial veil and an often crooked stipe base, and usually grows in urban habitats.

Agaricus "deardorffensis"
Kerrigan nom. prov.

CALIFORNIA FLAT-TOP AGARICUS

CAP: 5–15 cm across, buttons tightly convex and rather square, becoming broadly convex to flat. Whitish, but covered with dull inky gray to brownish black scales and fibrils, darker at center. Surface dry, often cracking in dry weather. **GILLS:** Free, close to crowded. Pale pinkish when young, becoming pinkish tan and finally brown to dark brown at maturity. **STIPE:** 7–15 cm long, 1–4 cm thick, equal or with a club-shaped or irregularly lobed and contorted, bulbous base. Whitish or off-white with dingy, pale tan tones. Surface dry, smooth. **PARTIAL VEIL:** Thick, somewhat spongy or felty, flat-edged collarlike ring. **FLESH:** Thin to moderately thick, off-whitish, marbled with dingy pale tan areas, weakly yellow in all parts, sometimes more strongly so, especially in base of stipe, stained areas slowly oxidizing to reddish brown. **ODOR:** Phenolic or chemical. **TASTE:** Mild. **KOH:** Bright yellow. **SPORE DEPOSIT:** Dark chocolate brown. **MICROSCOPY:** Spores 3.5–7 x 3–5 μm, ellipsoid, smooth.

ECOLOGY: Solitary, scattered, or in troops, clusters, and arcs in disturbed forest settings, especially along roads and trails in redwood duff, also with other conifers and oaks. Widespread and very common. Fruiting from early fall into early winter, mostly absent in cold weather, but often fruiting again in spring with sufficiently warm and wet weather.

EDIBILITY: Poisonous for most people, as are most yellow-staining, phenol-smelling *Agaricus*.

COMMENTS: Conspicuous due to its large size and stately posture, this beautiful mushroom is quite common during the fall in most years. It can be recognized by its large size, gray-brown to grayish black scales, flattened or squarish caps, phenolic odor, thick partial veil, and weak yellow bruising that becomes a dingy wine brown. Rick Kerrigan has shown that *A. "deardorffensis"* belongs to a complex of at least three species including *A. berryessae*, which is smaller, has a paler cap and slightly larger scales, and quickly stains bright yellow. There is also a larger species that grows under redwood that remains undescribed. All are sometimes confused with *A. hondensis*, which is paler, with flattened, pinkish tan to light brown fibrils, more rounded caps, and a tendency to fruit in winter and early spring in less-disturbed habitats. *A. augustus* usually smells sweet (like almonds), is different colored (paler golden tan to tawny brownish), and has a scaly-shaggy lower stipe.

MISAPPLIED NAMES: *A. molleri, A. praeclaresquamosus.*

Agaricus albolutescens

Zeller

AMBER-STAINING AGARICUS

CAP: 3–15 cm across, buttons rounded and inrolled, expanding to broadly convex or plane. Silvery whitish with yellow-orange streaks, sometimes with brownish or pinkish orange discoloration. Readily staining golden yellow when scraped, then darkening to golden brown. Surface dry, rather smooth and shiny. **GILLS:** Free, close to crowded. Whitish or very pale grayish tan at first, becoming pinkish tan and then dark brown as spores mature. **STIPE:** 3–15 cm long, 1.5–4.5 cm thick, smooth, with a distinctly enlarged, abruptly bulbous and flat-bottomed base adorned with white rhizomorphs. Whitish or off-white with yellowish orange or pinkish tan tones. **PARTIAL VEIL:** Thin, white, membranous skirt, often tearing irregularly and sometimes disappearing. **FLESH:** White with pale tan or brownish discolorations, bruising yellowish. **ODOR:** Distinctly sweet, like almonds or anise. **TASTE:** Mild. **KOH:** Yellow. **SPORE DEPOSIT:** Dark chocolate brown. **MICROSCOPY:** Spores (4) 5.5–6.5 (7.5) x 3.5–5 μm, ellipsoid, smooth.

ECOLOGY: Solitary or scattered, sometimes clustered in duff or humus. Found in a wide variety of forest habitats, but seems to prefer live oak duff. Although it can be locally common, it is infrequent over most of the California coast. Fruiting from midfall into early winter, occasionally in spring.

EDIBILITY: Edible and considered excellent by many who appreciate its firm texture and almond flavor.

COMMENTS: The smooth white cap that stains golden yellow, bulbous, flat-bottomed stipe, and sweet almond odor are helpful in distinguishing this species from other yellowing *Agaricus*. The toxic *A. xanthodermus* has an unpleasant phenolic odor and quickly stains yellow, then fades to brown, whereas *A. albolutescens* stains yellow and turns amber. *A. fissuratus* has a smaller partial veil and often has rings of small scales around the stipe base; it grows in urban areas. Other robust, yellowing *Agaricus* in our area have more darkly pigmented and/or scalier or more fibrillose caps. *A. summensis* is a larger, tawny-capped species that also stains yellow to ocher; it also grows in forested settings.

MISAPPLIED NAME: Long referred to as *A. abruptibulbus*, with the name *A. albolutescens* applied to a similar montane species. It is now understood that the coastal species is the "true" *A. albolutescens*; the montane species has been named *A. moronii*.

Agaricus crocodilinus

Murrill

CROCODILE AGARICUS

CAP: 5–12 cm across in button stage, 20–40 cm across when expanded. Buttons rounded or somewhat marshmallow shaped, becoming hemispheric to broadly convex and then nearly flat or shallowly uplifted with a slightly wavy margin in age. Whitish at first, soon dull creamy, developing blotches of light yellowish. Bruising yellowish ocher when scraped or rubbed or in age. Surface dry, nearly smooth or with small, thin scales near margin when very young. In dry conditions, caps break up into large, angular warts, giving them a "crocodile skin" appearance. **GILLS:** Free, close to crowded. Pale tan to dingy light pinkish gray when young, reddish brown to blackish brown in age. **STIPE:** 6–20 cm long, 2–4 cm thick, up to 6 cm thick at base. Shape variable, sometimes spindle shaped (with a tapered base and apex) or enlarged and club shaped at base. Whitish to off-white, with some yellowish gold to dingy ocher tones in age. Upper parts smooth, with rings of small scales below partial veil (becoming obscure in age). **PARTIAL VEIL:** A thin, whitish, membranous skirt with floccose patches on underside, persisting as a hanging skirt. **FLESH:** Thick, firm at first, soft in age. White. Bruising yellowish gold to ocher, then oxidizing to ocher-brown. **ODOR:** Sweet, like almonds, although sometimes faint. **TASTE:** Sweet or mild. **KOH:** Yellow on cap. **SPORE DEPOSIT:** Dark chocolate brown. **MICROSCOPY:** Spores 9–11 (14) × 6–7.5 μm, long ellipsoid, smooth.

ECOLOGY: Scattered or in large arcs and rings in coastal prairies with short, sparse vegetation, as well as in more lush, grassy livestock pastures. Fruiting primarily in fall and especially late spring. Uncommon and local in our area, known primarily from coastal areas around Point Reyes in Marin County.

EDIBILITY: Edible and by all accounts delicious. Sweetish when young, richer in age, with a firm, meaty texture.

COMMENTS: The large to enormous white caps, sweet odor, growth in open grassy areas, and large spores set this *Agaricus* apart. The species long called *A. osecanus* in our area (but now known more appropriately as *A. arvensis*) is very similar, but with smaller spores (6–6.5 x 4.5–5 μm), and perhaps less of a tendency to crack into large scales. Both *A. crocodilinus* and *A. arvensis* were historically common and even annually abundant in California's coastal pasturelands. Both were frequently collected for food by a small, enthusiastic group of mushroomers from at least the 1960s through the late 1980s, but they appear to have vanished entirely at some point in the past three decades. It is unknown whether climate change, changes in cattle husbandry practices, invasion of nonnative soil organisms, or some combination of all these (or other) factors is to blame.

Agaricus fissuratus

F. H. Møller

CAP: 4–12 cm across, young buttons egg or marshmallow shaped, expanding to broadly convex and then flat. Whitish to creamy beige or washed with yellow. Bruising warm yellow when rubbed or scraped. Surface dry, mostly smooth but with scattered, small, thin, flat scales. **GILLS:** Free, close to crowded. Pale when young, becoming pinkish tan and finally brown to blackish brown. **STIPE:** 4–13 cm long, 1–2.5 cm thick at apex, cylindrical, but often narrowed toward base and then enlarged again at extreme base. Whitish to creamy with pale golden and tan tones. Surface dry, smooth, sometimes with small cottony chunks or patches below veil. **PARTIAL VEIL:** A whitish membrane usually adorned with a striking pattern of felty patches (like a cogwheel) on underside, then forming a large, hanging skirt, occasionally disappearing in age. **FLESH:** Thick, firm when young, becoming soft. Stipe often stuffed with a cottony pith. White, bruising slowly yellow. **ODOR:** Sweet and slightly almondy when young, becoming musty and unpleasant in age. **TASTE:** Mild. **KOH:** Yellow. **SPORE DEPOSIT:** Dark chocolate brown. **MICROSCOPY:** Spores (5.5) 7.5–8.0 (9.5) x 5–5.5 (7) μm, ellipsoid, smooth.

ECOLOGY: Scattered in small groups, mostly in grassy areas or under trees in urban or disturbed areas. Widely distributed throughout our area, but patches seem few and far between. A single patch can fruit continuously throughout the mushroom season, given enough rain.

EDIBILITY: Edible and good when young; beware those growing in areas treated with pesticides.

COMMENTS: The tawny whitish beige to dull yellowish cap, yellow staining, sweet odor, cogwheel pattern on the underside of the veil, and medium-size caps help distinguish this species. *A. albolutescens* is quite similar but grows in forest habitats, as does *A. summensis*, which is larger, with an inherently tawny to golden beige cap and slightly smaller spores (6.5–7 x 4–5 μm); it is rare, known from woodland habitats around San Francisco Bay. *A. crocodilinus* and *A. arvensis* are usually much larger and grow in coastal pastures.

MISAPPLIED NAME: The species that has most commonly been identified as *A. arvensis* in California is more appropriately called *A. fissuratus*.

Agaricus californicus

Peck

CALIFORNIA AGARICUS

CAP: 2–11 cm across, buttons convex, becoming broadly convex to plane. Dull whitish to pale gray, often pale tan-brown to grayish over center. Not staining or weakly yellow when bruised. Surface dry and smooth, with scales around center of cap in age. **GILLS:** Free, close to crowded. Whitish at first, soon pink, becoming brown and then dark brown. **STIPE:** 3.5–9 cm long, 0.7–2 cm thick, equal or slightly enlarged, club-shaped base. Whitish or off-white with grayish tones. Surface smooth, dry. **PARTIAL VEIL:** A thick-edged, white, membranous skirt, often flaring outward or hanging shallowly, collapsing in age. **FLESH:** Thin, soft, whitish, staining absent or weakly pale yellow. **ODOR:** Mild or weakly phenolic. **TASTE:** Mild. **KOH:** Bright yellow on cap. **SPORE DEPOSIT:** Dark chocolate brown. **MICROSCOPY:** Spores (4.5) 5.5–6 (9) x 4–5 (6) μm, ellipsoid to broadly ellipsoid, smooth.

ECOLOGY: Solitary, scattered, or in loose groups in a variety of urban habitats (grass, landscaping), also around meadow edges and forest settings. Locally common, but rarely abundant. Appears to have been replaced in many areas by *A. xanthodermus*. Fruiting in late summer into late fall, then again in spring.

EDIBILITY: Mildly poisonous for most people.

COMMENTS: The small to medium size, whitish to pale grayish smooth cap, thick-edged ring, yellow KOH reaction, and slight phenolic odor are helpful in identification. It is often confused with *A. xanthodermus*, which is also whitish overall and yellows in KOH, but shows much stronger yellow bruising (especially on the cap and stipe base), has a larger, skirtlike ring and a stronger phenol odor, as well as slightly wider spores. However, some collections appear anomalous (lacking an odor but showing strong yellowing, or vice versa), and closer study is needed to understand the differences in morphology and distribution of these species. Both are poisonous for most people.

Agaricus xanthodermus

Genev.

YELLOW STAINER

CAP: 3–15 cm across, buttons rounded and strongly inrolled, often lobed, becoming broadly convex or flat in age. Pure white to dull whitish, when exposed to sun or in dry conditions becoming light, dull grayish brown to brown, often shiny and metallic in such conditions. Quickly and strongly staining yellow when rubbed or scraped. Surface dry, smooth, sometimes becoming distinctly cracked. **GILLS:** Free, close to crowded. Pale whitish gray when young, becoming light pink and then brown, dark brown to nearly black. **STIPE:** 6–17 cm long, 1–4 cm thick, usually equal and cylindrical but with a distinctly crooked, enlarged base, often lobed and with rhizomorphs. Whitish to grayish white or with light brownish tones, base readily staining yellow when damaged. Surface dry, smooth or wavy. **PARTIAL VEIL:** A prominent, white, membranous skirt, often with two layers, one thicker and blocky, the other thin and fragile. **FLESH:** Thin to moderately thick, whitish with dingy areas, staining yellow, especially in stipe base. **ODOR:** Strongly chemical or phenolic, although some people can't detect this odor. **TASTE:** Mild. **KOH:** Bright yellow. **SPORE DEPOSIT:** Dark chocolate brown. **MICROSCOPY:** Spores (4) 5–5.5 (6.5) × 3–4.5 µm, broadly ellipsoid to ellipsoid, smooth.

ECOLOGY: Scattered, sometimes solitary or clustered, more often in large troops, generally in urban habitats, grass lawns, wood chips, and gardens. Also occurs in a number of native habitats including live oak woodlands and riparian sycamore duff. Very common throughout the entire year, most abundant in fall and spring.

EDIBILITY: Poisonous! Typical of the phenol-smelling, yellow-staining *Agaricus*, this species causes relatively brief, but very unpleasant gastrointestinal disturbance in most people who consume it.

COMMENTS: Although data on the presence and abundance of this species are very sparse prior to the 1960s, the ruderal habitat preferences of the species strongly suggest that it is introduced and invasive. Although the date of introduction is unknown, *A. xanthodermus* was apparently common in central California by the 1960s. The intense yellowing reactions, whitish color overall, and phenol odor distinguish it from most *Agaricus*. *A. californicus* is very similar; it is distinguished by its weak yellowing (if at all), faint phenol odor, and longer, wider spores. Other phenolic-smelling *Agaricus* are larger and more pigmented, and do not stain yellow as strongly (*A. "deardorffensis"* group, *A. hondensis*). Some other strongly yellowing *Agaricus* smell sweet and almondy rather than phenolic (*A. albolutescens*).

Agaricus campestris group
MEADOW MUSHROOM

CAP: 3.5–10 cm across, buttons rounded and inrolled, becoming broadly convex and eventually flat. Margin at first with a ragged band of cottony tissue hanging past edge of gills, which disappears with age. Bright whitish to duller white, sometimes with a few tan brownish scales. Surface dry, sometimes shiny in dry weather. **GILLS:** Free, close to crowded. Bright bubblegum pink when young (even when cap is still closed), becoming pinkish brown and then deep chocolate brown when mature. **STIPE:** 2.5–8.5 cm long, 1–3 cm thick, usually fairly short and slender, straight to slightly curved, but almost always with a pointed or tapered stipe base. Whitish. Surface smooth or slightly scaly. **PARTIAL VEIL:** Ragged, thin, cottony, attaching to cap margin and forming a flaring ring, soon disappearing or becoming an inconspicuous ragged zone. **FLESH:** Thin, soft, white. **ODOR:** Like white Button Mushroom. **TASTE:** Mild. **KOH:** No reaction. **SPORE DEPOSIT:** Dark chocolate brown. **MICROSCOPY:** Spores 6–8 (9) x 4.5–6 µm, ellipsoid, smooth.

ECOLOGY: Solitary or scattered in troops, arcs, or rings in grassy areas, or in areas with bare soil and scattered short, weedy herbaceous plants. Infrequent, fruiting in early fall through late fall and again in spring.

EDIBILITY: Edible and very popular, especially the younger, firmer fruitbodies.

COMMENTS: Not particularly common in recent years, it was apparently once quite frequent in coastal pastures and other grassy areas during moist spells in fall and spring. It is distinguished from other *Agaricus* by its white color, tapered stipe base, ragged or inconspicuous partial veil, and lack of staining reactions. Among similar species, *A. bisporus* is stockier, lacks a tapered stipe base, and is usually not found in open meadows. *A. bernardii* is much firmer and larger, usually stains reddish, and more often grows in hard-packed soil or cypress duff. Older specimens may resemble *A. "incultorum"*, but that species is even more fragile and has a gray-brown, scruffy-felty cap.

Agaricus "incultorum"

Kerrigan nom. prov.

BROWN MEADOW MUSHROOM

CAP: 3–6 (8) cm across, buttons rounded and inrolled, becoming broadly convex to plane. Margin often with a band of sterile tissue. Dull grayish or brownish gray to evenly dull brown. Surface dry, distinctly shaggy or velvety-fibrillose, less commonly with appressed scales. **GILLS:** Free, close to crowded. Pinkish gray when young, becoming grayish brown and finally dark brown. **STIPE:** 2–5 cm long, 0.7–2 cm thick, equal or with a tapered base. White to dull whitish tan. Surface dry, shaggy or scurfy, especially when young and below the partial veil. **PARTIAL VEIL:** Ragged or shaggy, upward-flaring ring of fibers near middle of stipe, often disappearing in age. **FLESH:** Thin, rather soft and fragile, fruitbodies easily broken, whitish, not staining. **ODOR:** Like white Button Mushroom. **TASTE:** Mild. **KOH:** No reaction. **SPORE DEPOSIT:** Dark chocolate brown. **MICROSCOPY:** Spores (6) 7.5–9 (10) x 5–6 (7) µm, ellipsoid, smooth.

ECOLOGY: Solitary or scattered in grassy areas or in areas of poor, exposed soil with short weedy plant cover. Uncommon, known from the San Francisco Bay Area southward on coast, occurring farther north inland. Primarily fruiting in early to late fall, occasionally in winter and spring.

EDIBILITY: Edible but rather soft fleshed, considered inferior to *A. campestris*.

COMMENTS: *A. "incultorum"* is similar to *A. campestris* by virtue of its weak partial veil, pink gills when young, lack of staining reactions or strong odor, and growth in poor soil or grassy areas, but is easily distinguished by its shaggy, grayish to brownish cap.

MISAPPLIED NAME: Long identified as *A. cupreobrunneus*, which is now known to be a distinct European species.

Agaricus bisporus

(J. E. Lange) Imbach

BUTTON MUSHROOM

CAP: 3–10 cm across, buttons rounded and inrolled, becoming broadly convex to flat. Margin with a fragile band of sterile tissue. Whitish, dingy whitish gray to tan, sometimes with tightly appressed tan to light brown scales. Surface dry. **GILLS:** Free, close to crowded. Pale pinkish when young, becoming pinkish tan and finally brown to dark brown. **STIPE:** 3–7 cm long, 1.5–4 cm thick, equal or with an enlarged base. Whitish or off-white with tan tones, sometimes washed pinkish or brown above veil. **PARTIAL VEIL:** A thin, white, membranous band or flaring skirt with a spiky edge. **FLESH:** Thin, soft in cap, fibrous in stipe. White, rarely staining much when cut. **ODOR:** Like store-bought Button Mushroom (they are the same species after all), occasionally somewhat fishy in age. **TASTE:** Mild. **KOH:** No reaction. **SPORE DEPOSIT:** Dark brown. **MICROSCOPY:** Spores (5) 6–7.5 (9) x 4.5–6 (7) µm, broadly ellipsoid to ellipsoid, smooth, basidia 2-spored.

ECOLOGY: Scattered in arcs or troops, fruiting in a variety of urban and disturbed landscapes. Distribution is patchy in our area, but can be locally abundant in fall and spring in Monterey Cypress duff; also under other trees including eucalyptus, as well as in grassy lawns or compost piles.

EDIBILITY: Edible, but mediocre compared to most edible wild mushrooms. The cultivated form is the most widely consumed mushroom in the United States.

COMMENTS: A wild form of the common cultivated Button Mushroom, this species can be recognized by the whitish to tan cap, bandlike ring, and lack of a distinct staining reaction. The species name refers to the 2-spored basidia, which help separate it from the very similar *A. agrinferus*. The latter species usually also has a more club-shaped stipe and fairly pronounced orange to reddish staining when cut. *A. campestris* has brighter pink gills and a tapered stipe base and grows in grass. The very rare *Lepiota rhodophylla* occurs under Monterey Cypress. Because of its pink gills, robust stature, and pinkish beige to brownish cap, it could be mistaken for this or another *Agaricus*, but it produces a white spore print.

Agaricus agrinferus

Kerrigan & Callac

CAP: 3–7 cm across, bun shaped to round-convex at first, expanding to broadly convex, sometimes nearly flat, often with a narrow band of tissue hanging over edge of gills. Whitish to dingy tan or light fawn brown or with pinkish tan tones. Surface dry, innately fibrillose to nearly smooth. **GILLS:** Free, crowded. Dull grayish pink to pinkish brown when young, becoming reddish brown to dark brown in age. **STIPE:** 3.5–8 cm long, 1–2.5 cm thick, cylindrical or more often tapered upward with a distinctly swollen, club-shaped base. Whitish at first, soon dingy, often developing a striate appearance and murky grayish or dark brown tones near attachment to cap. Surface smooth or decorated with flat chevrons of silvery or white tissue. **PARTIAL VEIL:** A white membrane attached near midpoint of stipe, sometimes a flaring spiky-edged skirt, often collapsing against stipe but rarely disappearing. **FLESH:** Thin, soft in cap, fibrous, solid or with a small central hollow in stipe. White, bruising orangey pink to light reddish in cap and stipe, especially strong above gills. **ODOR:** Like white Button Mushroom. **TASTE:** Mild. **KOH:** No reaction. **SPORE DEPOSIT:** Dark chocolate brown. **MICROSCOPY:** Spores 5.5–6.5 x 4–4.5 µm, ellipsoid, smooth, basidia 4-spored.

ECOLOGY: Solitary or more often scattered in small troops in rich duff of Monterey Cypress, usually in disturbed areas, especially along roads and trails. Generally uncommon, but can be locally abundant. Primarily coastal, most records from south of Sonoma County. Fruiting from late fall into spring.

EDIBILITY: Edible.

COMMENTS: This *Agaricus* is tough to recognize; it is most likely to be confused with the extremely similar *A. bisporus*, but is usually slightly taller, has a more club-shaped stipe, has a consistently pinkish orange to light reddish staining reaction in the flesh when cut, and has predominantly 4-spored basidia. *A. benesii* is whitish overall, stains red much more strongly, and has a more shaggy stipe. Most other whitish *Agaricus* in our area stain yellow and have more pronounced almondy or phenolic odors.

Agaricus bitorquis

(Quél.) Sacc.

TORKS

CAP: 4–20 cm across, convex to broadly convex, sometimes finally becoming flat. Margin tightly inrolled and usually remaining so to some degree in age. White or off-white, sometimes with dull brownish discolorations. Surface dry, smooth, often cracked in dry weather or more rarely with broad, irregular flat scales. **GILLS:** Free, close to crowded. Pale pinkish to pink, becoming pinkish brown and finally dark brown. **STIPE:** 3–10 cm long, 3–5.5 cm thick, robust, stout, usually tapered to a point at base, occasionally equal. Whitish overall, sometimes with brownish discolorations. **PARTIAL VEIL:** A sheathing, often doubled, upward-flaring skirt, usually very low on stipe and thus sometimes confused for a volva. **FLESH:** Thick, rather hard when young, often very firm into age. White, bruising slowly and irregularly brownish in stipe and cap flesh. **ODOR:** Like portobello or cultivated Button Mushroom. **TASTE:** Mild. **KOH:** No reaction. **SPORE DEPOSIT:** Dark brown. **MICROSCOPY:** Spores (4.5) 5.5–6 x 4–5 (6) µm, broadly ellipsoid, smooth.

ECOLOGY: Solitary, scattered, or in clusters arranged in long rows, arcs, or "veins." Usually in loosely consolidated silt, in nutrient-poor soil, along roads, sometimes in agricultural soil. Often completely or partially underground, erupting in age, often best detected by investigating cracks in top layer of soil. Widespread, but rather uncommon. This species seems to have declined in our area in the last 30 years. Fruiting in winter and early spring.

EDIBILITY: Edible. Enjoyed by many as an excellent mushroom for the table, but it causes allergic reactions in some people.

COMMENTS: This species can be recognized by the white colors, low-set, upward-sheathing partial veil, lack of yellow or red staining, and squat, robust stature and firm texture. It is most easily confused with *A. bernardii*, which stains pinkish red and has a briny odor. It is probably more common inland than *A. bernardii*, but the two have probably been confused in the past, clouding our understanding of their distributions. *A. vinaceovirens* is a bit taller, lacks a sheathing veil, and has an unpleasant odor; it is mostly found under cypress.

Agaricus bernardii

Quél.

SALT-LOVING AGARICUS

CAP: 4–20 cm across, broad, convex with an inrolled cottony margin, becoming broadly convex to plane in age. Margin often persistently inrolled well into maturely. White, sometimes with gray or brown tones. Surface dry, innately fibrillose to smooth, sometimes with broad, flattened scales, coarsely cracking in dry weather. **GILLS:** Free, crowded, often rather thick edged. Pinkish at first, then dull brownish pink, finally dark brown. **STIPE:** 3–9 (13) cm long, 1.5–5 (8) cm thick, most often with a distinctly tapered base. White or whitish with gray or brownish tones, eventually staining reddish brown when scraped. Surface more or less smooth. **PARTIAL VEIL:** A thin, upward-flaring skirt, usually thin and ragged at margin, collapsing in age. **FLESH:** Thick, often firm and rubbery, whitish to tan, usually bruising distinctly pinkish to reddish when cut and rubbed, especially at juncture of cap and stipe. **ODOR:** Rather strong, briny, salty, "like low tide." **TASTE:** Mild or briny. **KOH:** No reaction. **SPORE DEPOSIT:** Chocolate brown. **MICROSCOPY:** Spores (5) 6–8 x 5–6.5 µm, ellipsoid, smooth.

ECOLOGY: Solitary, scattered, or in loose clusters, generally under cypress, on lawns or other urban areas, and in nutrient-poor, hard-packed soil, usually within a few miles of the coast. Common from the San Francisco Bay Area southward. Fruiting in fall and early winter, sometimes abundant in spring.

EDIBILITY: Edible and considered excellent by some; others consider it unpleasantly salty and tough.

COMMENTS: The stout stature, dense and often large fruitbodies, whitish colors, pinkish red–staining flesh, and briny odor are distinctive. The similar *A. bitorquis* does not typically bruise red (although this may be absent in some fruitbodies of *A. bernardii*), has a more sheathing veil, and has a mild odor. The rare and poorly known *A. vinaceovirens* is similar, but has a taller stature, a thinner, flocculent partial veil, and less red-staining flesh, and smells weirdly foul—to quote Noah Siegel, "like a wet dog who ran through a salt marsh and then rolled in dead fish." That being said, it is edible. The latter species is mostly found under cypress in our area.

5 • Dark-Spored Mushrooms

The mushrooms in this section share dark purplish gray to black spore prints, but exhibit quite a range in fruitbody size, stature, and coloration. The inky caps (whose gills dissolve into an inky, spore-laden liquid in a process called deliquescence) comprise two unrelated lineages: The few species of "true" *Coprinus* (based around the type species, *C. comatus*) are in the family Agaricaceae (related to *Agaricus* and *Lepiota*), while the majority of the inky caps have been assigned to other genera (*Coprinellus*, *Coprinopsis*, *Parasola*, and *Psathyrella*) in the family Psathyrellaceae. However, all of the generic concepts in the latter groups are being heavily reworked; it is likely that many of today's *Psathyrella* will be tomorrow's *Coprinopsis*, and vice versa, and you can expect a few new genera as well!

Dark-spored mushrooms can be very difficult or impossible to identify given current knowledge. A microscope is almost always required to examine details of spore size, cystidia, and other features.

Of the deliquescent species, *Coprinus* contains only one common species in our area—the distinctive Shaggy Mane. The other inky caps have small, fragile fruitbodies with caps ranging from scruffy-shaggy to smooth or finely flaky (*Coprinellus*, *Coprinopsis*) or pleated-sulcate (*Parasola*). However, these are rough generalizations. *Lacrymaria* is a bit of an oddball genus, with rather stout fruitbodies and strongly mottled gills that "weep" clear droplets of liquid when young. *Psathyrella* species tend to have thin, white, tall, brittle stipes (breaking when bent) and small caps. They often grow in troops on wood chips or in disturbed areas. *Panaeolus* are similar to the latter genus, but tend to be slightly less brittle, have strongly mottled gills, and are more likely to be found on dung or in grass. *Panaeolina* have very dark brown (but not truly black) spores, but otherwise look very similar to *Panaeolus*.

The following genera tend to have fruitbodies with slightly thicker, less fragile flesh and frequently have viscid or slightly gelatinized cap surfaces. *Protostropharia* grow on dung and are slimy on both cap and stipe. *Deconica* have small fruitbodies that could be confused with many of the other black-spored genera. *Psilocybe* tend to stain blue quite markedly when bruised, handled, or in age. *Hypholoma* mostly grow in clusters out of wood or on woody debris. The two *Leratiomyces* species are extremely common on wood chips. *Stropharia* have the largest, fleshiest fruitbodies of any of the genera listed here and usually have a distinct annulus and viscid caps.

Coprinus comatus

(O. F. Müll.) Pers

SHAGGY MANE

CAP: 2–6 cm across (unexpanded) to 12 cm across (in age). Egg shaped or cylindrical when young, elongating as stipe grows, becoming flatter and curling as cap and gills dissolve, sometimes with an umbo, edge often extensively curled up or ragged, or hanging down in black "drips." In age sometimes only a small disc of cap tissue is left on top of stipe. Mostly ocher-tan to light brown when very young, but soon cracking, becoming extensively white with a brown patch at center and many scattered flat, tan brownish patches, turning pinkish red upward from margin as gills dissolve. Dry, smooth when very young, soon fibrillose, with shaggy fibrils clumped into scales. **GILLS:** Free or narrowly attached, extremely crowded (often not individually distinguishable). White at first, becoming pink, then grayish, and finally black (often exuding red droplets). In age dissolving into a slightly viscous black liquid. Color change progressing upward from edge of cap, often with a distinct line between mature and immature areas. **STIPE:** 4–15 (25+) cm long, 1.5–2.5 (6) cm thick, tall, quite evenly cylindrical with slightly enlarged base. White, smooth to fibrillose. **PARTIAL VEIL:** Whitish band with brown edge around cap margin when very young, soon detached and sliding as a free ring, often settling at extreme base of stipe. **FLESH:** Stringy, soft, white. Stipe hollow with thin string running down center. **ODOR:** Indistinct or unpleasant in age.

TASTE: Mild. **SPORE DEPOSIT:** Black, inky. **MICROSCOPY:** Spores 10.5–14 (17.5) x 5.5–7.5 (10) μm, long ellipsoid to nearly cylindrical, smooth and dull, dark grayish black in KOH.

ECOLOGY: Solitary, clumped, or in dense troops in wood chips, lawns, and disturbed soils along trails and roads. Can fruit any time substrate is wet enough. Common throughout our range.

EDIBILITY: Edible and popular. However, it often grows in polluted areas (roadsides and lawns). Since it dissolves so quickly, only young specimens should be picked and cooked the same day. Immersing them in salt water and refrigerating them will slow deliquescence.

COMMENTS: The cylindrical, whitish caps with brownish scales, liquefying gills, and loose ring on the stipe distinguish this species. The only species likely to cause confusion are uncommon to rare in our area: *C. calyptratus* has a star-shaped patch at the center of the cap, and larger spores with an eccentric germ pore. *Podaxis pistillaris* is a desert-dwelling species with a cap that never expands, and with gills that dry out and split irregularly rather than turning to ink; the fruitbodies can persist for months. *C. atramentaria* usually grows in clusters, and has an egg- to bell-shaped cap that is smooth and gray in color.

Coprinopsis atramentaria

(Bull.) Redhead, Vilgalys & Moncalvo

ALCOHOL INKY CAP

CAP: 2–8 cm across, egg-shaped, ovoid to cylindrical at first, soon bell shaped, convex to plane, often deliquescing while expanding. Light to dark gray, sometimes with grayish brown disc, darkening in age. Surface moist to dry, mostly smooth, with a few scattered brownish scales over disc, and whitish hairs on margin when young, occasionally cracking when dry.
GILLS: Free, very crowded. Whitish at first, soon gray, then black and deliquescing. **STIPE:** 3–15 cm long, 0.3–1.5 cm thick, equal or enlarged toward, or tapering at, base. White, at times with brownish fibrillose scales at base. Surface dry, smooth, occasionally with veil remnants on lower stipe, with fibrillose scales at base. **FLESH:** Relatively thin and fragile, stipe hollow.
ODOR: Indistinct. **TASTE:** Indistinct. **SPORE DEPOSIT:** Black.
MICROSCOPY: Spores 6.5–10.5 x 4–6.5 μm, ellipsoid to ovoid, with a broad germ pore, smooth, dark reddish black in KOH. Pleurocystidia and cheilocystidia narrowly cylindrical or constricted with a rounded head.

ECOLOGY: Clustered or in troops, occasionally in small tufts, rarely solitary, usually on buried woody debris in grassy areas, on dead tree roots, or around stumps or dead trees. Widespread but with a patchy distribution throughout our range. Fruiting from early fall well into spring, or during summer in irrigated areas.

EDIBILITY: Edible, with caution! Consumption of alcohol within a few days of eating this mushroom could lead to sweating, light-headedness, severe headaches, and swelling and tingling in the extremities, as well as nausea and vomiting. In more severe cases, extremely low blood pressure and even fainting can occur. *C. atramentaria* contains the compound coprine, a disulfram-like chemical (the compound in Antabuse), which inhibits the body's ability to metabolize alcohol. These inky caps have a rich, mushroomy flavor well suited for use in sauces, but avoid those found in urban areas or near roadsides.

COMMENTS: The grayish, ovoid to cylindrical cap that expands and deliquesces as it ages, the clustered growth, and the somewhat large size are helpful in identification. Two other species are very similar in appearance. *C. acuminata* is typically smaller and has a more conical cap with a small, rounded umbo and smaller spores; it is more often found in forested settings. *C. romagnesiana* has a slightly scaly cap (at least in age) and small blackish scales on the lower stipe. Both species should be treated like *C. atramentaria* in terms of edibility. *Coprinus comatus* has a cylindrical, whitish cap that is usually scaly or shaggy.

SYNONYM: *Coprinus atramentarius* (Bull.) Fr.

Coprinopsis lagopus group
RABBIT'S-FOOT INKY CAP

CAP: 1–5 cm across, ovoid to cylindrical at first, soon expanding to plane. Margin often becoming upturned or incurled, often ragged, splitting, or dissolving when wet. Buff to grayish at first, soon gray with a buff disc, to all gray. Surface finely pleated or grooved, covered with short whitish to pale gray spiky to matted hairs, breaking up into irregular patches as cap expands, occasionally scuffing off completely. **GILLS:** Narrowly attached to free, close, narrow, often deliquescing, at least partly in age. Pallid at first, soon gray, then black. **STIPE:** 3–9 (12) cm long, 0.2–0.6 cm thick, equal or enlarged downward, occasionally with a bulbous base. White. Surface dry, finely floccose at first, to smooth. **FLESH:** Thin and fragile, stipe hollow. **ODOR:** Indistinct. **TASTE:** Indistinct. **SPORE DEPOSIT:** Black. **MICROSCOPY:** Spores 10–14 x 6–8.5 μm, ellipsoid to ovoid, with a broad germ pore, smooth, dark reddish black in KOH. Pleurocystidia and cheilocystidia oblong, or slightly constricted below rounded head. Veil consists of elongated, sausagelike cells.

ECOLOGY: Often in troops, tufts, or clusters, rarely solitary, usually in wood chips or on woody debris, also on straw or compost piles. Very common, most often fruiting in fall and spring (or in humid summers), less common in winter months.

EDIBILITY: Unknown, likely nontoxic.

COMMENTS: The fuzzy, grayish young caps, deliquescing gills, and growth in wood chips are good identifying features. Recent genetic work has found a number of species in this complex, but which or how many occur in California is still unknown. *C. jonesii* (=*C. lagopides*) is very similar, but often grows in mixed straw and dung, or compost piles. It is generally stockier and slightly browner. *C. friesii* is a very small grass-inhabiting species with scattered, flocculent-granular scales on the young caps.

SYNONYM: *Coprinus lagopus* (Fr.) Fr.

Coprinellus flocculosus

(DC.) Vilgalys, Hopple & Jacq. Johnson

CAP: 1.5–5 (7) cm across, cylindrical to oval at first, becoming bell shaped and ragged, with a deliquescing margin, expanding to plane, but dissolving while doing so. Buff to honey brown with whitish to pale buff veil patches at first, soon beige to gray, dissolving to blackish. Surface grooved, irregularly covered with flocculent-felty patches and specks of tissue, readily becoming bald in age. **GILLS:** Narrowly attached to free, very close, broad. Pallid to gray with whitish to pale gray fringed edges, soon deliquescing black. **STIPE:** 3–9 cm long, 0.2–0.6 cm thick, equal or enlarged slightly lower, and often with a bulbous base, with a collarlike volva when young, very fragile. Whitish. Surface dry, finely pruinose to smooth. **FLESH:** Very thin and fragile, stipe hollow. **ODOR:** Indistinct. **TASTE:** Indistinct. **SPORE DEPOSIT:** Black. **MICROSCOPY:** Spores 9.5–16.5 x 6.5–10 μm, ellipsoid to ovoid, smooth, thick walled with a germ pore, blackish in KOH. Pleurocystidia ellipsoid, ovoid, with a swollen base or cylindrical, 50–100 x 30–70 μm. Cheilocystidia smaller, more globose to ovoid. Pileocystidia absent.

ECOLOGY: Scattered in troops, small clusters, or occasionally solitary on wood chips. Common in urban habitats throughout our range. Fruiting throughout wet season, but generally in larger flushes in spring or early fall.

EDIBILITY: Nontoxic.

COMMENTS: This wood-chip-loving species can be recognized by the honey-colored to brownish gray, cylindrical to bell-shaped, deliquescing caps with small, flocculent-felty scales. Members of the *C. micaceus* group are often clustered and have pale granular specks (not felty scales) on the young caps. *C. domesticus* has persistent, flocculent-granular specks on the cap and grows from an orange, coarsely hairy mat. *Coprinopsis atramentaria* has a gray cap that is smooth or with scattered brownish scales.

Coprinellus domesticus

(Bolton) Vilgalys, Hopple & Jacq. Johnson

CAP: 1.5–5 (7) cm across, oval-conical to bell shaped at first, becoming convex to plane, margin often uncurled in age. Buff to honey brown at first, more rarely reddish brown, often darker on the disc, margin soon gray. Surface pleated and grooved, densely covered with pale flocculent-granular specks, rarely becoming bald in age. **GILLS:** Narrowly attached, often becoming free, close, narrow, edges often fringed when young, slowly deliquescing. Creamy at first, soon creamy gray, then blackish when spores mature. **STIPE:** 3–7 cm long, 0.2–0.4 cm thick, equal or enlarged slightly toward base. Whitish. Surface dry, finely pruinose to smooth, occasionally with veil tissue on lower portion. Base often with whitish hairs. **FLESH:** Very thin and fragile, stipe hollow. **ODOR:** Indistinct. **TASTE:** Indistinct. **SPORE DEPOSIT:** Black. **MICROSCOPY:** Spores 6–9 x 3.5–5 μm, ovoid to ellipsoid with a truncate end, smooth, thick walled with a germ pore, purplish black in KOH. Cheilocystidia globose to ovoid, 30–100 x 30–60 μm. Pileocystidia with a rounded base and a long, narrow neck.

ECOLOGY: Scattered in troops or in small clusters, occasionally solitary, fruiting on or near rotting hardwood logs, stumps, or woody debris in early fall or spring. It seems to have a preference for alder in our area. The substrate around the fruitbodies is usually covered with an orangish to dark orange-brown mat of spiky fibrils called an ozonium, which may be inconspicuous at times. Although it commonly fruits in wet basements on the East Coast, there are no such reports from California. Widespread but uncommon.

EDIBILITY: Nontoxic.

COMMENTS: This species resembles a few other inky caps, but can be told apart by the orangish mat of coarse, spiky hairs covering the substrate around the stipe base. If that feature is overlooked, the persistent, flocculent-granular specks on the cap are a helpful distinguishing feature. Members of the *C. micaceus* group are similar, but have pale granular specks on the cap when young and are often larger. *C. flocculosus* has flocculent-felty warts and patches that readily scuff off, often leaving the cap bald in age.

Coprinellus micaceus

(Bull.) Vilgalys, Hopple & Jacq. Johnson

MICA CAP

CAP: 1.5–4 (5) cm across, ovoid at first, becoming bell shaped to convex, margin sulcate, becoming ragged, often splitting, and deliquescing when wet. Yellow-brown, honey brown with a paler margin, occasionally more ocher-brown to brown on disc, becoming gray, at least on margin in age. Surface scattered with glistening whitish specks, which often disappear in age. **GILLS:** Narrowly attached to free, crowded, narrow, partially to completely deliquescing. Creamy buff at first, soon grayish brown to black. **STIPE:** 3–7 (11) cm long, 0.2–0.5 cm thick, equal or enlarged slightly toward base, fragile. White to pale buff. Surface dry, pruinose to smooth. **FLESH:** Thin and fragile, stipe hollow. **ODOR:** Indistinct. **TASTE:** Mild. **SPORE DEPOSIT:** Black. **MICROSCOPY:** Spores 7–10 x 5–7 µm, ellipsoid or ovoid, with a conical base and truncate apex, smooth, thick walled with a broad germ pore, dark reddish black in KOH. Pleurocystidia oblong to ovoid. Cheilocystidia similar or more rounded. Stipe covered in caulocystidia with swollen bases and tapered necks.

ECOLOGY: Generally clustered, scattered in small to large troops, in tufts, or occasionally solitary around stumps, dead trees, rotted logs, or buried woody debris. Very common, occurring throughout our range. Fruiting from early fall well into spring.

EDIBILITY: Edible and flavorful, but practically fleshless. Probably best used for gravy and sauces.

COMMENTS: The clustered growth, glistening whitish specks on honey yellow caps, and deliquescing gills are helpful identifying features. However, similar species in this group require microscopic examination to identify; keep in mind that no serious taxonomic work has been done on California's inky caps, and undoubtedly much is unknown about the diversity of these mushrooms. The following species lack caulocystidia: *C. truncorum* has slightly smaller, more rounded-ellipsoid spores (6.5–9 x 4.5–6.5 µm). *C. saccharinus* has spores measuring 6–9.5 x 5–6.5 µm, but with a somewhat conical base and rounded or truncate apex. *"Coprinus" rufopruinatus* has pinkish veil specks on the cap. Reports of the latter two species in California have yet to be confirmed. Also compare *C. domesticus*, which has granular specks on the cap and emerges from an orangish, hairy mat.

Coprinellus impatiens

(Fr.) J. E. Lange

CAP: 0.7–2.5 cm across, broadly egg shaped or cylindrical at first, soon expanding to broadly convex to plane, often with an umbo. Margin distinctly pleated or grooved. Yellow-brown to ocher-brown when young, very soon gray with an ocher-brown disc, to pale grayish white overall. **GILLS:** Narrowly attached, moderately spaced, edges often fringed. Pallid creamy gray when young, becoming gray and dark splotched as spores start maturing, slowly deliquescing and becoming dark grayish black in age. **STIPE:** 2–7 cm long, 0.1–0.3 cm thick, equal, or enlarged toward base. White to pale grayish yellow. Surface dry, pruinose to minutely hairy, often longitudinally lined near apex. **FLESH:** Very thin and fragile, stipe hollow. **ODOR:** Indistinct. **TASTE:** Indistinct. **SPORE DEPOSIT:** Black. **MICROSCOPY:** Spores 8.5–11 x 5–6 (7) µm, ovoid to egg shaped, smooth, thick walled with a germ pore, ocher-brown in KOH. Cheilocystidia abundant, bottle shaped, thin walls, colorless. Pileocystidia scattered, similar to cheilocystidia but with a tawny base. Pleurocystidia absent.

ECOLOGY: Solitary or scattered in leaf litter or grassy woodlands, generally under oaks. Locally common. Often fruiting after soaking rains in fall and winter.

EDIBILITY: Nontoxic.

COMMENTS: The yellow-brown cap with distinct pleats around the margin, slender whitish stipe, and black spores are helpful identifying features. *C. angulatus* is very similar, but grows on burnt ground and has spores shaped like a bishop's hat ("mitriform"). *Parasola lactea* is also very similar, but is often smaller and more finely pleated on the cap, with grooves going all the way to the center (which is often centrally depressed). *C. disseminatus* is smaller and often grows in large clusters or troops. Fruitbodies of species in the *C. micaceus* group are generally more colorful (warm yellowish browns) and have micalike specks on the caps when young.

Parasola auricoma

(Pat.) Redhead, Vilgalys & Hopple

CAP: 0.5–3.5 (5) cm across, ovoid at first, soon broadly convex to plane, slightly umbonate, but becoming depressed at disc in age. Margin pleated, ragged, and often splitting in age. Ocher-brown to orangish brown (less often dark brown) at first, fading to grayish brown, to all gray with a brownish spot on the disk. Surface striate at first, becoming grooved and pleated, moist to dry. **GILLS:** Narrowly attached, or becoming free, crowded to close, very narrow. Buff to light brown with a paler edge at first, soon darkening; brown to black. **STIPE:** 3–7 cm long, 0.2–0.4 cm thick, equal or enlarged slightly toward base. White to pale buff. Surface dry, smooth. **FLESH:** Very thin and fragile, stipe hollow. **ODOR:** Indistinct. **TASTE:** Indistinct. **SPORE DEPOSIT:** Black. **MICROSCOPY:** Spores 10–14.5 x 6–8 µm, ellipsoid to oblong, rounded at base and apex, with a germ pore, smooth, dark reddish black in KOH. Pleurocystidia nearly cylindrical or constricted below a rounded head. Cheilocystidia similar, smaller. Pileipellis and stipe base with thick-walled, long brown hairs (setae).

ECOLOGY: Solitary, scattered, or clustered in wood chips, grass, or disturbed soil. Common in urban areas, scattered elsewhere. Fruiting throughout wet season, but most common in early fall and spring.

EDIBILITY: Nontoxic.

COMMENTS: The ocher-brown to orangish brown colors of the young cap, pleated surface that lacks veil remnants, smallish size, and black spores are distinctive. *P. lactea* is paler, buff to grayish, and has distinctly free gills forming a collar around the stipe. *Coprinellus micaceus* (and others) are larger and have glistening specks of tissue on the cap when young. See comments under *Coprinopsis stercorea* for other look-alikes.

Parasola lactea

(Pat.) Redhead, Vilgalys & Hopple

UMBRELLA INKY CAP

CAP: 0.5–2.5 (3.5) cm across, ovoid at first, soon broadly convex to plane, often becoming depressed at disc in age. Buff to light brown, often with a grayish margin at first, soon gray with a buff to brown disc, to all gray in age. Surface dry, compressed with narrow groves at first, pleated, expanding with groves to disc. **GILLS:** Free, with a distinct collar around stipe, crowded to close, very narrow. Buff to light brown with a paler edge at first, soon darkening; brown to black. **STIPE:** 2–6 cm long, 0.1–0.2 cm thick, equal or enlarged slightly toward base. Whitish to pale buff. Surface dry, smooth. **FLESH:** Very thin and fragile, stipe hollow. **ODOR:** Indistinct. **TASTE:** Indistinct. **SPORE DEPOSIT:** Black. **MICROSCOPY:** Spores 8–12 x 7–10.5 x 5.5–7 μm, heart shaped to angular, with five sides, to lens shaped, with an eccentric germ pore, smooth, dark reddish black in KOH. Pleurocystidia oblong, ovoid, or constricted below a rounded head. Cheilocystidia similar or with a swollen base and slender neck.

ECOLOGY: Solitary, scattered, or in small tufts in grassy areas, often in disturbed soil, occasionally in wood chips. Locally common. Fruiting in fall and spring, or summer in irrigated lawns.

EDIBILITY: Nontoxic.

COMMENTS: This dainty inky cap can be recognized by its small size; buff-colored cap that lacks veil remnants becoming pleated and grayish in age; and distinctly free gills. *P. plicatilis* is very similar, but generally is slightly smaller, appears to be restricted to grassy lawns, and has longer spores (10–14.5 μm). *P. auricoma* has more ocher-brown to orangish brown when the cap is young, has narrowly attached gills, and is most commonly found on (but is not restricted to) wood chips. Also see comments under *Coprinopsis stercorea* for similar dung-dwelling species.

SYNONYM: *Parasola leiocephala* (P. D. Orton) Redhead, Vilgalys & Hopple.

Coprinellus disseminatus

(Pers.) J. E. Lange

FAIRY HELMETS

CAP: 0.4–1 cm across, egg shaped at first, soon bell shaped, or occasionally convex. Whitish to pale honey brown at first, often with a darker, ocher-brown disc, becoming more grayish brown and often dark gray to blackish from spores. Surface pleated and grooved, covered with minute glistening hairs when young. **GILLS:** Narrowly attached to free, moderately spaced, edges often fringed. Pallid creamy gray when young, becoming gray to blackish when spores mature. **STIPE:** 1.5–4 cm long, 0.1–0.2 cm thick, equal, or enlarged slightly toward base. Whitish to creamy buff, occasionally becoming pale gray in age. Surface dry, pruinose or finely hairy over entire portion. **FLESH:** Very thin and fragile, stipe hollow. **ODOR:** Indistinct. **TASTE:** Indistinct. **SPORE DEPOSIT:** Black. **MICROSCOPY:** Spores 7–10 x 4–4.5 μm, ellipsoid to egg shaped with a truncate end, smooth, thick walled with a germ pore, purplish black in KOH. Pileocystidia with a bulbous base and a long neck, occasionally continuing on gills around cap margin.

ECOLOGY: Usually in large troops or clusters, on or around dead trees, stumps, or woody debris (sometimes buried). Fruiting throughout wet season, but more common in fall and spring.

EDIBILITY: Flavorful, but tiny and practically fleshless.

COMMENTS: This tiny mushroom makes up for its small size by fruiting in troops that often number in the hundreds of fruitbodies. Besides the trooping or clustered fruiting habits, the diminutive size, pale, pleated cap with glistening hairs, and black spores are useful identifying features. *C. impatiens* is similar, but is slightly larger and doesn't grow in such large troops and clusters. *Parasola* species can be similar but have bare caps.

Coprinopsis stercorea

(Fr.) Redhead, Vilgalys & Moncalvo

DUNG-LOVING INKY CAP

CAP: 0.3–1.2 cm across, egg shaped to ovoid at first, becoming broadly convex to plane, margin pleated, soon deliquescing, occasionally becoming up-curled in age. Covered in whitish to pale gray hairy-flocculent to granular scales, becoming gray in age. **GILLS:** Narrowly attached, close to subdistant, narrow, soon deliquescing. Grayish at first, soon black. **STIPE:** 1–3.5 cm long, 0.05–0.1 cm thick, equal. Translucent to watery white. Surface dry, with scattered fine whitish specks or hairs. **FLESH:** Extremely thin and fragile, stipe hollow. **ODOR:** Pungent. **TASTE:** Indistinct. **SPORE DEPOSIT:** Black. **MICROSCOPY:** Spores 5.5–7.5 x 3–4.5 µm, broadly cylindrical to ellipsoid, smooth, thick walled with a broad germ pore, dark reddish black in KOH. Pleurocystidia oblong or with a slightly constricted neck, 25–65 x 12–23 µm. Cheilocystidia similar or more rounded. Veil made up of warty round cells.

ECOLOGY: Solitary, scattered, or in small troops on relatively fresh dung. Mostly occurring on dung of grazing animals, but is not picky; also found on bear and dog scat. Fruiting throughout wet season, generally soon after soaking rains following a dry period. Widespread and common, but fleeting—fruitbodies can come and go in less than a day.

EDIBILITY: Unknown.

COMMENTS: This tiny inky cap in one of a handful of very similar species that occur on dung; almost all require microscopic examination to identify. As a group, they can be recognized by their fragile fruitbodies, orangey brown colors, and deliquescing gills. *C. stercorea* is the most commonly encountered species in this group; it generally has a cap less than 1 cm wide and covered with hairy-flocculent or granular scales when young, and has small, ellipsoid spores. *Coprinellus ephemeroides* is a tiny species with a partial veil that leaves a flaring ring on the stipe. *Coprinellus heptemerus* is similar, but has a minutely spiky cap when young. Still smaller (caps 0.2–0.7 cm across) are *Coprinellus pellucidus* and *Coprinellus heterosetulosus,* which have finely granular to smooth, brownish caps that soon turn grayish. Microscopically, *Coprinellus heterosetulosus* has thick-walled cystidia on the cap cuticle (absent in *Coprinellus pellucidus*). Common, slightly larger, dung-inhabiting species include *Coprinopsis radiata,* with an egg-shaped cap at first covered with whitish to pale gray hairs (similar to *Coprinopsis lagopus*), and a rooting stipe, and *Coprinopsis macrocephala,* nearly identical but with wider spores (8–10 µm vs. 7.5–8.5 µm for *Coprinopsis radiata*). *Coprinopsis cinerea* is very similar to *Coprinopsis macrocephala,* but often has a slightly bulbous stipe base; it often grows in compost piles or in straw mixed with manure (rarely on pure dung). *Parasola schroeteri* (a *P. lactea*–like species) and *P. miser* (a tiny, honey brown to gray-capped species) are also common on dung.

Coprinopsis nivea

(Pers.) Redhead, Vilgalys & Moncalvo

SNOWY INKY CAP

CAP: 1–4 cm across, egg shaped to ovoid at first, becoming bell shaped, occasionally to plane. Margin often slowly dissolving after it opens, occasionally up-curled in age. Bright snowy white to pale grayish white in age. Surface dry, completely covered in flocculent-granular powder, persisting in age or sometimes washing off with rain. **GILLS:** Narrowly attached, often becoming free, close, narrow, edges often fringed when young, soon deliquescing. Whitish at first, soon gray, then black. **STIPE:** 3–7 cm long, 0.2–0.4 cm thick, equal or enlarged slightly toward base, fragile. White. Surface dry, finely powdery to smooth. **FLESH:** Very thin and fragile, stipe hollow. **ODOR:** Indistinct. **TASTE:** Indistinct. **SPORE DEPOSIT:** Black. **MICROSCOPY:** Spores 12–19 x 11–15.5 µm, lens shaped to lemon shaped in face view, ellipsoid in side view, smooth, thick walled with a germ pore, dark reddish black in KOH. Pleurocystidia bladder shaped, ellipsoid to nearly cylindrical, 50–150 x 25–60 µm. Cheilocystidia similar, smaller, up to 80 µm long. Veil made up of round cells, up to 100 µm.

ECOLOGY: Solitary, scattered, or in small clusters on dung. Fruitbodies rarely last more than a day or two. Fruiting throughout wet season, generally soon after soaking rains. Widespread and common.

EDIBILITY: Nontoxic.

COMMENTS: The white colors, powdery cap, dissolving gills, and growth on dung are distinctive. The species listed as *Psathyrella/Coprinopsis* "White" has a silky-fibrillose cap and grows in wet twig and leaf litter or on woody debris. Of the number of other inky caps on dung, none have the striking white fruitbodies of this species. *C. cinerea* has a cylindrical to bell-shaped gray cap that is covered with pale, shaggy hairs when young (similar to *C. lagopus*). Most other dung-dwelling species are smaller and are listed in the comments section under *C. stercorea*. *Panaeolus antillarum* has a smooth, whitish cap and nondeliquescing gills.

Psathyrella/Coprinopsis "White"

CAP: 0.5–3 cm across, egg shaped, rounded-conical or bell shaped at first, becoming broadly convex in age. Pure white at first, remaining so or becoming slightly grayish in age. Surface dry, silky-fibrillose to nearly smooth, not translucent-striate. Margin sometimes with small teeth of clumped pale fibrils. **GILLS:** Obscurely attached, close to subdistant. White when young, becoming mottled gray and black in age, occasionally with sterile areas that remain pale. Edges finely fringed in white. **STIPE:** 3.5–8 cm long, 0.2–0.6 cm thick, straight to curved, slightly enlarged at base. White. Surface floccose-scaly when very young, soon silky-hairy, nearly smooth in age. **FLESH:** Thin and fragile, white to grayish. Stipe hollow. **ODOR:** Indistinct. **TASTE:** Indistinct. **SPORE DEPOSIT:** Black. **MICROSCOPY:** Spores 5.5–7.5 x 3–4.5 µm, broadly cylindrical to ellipsoid, smooth, thick walled with a broad germ pore, dark in KOH. Pleurocystidia swollen with a slightly constricted neck. Cheilocystidia similar or more rounded. Veil made up of round cells up to 100 µm across.

ECOLOGY: Solitary, in pairs or small clusters in fine woody debris, especially well-rotted twig and branch litter in streambeds or ditches, as well as on rotting berry canes or in wood chips. Fairly common, known from Santa Cruz County to Sonoma County, but probably more widespread. Primarily fruiting in spring, smaller numbers found in late fall and winter.

EDIBILITY: Unknown.

COMMENTS: This distinctive species has somehow remained undescribed despite being well known to the mycological community for decades. Currently, it is unkown if this mushroom belongs in the genus *Psathyrella*, or if it will end up in *Coprinopsis*. Very few other mushrooms in our area show the combination of silvery white cap and stipe, and grayish to black gills in age. *Coprinopsis nivea* grows on dung, has a powdery cap surface, and is generally larger. Most *Coprinellus* in our area are not as white and have a more cylindrical cap shape. A few *Panaeolus* can look vaguely similar when faded but are much less fragile. Many *Psathyrella* are similar in stature and fragility, but none in our area are silvery white overall.

Psathyrella frustulenta

(Fr.) A. H. Sm.

FLUFFY PSATHYRELLA

CAP: 0.8–2 (3) cm across, oval-conical to bell shaped at first, becoming broadly bell shaped to convex. Honey brown, brown to grayish brown, covered with white, silky-flocculent hairs and tufts, becoming more gray with age, retaining whitish hairs around margin, or occasionally bald in age. **GILLS:** Attached, or seceding in age, close, narrow. Pale cinnamon buff at first, soon gray, then blackish when spores mature. Occasionally with paler, sterile patches. **STIPE:** 3–7 cm long, 0.2–0.4 cm thick, equal, very fragile. White to pale buff in age. Surface dry, covered with silky white fibrils or tufts with a pruinose apex, to nearly smooth in age. **PARTIAL VEIL:** Leaving remnants on cap margin. **FLESH:** Very thin and fragile, stipe hollow. **ODOR:** Indistinct. **TASTE:** Indistinct. **SPORE DEPOSIT:** Deep reddish black. **MICROSCOPY:** Spores 7–8.5 x 4.5–5.5 μm, ellipsoid, oblong to ovate smooth, bright ochraceous to pale reddish cinnamon in KOH. Pleurocystidia and cheilocystidia numerous, narrowly to broadly spindle shaped, to nearly cylindrical, tips pointed or slightly rounded.

ECOLOGY: Scattered in small troops or solitary in coniferous duff or moss. Fairly common, but often overlooked. Fruiting in fall and early winter on the North Coast.

EDIBILITY: Nontoxic.

COMMENTS: The honey brown to grayish cap covered with striking silky-flocculent hairs and tufts; small size; and fragile fruitbody are helpful clues to the identity of this *Psathyrella*. A nearly identical species, *P. hirta*, has a slightly darker brown cap and grows on dung. *P. subcaespitosa* is also similar, but often grows in clusters and has whitish to brown gills.

Psathyrella corrugis group

GRACEFUL PSATHYRELLA

CAP: 0.7–2 cm across, bell shaped to convex, becoming more broadly convex at maturity. Brown to beige, becoming paler as it loses moisture. Margin translucent-striate. Surface smooth, often with silvery whitish fibrils scattered over surface when young. **GILLS:** Broadly or obscurely attached, close. Pale beige to grayish tan, often developing a distinct pinkish or light reddish wash, becoming mottled with darker gray in age. **STIPE:** 3–7 cm long, 0.1–0.25 cm thick, cylindrical, very straight and upright, base often narrowed and elongated into tapered pseudorhiza (root) covered in spiky white tomentum (easily broken off or overlooked). Whitish to cream or pale dingy yellowish. Surface obscurely finely scurfy. **FLESH:** Very thin, fragile. **ODOR:** Indistinct. **TASTE:** Indistinct. **SPORE DEPOSIT:** Dark purplish brown to nearly black. **MICROSCOPY:** Spores 11–14 x 6.5–8 μm, enlongate-ellipsoid, with a pronounced germ pore, smooth. Pleurocystidia needle- to narrowly spindle-shaped, with a long, pointed apex. Cheilocystidia similar.

ECOLOGY: Solitary or in large troops on wood chips and in other disturbed areas. Common throughout our range, at least when considered as a species complex. Fruiting from fall through early spring.

EDIBILITY: Nontoxic.

COMMENTS: This species complex includes the small-capped *Psathyrella* with very thin, tall, straight stipes. Although often cited as a diagnostic feature, the pinkish wash to the mature gills seems to be present in a number of species of *Psathyrella*. A taxon often referred to as *P. prona* in our area becomes extensively pinkish on the cap in age, and the mature gills have pinkish red edges; its basidia often bear only two spores (on average measuring 12.5–17 x 6.5–8 μm), and it has pointed pleurocystidia that are occasionally forked.

SYNONYM: *Psathyrella gracilis* (Fr.) Quél.

Psathyrella atrospora

A. H. Sm.

CAP: 0.5–3 cm across, hemispheric to rounded conical, occasionally bell shaped. Bright bay red to reddish brown, soon ocher-brown fading outward to beige or cream, translucent-striate at maturity. Surface smooth, moist to dry (finely hairy when young if viewed with a hand lens). **GILLS:** Narrowly attached, close. Pallid grayish beige when young, soon mottled grayish brown to dark grayish with lilac-brown tones. **STIPE:** 3–8 cm long, 0.2–0.4 cm thick, cylindrical, quite straight. Pale cream or whitish when young, most often dull yellowish in age. **FLESH:** Very thin, fragile. Cap brownish, stipe whitish. **ODOR:** Indistinct. **TASTE:** Indistinct. **SPORE DEPOSIT:** Dark gray to nearly black. **MICROSCOPY:** Spores 15–20 x 7–9.5 µm, ellipsoid to subcylindrical, smooth. Pileipellis with long, rusty-colored setae (thick-walled pointed hairs).

ECOLOGY: In troops of solitary fruitbodies, occasionally scattered or in small clusters. Most abundant on beds of wood chips, but also on nutrient-rich soil in disturbed areas. Common in urban areas throughout our range, fruiting from late fall into spring.

EDIBILITY: Unknown, likely nontoxic.

COMMENTS: The rounded-conical cap with bay reddish tones when young, dull yellowish stipe, straight and tall stature, and lack of any veil tissue help distinguish this species. Based on genetic data, this species will likely be moved to *Parasola* (or at least out of *Psathyrella*). Very young fruitbodies can sometimes be found when still covered in spiky, bright yellow hairs. Mushrooms fitting the description above have often been lumped under the name *Parasola* (=*Psathyrella*) *conopilus*—a very similar species first described from Europe.. Some *Conocybe* species can look similar, but have rusty brown gills and spores.

Psathyrella bipellis

(Quél.) A. H. Sm.

CAP: 0.5–3.5 cm across, hemispheric at first, soon broadly convex to nearly plane. Dark burgundy-purple to wine red, occasionally bright red, distinctly translucent-striate when young and wet, often paler, creamy around margin at maturity. Strongly hygrophanous, becoming tan-brown, yellowish beige, or almost entirely whitish when dry. Surface thinly viscid when very wet, otherwise moist to dry, bumpy or wrinkled, sometimes sulcate. Fine patches of white veil fibrils clinging to edge of cap, sometimes in a narrow, continuous band. **GILLS:** Narrowly to broadly attached, close to well spaced. Pale gray or beige with a pinkish wash when young, soon dark mottled gray with lilac and dark pink tones. Edge often slightly paler. **STIPE:** 1.5–6.5 cm long, 0.3–0.6 cm thick, cylindrical, base slightly enlarged. Whitish, beige, or pinkish, usually paler toward base. Surface with scattered small streaks and chevrons of silvery tissue; apex densely covered in pale scurfs. **FLESH:** Thin, very fragile, pallid to pinkish, hollow in stipe. **ODOR:** Faintly to strongly fruity and pleasant, or unpleasant, especially in age, sometime reported as "like cat urine." **TASTE:** Mild. **SPORE DEPOSIT:** Blackish. **MICROSCOPY:** Spores 11–16.5 x 6.5–9 µm, ellipsoid to rounded-cylindrical, smooth. Pleurocystidia and cheilocystidia present.

ECOLOGY: Solitary or in clumps often in large troops in wood chips and leaf litter, almost always in urban or disturbed areas. Fruiting from winter through early spring. Locally common around the San Francisco Bay Area and in southern California, uncommon elsewhere.

EDIBILITY: Nontoxic.

COMMENTS: Few mushrooms in our area could be confused with this lovely little *Psathyrella*. The fragile, hygrophanous, purplish to wine red cap, gills that are mottled dark gray and pinkish at maturity, and growth in lignin-rich humus distinguish it. *P. prona* is sometimes pinkish, but usually much taller and thinner, and doesn't grow in clusters.

SYNONYMS: *P. barlae* (Bres.) A. H. Sm., *P. odorata* Sacc.

Psathyrella longipes

(Peck) A. H. Sm.

TALL PSATHYRELLA

CAP: 1.7–5 cm across, rounded-conical to nearly cylindrical when young, becoming bell shaped. Dull grayish tan or brown, fading outward to beige or cream as it loses moisture. Surface moist to dry, smooth, margin with hanging "teeth" of thin white veil tissue soon stained blackish by spores. **GILLS:** Obscurely attached or seceding, close to crowded, broad. Off-white to pallid gray, becoming dark brownish gray to nearly black. **STIPE:** 5–15 cm long, 0.2–0.7 cm thick, cylindrical with a slightly enlarged base, often tall, upright and slightly sinuous. White to creamy, often shiny. Surface smooth, or sometimes with chevrons of tissue around lower part, sometimes with extensive mycelium at base. **PARTIAL VEIL:** Thin sheet of white tissue, leaving hanging "teeth" from the cap margin. **FLESH:** Thin, fragile, beige in cap, stipe hollow, white. **ODOR:** Indistinct. **TASTE:** Indistinct. **SPORE DEPOSIT:** Dark purple brown. **MICROSCOPY:** Spores 7–13 x 7–9 µm, ellipsoid to subcylindrical, germ pore distinct, smooth. Cheilocystidia spindle shaped to narrowly club shaped. Pleurocystidia absent.

ECOLOGY: Solitary or in small groups in humus or well-decayed leaf litter, especially under live oak in wet, dark microhabitats. Common in the southern half of our range, rare on the Far North Coast. Fruiting throughout the wet season.

EDIBILITY: Edible.

COMMENTS: The slender stature, tall grayish brown cap with hanging patches of veil tissue, fairly large size (for a *Psathyrella*), and fragile white stipe help identify this species. A few other dark-spored genera can produce similar-looking fruitbodies, but they are less fragile and have viscid caps and/or brighter colors. *P. elwhaensis* is very similar, but has spores 12–15 µm long; it might just be a long-spored variant of *P. longipes*.

Psathyrella longistriata

(Murrill) A.H. Sm.

RINGED PSATHYRELLA

CAP: 2.5–8 (10) cm across, conical to almost bell shaped at first, becoming broadly convex to nearly plane, often with a low, broad umbo, and becoming wavy in age. Color variable, often reddish brown, brown to dark grayish brown, but can be pale creamy beige to yellow-brown. Hygrophanous, translucent-striate when wet, fading when dry. Surface moist to dry, with scattered, fine whitish hairs when young, otherwise smooth. **GILLS:** Narrowly attached to free, close, narrow, grayish buff, becoming grayish brown to dark purplish brown as the spores mature. **STIPE:** 4–12 cm long, 0.3–1 cm thick, often tall and rather straight, equal, with a swollen base. White to pale gray. Surface dry, covered with fine, appressed, floccose scales when young, to nearly smooth in age. **PARTIAL VEIL:** Membranous, often forming a skirtlike ring on stipe that collapses in age, occasionally breaking up and leaving remnants on cap margin. **FLESH:** Thin, very fragile buff to light brown in cap. Stipe fibrous but fragile, hollow, whitish. **ODOR:** Indistinct. **TASTE:** Indistinct. **SPORE DEPOSIT:** Dark purplish brown to purplish black. **MICROSCOPY:** Spores 6.5–9 x 4–5 µm, ellipsoid to ovoid, with a small, indistinct germ pore, smooth, moderately thick walled, inamyloid, dark brown in KOH.

ECOLOGY: Solitary or scattered, rarely in large numbers on well-decayed wood or duff, especially under alder and Bigleaf Maple. Although it occurs throughout much of our range, it is only common on the Far North Coast and is rather infrequent south of Mendocino County. Fruiting in fall and winter.

EDIBILITY: Unknown.

COMMENTS: The hygrophanous brownish to beige cap, dark spores, and fragile flesh and stipe help identify this species to genus level, and the membranous veil sets it apart from the horde of other brown *Psathyrella*. A similar-looking (but not closely related) species, *Stropharia albivelata*, has an umbonate, vinaceous brown to yellow-brown cap, a flaring partial veil, and ocher-brown spores; it is a rare species, occurring on debris in coniferous forest.

Psathyrella candolleana

(Fr.) Maire

COMMON PSATHYRELLA

CAP: 1.5–7 cm across, rounded to ovoid at first, becoming convex to nearly flat or uplifted, margin wavy and often splitting in age. Caramel brown to ocher-tan at first (but often with white patches of veil), strongly hygrophanous and soon fading to yellowish beige to nearly white. Surface smooth, dry, often extensively covered in patches of whitish veil tissue when young, margin frequently with hanging "teeth" of veil remnants even in age. **GILLS:** Narrowly to broadly attached, close to crowded. Whitish to pale grayish beige when young, becoming lilac-gray to dark brownish purple in age. Margin often paler whitish. **STIPE:** 2–9 cm long, 0.4–0.9 cm thick, cylindrical, often splitting. White. **FLESH:** Thin, fragile, often splitting and breaking, stipe hollow. **ODOR:** Indistinct. **TASTE:** Indistinct. **SPORE DEPOSIT:** Dark purplish brown. **MICROSCOPY:** Spores 7–8 x 4–5 µm, ellipsoid to subcylindrical. Pleurocystidia absent, cheilocystidia present.

ECOLOGY: In clusters, clumps, and troops, rarely solitary. Common in urban areas, usually in disturbed, nutrient-rich soil or near woody debris (including roots of dying trees). Most abundant in spring and fall, but can fruit at any time of the year.

EDIBILITY: Edible.

COMMENTS: The fragile but fairly large fruitbodies, hanging veil remnants around the edge of the cap, caramel to beige colors, and dark purplish gray gills identify this species. Most other *Psathyrella* are smaller or grow in forest settings. Other similar dark-spored genera (*Psilocybe*, *Stropharia*) have more robust, less fragile fruitbodies. *P. rogueiana* is a similar species described from southern Oregon, differing by having pinkish buff veil remnants (yellow in KOH under the microscope) and spores 8–11 µm long.

Psathyrella spadicea group

CAP: 3–10 cm across, convex when young, soon becoming flatter, sometimes with an indistinct broad umbo, slightly wavy in age. Pinkish tan to fawn brown, often paler to beige or ocher at center, outer portion often with a darker grayish brown band. Surface smooth to wrinkled, moist to dry, margin opaque. **GILLS:** Broadly attached with a fine decurrent tooth, close to crowded. Pallid beige at first, soon tan or with pinkish gray tones, darkening to brown or light purplish brown in age. **STIPE:** 4–12 cm long, 0.8–2.5 cm thick, cylindrical, often curved. Creamy white to beige, becoming tan with yellowish areas, often vertically striate with fine, dark lines. Surface dry, smooth or roughened, finely fibrillose. **FLESH:** Moderately thin, fairly firm but brittle in cap, fibrous, solid in stipe. **ODOR:** Indistinct. **TASTE:** Indistinct. **SPORE DEPOSIT:** Dark reddish brown to medium purplish brown. **MICROSCOPY:** Spores 6.5–11 x 4–5.5 µm, ellipsoid to oblong, smooth, germ pore quite inconspicuous or absent, pale brown. Cheilocystidia bottle shaped, with thick walls and crystals at apex.

ECOLOGY: Clustered at base of dead or dying trees, especially Grand Fir. Uncommon. Fruiting in late fall and winter from Sonoma County northward.

EDIBILITY: Edible.

COMMENTS: The large, pale caps, robust solid stipes, lack of veil tissue, and clustered growth at the base of trees help identify this species. *P. piluliformis* fruitbodies are smaller and have darker brown caps and white veil tissue. *P. sublateritia* is a similar species with more reddish brown spores; better known from montane habitats, it might occur in our area on cottonwoods. *Psathyrella spadicea* and allies were recently placed in the genus *Homophron*; like many species in the family, more work is needed to fully comprehend the correct genus placement on Californian species.

Psathyrella piluliformis
(Bull.) P. D. Orton
CLUSTERED PSATHYRELLA

CAP: 1.5–6 cm across, ovoid or dome shaped, becoming convex or round conical, eventually flattening in age. Reddish brown to brown, becoming brownish tan as it dries out, often two-toned with a pale beige center as it loses moisture. Surface often with a conspicuously contrasting band of silky white fibrils around edge, moist to dry, smooth or radially wrinkled near edge. **GILLS:** Finely attached, occasionally notched, close to crowded. Pallid beige to pale pinkish tan, becoming darker as spores mature. **STIPE:** 3–10 cm long, 0.2–0.7 cm thick, cylindrical, usually curved. White to beige or pale yellowish tan over lower part. Surface smooth to silky, with chevrons or patches of veil tissue. **PARTIAL VEIL:** Thin white cortina, leaving remnants around cap margin, becoming covered with dark spores. **FLESH:** Thin, fragile, pale. Stipe hollow at least in age. **ODOR:** Indistinct. **TASTE:** Indistinct. **SPORE DEPOSIT:** Dark purplish brown. **MICROSCOPY:** Spores 4–6 (7) x 3–4 µm, ellipsoid to ovoid, smooth, germ pore sometimes obscure, brown in KOH. Pleurocystidia present, with an elongated narrow apex. Cheilocystidia similar.

ECOLOGY: In small to large clusters of densely packed caps, rarely in small tufts or solitary. Usually found near well-decayed logs or stumps or decaying roots of hardwoods. Common from early fall through winter in the southern part of our range, less frequent on the Far North Coast.

EDIBILITY: Edible.

COMMENTS: The tightly clustered growth habit, rich brown caps with strongly contrasting white bands of veil tissue around the margins when young, and occurrence near rotting wood help identify this species. *P. spadicea* and its relatives also occur in large clusters near rotting wood, but are usually paler, with fleshier stipes, and lack the pronounced veil tissue.

SYNONYM: *Psathyrella hydrophila* (Bull.) Maire.

Psathyrella ammophila
(Durieu & Lév.) P. D. Orton
SAND-LOVING PSATHYRELLA

CAP: 1.5–6 across, convex with an incurved margin at first, becoming broadly convex to plane. Reddish brown to brown, fading to light brown when dry. Surface smooth, moist to dry. **GILLS:** Broadly attached with a notch, close, broad. Light brown at first to dark brown when spores mature. **STIPE:** 3–9 cm long, 0.4–1 cm thick, equal, or tapered downward, to a thicker base with adhering sand. Pinkish buff to light brown, darkening to reddish brown from base up in age. Surface dry, finely pruinose at first, longitudinally striate on upper portion. **FLESH:** Thin to moderately thick, firm but brittle, light brown. **ODOR:** Indistinct. **TASTE:** Indistinct. **SPORE DEPOSIT:** Dark brownish black. **MICROSCOPY:** Spores 9–14 x 6–7 µm, broadly ellipsoid to ovoid, smooth, moderately thick walled, obscurely truncate from germ pore, dark chocolate brown in KOH. Pleurocystidia scattered, spindle shaped, often with swollen middle and pointed tip. Cheilocystidia similar or more club shaped.

ECOLOGY: Scattered in small troops, solitary, or in small clusters in coastal sand dunes around dune grass (*Ammophila* spp.). Common in appropriate habitat. Fruiting in late winter and spring.

EDIBILITY: Nontoxic.

COMMENTS: The growth in sand dunes around the roots of dune grass, reddish brown colors, and dark spores readily identify this mushroom. An undescribed species of *Agrocybe* occurs in the same habitat, but has more yellow-brown colors and paler brown spores.

Coprinopsis uliginicola (sensu CA)

CAP: 2–8 cm across, egg shaped to domed when young, becoming convex to nearly flat, sometimes with a broad, round umbo, occasionally slightly uplifted and wavy in age. Usually distinctly gray, but also light grayish brown, grayish tan, or light brown when faded. Surface dry, innately fibrillose but nearly smooth, sometimes with matted fibrils or dense patches of hairs near center. Often rather virgate looking and in age sometimes radially wrinkled. Margin opaque, not translucent-striate. Whitish veil tissue present around margin as a narrow band or many small patches when young, soon disappearing. **GILLS:** Narrowly attached, close. Pale gray to dingy tan when young, becoming light brownish tan in age. **STIPE:** 3–8 cm long, 0.7–2 cm thick, cylindrical, fairly often tapered toward base. White to grayish, creamy or beige when handled or in age. Surface finely scurfy or silky to nearly smooth, usually with distinct vertical lines at apex. **FLESH:** Quite fragile, whitish to grayish brown in cap. Stipe stuffed at first, hollow in age. **ODOR:** Indistinct. **TASTE:** Indistinct. **SPORE DEPOSIT:** Dull medium brown. **MICROSCOPY:** Spores 10–12 x 5–6 μm, ellipsoid to oblong, smooth, germ pore inconspicuous, pallid in KOH. Cheilocystidia abundant, cylindrical to club shaped. Pleurocystidia absent.

ECOLOGY: Solitary, in small groups, or occasionally clustered on well-decayed wood. Most abundant under live oak where rotten stumps or logs are present. Can be locally abundant from late fall through early spring in the southern part of our range. Largely absent north of Marin County.

EDIBILITY: Unknown, likely nontoxic.

COMMENTS: This odd species doesn't fit neatly into common generic concepts. It is larger and more robust than most other *Psathyrella*, more closely resembling *Lacrymaria*. However, the dull brown spore deposit of this species is atypical for either of those genera. Young specimens with pale gills could even be confused with gray *Tricholoma* species, but the texture is more fragile and the gills soon turn brownish with spores.

Lacrymaria velutina
(Bull.) Pat.

WEEPING WIDOW

CAP: 3–9 cm across, egg shaped or tightly hemispheric at first, sometimes with a rounded umbo, becoming broadly convex to nearly plane. Margin occasionally uplifted and wavy in age. Color quite variable, ocher-beige to warm orangey brown, becoming tan to creamy beige overall. Surface dry to moist, radially silky-fibrillose to nearly smooth or slightly wrinkly when wet or in age; sometimes with hairs bunched into scales. Margin opaque and often with flattened or hanging patches of veil tissue. **GILLS:** Notched to broadly attached, close; usually adorned with many clear to milky droplets when young. Beige to pinkish tan, dark grayish brown mottled with black, often with a reddish brown hue in age. Edges always whitish fringed. **STIPE:** 4–10 cm long, 0.7–2 cm thick, cylindrical, slightly enlarged at base. Whitish beige to grayish tan, ocher, or with yellowish or orangey-brown tones in age. Surface silky-fibrillose to scurfy near apex, often with a ring zone, less often with a distinct annulus. Lower portions fairly smooth or fibrillose, sometimes with shaggy, pale chevrons. **PARTIAL VEIL:** Cottony, cobwebby, or spiky ring zone near stipe apex, soon collapsed and obscure. **FLESH:** Cap fragile, easily broken, stipe usually hollow at maturity; light beige, becoming brownish or with reddish tones in age. **ODOR:** Indistinct or musty. **TASTE:** Indistinct. **SPORE DEPOSIT:** Dark purplish brown. **MICROSCOPY:** Spores 10–12 X 6.5–8 µm, lemon shaped with a protruding chimneylike germ pore, roughened with dark warts, blackish brown. Pleurocystidia and cheilocystidia clustered, narrowly club shaped with rounded heads.

ECOLOGY: In small clusters and troops on wood chips, in nutrient-rich soil, or in disturbed areas with woody debris, such as logging roads. Rare in southern part of range, occasional in the north. Can be locally common in urban habitats in the San Francisco Bay Area. Fruiting from fall to spring, but commonly occurs in summer in irrigated areas as well.

EDIBILITY: Nontoxic.

COMMENTS: Distinguished from similar genera (*Coprinopsis*, *Psathyrella*) by a less fragile texture and by the droplets on the young gills. *Stropharia kauffmanii* can look quite similar but is usually more squat, with a firmer texture and without droplets on the gills. Other *Lacrymaria* are more difficult to tell apart: *L. echiniceps* has a much scalier-looking cap and stipe covered in dark, often pointed clumps of fibrils; its spores average less than 9.5 µm long and are weakly ornamented to nearly smooth. *Psathyrella* (*Lacrymaria*) *rigidipes* is smaller and more slender, and has spores that are greater than 9.5 µm in length with less pronounced warts, and without a protruding germ pore.

SYNONYM: *Psathyrella velutina* (Pers.) Singer. Some authors treat *L. lacrymabunda* as a synonym; however, it is described as having smaller and less strongly ornamented spores; such specimens may be consistently genetically distinct.

Panaeolina foenisecii
(Pers.) Maire
LAWNMOWER'S MUSHROOM

CAP: 1–3 (4) cm across, hemispheric at first, becoming broadly convex. Dark brown to reddish brown when wet; hygrophanous, fading from disc often with a darker band at margin, becoming tan to grayish beige when dry. Surface moist to dry, smooth, or with fine cracks in dry weather. **GILLS:** Broadly to narrowly attached, close, broad. Light brown at first, becoming mottled and darker to dark brown in age, edges paler. **STIPE:** 3–9 cm long, 0.2–0.4 cm thick, more or less equal. Whitish buff to light brown, darkening from base up in age. Surface dry, pruinose to smooth. **FLESH:** Moderately thin, fragile in cap, stipe fibrous. **ODOR:** Indistinct to musty. **TASTE:** Indistinct. **SPORE DEPOSIT:** Dark brown. **MICROSCOPY:** Spores 14–17 (22) x 7.5–11 μm, lemon shaped, wrinkled-roughened, with a projecting germ pore, dark golden brown in KOH. Pleurocystidia bottle shaped, occasionally with a rounded head.

ECOLOGY: Solitary or scattered in troops in lawns. Very common and widespread, occurring throughout our range. Fruiting throughout the warmer months, especially in watered lawns.

EDIBILITY: Nontoxic. Reports of it being mildly hallucinogenic are unsubstantiated and may have been the result of confusion with *Panaeolus cinctulus*.

COMMENTS: The hygrophanous brown to tan cap, often with a different color band around the margin; mottled gills; dark brown spores; and growth in manicured lawns are helpful identifying features. *Panaeolina castaneifolius* is nearly identical macroscopically, but has narrower spores (rarely more than 9.5 μm wide). *Panaeolus cinctulus* is also very similar, but has black spores.

SYNONYM: *Panaeolus foenisecii* (Fr.) J. Schroet.

Panaeolus cinctulus
(Bolton) Sacc.
BELTED PANAEOLUS

CAP: 1.5–6 cm across, broadly conical to hemispheric at first, becoming broadly convex. Generally brown to cinnamon brown, occasionally darker, reddish brown when wet; hygrophanous, fading from disc, often with a darker band at margin, tan, gray, to grayish beige when dry. Surface moist to dry, smooth, or with fine cracks in dry weather. **GILLS:** Broadly to narrowly attached, close, broad. Grayish brown at first, becoming mottled darker, to blackish in age, edges paler. **STIPE:** 3–10 cm long, 0.2–0.5 (0.8) cm thick, more or less equal. Buff to light reddish brown, covered with whitish powder, darkening in age. Base often staining pale bluish. Surface dry, pruinose at first, becoming smooth. **FLESH:** Moderately thin, fragile in cap, fibrous in stipe. **ODOR:** Earthy. **TASTE:** Mild to slightly farinaceous. **SPORE DEPOSIT:** Black. **MICROSCOPY:** Spores 11.5–14 (16) x 7.5–9.5 μm, lemon shaped to ellipsoid, with a germ pore, smooth, thick walled, dark brown to blackish brown in KOH. Cheilocystidia swollen at base with variably shaped necks. Pleurocystidia absent.

ECOLOGY: Clustered, in scattered troops, or solitary in a variety of habitats: gardens, rich soil, compost piles, dung mixed with straw, or grassy lawns. Locally common from the San Francisco Bay Area south, rare on the North Coast. Fruiting from early spring until fall, especially in watered lawns.

EDIBILITY: Mildly hallucinogenic.

COMMENTS: The differences between this species and *Panaeolina foenisecii* are subtle; slightly "cooler" colors, a stockier stature, and black spores are indicative of *Panaeolus cinctulus*. Although it does grow in lawns, it's not restricted to this habitat, as *Panaeolina foenisecii* seems to be. Another grassland species, *Panaeolus fimicola*, has a dark grayish brown to black, convex to nearly flat cap and a dark stipe with a paler pruinose coating. *P. acuminatus* has a conical to bell-shaped cap, is generally dark reddish brown to dark brown at first, and grows on dung or in grassy areas. *P. papilionaceus* has a bell-shaped cap with small white teeth of veil tissue around the margin; it grows on dung.

SYNONYM: *Panaeolus subbalteatus* (Berk. & Broome) Sacc.

Panaeolus papilionaceus

(Fr.) Quél.

BELL-CAPPED PANAEOLUS

CAP: 1–4 cm across, conical at first, becoming bell shaped, occasionally convex in age. Margin with a partial veil–like band of sterile tissue, breaking up into toothlike remnants hanging from margin. Color variable, generally gray to grayish brown, at times dark gray, olivaceous gray, pinkish gray, or even pale grayish beige when dry. Surface moist to dry, smooth, or with fine cracks in dry weather. **GILLS:** Broadly to narrowly attached, close to subdistant, broad. Grayish at first, becoming mottled with dark grayish spots, to blackish in age, edges paler. **STIPE:** 4–12 cm long, 0.2–0.6 cm thick, more or less equal, often quite tall and slender. Buff to light reddish brown, covered with powder, appearing whitish, darkening when handled and in age. Surface pruinose at first, becoming smooth, occasionally with clear droplets when young. **FLESH:** Thin, fragile in cap, fibrous in stipe. **ODOR:** Earthy. **TASTE:** Mild to earthy. **SPORE DEPOSIT:** Black. **MICROSCOPY:** Spores 11–18 x (7.5) 8.5–12 μm, lemon shaped to ellipsoid, with a germ pore, smooth, thick walled, dark brown to blackish brown in KOH. Cheilocystidia club shaped or cylindrical, often with an S-curved neck. Pleurocystidia absent.

ECOLOGY: Scattered in troops or solitary on dung, most often that of cows and horses. Common, occurring throughout our range. Fruiting throughout the wet season.

EDIBILITY: Nontoxic, but similar species are mildly hallucinogenic.

COMMENTS: This species can be readily identified by the conical to bell-shaped cap with small "teeth" around the margin, mottled gills, and black spores. Another common dung dweller is *P. semiovatus*, which is generally stockier and has a partial veil that leaves a ring on the stipe. *P. acuminatus* also grows on dung or in grassy areas; it lacks the toothed cap margin. *P. cinctulus* occurs in similar habitat, but generally has a dark band around the cap margin and has a more convex cap; it's mildly hallucinogenic. *P. cyanescens* (also hallucinogenic) has a pale gray to whitish cap and stains blue when handled. It's rare in California, known from southern California, but it might turn up on dung-rich compost elsewhere in the state during warm, humid weather.

SYNONYMS: Many taxonomists (ourselves included) consider *P. campanulatus*, *P. retirugus*, and *P. sphinctrinus* to be synonyms of *P. papilionaceus*.

Protostropharia dorsipora

(Esteve-Rav. & Barassa) Redhead

SLIMY DUNG DOME

CAP: 1.5–7 cm across, hemispheric to broadly convex, occasionally becoming plane, at times with a low umbo. Pale golden yellow, yellowish beige to creamy yellow. In shaded areas becoming darker golden brown, or bleached to creamy white in sunny areas, occasionally with purplish brown spores sticking to cap. Surface smooth, viscid to tacky when wet, shiny when dry. **GILLS:** Broadly attached, often with a shallow notch at stipe, close, broad. Pallid at first, soon gray to lilac-gray, often with a slight olive tone, to dark gray then purplish black in age. **STIPE:** 4–11 (15) cm long, 0.2–0.8 cm thick, tall, slender, more or less equal with an enlarged or bulbous base. Pale creamy yellow to yellowish gold, often darker toward base, discoloring golden when handled or in age. Surface smooth, viscid to tacky when wet, shiny when dry. **PARTIAL VEIL:** Thin, membranous, often slime covered when young, forming a flaring ring at first, becoming purplish brown from spore drop, constricting to a faint ring on upper stipe in age. **FLESH:** Thin, somewhat firm, creamy yellow. **ODOR:** Faint, musty-earthy. **TASTE:** Mild to slightly farinaceous. **KOH:** Orange-brown on cap. **SPORE DEPOSIT:** Dark purple-brown to purplish black. **MICROSCOPY:** Spores 12–21.5 x 7–10 µm, ovoid to ellipsoid, smooth, thick walled with a small, eccentric germ pore, dark brown in KOH. Pleurocystidia club shaped, often pointed at tips, with a yellowish mass inside cell. Cheilocystidia club shaped, or with a swollen base and a long neck, often with a rounded head, with a yellowish mass inside cell.

ECOLOGY: Solitary, scattered, or in small clusters or troops on dung, especially cow patties. Widespread and common. Fruiting throughout the wet season, with primary flushes soon after first rains in early fall and in spring.

EDIBILITY: Nontoxic.

COMMENTS: The hemispheric, viscid, creamy yellow cap; long, viscid stipe with a faint ring; purplish black spores; and growth on dung are distinguishing features for this group of species. *Protostropharia semiglobata* (also recently transferred from *Stropharia*) is a very similar but far less common species in our area. The only way to reliably distinguish them is to look at the spores: *P. dorsipora* has a small, off-center germ pore, whereas *P. semiglobata* has a wide germ pore centrally positioned at the top of the spore. *P. semiglobata* is generally slightly larger, but this doesn't seem to be a reliable feature to separate them. Other dung-dwelling species include *Deconica merdaria*, which looks very similar but has a dry to thinly viscid cap that is often slightly browner, and a dry stipe. *Panaeolus* have dry caps with more gray tones, and black spores. The dung-loving *Agrocybe semiorbicularis* has a golden brown, thinly viscid to dry cap, and brown spores.

SYNONYM: *Stropharia dorsipora* Esteve-Rav. & Barrasa.

Deconica coprophila

(Bull.) P. Kumm.

DUNG-LOVING DECONICA

CAP: 0.5–4 cm across, tightly hemispheric or egg shaped when young, soon bell shaped to convex, flattening only in age. Strikingly bright reddish orange when young, but very soon dingier brick red to dull dark brown, then fading as it loses moisture, becoming beige to tan. Margin translucent-striate, often decorated with fine white patches, a narrow band, or triangular "teeth" of veil tissue. Surface smooth, thinly viscid at first, then greasy to moist. **GILLS:** Broadly attached or slightly decurrent, fairly widely spaced. Beige when young, then ocher-brown, soon mottled dark brown with purplish gray tones as spores mature; edge contrastingly fringed whitish. **STIPE:** 1–5 cm long, 0.2–0.5 cm thick, usually curved, often clustered and with immature fruitbodies at base. White to beige or tan, darkening grayish to blackish upward from base. Surface covered in pale silvery veil tissue, breaking up into chevrons in age; apex slightly scurfy. **FLESH:** Thin and fragile, watery, dark. **ODOR:** Indistinct. **TASTE:** Indistinct. **SPORE DEPOSIT:** Dark purplish brown to nearly black. **MICROSCOPY:** Spores 7–8.5 (10) x 4.5–5.5 μm, slightly angular to nearly ellipsoid, with a broad germ pore, thick walled, smooth, dark brown in KOH. Cheilocystidia with a swollen base and elongated neck. Pleurocystidia absent.

ECOLOGY: Solitary or more often in clumps and troops on dung of grazing animals, primarily that of cows and horses. Very common in appropriate habitat throughout our area.

Generally fruiting in spring, but can fruit whenever substrate is moistened thoroughly.

EDIBILITY: Nontoxic, not hallucinogenic. Bacterial contamination from the dung habitat can cause gastrointestinal distress.

COMMENTS: The small size, reddish brown cap, slightly decurrent gills that are quite dark in age, and dung habitat are helpful clues in identification. Unfortunately, these features stop just short of identifying the species; the very similar *D. subcoprophila* has larger (15–20 μm long) spores that are ellipsoid or almond shaped (not angular). Two other species fruiting on dung produce slightly larger fruitbodies with paler (less reddish) caps and often more pronounced ring zones on the stipe: *D. merdicola* has large (greater than 12 μm) ellipsoid spores, while *D. merdaria* has more angled (hexagonal), smaller spores (less than 13 μm long). Morphological variation has left species concepts in *Deconica* quite vague, and the situation is complicated by the occurrence of European taxa alongside (perhaps undescribed) native taxa due to widespread introduction with grazing animals.

SYNONYM: *Psilocybe coprophila* (Bull.) P. Kumm.

Deconica montana

(Pers.) P. D. Orton

MOSS-LOVING DECONICA

CAP: 0.5–2 cm across, hemispheric to dome shaped at first, becoming broadly convex. Dark brown to dark reddish brown when young, lighter brownish tan in age; fading to beige or paler outward from center as it loses moisture. Margin translucent-striate, sometimes with scattered silvery fibrils of veil tissue. Surface smooth, thinly viscid to moist, drying out in age. Cap cuticle gelatinous, easily separated. **GILLS:** Broadly attached with a decurrent tooth or slightly decurrent, moderately spaced. Pale beige-tan becoming warm brown, then dark purplish brown as spores mature. Edge whitish fringed. **STIPE:** 1–5 cm long, 0.2–0.4 cm thick, cylindrical. Dull beige or tan, soon browner overall and darkening from reddish brown or grayish to nearly black upward from base. Surface fibrous, vertically striate, surface covered in pale silvery streaks and chevrons of veil tissue (especially near base). **FLESH:** Thin, fibrous, fairly dark. **ODOR:** Indistinct. **TASTE:** Indistinct. **SPORE DEPOSIT:** Dark purplish brown. **MICROSCOPY:** Spores 7–8.5 (10) x 4.5–5.5 µm, slightly angular to nearly ellipsoid, with a broad germ pore, thick walled, smooth, dark brown in KOH. Cheilocystidia with a swollen base and elongated neck. Pleurocystidia absent.

ECOLOGY: Solitary or in unfused clusters and scattered troops in moss beds. Generally in higher elevations in late fall through winter. Infrequent in our range, but can be locally common in mountains.

EDIBILITY: Probably nontoxic.

COMMENTS: The rather dark reddish brown young caps, upward-darkening stipe covered in silvery veil remnants, and growth in moss beds help distinguish this species. Other *Deconica* are quite similar and probably best definitively told apart by microscopic details (especially spore shape and size). *D. phyllogena, D. rhomboidospora,* and *D. subviscida* should be compared; all often grow on woody debris or in grass (not moss) and have an angular, four-sided look to the spores ("rhomboid").

SYNONYM: *Psilocybe montana* (Pers.) P. Kumm.

Deconica inquilina

(Fr.) Bres.

CAP: 0.5–2.5 cm across, egg shaped to hemispheric when young, soon domed to bell shaped with a broad, round umbo, to convex in age. Reddish brown to ocher brown, darkest when young, soon becoming butterscotch or caramel colored. Losing color as it dries out, to pale beige. Margin translucent-striate. Surface smooth, thinly viscid to greasy when wet. **GILLS:** Broadly attached, occasionally with a small decurrent tooth, moderately spaced. Dull cream to grayish beige when young, then ocher-brown, darkening to dark purplish brown. Edge whitish fringed. **STIPE:** 2–8 cm long, 0.2–0.4 cm thick, cylindrical, often curved and narrower downward. Creamy to beige, becoming vertically striate, darkening to ocher, reddish brown and dark brown upward from base. Lower part often sheathed in silky-silvery tissue. **FLESH:** Thin, stringy. Base of stipe quite tough and difficult to remove from substrate. **ODOR:** Indistinct. **TASTE:** Indistinct. **SPORE DEPOSIT:** Dark purplish brown. **MICROSCOPY:** Spores 7–9 (14) x 4.5–6 µm, quite variable in size, ovoid or almond shaped, sometimes weakly angled and with 4 sides, smooth, dark brown in KOH. Cheilocystidia with enlarged bases, and short to moderately elongate cylindrical necks. Pileipellis a gelatinized cutis.

ECOLOGY: In small clumps and scattered troops on grassy lawns, especially those with wet, dead thatch of older blades. Perhaps occasionally on very rotten woody debris as well (see comments). Uncommon throughout our area. Fruiting when lawns are very wet and weather is not too cold.

EDIBILITY: Nontoxic.

COMMENTS: The small, chestnut to ocher-brown caps, dark purplish brown gills, slender stipes, and growth in lawns are helpful identification features. The apparent lack of veil tissue helps distinguish *D. inquilina* from some other *Deconica* (especially those that grow on dung, on rotten wood, or in moss). *D. crobula* has slightly smaller spores (5.5–8 x 3.5–5 µm) and more elongated cheilocystidia. *D. subviscida* is very similar and has whitish partial veil tissue adhering to the edge of the cap. However, both are perhaps more likely to be found on rotting wood and might be confused with *Psilocybe*, which have whiter stipes, larger fruitbodies, and blue bruising reactions.

SYNONYM: *Psilocybe inquilina* (Fr.) Bres.

Psilocybe cyanescens

Wakef.

WAVY CAPS

CAP: 1–6 (8) cm across, rounded-conical or bell shaped with an incurved margin when young, expanding to irregular convex to plane, sometimes with a low, round umbo. Margin usually strongly and neatly wavy when mature, sometimes lobed even when young. Orange-brown, caramel to chestnut brown or reddish brown and yellower toward margin. Strongly hygrophanous, usually fading outward from center to light brown to beige; soon developing blue-green blotches. Surface smooth, viscid to moist or dry. Margin translucent-striate in age or when wet, edges sometimes with wispy white veil remnants when young. **GILLS:** Broadly attached to notched or with a decurrent tooth. Pallid pinkish beige to yellowish brown, often with greenish stains, mottled to darker brown as spores mature. **STIPE:** 5–10 cm long, 0.4–0.7 cm thick, cylindrical with a slightly enlarged base. Usually bright whitish, in age sometimes dingy beige-tan; can be mottled with bluish bruising at any age. Surface silky-fibrillose to smooth. **PARTIAL VEIL:** Relatively thin, whitish cortina, leaving wispy hairs on stipe when young, soon disappearing. **FLESH:** Fibrous, somewhat rubbery in cap and tough in stipe. Whitish to dingy tan. Bruising slowly blue-green to dark dingy blue. **ODOR:** Mild, farinaceous or earthy. **TASTE:** Bitter to farinaceous. **SPORE DEPOSIT:** Dark purple-brown. **MICROSCOPY:** Spores (9) 11–13 (15.5) x 6–8 μm, elongate-ellipsoid, smooth, moderately thick walled, with a broad germ pore, brown in KOH. Cheilocystidia swollen with a narrowed, cylindrical tip, occasionally forked or branched.

ECOLOGY: In small clusters, troops, or occasionally solitary, most commonly in wood chips or on woody debris, less commonly in disturbed grassy areas. Also in grassy dune habitat on the North Coast. Found throughout our range, but uncommon south of the San Francisco Bay Area. Fruiting from late fall through winter.

EDIBILITY: Strongly hallucinogenic.

COMMENTS: This attractive species is distinguished by the tough white stipe, blue-green staining, dark purplish brown spores, and a caramel-colored, wavy-edged cap. The deadly poisonous *Galerina marginata* sometimes grows right alongside this species, but can be told apart by its hemispheric to broadly convex cap, orangish stipe that darkens from the base up in age, and rusty brown spores. A few other *Psilocybe* are quite similar: *P. allenii* is rarely wavy capped, but otherwise is difficult to distinguish. *P. azurescens* is very similar, but has a broadly conical or peaked-umbonate cap, while *P. stuntzii* and *P. ovoideocystidiata* have more pronounced partial veils and browner (less orange) colors.

Psilocybe allenii
Borov., Rockefeller & P. G. Werner
TURPID PSILLY

CAP: 1.5–7 (9) cm across, hemispheric to convex at first, soon broadly convex to plane. Margin plane, rarely slightly wavy, translucent-striate around edge. Orange-brown to caramel, often with a slightly paler margin, becoming more dingy colored in age. Strongly hygrophanous, usually fading outward from center when dry; yellowish brown, light brown to beige; soon developing blue-green blotches. Surface smooth, viscid to moist or dry. **GILLS:** Broadly attached, notched, or with a decurrent tooth. Creamy beige at first, to light brown, mottled to all purple-brown as spores mature. **STIPE:** 3–7 (9) cm long, 0.2–0.7 cm thick, cylindrical with a slightly enlarged base. Whitish at first, soon discoloring dingy brownish; can be mottled with bluish bruising at any age. Surface silky-fibrillose, with a pruinose apex to smooth, base with white, binding rhizomorphs. **PARTIAL VEIL:** Thin, whitish cortina, leaving wispy hairs on the stipe when young, soon disappearing. **FLESH:** Fibrous, rather firm, somewhat rubbery in cap, tough in stipe. Whitish to dingy tan. Bruising slowly blue-green to dark dingy blue. **ODOR:** Mild, farinaceous or earthy. **TASTE:** Farinaceous. **SPORE DEPOSIT:** Dark purple-brown. **MICROSCOPY:** Spores 11–14 x 6.5–8 µm, elongate-ellipsoid, smooth, moderately thick walled, with a broad germ pore, brown in KOH. Cheilocystidia swollen with a narrowed, cylindrical tip. Pleurocystidia similar.

ECOLOGY: In small clusters or in troops, occasionally solitary. Almost always in wood chips in urban areas. Also found on grassy sand dunes on the Far North Coast. Although it can occur throughout our range, it is only common in the greater San Francisco Bay Area and on the Far North Coast. Fruiting from late fall through winter.

EDIBILITY: Strongly hallucinogenic.

COMMENTS: This species is very close in appearance to *P. cyanescens* and can sometimes be difficult to distinguish; the primary morphological difference is the convex to flat cap that does not become strongly wavy as in *P. cyanescens*. The range of *P. allenii*, like that of most of the wood chip *Psilocybe*, is extending, especially since it is commonly "wild" cultivated; it is frequently planted and now naturalizing in wood chip beds across California.

Psilocybe stuntzii
Guzmán & J. Ott
BLUE RINGER

CAP: 1–3.5 (5) cm across, broadly conical to bell shaped when young, becoming convex to plane. Margin often wavy and occasionally uplifted in age. Color variable, olivaceous brown, dark brown to grayish brown, with a translucent-striate margin when wet. Hygrophanous, fading from center out, to ochraceous brown to yellowish brown when dry. Surface smooth, slightly viscid or lubricious. **GILLS:** Broadly attached, moderately spaced, pallid grayish at first, soon medium brown with a purplish cast and often covered in dark brown speckles as spores mature to dark purplish brown in age. **STIPE:** 2–7 cm long, 0.2–0.5 cm thick, equal or enlarged slightly toward base. Whitish, buff to grayish brown, darkening in age; occasionally with bluish stains. Surface finely fibrillose to smooth. **PARTIAL VEIL:** Membranous to cortina-like, leaving a flaring and often bluish ring when young, collapsing in age. **FLESH:** Whitish to pallid brown, weakly bluing when young. **ODOR:** Farinaceous. **TASTE:** Farinaceous. **SPORE DEPOSIT:** Dark purple brown. **MICROSCOPY:** Spores (9) 11–13 (15) x 6–8.5 µm elongate-ellipsoid, thick walled, smooth, with a distinct broad germ pore. Cheilocystidia abundant, spindle to flask shaped, with a long, thin neck, occasionally branched. Pleurocystidia spindle, flask to pear shaped with a pointed neck.

ECOLOGY: Often clustered, but also solitary or scattered in troops on wood chips, in grassy areas, and in garden soil. Fruiting in late fall and winter. Relatively rare in California, but common in the Pacific Northwest. Like many *Psilocybe*, it is often intentionally introduced to wood chip beds, which may lead to a larger range in California in the near future.

EDIBILITY: Mildly hallucinogenic.

COMMENTS: The olivaceous brown to grayish brown cap, flaring ring with bluish stains, and dark purple-brown spore deposit help identify this uncommon species. *P. ovoideocystidiata* is similar, but has a whitish, evanescent veil, readily develops greenish blue stains overall, and often fruits in late winter or spring. *P. cyanescens* has a caramel-colored cap and a nondescript partial veil that doesn't form a ring. The deadly poisonous *Galerina marginata* also has a ring, but has rusty brown spores, a stipe that darkens from the base up in age, and doesn't stain blue.

Psilocybe baeocystis

Singer & A. H. Sm.

BLUE BELLS

CAP: 1.5–3.5 (5.5) cm across, broadly conical to bell shaped and often irregular and somewhat knobby when young, becoming convex to plane. Margin sometimes wavy but rarely uplifted. Color variable, caramel brown to dark brown, gray-brown to gray; hygrophanous fading to yellowish beige or whitish when dry. Most often appearing caramel brown mottled heavily with bluish green stains, although sometimes appearing entirely dark blue-green even when young. Surface smooth, slightly viscid to dry. **GILLS:** Broadly attached, moderately spaced, grayish at first, soon medium brown with a purplish cast and often covered in dark brown speckles as spores mature. **STIPE:** 3–7 cm long, 0.2–0.4 cm thick, equal or enlarged slightly toward base, squatter when growing in wood chips, taller when growing in grass. Whitish to pale grayish or tan-brown, very soon becoming dingy, dark blue black to greenish blue in blotches. Surface smooth or twisted-fibrous. **PARTIAL VEIL:** Indistinct, consisting of a thin cortina when young. **FLESH:** Pallid brown. Bruising dark greenish blue to nearly black, especially where handled or in age. **ODOR:** Mild to farinaceous. **TASTE:** Farinaceous. **SPORE DEPOSIT:** Dark purple brown. **MICROSCOPY:** 8.5–14 (17) x 5–7 μm, ellipsoid, but asymmetrical and slightly curved near apiculus, often described as "mango shaped," thick walled, smooth, dark brown in KOH. Cheilocystidia abundant, spindle to flask shaped, with a long, thin neck, occasionally forking or irregularly branched. Pleurocystidia generally absent.

ECOLOGY: Solitary, scattered in groups, or in small clusters on wood chips or in grassy areas in late fall and winter, usually during cold, wet periods. Rare in California, but readily introduced and naturalized in wood chip beds.

EDIBILITY: Strongly hallucinogenic.

COMMENTS: The fruitbodies of this species are, for lack of a gentler word, ugly. Rarely does this mushroom appear fresh, even when it is! It has none of the stately grace of *P. cyanescens*, usually appearing dingy and distressed. The introduced *P. ovoideocystidiata* has a distinct cortina-like ring on the stipe, and *P. stuntzii* usually has a blue-stained, flaring, membranous veil. *P. cyanescens* and *P. azurescens* have more stately fruitbodies (with uplifted and wavy caps in former) that don't tend to become so completely stained and murky looking.

Psilocybe semilanceata

(Fr.) P. Kumm.

LIBERTY CAP

CAP: 0.5–2.5 (3.5) cm across, conical to bell shaped, often with a peaked umbo, rarely expanding. Dark brown to chestnut brown, generally with a paler, somewhat translucent umbo when wet, occasionally with olivaceous tones; strongly hygrophanous, fading to pale yellowish brown to tan when dry. More dingy colored in age, occasionally with bluish stains. Margin translucent-striate when wet, opaque when dry. Surface smooth, thinly viscid to dry. Cap skin gelatinous, peelable. **GILLS:** Broadly to narrowly attached, often notched, close, narrow. Pale grayish buff when young, darkening, grayish brown to purplish brown as spores mature, with a whitish edge. **STIPE:** 3–10 (14) cm long, 0.1–0.3 cm thick, often quite tall, slender and wavy. Whitish to pale buff at first, darkening to brownish in age. Staining blue when damaged, sometimes with dingy bluish stains in age. Surface dry, pruinose at apex, with sparse, silky hairs over lower portion when young, smooth in age. **PARTIAL VEIL:** Indistinct, consisting of sparse, silky hairs, not forming a ring. **FLESH:** Very thin, fragile in cap, fibrous in stipe, concolorous, staining blue when damaged. **ODOR:** Slightly farinaceous. **TASTE:** Mildly farinaceous. **SPORE DEPOSIT:** Purple-brown. **MICROSCOPY:** Spores 11–14 (18) x 7–9 µm, ellipsoid, occasionally slightly lemon shaped, smooth, moderately thick walled, with a broad germ pore, brown in KOH. Cheilocystidia abundant, swollen in lower portion, or bottle shaped with an elongated neck. Neck occasionally forking or branched.

ECOLOGY: Solitary or scattered in troops in rich grassy pastures, usually in clumps of taller grass, generally around but not on dung. Common in appropriate habitat on the Far North Coast, not known to occur south of Humboldt County. Fruiting in late fall into winter, with scattred flushes into spring.

EDIBILITY: Strongly hallucinogenic.

COMMENTS: The peaked, conical to bell-shaped cap, purplish brown mature gills, grassland habitat, and slender stipe that stains blue when damaged are distinctive. *P. pelliculosa* is fairly similar, but grows in forest settings. Some *Panaeolus* are frequently found fruiting alongside this species, but have bell-shaped to convex caps and black spores.

Psilocybe pelliculosa

(A. H. Sm.) Singer & A. H. Sm.

WOODLAND PSILLY

CAP: 0.5–2 (3) cm across, conical to bell shaped, often with a rounded umbo, occasionally expanding to convex. Yellowish brown to brownish orange, occasionally slightly darker, reddish brown; hygrophanous, fading to pale yellowish brown to tan when dry. Becoming more dingy in age, often with grayish or greenish blue tints. Margin translucent-striate when wet, opaque when dry. Surface smooth, thinly viscid to dry. **GILLS:** Broadly attached, often notched, close, narrow. Pale grayish buff when young, darkening, grayish brown to purplish brown as spores mature. **STIPE:** 3–10 cm long, 0.1–0.3 cm thick, often quite tall, slender and wavy, equal, or with an enlarged or slightly bulbous base. Whitish to pale buff with a brownish base, darkening in age. Base occasionally with bluish stains. Surface dry, pruinose at apex, with sparse, silky hairs over lower portion. **PARTIAL VEIL:** Indistinct, consisting of sparse, silky hairs, not forming a ring. **FLESH:** Very thin, fragile, concolorous, sometimes staining blue in lower stipe. **ODOR:** Slightly farinaceous. **TASTE:** Mild to slightly farinaceous. **SPORE DEPOSIT:** Purple-brown. **MICROSCOPY:** Spores (8.5) 9.5–11 (13) x 5–7 µm, ellipsoid to ovoid, smooth, moderately thick walled, with a broad germ pore, brown in KOH. Cheilocystidia abundant, spindle shaped, or more elongate, with a pointed tip.

ECOLOGY: Solitary, scattered, or in small clusters and troops in duff or soil, often in disturbed areas. Uncommon in California, restricted to the Far North Coast, generally on trailsides or in logged areas. Fruiting in fall and early winter.

EDIBILITY: Mildly hallucinogenic.

COMMENTS: Small and often overlooked, this mushroom can be recognized by the markedly translucent-striate, yellowish brown cap, purple-brown spores, and thin stipe that often has bluish stains at the base. *P. silvatica* is nearly identical, but has smaller spores [6.5–9.5 (11) x 4–5.5 µm]. The cap of *P. semilanceata* is often more peaked-conical; it grows in grasslands. Many other mushrooms with a similar stature also grow on coniferous duff; these include *Galerina*, with paler gills and rusty brown spores; *Mycena*, with fragile, translucent stipes; and *Hypholoma dispersum*, which is generally slightly larger, often has a more domed cap that isn't as strongly translucent-striate, and has silvery chevrons covering the stipe.

Hypholoma dispersum

(Fr.) Quél.

CAP: 1–5 cm across, domed to bell shaped, becoming convex, occasionally with a rounded umbo when young. Ocher-brown to pinkish tan or with yellowish brown colors, paler to beige or white near margin. Fading as it dries out, extensively pale beige in age. Surface smooth, moist to dry. Margin opaque, often with silky patches or a narrow ring of whitish veil fibrils. **GILLS:** Obscurely attached with a small notch, close to crowded. Pale cream when young, then light yellowish, soon mottled with gray and becoming brownish gray in age, occasionally with an olive cast. **STIPE:** 3–10 cm long, 0.2–0.5 cm thick, cylindrical, base slightly enlarged; usually rather straight and upright. Whitish overall, soon light tan near apex, brownish toward base. Covered in shiny chevrons of silky veil tissue, often making a striking snakeskin pattern most distinct against dark lower stipe. **FLESH:** Thin, tough in stipe. **ODOR:** Indistinct. **TASTE:** Mild to bitter. **SPORE DEPOSIT:** Very dark grayish with purple-brown tones. **MICROSCOPY:** Spores 7–10 x 4–5 μm, ellipsoid, with a central germ pore, smooth, dark brown in KOH. Pleurocystidia and cheilocystidia present, some yellow in KOH.

ECOLOGY: Solitary, in troops, small clusters, or arcs in deep needle litter and moss beds, or among woody debris. Fairly common during late fall and winter in the northern part of our range, absent south of Sonoma County.

EDIBILITY: Unknown.

COMMENTS: Although *H. dispersum* is definitely a little brown mushroom, the domed ocher-brown cap, rather upright stature, and snakeskin pattern of silvery chevrons on the stipe distinguish this species. *H. myosotis* is smaller, with a yellowish to ocher, slightly viscid cap and a more sinuous stipe. *H. udum* is very similar to the latter species, but lacks a viscid cap and has more purplish brown gills and bigger, almond-shaped, finely roughened spores; its segregation into the genus *Bogbodia* is not well supported. *Psilocybe pelliculosa* has a distinctly translucent-striate cap. Some species of *Galerina* have silvery bands on the stipe, but usually less extensively. They also tend to have translucent-striate caps and rusty brown spore deposits.

SYNONYM: *Hypholoma marginatum* J. Schröt.

Hypholoma capnoides

(Fr.) P. Kumm.

CONIFER TUFT

CAP: 1.5–5 (7) cm across, convex to broadly convex with a low umbo, to plane in age, sometimes wavy. Dull or bright yellow-orange, or honey tan, ocher-brown to nearly reddish brown. Surface smooth, viscid when wet, otherwise dry, hygrophanous, not striate. Margin often with thin patches of adhering cobwebby veil. **GILLS:** Broadly attached to slightly decurrent, close, narrow. Creamy to grayish beige, very soon smoky gray, to dark smoky gray as spores mature. **STIPE:** 4–12 cm long, 0.3–1.2 cm thick, cylindrical, somewhat tapered to base, often loosely fused or clustered. Whitish to creamy beige or dingy brownish tan (especially lower parts). Surface dry, covered in inconspicuous pale chevrons or fibrils. **PARTIAL VEIL:** Thin cortina, or cobwebby and nearly membranous when young, soon splitting, leaving remnants on cap margin, sometimes collapsing on upper stipe and collecting dark spores, but often disappearing entirely. **FLESH:** Thin, tough in stipe. **ODOR:** Indistinct. **TASTE:** Mild. **KOH:** Slight reaction, orange-brown on cap. **SPORE DEPOSIT:** Dark grayish purple to purplish black. **MICROSCOPY:** Spores 6–7.5 x 3.5–5 μm, ellipsoid, smooth, with a small germ pore. Pleurocystidia with a golden yellow mass in KOH. Cheilocystidia present.

ECOLOGY: In clusters and troops on decaying coniferous logs and stumps from late fall into spring. One of the few species that will grow on Coast Redwood. Common and widespread, occurring throughout our range.

EDIBILITY: Edible, but infrequently collected. Be careful not to mistake it with the toxic *Hypholoma fasciculare*.

COMMENTS: Only slightly less common than its brighter yellow, bitter-tasting cousin *H. fasciculare*, this species is nonetheless a familiar sight on conifer logs, particularly in the coastal forests north of the San Francisco Bay. It can be recognized by the yellow-brown to ocher-brown cap, smoky gray gills, growth on coniferous wood, and dark grayish purple spores. *Pholiota astragalina* is somewhat similar but has a peachy pink cap, tastes bitter, has brown spores, and sometimes slowly bruises ash gray.

Hypholoma fasciculare

(Huds.) P. Kumm.

SULFUR TUFT

CAP: 1–6 (10) cm across, rounded-conical to hemispheric when young, soon broadly convex, often wavy and uplifted in age. Pale yellowish green to lemon-yellow, orange-yellow, or ocher, sometimes nearly reddish brown, especially over disc or when young. Surface smooth, dry, often with cobwebby veil remnants clinging to margin in age. **GILLS:** Attached, fairly close. Greenish yellow to bright yellow or pale, dull yellowish beige when young, maturing to brownish or purplish brown or grayish black, but often retaining an underlying greenish yellow hue. **STIPE:** 3–10 cm long, 0.4–0.7 cm thick, mostly cylindrical, often tapered to base and fused into moderate-size clusters. Pale yellow, greenish yellow to ocher brown, especially browner near base. **PARTIAL VEIL:** Cobwebby, sometimes forming a nearly membranous-fibrillose sheet over young gills, often leaving a dark ring zone near apex of stipe collecting purplish black spores. **FLESH:** Thin, fibrillose, yellowish in cap, often with brownish stains in stipe. **ODOR:** Indistinct. **TASTE:** Quite bitter. **KOH:** No reaction. **SPORE DEPOSIT:** Variable, dark purple-brown to dark purplish black. **MICROSCOPY:** Spores 6–8 x 3.5–5 μm, ellipsoid, with a germ pore. Pleurocystidia with swollen bases and elongated necks, with golden yellow content in KOH. Cheilocystidia present.

ECOLOGY: In small groups, arcs, and clumps or in tremendous clusters and rows on fallen conifer and hardwood logs or in wood chips. Extremely common and cosmopolitan. Fruiting from early fall through late winter.

EDIBILITY: Toxic, causing severe gastrointestinal distress. Luckily, the bitter taste is a deterrent.

COMMENTS: The yellow to orange cap, yellowish green gills that become grayish purple as spores mature, clustered growth on wood, and bitter taste set this species apart. *H. capnoides* is quite similar but lacks the yellow-green tones in the gills (and is usually less bright overall), tastes mild, and grows only on conifer wood. A few smooth-capped yellow *Pholiota* are similar, but taste mild and have brown spores.

Leratiomyces ceres

(Cooke & Massee) Spooner & Bridge

CHIP CHERRIES

CAP: 2–7.5 cm across, bell shaped when young, becoming convex to broadly convex, often with a round umbo, margin often uplifted or wavy in age. Dark red to orange-red, dingier in age. Margin opaque, often with small patches of whitish veil fibrils. Surface smooth, slightly viscid when young, soon dry. **GILLS:** Broadly attached or with a slight decurrent tooth. White when young, becoming dingy gray-tan, then cool gray mottled with dark brown; can be spotted or extensively blushed with orange to red at any age. **STIPE:** 3–8 (11) cm long, 0.5–0.8 cm thick, cylindrical, straight to sinuous, base slightly enlarged. White at first, becoming dingy cream, reddish at base and becoming orange-red streaked upward in age. Surface slightly scaly or covered in chevrons over lower portion, smooth in age. Extreme base with white rhizomorphs. **PARTIAL VEIL:** Cottony, white and pronounced when young, but soon collapsing to an indistinct fibrillose zone that becomes darkened with spores in age. **FLESH:** Fibrous, pale, fairly tough in stipe. **ODOR:** Indistinct. **TASTE:** Indistinct. **KOH:** Orangish on cap. **SPORE DEPOSIT:** Dark purplish brown. **MICROSCOPY:** Spores 10–13 x 6–7.5 µm, ellipsoid, smooth, dark brown in KOH. Cystidia on gill edges with yellow contents in KOH.

ECOLOGY: Scattered, in troops or small clusters on wood chips and nutrient-rich soil in landscaped areas throughout our range. Very common; probably our most abundant member of the wood chip fungal flora. Fruiting in wet weather at any time of year.

EDIBILITY: Unknown, possibly toxic.

COMMENTS: The red cap, clustered or trooping growth, and frequently red-streaked stipe identify this species. It sometimes fruits by the thousands in wood chip beds, often in the company of *L. percevalii*, as well as various species of *Agrocybe*, *Psilocybe*, and *Psathyrella*. Thought to be native to Australia, it is expanding throughout North America. *Hypholoma tuberosum* is similar but quite rare; it also grows on wood chips, but has a paler red to orange-brown cap, which is conical to convex (often with a pronounced umbo), and grows in tufts or clusters emerging from hard, brownish to black underground sclerotia.

SYNONYMS: *Hypholoma aurantiacum* (Cooke) Faus, *Stropharia aurantiacum* (Cooke) Ryman.

Leratiomyces percevalii

(Berk. & Broome) Bridge & Spooner

MULCH MAIDS

CAP: 2–7 (12) cm across, bell shaped to convex with a round or slightly conical umbo (sometimes rather pointed), flattening out, or occasionally wavy in age. Light beige to pale golden yellow, sometimes ocher or darker tan at center, often shiny in dry weather. Surface moist to dry, sometimes finely scaly. Margin often with patches or hanging "teeth" of beige-yellow or whitish veil tissue. **GILLS:** Broadly attached, often with a decurrent tooth. Pale cream at first, soon dingy grayish beige, mottled with lilac-gray, to dark gray in age, edges whitish fringed. **STIPE:** 5–12 cm long, 0.3–1 cm thick, cylindrical, slightly enlarged at base, often quite curvy. Light whitish cream to yellowish beige, dingier and darker in age, developing orangey brown tones. Surface smooth to finely scaly over lower half. **PARTIAL VEIL:** White, fibrillose to slightly membranous. Usually only present as a spiky or narrow flattened ring zone darkened with grayish spores. **FLESH:** Rather tough, stipe fibrous, beige to dingy ocher-brown. **ODOR:** Indistinct. **TASTE:** Indistinct. **KOH:** Orangish on cap. **SPORE DEPOSIT:** Dark purplish gray. **MICROSCOPY:** Spores 12–14.5 x 7–9 μm, ellipsoid to almond shaped, smooth, brown in KOH. Cystidia yellow in KOH, at least in young specimens.

ECOLOGY: Solitary, scattered, in small clusters and troops on wood chips. Very common in urban areas or along roads and trails throughout most of our area, but rare on the Far North Coast. Fruiting throughout the year (at least in irrigated areas); most abundant from late fall into spring.

EDIBILITY: Nontoxic. A few people have reported a good texture but poor taste; it often grows in contaminated areas.

COMMENTS: This is a fairly distinctive species by virtue of the beige to golden cap, broadly attached lilac-gray gills, tough, pale stipe with indistinct annular zone, and growth in wood chips. *Agrocybe* species often grow alongside this species and can look similar, but have brown spores and gills and either have skirtlike rings on the stipe or lack the shaggy annular zone of this species. *Stropharia ambigua* is larger, with a viscid cap when young, a shaggier white stipe, and a floccose partial veil.

NOMENCLATURE NOTE: Some people use the name *L. riparius* (described from Washington state) for our common tan *Leratiomyces* found in wood chips. However, we believe it represents *L. percevalii*, which, although described from England, now appears to have a cosmopolitan distribution due to introduction and spread on wood chips.

Stropharia ambigua

(Peck) Zeller

AMBIGUOUS STROPHARIA

CAP: 3–10 (15) cm across, rounded-conical to convex at first, expanding to plane or wavy in age. Margin incurved at first, expanding and covered with hanging, white veil, becoming bald and occasionally lightly uplifted in age. Dull yellow to yellow-brown, often fading to creamy beige in age. Young caps can also be more ocher-yellow to butterscotch yellow; older caps can become purplish brown from spores, especially when clustered. Surface thinly viscid to lubricious, smooth, except for white veil remnants near margin when young. **GILLS:** Broadly attached, often with a shallow notch and a decurrent tooth, close to crowded, partial gills numerous. Gray at first, soon darkening to purplish gray to purplish black as spores mature. **STIPE:** 5–20 cm long, 0.5–2 cm thick, club shaped at first, becoming more or less equal, tall and slender in age. White, occasionally with buff to orangish brown stains over lower portion. Surface dry, covered with fine white tufts and scales when young, to nearly smooth in age. Base with binding white rhizomorphs. **PARTIAL VEIL:** Flocculent-membranous, adhering to cap margin, with hanging triangular pieces when young, often disappearing in age. Rarely leaving a ring on the stipe. **FLESH:** Moderately thin to thick, soft, whitish. **ODOR:** Mild to earthy. **TASTE:** Earthy. **SPORE DEPOSIT:** Dark purple-brown to purplish black. **MICROSCOPY:** Spores 11–14 x 6–7.5 μm, ellipsoid, appearing truncate from an apical germ pore, smooth, thick walled, grayish brown in KOH. Cheilocystidia abundant, bottle shaped with a rounded

head, cylindrical to club shaped. Pleurocystidia embedded, inconspicuous, broadly spindle shaped with pointed tips, and with yellow contents in KOH.

ECOLOGY: Solitary or scattered in small troops when fruiting in leaf litter and duff in natural habitats. Patches growing in urban areas often fruit more profusely in large troops and clusters, especially in wood chips. Very common throughout our area, with large flushes in late fall and winter, with scattered fruitings throughout wet season.

EDIBILITY: Edible, but (very) poor from a culinary perspective.

COMMENTS: The thinly viscid to slimy yellowish caps, fluffy white veil tissue (often disappearing), tall, scaly to nearly smooth white stipes, and dark purple-brown spores make this ubiquitous species easy to identify. Older specimens with spores covering the caps might look similar to *S. hornemannii*, but that species has a skirtlike ring and a more distinctly scaly stipe. *Leratiomyces percevalii* is a somewhat similar species commonly found on wood chips but has smaller fruitbodies with fibrillose partial veils and dry caps.

Stropharia hornemannii

(Fr.) S. Lundell & Nannf.

CAP: 4–12 (15) cm across, rounded-conical at first, becoming convex to plane with a broad umbo, margin incurved and often adorned with whitish specks when young, often uplifted and wavy in age. Color variable, usually lilac-brown to reddish brown at first, fading to grayish brown or smoky buff in age. Surface viscid to lubricious, with veil remnants on margin when young, soon smooth. **GILLS:** Broadly attached, often with a decurrent tooth, close, broad, partial gills numerous. Pallid at first, soon gray to vinaceous gray, becoming deep vinaceous gray to blackish when spores mature. **STIPE:** 6–11 (15) cm long, 0.5–2 cm thick, equal or club shaped with an enlarged base. Whitish to buff at first, darkening slightly in age. Surface dry, covered with white flocculent scales below veil, pruinose above. **PARTIAL VEIL:** Thick, membranous, forming flaring skirtlike ring high on stipe, collapsing in age. **FLESH:** Thick, soft, whitish to buff, discoloring ocher-brown around larva tunnels and in age. **ODOR:** Mild to unpleasant. **TASTE:** Disagreeable. **SPORE DEPOSIT:** Dark purple-brown to purplish black. **MICROSCOPY:** Spores 11–13 x 6–6.5 μm, ellipsoid to almond shaped, with a broad germ pore, smooth, thick walled, grayish brown in KOH. Pleurocystidia with a swollen base and slender necks. Cheilocystidia more club shaped.

ECOLOGY: Solitary, scattered, or in small clusters near well-rotted wood. Uncommon in California, primarily found on the Far North Coast and in the northern mountains. Fruiting in fall or early winter.

EDIBILITY: Unknown.

COMMENTS: This large *Stropharia* can be recognized by the chestnut brown cap (soon fading to brownish buff), cool gray gills, shaggy white stipe with a skirtlike ring, and dark purple-brown spores. *S. ambigua* has a yellowish cap and a fluffy partial veil that leaves ragged remnants on the cap margin. *S. rugosoannulata* has a wine red, slightly viscid to dry cap that fades to brownish and a thick, cogwheel-shaped partial veil. It is a commonly cultivated species that has become naturalized in Pacific Northwest wood chip habitats, but has yet to do so in California (to our knowledge).

Stropharia coronilla

(Bull.) Quél.

GARLAND STROPHARIA

CAP: 2–6 cm across, convex to plane, occasionally slightly uplifted in age. Golden yellow at first, soon straw yellow, creamy yellow to beige. Surface lubricious to dry, generally smooth, occasionally cracking to form small scales. **GILLS:** Broadly to narrowly attached, often notched, close to subdistant. Whitish at first, soon with grayish purple tones, to dark purple-brown to purplish black in age. **STIPE:** 2–5 (7) cm long, 0.3–1 cm thick, equal or enlarged lower, often fairly stout, although slender forms can occur. White to dingy whitish in age. Surface dry, finely fibrillose to smooth. **PARTIAL VEIL:** Thick, tight white ring, usually grooved on upper surface (and darkened from spore drop). **FLESH:** Thin to moderately thick, firm when young, becoming soft, white. **ODOR:** Mild to slightly pungent. **TASTE:** Indistinct. **SPORE DEPOSIT:** Dark purple-brown to purple-black. **MICROSCOPY:** Spores 7–9 (11) x 4.5–5 (6) μm, ovoid to broadly ellipsoid, smooth, thick walled, without distinct germ pore, cinnamon brown in KOH. Pleurocystidia and cheilocystidia club shaped, with a swollen base and pointed tips.

ECOLOGY: Solitary or scattered in small troops in grassy areas. Widely distributed, but only locally common. Fruiting throughout the wet season, more abundant in late fall and winter.

EDIBILITY: Reportedly toxic.

COMMENTS: This stocky grassland species can be recognized by the yellowish cap, dark mature gills, grooved, membranous partial veil, and dark purple-brown spores. *Agrocybe praecox* is somewhat similar, but is usually taller and has brown spores. Grassland *Agaricus* species can be quite similar, but generally have whiter caps, free gills, and dark chocolate brown spores. *Leratiomyces percevalii* generally has a taller stature, specks of veil remnants on the cap, and a faint ring on the stipe, and grows in wood chips.

Stropharia caerulea

Kreisel

BLUE STROPHARIA

CAP: 1–4 (6) cm across, rounded-conical to broadly conical at first, becoming convex to plane, occasionally umbonate. Turquoise blue to bluish green at first, soon fading from the center out; paler blue to creamy yellow with a bluish margin, occasionally all yellowish beige in age. Surface thinly viscid to lubricious, with whitish specks on margin when young, soon smooth. **GILLS:** Broadly attached, close to subdistant. Pallid at first, developing vinaceous brown tones as spores start maturing, to brown in age. **STIPE:** 3–7 (10) cm long, 0.3–1 cm thick, equal or tapering downward, often tall and slender. Bluish at first, fading to whitish, then buff in age. Surface moist to dry, with whitish fibrillose scales below veil, pruinose above, to nearly smooth in age. **PARTIAL VEIL:** Fleeting, forming a thin ring and adhering to cap margin when young, disappearing in age. **FLESH:** Moderately thin, soft, bluish when young, to buff in age. **ODOR:** Indistinct. **TASTE:** Mild. **SPORE DEPOSIT:** Dark purple-brown to purple-black. **MICROSCOPY:** Spores 7–9.5 x 4.5–6 μm, ellipsoid, smooth, thick walled, grayish brown in KOH. Pleurocystidia and cheilocystidia abundant, with a swollen midsection and short, slender necks, to bottle shaped.

ECOLOGY: Solitary, scattered, or in small clusters in grassy areas, leaf litter, or occasionally rich soil at garden edges or in wood chips. Fruiting in winter. Rare throughout our range.

EDIBILITY: Unknown.

COMMENTS: Although species in this group are easily recognized by their beautiful turquoise colors and purplish brown to black spores, differentiating the three species in our area can be very difficult: *S. caerulea* is small to medium size, has blue colors that slowly fade to creamy yellow, and has an evanescent partial veil. *S. aeruginosa* is generally larger, often grows in wood chips, retains the blue color longer, and has a more membranous partial veil and whitish fringed gill edges (from the abundant cylindrical and capitate cheilocystidia). *S. pseudocyanea* is smaller and has a conical to distinctly umbonate cap, and its colors fade more quickly than in the other two. All are infrequently encountered in California and are a treat to find. *Psilocybe* species are more slender and are never as brightly colored (but can become bluish stained).

Stropharia kauffmanii

A. H. Sm.

CAP: 4–10 (15) cm across, convex to broadly convex, occasionally expanding to plane, margin incurved at first, often with whitish remnants. Buff to honey brown, covered with slightly darker brown or yellow-brown scales. Surface dry, covered with recurved scales becoming more matted-fibrillose in age. **GILLS:** Narrowly attached, often seceding, close to crowded, partial gills numerous. Gray to lilac-gray, becoming deep brown as spores mature. **STIPE:** 5–10 cm long, 1–3 cm thick, more or less equal. White to pale buff, lower portion often with pale orange scales, and discoloring orangish in age. Surface dry, covered with recurved scales below veil. Base often with white rhizomorphs. **PARTIAL VEIL:** Membranous, adhering to cap margin and forming a short skirt- to collar-like ring on stipe, collapsing or disappearing in age. **FLESH:** Thick, soft, whitish. **ODOR:** Mild to unpleasant, rancid. **TASTE:** Slightly disagreeable. **SPORE DEPOSIT:** Dark purple-brown. **MICROSCOPY:** Spores 6–8 x 4–4.5 μm, ellipsoid, smooth, thick walled, grayish brown in KOH. Cheilocystidia nearly cylindrical to spindle shaped.

ECOLOGY: Solitary or scattered in leaf litter of alder and Bigleaf Maple, more rarely in conifer duff or around well-rotted woody debris. Rare in California, known from the Far North Coast and mountains. Fruiting in fall.

EDIBILITY: Unknown.

COMMENTS: The dry, scaly cap, lilac-gray to brown gills, dark purple-brown spores, and often disagreeable odor make this rare mushroom distinctive. At first glance, one might mistake it for an *Agaricus*, but the gills are attached, not free, and the purple-brown spores help set it apart.

6 • Brown-Spored Decomposers

The various genera in this group are not all closely related, as you might guess upon seeing the wide range of fruitbody morphologies and ecological strategies of the species included here.

Pholiota are usually medium size, and most often grow in troops or clusters, usually (but not always) on wood or woody debris; their spore deposits are dull brown. Many have scaly caps and/or stipe surfaces, but a few are smooth. The stipes of some species have a partial veil that can be slimy, weakly membranous, or evanescent and cortina-like.

Gymnopilus are distinguished by their rusty orange-brown spore deposits, usually bitter taste, and comparatively large, fleshy fruitbodies. They grow on lignin-rich humus or more often directly on dead wood.

Galerina also have rusty spore deposits, but produce tiny to small fruitbodies, usually with bell-shaped or conical, translucent-striate caps; they are particularly common in moss beds. The deadly toxic G. *marginata* is a bit different from the rest; it produces larger fruitbodies with rings on the stipe.

Tubaria have small, dull pinkish tan to brick brown or blood red to wine-colored fruitbodies with caps that are usually translucent-striate. *Agrocybe* have opaque, yellowish to tan or dull ocher caps. Only a few *Agrocybe* produce fruitbodies with truly membranous partial veils, but many species in these two genera show some form of partial veil tissue. Both genera produce dull brown spore deposits.

Conocybe, *Bolbitius*, and *Pholiotina* have small, fragile fruitbodies and have ocher spore deposits. They are most often encountered on lawns, in manure piles, and in urban habitats with wood chips, but some species in each genus also occur in forest settings. *Bolbitius* are easily distinguished from most *Conocybe* by their distinctly slimy caps. *Pholiotina* are much harder to tell apart from *Conocybe*, with the exception of P. *rugosa* and its relatives, which have membranous partial veils.

Simocybe, *Flammulaster*, and *Phaeomarasmius* produce very small fruitbodies with finely velvety, granular, or hairy-scaly cap textures, respectively. All have dull brown to beige spore deposits and are mostly found on dead branches and well-decayed logs in forest settings (rarely in urban habitats).

Identification of species within most of these genera is quite difficult and often requires examination of microscopic features. *Gymnopilus*, *Pholiota*, *Agrocybe*, and *Bolbitius* are common and conspicuous, and are slightly easier to put names to. *Galerina*, *Pholiotina*, and *Conocybe* are much harder to identify.

Pholiota terrestris

Overh.

TERRESTRIAL PHOLIOTA

CAP: 1–6 (10) cm across, conical to convex with an incurved margin at first, becoming broadly convex to plane, occasionally with an umbo. Light to dark brown scales completely covering the cap, soon breaking up, retaining brown color in the center and exposing a paler, yellowish brown to creamy beige base color. Occasionally losing all the scales and becoming yellowish brown to creamy beige overall. Surface with dry appressed, flattened scales over center and paler tufted scales around margin, breaking up upon cap expansion, exposing a slightly viscid, smooth skin. Becoming more appressed-scaly or almost smooth in age. **GILLS:** Broadly attached, or with a slight notch and a decurrent tooth, close to subdistant, narrow, partial gills numerous. Beige, tan to pale yellowish brown at first, darkening slightly in age. **STIPE:** 3–9 cm long, 0.3–1 cm thick, equal, or tapering toward the base. Whitish to tan, covered with brown tufts and scales, darkening slightly in age. Surface dry, covered with pointed tufts and scales below veil at first, becoming more sparse and matted in age. **PARTIAL VEIL:** Thin, membranous, leaving ragged remnants on cap margin, or occasionally an evanescent ring on stipe. **FLESH:** Moderately thin, firm, watery whitish, yellowish to light brown, darkening in age. Stipe fibrous, hollow in age. **ODOR:** Indistinct. **TASTE:** Indistinct. **KOH:** Pinkish brown on cap, orange-brown elsewhere. **SPORE DEPOSIT:** Brown. **MICROSCOPY:** Spores 4.5–7 x 3.5–4.5 μm, elliptical to nearly ovoid, with a small germ pore, smooth, moderately thick walled, inamyloid, cinnamon brown in KOH.

ECOLOGY: In small to large clusters, rarely solitary, often in disturbed ground, lawns, and roadsides, or in wood chips or on buried woody debris. Common and widespread. Fruiting in late fall and winter.

EDIBILITY: Reportedly edible, but not recommended.

COMMENTS: Although highly variable in size and color; this *Pholiota* can easily be recognized by its scaly brown cap, scaly stipe (at least when young), and tendency to grow in clusters in disturbed or urban areas. Older, rain-beaten specimens can be difficult to distinguish from other *Pholiota*, but the clustered growth on ground, (rarely directly on wood) is also helpful. *Lacrymaria lacrymabunda* and *L. velutina* have fibrillose-scaly caps and often grow in clusters, but are more fragile and have black spores. *Inocybe hirsuta* var. *maxima* has a dry, scaly, dark brown cap and stipe, stains reddish when rubbed, and has a bluish gray stipe base.

Pholiota carbonaria

A. H. Sm.

CAP: 2–5 cm across, convex with an incurved margin at first, becoming broadly convex to plane, occasionally with an umbo. Pale creamy beige to yellowish buff at first, darkening to orangish buff or cinnamon brown in age. Surface viscid to tacky, adorned with reddish orange specks of veil remnants around margin when young, smooth in age. **GILLS:** Broadly attached, or with a slight notch and a decurrent tooth. Close to subdistant, narrow. Whitish beige at first, darkening to tan, then brown as spores mature. **STIPE:** 3–6 cm long, 0.3–0.5 cm thick, relatively slender, equal, or enlarged slightly toward base. Whitish to pale yellow as first, covered with reddish ocher scales; apex smooth. **FLESH:** Thin, fibrous in stipe; watery yellow to yellow-brown. **ODOR:** Indistinct. **TASTE:** Indistinct. **KOH:** Orange-brown on cap. **SPORE DEPOSIT:** Brown. **MICROSCOPY:** Spores 5–8 x 3.5–4.5 µm, ovoid, with a small germ pore, smooth, thick walled, inamyloid, cinnamon in KOH.

ECOLOGY: In scattered troops or clusters on burnt ground or in charcoal. Often seen in small numbers in campfire pits or burnt brush piles, but can occur in huge numbers after forest fires. Common in its specialized habitat within a limited time frame (1–5 years postfire). Fruiting in late fall and winter, occasionally in spring.

EDIBILITY: Unknown.

COMMENTS: This is one of a handful of *Pholiota* species that thrive in burnt areas. After forest fires, one can find thousands of fruitbodies carpeting the ground in otherwise barren areas. Among the species found in burnt habitats, this one stands out by its combination of a yellowish buff cap with reddish ocher veil remnants on the margin, pale gills that darken in age, and a slender stipe covered with reddish ocher scales. *P. brunnescens* is a larger, stockier species, with a viscid, orange to brown cap, yellowish veil bands on the stipe, and a glutinous partial veil. It's fairly common in campfire pits on the North Coast and in burnt montane forest in the fall. *P. highlandensis* (sensu CA) is a very similar, spring-fruiting, montane species with a whitish stipe and pale veil bands. *Crassisporium funariophilum* (=*Pachylepyrium carbonicola*) is another vernal montane species that grows in large troops in burnt areas. It produces smaller fruitbodies with smooth, ocher to reddish brown or brown colors, and has distinctive, vaguely diamond-shaped spores. Globally, there is some confusion surrounding the name *Pholiota carbonaria*: a European species with the epithet *carbonaria* was transferred to *Pholiota* illegitimately (since it was done after Smith's name was already published). It since has been synonymized with Peck's *P. highlandensis* (which is distinct from the Californian species going by that name).

Pholiota flammans

(Batsch) P. Kumm.

FLAMING PHOLIOTA

CAP: 3–7 (10) cm across, convex with an incurved margin at first, becoming broadly convex to nearly plane, occasionally with an umbo. Brilliant golden yellow when young, developing orange tones at it ages. Surface dry, covered with tufts or curved fibrillose scales when young, occasionally washing off and becoming nearly smooth in age. **GILLS:** Broadly attached, at times with a shallow notch and a decurrent tooth, close, narrow. Bright yellow at first, becoming rusty yellow as spores mature. **STIPE:** 3–8 (10) cm long, 0.4–1 cm thick, equal, or enlarged downward, often with a pointed base. Bright yellow, discoloring only slightly in age. Surface covered with bands of curved and sometimes pointed scales, becoming more matted in age; smooth above the veil. **PARTIAL VEIL:** Forming a ragged ring zone on upper stipe and leaving remnants on the cap margin when young, nearly disappearing in age. **FLESH:** Moderately thick, firm, fibrous in stipe. Yellow to greenish yellow. **ODOR:** Indistinct. **TASTE:** Indistinct. **KOH:** Orange-brown. **SPORE DEPOSIT:** Brown. **MICROSCOPY:** Spores 4–5 x 2.5–3 μm, oblong to elliptical, smooth, thick walled, ocher in KOH. Pleurocystidia abundant, club shaped or pointed. Cheilocystidia similar or small and cylindrical.

ECOLOGY: Solitary, scattered, or in small clusters on decaying logs and stumps. Often on conifers, occasionally on alder as well. Uncommon in California, mostly restricted to the northern part of our range; not known south of Mendocino County. Often fruiting early in the season, soon after first soaking rains in fall, with scattered fruitings into winter.

EDIBILITY: Reportedly nontoxic; not recommended.

COMMENTS: The striking golden yellow colors, scaly cap and stipe, and growth on wood make this beautiful *Pholiota* easy to recognize. Most of the look-alike species are rare on the coast, but common in the mountains. These include *P. aurivella*, with a yellow to golden orange, viscid cap with tufted to appressed scales and large spores [7–9.5 (11) x 4.5–6 μm]; *P. limonella*, which is nearly identical to the latter but with smaller spores (6–7.5 x 4–5 μm); and *P. squarrosoides*, with a viscid, beige to orange cap covered with upright or curved orangish scales. *P. squarrosa* is also very similar, but has a dry, beige to brown, scaly cap and stipe.

Pholiota decorata

(Murrill) A. H. Sm. & Hesler

CAP: 2–8 cm across, convex at first, becoming convex to plane, occasionally with a low, broad umbo. Creamy beige with a brown center or extensively pinkish brown, radially streaked with darker fibrils. Surface viscid to dry, appressed-fibrillose at first, soon nearly smooth. **GILLS:** Broadly attached to slightly notched, close to subdistant, broad, partial gills numerous. Creamy beige to yellowish tan at first, darkening to yellowish gray or pallid clay colored. **STIPE:** 3–9 cm long, 0.3–1 cm thick, equal. Whitish as first, soon with brownish streaks and stains, darkening from base up. Surface dry, covered with scruffy white tufts and hairs, flocculent when young, appressed in age. Base often with fuzzy mycelium and rhizomorphs. **PARTIAL VEIL:** A thin, flocculent membrane, leaving tissue on cap margin and stipe. **FLESH:** Thin, fibrous in stipe, whitish to brown. **ODOR:** Indistinct. **TASTE:** Indistinct. **KOH:** Orange-brown on all parts. **SPORE DEPOSIT:** Dark brown. **MICROSCOPY:** Spores 5.5–8.5 x 3.5–4.5 μm, ovoid to elliptical in face view, or flattened on one side in side view, smooth, thick walled, rusty brown in KOH.

ECOLOGY: Solitary or scattered on small logs, branches, and surrounding duff of both conifers and hardwoods. Common in the northern part of our range, uncommon in the San Francisco Bay Area, not yet recorded south of San Francisco. Fruiting from late fall through winter.

EDIBILITY: Unknown.

COMMENTS: This is one of the more easily identified *Pholiota* in our area; the combination of a creamy beige cap with a brown, scaly center, white stipe with cottony scales, and growth on smallish dead branches is distinctive. *P. terrestris* has a scalier cap and darker scales on the stipe, and grows in clumps and clusters on the ground or in wood chips.

Pholiota spumosa group
SLENDER PHOLIOTA

CAP: 2–6 cm across, conical to convex with an incurved margin at first, becoming broadly convex to plane, often with an umbo. Color variable, generally yellow-brown with an ochraceous disc and an overall olive cast when young, becoming more ochraceous brown and dingy in age. Occasionally lacking ochraceous brown tones or more olive-yellow in color. Surface viscid to tacky, fibrillose-streaked to smooth. **GILLS:** Broadly attached, or with a slight notch and a decurrent tooth, close to subdistant, narrow, partial gills numerous. Yellowish green at first, soon beige, darkening to brown as spores mature. **STIPE:** 3–6 (10) cm long, 0.3–0.5 (0.7) cm thick, equal, or enlarged slightly toward base. Pale yellow to pale greenish yellow as first, developing stains and streaks and darkening to dingy orangish brown from base up in age. Surface dry, finely fibrous to smooth below veil, pruinose above. **PARTIAL VEIL:** Thin, cobwebby, adhering to cap margin when young, leaving sparse silky fibrils on stipe. **FLESH:** Thin, firm, stipe fibrous, watery greenish in cap. Stipe becoming hollow yellowish green at first, darkening in age. **ODOR:** Indistinct. **TASTE:** Indistinct. **KOH:** Pale orange-brown. **SPORE DEPOSIT:** Brown. **MICROSCOPY:** Spores 5.5–8 x 4–4.5 (5.5) µm, ovoid, with a small germ pore, smooth, moderately thick walled, inamyloid, golden brown in KOH.

ECOLOGY: Scattered in troops or small clusters, often on wood chips or on or around well-decayed logs and stumps of both conifers and hardwoods. Common and widespread. Fruiting in late fall and winter.

EDIBILITY: Unknown.

COMMENTS: The viscid, yellow-brown cap with an ochraceous disc, olive tones, light yellowish gills with a greenish cast when young, cobwebby partial veil, and growth on rotting wood (most often wood chips) are helpful identifying features of this *Pholiota. P. spumosa* is a European species that probably doesn't occur in California. There are a number of species in this complex described from North America, but modern taxonomic work (that is, sequencing type collections) is needed to help sort out the mess. Two West Coast species in this complex are *P. subflavida* and *P. vialis*; one of which may be the correct name for our western *P. "spumosa". P. malicola* var. *macropoda* often grows in clusters, has a smooth yellow-ocher cap, and a green-corn odor. *P. alnicola* is a similar but slightly smaller species, often grows in clusters on dead alder stumps and logs and has a smooth yellow cap when young that darkens slightly in age; pallid to straw yellow gills; and a fragrant odor. *Hypholoma fasciculare* has a yellowish cap (occasionally olive or ocher toned), yellowish green gills that turn gray, and purplish black spores. *H. capnoides* has pallid gills that turn gray and a smoky gray spore deposit. Similar *Gymnopilus* species have rusty orange spores and brighter orange colors, and most have a bitter taste.

Pholiota velaglutinosa

A. H. Sm. & Hesler

SLIMY-VEILED PHOLIOTA

CAP: 2–7 cm across, convex with an inrolled margin at first, becoming broadly convex to plane. Dark reddish brown to dark brown overall at first, fading around margin as it ages. Surface smooth or with a few whitish veil specks, covered with glutinous slime when wet, tacky and/or shiny if dry. Margin often with overhanging slime. **GILLS:** Broadly attached, or with a decurrent tooth, close to subdistant, broad, partial gills numerous. Dingy yellowish at first, to clay brown to brown. **STIPE:** 3–7 cm long, 0.3–0.8 cm thick, equal or enlarged slightly toward base. Whitish to pale yellow as first, developing orangish brown streaks and stains in age. Surface moist, covered with finely white fibrils, often in chevronlike scales, becoming more matted in age. **PARTIAL VEIL:** A translucent, glutinous membrane with embedded silky fibrils, soon breaking away from the stipe, leaving a roll of glutinous tissue on cap margin and an annular zone on stipe. **FLESH:** Moderately thin, firm, greenish yellow, and slowly staining rusty brown in cap. Stipe fibrous, yellowish buff to brownish, darkening in age. **ODOR:** Indistinct. **TASTE:** Indistinct. **KOH:** Orange-brown. **SPORE DEPOSIT:** Brown. **MICROSCOPY:** Spores 6–7.5 x 3.5–4.5 μm, ovoid to elliptical in face view, bean shaped in side view, with a small germ pore, smooth, thick walled, inamyloid, cinnamon in KOH.

ECOLOGY: Scattered in troops or clusters, rarely solitary, in duff, woody debris, or wood chips under pine. Common throughout our range. Fruiting in late fall and winter.

EDIBILITY: Unknown.

COMMENTS: This *Pholiota* can be readily identified when the glutinous veil is present; older specimens can be tough to tell apart from similar species. Other distinguishing features are the viscid to tacky, brown to reddish brown cap with a paler margin, moist to dry stipe covered with whitish fibrils, frequently yellow upper stipe, and growth in pine debris. A number of other *Pholiota* are similar, but lack the glutinous veil when young, are paler in color, or occur in different habitats.

Pholiota astragalina

(Fr.) Singer

PINK PHOLIOTA

CAP: 1–5 cm across, rounded-conical to convex and sometimes broadly umbonate at first, soon flattening or with wavy margin in age. Pinkish orange to peachy orange with a paler margin at first, fading to dull peachy with a whitish to buff margin, occasionally developing dingy stains in age. Surface dry to moist, appressed-fibrillose to smooth, often with veil remnants around margin when young. **GILLS:** Broadly attached to slightly notched, close to subdistant, broad. Pinkish to pale yellow at first, darkening to yellow-brown or light brown. **STIPE:** 4–12 cm long, 0.3–0.7 cm thick, equal, or rooted. Whitish as first, with a pinkish buff base color, becoming more orange in age. Surface dry, covered with fine scruffy white fibrils, becoming more matted in age. **PARTIAL VEIL:** Thin, whitish wispy fibrils, leaving patches on stipe and cap margin when young. **FLESH:** Thin, relatively fragile, orangish to brownish in cap. Stipe fibrous, hollow in age, whitish to pale orangish brown. All parts slowly turning grayish to blackish when damaged. **ODOR:** Indistinct. **TASTE:** Very bitter. **KOH:** Orange-brown on all parts. **SPORE DEPOSIT:** Brown. **MICROSCOPY:** Spores 5–7 x 3.5–4.5 μm, ovoid to elliptical in face view, flattened on one side in side view, with a small germ pore, smooth, thick walled, inamyloid, rusty brown in KOH.

ECOLOGY: Solitary, scattered, or in small clusters on rotting coniferous logs and stumps. Uncommon in the northern part of our range, often on Western Hemlock. Occurs at least as far south as Santa Cruz County on Douglas-fir, but generally quite rare south of Sonoma County. Fruiting from fall into winter.

EDIBILITY: Unknown.

COMMENTS: The peachy pink cap, pale pinkish or yellowish gills that become brown, pale stipe, and bitter taste identify this species. It is most likely to be confused for a *Hypholoma*: *H. fasciculare* has a yellower cap (occasionally orangey) and bright yellowish green to greenish gray gills. *H. capnoides* has a paler, dull orangey beige cap and light grayish gills. Both produce dark grayish purple spore deposits.

Gymnopilus luteofolius

Peck (Singer)

PURPLE GYM

CAP: 3–12 cm across, convex with an inrolled margin at first, to nearly plane in age. Purple or maroon-red at first, soon duller red to reddish pink or purplish pink (often with a contrasting yellow or whitish margin), then showing yellow-orange between scales, often predominantly orange overall at maturity; occasionally extensively creamy yellowish when young. Variably marked with blue-green dots or blotches. Dry to moist, densely covered with upright scales and tufts when young, in age becoming radially appressed fibrillose, often metallic looking. Edge often with whitish patches of veil, soon covered with rusty spores. **GILLS:** Broadly attached with a fine decurrent tooth, close. Greenish yellow when young, becoming bright orange to brownish orange in age. **STIPE:** 5–10 cm long, 1–2.5 cm thick, cylindrical with an enlarged base, often curved. Pinkish purple to yellowish orange, with a silky white sheen over entire surface when young, easily rubbed off when handled and then appearing vertically marbled or variegated. Smooth when young, soon vertically striate. **PARTIAL VEIL:** A pronounced white cortina (sometimes nearly membranous) set high on stipe, soon covered in rusty spores, sometimes present only as a sparse ring zone in age. **FLESH:** Firm, fibrous, yellowish orange, mottled with purplish red in cap. Bruising erratically blue-green on surfaces, not when cut. **ODOR:** Indistinct. **TASTE:** Very bitter. **KOH:** Orangish. **SPORE DEPOSIT:** Bright orange-brown. **MICROSCOPY:** Spores 5.5–8.5 x 3.5–4.5 μm, ellipsoid to nearly ovoid, flattened on one side in side view, finely warted, dextrinoid. Cheilocystidia bottle shaped to flask shaped, with a slender neck, with or without a round head.

ECOLOGY: In large clumps and troops on wood chips. Fairly common in urban areas, occasional elsewhere. Fruiting from late fall through winter.

EDIBILITY: Mildly hallucinogenic, very bitter.

COMMENTS: Several features are helpful in distinguishing this species: purple to maroon-red, scaly young caps, that soon become duller red with yellow-orange showing in the cracks, or entirely orange in age; greenish yellow young gills; a cobwebby whitish partial veil; orange-brown spores; and growth in clusters on wood chips. A number of other *Gymnopilus* species are similar: *G. thiersii* was described from San Francisco County and might be the appropriate name for the brightest purplish to red specimens found on pine mulch or logs in pine forests. An often smaller form (species?) with more purple tones and paler colors between the scales appears on native oak (rarely) and possibly also in disturbed habitats. *G. aeruginosus* may be a more appropriate name for the species profiled above, but until this assertion is backed by more conclusive evidence, we have chosen to continue using the more familiar name, *G. luteofolius*. Others to compare include *G. pulchrifolius*, *G. viridans*, and *G. braendlei*—good luck! *Tricholompsis rutilans* can look very similar and sometimes grows in wood chips, but its fruitbodies taste mild, lack a cobwebby partial veil, and produce a pale spore deposit.

Gymnopilus ventricosus

(Earle) Hesler

WESTERN JUMBO GYM

CAP: 5–30+ cm across, rounded with an inrolled margin at first, becoming convex to plane, sometimes wavy in age. Yellowish orange to orange, often with a pale virgate appearance, in age extensively streaked with darker reddish brown. Surface dry to moist, smooth, innately fibrillose, sometimes with scattered, pale, appressed-fibrillose patches. Margin often with adhering pale patches of partial veil. **GILLS:** Notched with a fine decurrent tooth, sometimes more broadly attached, close, edges irregular. Pale yellow when young, becoming bright ocher, sometimes with distinct pale edges; darker brown in age, bruising brownish when damaged. **STIPE:** 10–30 cm long, 2.5–10 cm thick. Curved, with a strongly swollen lower portions, tapering at base, especially when fruiting in dense clusters. Whitish to yellowish or with ocher-brown vertical streaks, bruising brownish when handled. Surface fibrillose, often ridged, and sometimes scaly near apex. **PARTIAL VEIL:** Conspicuous, a thick cortina or membranous. Whitish yellow when young, eventually collapsing and becoming covered in rusty spores. **FLESH:** Thick, rubbery-firm, fibrous. Marbled yellowish to light orange throughout. **ODOR:** Indistinct. **TASTE:** Very bitter. **KOH:** Dark red-brown on cap, orangey red on gills. **SPORE DEPOSIT:** Rusty orange-brown. **MICROSCOPY:** Spores 7.5–9 x 4–5.5 μm, ellipsoid, roughened, slowly dextrinoid, ocher-brown in KOH. Pleurocystidia swollen or bottle shaped. Cheilocystidia bottle shaped, capitate.

ECOLOGY: Sometimes in enormous clusters, rarely solitary, usually on pine stumps or at base of dead pines, occasionally higher up trunk. Regularly found on oak and eucalyptus in some areas (see comments). Common and conspicuous from fall through midwinter.

EDIBILITY: Inedible, very bitter. Similar species in eastern North America are hallucinogenic, but this one is not.

COMMENTS: The large, orange fruitbodies in clusters on wood, bitter taste, sinuate gills, rusty brown spores and partial veil distinguish this species. Although often referred to as *G. junonius* (=*G. spectabilis*) in the past; that species is a European taxon with larger spores and different gill cystidia. There appear to be additional species of Jumbo Gyms on the Far North Coast, but more work is needed to sort them out. *G. subspectabilis*, reported on oak in California, has larger spores (8.5–13 x 5–7.5 μm). *Omphalotus olivascens* very rarely grows on pine, lacks a partial veil, has decurrent gills, and whitish spores.

Gymnopilus aurantiophyllus

Hesler

CAP: 3–11 cm across, convex with an incurved margin, expanding to nearly plane. Orange to ocher at first, usually marbled and streaked with brick red, brownish red and yellowish, and whitish yellow toward margin. Dry to moist, edge opaque, often with adhering remnants of whitish partial veil, becoming covered in spores. **GILLS:** Narrowly attached with a distinct notch, close, narrow, edge even or irregularly saw-toothed in age. Dull pale beige to yellowish beige, becoming more orange to ocher brown in age; edge contrastingly pale, bruising brownish when damaged. **STIPE:** 5–13 cm long, 1–2.5 cm thick, cylindrical with an enlarged base, often curving. Pale yellowish white to beige, vertically streaked, darkening to ocher-brown or reddish brown upward in age or when handled. Base with whitish mycelium attached to woody substrate. Dry, longitudinally striate, fibrous or ridged. **PARTIAL VEIL:** Thick white cortina, sometimes nearly membranous when young, set high on stipe, soon covered in rusty spores, often collapsing and sparse in age. **FLESH:** Fibrous, marbled whitish yellow. **ODOR:** Indistinct to slightly farinaceous. **TASTE:** Bitter. **SPORE DEPOSIT:** Bright orange-brown. **MICROSCOPY:** Spores 7–9 x 4–5.5 µm, almond shaped, slightly roughened, slightly thick walled, dextrinoid. Cheilocystidia densely packed, vase shaped with a rounded head.

ECOLOGY: Solitary, scattered, or in clumps on wood chips, woody debris, and stumps of pines. Common in pine forest throughout our range. Fruiting from fall into winter, or spring in wet years.

EDIBILITY: Inedible; very bitter.

COMMENTS: Species in this group of *Gymnopilus* are recognized by their yellowish orange to ocher, medium-size fruitbodies with conspicuous partial veils, and growth on fairly well-decayed logs, wood chips, and other lignin-rich debris. However, identification to species is a real challenge, primarily due to vague and sometimes competing original concepts, but also due to genuine similarity to one another. Most prior authors used the name of a European taxon, *G. sapineus*, as a catchall for this group. From the perspective of convenience (to distinguish these species from the large and tiny *Gymnopilus*), this may be the best approach until a clearer understanding is reached. However, we think the species illustrated here is a good match for Lexemuel Hesler's concept of *G. aurantiophyllus* on the basis of its white veil, dextrinoid spores, incrusted cap cells, and dense bundles of cheilocystidia. A few other species (some undescribed) have smaller fruitbodies and/or more scaly caps, and at least one grows on oak. *G. luteocarneus* grows on conifers and has small fruitbodies with smooth caps, thin stipes, yellowish or buff flesh, and spores 7–8 x 4–4.5 µm.

Gymnopilus punctifolius

(Peck) Singer

BLUE-GREEN GYM

CAP: 3–10 cm across, rounded to broadly convex with an inrolled edge, often very irregular and wavy or contorted. Color variable within a distinctive range. Dull sky blue when young, becoming mottled and streaked with green, yellow, olive, and ocher-tan, often all on a single cap. Generally duller and more brownish in age, brighter greenish yellow to blue-gray when young. Surface dry, silky-fibrillose, sometimes cracking into small plaques in age; covered in a fine bloom that rubs off when handled. **GILLS:** Narrowly attached or finely notched. Bright yellow-green when young, olive-tan to ocher or brown in age; often with a pale and uneven edge, bruising brown when damaged. **STIPE:** 5–16 cm long, 0.5–3 cm thick, cylindrical, curved, often strongly sinuous with a tapered, rooting portion buried in wood. Bright greenish yellow to powdery blue-gray or dingy olive overall, extreme base with patches of lilac mycelium. Surface with a pale bloom (sometimes vertically striate), apex with small yellow scurfs. **FLESH:** Firm, fibrillose, tough. Yellowish olive, bruising brown. **ODOR:** Indistinct. **TASTE:** Bitter. **KOH:** No reaction. **SPORE DEPOSIT:** Rusty brown to darker ocher brown. **MICROSCOPY:** Spores 4–6.5 x 3.5–5 μm, ovoid to broadly elliptical, finely warted, dextrinoid, brownish in KOH. Pleurocystidia slender or swollen in the middle, with a swollen head. Cheilocystidia similar.

ECOLOGY: Solitary or in clumps or dispersed small groups inside hollowed cut ends and undersides of old-growth conifer logs, especially hemlock, redwood, and Douglas-fir. Known from southern Santa Cruz County northward, but uncommon south of Humboldt County, and rarely occurring outside old-growth forest. Fruiting in late fall and winter.

EDIBILITY: Unknown. Reports of psilocybin content are unsubstantiated.

COMMENTS: The bitter taste, growth on wood, and rusty spores place this species in *Gymnopilus*. No other species in this genus shows powdery blue or inherent, extensive greenish cap tones. Smaller, yellower forms could be confused with *Hypholoma fasciculare*, but that species produces a purplish gray spore deposit.

Gymnopilus rufescens

Hesler

CAP: 1.5–5 cm across, rounded to convex, sometimes with a broad umbo. Pale brownish orange to orange, with more yellow tones in age. Surface moist to dry, finely scurfy at first, soon nearly smooth. Margin faintly striate and with a narrow overhanging band of translucent tissue. **GILLS:** Finely attached, close, and rather narrow. Pale creamy yellow to yellow when young, soon dull orange-tan, rusty to brown as spores mature. **STIPE:** 2–6 cm long, 0.2–0.5 cm thick, curved with an enlarged base. Dull orange-brown with a pale yellowish white bloom (often visible only at apex), darker brick to reddish brown toward base. **PARTIAL VEIL:** Absent. **FLESH:** Very thin, orange-brown. **ODOR:** Indistinct. **TASTE:** Bitter. **KOH:** Very dark, blackish on cap. **SPORE DEPOSIT:** Rusty orange-brown. **MICROSCOPY:** Spores 4–5.5 x 2.5–3.5 μm, ellipsoid or like an apple seed ("pip shaped"), coarsely roughened to nearly warted, ocher-brown in KOH. Cheilocystidia bottle shaped with a slender neck, with or without a round head. Hyphae in flesh beneath pileipellis radially arranged.

ECOLOGY: Solitary or in small clusters on decaying conifer wood. Locally common especially in northern part of our range. Fruiting from late fall through early spring, especially in dark, wet areas.

EDIBILITY: Unknown.

COMMENTS: The small size, nearly smooth cap, light orange-brown colors, bitter taste, and lack of a partial veil help distinguish this species. A few other *Gymnopilus* are quite similar, including *G. oregonensis* and *G. picreus*. Both have much larger spores (7–9 x 4.5–5.5 μm, perhaps up to 10.5 μm long in *G. picreus*). The squamulose to markedly scurfy cap of *G. picreus* appears to set it apart from *G. oregonensis*, and both appear to be darker reddish brown than *G. rufescens*. Although *G. picreus* was described from Europe, it matches our material quite well, and we lack a better alternative. Every small *Gymnopilus* in California that we have examined shows a fine bloom on the stipe due to the presence of caulocystidia. This feature is evanescent and easily destroyed; thus we believe Lexemuel Hesler overlooked it in his original description of *G. rufescens*.

Galerina marginata group
DEADLY GALERINA

CAP: (0.5) 1.5–5 (8) cm across, hemispherical to convex at first, expanding to broadly convex to nearly plane, occasionally with a low, broad umbo. Color variable, strongly hygrophanous; when wet, generally ocher-orange to ocher-brown, or slightly paler; cinnamon brown, but can be dark reddish brown to brown as well. Fading around margin to pale cinnamon brown, to pale tan to pinkish beige. Surface smooth, moist to dry, margin translucent-striate when wet. **GILLS:** Broadly attached, occasionally with a decurrent tooth, close to subdistant, edges often fringed. Beige to buff at first, darkening to cinnamon brown to ocher-brown as spores mature, but often retaining a paler edge. **STIPE:** 2–8 cm long, 0.3–1 cm thick, equal or enlarged toward base. Beige to cinnamon brown, covered with whitish fibrils at first, becoming dark brown to reddish brown and then grayish to black from the base up in age. Surface dry, covered with fine, appressed-silky hairs below annular zone, pruinose above. **PARTIAL VEIL:** Membranous, and forming a flaring ring at first, very soon collapsing, leaving a ring of tissue on stipe that often turns rusty brown from dropping spores. **FLESH:** Thin, brownish in cap, cinnamon in stipe when young, dark brown to black in age. **ODOR:** Generally indistinct (however, we have had collections that smelled like mint–chocolate chip ice cream). **TASTE:** Indistinct. **KOH:** No reaction. **SPORE DEPOSIT:** Rusty brown. **MICROSCOPY:** Spores 8–11 x 5–6.5 µm, ovoid, wrinkled-roughened, lacking a germ pore, ocher-brown. Basidia 4-spored. Pleurocystidia sparse, spindle shaped, with wavy necks and pointed tips. Cheilocystidia more abundant, similar in shape but smaller.

ECOLOGY: Solitary, scattered, or in troops on rotten logs, stumps, or woody debris (rarely in duff or moss). Also occurs in wood chips, often in clusters. Very common and widespread throughout our range. Fruiting from fall into winter, occasionally in spring.

EDIBILITY: Deadly poisonous! Contains amatoxins (see p. 47). It's a good rule of thumb to avoid consuming little brown mushrooms. Most are difficult to identify, and a few are seriously toxic.

COMMENTS: The strongly hygrophanous, ocher-brown to tan cap, membranous partial veil (sometimes collapsed into a faint ring), growth on wood, and rusty brown spore deposit are important features. Make an effort to learn to recognize this mushroom since it is fairly common and one of our few deadly poisonous species. This is especially true for those seeking hallucinogenic *Psilocybe* species—the two often grow alongside one another in wood chips. *Psilocybe* differ by having dark purplish black spores and stain blue when damaged (the dark grayish stipe of *Galerina marginata* is often interpreted as bluing by overeager pickers). *G. marginata* appears to be a species complex with at least two common species in California; one usually grows on Douglas-fir (and other conifers) and has much larger fruitbodies with flatter caps.

SYNONYM: *Galerina autumnalis* (Peck) A. H. Sm. & Singer.

Galerina badipes

(Pers.) Kühner

CAP: 0.5–2 (2.5) cm across, bell shaped to convex at first, expanding to broadly convex or nearly plane with a low, broad umbo. Strongly hygrophanous and thus color variable, brown to ocher when young and wet, soon fading around margin to pale cinnamon brown, fading in a ring around margin and from center outward as it dries. Finally pale tan to pinkish beige. Surface smooth, moist to dry, margin translucent-striate when wet, often with pale veil remnants. **GILLS:** Broadly attached, occasionally with a decurrent tooth, close to subdistant, edges often fringed. Cinnamon brown to ocher-brown. **STIPE:** 2–6 cm long, 0.2–0.4 cm thick, more or less equal, often twisted. Pale cinnamon brown over much of length, with a reddish brown base, and covered with appressed whitish fibrils when young, losing some of the fibrils, and darkening from base up in age. Surface dry, covered with fine, appressed-silky hairs below annular zone, pruinose above. **PARTIAL VEIL:** Very faint, often overlooked, consisting of a few silky white hairs and silky remnants on stipe. **FLESH:** Very thin. **ODOR:** Indistinct. **TASTE:** Indistinct. **SPORE DEPOSIT:** Light rusty brown. **MICROSCOPY:** Spores 10–13 x 5–7 μm, ovoid, wrinkled-roughened, lacking a germ pore, pale ocher-brown. Basidia 2-spored. Pleurocystidia abundant, swollen with curving necks and pointed tips. Cheilocystidia also abundant, similar in shape, with a blunt end.

ECOLOGY: Solitary, scattered, or in troops in duff, soil, moss, or on small woody debris, or occasionally in wood chips. It has a preference for slightly disturbed areas, such as road- and trailsides, or old logging roads with a thin moss or duff covering. Very common and widespread throughout our range. Fruiting from fall into winter.

EDIBILITY: Unknown, closely related to the Deadly Galerina. Do not experiment!

COMMENTS: The strongly hygrophanous, ocher-brown to tan cap, cinnamon brown gills, and stipe that darkens from the base upward are helpful in identification. *Galerina marginata* generally grows on wood, has a membranous partial veil (when young), and is generally larger. *G. nana* is very similar macroscopically, but is usually slightly smaller. Microscopically, it differs from nearly all other *Galerina* by having cystidia with crystals at their tips. A number of *Deconica* are similar, but produce purplish black spore deposits.

Galerina "Smooth Spores"

CAP: 0.3–1 (1.5) cm across, distinctly bell shaped with a rounded, sometimes translucent umbo, margin often flaring in age. Light yellow brown, tan to buff, translucent-striate when wet, fading to pale buff or beige when dry. Surface smooth, moist to dry. **GILLS:** Broadly to narrowly attached, close to subdistant. Whitish to pale pinkish buff at first, to pale cinnamon brown as spores mature. **STIPE:** 2–6 cm long, 0.05–0.15 cm thick, tall, slightly enlarged toward base. Translucent pale ocher-buff when wet, pale buff when dry. Surface moist to dry, smooth over much of length, with scattered fine hairs on lower portion. **FLESH:** Very thin and fragile, concolorous. **ODOR:** Indistinct. **TASTE:** Indistinct. **SPORE DEPOSIT:** Light rusty brown. **MICROSCOPY:** Spores 11–12 x 6–7 μm, ellipsoid to almond shaped, appearing nearly smooth; due to a very tight perispore (a sheath covering the spore wall), strongly dextrinoid. Pleurocystidia absent. Cheilocystidia and caulocystidia narrowly bottle shaped to cylindrical.

ECOLOGY: Solitary, scattered, or in small troops in moss or on mossy logs. Locally common, known from the San Francisco Bay Area north to Mendocino County, but probably more widespread. Fruiting from fall into winter.

EDIBILITY: Unknown; do not experiment.

COMMENTS: At first glance, this tiny moss dweller could be mistaken for a *Mycena*, but the yellow-brown colors, smaller stature, and rusty spore deposit set it apart. Probably the most distinctive macrofeature is the bell-shaped cap with a rounded, sometimes translucent umbo. Douglas Smith, one of California's leading little brown mushroom aficionados, is in the process of describing this species and graciously supplied the microscopic data. *G. sahleri* has a similar cap shape and shares the same habitat, but differs microscopically in having calyptrate spores—the outer layer separates to form a partial bag, hood, or sheath around the spore.

Galerina vittiformis

(Fr.) Singer

CAP: 0.3–1 (1.5) cm across, conical to bell shaped when young, becoming convex with a rounded to pointed umbo. Ocher-buff to pale cinnamon brown, with a paler, almost white margin. Strongly translucent-striate, with dark lines most of the way to a translucent umbo. Surface smooth, moist to dry. **GILLS:** Broadly attached, close to subdistant. Whitish buff to pale cinnamon brown. **STIPE:** 2.5–6 (9) cm long, 0.07–0.2 cm thick, tall, thin, equal or with a slightly enlarged base. Reddish brown over lower portion, cinnamon brown in midportion and whitish to pale buff at apex when young; darkening from base up, to dark reddish brown to blackish brown, with a paler apex. Surface dry, covered with short, fine, hairlike projecting caulocystidia. **FLESH:** Very thin and fragile, concolorous. **ODOR:** Indistinct. **TASTE:** Indistinct. **SPORE DEPOSIT:** Light rusty brown. **MICROSCOPY:** Spores (8) 10–12.5 x 5–6.5 μm, ovoid to almond shaped, wrinkled-roughened, ocher-brown. Smooth area around apiculus (plage) sharply delimited, germ pore absent. Basidia 2-spored in var. *vittiformis*, 4-spored in var. *tetraspora*. Pleurocystidia and cheilocystidia present and often abundant, both with a narrow base, swollen midsection, and variably pointed or rounded tip. Cystidia absent from cap.

ECOLOGY: Solitary, scattered, or in troops in moss. Very common, especially on the Far North Coast. Fruiting from fall through winter, generally earlier in northern part of our range and later southward.

EDIBILITY: Unknown; do not experiment.

COMMENTS: This species is one of many tiny, moss-dwelling species that have cinnamon brown, translucent-striate caps and thin stipes that darken from the base upward. *G. badipes* has a reddish brown to grayish black lower stipe, but is larger, has a silky partial veil, and grows in duff (not moss). *G. sahleri* has more of a bell-shaped cap with a rounded umbo; it is almost always on moss-covered trunks or logs of Douglas-fir. *G. semilanceata* is taller, with a more conical cap and a stipe that doesn't darken from the base upward in age; it also has capitate cheilocystidia. There are plenty of other *Galerina* reliably separated only by microscopic examination. A number of *Mycena* species have the same general stature and habitat preference, but most aren't ocher-colored, and they usually have taller stipes that don't darken upward in age.

Galerina semilanceata

(Peck) A. H. Sm. & Singer

CAP: 0.4–2 (2.5) cm across, conical to bell shaped when young, margin flaring, becoming broadly convex with a rounded umbo in age. Yellowish to ocher when young, soon fading around margin to pale cinnamon buff with a darker center, or buff to tan when dry. Strongly hygrophanous, translucent-striate when wet, opaque when dry. **GILLS:** Broadly to narrowly attached, close to subdistant. Pallid at first, then cinnamon brown, often with darker rusty stains from mature spores. **STIPE:** 3–7 (10) cm long, 0.01–0.3 (0.5) cm thick, enlarging downward with a club-shaped base. Whitish to cinnamon brown, covered with white hairs. Surface dry, covered with fine, appressed-silky hairs over lower portion, pruinose over upper portion. **PARTIAL VEIL:** Very faint, easily overlooked. Consisting of a few silky white hairs leaving sparse remnants on stipe. **FLESH:** Very thin, pale. **ODOR:** Indistinct. **TASTE:** Indistinct. **SPORE DEPOSIT:** Light rusty brown. **MICROSCOPY:** Spores 8–10 x 5–6 μm, ovoid, wrinkled-roughened, lacking a germ pore, pale ocher-brown. Basidia 4-spored. Cheilocystidia abundant, ventricose, with slender necks and rounded heads.

ECOLOGY: Solitary, scattered, or in troops in moss. Very common and widespread, fruiting from fall into winter. Probably most commonly encountered *Galerina* in California, especially in open mossy forest on the Far North Coast.

EDIBILITY: Unknown; do not experiment.

COMMENTS: The bell-shaped, yellowish buff to ocher-orange, translucent-striate cap; pallid to cinnamon brown gills; stipe covered in whitish fuzz; rusty brown spore deposit; and growth in moss are helpful in identification. Microscopically, the capitate cheilocystidia are distinctive. It has also gone by *G. clavata*, which is actually a distinct, much less common species (in California) most easily distinguished by microscopy: its spores are larger, averaging 12 μm long. A number of *Conocybe* are similar, but generally darker, with straighter stipes and smooth spores.

MISAPPLIED NAME: *G. heterocystis.*

"Albogymnopilus nanus"

N. Siegel & C. F. Schwarz nom. prov.

CAP: 0.4–2 (2.5) cm across, rounded to convex with a tightly inrolled margin when young, soon broadly convex to nearly plane. Margin often ribbed when young, wavy and irregular in age. Chalky white at first, becoming creamy white and developing grayish to beige patches, occasionally with rusty stains in wet weather. Surface smooth, moist to dry. **GILLS:** Broadly to narrowly attached, close to subdistant, edges often fringed when young. Creamy at first, becoming light brown to ocher. **STIPE:** 0.7–3.3 cm long, 0.1–0.4 cm thick, widest at apex, often curved, tapering toward a pointed base. Whitish, ivory to pale buff. Surface dry, nearly smooth. **FLESH:** Fragile, thin, pale. **ODOR:** Indistinct. **TASTE:** Slightly bitter, astringent. **KOH:** No reaction. **SPORE DEPOSIT:** Light rusty brown to ocher-brown. **MICROSCOPY:** Spores (8) 9–10 x 5–6 μm, ellipsoid to almond shaped, broader near apiculus, surface barely roughened, with two internal oil droplets, distinctly dextrinoid, no germ pore. Pileipellis a loose, irregular cutis with some upright elements. These elements irregularly inflated and sometimes branched or cylindrical-capitate. Cheilocystidia irregularly cylindrical-capitate, sometimes with round or finger-like projections at the tip, occasionally inflated at base so as to be ventricose-capitate with moderately long, sometimes sinuous necks.

ECOLOGY: Scattered in troops, on small woody debris (sometimes buried), or on wood chips, often in disturbed areas, rarely on rotting logs or in spruce and hemlock duff. Widespread but rather uncommon; we have seen it from Santa Cruz County to central Oregon on the coast, and in the Sierra Foothills. Fruiting from late fall into spring.

EDIBILITY: Unknown.

COMMENTS: This small white character doesn't fit neatly into our common generic concepts (and genetic data suggest it may need a genus of its own). It can easily be recognized by its small opaque, chalky white caps that develop creamy buff tones in age, ocher gills, and short, curved stipes that taper downward.

Agrocybe praecox

(Pers.) Fayod

SPRING AGROCYBE

CAP: 2–8 (15) cm across, convex with an incurved margin at first, becoming broadly convex to plane, often with a low umbo or depressed at center, margin occasionally uplifted and/or splitting in age. Light brown, creamy brown to yellowish beige, fading or browning in age. Surface dry, smooth, often cracking in dry weather. **GILLS:** Broadly attached, close to crowded, broad. Whitish to pale clay colored when young, darkening to brown as spores mature. **STIPE:** 4–12 cm long, 0.5–2 cm thick, cylindrical or club shaped with an enlarged base. Whitish to tan, darkening slightly in age. Surface dry, fibrous to smooth, base often with white rhizomorphs. **PARTIAL VEIL:** Relatively thick, membranous, skirtlike. In age collapsing to form a band on stipe, or breaking up and adhering to cap margin, or falling off completely. **FLESH:** Thin to thick and firm, fibrous in stipe, pallid. **ODOR:** Farinaceous, or like "stale chocolate." **TASTE:** Earthy-farinaceous. **SPORE DEPOSIT:** Brown. **MICROSCOPY:** Spores 8–11 x 5–7 µm, ellipsoid, with an apical germ pore, smooth, thick walled, yellow-brown.

ECOLOGY: Solitary, scattered, or in small to large clusters in a wide variety of habitats. Generally in wood chips or grassy areas, also in coastal scrub, disturbed forest, or live oak woodlands. Very common, especially in urban areas in spring, very rarely in fall.

EDIBILITY: Not recommended.

COMMENTS: This *Agrocybe* is highly variable in size and stature; even single patches can have small, slender specimens next to large, stocky ones. It is easily recognized by its smooth, light brown to yellowish beige cap, brown mature gills, and membranous partial veil. When exposed to sun or dry weather, the cap often develops deep cracks. Although the name *A. dura* has sometimes been applied to larger, cracked specimens, or collections in which the ring remains attached to the cap margin, we believe these represent variations of *A. praecox*. *Cyclocybe erebia* (=*Agrocybe erebia*) has a darker brown, reddish brown to olive-brown cap and a persistent ring. It is uncommon in California, generally fruiting in the fall and spring near alders. *Cortinarius caperatus* is similar but has brighter ocher-brown spores, grows under conifers in forest settings, and is much less common in our area, found primarily in fall on the Far North Coast.

Agrocybe putaminum

(Maire) Singer

CAP: 2–7.5 cm across, irregular-conical to convex with a tightly inrolled and often ribbed margin at first, becoming broadly convex with a broad umbo to plane, or wavy and irregular in age. Pale ocher-brown to yellow-brown, often fading to light brown to tan. Sometimes darker to orange-brown in age. Surface dry, velvety to finely pubescent at first, often becoming smooth. **GILLS:** Broadly attached or with a narrow notch, close to crowded, broad. Whitish to pale clay colored when young, darkening to clay brown to brown as spores mature. **STIPE:** 2–8 cm long, 0.4–1 cm thick, club shaped with an enlarged base to nearly equal. Pale ocher-buff to yellowish brown, darkening slightly in age. Surface dry, velvety to finely pubescent, longitudinally grooved, at least over upper portion, base often with white rhizomorphs. **FLESH:** Thin to fairly thick and firm, fibrous in stipe, whitish to buff. **ODOR:** Musty. **TASTE:** Bitter-farinaceous. **SPORE DEPOSIT:** Brown. **MICROSCOPY:** Spores 9.5–15 x 6–8 μm, ellipsoid to oblong, with narrow apical germ pore, smooth, thick walled, yellow-brown.

ECOLOGY: Clustered or in scattered troops in wood chips. Common in urban areas. Fruiting in spring and early fall, occasionally in summer in irrigated areas.

EDIBILITY: Nontoxic.

COMMENTS: This *Agrocybe* can be recognized by its velvety, ocher-brown to yellow-brown cap, velvety stipe with longitudinal lines, and growth in wood chips. It was recently introduced to California, probably from Europe, and now is a common member of our wood chip fungal community. Although other *Agrocybe* are similarly colored, they lack the velvety cap and powdery lines on the stipe. *A. arvalis* is much smaller and more slender, and grows out of a large, dark sclerotium. Microscopically, it has swollen pleurocystidia with bunches of "fingers" at the top. *Hypholoma tuberosum* has a paler red to orange-brown cap, smooth stipe, and dark gray-brown spores; the base of its stipe emerges from a knobby, brown, hard sclerotium.

Agrocybe pediades

(Fr.) Fayod

HEMISPHERIC AGROCYBE

CAP: 1–3 cm across, rounded to hemispherical at first, becoming convex, margin incurved when young, often remaining so in age. Golden brown to pale ocher when young, fading to yellowish or creamy beige. Surface smooth, slightly viscid to dry, shiny, occasionally cracking in age or when dry. **GILLS:** Broadly attached, close to subdistant, broad. Yellowish buff when young, darkening to brown as spores mature, to dark brown in age. **STIPE:** 2–5 cm long, 0.1–0.5 cm thick, equal or enlarged slightly toward base. Whitish as first, darkening to yellowish buff. Surface dry, finely powdery when young, often smooth, fibers appearing twisted in age. **FLESH:** Thin, fibrous in stipe, whitish to pale buff. **ODOR:** Earthy-farinaceous. **TASTE:** Mild to farinaceous. **SPORE DEPOSIT:** Dull brown. **MICROSCOPY:** Spores 10.5–14 x 6.5–8 μm, broadly ellipsoid in face view, flattened on one side in side view, smooth, thick walled with an apical germ pore, brown. Basidia 4-spored.

ECOLOGY: Solitary to scattered in grassy habitats, often in lawns or fields, occasionally in areas of nutrient-poor soil, but always with grass present. Common throughout our range, especially in urban areas. Typically fruiting in spring, but scattered fruiting can occur throughout year.

EDIBILITY: Nontoxic.

COMMENTS: The small size, hemispherical and often shiny, golden brown to creamy beige cap, brown mature gills, and growth in grass help identify this common lawn mushroom. That being said, there are probably a few undescribed species passing under the name *A. pediades* in California. *A. semiorbicularis* is very similar, but is often slightly larger and darker ocher-brown when young; microscopically it differs by having mostly 2-spored basidia. It usually grows in nutrient-rich areas, often near herbivore dung. Another similar (apparently undescribed) *Agrocybe* is common on the roots of dune grass (*Ammophila* spp.) in spring on the Far North Coast. It is similarly colored, but is larger and stockier in stature. *Protostropharia* are similar in stature, but have viscid caps and stipes, purplish black spores, and grow on dung. Small specimens of *Leratiomyces percevalii* could also be mistaken for this *Agrocybe,* but they have grayish mature gills and purplish black spores.

Tubaria furfuracea

(Pers.) Gillet

FRINGED TUBARIA

CAP: 1–4 (5) cm across, convex when young, soon broadly convex to plane, often with a depressed center and wavy margin in age. Cap strongly hygrophanous, thus color quite variable: brown, reddish brown to cinnamon brown when wet, light brown, tan to whitish beige when dry. Surface with fine silky fibers when young, often with a ring of white veil flecks around margin, bare in age. Moist to dry, smooth, translucent-striate when wet. **GILLS:** Broadly attached, close to subdistant, moderately broad. Light tan to cinnamon brown. **STIPE:** 1–4 (6) cm long, 0.1–0.4 cm thick, equal or slightly enlarged toward base. Whitish cream to tan with a silky white sheen and white basal mycelium. Surface dry, often silky. **PARTIAL VEIL:** Faint, silky, generally leaving flecks of white tissue on cap margin, rarely a slight ring on stipe. **FLESH:** Very thin and fragile. Light brown to buff. **ODOR:** Indistinct. **TASTE:** Indistinct. **SPORE DEPOSIT:** Pale brown to cinnamon brown. **MICROSCOPY:** Spores 6.5–9.5 x 4–5.5 μm, ellipsoid to nearly cylindrical, smooth, light yellow, without a germ pore.

ECOLOGY: Solitary, in clusters, in large troops, generally in urban or disturbed areas, especially on wood chips, disturbed soil, or overgrown grassy areas. Very common throughout our range. Fruiting throughout wet season, primarily in fall, with scattered fruitings through winter and larger flushes in wet springs.

EDIBILITY: Nontoxic.

COMMENTS: This rather nondescript mushroom can be recognized by its translucent-striate cap that fades dramatically as it dries, whitish veil flecks around the margin, and cinnamon brown gills. *Tubaria confragosa* (in the California sense) is very similar, but is slightly darker reddish brown when young and has a more distinct ring on the stipe. *T. tenuis* (=*T. hiemalis*, sensu CA?) has a slightly viscid, translucent-striate, ocher-orange cap and strongly capitate cheilocystidia. It is quite common on small branches and twigs of live oak. *Galerina marginata* has a hemispherical to convex cap, a membranous partial veil when young, and rusty orange spores. *Meottomyces* species are similar but quite rare in California; they have browner fruitbodies and subdecurrent gills.

Tubaria punicea

(A. H. Sm. & Hesler) Ammirati, Matheny & P.-A. Moreau

MAROON MADRONE TUBARIA

CAP: 1–6 cm across, convex at first, becoming broadly convex to nearly plane, often with a slight umbo, margin becoming wavy in age. Deep red to burgundy or maroon when wet, duller when dry, developing some brownish tones in age. Surface finely fibrillose to smooth, moist to dry, hygrophanous, translucent-striate around margin when wet. **GILLS:** Broadly attached to slightly decurrent, close, broad. Deep wine red to burgundy when young, becoming brownish maroon in age. **STIPE:** 2–8 cm long 0.2–0.7 cm thick, equal or slightly enlarged at base. Wine red to pale orangish red with a whitish, silky fibrous sheen, and white basal mycelium. Surface dry, often with silky longitudinal fibers. **PARTIAL VEIL:** Silky, forming a faint whitish ring when young, disappearing completely in age. **FLESH:** Thin, fibrous, brittle. Reddish brown. **KOH:** No reaction. **ODOR:** Indistinct. **TASTE:** Indistinct. **SPORE DEPOSIT:** Pale brown to cinnamon brown. **MICROSCOPY:** Spores 6.5–9.5 x 4–5.5 μm, bean shaped to ellipsoid, smooth, thick walled without a germ pore, cinnamon.

ECOLOGY: Solitary, or in scattered groups or clusters on Madrone wood. Almost always fruiting on and in rotted-out bases of living or recently dead trees. Fairly common, but hard to see and often overlooked. Fruiting primarily in fall, also in smaller numbers well into spring

EDIBILITY: Nontoxic.

COMMENTS: The burgundy colors and growth on Madrone are usually enough to identify this striking *Tubaria*. It is common, but can be tough to find unless one makes an effort to closely inspect the inner parts of rotted-out Madrone trunks. *T. vinicolor* is practically indistinguishable except by habitat—it grows in urban areas, usually in gardens or landscaped areas, especially on the rotted trunks of nonnative trees. *Cortinarius neosanguineus* is also dark red, but usually grows on the ground under conifers and has a taller, thinner stature, ocher-yellow basal mycelium, and a purple KOH reaction. Red-colored *Hygrocybe* grow in duff or on the ground and have waxy gills and white spore deposits.

SYNONYM: *Pholiota punicea* A. H. Sm. & Hesler.

Flammulaster rhombosporus

(G. F. Atk.) Watling

CAP: 0.5–2 cm across, rounded with an incurved margin at first, soon convex to broadly convex. Beige to tan at first, staying so, or darkening slightly to buff or light brown. Surface dry, covered with granular powder or warts with interspersed, finely floccose hairs, often nearly smooth in age. **GILLS:** Broadly to narrowly attached, close to subdistant, broad, edges often fringed. Buff when young, to light cinnamon brown. **STIPE:** 1–4 cm long, 0.1–0.2 cm thick, equal. Buff to cinnamon brown. Surface dry, covered with granules and silky-floccose hairs, apex very finely powdery. **PARTIAL VEIL:** Thin, scruffy-floccose tissue when very young. **FLESH:** Very thin and fragile. **ODOR:** Indistinct. **TASTE:** Indistinct. **SPORE DEPOSIT:** Pale yellow-brown to brown. **MICROSCOPY:** Spores (6) 7–9 x 4–6 µm, ellipsoid to irregularly lemon shaped, smooth, thin walled, without a germ pore, nearly colorless. Cap surface with clumps of large, round encrusted cells.

ECOLOGY: Solitary or scattered on duff under hardwoods, often eucalyptus and Tanoak, especially on old, rotting catkins and acorn cups of Tanoak. Common, but often overlooked. Fruiting in wet periods from fall into spring.

EDIBILITY: Unknown.

COMMENTS: At first glance, this is just another little brown mushroom, but when seen close up, the young fruitbodies can actually be quite stunning: the granular powder on the cap interspersed with the silky white hairs makes a lovely combination. *Tubaria conspersa* is similar, but generally slightly darker pinkish tan (when wet) and lacks the truly granular coating on the cap (but it does have floccose clumps of fine hairs). *F. muricatus* is also similar, but has more ocher-brown to ocher-buff colors, and grows on very rotten wood.

Flammulaster muricatus

(Fr.) Watling

CAP: 0.5–2 cm across, rounded with an incurved margin at first, soon convex to broadly convex. Ocher-orange, with slightly darker granular powder at first, slowly fading to cinnamon, then pale ocher-buff in age. Surface dry, covered with granular to finely floccose powder, becoming less apparent in age, to nearly smooth. **GILLS:** Broadly to narrowly attached, close, broad, partial gills numerous, edges often fringed. Creamy buff when young, becoming pale ocher-buff to pale cinnamon brown. **STIPE:** 1–3 cm long, 0.1–0.3 cm thick, equal. Ocher-orange to ocher-buff. Surface dry, covered with scruffy-floccose powder. **PARTIAL VEIL:** Thin, scruffy-floccose tissue, more like a continuation of cap surface, leaving jagged remnants on cap margin. **FLESH:** Very thin and fragile, more fibrous in stipe. **ODOR:** Indistinct. **TASTE:** Indistinct. **SPORE DEPOSIT:** Rusty brown to dark brown. **MICROSCOPY:** Spores (6) 7–9 x 4–5 µm, ellipsoid to bean shaped, smooth, thin to moderately thick walled, with a germ pore, rusty orange. Cheilocystidia cylindrical with a wider apex.

ECOLOGY: Solitary or scattered on or inside decaying hardwood logs and stumps, especially those of Tanoak. Fruiting in wet periods from fall into spring. Fairly common, but often overlooked due to small size and habit of fruiting inside rotten stumps.

EDIBILITY: Unknown.

COMMENTS: The ocher-orange to buff-tan colors, granular to finely floccose powdery covering on the cap and stipe, and growth on rotting wood make this mushroom fairly easy to recognize. *Phaeomarasmius erinaceus* is darker and scaly or hairy (without a powdery granular coating) and often grows on small dead branches of live oak. *F. rhombosporus* is much paler, with more drab colors and a whitish, granular to finely floccose covering; it often grows in duff under eucalyptus or Tanoak. *Tubaria conspersa* is a common species in the northern part of our range; it has pale orangish buff to pinkish beige colors and fibrillose-scruffy (rather than granular) surfaces. It generally grows on small woody debris on the ground, not on logs.

Phaeomarasmius erinaceus

(Fr.) Romagn.

CAP: 0.5–2 cm across, rounded with an incurved margin at first, soon convex to broadly convex. Dark brown or dark reddish brown to orange-brown. Surface dry, covered with hairy scales, becoming appressed-scaly in age. **GILLS:** Broadly attached, occasionally with a decurrent tooth, well spaced, broad, edges often fringed. Creamy buff when young, brown with a paler edge in age. **STIPE:** 0.5–3 cm long, 0.1–0.3 cm thick, equal, often curved. Orange-brown with a whitish apex at first, darkening to blackish brown in age with a pale apex. Surface dry, covered with fuzzy hairs or fine scales over lower part, appressed-scaly in age. **PARTIAL VEIL:** Absent or obscure, spiky annular zone on stipe. **FLESH:** Very thin, somewhat tough, brownish. **ODOR:** Indistinct. **TASTE:** Indistinct. **SPORE DEPOSIT:** Brown. **MICROSCOPY:** Spores 9–10 x 6–6.5 μm, ellipsoid or slightly rhomboid, smooth, without a germ pore. Cheilocystidia highly contorted, irregular, knobby. Pileipellis with clamps, cells strongly incrusted.

ECOLOGY: Solitary or scattered on small branches of hardwoods, especially live oak, usually fruiting on dead branches still attached to tree. Uncommon, fruiting throughout wet season, but more common in late winter and spring.

EDIBILITY: Unknown.

COMMENTS: The small size, dark colors, scaly cap and stipe, and growth on hardwood branches are helpful identifying features of this uncommon mushroom. *P. rimulincola* is a tiny species (under 1 cm) with a pinkish brown, sulcate cap with scattered coarse hairs and widely spaced gills. It is common in the Sierra Foothills on the bark of Tanoak and Black Oak, but probably occurs on the coast as well. A similar *Pleuroflammula* species grows on the bark of oaks and eucalyptus; it is slightly paler and much more golden brown, with a very short, off-center to nearly lateral stipe. *Flammulaster muricatus* has warmer ocher-brown to ocher-buff colors and a granular to finely floccose powdery coating on the cap and stipe.

Simocybe centunculus group

CAP: 0.5–3 cm across, rounded or bell shaped when young, soon convex, sometimes umbonate, nearly flat in age. Dark hunter green or dark olive-brown, soon paler to olive drab or brownish tan, almost always with a pale sheen. Margin narrowly whitish when very young, extensively beige or ocher in age, often becoming translucent-striate when wet. Surface dry, at first distinctly finely velvety, but can be nearly smooth in age. **GILLS:** Finely attached, broad. Dull olive to pale brownish tan, sometimes dark brown in age, margin fringed with white beads, sometimes also with clear, milky, or brown droplets when young. **STIPE:** 1–6 cm long, 0.1–0.4 cm thick, cylindrical, often curved, slightly enlarged toward base. Olive to grayish tan or yellowish olive, base often with a conspicuous tuft of white mycelium. Surface entirely covered with tiny pale scurfs, smoother and often vertically striate in age. **FLESH:** Very thin, fibrous, brownish. **KOH:** Reddish or orange on cap. **ODOR:** Indistinct. **TASTE:** Indistinct or slightly bitter. **SPORE DEPOSIT:** Dull brown. **MICROSCOPY:** Spores 7–8 x 5.5–6 μm, bean shaped or broadly ellipsoid, lacking a germ pore. Cheilocystidia cylindrical, up to 60 μm long. Pileipellis with many upright urn- or bottle-shaped cells.

ECOLOGY: Solitary, scattered in small troops or clusters on decaying hardwood branches and logs, especially oak and Tanoak. Fruiting from early fall through early spring. Common on the Central Coast and northern inland zones, uncommon on the Far North Coast.

EDIBILITY: Unknown.

COMMENTS: *Simocybe* are recognized by their small size, finely velvety cap, olive colors, pale gill edges, and growth on wood. *Callistosporium luteo-olivaceum* also has olive colors, but is usually larger, has a smooth cap, and has white spores; it grows on conifer wood. Many other small brown-spored mushrooms share this habitat and fruit alongside this species, but have differently textured caps (more granular or smoother), lack the olive tones, have a ring on the stipe, and/or have less contrasting gill edges. *Simocybe centunculus* has been the catchall name applied to an unknown number of species in California. The species that produces tiny fruitbodies with distinctly translucent-striate caps is closer to the European concept of *S. haustellaria*. The larger and more common *S. sumptuosa* is often dingier olive-brown, with cheilocystidia that have broad, round heads and tapered bases.

Pholiotina rugosa

(Peck) Singer

DEADLY PHOLIOTINA

CAP: 0.5–3 cm across, conical to bell shaped at first, expanding to broadly convex or plane, margin occasionally slightly uplifted in age. Reddish brown, ocher-brown to orange at first, with a paler margin. Hygrophanous, fading to ocher-buff or creamy when dry. Translucent-striate when wet, opaque when dry. Surface wrinkled to smooth, moist to dry. **GILLS:** Narrowly attached, close. Buff when young, becoming cinnamon brown to brown as spores mature. **STIPE:** 1.5–5 (8) cm long, 0.1–0.4 cm thick, equal or slightly swollen toward base, rather tall and straight. Pallid or yellowish, with an orangey brown base when young, darkening upward in age. Surface dry, finely scruffy. **PARTIAL VEIL:** A rather thick, feltlike, flaring, circular skirt. Occasionally detaching and sliding down stipe, or falling off altogether. Whitish to cream colored, upper side often covered in rusty spores. **FLESH:** Very thin and fragile, pallid. **TASTE:** Not sampled. **ODOR:** Indistinct. **SPORE DEPOSIT:** Dark ocher-brown to chestnut brown. **MICROSCOPY:** Spores 8–12 x 4–6 μm, elongate-ellipsoid to almond shaped, smooth, thin to moderately thick walled, with a large germ pore, ocher-yellow to rusty orange.

ECOLOGY: Solitary, scattered, or in small clusters or troops in a wide variety of habitats. Often in wood chips or compost-rich soil, but also in duff or disturbed soil in woodland habitat, and commonly in soil of potted houseplants. Locally common on the Far North Coast, uncommon to occasional elsewhere. Fruiting can be found anytime during the rainy season, or throughout the entire year in cultivated areas.

EDIBILITY: Extremely poisonous and potentially deadly, due to the presence of amatoxins (the same toxins in the Death Cap, *Amanita phalloides*).

COMMENTS: This small mushroom can be recognized by the hygrophanous reddish brown cap that fades to ocher-buff, and the disc-like ring on the stipe. It is part of a species complex, a handful of which are represented in California (all are distinguished microscopically). There are also a number of *Pholiotina* that have thinner veils that leave remnants only at the cap margin. Still other *Pholiotina* and similar *Conocybe* species lack a partial veil entirely. Careful microscopic examination and a dedicated monograph are needed for even a chance of identifying most of these little brown mushrooms.

SYNONYMS: *Conocybe rugosa* (Peck) Watling, *C. filaris* (Fr.) Kühner, *Pholiotina filaris* (Fr.) Singer.

Conocybe aurea

(Jul. Schäff.) Hongo

GOLDEN DUNCE CAP

CAP: 1–4 (6) cm across, conical to bell shaped, occasionally pointed, or broadly convex with an umbo; margin often flaring, sometimes uplifted in age. Dark ocher-orange to ocher-yellow at first, sometimes dark reddish ocher when wet. Hygrophanous, becoming light golden beige when dry. Slightly translucent-striate when wet, opaque when dry. Surface smooth, thinly viscid when young, soon dry. **GILLS:** Narrowly attached, close to crowded, broad. Creamy tan when young, very soon light cinnamon brown to rusty brown as spores mature. **STIPE:** 2–7 cm long, 0.1–0.5 cm thick, equal, or widening toward a small basal bulb. Buff to light ocher-brown, darkening in age. Surface dry, finely pruinose, smooth if handled. **FLESH:** Very thin and fragile, stipe hollow in age. **ODOR:** Indistinct. **TASTE:** Indistinct. **SPORE DEPOSIT:** Dark ocher-brown to chestnut brown. **MICROSCOPY:** Spores 10–13 x 6–7 µm, broadly ellipsoid to ovoid, smooth, thick walled, with a large germ pore, dark ocher-brown. Cheilocystidia and caulocystidia bowling pin shaped.

ECOLOGY: Solitary, scattered, or in troops in compost-rich soil or occasionally wood chips. Locally common in urban areas. Fruiting from fall into spring, occasionally in summer in watered gardens.

EDIBILITY: Unknown; do not experiment. Related species are known to be toxic.

COMMENTS: The large caps and relatively short stature (for a *Conocybe*), ocher-orange to golden brown cap, rusty brown gills, and growth in compost-rich soil are identifying features. This is one among a few species that often show up in houseplants or in gardens. *C. rickenii* shares the same habitat, but is generally smaller and light brown, tan, or beige. A whole suite of similar-looking species are found on dung: *C. magnispora* is relatively small (caps usually less than 1 cm wide), with orange-brown to yellow-brown caps when young, and large spores (13.5–20.5 x 7.5–11 µm). *C. macrospora* (and others) are similar dung dwellers, but generally have larger fruitbodies (caps greater than 1 cm wide) and differ microscopically (primarily by 2-spored basidia and larger spores). Another dung species, *C. antipus*, has a rooting stipe base up to 4 cm long. The brown to ocher-capped grassland species are frequently lumped under *C. tenera*, but keep in mind that there are a number of species in that complex, and nobody has seriously studied *Conocybe* in California. Some *Psathyrella* are similar, but have darker, blackish brown spore deposits.

Conocybe apala

(Fr.) Arnolds

WHITE DUNCE CAP

CAP: 1–3.5 (5.5) cm across, conical to bell shaped, margin often flaring in age. Whitish to creamy white, often developing beige to pale buff colors in age. Surface smooth, dry. **GILLS:** Narrowly attached to free, close to crowded, narrow. Whitish to beige at first, soon darkening to light rusty brown. **STIPE:** 4–11 cm long, 0.1–0.2 cm thick, cylindrical with a bulbous base. Whitish to cream. Surface dry, finely scruffy to smooth. **FLESH:** Very thin and fragile, stipe hollow. **ODOR:** Mild to slightly garlicky. **TASTE:** Indistinct. **SPORE DEPOSIT:** Light rusty brown. **MICROSCOPY:** Spores 10–14 x 7.5–8.5 µm, broadly ellipsoid to ovoid, smooth, thick walled, with large germ pore, ocher-brown.

ECOLOGY: Solitary, scattered, or in troops in grass. Almost always fruiting in summer months in watered lawns. Common in the southern half of our range, rarer northward.

EDIBILITY: Potentially dangerous. Some collections have been found to contain phallotoxins. (Although not usually implicated in *Amanita* poisonings because they are not absorbed through the human gut, phallotoxins are highly toxic to liver cells in vitro.)

COMMENTS: The white to pale buff, conical to bell-shaped cap, ocher-brown spores, and growth in grass (usually in the warmer months) are distinctive features. These mushrooms often come up during the night and wilt as soon as the morning dew dries up and the sun starts to shine. Another summer-fruiting species in the same habitats is *Bolbitius lacteus*, which has a whitish to pale grayish, conical to bell-shaped cap that becomes plane or uplifted in age. *B. titubans* has a viscid yellow cap when young, but can fade to whitish buff in age; it fruits from fall through spring, often in similar habitats as the other species mentioned here.

SYNONYMS: *Conocybe albipes* Hauskn., *Conocybe lactea* (J. E. Lange) Métrod.

Bolbitius aleuriatus (sensu CA)

CAP: 1–5 cm broad, egg shaped when young, soon round-conical, then broadly convex to plane with a low umbo. Light gray with a dark disc or dingy tan-gray, occasionally with a violet tone when young, soon pallid dingy gray-brown with a dark center, occasionally with limited yellowish hues. Hygrophanous, drying to pale beige-tan. Surface wrinkled-rivulose but covered with a smooth viscid layer (slime layer can be very thick when young), soon expanding to become pleated and translucent-striate. **GILLS:** Narrowly to broadly attached, seceding in age. Pale gray-tan when young, soon with an ocher wash as spores begin to mature, to brighter ocher-brown to dull warm brown in age, edges paler, beaded with a whitish band of cheilocystidia. **STIPE:** 3–10 cm long, 0.3–0.8 cm thick, cylindrical, slightly enlarged toward base. Whitish to yellow, often brighter yellow near base, scales distinctly paler whitish yellow. Surface covered in small scurfy scales (fine and punctate near apex, forming coarser chevrons near base). **FLESH:** Extremely thin, fragile, hollow in stipe. **ODOR:** Indistinct. **TASTE:** Indistinct. **SPORE DEPOSIT:** Ocher-brown. **MICROSCOPY:** Spores 9–12 x 5–7 µm, ellipsoid to ovoid, smooth, thick walled with a germ pore, ocher-brown.

ECOLOGY: Solitary or in small troops on well-decayed hardwoods, especially Tanoak and live oak. Common on the Central Coast and in northern inland zones, uncommon on much of the conifer-dominated Far North Coast. Fruiting throughout mushroom season, most common in fall and spring.

EDIBILITY: Nontoxic.

COMMENTS: The thin, fragile fruitbodies with grayish caps, contrasting yellowish white stipes, ocher-brown mature gills, and growth on well-decayed wood are quite distinctive. The ocher-brown spore deposit helps distinguish it from small *Pluteus* species that often share the same habitat. The Californian species is not conspecific with the "true" European *B. aleuriatus* (based on genetic data). Macroscopically the two are very similar, and use of the name in our area persists since an alternative for the local taxon has not been published.

MISAPPLIED NAME: *Bolbitius reticulatus* (Pers.) Ricken.

Bolbitius titubans

(Bull.) Fr.

SUNNY SIDE UP

CAP: 0.7–2.5 cm wide when young, 2–7 (10) cm when expanded. At first egg shaped or cylindrical, soon expanding to round-conical, broadly convex, finally flat, margin soon translucent-striate, often pleated in age, sometimes curling inward. Color variable, bright lemon yellow to deep golden yellow when young, fading outward to pallid dingy grayish yellow or ocher, to dingy cream or whitish. Often olive or with orange to red tones in var. *olivaceus.* Surface viscid at first, sometimes becoming dry, faintly translucent-striate when wet, often sulcate in age. In *B. titubans* var. *olivaceus,* strongly netted at center of cap. **GILLS:** Narrowly attached, often seceding, rather close, edges occasionally fringed with clusters of cheilocystidia. Pallid creamy yellow when young, becoming ocher-brown as spores mature. **STIPE:** 3–15 cm long, 0.3–1.5 cm thick, cylindrical, straight or curved, thickest toward base. Creamy whitish to pallid yellow, covered with fine whitish flocculose tufts of caulocystidia, coarsest toward base. **FLESH:** Very thin and fragile, stipe hollow. **ODOR:** Indistinct. **TASTE:** Indistinct. **SPORE DEPOSIT:** Ocher-brown. **MICROSCOPY:** Spores 12–13 (15) x 6–7 μm, ellipsoid, smooth, thick walled with a germ pore, ocher-brown.

ECOLOGY: Solitary, scattered, or clustered in grassy areas, straw, compost piles, or wood chips, or occasionally on dung. Common, occurring throughout our range. Fruiting after wet periods from fall until spring.

EDIBILITY: Nontoxic.

COMMENTS: The fragile fruitbodies of this species are immediately recognizable by their bright yellow, viscid caps, pale stipes, and preference for nutrient-rich, disturbed, or urban settings. Fruitbodies growing in grass can be very slender, thin fleshed, and flimsy, while those in dung-rich habitat or in wood chips are often more robust and stocky. *B. aleuriatus* grows on well-decayed oak and Tanoak wood in forests and has a grayish cap. The rarer *B. coprophilus* has a pinkish cap that becomes grayish; it is found primarily on dung. *B. lacteus* shares a similar stature, but has a whitish cap and grows in irrigated lawns in the summer. *Conocybe apala* is very similar to the last species in appearance and habitat, but has a dry, conical to bell-shaped cap.

SYNONYM: *Bolbitius vitellinus* (Pers.) Fr.

7 • Mycorrhizal Brown-Spored Mushrooms

This group contains some of the most difficult-to-identify and anonymous-looking mushrooms (*Inocybe*), as well as some of the most interesting and characteristic mushrooms of the California conifer rainforest (*Phaeocollybia*). All have dull brownish spore deposits and are mycorrhizal. Although *Cortinarius* are also mycorrhizal, they are treated in a separate section because of their distinctly rusty spore deposits, cobwebby cortina (partial veil), and exceptional diversity.

Alnicola—Fruitbodies small, drab colored (browns, tans, beiges) with hygrophanous caps, faint cortina-like partial veil and medium brown spores. Always with alder, which typically inhabits wet areas.

Inocybe—Fruitbodies usually slender and small, but can be quite tall (a few species are more robust or squat). Frequently with distinctive odors, most commonly spermatic or like green corn. The caps are usually distinctly radially fibrillose, silky, or finely hairy-scaly. The stipe surfaces often have parallel lines of pale cystidia, and there is sometimes a small bulb at the base of the stipe.

Hebeloma—Fruitbodies usually medium size to large, but a few species are small. The caps are always smooth and often viscid when fresh, but can dry out and appear shiny. The frequently radishlike odor and gills often adorned with clear, milky, or brownish droplets when young are features highly suggestive of this genus.

Phaeocollybia—The distinctive cartilaginous stipe that usually tapers strongly and roots deeply into the soil is very distinctive among brown-spored mushrooms. The caps are often viscid, and the fruitbodies of many species are rather colorful. Most are restricted to older-growth conifer forests, and some are rare and likely endangered.

Alnicola escharioides group
BROWN ALDER MUSHROOM

CAP: 1–4 cm across, convex at first, becoming broadly convex to plane, margin occasionally uplifted, wavy and splitting in age. Dark to light brown, beige with an olive cast, or orangish brown, hygrophanous, fading when dry. Surface moist to dry, silky looking when young, appressed-fibrillose to smooth in age. **GILLS:** Broadly attached or with a decurrent tooth, close to subdistant. Beige to brown. **STIPE:** 1.5–4 cm long, 0.1–0.4 cm thick, equal or slightly enlarged and often twisted or curved toward base. Light brown to ocher-brown, darkening from base up in age. **PARTIAL VEIL:** Fleeting, silky cortina leaving irregular chevrons or patches on stipe (when young). **FLESH:** Concolorous, thin, fragile, fibrous in stipe. **ODOR:** Indistinct. **TASTE:** Mild to slightly bitter. **SPORE DEPOSIT:** Brown. **MICROSCOPY:** Spores 9.5–13 x 5–7 μm, almond shaped to broadly ellipsoid, finely roughened, inamyloid, yellow-brown.

ECOLOGY: Solitary, scattered, or in small clusters on ground. Mycorrhizal with alder trees, which are usually found in swampy areas and riparian zones. Very common in appropriate habitat throughout our range. Fruiting in fall or early winter in northern areas, more common in spring southward.

EDIBILITY: Unknown; do not experiment.

COMMENTS: This group of drab little brown mushrooms can be recognized by their small size, relatively smooth caps, thin and often twisted stipes with silky veil remnants, and growth under alder. There appears to be a large number of species in California going by too few (European) names; this group is in serious need of taxonomic work. A number of *Inocybe* species are similar, but have a different "look"—generally with more radially fibrillose or with scaly caps and stipes, and/or a stockier stature. Similar small *Cortinarius* generally have a strongly hygrophanous cap and a more prominent cobwebby veil when young.

SYNONYM: *Naucoria escharioides* (Fr.) P. Kumm.

Inocybe hirsuta var. *maxima*
A. H. Sm.
LARGE SCALY FIBERHEAD

CAP: 2–8 (11) cm across, conical to bell shaped at first, becoming broadly convex to nearly plane, often with a low, broad umbo. Margin incurved and tomentose at first, becoming wavy, uplifted, and often splitting in age. Brown to reddish brown, covered with reddish brown scales, darkening in age. Staining red when damaged and occasionally showing bluish gray stains. Surface dry, covered with fibrillose tufts and scales. **GILLS:** Narrowly to broadly attached, close, narrow. Grayish at first, becoming grayish buff to brown. **STIPE:** 4–15 cm long, 0.4–2.5 cm thick, more or less equal or with a swollen base. Pallid at first, soon light brown to grayish brown, often with reddish brown scales, and a bluish lower portion. Staining red when damaged. Surface dry, scaly, fibrous with a pruinose apex. **PARTIAL VEIL:** Very scant or absent. **FLESH:** Moderately thin to thick, fibrous, somewhat fragile in cap, tough in stipe. Whitish to light brown in cap, whitish in stipe, with some bluish gray around base; slowly staining red when cut. **ODOR:** Mild to fishy when young, strong, disagreeable, like rotting fish in age. **TASTE:** Indistinct. **SPORE DEPOSIT:** Brown. **MICROSCOPY:** Spores 8–12 x 5–6 μm, elliptical to ovoid, smooth, thick walled, brownish. Cheilocystidia subcylindric to club shaped, frequently with rounded heads. Pleurocystidia absent.

ECOLOGY: Solitary, scattered, or in troops in duff or moss under conifers, especially Sitka Spruce. Very common on the Far North Coast, rare elsewhere. Fruiting from midfall into winter.

EDIBILITY: Unknown, possibly poisonous.

COMMENTS: This large, scaly *Inocybe* can be easily recognized by the reddish brown cap, pallid to brown gills, bluish gray stipe base, fishy odor, and red staining when damaged. *I. calamistrata* (sensu CA) is generally much smaller and has a tall, slender stature, a dark brown cap, and an odor like the skin of fresh fish. It's also common on the Far North Coast, usually growing in wet areas with alder. *I. lanuginosa* also has a dry, scaly brown cap, but lacks the bluish stipe base, often grows on well-rotted wood, and lacks a distinctive odor.

Inocybe griseolilacina

J. E. Lange

LILAC-STEMMED FIBERHEAD

CAP: 2–5 cm across, rounded-conical with an incurved margin at first, soon bell shaped to convex, expanding to nearly flat, sometimes with a low umbo. Pale tan-brown to brown or dull red-brown at center, paler toward edge, nearly white or lilac at margin when young from adhering veil tissue. Surface dry, shaggy-scaly with radially arranged flattened fibrillose scales. **GILLS:** Narrowly attached or notched, close, narrow. Pale whitish gray with a lilac wash when young, dingy tan to brown in age; often with irregular, contrasting lighter orange zones or bands, Edge contrastingly pale, finely beaded. **STIPE:** 5–9 cm long, 0.5–1.3 cm thick, cylindrical or with a slightly enlarged base. Silvery whitish with lilac to purple tones (especially over lower half) when young, soon pinkish tan to brownish with fine, pale shaggy chevrons. **PARTIAL VEIL:** Relatively pronounced whitish to lilac-purple cortina, usually adhering to edge of cap and not forming ring zone on stipe. **FLESH:** Fibrous, whitish. **ODOR:** Faintly spermatic or mild. **TASTE:** Indistinct. **SPORE DEPOSIT:** Dull brown. **MICROSCOPY:** Spores 7–9.5 x 4.5–6 µm, elliptical or bean shaped, smooth, thick walled. Pleurocystidia abundant, variable in shape, capitate to bottle shaped, without crystals. Cheilocystidia narrowly club shaped, thin walled, with small crystals encrusting tips.

ECOLOGY: Scattered in small troops on ground under conifers, especially Monterey Pine in the late fall and winter. Locally common in appropriate habitat.

EDIBILITY: Likely toxic. Many *Inocybe* contain muscarine.

COMMENTS: This *Inocybe* is distinct from most other species by virtue of the scaly cap texture, silvery gray stipe with purple tones, and growth with pine. *I. pusio* is quite similar and also grows with pine, but has a slightly smoother cap and purple tones primarily at the apex of the stipe, which has a more powdered appearance. *I. lilacina* is more evenly lilac overall when young (paler to whitish lilac in age) and has a smoother cap surface.

Inocybe lanatodisca

Kauffman

CAP: 4–10 cm across, bell shaped to convex, expanding to nearly flat, usually with a distinct rounded umbo. Yellowish ocher to orangey tan with brown radial fibers, creamy whitish near margin when young; center of cap often with patches of white cottony material. Surface moist to dry, innately radially fibrillose, edge often splitting or eroded. **GILLS:** Very finely attached, fairly close, edges irregular to eroded. Whitish cream when young, becoming dingy tan to light brown. **STIPE:** 6–12 cm long, 0.6–2 cm thick, cylindrical, base often enlarged and curved. White to pale creamy or light tan. Surface smooth (lacking a bloom or lines of pale cystidia). **PARTIAL VEIL:** Very scant or absent. **FLESH:** Stringy, white, solid throughout but fairly fragile. **ODOR:** Variably floral, spermatic, or like green corn. **TASTE:** Indistinct. **SPORE DEPOSIT:** Dull brown. **MICROSCOPY:** Spores 8.0–11.5 x 4.5–5.5 µm, elliptical or bean shaped. Cheilocystidia variable in size and shape, tips rounded and without crystal encrustations. Pleurocystidia absent.

ECOLOGY: Solitary, in small clumps, or scattered in troops in duff. On the Redwood Coast we have only seen it with willow (*Salix* spp.) and possibly cottonwood, but it is known to occur with other hardwoods and even conifers outside our range. Fruiting from fall into winter.

EDIBILITY: Unknown, likely toxic due to presence of muscarine.

COMMENTS: The yellowish to ocher-tan cap with thin, cottony patches in combination with the pale, rather smooth stipe and growth with willow makes this *Inocybe* somewhat distinctive. Since its frequency and host range is unknown in California, definitive identification should be made with microscopic data.

Inocybe "Citrine Gills"

CAP: 1–3 (5) cm across, conical at first, soon convex to nearly plane, with low umbo. Yellow-brown, ocher-brown to reddish brown, often with a darker center. Surface dry, covered with appressed fibrils radiating out from center, occasionally lifting up into small scales. **GILLS:** Narrowly attached with a deep notch, or free in age, close, narrow. Citrine yellow when young, becoming yellowish to dingy yellow-brown as spores mature. **STIPE:** 4–8 cm long, 0.3–0.8 cm thick, equal or enlarged toward base. Yellowish at first, developing ocher stains in age. Surface dry, pruinose over entire length, often with sparse fibrils in age. **PARTIAL VEIL:** Indistinct, consisting of sparse wispy hairs adhering to cap margin, disappearing soon after button stage. **FLESH:** Thin, fragile in cap, fibrous in stipe, yellowish. **ODOR:** Mildly spermatic. **TASTE:** Mild to unpleasant. **SPORE DEPOSIT:** Dull brown in mass, to grayish brown from prints. **MICROSCOPY:** Spores (7.5) 9–11.5 x 5–6 (7) µm, ovoid to broadly elliptical, smooth, thick walled, inamyloid, brownish. Cheilocystidia abundant, spindle shaped, 65–85 µm long, moderately thick walled, ends with encrusted crystals. Pleurocystidia similar.

ECOLOGY: Solitary, scattered, or in troops in moss or duff of mixed Tanoak and Douglas-fir forest. Locally common, known from Santa Cruz County to northern Mendocino County, but probably occurs farther north. Fruiting from late fall into spring.

EDIBILITY: Toxic.

COMMENTS: The brown, slightly shaggy cap, slender, silky-pruinose stipe, and brown spores identify it as an *Inocybe*, but the citrine-yellow gills set it apart. The gills of *I. cinnamomea* are also bright, but are much more orange cinnamon. A few small *Cortinarius* have yellow gills, but they have smoother caps and rusty brown spores.

Inocybe cinnamomea

A. H. Sm.

CINNAMON FIBERHEAD

CAP: 0.8–2.5 cm across, conical at first, soon convex to nearly plane, with a rounded to pointed umbo. Brown, ocher-brown to reddish brown, often but not always lighter toward margin. Surface moist to dry, covered with densely matted fibrils. **GILLS:** Narrowly attached with a deep notch or free in age; close, narrow. Cinnamon orange when young, to cinnamon brown in age. **STIPE:** 4–8 cm long, 0.2–0.6 cm thick, equal or enlarged toward base. Cinnamon base color, covered with whitish silky fibrils, darkening when handled and in age. Surface dry, silky-flocculent to finely tomentose with a pruinose apex. Often becoming smooth, or nearly so in age. **PARTIAL VEIL:** Cortina-like, consisting of a few sparse silky hairs, disappearing soon after button stage. **FLESH:** Thin, fragile, fibrous, pale cinnamon. **ODOR:** Indistinct. **TASTE:** Unpleasant. **SPORE DEPOSIT:** Dull brown in mass; grayish brown from sparse prints. **MICROSCOPY:** Spores 7–9 x 4.5–5 µm, ovoid to broadly elliptical, smooth, thick walled, inamyloid, brownish. Cheilocystidia club shaped, thin walled.

ECOLOGY: Solitary, scattered, or in troops in moss or duff under conifers, especially Sitka Spruce and Douglas-fir. Common on the Far North Coast, locally common elsewhere, occurring throughout our range. Fruiting from late fall well into winter or early spring.

EDIBILITY: Toxic.

COMMENTS: The bright orange cinnamon gills are unusual for an *Inocybe*. The matted-hairy cap and dull brown spore deposit help distinguish it from the orange-gilled *Cortinarius cinnamomeus* group, which have rusty brown spores and smoother caps. *Inocybe* "Citrine Gills" has a brown to golden brown cap, citrine-yellow gills, and a finely powdery stipe.

Inocybe albodisca group
WHITE-DISC FIBERHEAD

CAP: 1.5–2.5 (3.5) cm across, conical at first, becoming broadly conical to nearly plane, with a rounded to pointed umbo. Straw brown, pale ocher-brown to brown, with a whitish center, and often a slightly paler margin, and whitish fibrils streaked over cap. Surface dry, with appressed fibrils radiating from center, occasionally lifting up into small scales. **GILLS:** Narrowly attached with a deep notch to free, close, narrow, often ragged in age. Pallid gray to grayish beige, darkening to grayish brown to brown as spores mature. **STIPE:** 3–6 cm long, 0.3–0.5 cm thick, equal with an abrupt rounded bulb at base. Whitish to beige, developing orangish stains in age. Surface dry, pruinose over upper portion when young, to smooth in age. **PARTIAL VEIL:** Sparse, cortina-like. Fringing the cap margin when young, soon disappearing. **FLESH:** Whitish to pale buff, moderately thin, fibrous, but fragile in cap. Fibrous, somewhat tough in stipe. **ODOR:** Strongly spermatic. **TASTE:** Unpleasant. **SPORE DEPOSIT:** Dark brown in mass, more grayish brown from prints. **MICROSCOPY:** Spores 6–8 x 5–6 µm, irregularly broadly ellipsoid, obscurely nodulose (bumpy), thick walled, inamyloid, brownish. Pleurocystidia and cheilocystidia abundant, swollen, thick walled.

ECOLOGY: Solitary, scattered, or in troops on ground or in duff under oaks. Common and widespread. Fruiting from fall into early winter.

EDIBILITY: Toxic.

COMMENTS: This species can be recognized by the distinctly white center to the cap, pallid gills, powdery stipe apex, and bulb at the base of the stipe. *I. brunnescens* is generally larger, has a browner cap with a whitish patch of tissue over the center, lacks a distinct bulbous base to the stipe, and has a mild odor and taste, and larger elliptical spores. It grows with oaks, especially in the southern part of our area. *I. grammata* (sensu western NA) has a pinkish to lilac grayish cap with a white center, pallid gills, and a spermatic odor, and grows with conifers. *I. corydalina* has a brownish cap with dingy greenish blue blotches.

Inocybe fuscodisca
(Peck) Massee
DARK-EYED FIBERHEAD

CAP: 1.5–2.5 cm across, conical at first, becoming broadly conical to nearly plane, with a rounded umbo. Umbo dark brown, with darker appressed fibrils radiating over a grayish brown to pale gray base color, often creaking in age, exposing the whitish flesh. Surface moist to dry, covered with appressed fibrils. **GILLS:** Narrowly attached with a deep notch, or free in age, close, narrow. Whitish gray at first, to grayish buff, then to gray-brown as spores mature. **STIPE:** 3–8 cm long, 0.2–0.4 cm thick, equal or enlarged lower, often with a swollen bulb at base. Whitish base color, sheathed with dark brown fibrils and scales to veil, whitish above. Darkening to grayish or dingy grayish brown in age. Surface dry, pruinose above veil, appressed-fibrillose below. **PARTIAL VEIL:** Cortina-like, leaving sparse, silky fibrils on stipe when young, disappearing altogether in age. **FLESH:** Thin, fibrous, whitish to pale buff. **ODOR:** Strongly spermatic. **TASTE:** Mild to slightly unpleasant. **SPORE DEPOSIT:** Dull brown in mass, more grayish brown from prints. **MICROSCOPY:** Spores 8–10 x 4.5–5.5 µm (or up to 13 µm long from 2-spored basidia), almond shaped, thick walled, inamyloid, brownish. Pleurocystidia swollen in middle, cylindrical or with a slightly tapering neck. Cheilocystidia like pleurocystidia or club shaped.

ECOLOGY: Solitary, scattered, or in troops under a wide variety of conifers. Common on the North Coast. Fruiting from early fall in the north, well into early winter in the south.

EDIBILITY: Toxic.

COMMENTS: This is one of our more distinctive *Inocybe*—the dark brown cap center and sheath of dark fibrils on the stipe are distinctive. Perhaps most similar is *Hygrophorus olivaceoalbus*, which has the same color scheme, but has a viscid cap, thick, waxy white gills, and white spores.

Inocybe geophylla group
WHITE FIBERHEAD

CAP: 1–3.5 (5) cm across, conical at first, soon convex to nearly plane, with a rounded to pointed umbo. White to creamy, becoming dingy white to pale grayish in age. Surface dry, silky-fibrillose to smooth. **GILLS:** Broadly to narrowly attached, close, narrow. Pallid when young, becoming grayish to grayish brown, edges often paler. **STIPE:** 2–7 cm long, 0.2–0.5 cm thick, equal or enlarged downward, occasionally with a slightly bulbous base. White to off-white. Surface dry, silky-fibrillose to smooth. **PARTIAL VEIL:** Cortina-like, relatively sparse, leaving silky hairs on stipe and cap margin, disappearing in age. **FLESH:** Thin, firm to fragile, fibrous, white. **ODOR:** Mildly to strongly spermatic. **TASTE:** Mild to unpleasant. **SPORE DEPOSIT:** Brown in mass, more grayish brown from prints. **MICROSCOPY:** Spores 7.5–9.5 x 4.5–5.5 μm, elliptical to ovoid, smooth, thick walled, brownish. Cheilocystidia and pleurocystidia abundant, narrowly spindle shaped to nearly cylindrical, with crystals at apex.

ECOLOGY: Solitary, scattered, or in small clusters and troops on ground, in moss or duff under conifers and hardwoods. Like many *Inocybe*, it has a preference for younger forest. Very common throughout our range. Fruiting from fall through winter.

EDIBILITY: Toxic.

COMMENTS: The dry, white, silky-fibrillose cap, grayish to brown gills, scant, cortina-like partial veil, and spermatic odor are distinctive. The *Inocybe geophylla* complex consists of a couple of undescribed species in California, as well as one described: *I. insinuata*, a chalk white species with a stockier stature found under pine and live oak (occasionally in clusters) throughout the southern half of our range. *I. lilacina* is very similar but has lilac colors (at least when young); it can fade to pale whitish or gray and is often confused with members of the *I. geophylla* group. Similar *Hebeloma* species usually have a slightly viscid cap and radishlike odor.

Inocybe lilacina (sensu CA)
LILAC FIBERHEAD

CAP: 1–5 cm across, conical at first, soon convex to nearly plane, with a rounded to pointed umbo. Color variable, often lilac-purple when young, more grayish lilac, with an ocher tone to the center; more grayish to dingy whitish in age. Surface dry, silky-fibrillose to smooth. **GILLS:** Broadly to narrowly attached, close, narrow. Pale lilac-gray when young, becoming grayish to grayish brown, edges often paler. **STIPE:** 2–7 cm long, 0.3–0.8 cm thick, equal or enlarged downward, occasionally with a slightly bulbous base. Pale lilac-gray with a whitish base. Surface dry, silky-fibrillose to smooth. **PARTIAL VEIL:** Cortina-like, relatively sparse, leaving silky hairs on stipe and cap margin, disappearing in age. **FLESH:** Thin to moderately thick, firm, fibrous, white to pale lilac. **ODOR:** Spermatic. **TASTE:** Unpleasant. **SPORE DEPOSIT:** Brown in mass, more grayish brown from prints. **MICROSCOPY:** Spores 7–9 x 4.5–5.5 μm, elliptical to ovoid, smooth, thick walled, inamyloid, brownish. Cheilocystidia and pleurocystidia abundant, fusoid-ventricose, thick walled.

ECOLOGY: Solitary, scattered, or in small to large troops on ground or in moss or duff. Usually found under conifers, especially Douglas-fir, occasionally in hardwood forest as well. Fairly common throughout our range. Fruiting from late fall through winter.

EDIBILITY: Toxic.

COMMENTS: The combination of lilac colors (most apparent when young), pallid to brown gills, and faint cortina is distinctive. Members of the *I. geophylla* group can be very similar, but are white when young; faded specimens of *I. lilacina* can be difficult to distinguish. Small, lilac-colored *Cortinarius* can be differentiated by their brighter rusty brown spores and lack of a spermatic odor. The western North American representative of this group is known to be genetically distinct from similar species in eastern North America and Europe, but has yet to be given its own name.

SYNONYM: *I. geophylla* var. *lilacina* (Peck) Gillet.

Inocybe pudica

Kühner

BLUSHING FIBERHEAD

CAP: 2–8 cm across, conical to bell shaped at first, becoming broadly conical to nearly plane, often with a rounded umbo. Whitish to pale pinkish, becoming blushed with pinkish salmon to peachy orange streaks, spots, and stains; often extensively dingy pinkish buff in age. Surface dry to moist, smooth at first, except for silky veil remnants on margin, soon with appressed fibrils radiating from the center, margin occasionally splitting in age. **GILLS:** Narrowly attached to free, close, narrow. Whitish to pale pinkish salmon when young, becoming grayish to grayish brown; staining pinkish salmon when damaged. **STIPE:** 2.5–8 cm long, 0.3–1 cm thick, equal or enlarged downward, not bulbous. Whitish to pale pinkish, staining or discoloring pinkish salmon to peachy orange. Surface dry, silky-flocculent to smooth. **PARTIAL VEIL:** Cortina-like, consisting of silky white fibrils, leaving remnants on cap margin and a few silky threads on stipe when young, disappearing in age. **FLESH:** Moderately thin to thick, fibrous but somewhat fragile in cap; fibrous, tough in stipe. Whitish, slowly staining pinkish salmon to peachy orange. **ODOR:** Spermatic. **TASTE:** Unpleasant. **SPORE DEPOSIT:** Brown. **MICROSCOPY:** Spores 8.5–10 x 5–5.6 μm, elliptical, thick walled, inamyloid, brownish. Pleurocystidia abundant to sparse, rather narrow, spindle shaped with crystals at apex, walls thick, yellow in color. Cheilocystidia abundant, similar to pleurocystidia, walls bright yellow.

ECOLOGY: Solitary, scattered, or in troops on ground or in duff under conifers. Very common and widespread. Fruiting in fall and winter, into spring in wet years.

EDIBILITY: Toxic.

COMMENTS: The pallid colors with pinkish orange staining and growth with conifers make this *Inocybe* easy to recognize. Members of the *I. geophylla* group also have white to off-white colors, but don't develop the pinkish blush.

NOMENCLATURE NOTE: Although the name *I. pudica* has been synonymized with *I. whitei*, the better-known "placeholder" name is used here, since the species described above is likely distinct from European and eastern North American taxa.

Inocybe corydalina (sensu CA)

BLUING FIBERHEAD

CAP: 3–7 cm across, conical to bell shaped at first, becoming broadly convex to nearly plane, often with a low, broad umbo. Whitish, blotchy light orangish brown to light brown at first, with irregular bluish to bluish gray patches, to more dingy brown in age. Surface dry to moist, with random appressed-fibrillose patches. **GILLS:** Narrowly attached, close, narrow. Whitish at first, becoming grayish buff to brown. **STIPE:** 3–10 cm long, 0.4–1 (1.5) cm thick, more or less equal or with a swollen base. Whitish at first, soon buff to grayish brown, often with bluish stains over lower portion. Surface dry, longitudinally striate and fibrous. **PARTIAL VEIL:** Scant, soon disappearing. **FLESH:** Moderately thin to thick, fibrous, somewhat fragile in cap, tough in stipe. Whitish to pale bluish gray, especially in stipe base. **ODOR:** Like cinnamon, spicy-musty or slightly unpleasant; similar to the Matsutake (*Tricholoma magnivelare*), but usually with other disagreeable components. **TASTE:** Indistinct. **SPORE DEPOSIT:** Brown. **MICROSCOPY:** Spores 7.5–11 x 5–6 μm, elliptical to almond shaped, smooth, thick walled, inamyloid, brownish. Pleurocystidia and cheilocystidia sparse, spindle shaped with crystals at apex, walls thick.

ECOLOGY: Solitary, scattered, or in troops on ground or in duff under live oak and deciduous oaks. Rare on the Redwood Coast, mostly occurring in drier woodlands. Common in southern California and under deciduous oaks in the Central Valley. Fruiting from late fall into spring.

EDIBILITY: Possibly poisonous. It is known to contain the hallucinogenic toxin psilocybin and lacks muscarine (the typical *Inocybe* toxin); however, its chemistry is still dubious, and it may contain other toxins. Many look-alike species contain muscarine and are very toxic.

COMMENTS: The large size (for an *Inocybe*), stocky stature, whitish to brownish cap that almost always shows a bluish gray patch, and Matsutake-like (cinnamon-spicy) odor are distinctive. *I. fraudans* has a similar odor, but lacks the bluish gray color and stains reddish when damaged. This species is known to be different from the true *I. corydalina* of Europe, but has yet to be given its own name.

Inocybe fraudans (sensu CA)
MATSUTAKE FIBERHEAD

CAP: 2–8 cm across, conical at first, becoming broadly convex to nearly plane, often with a broad umbo and becoming wavy in age. Whitish to light brown at first, soon with dingy ocher reddish streaks, spots, and stains, to dingy ocher-brown or brown in age. Surface dry to moist, with appressed-fibrillose patches. **GILLS:** Finely attached, close, narrow. Pallid, buff to light brown at first, brown to rusty brown as spores mature. **STIPE:** 3–10 cm long, 0.5–1 (1.5) cm thick, more or less equal or with a swollen base. Whitish to buff, often discoloring pale reddish brown, staining pinkish to reddish brown when damaged. Surface dry, slightly fibrous to smooth. **PARTIAL VEIL:** Cortina-like, consisting of silky white fibrils, leaving remnants on cap margin and a few silky threads on stipe when young, disappearing altogether in age. **FLESH:** Moderately thin to thick, fibrous, somewhat fragile in cap, tough in stipe. Whitish, slowly staining pinkish to reddish. **ODOR:** A mix of cinnamon spicy and musty; remarkably similar to the odor of Matsutake (*Tricholoma magnivelare*). **TASTE:** Indistinct or pleasant. **SPORE DEPOSIT:** Brown. **MICROSCOPY:** Spores 9–11.5 x 6–8 µm, almond shaped, smooth, thick walled, inamyloid, brownish. Pleurocystidia and cheilocystidia sparse, spindle shaped, thick walled, with crystals at the apex.

ECOLOGY: Solitary, scattered, or in troops in duff under live oak and deciduous oaks. Uncommon on much of the Redwood Coast, common in southern California. Fruiting from late fall into spring (later in wet years).

EDIBILITY: Sniff but don't eat. Despite the tempting odor of this species, it's still an *Inocybe* and likely contains muscarine, a serious toxin.

COMMENTS: This large, stocky *Inocybe* can be recognized by the whitish to dingy brown cap, whitish stipe that stains pinkish red when damaged, and cinnamon-spicy odor. Another stocky *Inocybe* that often grows in the same areas is *I. corydalina*. It has a similar odor, but differs in having bluish gray stains on the cap and lower stipe. *I. brunnescens* is a very similar oak associate, but generally has a darker brown cap with a whitish patch of tissue over the center, and a mild odor and taste.

Inocybe "Dingy Red"

CAP: 5–10 (12) cm across, bell shaped to convex at first, expanding to nearly flat with a pronounced round umbo, edge often wavy and splitting or eroded in age. Beige to fawn brown or grayish brown, with extensively pinkish red to wine red blotches, especially in age. Often radially streaked with paler and darker fibrils. Surface dry to moist, smooth, innately fibrillose. **GILLS:** Narrowly attached to nearly free, sometimes with a small sinuate curve. Whitish to pale grayish cream, soon yellowish gray to dingy brownish tan with a contrasting pale edge. **STIPE:** 7–12 cm long, 0.8–2 (3) cm thick, equal with a tapered or pointed base. White when young, soon grayish cream to dingy tan, developing pink to red tones upward in age, base of stipe usually strongly blushed blood red or wine red. Surface smooth to vertically striate or with fine silvery chevrons. **PARTIAL VEIL:** Very weak cortina, usually not visible. **FLESH:** Fibrous, mostly whitish, pinker in stipe base, sometimes grayish pink in cap. **ODOR:** Mild to slightly farinaceous. **TASTE:** Mild to slightly farinaceous. **SPORE DEPOSIT:** Dull medium brown. **MICROSCOPY:** Spores 9–12 x 6–7.5 µm, ellipsoid, smooth, thick walled, brownish. Cheilocystidia cylindrical to club shaped. Pleurocystidia absent.

ECOLOGY: Solitary or in small groups on ground under live oak, Valley Oak, and perhaps other deciduous oaks. Irregularly fruiting but can be locally common in the southern part of our range, otherwise quite uncommon.

EDIBILITY: Unknown, likely toxic.

COMMENTS: This species is easy to recognize by the large size, radially streaked beige to dingy brown cap that soon blushes extensively pinkish red, and dingy grayish brown mature gills. *I. pudica* is generally smaller and more whitish, and stains peachy orange to pinkish. Stout forms with whitish young gills could be mistaken for *Hygrophorus russula*, but that species has greasy feeling, widely spaced gills and a white spore deposit.

MISAPPLIED NAME: *Inocybe adaequata* (Britzelm.) Sacc.

Inocybe sororia

Kauffman

CORN-SILK FIBERHEAD

CAP: 3.5–11 cm across, tall-conical to broadly conical, usually sharply peaked but sometimes rounded, expanding in age to nearly flat but retaining pointed central umbo. Margin incurved at first, but soon radially splitting, especially in age. Light yellowish beige to pallid gold, sometimes with a pale greenish wash and metallic looking in dry weather, older caps with brownish ocher blotches. Surface dry, innately fibrillose and soon splitting radially, appearing like thatched straw. **GILLS:** Narrowly attached or weakly sinuate, close. Pale beige to yellowish, becoming grayish tan to light brownish. Edges beaded, contrastingly pale. **STIPE:** 6–12 (15) cm long, 1–2 cm thick, cylindrical with a slightly enlarged base, rather tall and straight. Whitish to pale tan, with silvery white chevrons over most of surface, finely powdery near apex. **PARTIAL VEIL:** Thick whitish cortina adhering to cap edge, not forming a ring zone on stipe. **FLESH:** Stringy, whitish. **ODOR:** Strongly like green corn or corn silk, especially when crushed. **TASTE:** Indistinct. **SPORE DEPOSIT:** Dull brownish. **MICROSCOPY:** Spores 9–12 x 5.5–6.5 (7) μm, smooth, ellipsoid to bean shaped. Cheilocystidia large, swollen with blunt or slightly squared-off ends, without conspicuous crystal incrustation. Pleurocystidia absent.

ECOLOGY: Solitary or scattered in troops on ground under conifers and hardwoods. Widely distributed throughout western North America and common throughout our range. Fruiting from early fall into winter. Usually one of the first *Inocybe* to appear in fall.

EDIBILITY: Likely toxic, containing muscarine.

COMMENTS: The large size (for an *Inocybe*), peaked cap, with extensive radial splitting, often tall straight stipe, pallid yellowy colors, and strong green-corn odor render this species practically unmistakable in our area. *I. aestiva* is a summer-fruiting interior-montane species that grows with conifers.

Hebeloma mesophaeum group

CAP: 1.5–4 cm across, domed when young, expanding to broadly convex, often with a low, round umbo, sometimes uplifted in age. Pinkish brown to ocher-brown at center, tan-beige outward, creamy at margin. Surface thinly viscid when wet, soon dry, smooth. **GILLS:** Attached with a small notch, close, broad. White to yellow at first, becoming dingy tan. **STIPE:** 3–6 cm long, 0.3–0.6 cm thick, cylindrical. White to beige, dingy tan in age. Surface dry, fibrillose. **PARTIAL VEIL:** White, fibrillose or cobwebby, leaving a sparse silky annular zone on stipe, occasionally thicker and persistent. **FLESH:** Thin, often hollow in stipe; white to beige. **ODOR:** Radishlike. **TASTE:** Earthy. **SPORE DEPOSIT:** Dull brown. **MICROSCOPY:** Spores 8–10.5 x 4.5–6.5 µm, ellipsoid, nearly smooth. Cheilocystidia bottle shaped.

ECOLOGY: Scattered in small groups, especially under conifers (some variants under hardwoods), frequently encountered in urban areas with planted trees but some also in forests. Found throughout our range, can be locally quite common. Fruiting from early fall into spring, but most abundant in mid winter.

EDIBILITY: Unknown; likely toxic.

COMMENTS: This name is used here rather loosely to refer to a number of *Hebeloma* with small to medium-size fruitbodies, smooth, pinkish tan to brown and beige caps, and a cobwebby partial veil when young. One species in the group is common in yards under planted birch; another primarily grows with introduced European pines. A few similar species occur in native forest habitats, but names for taxa in this complex are lacking and/or very difficult to sort out. Some *Inocybe* are similar but have more radially fibrillose caps and/or bulbous stipe bases and are microscopically very different. *Pholiota* species tend to grow directly on wood and have smooth spores with a (sometimes inconspicuous) germ pore. Other *Hebeloma* in our area are larger, differ in color (paler capped in the *H. crustuliniforme* group, darker brown in the *H. theobrominum* group), and lack the cobwebby veil.

Hebeloma theobrominum group

CAP: 3–10 cm across, domed at first, becoming broadly convex, margin sometimes wavy. Dark reddish brown to pinkish brown or chestnut at first, becoming lighter pinkish brown to ocher, covered in a grayish white bloom when young, often remaining dusty whitish at margin. Surface slightly viscid to greasy, becoming moist or dry, smooth or radially wrinkly. **GILLS:** Attached, notched, edges often eroded. Pale grayish white when young, soon clay tan, becoming dull brown to warm brown in age, edges with a white fringe and often beaded with clear or brown droplets when young. **STIPE:** 5–8.5 cm long, 1–2.5 cm thick, cylindrical, stout, straight to curved, base equal or enlarged. Whitish to creamy, developing yellowish tan tones in age. Surface fibrillose, nearly smooth or more often with small shaggy scales and obscure chevrons overall, more densely scurfy at apex. **FLESH:** Fibrous, solid. White, bruising slowly olive brownish where injured. **ODOR:** Musty, like stale chocolate, or slightly mineral, like gravel or wet asphalt. **TASTE:** Indistinct. **SPORE DEPOSIT:** Dull brown. **MICROSCOPY:** Spores 8–11 x 4–5.5 µm, almond shaped, coarsely wrinkled, dextrinoid. Cheilocystidia club shaped and somewhat elongate.

ECOLOGY: Solitary, in clumps, or scattered in loose troops. Most commonly encountered in thick needle duff under Monterey Pine, also found with other pines and even hardwoods. Fruiting from winter into early spring.

EDIBILITY: Unknown; do not experiment, as many *Hebeloma* are toxic.

COMMENTS: This species can be recognized by the reddish brown to chestnut-colored caps with a pale bloom, stout stipes, and odor of stale chocolate (*theobrominum* refers to the genus of the cacao tree). Other *Hebeloma* in our area are mostly smaller and/or paler in color. The most similar *Cortinarius* species are usually strongly hygrophanous, have a cobwebby cortina and brighter rusty brown spores, and never have clear droplets on the edges of the young gills.

Hebeloma crustuliniforme group
POISON PIE

CAP: 2–8 cm across, domed with an inrolled and often finely ribbed margin when young, expanding to broadly convex or flat but often retaining finely inrolled margin. Light tan creamy to warm beige, extensively creamy outward, whitish at margin. Surface smooth, viscid when wet, becoming dry, often with adhering debris, often shiny in dry weather. **GILLS:** Finely attached, notched. White to creamy at first, soon gray-tan and then dingy brown in age. Edges often beaded with many tiny, clear or milky droplets when fresh. Droplets soon dry out and leave brown spots. **STIPE:** 4–10 cm long, 0.7–2 cm thick, cylindrical, fairly straight, base often enlarged or club shaped. White to cream, dingier tan in age, often covered in brown spores. Surface mostly smooth or fibrillose except for dense powdery-scurfs near apex. **FLESH:** White, fibrous, solid. **ODOR:** Strong, radishy or earthy. **TASTE:** Indistinct to unpleasant, radishlike. **SPORE DEPOSIT:** Dull brown. **MICROSCOPY:** Spores 7–11.5 x 5.5–7 μm, almond shaped, finely roughened (can be hard to see). Cheilocystidia abundant, cylindrical with wavy walls and a swollen, round head.

ECOLOGY: Solitary or scattered in loose troops, often in dense clumps forming arcs and rings around host tree in urban areas. Very common and abundant in late fall and winter around live oak and Monterey Pine, also around nonnative oaks, pines, and especially Strawberry Tree (*Arbutus unedo*). Less common in forested settings, mostly with live oak and pines. A few variants seem to be specific to conifers or willow (*Salix* spp.).

EDIBILITY: Quite toxic, causing gastrointestinal distress lasting a day or two.

COMMENTS: The most commonly encountered member of this group has a pale creamy cap with an inrolled margin, a slightly scurfy stipe (without a cobwebby veil), and, most importantly, clear to milky or brown droplets and spots on the gills. However, this "group name" is rather shamelessly used here for what could be a dozen different species in our area. Current species concepts in this genus are weak and/or based on European specimens (keep in mind that some *Hebeloma* seem to be introduced from Europe!); this is partly due to the inherent tendency of *Hebeloma* to be rather featureless and similar to one another. Nevertheless, a few are quite distinct, including a Shore Pine associate that grows in sand dunes on the Far North Coast and a lovely pink-capped species that grows in mossy spruce forests. Modern studies of *Hebeloma* are under way. Expect many new names in the near future.

Hebeloma "Big Scaly"

CAP: 6–18 (22) cm across, domed with a strongly inrolled and often lobed and distinctly ribbed margin when young, becoming broadly convex to nearly flat but remaining inrolled at the margin. Dingy reddish brown to pinkish brown at center with radiating brown streaks, extensively beige near margin, inrolled part often with a whitish bloom or pale fibrils. Surface thinly viscid to dry, smooth, often with leaves and debris stuck on when dry. **GILLS:** Broadly attached, notched, close to crowded, edges finely fringed or serrate and eroded. Light, dull grayish beige at first, becoming tan and then brown in age. Edges strikingly covered with clear to brown droplets when young and fresh. **STIPE:** 6–16 cm long, 2–3.5 cm thick, cylindrical, upright, club shaped or distinctly swollen at base (up to 5 cm thick). Whitish to cream, becoming beige to tan. Covered in large shaggy or pointed scales, especially dense near apex, lower parts smooth and vertically fibrillose. Base often with strong white rhizomorphs. **FLESH:** Thin, firm, fibrous. White marbled with tan, bruising slowly olive brownish where injured. **ODOR:** Radishy. **TASTE:** Indistinct. **SPORE DEPOSIT:** Dull brown. **MICROSCOPY:** Spores 10–12 x 6–7 μm, elongate almond shaped, thick walled, coarsely roughened.

ECOLOGY: Solitary or in small groups in forests of mixed hardwoods, especially Madrone, which appears be its preferred partner. Uncommon and local throughout our range. Fruiting in late fall and winter.

EDIBILITY: Unknown; do not experiment. Closely related species are known to be toxic.

COMMENTS: This species is recognizable by the hefty fruitbodies, strongly scaly stipe with a bulbous base, rosy brown to beige cap, and droplets on the gills. *Tricholoma dryophilum* can look somewhat similar, but has a smooth stipe and white spores. Some *Cortinarius* are similar, but lack the droplets on the gills and have brighter rusty brown spores. All other *Hebeloma* in our area are significantly smaller. Because of the scaly stipe, this apparently undescribed taxon has sometimes been identified as *H. sinapizans* (which is smaller and paler capped, and lacks the pale bloom) and as *H. insigne* [which is darker capped and associated with Quaking Aspen (*Populus tremuloides*) in the Rocky Mountains].

Phaeocollybia kauffmanii

(A. H. Sm.) Singer

CAP: 5–15 (20) cm across, conical with an inrolled margin at first, becoming broadly conical to convex with a broad umbo. Margin becoming wavy and sometimes slightly uplifted in age. Orange-brown, ochraceous brown to warm brown, often with a slightly darker center, developing darker stains or blotches when damaged or in age. Surface smooth, shiny, viscid, and covered with thick slime when wet, dull but moist in dry weather. **GILLS:** Narrowly attached or appearing free, close to crowded. Buff to tan when young, becoming dingy golden brown in age. **STIPE:** 4–12 cm long aboveground, 20–30 cm (up to 40 cm) long overall, 1–3.5 cm thick, equal, swollen in middle or enlarged toward lower stipe, with a tapering root underground. Pinkish buff to cinnamon buff on upper stipe, orange-brown lower when young. Whole stipe darkens in age; pale cinnamon brown to reddish brown at apex, dark orange-brown to reddish brown spreading up from base. Surface moist to dry, appressed-fibrillose to smooth, often cracking longitudinally in age or dry weather. **FLESH:** Thin to moderately thick, watery brown when wet, pinkish buff when dry in cap. Stipe stuffed with pinkish ivory pith that slowly stains orangish brown and a fibrous to cartilaginous pale pinkish brown to orangish brown rindlike skin. **ODOR:** Slightly farinaceous. **TASTE:** Strongly farinaceous to slightly bitter. **KOH:** Brownish. **SPORE DEPOSIT:** Cinnamon brown. **MICROSCOPY:** Spores 7.5–10 x 4–6 μm, averaging 8.8 x 5.2 μm, almond shaped or lemon shaped in side view, roughened,

but with a smooth beak. Cheilocystidia narrowly club shaped, thin walled.

ECOLOGY: Solitary, scattered, or in small clusters, often in rows and arcs under coastal conifers, especially in mature and old-growth forests. Widespread throughout most of our area, but rare south of Sonoma County; has been recorded as far south as the Santa Cruz Mountains. Fruiting from early fall into winter.

EDIBILITY: Unknown, possibly poisonous.

COMMENTS: The large, viscid, orange-brown to warm brown cap, solid, rooting stipe, and farinaceous taste help identify this *Phaeocollybia*. The *P. kauffmanii* complex includes many similar species that can be very tough to tell apart. Some can be locally common, but restricted to the North Coast conifer forests, and are quite rare or absent south of Mendocino County. *P. ammiratii* is generally smaller and more slender, and has a brownish yellow to butterscotch-colored cap, orangish cream gills when young, and a distinctive burgundy blush to the lower stipe. Microscopically, it differs from others in this group by having easily observed clamp connections. *P. ochraceocana* differs by having a tawny ocher-orange, dry to moist (not viscid) cap, creamy to creamy buff gills, and a pale creamy buff to pale ochraceous salmon stipe; its spores measure 5.8–8.9 x 3.8–5.6 μm. *P. luteosquamulosa* is another ocher-yellow species that looks almost identical to *P. ochraceocana*, but can be easily told apart by its larger spores (8.7–12 x 5–7.2 μm). Also see *P. redheadii* (p. 161) for large, darker-capped species.

Phaeocollybia redheadii

Norvell

SCOTT'S PHAEOCOLLYBIA

CAP: 6–20 (30) cm across, conical with an inrolled margin at first, becoming broadly conical to convex with a broad umbo. Margin becoming wavy and sometimes slightly uplifted in age. Dark orange-brown to red-brown overall at first, becoming unevenly dark brown, often with darker spots or lighter blotches. Surface smooth, shiny, viscid, and covered with thick slime when wet, shiny and moist in dry weather. **GILLS:** Narrowly attached, or appearing free, close to crowded. Dingy whitish at first, becoming buff to dingy yellowish brown in age. **STIPE:** 4–15 cm long aboveground, 10–30 (40) cm long overall, 1.5–3.5 cm thick at the apex. Equal, swollen in middle or enlarged toward lower stipe with a tapering root underground. Pinkish buff to cinnamon buff over upper portion, orange-brown lower when young. Whole stipe darkens in age; cinnamon brown to reddish brown at apex, dark reddish brown to vinaceous brown from base up in age. Surface moist to dry, appressed-fibrillose to smooth, often cracking longitudinally in age or dry weather. **FLESH:** Moderately thick, firm, creamy to pale yellowish, staining orangish brown. Stipe solid, stuffed with creamy buff pith that slowly stains orangish brown and a fibrous to cartilaginous brownish rindlike skin. **ODOR:** Strongly farinaceous, at times with a floral element. **TASTE:** Strongly farinaceous, cap slime bitter. **KOH:** Brownish. **SPORE DEPOSIT:** Cinnamon brown. **MICROSCOPY:** Spores 8–13 x 4.9–7.3 µm, averaging 10.5 x 6 µm, almond shaped in side view, elongate-ellipsoid in face view, roughened, but with a smooth beak. Cheilocystidia abundant, broadly club shaped or with rounded heads, thin walled.

ECOLOGY: Often scattered in small clusters, or in rows and arcs in coniferous forest. Locally common, occurring from Mendocino County northward, but seemingly restricted to mature or old-growth forest. Fruiting from early fall well into winter.

EDIBILITY: Unknown, possibly poisonous.

COMMENTS: This is our largest *Phaeocollybia*; caps upward of 30 cm across can be found (more typically 12–15 cm). The combination of the large size, reddish brown to dark brown colors, long rooting stipe stuffed with creamy pith, and large spores distinguishes this species. *P. benzokauffmanii* is a similar large species, but differs in having a smoky violet to pinkish brown cap when young and becoming dark vinaceous brown to dark brown in age. It also has smaller spores (9 x 5.5 µm). *P. spadicea* has a dark brown to gray-brown cap that develops vinaceous tones in age, but is generally smaller (cap 3–12 cm across). Also see *P. kauffmanii* (p. 160) for large, paler species.

Phaeocollybia californica

A. H. Sm.

CALIFORNIA PHAEOCOLLYBIA

CAP: 3–5 (9) cm across, conical to broadly conical and often umbonate when young, becoming bell shaped to convex, with a low broad umbo. Margin downcurved at first, becoming uplifted in age. Orange-brown, reddish orange to yellowish brown when young, becoming dingy reddish brown to yellowish brown, occasionally with darker spots or stains in age. Surface smooth, viscid when wet, shiny when dry. **GILLS:** Narrowly attached or appearing free, close to moderately crowded. Whitish to pale buff at first, becoming cinnamon buff to cinnamon orange, to rusty brown in age. **STIPE:** 3–9 cm long aboveground, up to 20 cm long overall (but tough to extract from ground without breaking), 0.5–1.5 cm thick, more or less equal aboveground, tapering to a long, fragile, root underground. Buff to tan at first, darkening to orangish brown and gradually becoming dark brown from base up in age. **FLESH:** Thin, watery brown, orangish brown to tan, slowly staining reddish. Stipe hollow, with silky whitish fibrils when young. Cartilaginous outer skin often concolorous with the outer stipe. **ODOR:** Earthy when young; stronger, often unpleasant, chlorinelike in age. **TASTE:** Mild to slightly bitter, or radishlike. **SPORE DEPOSIT:** Reddish brown. **MICROSCOPY:** Spores 8–11 x 5–6 (7) µm, average 9 x 5.4 µm, broadly lemon shaped with a beak in side view, ovoid in face view. Surface strongly roughened, warted, apiculus and beak smooth. Cheilocystidia sparse, bottle shaped or cylindrical with an abrupt rounded head.

ECOLOGY: Often growing in small to large clusters, arcs, or troops, rarely solitary, usually in mixed conifer-Tanoak forest. Generally rare throughout our range, but can be locally common. Fruiting from early fall in north into early winter southward.

EDIBILITY: Unknown, possibly poisonous.

COMMENTS: The orange- to red-brown cap that fades to yellowish brown when dry, pallid young gills, hollow, rooting stipe that darkens from the base up in age, clustered growth, and medium size help distinguish this species. *P. scatesiae* produces medium-size, clustered fruitbodies that that differ in having glutinous, tawny yellow-brown to dark brown caps, and abundant cheilocystidia; it is rare in California, known only from the Far North Coast. *P. attenuata* is smaller and has an orange-brown to clay brown cap and a thin, wiry, rooting stipe. *P. sipei* is larger and rarely grows in large clusters; it has a red-brown to orange-brown cap, a tall, fragile, hollow stipe, and small spores (5.8–7.5 x 3.3–5 µm). *P. dissiliens* is a small- to medium-size species with a bright orange to butterscotch yellow, conical to bell-shaped cap with a pointed umbo, a fragile, hollow stipe, and smaller spores (6–8 x 3.7–4.8 µm).

SYNONYM: Redder-capped specimens were described as *P. rufotubulina*, but genetic data show them to be a color varient of *P. californica*.

Phaeocollybia attenuata
(A. H. Sm.) Singer
LITTLE PHAEOCOLLYBIA

CAP: 1–5 cm across, conical to broadly conical, often with a pointed umbo when young, becoming bell shaped to convex, occasionally with a low broad to pointed umbo. Dull orange-brown to clay brown or yellowish brown, occasionally with a slightly darker or lighter margin if in the process of drying out, sometimes developing reddish brown spots or stains in age. Surface smooth, moist to slightly viscid if wet. **GILLS:** Narrowly attached or appearing free at times, close to moderately crowded. Light warm buff to pinkish buff when young, soon darkening to buff-brown, becoming dingy orange-brown in age. **STIPE:** 2–6 (8) cm long aboveground, 20+ cm long overall (but tough to extract from ground without breaking), 0.2–0.5 cm thick, more or less equal aboveground, tapering to a long, brittle wirelike root underground. Clay brown to yellowish brown at first, darkening to orangish brown and gradually becoming dark brown to blackish brown from base up in age. **FLESH:** Very thin, creamy ivory when young, darkening in age. Stipe loosely stuffed with whitish fibrous pith at first, soon becoming hollow. **ODOR:** Indistinct to strong; often radishlike with sweet floral undertones and disagreeable burnt fish odor. **TASTE:** Mildly to strongly unpleasant and often bitter. **KOH:** Dark brown on all parts. **SPORE DEPOSIT:** Reddish brown. **MICROSCOPY:** Spores 7–10 x 4.5–7 μm, average 9 x 7.8 μm, lemon shaped or almond shaped with a long beak. Surface strongly roughened, warted, apiculus and beak smooth. Cheilocystidia numerous, thread like or narrowly club shaped.

ECOLOGY: Scattered in small to large troops (rarely solitary or clustered) on ground in mixed conifer forest on the Far North Coast and in mixed Douglas-fir and Tanoak forests south to the greater San Francisco Bay Area. Probably most common *Phaeocollybia* in California. Fruiting from early fall in the north well into winter southward.

EDIBILITY: Unknown, possibly poisonous.

COMMENTS: The clay brown to orange-brown cap, slender stipe that darkens from the base up, and very small size (for a *Phaeocollybia*) help identify this mushroom. Other similarly colored small *Phaeocollybia* in California include *P. radicata*, which generally has a brighter yellow-orange cap and pale orange to cinnamon buff gills when young, and lacks the dark wiry root to the stipe; it's a rare species. *P. pleurocystidiata* has an ocher-yellow to tawny yellow cap, yellowish buff gills, and a salmon-colored, deeply rooting stipe, and only fruits in spring. *P. piceae* is slightly larger and more orange in color, and has a solid stipe stuffed with whitish pith. Also compare with the medium-size, orange-brown *P. californica* and its kin.

Phaeocollybia olivacea

A. H. Sm.

OLIVE PHAEOCOLLYBIA

CAP: 3.5–10 (12) cm across, conical with an incurved margin at first, becoming broadly conical to convex, often with a low umbo. Margin becoming wavy and slightly uplifted in age. Color variable and hygrophanous, dark olive to olive-green or forest green when young, becoming olive-green, orangish green to brownish olive or even reddish olive in age. Surface smooth to slightly wrinkled, viscid and shiny when wet, becoming dull when dry. **GILLS:** Narrowly attached or appearing free at times, close to crowded. Dingy yellow to pale greenish yellow when young, often with reddish spots and stains, becoming yellowish brown to dirty reddish brown in age. **STIPE:** 7–14 cm long aboveground, up to 22 cm long overall, 0.8–2 cm thick, equal or slightly enlarged toward lower stipe and tapering to a long root underground. Drab yellowish to greenish buff on upper portion, darker, orange-red on lower stipe when young, becoming reddish to orange-brown from base up in age, but often staying paler buff-brown at apex. **FLESH:** Thin, greenish in cap. Stipe stuffed with solid white to ivory silky-fibrous pith that stains reddish brown, with a fibrous to cartilaginous reddish to watery gray rindlike skin. **ODOR:** Farinaceous to cucumber-like. **TASTE:** Indistinct. **KOH:** Orange on cap, reddish brown on flesh. **SPORE DEPOSIT:** Rusty brown. **MICROSCOPY:** Spores 5–7 x 8–11 μm, average 6 x 10 μm, broadly ovoid with an abruptly projecting snout in face view, flattened on one side in side view, roughened or warted with dark ornamentation with a smooth apical beak. Cheilocystidia thin walled, variable, club shaped, often with threadlike extensions in age.

ECOLOGY: Often in clusters or in large arcs, occasionally solitary or scattered in small groups on ground under conifers. Especially frequent with Sitka Spruce and Western Hemlock from Mendocino County northward, rarer with Douglas-fir and occasionally pines southward. Fruiting from midfall in the north, well into winter in the south.

EDIBILITY: Unknown, possibly poisonous.

COMMENTS: The conical greenish cap, dingy yellowish gills, and solid, rooting stipe help identify this mushroom. *P. fallax* is a common green-capped species distinguished by its beautiful lilac gills when young; however, the gills lose their distinctive color as they age, and mature fruitbodies of the two species can be very tough to tell apart, especially since the they often grow alongside one another! However, *P. fallax* is usually smaller and has a hollow stipe, especially in age. *P. pseudofestiva* also has a greenish cap and drab yellowish gills, but is usually much smaller and has a green stipe apex when young; microscopically, it has relatively thick-walled cheilocystidia that are bottle shaped or cylindrical with abruptly rounded heads.

Phaeocollybia fallax

A. H. Sm.

CAP: 1.5–6 (9) cm across, conical to broadly conical with an incurved margin when young, becoming broadly convex to nearly plane, often with a pronounced pointed umbo. Margin becoming wavy and occasionally slightly uplifted in age. Color variable, hygrophanous; fading when dry. Dark olive, olive-green to forest green, but often with paler blotches when young, becoming dingy olive to olive-brown in age. Surface smooth, shiny, viscid, margin translucent-striate when wet; become dull when dry. **GILLS:** Narrowly attached or appearing free, close to crowded. Bluish lilac to violet when young, soon fading to grayish lilac and developing orangish spots and stains, becoming dingy orange-brown, but often retaining some grayish color in age. **STIPE:** 4–8 cm long aboveground, up to 25 cm long overall (but tough to extract from ground without breaking), 0.3–1 cm thick. Equal, or occasionally slightly tapered toward the base, with a short and abrupt to long tapering root underground. Drab gray to pale lilac-gray on upper portion, darker, reddish to orange-red on lower stipe when young, then reddish brown to dark brown from base up in age, but often staying paler dingy gray to olive-gray at apex. **FLESH:** Very thin, fragile, watery green to grayish in cap. Stipe stuffed with whitish silky-fibrillose pith when young, soon hollow, with a fibrous to cartilaginous outer skin. **ODOR:** Generally mild; mix of radish and sweet floral scent with an unpleasant undertone. **TASTE:** Mild, slightly radishlike or occasionally bitter. **KOH:** Dingy orangish brown on cap. **SPORE DEPOSIT:** Pinkish cinnamon to pale rusty brown.

MICROSCOPY: Spores 7–10 x 4–6 μm, average 9 x 5.3 μm, lemon to almond shaped in side view, ovoid with an elongated beak and a pointed apiculus in face view, surface marbled and roughened. Cheilocystidia numerous, thin walled, with small, swollen heads.

ECOLOGY: Solitary, scattered, or in small clusters on ground under conifers. Common under Sitka Spruce, Western Hemlock, and Grand Fir from Mendocino County northward. Much rarer (occasionally locally common) in mixed Douglas-fir and Tanoak forests south to the greater San Francisco Bay Area. Fruiting from late fall into early winter.

EDIBILITY: Unknown, possibly poisonous.

COMMENTS: The small size, conical green cap, beautiful bluish lilac gills, and rooting stipe make this *Phaeocollybia* unmistakable when young. However, be sure to check each fruitbody carefully, as two other green-capped *Phaeocollybia* often grow intermingled with *P. fallax*. The rare *P. pseudofestiva* shares the same stature, size, and cap color as *P. fallax*, but has drab yellowish gills and a green stipe apex when young. *P. olivacea* can be separated by its larger size, yellowish to drab gills, and solid stipe (stuffed with firm white pith); it is fairly common. *P. rifflipes* and *P. lilacifolia* also have lilac to violet gills when young, but have dark brown to gray-brown caps. *P. rifflipes* is the smaller and paler of the two, and has smaller spores (7–7.5 x 4.4–5 μm) than those of *P. lilacifolia* (7–8.5 x 5–5.5 μm). Both primarily occur on the Far North Coast.

8 • *Cortinarius*

This is the most species-rich genus of gilled mushrooms on Earth, with thousands currently recognized (and many more still undescribed). *Cortinarius* are often difficult to identify, but many are very beautiful, and the group can be very satisfying to work with once you become familiar with it. There is a great diversity in the appearance of mushrooms in this genus: some are tiny, others are enormous, a great many are dull, and many others are brilliantly colored. All are ectomycorrhizal and of great ecological importance in temperate forests around the world. The forests of the Pacific Coast in particular are a global hotspot for *Cortinarius* biodiversity.

Most of the *Cortinarius* in our area can be recognized by their cortinas (a partial veil made of fine fibrils); ocher-brown to rusty brown to bright orange-brown spore deposit; growth in soil or humus (none consistently fruit from wood); and attached gills. Keep in mind that the cortina can be very inconspicuous, covered in a layer of slime, or even, in one case, membranous.

Important characters to make note of when identifying *Cortinarius* include the following: coloration of cap and gills when young as well as in age; texture of the surfaces (dry or viscid? smooth or scaly?); shape of the stipe base; odor; KOH reactions on all parts; bruising reactions; and nearby tree species. Important microscopic features include size and shape of spores; coarseness of warts, bumps, and wrinkles on the spore surface; and the presence or absence of clamp connections in the hyphae of the cap.

Although *C. caperatus* and *C. ponderosus* are frequently collected for food in our area, the edibility of most *Cortinarius* remains unknown. It is likely that in such a large and chemically varied genus, some contain gastrointestinal irritants; members of the *C. rubellus* group are potentially lethally toxic due to the presence of orellanine, a serious kidney toxin. We advise that you do not experiment with the edibility of *Cortinarius*!

Dimitar Bojantchev has pushed our understanding of California's *Cortinarius* forward substantially in the past decade. Much can be learned about this group by exploring his website: www.mushroomhobby.com.

To make the task of identifying a *Cortinarius* less overwhelming, it's extremely useful to learn to recognize the subgroups below. Keep in mind that these groups are not necessarily "natural" (phylogenetically supported), and not every species you encounter will fit neatly into one of these groups.

Leprocybe—Species with scaly or scurfy caps and with dull yellowish to brown pigments. KOH reactions usually lacking. Almost always brilliantly fluorescent under UV light. Some are deadly toxic.

Dermocybe—Yellow, orange, red, or brownish pigmented species (gills usually brightest in color) with thin stipes. Caps not hygrophanous. KOH reactions intense and distinctive. Sometimes treated as a full-fledged genus. Members of this subgroup are relatively distinctive, but can be confused with *Leprocybe* or brightly colored *Telamonia*.

Telamonia—Cap and stipe dry. Surfaces usually smooth to silky-fibrillose. Caps usually hygrophanous. Most are brownish, but a few have bright red colors, leading to confusion with *Dermocybe*. KOH reactions usually indistinct except in those species with strong red pigmentation.

Cortinarius—All surfaces dry, intense, deep purplish black in color, cap scaly. A very small subgroup, represented in our area only by *C. violaceus*.

Sericeocybe—Dry, silky-capped species with silvery, pale blue, lavender, or lilac tones. Not strongly hygrophanous, if at all. KOH reactions usually indistinct.

Phlegmacium—Cap viscid, stipe dry, often brightly colored. Some have very distinctive KOH reactions (pink, yellow, red, orange). The "Bulbopodiums" are those *Phlegmaciums* that have an abruptly swollen, often rimmed bulb at the base of the stipe.

Myxacium—Cap and stipe both viscid (sometimes only weakly). Often brightly colored. KOH reactions usually indistinct.

"Rozites"—One species (*C. caperatus*) in our area has a membranous partial veil rather than a cortina; despite this, the medium-large fruitbodies, growth on soil (rather than wood), attached gills, and rusty spore print help identify this species as a *Cortinarius*.

Cortinarius clandestinus group

CAP: 2–6 (9) cm across, rounded to convex with an inrolled margin when young, becoming broadly convex to plane, margin becoming wavy and often slightly uplifted. Olive-brown with darker brown to blackish scales, often blackish olive over center, becoming ocher-olive to rusty in age. Surface dry, appressed-fibrillose, breaking up into fine scales. **GILLS:** Broadly to narrowly attached, often notched, close to subdistant. Straw yellow to pale olivaceous brown at first, soon orange-brown as spores mature. **STIPE:** 4–9 cm long, 0.8–2 cm thick at apex, club shaped with an enlarged base. Pale, dingy greenish yellow to yellowish brown, often with a narrow, dark annular zone. Surface dry, covered with longitudinal fibrils and silky cortina remnants. **PARTIAL VEIL:** Sparse cortina, leaving an annular zone of yellowish to dark olive fibrils on stipe, soon covered in rusty brown spores. **FLESH:** Thin to moderately thick, firm, pale olivaceous yellow to dingy yellowish brown. **ODOR:** Radishlike. **TASTE:** Indistinct. **KOH:** Indistinct to dark orange-brown. **SPORE DEPOSIT:** Rusty brown. **MICROSCOPY:** Spores 5.5–7 x 4.5–6 µm, subglobose, slightly roughened to nearly smooth.

ECOLOGY: Solitary or scattered under Tanoak. Generally infrequent, but can be locally common in the southern part of our range. Fruiting from late fall into spring.

EDIBILITY: Unknown, possibly toxic.

COMMENTS: *Leprocybe* in California are very poorly known, and most of our species remain undescribed. As a group they can be recognized by their dry, often scaly caps; mustard, olive, brown, or rusty orange colors; and, frequently, dark bands on the lower stipe. Most fluoresce brightly in UV light. The description and image above represent our most common *C. clandestinus*–like species (with Tanoak). *C. "subalpinus"* is a similar (less olive-toned) spring-fruiting species that occurs in the high mountains under conifers. Another common undescribed species in this group is more slender, with an olive-green cap; it grows on the North Coast under conifers.

Cortinarius limonius group

CAP: 2–6 (8) cm across, conical to irregularly bell shaped, expanding to plane, often wavy, irregular or lobed, often with a rounded umbo. Dark ocher-orange to orange-brown, fading slightly in age. Surface moist to dry, smooth to wrinkled or slightly roughend. **GILLS:** Narrowly attached with a deep notch, close to subdistant. Straw yellow at first, very soon ocher-orange to rusty orangish brown. **STIPE:** 3–10 (13) cm long, 0.5–1.5 cm thick, tall, often twisted or curved. Yellowish ocher to orange-brown, covered with pale yellowish fibrils, staining brown when handled. Surface moist to dry, with longitudinal fibrils. **PARTIAL VEIL:** Sparse yellowish cortina, leaving silky remnants on stipe that soon disappear. **FLESH:** Moderately thin, fragile in cap, fibrous in stipe, becoming hollow. Orangish, slowly staining ocher-brown. **ODOR:** Mild to radishlike. **TASTE:** Indistinct. **KOH:** Turning slightly orangish. **SPORE DEPOSIT:** Rusty brown. **MICROSCOPY:** Spores 6.5–10 x 5–7.5 µm, broadly ellipsoid, coarsely roughened.

ECOLOGY: Solitary, scattered, or in small troops in coniferous forest on the Far North Coast. Common in Sitka Spruce forest, rare elsewhere. Fruiting from early fall into winter.

EDIBILITY: Unknown; do not experiment! Similar species are deadly poisonous, containing the toxin orellanine.

COMMENTS: The irregularly bell-shaped to flat cap, ocher-orange colors, tall stature, and evanescent yellowish cortina are helpful identifying features. The deadly poisonous *C. rubellus* is very similar but slightly duller in color, and has more widely spaced gills and a slightly scaly cap; it is rare in California, known only from a few locations on the Far North Coast. *C. callisteus* has a paler yellow-orange cap with small scales, smaller spores, and a distinctive odor described variously as "motor-oil" or "a hot iron scorching linen." *C. gentilis* is a smaller, more slender species with a hygrophanous cap that fades from warm brown to cinnamon.

NOMENCLATURE NOTE: The newly described *C. kroegeri* likely represents the common Californian specied in this complex.

Cortinarius idahoensis
Ammirati & A. H. Sm.

CAP: 1–5 cm across, broadly conical to convex, becoming broadly convex to plane, at times with a low broad umbo. Dark brown to cinnamon brown, but often with some olive tones when wet, fading slightly and developing darker olive tones when dry. Surface smooth to finely fibrillose, moist to dry. **GILLS:** Broadly attached, or with a slight notch and a decurrent tooth, closely spaced. Yellowish brown to ocher-brown when young, becoming more ocher-orange as spores mature. **STIPE:** 3–8 cm long, 0.5–1 cm thick, more or less equal, or with a slightly enlarged base. Yellowish to yellowish buff with darker, ocher-brown fibrils, darkening more toward ocher-brown as it ages, basal mycelium pale creamy yellowish, staining dingy yellowish when handled. Surface dry, covered with silky longitudinal fibers. **PARTIAL VEIL:** Sparse cortina of pale ocher-brown fibrils, leaving silky fibrils on the stipe, or disappearing altogether. **FLESH:** Thin, fibrous. Brown in cap, paler buff-brown to ocher-brown in stipe. Stipe stuffed with pith when young, soon becoming hollow. **ODOR:** Indistinct. **TASTE:** Indistinct. **KOH:** Dark reddish brown on cap. **SPORE DEPOSIT:** Rusty brown. **MICROSCOPY:** Spores 7–9 × 4.5–5.5 μm, ellipsoid, minutely roughened. Cells of cap surface with bluish particles in KOH.

ECOLOGY: Solitary or scattered on ground or on well-rotted wood under pine, but reported with other conifers as well. Locally common, easily overlooked because of small size and dull colors. Known from the San Francisco Bay Area north. Fruiting in late fall or winter.

EDIBILITY: Unknown.

COMMENTS: This drab *Dermocybe* is one among a fairly large group of difficult-to-distinguish species. The dark brown to cinnamon brown cap with olive tones, ocher-brown gills, and reddish brown KOH reaction are helpful distinguishing features. Under the microscope, look for the presence of bluish pigment particles in the cap skin (when mounted in KOH). *C. humboldtensis* has an olive-yellow to cinnamon brown cap, olive-yellow to ocher-yellow gills, and a dark inky violet KOH reaction on the cap. *C. croceus* has a warm cinnamon brown cap, yellowish gills with a greenish hue when young, and dingy olive-yellow flesh. Members of the *C. cinnamomeus* group have bright orange gills from the start.

Cortinarius cinnamomeus group
CINNAMON CORT

CAP: 1.5–5 cm across, broadly conical to convex when young, becoming broadly convex to plane or wavy in age, often with a low broad umbo. Orange-brown to cinnamon brown, occasionally darker, reddish brown to brown, or with an olive cast. Surface smooth to finely fibrillose, moist to dry. **GILLS:** Broadly attached, or with a slight notch and a decurrent tooth, close to subdistant. Bright orange to cinnamon orange when young, becoming rusty orange as spores mature. **STIPE:** 3–10 cm long, 0.5–1 cm thick, more or less equal, or enlarged slightly toward base. Yellow to dingy yellow, developing sordid brownish to olive-orange discoloration. Basal mycelium whitish to creamy yellow. Surface dry, covered with brownish, silky longitudinal fibers. **PARTIAL VEIL:** Sparse, cortina-like, leaving some silky fibrils on stipe. Yellowish at first, becoming rusty with spores. **FLESH:** Moderately thin, fibrous. Pale buff to yellowish buff in cap, yellowish on outer stipe, buff on hollow interior. **ODOR:** Indistinct to slightly radishlike. **TASTE:** Indistinct. **KOH:** Reddish brown on cap and flesh, reddish on gills. **SPORE DEPOSIT:** Rusty brown. **MICROSCOPY:** Spores 6–8 × 4–5 μm, ellipsoid, minutely roughened.

ECOLOGY: Scattered, in small clusters or troops on ground under pine. Common throughout our range. Fruiting from late fall into early spring.

EDIBILITY: Unknown.

COMMENTS: The orange-brown cap, bright orange young gills, pale flesh, and rusty brown spores are helpful identifying features. *C. malicorius* is very similar, but generally has brighter orange gills, darker olive flesh, and slightly smaller spores (6 x 4 μm). *C. aurantiobasis* has a dingy orange-red lower stipe and bright yellow-orange stipe flesh; it appears to be restricted to the Far North Coast under Sitka Spruce and Shore Pine. *C. croceus* has mustard yellow young gills that become rusty orange as the spores mature. Additional minutely different, undescribed species are also commonly encountered. Probably most common is one with a distinctly olive cap found under Shore Pine. *Inocybe cinnamomea* looks somewhat similar, but has an innately radially fibrillose to felty cap and duller brown spores.

SYNONYM: *Dermocybe cinnamomea* (Schaeff.) M. M. Moser.

Cortinarius croceus

(Schaeff.) Gray

MUSTARD-GILLED CORT

CAP: 1.5–6 cm across, broadly conical to convex when young, becoming broadly convex to plane or wavy in age, often with a low broad umbo. Color variable, cinnamon brown, orange-brown, or medium olive-brown, fading slightly, often more reddish brown in age. Surface smooth to finely fibrillose, moist to dry. **GILLS:** Broadly attached or with a slight notch and a decurrent tooth, close to subdistant. Mustard yellow to greenish yellow when young, quickly becoming more orange-yellow, finally orange to rusty orange as spores mature. **STIPE:** 3–8 cm long, 0.5–1 cm thick, more or less equal or enlarged slightly toward base. Yellow to mustard yellow at first, becoming dingy yellow and developing orangish brown discoloration. Basal mycelium creamy yellow to pale yellow. Surface dry, covered with brownish, silky longitudinal fibers. **PARTIAL VEIL:** Faint, cortina-like. Yellow at first, remnants on stipe becoming rusty with spores. **FLESH:** Thin, fibrous. Pale watery olive-yellow in cap, yellowish, but often with an olive tinge in stipe. Stipe stuffed to hollow. **ODOR:** Indistinct to slightly radishlike. **TASTE:** Indistinct. **KOH:** Reddish brown on cap, reddish on gills. **SPORE DEPOSIT:** Rusty brown. **MICROSCOPY:** Spores 6.5–8 × 4.5–5 μm, ellipsoid, minutely roughened.

ECOLOGY: Scattered, in small clusters or troops on ground under conifers, especially pine. Very common throughout our range. Fruiting from late fall into early spring.

EDIBILITY: Unknown.

COMMENTS: This widespread and common *Dermocybe* is fairly easy to identify when you have a large collection of young to old specimens; older specimens by themselves can be nearly indistinguishable from similar species. The cap color is generally cinnamon brown to orange-brown, but can be dark brown to olive-brown or even reddish brown in age. The mustard yellow to greenish yellow color of the young gills (often best observed when still in button stage) is the most helpful identifying feature. However, the gills quickly become more orange in age, leading to confusion with the very similar *C. cinnamomeus* group. *C. idahoensis* has a dark brown to cinnamon brown cap with olive tones, and slightly darker ocher-brown gills that don't become as bright orange as in *C. croceus*. *C. humboldtensis* has an olive-yellow to cinnamon brown cap, olive-yellow to ocher-yellow gills, and a dark inky violet KOH reaction on the cap. *C. aurantiobasis* has brighter yellowish to orange gills, an orange-red stipe base, and bright yellow basal mycelium. *C. thiersii* is a very similar, primarily montane species (it may occur on the coast as well) with larger spores [8–10 (12) × 5–6 μm].

SYNONYM: *Dermocybe crocea* (Schaeff.) M. M. Moser.

Cortinarius cinnamomeoluteus
P. D. Orton

CAP: 1.5–5 cm across, rounded to broadly conical when young, becoming broadly convex or nearly plane in age, with a low umbo at times. Cinnamon brown to orange-brown with a yellow-brown margin when young, becoming darker brown at center, especially in age. Surface smooth to finely fibrillose, moist to dry. **GILLS:** Attached with a distinct notch, close. Mustard yellow to pale ocher-yellow when young, becoming duller, more dingy ocher yellow to rusty yellow as spores mature. **STIPE:** 3–7 cm long, 0.3–0.8 cm thick, more or less equal, or slightly swollen toward base. Mustard yellow to golden yellow with orangish veil remnants and an orangish brown base at first, developing orangish brown discoloration as it ages. Basal mycelium whitish. Surface dry to moist, covered with matted orange, silky longitudinal fibers. **PARTIAL VEIL:** Faint, cortina-like, leaving an annular zone of silky fibrils near apex. Ocher-orange at first, becoming rusty with spores. **FLESH:** Thin, fibrous. Dull mustard yellow in cap, bright golden yellow in stipe. Stipe stuffed when young, becoming hollow in age. **ODOR:** Radishlike to earthy. **TASTE:** Radishlike. **KOH:** Bright blood red on cap and stipe, orange on flesh. **SPORE DEPOSIT:** Rusty brown. **MICROSCOPY:** Spores 7–11 x 4–6.5 µm, ellipsoid to nearly almond shaped, minutely roughened.

ECOLOGY: Scattered, in small clusters or troops on ground in wet swampy areas with willow (*Salix* spp.). Fruiting in fall and early winter. Rare, or at least rarely reported in California, as mushroom hunters rarely spend time in this habitat.

EDIBILITY: Unknown.

COMMENTS: This orange-brown *Cortinarius* in the subgenus *Dermocybe* looks like lot of other species and would be tough to tell apart were it not for its unique habitat (all other Californian *Dermocybe* appear to be conifer associates). The bright red KOH reaction also helps distinguish it. *C. aurantiobasis* has brighter yellowish to ocher-orange gills, an orange-red stipe base, a bright yellow basal mycelium, and a reddish brown KOH reaction on the cap. It is common under Shore Pine and spruce (note that this habitat sometimes is intermixed with willow on the North Coast dunes!).

Cortinarius neosanguineus
Ammirati, Liimat. & Niskanen
WESTERN BLOOD-RED CORT

CAP: 1–5 cm across, broadly conical to convex, becoming broadly convex. Deep red to purple-red when wet, burgundy to deep reddish brown when dry. Surface smooth to finely fibrillose, moist to dry. **GILLS:** Broadly attached, often with a decurrent tooth, close to subdistant. Deep red to purple-red, becoming rusty red as spores mature. **STIPE:** 3–8 cm long, 0.2–0.8 cm thick, equal or with a slightly enlarged base. Deep red, purple-red or reddish brown, basal mycelium ocher-yellow, staining purplish brown when handled. Surface dry, often with silky longitudinal fibers. **PARTIAL VEIL:** Faint, cortina-like, leaving a few silky fibrils or disappearing. **FLESH:** Thin, fibrous, deep reddish purple. **ODOR:** Slightly fragrant, cedarlike. **TASTE:** Radishlike. **KOH:** Dark vinaceous on all parts. **SPORE DEPOSIT:** Rusty brown. **MICROSCOPY:** Spores 7–9 x 4.5–5.5 µm, ellipsoid to almond shaped, slightly roughened.

ECOLOGY: Solitary or scattered on ground, in moss or at the base of moss-covered logs and stumps under conifers, especially Western Hemlock. Uncommon to rare in California, occurring from Mendocino County north along the coast. Fruiting from late fall through winter.

EDIBILITY: Unknown.

COMMENTS: This beautiful *Cortinarius* can be identified by the deep red color of the cap, gills, and stipe, slender stature, ocher-yellow basal mycelium, and growth under conifers. It's a wonderful dye mushroom, giving red, orange, and purple colors, but is rarely abundant enough in California to collect. *Tubaria punicea* is similarly colored but grows on woody debris near the base of rotting Madrones and has whitish or pinkish basal mycelium and a strongly hygrophanous cap, and does not turn vinaceous in KOH. *C. sierraensis* is a closely related species that is more orange-red; it's known only from Lodgepole Pine (*Pinus contorta* subsp. *murrayana*) forests in the mountains. *C. smithii* is stockier, with a deep red cap and yellow stipe. *C. californicus* is larger and deep reddish orange when young, and has a bright purple KOH reaction.

SYNONYM: Formerly lumped in with *C. sanguineus* (=*Dermocybe sanguinea*), a genetically distinct species occurring in eastern North America and Europe.

Cortinarius smithii

Ammirati, Niskanen & Liimat.

WESTERN RED DYE

CAP: 2–8 (12) cm across, broadly conical to convex, becoming broadly convex with a low broad umbo. Deep burgundy-red, purple-red, or maroon when wet, burgundy-red to reddish brown when dry. Occasional forms with mustard yellow to honey brown caps also occur. Surface smooth to finely fibrillose, moist to dry. **GILLS:** Broadly attached, often with a decurrent tooth, close to subdistant. Deep burgundy-red to maroon when young, remaining burgundy-red or becoming deep orangish red to rusty orange as spores mature. **STIPE:** 3–15 cm long, 0.5–2 cm thick, equal, or with a slightly enlarged base. Yellow, yellowish orange to yellowish beige, becoming paler, more dingy yellowish brown in age, basal mycelium pale yellowish white, staining pale pinkish orange when handled. Surface dry, often with silky longitudinal fibers. **PARTIAL VEIL:** Faint, cortina-like, leaving some silky fibrils or disappearing altogether. Bright yellow at first, soon fading. **FLESH:** Thin to moderately thick, fibrous. Reddish under cap skin, more reddish brown in cap flesh, transitioning to yellow or yellow-brown in stipe, dingy yellow-brown in age. Stipe stuffed with yellowish pith when young, sometimes becoming hollow. **ODOR:** Indistinct. **TASTE:** Indistinct. **KOH:** Purple-black on cap. **SPORE DEPOSIT:** Rusty brown. **MICROSCOPY:** Spores 6.5–8 x 4–5 μm, ellipsoid, minutely roughened.

ECOLOGY: Solitary, scattered, or in small clusters in duff or moss. Usually under conifers, especially pine, but occasionally under manzanita or Madrone, especially in southern part of range. Known from as far south as the Santa Cruz Mountains, but rarely found south of the San Francisco Bay Area. Fruiting from late fall into spring, peaking in midwinter.

EDIBILITY: Unknown.

COMMENTS: Our most common brightly colored *Dermocybe* is readily identified by the deep burgundy-red cap and gills, yellowish stipe, and rusty orange spores. Like many *Dermocybe*, it is a wonderful dye mushroom, yielding vibrant, colorfast, red, orange, and purple colors on wool and silk. Although scattered fruitings occur throughout the mushroom season, it can be collected in abundance from late winter through early spring under pines on the Sonoma and Mendocino coast. *C. californicus* can look similar when young, but all parts are deep reddish orange when young, and it has a brighter purple KOH reaction. *C. neosanguineus* is a small, slender species that is entirely deep red, with dark purplish red flesh.

SYNONYM: Previously called *Cortinarius phoeniceus* var. *occidentalis* (=*Dermocybe phoenicea* var. *occidentalis*). Following most authorities, we have chosen to lump *Dermocybe* as a subgenus of *Cortinarius*. California collections with brown caps are often referred to *C. semisanguineus*, but genetic evidence suggests they are color variants of *C. smithii*.

Cortinarius californicus

A. H. Sm.

CALIFORNIA RED DYE

CAP: 2–10 cm across, rounded or conical when young, becoming broadly convex to plane, or somewhat wavy and irregular in age, often with a low, broad umbo. Deep burgundy-red to deep brownish red when young and wet, duller and more orange-red when dry, becoming more orange-brown in age. Surface dull, moist to dry, finely fibrillose around margin at first, becoming smooth and metallic looking. Strongly hygrophanous, fading as it dries. **GILLS:** Broadly attached, or with a shallow notch, close to subdistant. Intensely red to duller red or orange-red when young, becoming bright orange and finally rusty orange with mature spores. **STIPE:** 5–20 cm long, 0.5–2.5 cm thick, equal, or swollen toward the lower portion, but with a pointed, somewhat rooting base. Orangish red with a pinkish sheen when young, becoming orange to orange-brown in age. Surface dry, covered with orangish, silky fibrils when young, becoming nearly smooth in age. Basal mycelium orange-red. **PARTIAL VEIL:** Cortina-like, often scant or leaving silky fibrils on cap margin and stipe. Orangish in color. **FLESH:** Thin to moderately thick, fibrous or brittle. Marbled pinkish beige, reddish to orangish, brighter red or orange in cap. **ODOR:** Earthy to slightly radishlike. **TASTE:** Indistinct to slightly radishlike. **KOH:** Bright purple to inky purple on all parts. **SPORE DEPOSIT:** Rusty brown. **MICROSCOPY:** Spores 7.5–9.5 (11) x 4.6–5.8 (7.0) μm, ellipsoid to spindle shaped, roughened.

ECOLOGY: Solitary or scattered in small troops on ground, common throughout our range in a wide variety of forest types. Tanoak seems to be a preferred associate, but it occurs with most conifers. Fruiting from fall well into winter.

EDIBILITY: Unknown.

COMMENTS: The deep burgundy red cap that fades to orange, bright reddish to orange gills, and bright purple KOH reaction are distinctive features of this brightly colored *Telamonia*. It is often mistaken for a *Dermocybe*, but the hygrophanous cap, larger size, and stockier stature help separate it. This species produces a wonderful wool dye (best extracted in an alkaline dye bath, using alum as a mordant). *C. smithii* looks somewhat similar, but has a deep red cap and gills and a yellow stipe. *C. neosanguineus* is a smaller, slender species with a deep red cap, gills, and stipe, and ocher-yellow basal mycelium. Like most *Cortinarius*, this species "browns out" in age, losing its distinctive colors and becoming dingy orange-brown overall.

Cortinarius anthracinus (sensu CA)

CAP: 1–3 cm across, conical when young, remaining conical or becoming bell shaped or convex in age, usually with a well-defined umbo at all stages. Dark purplish brown to reddish brown with olive-green tones and a paler, salmon pinkish margin. Surface radially fibrillose, covered with buff, silky fibrils, occasionally finely scaly. Hygrophanous, fading to light brown or buff-brown as it dries. **GILLS:** Narrowly attached and with a distinct notch. Salmon pink to salmon buff when young, becoming salmon ocher with dark purplish color "bleeding" in from cap in age. **STIPE:** 3–8 cm long, 0.2–0.5 cm thick, tall, slender, more or less equal, or slightly swollen toward base. Purplish pink with a silvery sheen, covered with ocher veil remnants when young, becoming more purple-brown in age, handled areas deeper purple. Surface dry to moist, with faint longitudinal lines and fibrils and a matted-tomentose purplish orange base. **PARTIAL VEIL:** Sparse cortina, leaving a few silky fibrils on stipe and cap margin, ocher salmon in color. **FLESH:** Very thin, brittle, dark purplish brown in cap, in stipe fibrous, purplish to grayish purple. **ODOR:** Indistinct to slightly radishlike. **TASTE:** Indistinct. **KOH:** Deep vinaceous purple to inky purple on all parts. **SPORE DEPOSIT:** Often scant, rusty brown. **MICROSCOPY:** Spores 6–7 x 4–5 μm, narrowly ellipsoid, finely roughened to nearly smooth.

ECOLOGY: Solitary or scattered in small groups on ground in mossy northern conifer forest. Found in mixed Sitka Spruce, Western Hemlock, Grand Fir, and Douglas-fir forest on the Far North Coast. Rare in California. Fruiting from midfall into midwinter.

EDIBILITY: Unknown.

COMMENTS: The small size, conical to bell-shaped cap with purplish, reddish, or olive colors, salmon-toned gills, and slender pinkish purple stipe make this rare *Cortinarius* quite distinctive. It is slightly different from the "true" European *C. anthracinus*, but no alternative names have been proposed. *C. bibulus* is a small species with a more royal purple to purple-brown cap, and grows with alder trees. The *C. flexipes* group and *C. hemitrichus* can be similar in stature, but lack the deep purplish cap colors and salmon gills.

Cortinarius obtusus

(Fr.) Fr.

CAP: 1–4 (5) cm across, conical to bell shaped, expanding to plane with a well-defined pointed or rounded umbo at all times. Warm brown to orangish brown when wet; hygrophanous, fading to cinnamon to beige when dry. Surface smooth, moist to dry, often with whitish veil remnants around margin when young. Margin translucent-striate. **GILLS:** Narrowly attached, often with a broad notch, close to subdistant. Cinnamon buff at first, orange-brown in age. **STIPE:** 4–10 cm long, 0.2–0.7 cm thick, tall, slender, equal with a tapered and often pointed base. Pale buff to light cinnamon brown, covered with silky white fibrils. **PARTIAL VEIL:** Thick, white cortina, leaving remnants on cap margin and silky fibrils on stipe. **FLESH:** Thin, fragile, fibrous in stipe but becoming hollow, pale buff to light brown. **ODOR:** Mild to radishlike. **TASTE:** Indistinct. **KOH:** No reaction. **SPORE DEPOSIT:** Rusty brown. **MICROSCOPY:** Spores 6–8 x 4–5 μm, broadly ellipsoid, minutely roughened.

ECOLOGY: Often in loose clustered or scattered troops, occasionally solitary in duff under conifers, especially pine. Common on the Far North Coast, occasional elsewhere. Fruiting from midfall into winter.

EDIBILITY: Unknown.

COMMENTS: The conical-umbonate ocher caps, white cortina, and slender stature are distinctive. The very similar *C. acutus* is smaller, with a pointed and often semitranslucent umbo; it grows with spruce. *C. gualalaensis* is larger, grows in clusters under pine, and has a darker, rounded cap with flesh that bruises purplish brown. *C. flexipes* has a darker brown, finely scaly cap with a pointed umbo, a shorter stature, and often a sweet odor.

Cortinarius acutus

(Pers.) Fr.

ACUTE CORT

CAP: 0.4–1.8 cm across, conical when young, remaining conical or becoming bell shaped or broadly conical in age, almost always with an acute (pointed), semitranslucent umbo at all stages. Warm light brown, cinnamon tan to cinnamon brown when wet, umbo translucent cinnamon tan to buff; hygrophanous, becoming paler to whitish tan when dry. Surface covered with silky white fibrils (most visible when dry), margin translucent-striate when wet. **GILLS:** Broadly attached, moderately close. Pale cinnamon to cinnamon tan when young, becoming cinnamon orange in age. **STIPE:** 1–5 cm long, 0.1–0.3 cm thick, more or less equal, or pointed slightly toward base. Pale cinnamon buff base color, covered with whitish fibrils, becoming more cinnamon colored as fibrils wear off. Surface dry, covered with silky longitudinal fibrils. **PARTIAL VEIL:** Cortina-like, often sheathing to cap margin, leaving white fibrils on stipe and cap margin. **FLESH:** Very thin, brittle, fibrous in stipe. Pale cinnamon tan. **TASTE:** Indistinct to slightly radishlike. **ODOR:** Indistinct to slightly radishlike. **KOH:** No reaction. **SPORE DEPOSIT:** Scant, rusty brown. **MICROSCOPY:** Spores 6.5–8 (9) x 4–5 μm, ellipsoid, finely roughened to nearly smooth.

ECOLOGY: Solitary or scattered in troops on mossy humus under Sitka Spruce. Locally very common on the Far North Coast, absent elsewhere. Fruiting from late fall into winter.

EDIBILITY: Unknown.

COMMENTS: The small size, cinnamon-colored, acutely conical cap with a semitranslucent pointed umbo, and rusty orange spores help identify this dainty *Cortinarius*. It can carpet the floor of spruce forests, but is easily overlooked among swarms of *Mycena* and *Galerina*. Compare with *C. obtusus*, which is quite similar but slightly larger and stockier, and with a broader umbo. Some *Galerina* species (such as *G. vittiformis*) resemble *C. acutus*, but are usually smaller, with rounded-umbonate caps, without the white cortina when young, and with stipes that darken from the base up.

Cortinarius laniger group

CAP: 3–9 cm across, rounded to conical when young, expanding to broadly convex, nearly plane or wavy in age; often with a low, broad umbo. Cinnamon brown to ocher-brown or light reddish brown when wet, often with a whitish band around margin when young, in age developing darker streaks overall. Hygrophanous, fading to beige-tan when dry, then "browning out" in age. Surface dry, smooth or nearly so, often with small, matted fibrils when young. **GILLS:** Narrowly attached with a distinct notch, close to subdistant. Cinnamon buff at first, becoming bright orange-tan to ocher, then browner as spores mature. **STIPE:** 4–12 cm long, 0.7–2 cm thick at apex, equal or with a swollen to distinctly bulbous base. Usually extensively whitish, becoming pale cinnamon, bruising cinnamon tan when handled and developing grayish tan discolorations in age. Surface dry, covered in silky fibrils below veil. **PARTIAL VEIL:** Scant, cortina-like, leaving a faint, evanescent annular zone of silky whitish fibrils on stipe and cap margin. **FLESH:** Thick to moderately thin, firm, fleshy in cap, fibrous in stipe. Cinnamon brown, light brown to pale tan. **ODOR:** Faint, sweet, earthy, or radishlike. **TASTE:** Mild to earthy. **KOH:** Indistinct to faintly greenish gray in some forms. **SPORE DEPOSIT:** Rusty brown. **MICROSCOPY:** Spores 8.5–10.5 (11.5) x 5–6.5 μm, ellipsoid to broadly ellipsoid, moderately roughened.

ECOLOGY: Solitary or scattered on ground in mixed coastal forest. Prefers conifers, especially pine and Douglas-fir, but might be associated with Tanoak as well. Widespread throughout our range, but generally uncommon. Fruiting in late fall and winter.

EDIBILITY: Unknown.

COMMENTS: The large size, cinnamon tan cap, rather bright ocher cinnamon gills, and whitish stipe help identify this species group. A few very similar species can be differentiated by such subtleties such as young gill color and spore characters. The name *C. solis-occasus* has been used for specimens with lilac-purple cortina fibers. *C. subbalaustinus* has a shorter stature and a lightly streaked cap; it grows with a number of ectomycorrhizal trees in urban areas. *C. bivelus* has been introduced from Europe to the Pacific Northwest, where it grows with planted birch; it is not yet known from California, but can be expected to occur here eventually. This group is badly in need of modern taxonomic treatment and for now is referred to as *C. laniger*.

Cortinarius rubicundulus

(Rea) A. Pearson
YELLOW-STAINING CORT

CAP: 3–9 cm across, rounded to broadly bell shaped when young, becoming broadly convex to plane and irregular or uplifted in age. Creamy beige to yellowish beige at first, developing reddish streaks and stains, becoming ocher to orange-brown overall in age. Bruising bright yellow-orange, slowly darkening to ocher-red when damaged. Surface dry to thinly viscid, slightly fibrillose, smooth or with reddish appressed fibrils. **GILLS:** Broadly attached, close to subdistant. Beige at first, darkening to tan or light brown, developing orangish browns stains and becoming orange-brown to rusty brown as spores mature. **STIPE:** 4–8 (12) cm long, 0.5–2 cm thick, often club shaped, enlarging downward to a swollen or bulbous lower portion, often with a tapered, rooting base below that. Off-white to creamy beige at first, developing ocher-orange streaks and discoloration from base up, becoming ocher-orange to orange-brown in age. Bruising like cap. Surface dry, slightly fibrillose. **PARTIAL VEIL:** Scant, cortina-like, leaving a annular zone of silky whitish to pale ocher fibrils on stipe that become rusty brown with spores. **FLESH:** Thin to moderately thick, fleshy in cap, fibrous in stipe. Whitish to buff. Bruising bright yellow-orange, slowly becoming ocher-red to reddish orange on all parts. **ODOR:** Indistinct. **TASTE:** Indistinct. **KOH:** No reaction. **SPORE DEPOSIT:** Rusty brown. **MICROSCOPY:** Spores 6–8 x 3.5–4 μm, almond shaped to ellipsoid, slightly roughened.

ECOLOGY: Solitary, scattered, or in small clusters on the ground or in mossy forest throughout our range. Often the most prolific *Cortinarius* in Sitka Spruce forests on the Far North Coast; elsewhere fairly common under Douglas-fir, Tanoak, and pine (uncommon south of Sonoma County). Fruiting from early fall into early winter in north, late fall through winter southward.

EDIBILITY: Unknown.

COMMENTS: The best features to identify this mushroom are the pale creamy beige cap and bright yellow-orange to ocher-red or reddish staining. Although it superficially resembles a number of other *Cortinarius* in California, none show this color-staining reaction.

Cortinarius infractus group

BITTER OLIVE CORT

CAP: 3–10 cm across, rounded to convex when young, expanding to broadly convex to nearly plane; often becoming wavy in age. Dark grayish olive, or dark grayish blue in variety *obscurocyaneus* (see comments), becoming paler olive-brown to gray-brown, to gray or yellowish brown in age, often with darker streaks and with inconspicuous concentric zones and darker spots around margin. Surface slightly viscid to dry, smooth or with appressed fibrils. **GILLS:** Narrowly attached, with a distinct notch, close to moderately crowded. Olive-gray to dark olive, or olivaceous gray with violet tones, becoming more cinnamon brown to rust brown as the spores mature, but usually retaining some olive tones well into maturity. **STIPE:** 3–10 cm long, 0.7–1.5 cm thick at apex, more or less equal, or with a swollen base, but occasionally tapering or pointed at base. Whitish with olivaceous tinge and a bright white base when young, becoming dingy grayish to brownish, but often staying white at base. Surface dry, covered with silky fibrils, below a silky cortina. **PARTIAL VEIL:** Scant, leaving a faint, evanescent annular zone of fibrils on stipe. **FLESH:** Fairly firm, fleshy to fibrous. Off-white but often tinged with olivaceous tones (or bluish to violet-gray in var. *obscurocyaneus*) at stipe apex, becoming grayish. **ODOR:** Indistinct. **TASTE:** Bitter. **KOH:** Indistinct or darkening slightly to grayish. **SPORE DEPOSIT:** Rusty brown. **MICROSCOPY:** Spores 6–8 (9) x 5–6.5 (7.5) μm, subglobose to nearly globose, roughened.

ECOLOGY: Solitary, scattered, or in small clusters on ground in conifer and hardwood forests, especially under Douglas-fir and Tanoak. Widespread and common. Fruiting in fall in the north, well into winter in the southern part of our range.

EDIBILITY: Unknown.

COMMENTS: Members of this species complex can be identified by the combination of grayish olive colors, bitter taste, and rusty brown spore deposit. There are likely at least three different species in California. Fruitbodies showing bluish or violet tones on the gills, cap, or upper stipe have often been referred to as var. *obscurocyaneus* (one of the many named varieties of this species); such specimens are especially common under Tanoak in the southern half of our range. Some members of the *C. anomalus* group can look similar, but are usually smaller and lack the bitter taste.

Cortinarius regalis

A. H. Sm. & P. M. Rea

COTTON-CAPPED CORT

CAP: 4–15 cm across, convex to broadly convex with a downcurved margin, then plane, to uplifted and wavy in age. Dingy whitish to gray-brown or beige-brown, generally darker in age. Surface dry, innately fibrillose, adorned with patches of adhering bright white cottony tissue that often disappears in age or when rain washed. **GILLS:** Broadly attached, dull tan when young, dingy to ocher-brown in age. **STIPE:** 5–12 cm long, 1.5–3 cm thick at apex, cylindrical, base ranging from merely club shaped to strongly bulbous (up to 6 cm across), generally more markedly lobed or rimmed in young buttons. White, usually remaining so or becoming slightly dingier with brownish tones. **PARTIAL VEIL:** White cobwebby cortina, soon becoming very scant or disappearing in most specimens, some retaining a more noticeable cortina in age. **FLESH:** Thick, firm, whitish, usually finely marbled with dirty beige or dingy pallid ocher-brown, usually darker toward base. **ODOR:** Indistinct. **TASTE:** Indistinct. **KOH:** No reaction. **SPORE DEPOSIT:** Rusty brown. **MICROSCOPY:** Spores 10–12 x 6–7 μm, bean shaped to broadly ellipsoid, moderately roughened.

ECOLOGY: Solitary or in troops under live oak. Fairly common from the San Francisco Bay Area south, but fruitings are irregular (abundant in some years, nearly absent in others). Fruiting from early fall into winter.

EDIBILITY: Unknown.

COMMENTS: Among *Cortinarius*, this species stands out as something of an anomaly: the stipe base is often swollen and rimmed like a *Phlegmacium*, but the cap is totally dry, and the colors are brownish overall like many *Telamonia*.

Cortinarius cisqhale

Bojantchev

CAP: 3–12 cm across, rounded to convex when young, becoming broadly convex to nearly plane, often with a low, rounded umbo. Color extremely variable, dark brown to dark gray-brown, often with olive tones, occasionally with obscure bluish tones when young, becoming paler and developing more olive or rusty tones in age. Surface moist to dry, often with a narrow and strongly contrasting pale band of partial veil remnants around margin. Strongly hygrophanous, often fading in concentric zones and often shiny when dry. **GILLS:** Broadly attached, often with a small notch, close to subdistant. Gray-brown to clay brown, occasionally with a bluish cast when young, becoming dark clay brown to rusty brown as spores mature. **STIPE:** 5–12 cm long, 1–4 cm thick, more or less equal or club shaped. Pale grayish brown to dark brown with whitish to pale gray veil tissue. Surface dry, fibrillose-silky. **PARTIAL VEIL:** Cortina-like, leaving an annular zone of silky whitish fibrils on stipe, becoming matted or disappearing altogether in age. **FLESH:** Whitish to gray marbled with brownish spots, often with violet to bluish coloration in stipe apex when young. **ODOR:** Faint, earthy. **TASTE:** Earthy. **KOH:** No reaction. **SPORE DEPOSIT:** Rusty brown. **MICROSCOPY:** Spores 6.5–9 x 4–5 μm, averaging 7.5 x 4.4 μm, almond shaped to ellipsoid, finely roughened.

ECOLOGY: Solitary, scattered, or in clusters in duff. Common under Tanoak throughout our range, but also grows with live oaks. Fruiting from midfall through early winter.

EDIBILITY: Unknown.

COMMENTS: This ubiquitous *Telamonia* can occur by the hundreds in Tanoak forests. Like the caps of many of its brethren, the color is extremely variable, depending on age and moisture content. Typical forms can be identified by the rather dark brown, hygrophanous cap (often with olive tones), whitish band around the edge of the cap (at least when young), and lack of pronounced violet tones. *C. athabascus* generally has violet tones (at least when young) and grows with conifers. *C. ohlone* is a smaller, often clustered species with a white stipe that grows primarily under live oak.

Cortinarius ohlone

Bojantchev

CAP: 3–8 cm across, convex when young, becoming broadly convex to plane or irregular and slightly uplifted in age, occasionally with a low, broad umbo. Dark brown to gray-brown with a bluish gray cast and a silvery sheen of pale fibrils. Duller brown to cinnamon brown in age. Surface dry, smooth or nearly so. **GILLS:** Broadly attached, often with a small notch, close to subdistant. Grayish brown to clay brown at first, becoming orangish brown to rusty brown as spores mature. **STIPE:** 3–10 (12) cm long, 0.5–2 cm thick, equal or tapering at base, especially when clustered. White to off-white, occasionally with grayish tan discoloration. Surface dry, silky-fibrillose below a well-defined ring when young, becoming smooth or nearly so in age. **PARTIAL VEIL:** Cortina-like, fairly thick, leaving an often upward-flaring annular zone of fibrils on stipe, soon matted and covered with rusty spores or disappearing in age. **FLESH:** Fairly thick, fleshy in cap, fibrous in stipe. Marbled brown in cap, transitioning to whitish in stipe. **ODOR:** Mildly spicy, similar to cedarwood. **TASTE:** Indistinct. **KOH:** Yellow on stipe, trivial darkening on cap. **SPORE DEPOSIT:** Rusty brown. **MICROSCOPY:** Spores 7–9 x 5–6 µm, averaging 7.8 x 5.2 µm, almond shaped to ellipsoid, finely roughened.

ECOLOGY: Often in clusters, but can be scattered or solitary on ground under oak, especially Coast Live Oak, from late fall into early spring. Very common from the greater San Francisco Bay Area south (northern extent of range unknown).

EDIBILITY: Unknown.

COMMENTS: When young, this *Telamonia* can be identified by the cool brown cap with a silvery sheen, bright white stipe, clustered growth under live oak, and spicy odor. As it matures, it loses its distinctive color and "browns out," making it really tough to separate from other *Telamonia*. According to California's *Cortinarius* expert Dimitar Bojantchev, the slightly spicy cedar-wood odor is a feature not present in any similar species growing with live oak in our area. *C. gualalaensis* is quite similar, but has flesh that bruises purplish brown and grows under pines north of the San Francisco Bay Area.

Cortinarius athabascus

Bojantchev

CAP: 3–9 cm across, conical to bell shaped when young, becoming broadly bell shaped to convex with a low, rounded umbo and often with concentrically raised or depressed steps or rings. Color extremely variable, often dark brown to dark gray-brown with violet or bluish tones when young, losing the violet tones and fading to brown in age. Paler forms show grayish lilac to grayish buff caps. Surface moist to dry, often with a silky sheen of appressed fibrils and a band of partial veil remnants around margin when young. Strongly hygrophanous, often fading in concentric zones, dull when wet, often shiny when dry. **GILLS:** Broadly attached, but often with a small notch, close to subdistant. Lilac-gray to gray-tan when young, becoming clay brown and finally rusty brown as spores mature. **STIPE:** 5–10 (15) long, 0.5–2.5 cm thick, more or less equal or club shaped, swollen slightly at base. Pale grayish brown with violet tinge at apex and covered with felty-fibrillose white veil remnants when young, becoming light dingy brown in age. Surface dry, felty-fibrillose base below a silky annular zone. **PARTIAL VEIL:** Cortina-like, leaving an annular zone of silky whitish fibrils on stipe, become matted and rusty brown with spores and often disappearing altogether in age. **FLESH:** Moderately thin, fleshy in cap, fibrous in stipe. Whitish, marbled or streaked with purplish or light brown spots in cap, pronounced violet to bluish coloration at stipe apex and often with orange-brown spots at base. **ODOR:** Mild, earthy. **TASTE:** Indistinct to slightly earthy. **KOH:** Indistinct or darking slightly. **SPORE DEPOSIT:** Rusty brown. **MICROSCOPY:** Spores 8–10.5 x 5–6 μm, averaging 9.2 x 5.6 μm, almond shaped to ellipsoid, finely roughened.

ECOLOGY: Solitary, scattered, or in clusters on ground in mixed coastal forests. Fruiting from fall through early winter. Known from the greater San Francisco Bay Area north, may occur in the Santa Cruz Mountains as well. Appears to associate with many different conifers, particularly Douglas-fir in the southern part of range, Sitka Spruce and Western Hemlock in the north.

EDIBILITY: Unknown.

COMMENTS: The cap color is highly variable, but usually some shade of brown to gray-brown, with pronounced violet or bluish tones when young (these are often quickly lost). The flesh in the upper stipe retains the bluish violet well into maturity, which helps separate it from the many similar species. *C. evernius* has a reddish brown cap with violet hues when young, a pale bluish violet stipe, violet flesh, and longer spores (9–11.5 x 5–6.5 μm). *C. biformis* has a slightly glutinous cap with short striations at the margin when wet, a thicker veil, and slightly smaller spores (7–9 x 5–6 μm). *C. cisqhale* usually has a darker brown cap with olive tones and a narrow but strongly contrasting whitish margin; it grows with Tanoak and live oak.

Cortinarius hemitrichus

(Pers.) Fr.

CAP: 1–4 cm across, conical when young, becoming broadly convex to plane, often with an umbo. Dark grayish to brownish gray at first, becoming browner in age; hygrophanous, fading to beige as it dries out. Entirely covered with many fine silvery to pale gray scales (contrasting most when young). Surface dry, finely scaly overall, occasionally nearly smooth. **GILLS:** Attached, often slightly notched, close to crowded. Dark gray-brown to gray-buff when young, becoming rusty grayish to rusty brown as the spores mature. **STIPE:** 3–7 cm long, 0.4–1 cm thick, equal, or slightly swollen toward base. Nearly smooth, covered with silvery fibrils when young. Grayish to violet-gray above the veil. Silky fibrils wearing off or become matted in age, exposing a grayish to brownish color. **PARTIAL VEIL:** Fairly heavy silvery cortina that leaves a few silky fibrils on cap margin and a silky annular zone on stipe in age. **FLESH:** Thin, brittle in cap, fibrous in stipe, tan to brownish. **ODOR:** Indistinct. **TASTE:** Indistinct. **KOH:** Dingy dark reddish or dark vinaceous purple. **SPORE DEPOSIT:** Often scant, rusty brown. **MICROSCOPY:** Spores 7–8.5 x 4–5.5 μm, ellipsoid, slightly roughened.

ECOLOGY: Solitary or scattered in small groups or clusters on ground under introduced birch trees, often in lawns or grassy areas. Uncommon and easily overlooked. Generally fruiting in late fall and winter, often with a second flush in late spring (but can occur any time of year).

EDIBILITY: Unknown.

COMMENTS: The small size, often somewhat conical cap covered with whitish tufts or scales, grayish gills, and growth under introduced birch trees are the primary identification features for this small Cort. It belongs to a confusing group of *Telamonia*; most of the Californian representatives remain undescribed. Members of the *C. flexipes* group are very similar, but generally have some purplish tones to the stipe and a sweet odor, and grow with conifers.

Cortinarius anomalovelatus

Ammirati, Berbee, Harrower, Liimat. & Niskanen

CAP: 0.5–3 cm across, rounded-conical when young, becoming convex to broadly convex. Silvery blue to silvery lilac base color, covered with silvery white fibrils; often developing an orangish center in age. Surface dry, covered with matted silky fibrils. **GILLS:** Broadly attached, close to moderately crowded. Bright bluish to lilac when young, developing gray tones and then becoming rusty orange as spores mature. **STIPE:** 3–10 cm long, 0.3–0.7 cm thick, long, slender, enlarged slightly downward to a swollen base. Silvery whitish, often more lilac near apex. Surface dry, with tufts or bands of whitish to pale buff, silky or felted fibrils. **PARTIAL VEIL:** Nearly membranous white cortina, leaving bands or patches on stipe as well as around cap margin. **FLESH:** Thin, brittle in cap, fibrous in stipe. Bluish gray to grayish lilac in cap, bright bluish purple to purplish gray in upper stipe, transitioning to whitish or pale buff toward base. **ODOR:** Indistinct. **TASTE:** Indistinct. **KOH:** No reaction. **SPORE DEPOSIT:** Rusty brown. **MICROSCOPY:** Spores 8–8.5 x 5–7 μm, round-ellipsoid to subglobose, moderately roughened.

ECOLOGY: Scattered or in small groups and clusters of two to three fruitbodies on ground or on very decayed mossy logs and stumps under Sitka Spruce and Western Hemlock. Uncommon on the Far North Coast, known from Humboldt County northward. First fruiting early in fall, often peaking in late fall or early winter, then in smaller numbers into spring.

EDIBILITY: Unknown.

COMMENTS: The small size, slender stature, slivery blue cap, blue to lilac-gray gills, and silky-felty partial veil remnants forming bands on the stipe help identify this beautiful, uncommon *Cortinarius*. It is a member of the *C. anomalus* group, a confusing complex with at least five different species in California, most of which appear to be undescribed. As a group, they can be told by their relatively small size, silvery bluish to lilac or beige colors, and subglobose spores.

Cortinarius anomalus group

CAP: 2–8 cm across, convex to broadly convex (sometimes slightly umbonate), nearly flat in age, occasionally wavy. Silvery gray, bluish or lilac-gray, brownish, tan, or pale beige, usually palest at margin. Most species becoming duller in age, often with a radially streaked appearance. Surface dry to moist, or in some species slightly viscid, innately silky-fibrillose (sometimes finely scaly) to nearly smooth in age. Margin opaque, sometimes with clinging wispy fibrils of cortina. **GILLS:** Finely attached or notched, close. Pallid to grayish blue or light sky blue when young, becoming tan to dull brown, often with ocher blotches, usually maturing to rusty orange. **STIPE:** 5–10 cm long, 0.4–1.5 cm thick, straight to slightly curved, base often swollen. Bluish lilac, whitish, beige or tan, becoming dingier in age. Surface silky-fibrillose, a few species with bands of ocher or brownish tissue. **PARTIAL VEIL:** Inconspicuous cortina usually set high on stipe, mostly disappearing in age. **FLESH:** Thin, fibrous, pallid. **ODOR:** Rarely distinctive. **TASTE:** Indistinct. **KOH:** Rarely producing distinctive reactions, sometimes faintly grayish olive. **SPORE DEPOSIT:** Rusty brown. **MICROSCOPY:** Spores rarely longer than 7.5 µm, often less than 6 µm long, round to globose-ellipsoid, weakly roughened (occasionally coarsely warted).

ECOLOGY: Solitary or in small clumps in forested settings. Some species restricted to conifer habitats, others primarily found with hardwoods. Fruiting primarily in late fall through winter.

EDIBILITY: Probably nontoxic, although one should not experiment with *Cortinarius*.

COMMENTS: This subgroup of *Cortinarius* is characterized by fairly small, slender fruitbodies and bluish or lilac tones of the gills and upper stipe when young. The stipes are never viscid, but the caps of some species are. A few species lack blue tones almost entirely, but their small fruitbodies and smallish, round spores indicate their relation to this group. Western species concepts for these species are weak and/or complex and difficult to sort out. There appear to be at least five species in California, all of which remain undescribed. Also see *C. anomalovelatus* (p. 180), a distinctive and recently named species in this group. A species to look out for on the Far North Coast is *C. caesiifolius*—it has slightly more elongate spores, tan girdles around the stipe, and an unpleasant smell. Less similar species include *C. alboviolaceus* and others in the *Sericeocybe* group, but these usually have more prominent cobwebby cortinas and larger fruitbodies, and often distinctive strong odors.

Cortinarius "Winter Blues"

CAP: 2–8 cm across, conical when young, becoming broadly conical to convex, often with a low, broad umbo. Margin inrolled, often squared off when young, becoming downcurved to straight, then wavy and occasionally uplifted in age. Strongly hygrophanous, color extremely variable. When wet: deep purple, becoming silvery purple and finally purplish tan to purple-brown. In dry conditions silvery blue to bluish purple with appressed silvery fibrils, becoming bluish tan to purplish tan and finally grayish tan. Surface smooth, shiny, moist to slightly viscid when wet; dull, covered with appressed fibrils when dry. **GILLS:** Broadly attached or with a small notch, close to moderately crowded. Beige at first, becoming cinnamon tan and then darker ocher-brown as spores mature. **STIPE:** 4–14 cm long, 0.5–1.5 cm thick, equal or slightly enlarged downward, then with a tapered, "rooting" base. Whitish with a band of violet or bluish color below partial veil, developing dingy tan stains in age. Surface dry to slightly moist, covered with silky fibrils when young, smooth in age. **PARTIAL VEIL:** Thin, sparse, cortina-like, binding to cap margin and leaving a silky annular zone of whitish fibrils on stipe. **FLESH:** Thin, brittle, purplish to purple-brown in cap; fibrous, solid to stuffed, tan in stipe. **ODOR:** Indistinct. **TASTE:** Indistinct. **KOH:** No reaction. **SPORE DEPOSIT:** Rusty brown. **MICROSCOPY:** Spores 8.5–10.5 x 5–6 μm, almond shaped to lemon shaped (with small knobs at either end), weakly roughened to nearly smooth.

ECOLOGY: Solitary or scattered on ground in colder winter months. Rare, only known from a few locations on the North Coast. Most frequently encountered in coastal Sitka Spruce forest in Humboldt County, but also in mixed Western Hemlock and Grand Fir forest in Mendocino County.

EDIBILITY: Unknown.

COMMENTS: This beautiful winter-fruiting *Cortinarius* can look like two different species depending on the dryness of the fruiting conditions—markedly purple when wet, silvery blue when dry. In addition to the variable cap color, this species can be recognized by the pallid gills, somewhat rooting stipe with purplish blushes, and preference for fruiting late in the season. In Humboldt County, it fruits about a month after most other *Cortinarius* have gone by. Although it appears to be rare and quite range restricted, Joann Olson, a *Cortinarius* expert from Humboldt County, finds it every winter. It bears a resemblance to the members of the *C. evernius* and *C. biformis* groups, but has a deeper purplish blue color and lacks thick veil bands on the stipe. The much smaller, more slender *C. bibulus* has a deep purple to purplish brown, umbonate cap and grows in wet areas with alder. *C. iodeoides* has a deep bluish purple cap, is viscid on both cap and stipe, and has a decidedly bitter taste.

Cortinarius violaceus

(L.) Gray

VIOLET CORT

CAP: 4–10 (13) cm across, rounded to broadly conical when young, becoming broadly convex to nearly plane. Deep blackish violet to deep purple, becoming paler royal purple or graying slightly in age. Surface dry, somewhat velvety, covered in upturned tufts of hairs that become matted in age. Occasionally cracking when dry. **GILLS:** Broadly attached with a distinct notch, moderately close at first, becoming well spaced. Deep blackish violet to deep purple when young, becoming grayish purple and finally dusted with rusty brown spores when mature. **STIPE:** 6–15 (18) cm long, 1–2.5 cm thick at apex, often club shaped with a swollen or bulbous base (up to 4 cm across). Deep blackish violet to deep purple with a sheen of paler grayish violet fibrils, darker purple where handled. Surface dry, covered with longitudinal fibrils and a silky annular zone. Basal mycelium bright violet to whitish. **PARTIAL VEIL:** Cobwebby cortina, leaving a prominent zone of fibrils on stipe that becomes covered in rusty brown spores. **FLESH:** Thick, fibrous to somewhat brittle in stipe. Dark violet to violet-gray, becoming more gray in age, staining deep purple, especially in stipe when young. **TASTE:** Indistinct. **ODOR:** Faint, somewhat cedarlike. **KOH:** Deep blood red on all parts. **SPORE DEPOSIT:** Reddish brown to orange-brown. **MICROSCOPY:** Spores 13–17 x 7–10 μm, ellipsoid to almond shaped, distinctly roughened. Pleurocystidia and cheilocystidia abundant, bottle shaped with elongated necks.

ECOLOGY: Solitary or scattered, but rarely in large numbers, in humus in a variety of forest types: on the Far North Coast usually with Sitka Spruce or Western Hemlock, in the central part of our range with Douglas-fir or Tanoak, occasionally with Coast Live Oak and pine in the Santa Cruz Mountains. Occasional from Humboldt County north, uncommon to south. Fruiting from fall into winter.

EDIBILITY: Edible; reportedly "earthy" or with a slight bitter metallic flavor.

COMMENTS: This beautiful *Cortinarius* can easily be recognized by the dry, velvety to coarsely hairy cap, blackish violet to deep purple coloration, and rusty brown spores covering a cobwebby cortina. Some *Leptonia* can have similar colors and cap texture, but are smaller and have pinkish buff spores and paler gills. Recent findings suggest that *C. violaceus* is a widespread and variable species that encompasses the larger conifer-associated forms often identified as *C. hercynicus* (=*C. violaceus* var. *hercynicus*).

Cortinarius camphoratus

(Fr.) Fr.

STINKY CORT

CAP: 5–10 (12) cm across, rounded to convex when young, becoming broadly convex to nearly plane. Margin downcurved or incurved and often lobed at first, often wavy and irregular in age. Grayish lilac, silvery violet to silvery gray, rarely entirely violet when young; slowly losing lilac and violet color as it ages, becoming silver to gray and developing orangish tones at center. Surface viscid to dry, smooth, or occasionally with appressed fibrils or tiny scales. **GILLS:** Broadly to narrowly attached, often with a distinct notch, close to crowded. Violet-purple at first, becoming violet-gray to lilac-gray and developing rusty brown tones as spores mature. **STIPE:** 5–10 (15) cm long, 1–2 cm thick at the apex, 1–3.5 cm thick at base, variable in shape, often club shaped with a swollen base, but can be distinctly bulbous or even a tapered to a pointed base. Silvery violet to violet, covered with longitudinal whitish fibrils and bruising violet when handled, becoming dingy grayish violet and developing orangish brown stains in age. Surface dry, covered with silky fibrils that become matted in age. **PARTIAL VEIL:** Cortina-like, leaving an annular zone of whitish to lilac fibrils on the stipe (stained rusty brown with spores in age), and often a band or patches of silky fibrils on cap margin. **FLESH:** Thick, firm. Violet to violet-gray marbled with paler whitish, lower parts of stipe developing tan to orangish brown colors in age. **ODOR:** Strong, distinct; putrid and unpleasant, especially in age. Often compared to rotting potatoes. **TASTE:** Unpleasant. **KOH:** No reaction. **SPORE DEPOSIT:** Rusty brown.

MICROSCOPY: Spores (9) 9.5–10.5 x (5.5) 6–6.5 μm, ellipsoid to almond shaped, slightly roughened.

ECOLOGY: Solitary, scattered, or in small clusters on ground in coniferous forests from late fall into winter. Known from Sonoma County north, uncommon south of central Mendocino County.

EDIBILITY: Likely nontoxic.

COMMENTS: The grayish lilac cap, violet-purple gills that fade to grayish lilac, rusty brown spores, marbled violet-gray flesh, and putrid odor make this *Cortinarius* distinctive. The memorable odor has been described in many different ways, from "burnt flesh or burnt horn" (Brandrud) to "cold mashed potatoes and goat cheese" (Courtecuisse), to simply "smelling like goats." It can become so nauseating that we have thrown collections out of the car trunk in disgust. The fairly similar *C. traganus* has a pearlike, fruity odor when young that becomes more unpleasant in age; it is readily told apart by its light brown to cinnamon brown gills and marbled rusty brown flesh. *C. alboviolaceus* has a silvery violet cap, lilac-gray to brownish gills, and pale to lilac-gray marbled flesh, and lacks a distinctive odor.

NOMENCLATURE NOTE: The recently described and poorly known *C. putorius* supposedly has more purple colors when young, white veil tissue, and slightly smaller spores.

Cortinarius traganus

(Fr.) Fr.

LILAC CORT

CAP: 4–10 (12) cm across, rounded to convex when young, becoming broadly convex to nearly plane, occasionally wavy in age. Lilac-purple to bluish lilac when young, fading to lilac-gray, occasionally grayish overall, or at times extensively ocher. Surface dry, with silky, matted fibrils when young, often cracking into irregular patches or upturned scales over center in age or when dry. **GILLS:** Broadly to narrowly attached, often with a distinct notch, close to crowded. Light brown to cinnamon brown at first, becoming rusty brown. **STIPE:** 4–10 cm long, 0.7–2 cm thick at apex, 1–3.5 cm thick at base, club shaped, with a swollen or distinctly bulbous base. Lilac-purple to grayish lilac, covered with irregular belts or patches of white to lilac fibrils on lower portion, developing orangish discolorations from base up in age. Surface dry, covered with silky to flocculent fibrils that become matted in age, below a silky cortina. **PARTIAL VEIL:** Thick cortina, leaving an annular zone of whitish to lilac fibrils on stipe, soon covered in rusty brown spores; often leaving a band of silky fibrils on cap margin. **FLESH:** Thick, firm, fleshy to fibrous. Rusty brown to orange-brown, with creamy or tan marbling. **ODOR:** Usually strong; a pleasant, sweet, fruity odor often described as pearlike when young, becoming stronger and disagreeable in age. **TASTE:** Earthy. **KOH:** No reaction. **SPORE DEPOSIT:** Rusty brown. **MICROSCOPY:** Spores 7–10 x 5–6 μm, ellipsoid to almond shaped, roughened.

ECOLOGY: Solitary or scattered on ground in a wide variety of forest types. Like a lot of *Cortinarius*, it seems to prefer Tanoak in southern part of range, but becomes more common under conifers, especially Western Hemlock and Sitka Spruce, in the northern part of California. Common and widespread. Fruiting from fall through winter.

EDIBILITY: Likely nontoxic.

COMMENTS: The dry, lilac-purple cap, marbled orange-brown flesh, and pearlike fruity odor are distinctive. Old *C. traganus* can have an unpleasant odor similar to *C. camphoratus*, but the odor is usually still somewhat sweet and slightly less disgusting. *C. camphoratus* also differs by having violet-purple gills and violet-gray marbled flesh in the cap (sometimes marbled rusty brown, but only in the lower stipe). *C. alboviolaceus* has a silvery violet cap, violet to lilac-gray gills, and violet marbled flesh, and lacks a distinctive odor. Care should be taken not to mistake these violet *Cortinarius* with the edible *Clitocybe nuda*, which has a smooth, greasy cap, and pale pinkish buff to lilac-buff (not rusty brown) spores, and lacks a cortina.

Cortinarius alboviolaceus

(Pers.) Fr.

SILVERY-VIOLET CORT

CAP: 3–7 (10) cm across, rounded to convex when young, becoming broadly convex to nearly plane or occasionally wavy in age. Silvery lilac to silvery white with a slight bluish violet cast when young, becoming silvery gray to grayish lilac, developing some blotchy orangish coloration in age. Surface dry, silky and often with small tufts of fibrils when young, more appressed-fibrillose to smooth in age. **GILLS:** Broadly to narrowly attached, often with a distinct notch, crowded to subdistant. Lilac or brownish at first, becoming lilac-gray and developing rusty tones as spores mature. **STIPE:** 5–10 cm long, 1–2 cm thick at the apex, enlarging downward to a swollen base. Silvery violet to pale lilac, covered with silvery white, silky to felty fibrils on lower portion. Surface dry. **PARTIAL VEIL:** Cortina-like, leaving a zone of whitish to grayish fibrils on stipe, which become covered in rusty brown spores; often leaving a band of silky fibrils around cap margin. **FLESH:** Thick, firm to soft, whitish gray marbled with lilac, especially in stipe apex. Occasionally becoming orangish in lower stipe. **ODOR:** Indistinct. **TASTE:** Indistinct. **KOH:** No reaction. **SPORE DEPOSIT:** Rusty brown. **MICROSCOPY:** Spores 7–10 x 4.5–6.5 µm, ellipsoid to almond shaped, slightly roughened.

ECOLOGY: Solitary or scattered in a wide variety of forest types. Locally common under Tanoak and Chinquapin in the southern half of our range. Prefer conifers, especially Western Hemlock and Sitka Spruce, in the northern part. Fruiting from fall into early winter in north, late fall until late winter in south.

EDIBILITY: Unknown; likely nontoxic.

COMMENTS: The dry, silvery violet cap, lilac-gray to brownish gills, rusty brown spores, lack of an odor, and whitish to lilac-marbled flesh help identify this beautiful species. *C. camphoratus* is quite similar, but can easily be told apart by its awful odor of rotting potatoes. *C. traganus* is also quite similar, but usually has a richer violet-purple cap, marbled tan to orange flesh, and a sweetish odor. Another *Cortinarius* in our area (apparently closely related to the European *C. emunctus*) also has silvery violet colors, but has a viscid cap and stipe. *C. occidentalis* has a darker, thinly viscid cap, often streaked with purplish fibrils, and bluish violet flesh that stains purple.

Cortinarius "Casper Blue"

CAP: 3.5–8 (10) cm across, rounded to convex when young, becoming broadly convex, margin incurved to downcurved well into maturity. Rich blue to bluish violet, often with silvery whitish patches or fibrils. Surface thinly viscid to dry, appressed-fibrillose to smooth, often with whitish cottony veil tissue over center and partial veil remnants on margin. **GILLS:** Broadly attached or with a small notch, close to crowded. Distinctly royal blue to grayish blue when young, fading slightly and developing rusty tones as spores mature. **STIPE:** 5–10 cm long, 1–2 cm thick at apex, enlarging downward to an often abrupt, rounded to slightly elongate bulb, 2–4 cm across. White, at times with a faint bluish cast. Surface dry to slightly moist, with longitudinal fibrils and silky cortina remnants. **PARTIAL VEIL:** Rather thick, sheathing white cortina, leaving fibrils on cap margin and an annular zone of white fibrils on stipe, soon covered in rusty spores. **FLESH:** Thick, firm, white. **ODOR:** Indistinct. **TASTE:** Indistinct. **KOH:** Yellowish ocher on flesh in stipe base, no reaction elsewhere. **SPORE DEPOSIT:** Rusty brown. **MICROSCOPY:** Spores 7.5–9.5 x 4.5–6.5 µm, almond shaped to lemon shaped, distinctly roughened.

ECOLOGY: Solitary or scattered in small troops under conifers in the northern part of our range. Locally common. Fruiting from midfall into winter.

EDIBILITY: Unknown.

COMMENTS: The distinctive bluish cap, often with adhering whitish tissue; beautiful blue gills; and white stipe and flesh help distinguish this species. At first it was thought that the name *C. calyptratus* might refer to this species, but is now known to belong to a very similar mushroom with a violet-purple cap and paler gills. Because this beauty lacks a good name, the nickname used here is Casper Blue, as it can be very abundant in the northern coniferous forest around Casper, California. Genetic sequences have shown it to be fairly close to *C. caerulescens*, a European species. A fairly similar local species, *C. subfoetidus*, also has a bluish purple cap, but has pale whitish gray to tan gills. Another common, undescribed species on the Far North Coast, given the unwieldy nickname Blue-Gilled Green Corn, has a strong green-corn odor, lilac-blue young gills, and a silvery cap that quickly becomes ocher-beige in age. Members of the *C. glaucopus* group can be similar, but they have darker, often mottled flesh and a brownish KOH reaction on the cap.

Cortinarius cyanites (sensu CA)

CAP: 6–15 cm across, rounded to convex when young, expanding to broadly convex, plane or wavy. Margin downcurved when young, becoming wavy and slightly uplifted in age. Color variable and changing rapidly. Bluish in button stage, becoming bluish purple and developing darker streaks made up of appressed fibrils. Grayish and brown tones developing as it expands, then becoming purplish brown in age. Surface dry to slightly viscid when wet, covered with radiating streaks of appressed fibrils. **GILLS:** Broadly to narrowly attached, often with a distinct notch, close to crowded. Bluish to bluish violet at first, becoming lilac-gray and eventually darkening to deep lilac-gray to gray-brown. Staining vinaceous red slowly where damaged. **STIPE:** 6–15 (18) cm long, 1–2.5 cm thick at apex, club shaped or swollen with an elongated bulbous base, 1.5–4.5 cm thick. Bluish lilac to pale violet at first, fading to pale grayish lilac, to pale gray and then darkening to dingy gray to grayish lilac and developing inky brown color from base up. Staining bright violet when handled, then slowly becoming vinaceous red to deep purple. Surface dry, covered with silky fibrils that become matted in age. **PARTIAL VEIL:** Cortina-like, leaving an annular zone of lilac fibrils on stipe, which become grayish and then rusty brown from spores. **FLESH:** Thick, firm, fleshy. Bluish lilac at first, fading to grayish lilac, staining red to vinaceous red, especially in stipe base. **ODOR:** Indistinct. **TASTE:** Indistinct to slightly bitter. **KOH:** No reaction. **SPORE DEPOSIT:** Rusty brown. **MICROSCOPY:** Spores 8.5–11.5 x 5–7 µm, ellipsoid to almond shaped, roughened.

ECOLOGY: Solitary, scattered, or in small groups on ground in northern coniferous forests. Found in mixed Sitka Spruce, Western Hemlock, and Grand Fir forest on the Far North Caost. Rare in California, known from Mendocino County north. Fruiting from early fall into winter.

EDIBILITY: Unknown.

COMMENTS: This beautiful and highly variable *Cortinarius* can be recognized by the large size; bluish lilac cap that develops dark streaks and darkens in age; bluish lilac to grayish brown gills; and reddish vinaceous staining flesh. No other *Cortinarius* in California stains quite like this one. *C. occidentalis* (=*C. mutabilis*) is grayish purple to lilac capped, and stains purple or pinkish magenta when cut and rubbed. It also shows a purplish red reaction to iodine, especially in the stipe base, whereas *C. cyanites* has no reaction. Also see the *C. purpurascens* group (p. 191) for other purple stainers.

Cortinarius glaucopus group

CAP: 5–12 cm across, convex with an inrolled margin when young, becoming broadly convex to plain or slightly wavy and lobed in age, often retaining a downcurved margin. Color extremely variable within a characteristic range. Usually some blend of olive, bluish, purple, and grayish tan or brownish. Most colorful and extensively bluish green when young, fading to duller beige, brownish or with ocher-tan areas in age. Margin often retaining bluish or lilac tones in age. In some variants, extensively orangey or ocher even when young, most often with a streaked appearance. Surface smooth but innately fibrillose, sometimes radially wrinkled, viscid when wet but often drying out. Sometimes with scattered whitish patches of flat or cottony tissue. **GILLS:** Attached, closely spaced. Variably colored; usually lilac to sky blue or royal purple when young, becoming grayish to tan, eventually more ocher to dull brown in ages. **STIPE:** 5–11 cm long, 2–3.5 cm thick at apex, base larger, to 6 cm thick. Stout and cylindrical over most of length but strongly swollen (usually abruptly enlarged) at base. Bluish white to lilac (often strongly so at apex), overall silvery streaked, becoming creamy to yellowish tan. **PARTIAL VEIL:** Prominent cobwebby cortina broadly attached over most of lower stipe, soon collapsing on stipe and becoming rusty brown in age. Occasionally rather scant and evanescent. **FLESH:** Thick, firm, with a distinctive mottled appearance. Creamy whitish mottled with bluish or lilac (especially at midpoint of stipe), usually paler in cap and yellowish or ocher in flesh of bulbous base. **ODOR:** Indistinct. **TASTE:** Indistinct. **KOH:** Dingy reddish brown or ocher on cap surface and flesh. **SPORE DEPOSIT:** Rusty brown. **MICROSCOPY:** Spores 6–8 x 4–4.5 μm, ellipsoid to almond shaped, slightly roughened to nearly smooth.

ECOLOGY: Solitary, in clusters, or scattered in arcs, primarily under live oak and Tanoak in midfall into early winter. Most species in this group occur throughout our range wherever hardwoods are present in the forest. A greenish, conifer-associated species occurs on the Far North Coast.

EDIBILITY: Likely nontoxic.

COMMENTS: Genetic data show that at least five species in this group are found in our area, with one variable-looking species being substantially the most common. As a group, they can be recognized by their viscid caps (when wet), stout stipes with a distinct bulb, olive to bluish purple cap, lilac to purple young gills, and reddish brown to ocher KOH reactions. *C. glaucocephalus* is an intensely bluish species with slightly longer, lemon-shaped to snowshoe-shaped spores. The *C. purpurascens* group shows a darker reddish brown cap without olive or bluish lilac tones, and the gills bruise purplish when scratched or crushed; they are more often found under conifers. Also compare with *Cortinarius* "Casper Blue" (p. 186).

Cortinarius perplexus

Bojantchev, Ammirati & N. Siegel

CAP: 3–12 cm across, rounded to convex when young, becoming broadly convex, margin inrolled to downcurved well into maturity. Creamy ocher to pale ocher, occasionally with a violet cast, or more extensively mottled violet. Surface thinly viscid to dry, at times with sparse whitish cottony veil tissue. **GILLS:** Broadly attached or with a small notch, close to moderately crowded. Pale, lilac-gray at first, becoming more purple with age. **STIPE:** 4–9 cm long 1–2 cm thick at apex, equal or enlarging downward to an abrupt, triangular bulb. Whitish to pale violet. Surface dry, covered with longitudinal fibrils and silky cortina remnants. **PARTIAL VEIL:** Rather thick, sheathing white cortina, leaving an annular zone of white fibrils on stipe, soon becoming rusty brown from spores. **FLESH:** Thick, firm, whitish to pale ocher-buff. **ODOR:** Indistinct. **TASTE:** Indistinct. **KOH:** Bright rosy pink on cap and on exterior of bulb, dingy pink on stipe, no reaction on flesh. **SPORE DEPOSIT:** Rusty brown. **MICROSCOPY:** Spores 9.5–11 (12) x 5.5–7 µm, almond shaped, moderately to coarsely roughened.

ECOLOGY: Solitary or scattered in small troops under both evergreen and deciduous oaks. Rare, known from the greater San Francisco Bay Area into the Pacific Northwest. Fruiting from midfall into winter.

EDIBILITY: Unknown.

COMMENTS: The pale creamy ocher cap, rosy pink KOH reaction, pale grayish lilac gills that develop purple colors in age, and growth with oaks are important identifying features. Most *Cortinarius* lose purple gill colors as they age, making this a rare exception. *C. intricatus* is extremely similar, but generally has a slightly darker ocher-beige cap, less purple tones to mature gills, an orange- to reddish brown KOH reaction on the cap, and slightly wider, distinctly roughened spores, 10–12 x 6.5–7 (8) µm.

Cortinarius lilacinocolossus

M. M. Moser

CAP: 4–15 cm across, rounded-convex with an inrolled margin, expanding to plane or slightly wavy in age, often retaining a partially inrolled margin. Lilac to lilac grayish, sometimes dull brownish with a purplish wash, margin whitish. Surface thinly viscid, soon dry, smooth or slightly fibrillose, sometimes with whitish patches or a faint bloom. **GILLS:** Attached, close to crowded. Pale creamy white to very light grayish white, remaining pale for a long time and then becoming dingy brown as spores mature. **STIPE:** 3.5–9 cm long, 2.5–5 cm thick, often rather squat or stout, cylindrical or more often club-shaped, with a swollen base. White, or with a slight lilac blush. Surface finely fibrous to smooth, dry. **PARTIAL VEIL:** Sparse lilac to whitish cortina leaving an evanescent annular zone. **FLESH:** Thick, firm, whitish, marbled with lilac above gills and below cap skin. **ODOR:** Indistinct. **TASTE:** Indistinct. **KOH:** Yellow to light ocher on cap, bright yellow on stipe and flesh. **SPORE DEPOSIT:** Rusty brown. **MICROSCOPY:** Spores 10.5–12.5 x 4.5–7.5 µm, ellipsoid to almond shaped, finely roughened.

ECOLOGY: Solitary or in small groups under live oak. Rather uncommon, but locally abundant in some places in the Santa Cruz Mountains. Fruiting from fall into winter.

EDIBILITY: Unknown, unlikely to be poisonous.

COMMENTS: This uncommon *Phlegmacium* can be told apart by the lilac cap, stout stature, white flesh and stipe, yellow KOH reaction, and growth under live oak. The yellow KOH reaction is rare among *Cortinarius* (especially for a *Phlegmacium*). Most similar species are more slender, have more abruptly bulbous stipe bases, and/or grow in different habitats. A similar undescribed species in the *C. balteatocumatilis* group also occurs here, but is generally slightly darker lilac-purple and probably grows with conifers.

Cortinarius fuligineofolius

(M. M. Moser) M. M. Moser & Peintner

CAP: 3–10 cm across, rounded to convex when young, becoming broadly convex to plane or wavy in age, often with a low, rounded umbo. Often distinctly two toned, with a buff, light yellowish tan to orange-brown center and darker watery brown tones or with spots on outer part. Extreme margin sometimes shows violet to lilac-gray tones, often with a greenish or olive cast to entire cap. Hygrophanous, becoming paler and more evenly colored in age or when dry. Surface smooth, viscid to dry. **GILLS:** Broadly attached, or with a slight notch, close to crowded. Pale greenish yellow to pale olive-gray with a violet cast when young, soon developing buff tones before becoming rusty orange to brownish. **STIPE:** 4–12 cm long, 1–2 cm thick, equal or enlarged downward toward a rather bulbous base. Whitish to pale violet, becoming dingier in age. Basal mycelium bright greenish yellow when first collected, but soon fading. Surface dry, covered with brownish, silky longitudinal fibers. **PARTIAL VEIL:** Quite sparse, cortina-like, leaving evanescent silky fibrils on stipe. **FLESH:** Thick, firm, fibrous. Grayish to grayish buff in cap, often marbled with, or purple in stipe, paler in lower stipe. Developing ocher stains around larva tunnels and in age. **ODOR:** Indistinct to slightly radishlike. **TASTE:** Indistinct. **KOH:** Brownish on cap, violet on flesh. **SPORE DEPOSIT:** Rusty brown. **MICROSCOPY:** Spores 8–11 x 5–7 µm, broadly ellipsoid, finely roughened to nearly smooth.

ECOLOGY: Scattered or in small troops on ground, in moss of duff under conifers. Especially common under Sitka Spruce on the Far North Coast and in mixed forest south to Sonoma County. Fruiting from early fall into early winter, occasionally again in spring.

EDIBILITY: Unknown.

COMMENTS: The cap of this mushroom is hygrophanous, viscid, and oddly colored: orange-brown fading to buff, with greenish and/or violet tones. These features are distinctive when considered in combination with the yellowish mycelium around the stipe base. Other useful features are the pale violet KOH reaction on the cap and wine red iodine reaction on the flesh, as well as the long ellipsoid, relatively smooth spores. *C. montanus* is very similar and quite common, but it is primarily montane and usually has a darker cap. *C. napus* has an orange-brown to red-brown cap, and white flesh that does not change color in iodine solutions. Both latter species have bright greenish yellow mycelium around the bulb, unlike the somewhat similar (but usually more purple gilled) *C. subpurpureophyllus*, an occasional fruiter on the Far North Coast. Members of the *C. glaucopus* group are generally more stout, lack the violet KOH reaction, and usually exhibit more bluish coloration when young.

MISAPPLIED NAMES: *Cortinarius scaurus* (Fr.) Fr., *C. herpeticus* Fr.

Cortinarius purpurascens group
PURPLE-BLUSHING CORT

CAP: 4–10 cm across, domed with an inrolled margin at first, soon broadly convex to nearly plane, sometimes lobed and wavy in age. Color variable (distinct forms may represent different species). Specimens from the southern half of range generally show dingy tan to beige-brown caps mixed with olive and grayish tones; caps of fruitbodies farther north are more often rich orange-brown to reddish brown (there's much overlap, especially in northern inland areas). Surface slightly viscid to greasy or dry; innately fibrillose to nearly smooth. **GILLS:** Notched to narrowly attached, moderately close. Light grayish, becoming more lilac, maturing to dingy yellowish with brownish or ocher blotches. Bruising amethyst to purple when damaged. **STIPE:** 5–12 cm long, 1.5–3 cm thick at apex, cylindrical, base slightly enlarged or club shaped, sometimes more abruptly swollen. Whitish to tan or extensively lilac to purple, often vertically striate with a silvery sheen. Bruising purple when scraped. **PARTIAL VEIL:** Prominent cobwebby cortina of whitish to lilac fibers, often collapsed or sparse in age, becoming covered in rusty brown spores. **FLESH:** Thick, firm, whitish to pale lilac-grayish, sometimes yellowish in stipe base. **ODOR:** Indistinct to slightly peppery. **TASTE:** Mild. **KOH:** Dingy ocher-brown to slowly reddish brown on cap. Iodine solutions producing intense wine red to purple color on flesh. **SPORE DEPOSIT:** Rusty brown. **MICROSCOPY:** Spores 7–8.5 x 4–5 μm, almond shaped, distinctly roughened.

ECOLOGY: Solitary or in small groups in forested habitats from late fall through midwinter. Fairly common under live oak and pines throughout most of our area, more often with spruce, fir, and other conifers on the Far North Coast.

EDIBILITY: Unknown.

COMMENTS: The purplish bruising gills and stipe, often dingy or ocher-brown cap color, and lack of a rimmed bulb help distinguish this species. When you are first becoming familiar with this taxon, confirmation (at least to species group) by applying iodine to the flesh can be very helpful. Although there are likely two or three species in this complex in our area, boundaries between them remain unclear, and use of this European name continues. A related species, *C. occidentalis*, is entirely silvery lilac to bluish purple.

Cortinarius lilaciotinctus
Garnica & Ammirati

CAP: 4–10 cm across, domed when young, becoming broadly convex, lobed and sometimes slightly wavy in age. Reddish brown to ocher or orange-tan. Surface smooth, viscid when wet, drying out in dry weather. **GILLS:** Narrowly attached, moderately close. Pale grayish to lilac, becoming more extensively purplish, then ocher-brown as spores mature. **STIPE:** 5–10 cm long, 1.5–2.5 cm thick at apex, swollen base up to 5.5 cm thick. Stout, cylindrical, bulb strongly swollen with an abrupt margin. Pale or rich lilac (often darker purplish near apex) becoming dingy, pale ocher-brown, usually retaining at least some lilac tones. Surface dry, often silky looking when young. **PARTIAL VEIL:** Cobwebby cortina, leaving silky remnants on stipe, soon covered in rusty brown spores, sometimes disappearing in age. **FLESH:** Thick, firm, solid, whitish or light yellowish to pale lilac in cap and above gills. **ODOR:** Indistinct. **TASTE:** Bitter. **KOH:** Wine red to dark rosy purple on cap, pink to reddish magenta on stipe, neon pink on flesh. **SPORE DEPOSIT:** Rusty brown. **MICROSCOPY:** Spores 9.5–11 x 5–6.5 μm, broadly almond shaped to lemon shaped, roughened.

ECOLOGY: Solitary, in small clumps or troops under live oak and Tanoak. Uncommon, known from Santa Cruz County to Mendocino County. Fruiting from late fall through midwinter.

EDIBILITY: Unknown.

COMMENTS: This lovely mushroom can be recognized by the orangey brown cap, lilac tones of the gills and stipe, pink or wine red KOH reactions, and association with live oak or Tanoak. *C. fulvo-arcuatorum* is very similar, but may be duller capped on average (brownish to light brick) and more strongly lilac-purple on the stipe, with slightly larger spores (10–12 x 6–7 μm). *C. olympianus* has a pale violet to whitish lilac cap when young, stout stature, and rosy pink KOH reaction. Rare in California, it occurs under conifers on the Far North Coast and interior mountains. *C. perplexus* has a pale creamy ocher cap, pallid gills that become violet, before darkening and maturing spores. The cap turns rosy pink KOH, while the flesh does not react. *C. mikedavisii* is a rare conifer associate (likely with Sitka Spruce); it has a deep red to reddish brown cap with a paler yellowish margin when young, and yellowish violet gills. KOH produces a blood red reaction on the cap and vinaceous reaction on the flesh.

Cortinarius adonis

Bojantchev, Garnica & Ammirati

ADONIS' CORT

CAP: 4.5–10 cm across, domed to broadly convex with a narrowly inrolled margin, expanding to plane with a downcurved margin or becoming slightly uplifted and wavy in age. Color variable, ranging from light mustard yellow with a greenish wash when young to blotchy ocher-brown in age; usually paler (sometimes with a faint lilac or greenish wash) at margin. Surface smooth, viscid when wet. **GILLS:** Attached, close to crowded, edges slightly irregular. Light mustard yellow to ocher-yellow when young, to orange-brown in age. Faces and deeper parts of gills lilac to amethyst purple (most easily seen when crushed or cut), edges yellow at all ages. **STIPE:** 5–12 cm long, 1–2.5 cm thick at apex, up to 5 cm thick at bulbous base. Nearly cylindrical or more often club shaped toward abruptly bulbous, swollen, and then tapered-downward base. Surface dry, fibrillose to nearly smooth above cortina zone. Lower parts covered in matted fibrils of dingy, light grayish green or yellowish cortina, obscuring purplish lilac beneath at first, soon with collapsed cortina remnants covered in orangey-brown spores. Amethyst purple of stipe surface primarily visible at apex or more dramatically through cortina fibers where they are scraped away. **PARTIAL VEIL:** Cobwebby cortina of yellowish to greenish fibers, broadly covering stipe, soon collapsing and leaving a high-set annular zone and becoming covered in rusty brown spores. **FLESH:** Firm, solid. Yellowish green mottled with lilac in cap, purplish immediately above gills, strongly amethyst purple through much of stipe, yellowish to dingy ocher-brown in base. **ODOR:** Indistinct. **TASTE:** Indistinct. **KOH:** Blood red to wine red on cap, light wine red to raspberry or deep magenta-purple on yellow parts of flesh (no reaction on purple parts), as well as on outer surface of stipe base. **SPORE DEPOSIT:** Rusty brown. **MICROSCOPY:** Spores 11–16 x 6–9.5 μm, averaging 13.1 x 7.5 μm, lemon shaped, distinctly roughened.

ECOLOGY: Solitary or scattered in small patches; rarely more than a few fruitbodies found at once. Fruiting from late fall through winter under live oak and Tanoak. Rather rare; known from the Santa Cruz Mountains as well as Mendocino County, but patchy and fruiting irregularly.

EDIBILITY: Unknown.

COMMENTS: The striking combination of yellow and amethyst on the gills, yellowish green tones on the cap, yellowish green to purple stipe, purple flesh, and intense KOH reaction helps set this *Phlegmacium* apart. Other species with bright colors don't have such a pronounced mix of amethyst flesh and yellowish green colors elsewhere, and/or lack the raspberry KOH reaction seen in the photo above. *C. mikedavisii* has a darker red to red-brown cap, grows with conifers, and has a blood red KOH reaction on the cap and vinaccous KOH reaction on the flesh.

Cortinarius viridirubescens

M. M. Moser & Ammirati

YELLOW-GREEN CORT

CAP: 3.5–12 cm across, convex to broadly convex, nearly flat in age, occasionally wavy. Grass green to yellow-green to nearly chartreuse, sometimes with duller olive or brownish areas, and often developing rusty orange stains in age. Surface smooth, viscid when wet, metallic in dry weather. **GILLS:** Attached, moderately close. Whitish to pallid clay gray when young (sometimes with a lilac wash), maturing to dingy yellowish with brownish or ocher blotches. **STIPE:** 5–12 cm long, 1.3–3 cm thick at apex, swollen base up to 6.5 cm thick. Stout and cylindrical, usually with an abruptly enlarged bulb, but then tapering to extreme base. Pale whitish to yellowish, yellow-green to tan in age, exterior of base covered in felty, bright yellow-green mycelium surface. Dry or slightly viscid near base. **PARTIAL VEIL:** Prominent cobwebby cortina of whitish yellow to light greenish yellow fibers over much of cap and stipe when young, ephemeral and often absent at maturity. **FLESH:** Thick, firm, pale whitish yellow. Bruising slowly reddish brown to brighter red, at least when young, especially around maggot tunnels and on lower stipe when cut or scraped. **ODOR:** Indistinct. **TASTE:** Indistinct. **KOH:** Orange-red on cap and stipe base. **SPORE DEPOSIT:** Rusty brown. **MICROSCOPY:** Spores 8.5–11.5 x 5–6 μm, almond to lemon shaped (with small knobs at either end), moderately roughened.

ECOLOGY: Solitary or scattered in troops under oak and Tanoak. Prefers Coast Live Oak throughout much of the southern portion of our range, but occurs with Tanoak and deciduous oaks to the north. Fruiting from late fall into winter. Observed from Del Norte County south to San Luis Obispo County. Fruits prolifically in some years, seemingly absent in others.

EDIBILITY: Unknown.

COMMENTS: This is one of the most distinctive of all our Cortinarius. Few mushrooms of any kind rival the bright yellow-green to grass green colors of this species. Additionally, the tendency to bruise reddish brown to brighter red, orange-red KOH reaction on the cap, and bright chartreuse stipe base help confirm the identification.

Cortinarius xanthodryophilus

Bojantchev & R. M. Davis

CAP: 5–10 (15) cm across, convex to broadly convex, expanding to plane, occasionally uplifted and wavy in age. Variably dull yellow to straw yellow, becoming brighter yellow with ocher-brown or orangish areas in age. Surface smooth or occasionally cracked at center in age, viscid when wet, often with stuck-on debris in dry weather. **GILLS:** Attached, often with a distinct notch, close to crowded. Dull cream yellowish to yellowish beige when young, becoming ocher and eventually rusty brown overall. **STIPE:** 5–10 cm long, 1.5–3 cm thick at apex, 3–5 cm thick at base, generally equal above an enlarged to abruptly bulbous base. White, occasionally with a slight violet blush at apex when young, becoming dingy beige in age, with an ocher band of spore deposit on collapsed cortina. Surface dry, base often with adhering debris. **PARTIAL VEIL:** Cobwebby cortina leaving a distinct annular zone of pale yellowish beige to off-white fibrils on stipe, soon covered in rusty brown spores. **FLESH:** Firm, fleshy, whitish, occasionally bluish violet in stipe apex, slowly staining brownish in stipe. **ODOR:** Indistinct to slightly earthy. **TASTE:** Indistinct. **KOH:** Light brown to reddish brown on cap, slight yellowing on cap flesh, yellow-brown on flesh of lower stipe. **SPORE DEPOSIT:** Deep rusty brown. **MICROSCOPY:** Spores 9.5–13 x 5–7.5 μm, averaging 11.2 x 6.3 μm, almond to lemon shaped, distinctly roughened.

ECOLOGY: Solitary or scattered in troops under oak (especially Coast Live Oak), occasionally with deciduous oaks or Tanoak. Locally common throughout much of our area, but rare outside range of Coast Live Oak. Fruiting from midfall well into winter.

EDIBILITY: Unknown.

COMMENTS: The dull yellow colors, relatively weak KOH reactions, and association with oaks help distinguish this Cortinarius. C. vellingae has a deeper golden to ocher-yellow cap, brighter yellow gills, and a blood-red KOH reaction. Young fruitbodies of C. amabilis have cooler yellow caps, lilac tones to the gills, and a red KOH reaction. C. elegantio-occidentalis has an ocher yellowish cap, rimmed bulbous stipe base, and red KOH reaction on the flesh in stipe base, and grows with northern conifers.

Cortinarius vellingae

Bojantchev, Garnica & Ammirati

ELSE'S CORT

CAP: 4–12 cm across, rounded to convex with an inrolled margin, becoming broadly convex, margin often remaining downcurved, or occasionally wavy in age. Yellowish or dull gold at first, soon developing orange to ocher-brown blotches, becoming rusty brown in age. Surface smooth, viscid when wet, dry and metallic looking in dry weather. **GILLS:** Notched or narrowly attached. Distinctly (but sometimes light) yellow when young, becoming yellowish to dingy tan, then ocher-brown in age. **STIPE:** 5–10 cm long, 2–3 cm thick at apex, base up to 6 cm thick. Rather stout, with a swollen bulb at base (often with an abrupt margin). Whitish to cream or light yellowish. **PARTIAL VEIL:** Prominent cobwebby cortina broadly situated over middle of stipe, becoming reddish brown with spores. **FLESH:** Thick, firm, pale whitish yellow. Bruising slowly light brownish when cut. **ODOR:** Indistinct. **TASTE:** Mild. **KOH:** Blood red to reddish brown on cap, pinkish red to red on stipe flesh. **SPORE DEPOSIT:** Rusty brown. **MICROSCOPY:** Spores 9–10.5 x 4.9–6 µm, broadly almond shaped to lemon shaped (with small knobs at either end), roughened.

ECOLOGY: Solitary or in small groups under live oak, possibly Tanoak, and likely continuing much farther north with Garry Oak (*Quercus garryana*). Locally common and widespread; generally in drier, inland live oak woodlands, fruiting from midfall through midwinter.

EDIBILITY: Unknown, likely nontoxic.

COMMENTS: This *Cortinarius* was named to honor Else Vellinga, a mentor and inspiration to many in the mycological community. It is part of a complicated group of viscid-capped, bulky *Cortinarius* with bulbous stipe bases. Most are brightly colored, and the majority in our area grow with live oak and Tanoak. *C. amabilis* is extremely similar, but has lilac-beige gills when young (soon fading) and more distinctly roughened spores that are slightly larger (10–11.5 x 6.8–7.5 µm). *C. xanthodryophilus* has a duller straw yellow to yellowish beige cap, pale beige grayish gills when young, and more elongate, lemon shaped spores (10–12 x 5.5–7 µm). *C. elegantio-occidentalis* is similar in many respects, but has a more northern distribution, grows with conifers (spruce and Douglas-fir), and has larger spores, 12.5–15.5 (18!) x 7.5–9.5 µm; the cap surface and flesh turn red in KOH.

Cortinarius ponderosus

A. H. Sm.

PONDEROUS CORT

CAP: 6–30 (40) cm across, convex with a tightly inrolled margin when young, becoming broadly convex to plane, margin often incurved well into maturity, can become uplifted and wavy in age. Golden yellow with warm ocher-brown appressed scales when young, often appearing streaked, sometimes paler beige or whitish near margin, becoming more evenly orange-brown to medium brown in age. Fruitbodies of any age can show random olive blotches or streaks. Surface slightly viscid or tacky to dry, covered in small appressed-fibrillose scales. **GILLS:** Broadly attached to slightly notched, fairly close. Color variable, deep lilac or grayish with a light lilac wash, entirely grayish tan or whitish with no lilac tones when young, soon grayish tan, developing orange-brown spots and splotching as spores mature, becoming warm ocher-brown to rusty brown with a splotchy appearance in age. **STIPE:** 5–20 cm long, 3–10 cm thick, cylindrical, club shaped or bulbous, often with a tapered, slightly rooting base. Whitish at apex, orange to brownish near base, often streaky or blotched, darkening to rusty brown or ocher-brown overall in age. **PARTIAL VEIL:** Heavy white cortina when young, usually persistent into age as a collapsed webby skirt of fibrils stained rusty brown by spores, sometimes disappearing entirely in older specimens. **FLESH:** Very firm and thick, often hard and somewhat rubbery. Whitish, marbled with grayish to pale lilac-gray spots when young, developing ocher stains, especially around larva tunnels. Bruising slowly ocher to brownish. **ODOR:** Often

sour, or like slightly rancid vegetables. **TASTE:** Mild to slightly sour. **KOH:** No reaction. **SPORE DEPOSIT:** Rusty ocher-brown. **MICROSCOPY:** Spores 7–11 x 5–6 μm, ellipsoid to almond shaped, finely roughened.

ECOLOGY: Solitary, scattered, in clusters, or in troops in forested settings. Common especially under Tanoak and Douglas-fir. Fruiting in late fall, but will continue into late winter in the southern part of our range.

EDIBILITY: Edible, but not uniformly appreciated. Some people report a sour or "muddy" taste; others find it to have a pleasant flavor and pleasingly firm texture. Perhaps best used to flavor stock.

COMMENTS: Although the size alone is enough to set this species apart from most other *Cortinarius*, the orange-brown, innately scaly caps and purplish to light grayish gills when young further distinguish it. The difficult-to-describe sour odor is also unique among its brethren. *Tricholoma focale* is perhaps more likely to be confused for this species than any other *Cortinarius*. While it can be large and similarly colored, it usually has white gills (sometimes extensively spotted or stained orangish in age), a shorter, more strongly pointed stipe base with tufts of pale mycelium, and a white spore deposit.

Cortinarius callimorphus

Bojantchev & R. M. Davis

CAP: 5–12 cm across, rounded to convex when young, becoming broadly convex, margin incurved to downcurved well into maturity. Often uniformly ocher-orange, or with a paler yellowish margin, developing brown tones and often with rusty stains in age. Surface viscid when wet, dull glossy, but often with debris stuck to it when dry, smooth, or slightly wrinkled around margin in age. **GILLS:** Broadly attached or with a small notch, close to moderately crowded. Off-white to pale grayish white when young, becoming pale orangish brown to warm brown, sometimes with rusty stains as spores mature. **STIPE:** 6–12 (14) cm long, 1–2 cm thick at apex, enlarging downward to a large, often abrupt, rounded to slightly elongate bulb, 3–5 cm across. White to whitish when young, developing rusty ocher stains. Surface dry to slightly moist, with longitudinal fibrils and silky cortina remnants. **PARTIAL VEIL:** Relatively sparse cortina, often with a viscid covering when young, leaving an annular zone of white fibrils on stipe, soon rusty brown from spores, and occasionally fibrils on cap margin. **FLESH:** Thick, firm. Whitish, slowly staining orange-brown, especially around larva tunnels. **ODOR:** Indistinct. **TASTE:** Indistinct. **KOH:** Carmine red on cap (sometimes weakly), pale yellowish brown on flesh. **SPORE DEPOSIT:** Rusty brown. **MICROSCOPY:** Spores 8.5–10.5 x 5–6 μm, distinctly lemon shaped, moderately roughened.

ECOLOGY: Scattered in small to large troops under Sitka Spruce from late fall into winter. Common on the Far North Coast. Currently known from Mendocino County into southwest Oregon, but this is a recently described species, and it's still unknown how far north range extends; it may extend over much of the Sitka Spruce range.

EDIBILITY: Unknown.

COMMENTS: This common species of the North Coast is identified by the ocher-orange, viscid cap, fairly pallid gills, stipe, and flesh that can all develop rusty stains; the carmine red KOH reaction on the cap; and the occurrence with Sitka Spruce. Microscopically, the lemon-shaped spores set it apart from some similar-looking species in California. *C. largentii* shares the same habitat and can look very similar from the top, but has beautiful lilac-toned gills when young. *C. multiformis* is also very similar looking, but has a duller ocher-tan cap, often streaked with appressed whitish fibrils (especially when dry), an indistinct to orangish brown KOH reaction, and slightly smaller, ellipsoidal to almond-shaped spores. *C. talus* has a creamy to pale yellowish cap, pallid gills, an orange-brown KOH reaction, and small, relatively smooth spores.

MISAPPLIED NAME: This species was recently described by Bojantchev and Davis. In the past, it was probably lumped with *C. multiformis* in the species lists of fairs and forays.

Cortinarius albofragrans

Ammirati & M. M. Moser

PALE FRAGRANT CORT

CAP: 3–9 cm across, rounded to convex when young, becoming broadly convex, or wavy in age. Margin inrolled when young and with a sterile flap of tissue, becoming slightly uplifted in age. Creamy white or ivory to pale yellowish when young, becoming creamy buff to ochraceous buff at center. Surface smooth, glutinous to viscid when wet, dull glossy, but often with debris stuck to it when dry. **GILLS:** Broadly to narrowly attached, often with a distinct notch, close to moderately crowded. Off-white to grayish white at first, becoming pale buff to cinnamon, browner as spores mature. **STIPE:** 5–10 cm long, 1–2 cm thick at apex, enlarging downward to a swollen or slightly bulbous base, 1.5–3.5 cm across. White to whitish, at times with yellowish buff bands or patches on lower portion when young. Surface dry to slightly tacky, with silky longitudinal fibrils and a silky cortina. **PARTIAL VEIL:** Cortina-like, leaving an annular zone of whitish fibrils on stipe and occasionally fibrils on cap margin. **FLESH:** Thick, firm, fleshy in cap, fibrous in stipe. Whitish to pale whitish tan. **ODOR:** Faint to strong, fruity chemical or spicy fruity, usually with some trace of pleasant tropical fruit odor. **TASTE:** Mild to somewhat unpleasant. **KOH:** No reaction. **SPORE DEPOSIT:** Rusty brown. **MICROSCOPY:** Spores 10–12.5 x 5.5–7 µm, averaging 10.8 x 6 µm, almond shaped, roughened.

ECOLOGY: Solitary or scattered on ground under both live and deciduous oaks as well as Tanoak. Very common under Coast Live Oak in the southern part of our range, continuing well into southern California. Less common north of the greater San Francisco Bay Area, primarily with Canyon Live Oak and Tanoak. Fruiting from midfall into early winter in the north, through winter in the south.

EDIBILITY: Unknown.

COMMENTS: The creamy white to buff, viscid cap, pale young gills, rusty spores, and sweet-spicy odor help identify this common *Cortinarius* of oak woodlands. The distinctive odor reminds us of a cheap attempt at mango-scented hand lotion. *C. citrinifolius* has a similar odor, but can be readily separated by the darker creamy yellow to ocher-buff cap and bright yellow flesh, as well as by the growth with conifers (although it may occasionally occur with Tanoak). *C. multiformis* has a tan to light brown cap, but lacks the distinctive odor (although it can have a honeylike odor). Other species of *Cortinarius* (some undescribed) look similar, but all lack the distinctive odor.

MISAPPLIED NAME: *Cortinarius luteoarmillatus*, which refers to a similar eastern North American species.

Cortinarius citrinifolius

A. H. Sm.

CITRINE FRAGRANT CORT

CAP: 4–9 cm across, rounded to convex when young, becoming broadly convex to plane, margin incurved at first, becoming wavy and occasionally slightly uplifted in age. Creamy yellow with a greenish tinge, yellowish to ocher-yellow when young, becoming ocher-orange to ocher-buff in age. Surface smooth, viscid when wet, dull glossy when dry. **GILLS:** Broadly to narrowly attached, often with a distinct notch, close to moderately crowded. Greenish yellow to yellowish when young, becoming yellowish buff, and finally buff-brown as spores mature. **STIPE:** 4–12 cm long, 0.7–2 cm thick at apex, enlarging slightly downward to a swollen base. Yellowish white to greenish yellow at first, becoming pale yellow, yellowish buff to dingy ocher-yellow. Surface dry to slightly tacky, with silky longitudinal fibrils. **PARTIAL VEIL:** Cortina-like, leaving an annular zone of whitish to pale yellow fibrils on stipe that become rusty brown with spores, occasionally leaving fibrils on cap margin. **FLESH:** Thick, firm, yellow to greenish yellow, fading slightly in age. **ODOR:** Mild to strong, a spicy mix of pleasant tropical fruit, lemon, and often some chemical component. **TASTE:** Mild to somewhat unpleasant. **KOH:** Red on flesh of stipe base, dingy red on cap flesh. **SPORE DEPOSIT:** Rusty brown. **MICROSCOPY:** Spores 10–12 x 6–7 μm, almond to lemon shaped, roughened.

ECOLOGY: Scattered or in small to large clusters on ground under conifers, especially Grand Fir and Sitka Spruce, occasionally with Tanoak. Occurring throughout the central and northern part of our range, but generally uncommon. Fruiting from mid fall into winter.

EDIBILITY: Unknown.

COMMENTS: The greenish yellow to yellow-ocher viscid cap that darkens to ocher-orange; greenish yellow young gills; bright yellow flesh; and sweet, spicy odor help identify this *Cortinarius*. *C. albofragrans* is an oak-associated species with a similar odor, but has creamy white to creamy buff cap, pale gills, and whitish flesh. *C. superbus* has a darker cap, brownish veil bands on the stipe, and a green-corn odor.

MISAPPLIED NAME: The similar European species *Cortinarius percomis* is genetically distinct.

Cortinarius vibratilis

(Fr.) Fr.

SLIMY-BITTER CORT

CAP: 1.5–5 cm across, rounded to weakly round-conical when young, becoming broadly convex to plane or with an umbo. Warm ocher-brown, orange-brown or yellow-brown to yellowish beige at center, paler to nearly whitish around margin, duller and paler overall when dry. Surface viscid to dry, usually shiny, smooth or finely pebbly textured. **GILLS:** Narrowly attached, often with a distinct notch, close. Whitish when young, becoming rusty brown as spores mature. **STIPE:** 3–7 cm long, 0.4–1 cm thick, more or less equal or tapered below a swollen midpoint. White at first, often becoming dingy white and developing a rusty band of spores on veil zone. Surface thinly or distinctly viscid below an annular zone, dry to moist above. **PARTIAL VEIL:** Cortina-like, leaving a annular zone on stipe that becomes covered in rusty brown spores. **FLESH:** Thin, fibrous to fleshy. Whitish in color. **ODOR:** Indistinct. **TASTE:** Very bitter, especially slime of cap. **KOH:** No reaction. **SPORE DEPOSIT:** Rusty brown. **MICROSCOPY:** Spores 6.5–8.5 x 4.5–5.5 μm, ellipsoid, moderately roughened.

ECOLOGY: Solitary or scattered in moss or duff under conifers and occasionally Tanoak. Common throughout the northern part of our range, uncommon to rare south of Sonoma County. Fruiting from fall into early winter.

EDIBILITY: Unknown.

COMMENTS: The ocher-brown to yellowish, viscid cap, viscid white stipe, and bitter taste make this smallish *Cortinarius* fairly easy to identify. *C. pluvius* is very similar but is usually brighter orange to ocher-yellow in color; it is common on the Far North Coast, especially with Shore Pine.

Cortinarius "glutinosoarmillatus"

Bojantchev et al. nom. prov.

SLIMY-GIRDLED CORT

CAP: 3–10 cm across, rounded-convex with an inrolled margin when young, becoming broadly convex or plane, sometimes wavy in age. Color extremely variable, whitish beige, ocher-beige, dull brownish, olive, or even with obscure blue-gray tones (or mottled with any combination thereof). Surface smooth, extremely viscid, covered with a thick gelatinous slime layer when wet, and usually somewhat viscid or elastic even in dry weather. Slime layer so thick that it usually hangs well past edge of cap. **GILLS:** Narrowly to broadly attached, or occasionally notched. Color variable, lilac-gray, grayish tan, beige, or occasionally deep amethyst purple when young, but often fading rapidly, soon becoming dingy grayish tan to dingy beige, then ocher-brown to rusty as spores mature. **STIPE:** 5–12 (15) cm long, 1–3 cm thick, equal or slightly tapered toward base. Dingy whitish or grayish beige, sometimes purplish in young fruitbodies near apex, often developing orangish discolorations from base up in age. Surface viscid, usually with pronounced rings of slime (girdles) or chevrons of dried gelatin in age. **PARTIAL VEIL:** Whitish cortina embedded in a layer of slime, upward flaring when young, soon drying into a membranous-looking ring that accumulates rusty brown spores and becomes darker reddish brown in age. **FLESH:** Firm, rubbery-fleshy, pale whitish beige. Stipe fibrous-fleshy, whitish beige, but becoming orangey beige toward base. **ODOR:** Indistinct.

TASTE: Indistinct. **KOH:** No reaction. **SPORE DEPOSIT:** Rusty orange-brown. **MICROSCOPY:** Spores 11.5–14.5 x 7.5–9 μm, elongate-ellipsoid, distinctly roughened.

ECOLOGY: Solitary or scattered in troops or small clusters under live oak. Very common from Sonoma County southward. Fruiting from late fall into winter or even early spring some years.

EDIBILITY: Unknown.

COMMENTS: The belts of slime on the stipe, dingy but strangely attractive cap colors, grayish to purple gills, and growth with live oak make this *Cortinarius* unique in our area. Other *Myxacium* rarely have such discrete bands or belts of slime on the stipe, and are differently colored.

MISSAPPLIED NAME: Long called *Cortinarius trivialis*, the Californian species is distinct from the European entity.

Cortinarius vanduzerensis group
SLIMY PURPLE-STEMMED CORT

CAP: 3–10 (15) cm across, tightly domed, often with a crimped margin at first, soon egg shaped or bell shaped, expanding to conical-convex or nearly plane, usually with an umbo. In most common form, dark rich chestnut brown to dark brown at first, paler toward margin, becoming cinnamon to beige in age. Pale forms; brownish beige to brownish tan even when young, are less frequently encountered. Surface strongly wrinkled, but covered in a smooth layer of glutinous slime when wet, otherwise viscid or tacky. **GILLS:** Narrowly attached, often with a broad notch, close to crowded. Pale grayish beige to pale tan at first, becoming cinnamon brown to light rusty brown. **STIPE:** 5–15 (20) cm long, 1–2 cm thick, equal, swollen at middle, tapering toward base. Light to dark violet, apex and base often whitish. Occasionally losing violet and becoming whitish overall in age. Surface covered in a glutinous slime layer below annular zone, dry above. Variably smooth or broken into chevrons. **PARTIAL VEIL:** Mostly present as a termination of slime layer, inconspicuous when dry. **FLESH:** Thick, firm, solid, fibrous. Whitish to pale buff, often developing ocher stains in lower stipe. **ODOR:** Mild to earthy. **TASTE:** Indistinct. **KOH:** No reaction. **SPORE DEPOSIT:** Rusty brown. **MICROSCOPY:** Spores 11–15 x 7–9 µm, broadly ellipsoid, roughened.

ECOLOGY: Solitary, scattered, or in troops on ground, in moss or duff under northern conifers. Especially common under Sitka Spruce on the Far North Coast; also with Western Hemlock and Grand Fir south to Sonoma County. Fruiting from midfall in northern part of range, continuing well into winter southward.

EDIBILITY: Unknown.

COMMENTS: The combination of the brown cap with wrinkles visible under the viscid layer, slimy violet stipe, and pallid gills are distinctive of the group. Recent work in the Pacific Northwest has shown that *C. vanduzerensis* is rather rare, known from the Central Oregon Coast. A new widespread species (which has been called *C. vanduzerensis*) was given the name *C. seidliae*. However, California collections have not been studied closely, and it is likely we have additional undescribed species in this complex. *C. cylindripes* is similarly slimy, but has a violet cap when young and grows with oaks, Madrone, and manzanita. *C. "glutinosoarmillatus"* has a paler, grayish beige cap and distinct girdles of slime on the stipe; it grows with oaks and Tanoak. *C. mucosus* has an orange-brown cap and whitish stipe.

Cortinarius cylindripes (sensu CA)

CAP: 3–12 cm across, rounded-convex when young, becoming convex to plane or slightly wavy. Dull amethyst purple or bluish purple at first (some fruitbodies with paler silvery lilac caps), soon fading from center outward to dingy tan-brown or yellowish, but usually retaining some lilac mottling into age. Surface smooth, very slimy when wet and usually remaining viscid or at least gooey tacky even in drier weather. **GILLS:** Narrowly to broadly attached, close to crowded, edges pale and slightly serrate or heavily eroded in age. Pallid grayish white when very young, soon grayish tan, then becoming dingy ocher or warm brown as spores mature. **STIPE:** 4–15 cm long, 1.5–3 cm thick, fairly slender, usually gradually tapered toward base. Whitish to silvery, lower portions usually showing amethyst, bluish, or lilac tones. Surface dry and somewhat scurfy above veil, thinly viscid, often with slimy chevrons below. **PARTIAL VEIL:** Inconspicuous, embedded in a layer of slime spanning from cap margin to stipe when very young, soon more or less obliterated, with only a faint "veil zone" around midstipe where viscid layer stops or becomes markedly thinner. **FLESH:** Thick, firm, fleshy-fibrous in stipe, whitish to yellowish with bluish purple areas. **ODOR:** Indistinct. **TASTE:** Indistinct. **KOH:** No reaction. **SPORE DEPOSIT:** Rusty brown. **MICROSCOPY:** Spores 12–15 x 6.5–8 μm, ellipsoid, roughened.

ECOLOGY: Scattered or in troops in mixed forest, generally with hardwoods (especially where ericaceous shrubs or trees are present). Widespread but only locally common. Fruiting in late fall and winter.

EDIBILITY: Unknown, likely nontoxic.

COMMENTS: This beautiful species is superficially similar to a number of other *Cortinarius*, but once you learn to recognize the subtle amethyst color, it's actually quite distinctive, especially in combination with the often slightly girdled appearance of the stipe. Similar *Myxacium* include *C. vanduzerensis*, which has a yellowish tan or dark brown cap (but the stipe is similar in color and texture) and grows with North Coast conifers; the rare *C. iodeoides*, which has a more royal purple coloration over the cap and stem and a bitter-tasting cap surface; and *C. salor*, which has a smoky lilac to pale purple cap and whitish stipe, and is restricted to coniferous forests on the Far North Coast.

Cortinarius mucosus

(Bull.) J. Kickx f.

CAP: 3–10 cm across, rounded at first, becoming broadly convex to nearly plane. Dark orange-brown, ocher-brown to reddish brown, slightly paler when dry. Surface smooth, viscid to tacky, metallic in dry weather. **GILLS:** Attached, close to subdistant. Tan to light clay brown at first, becoming orange-brown. **STIPE:** 4–15 cm long, 1–2.5 cm thick, cylindrical, often very tall, straight. White, developing ocher-brown stains in age. Surface viscid, covered with glutinous, slimy chevrons, frequently clumped into girdles or bands. **PARTIAL VEIL:** Silky cortina embedded in glutinous slime, soon covered in rusty spores. **FLESH:** Moderately thick, stipe cartilaginous-fibrous and quite firm. Whitish, mottled with pale ocher-brown, darker in age and around larva tunnels. **ODOR:** Indistinct. **TASTE:** Mild. **KOH:** No reaction. **SPORE DEPOSIT:** Rusty brown. **MICROSCOPY:** Spores (10.5) 12–16 x 6–8 μm, elongate-ellipsoid, coarsely roughened.

ECOLOGY: Solitary, scattered, or in troops in duff under pines on the North Coast. Fairly common in sandy coastal forests of Shore Pine and Bishop Pine. Fruiting from late fall into winter.

EDIBILITY: Unknown, likely nontoxic.

COMMENTS: The viscid, orange- to reddish brown cap, viscid white stipe, and growth with pines are distinctive. The lack of violet colors distinguishes it from *C. vanduzerensis*, while the growth under pines and the warmer colors separate it from the oak-associated *C. "glutinosoarmillatus"*.

Cortinarius caperatus

(Pers.) Fr.

THE GYPSY

CAP: 4–12 (15) cm across, rounded, egg shaped or rounded-conical when young, becoming broadly convex with a low, broad umbo, often wavy and irregular in age. Tan to light brown with a whitish sheen or bloom when young, remaining so or becoming honey brown to ocher-brown at center. Surface dry to moist, dull, often wrinkled. **GILLS:** Broadly attached or with a small notch, close to crowded. Light beige to tan, becoming more rusty brown. **STIPE:** 5–12 (15) cm long, 1–2.5 cm thick, equal, enlarged slightly toward base or with a club-shaped base. White to dingy whitish, developing pale orangish tan stains at base. Surface dry, with longitudinal striations, slightly silky to smooth. **PARTIAL VEIL:** Thick and membranous, leaving a rimmed collarlike ring near midpoint of stipe. **FLESH:** Moderately thick to thin, firm but brittle in cap, fibrous and sometimes tough in stipe. Whitish to tan, or orangish tan around larva tunnels. **ODOR:** Indistinct. **TASTE:** Mild to slightly unpleasant. **KOH:** No reaction. **SPORE DEPOSIT:** Ocher-brown. **MICROSCOPY:** Spores 10–14 x 7–9 μm, ellipsoid to almond shaped, moderately roughened.

ECOLOGY: Scattered in small to large troops under conifers, especially Sitka Spruce. Fruiting from late fall into winter. Rare to uncommon on the California coast, known from as far south as Mendocino County.

EDIBILITY: Edible and good, one of the few widely consumed *Cortinarius*, though some California collections have a slightly bitter metallic taste.

COMMENTS: This atypical *Cortinarius* can be identified by the tan to light brown wrinkled cap with a whitish bloom, pale to rusty gills, and thick membranous partial veil that leaves a collarlike ring or skirt on the stipe. No other *Cortinarius* in the western United States has such a thick membranous partial veil like *C. caperatus*. Members of the *C. multiformis* group superficially resemble it, but have a cobwebby partial veil.

SYNONYM: *Rozites caperata* (Pers.) P. Karst. Segregation of *Rozites* was based on the membranous partial veil (differing from the typical cortina of most *Cortinarius*). Genetic studies have supported treating these species as *Cortinarius*, as the membranous veil appears to have evolved multiple times.

9 • Entoloma and Allies

The family Entolomataceae is a natural (phylogenetially supported) grouping represented by a great diversity of species in our area. However, the similarity in appearance of many of these species makes it a very challenging group to work with, and the group remains quite poorly known. There are many rare and endemic species in California, and many more remain undescribed.

Although Entolomataceae is clearly defined as a family, there has been much turmoil in the naming of the genera it contains. Based on recent phylogenetic work (Co-David et al. 2009, Morgado et al. 2013), European authors have taken a "lumping" approach, collapsing many smaller genera into *Entoloma*. However, we've chosen to retain the genera *Leptonia*, *Nolanea*, *Inocephalus*, and *Alboleptonia*. Although these might be more "artificial" groupings, they are very useful for field taxonomists.

Entoloma—Large, fleshy, and dull-colored fruitbodies (except for one blue species), often with a farinaceous odor.

Leptonia—Small, slender fruitbodies, ranging in color from pink to yellowish to brownish to more commonly blue, blue-black, grayish blue, or purple. Very diverse group in our area, especially in redwood forest.

Nolanea—Conical to bell-shaped caps that are usually translucent-striate and hygrophanous, stipe surfaces often silky, silvery, and twisted looking. Brown, tan, and beige colors predominate.

Alboleptonia have small fruitbodies that are white overall (except for the pinkish gills). *Pouzarella* have radially shaggy-fibrillose to silky cap surfaces. *Clitopilus prunulus* has medium-size fruitbodies with decurrent gills and a strong farinaceous odor. *Rhodophana* fruitbodies are small and orange, and have a cucumber-y odor. *Rhodocybe* have slightly decurrent gills that are often thick-edged and widely spaced. *Inocephalus* are oddballs and best learned individually.

When identifying *Entoloma* and allies, pay attention to texture of cap and stipe, subtle details of coloration, and size/stature. Definitive identification usually requires microscopic examination—the size and shape of the spores (as well as the number of angles), presence or absence of clamp connections, type and distribution of pigments, and size and shape of cheilocystidia (if present) are important features to note.

Clitopilus prunulus

(Scop.) P. Kumm.

THE SPY

CAP: 2–7 cm across, convex with an evenly inrolled margin when young, becoming broadly convex to irregularly uplifted and wavy in age, often retaining a finely inrolled margin. Dingy whitish gray to grayish tan, covered in a dusty white bloom when young (sometimes disappearing entirely in age). Surface dry, often somewhat lumpy, slightly velvety to smooth. **GILLS:** Slightly to strongly decurrent, closely spaced, short gills present. Pallid whitish to dingy tan, becoming pinkish tan as spores mature. **STIPE:** 3–10 cm long, 0.8–1.8 cm thick, overall cylindrical although often swollen near midpoint or enlarged lower, then with a tapered extreme base. Dingy whitish gray with tan areas. **FLESH:** Fibrillose-fleshy, whitish. **ODOR:** Strongly farinaceous. **TASTE:** Farinaceous. **SPORE DEPOSIT:** Dingy pinkish tan. **MICROSCOPY:** Spores 9–12 x 5–7 µm, long-ellipsoid to nearly cylindrical (but tapered at one end), longitudinally ridged (but these can be hard to see).

ECOLOGY: Solitary, in clumps, or scattered in arcs and troops in duff in forest settings where true *Boletus* species are present. Especially common in coastal pine and spruce forests in fall and early winter. Found throughout our area, often in very close proximity to fruitbodies of Porcini-group boletes (sometimes sharing practically the same spot in the ground!). The common name refers to its frequent habit of fruiting in same area and about a week earlier than these prized boletes, and thus acting as an indicator species. *C. prunulus* is likely involved in a symbiosis with these boletes (likely parasitic, but not well understood).

EDIBILITY: Edible and good, but caution is warranted, as *Clitocybe rivulosa* is a very toxic look-alike.

COMMENTS: Also known as the Sweetbread Mushroom, this distinctive species is fairly easy to recognize by its dry, velvety or smooth, grayish white cap with a whitish bloom; decurrent gills that eventually develop a pinkish tone; strong cucumber-farinaceous odor; and pinkish tan spores. Some *Rhodocybe* in our area are similarly colored, but usually have thinner stipes, and most lack the strong farinaceous odor. Perhaps most similar is the toxic *Clitocybe rivulosa*, which is smaller, grows in arcs and troops in grassy areas, and has a paler spore deposit. *Clitocybe nebularis* is much larger, with a very smooth (not suedelike) cap, and has a skunklike odor. *C. subconnexa* is larger, grows in clusters, has a whitish cap, and has buff-colored spores. Other *Clitocybe* can be similar but are differently textured or have less distinct odor and usually a paler (whitish) spore deposit.

Alboleptonia adnatifolia

(Murrill) Largent & R. G. Benedict

CAP: 1–6 cm across, domed to convex with an incurved and often scalloped margin, expanding to broadly convex; some specimens slightly umbilicate, others broadly conical with a small umbo, outer margin often narrowly uplifted even when young. Whitish to dull whitish beige or with yellowish tones near center (see comments). Surface dry, fibrillose-silky, sometimes with very small scales. **GILLS:** Broadly attached and often with a small decurrent tooth, fairly widely spaced to subdistant, margin usually appearing serrate or eroded. Whitish at first, soon creamy and then pinkish to salmon as spores mature. **STIPE:** Cylindrical with a slightly enlarged base. White to creamy or light dingy tan near base. Surface smooth to silky or with small, obscure chevrons of silvery fibrils near apex. **FLESH:** Thin, fragile, hollow in stipe, white. **ODOR:** Indistinct. **TASTE:** Indistinct. **SPORE DEPOSIT:** Pinkish to dull pinkish tan. **MICROSCOPY:** Spores 9–10 x 7–8 µm, with 5 or 6 sides, distinctly angled. Cheilocystidia scattered. Clamps present.

ECOLOGY: Solitary or scattered in small groups, rarely abundant. Often on gravelly or mossy soil, also in humus and needle litter. Fairly common from late fall through early spring throughout our range, especially under California Bay Laurel and redwood, also in other forested settings and occasionally in open areas in disturbed soil.

EDIBILITY: Unknown.

COMMENTS: The silky white cap and stipe and somewhat widely spaced gills that turn pinkish in age help identify this species. In our area, this species is probably most often confused for members of the *Inocybe geophylla* group, which have a silky white cortina when young and grayish tan to dull brown mature gills. Specimens with yellowish tones are often identified as *A. sericella* var. *lutescens*, which Largent also distinguished by its longer spores (on average greater than 10 µm). If found to be genetically distinct from *A. adnatifolia*, the latter variety may warrant elevation to species status since *sericella* is based on European type material. *A. ochracea* develops darker, dingy orange colors, has longer spores (up to 13 µm long) and a densely scurfy upper stipe (with abundant club-shaped caulocystidia 12–25 µm wide); it is known from the northern part of our range.

MISAPPLIED NAME: *Alboleptonia sericella*.

Entoloma acutipes

Largent

CAP: 2–5 cm across, convex with a weakly inrolled margin, becoming broadly convex or flat. Light grayish brown, tan to nearly whitish at the very margin; opaque when young, becoming translucent-striate near margin when moist; hygrophanous, becoming beige-tan and more opaque as it dries out. Surface moist to dry, smooth. **GILLS:** Broadly attached, often slightly sinuate. Edges even when young, becoming irregularly serrate. Ivory to cream, becoming pinkish as spores mature. **STIPE:** 4–7 cm long, 0.5–0.8 cm thick, cylindrical or slightly enlarged lower, then tapered or pointed at extreme base. Whitish to creamy, obscurely striate but nearly smooth. **FLESH:** Thin, fairly firm, stipe becoming hollow in age. **ODOR:** Slightly farinaceous. **TASTE:** Slightly farinaceous. **SPORE DEPOSIT:** Dull pinkish. **MICROSCOPY:** Spores 7–9 x 7–9 µm, nearly isodiametric, 5 or 6 sided, distinctly angled. Gill cystidia absent. Clamps present. Pileipellis pigmentation intracellular.

ECOLOGY: Solitary or in small clusters in mossy forest humus under Tanoak and redwood. Known from Santa Cruz County to Mendocino County, but status and distribution essentially unknown. Fruiting in fall and winter.

EDIBILITY: Unknown; do not experiment.

COMMENTS: The relatively slender stature, tan to light grayish brown caps with faintly translucent-striate margins, pink gills, and white stipes that are often pointed or narrowed at the base are initial clues that help identify this rather anonymous *Entoloma*. Most similar species are more robust, with larger fruitbodies and often different colors as well (darker or paler), or with a distinct bleach odor. The microscopic characters mentioned above should be checked in order to identify this species. Largent used *E. rhodopolium* (a European name) for a number of other variants (species?) in this group in the western United States. Given that most (if not all) of these taxa are actually distinct from their European counterparts, a modern taxonomic investigation will likely result in many new names for members of this group in California. One common member of the *E. rhodopolium* group fruits abundantly in large troops in coastal thickets of alder and willow (*Salix* spp.) in the northern part of our range.

Entoloma ferruginans

Peck

BLEACHY ENTOLOMA

CAP: 5–12 cm across, domed to convex with a slightly inrolled margin, expanding to broadly convex, occasionally nearly flat, margin often remaining inrolled, sometimes wavy. Color variable, from very dark blackish brown to grayish brown with an olive wash, grayish tan when wet; hygrophanous, fading to beige, often shiny and metallic when dry. Center of cap frequently extensively virgate with a whitish sheen. Surface thinly viscid, more often greasy to dry, slightly wrinkled to smooth. **GILLS:** Broadly attached to sinuate, edges even to weakly serrate or eroded. Pale dingy grayish tan, soon becoming pinkish to salmon, often darker grayish brown. **STIPE:** 6–14 cm long, 1–4 cm thick, cylindrical but enlarged toward base, extreme base often slightly tapered. Pale whitish to light beige or tan. Dry, vertically striate, fibrillose to smooth. **FLESH:** Solid or becoming hollow in stipe, fibrous, white. **ODOR:** Usually quite strong of bleach or like a swimming pool. **TASTE:** Mild or similar to odor. **SPORE DEPOSIT:** Pinkish to salmon tan. **MICROSCOPY:** Spores 7–9 x 6.5–8 µm, nearly isodiametric, 5 or 6 sided, distinctly angled. Gill cystidia absent. Pileipellis a cutis without incrustations. Clamps abundant.

ECOLOGY: Solitary, scattered, or in clusters, often in troops in leaf litter and humus. Almost always under live oak; fruiting from late fall through midwinter. Common from the San Francisco Bay Area southward. Apparently absent on the coast north of Sonoma County.

EDIBILITY: Apparently nontoxic, but the smell is not appetizing and can be confused for potentially dangerous species.

COMMENTS: The strong bleach odor, dark grayish tan cap, gills that turn salmon pink, and growth under live oak distinguish this species. *E. cinereolamellatum* is similar in many respects, the most obvious difference being its mild odor; it may also be on average smaller spored and paler capped. The range of variation within each species needs to be investigated; there is some difference in interpretation among mushroom identifiers in California. *E. subsaundersii* is larger, often has a paler grayish tan cap, grows in fused clusters, and lacks the strong bleachy odor. Members of the *E. lividoalbum* complex are often browner capped and are usually larger, with more robust stipes.

Entoloma subsaundersii

Largent

CAP: 5–13 (18) cm across, domed or rounded-conical with an inrolled and frequently slightly ribbed margin when young, expanding to broadly convex with an indistinct, low umbo, sometimes nearly flat in age, margin often wavy. Whitish when very young but very soon a mix of pinkish tan, pinkish brown, pale creamy beige, lilac-gray, or cool gray; occasionally predominantly one of these colors overall but usually blotchy and mixed, sometimes with a radially silky or finely streaked appearance. Center of cap usually markedly virgate with a whitish sheen. Surface slightly wrinkly, rough to nearly smooth, slightly viscid to greasy or dry. **GILLS:** Attached, often notched, edges usually slightly serrate to eroded. Whitish to creamy when young, sometimes with a light yellowish tone, becoming pinkish as spores mature. **STIPE:** 4–15 cm long, 1.2–4 (5) cm thick, stout, cylindrical or spindle shaped (with a swollen and then tapered base), with a club shaped base if fruiting singly, more often multiple stipe bases fused into a large mass of firm, rubbery tissue. Whitish to pale cream, slightly dingier in age. Surface dry, fibrillose to nearly smooth, apex often with obscure, small scurfy chevrons. **FLESH:** Thick, firm and often rubbery, fibrous, white, solid throughout until riddled with insect larvae. **ODOR:** Farinaceous, fairly strong. **TASTE:** Farinaceous. **SPORE DEPOSIT:** Pinkish to salmon tan. **MICROSCOPY:** Spores 7–8.5 x 7–9 µm, nearly isodiametric, 5 or 6 angled but fairly weakly. Clamps abundant.

ECOLOGY: Usually fruiting in clumps of two to six fruitbodies, occasionally solitary; the largest patches we have seen are in hard-packed soil or rich humus of Tanoak-Madrone forests. Generally infrequent but can be locally abundant from late fall through winter throughout our range.

EDIBILITY: Unknown; can be confused for species that may be toxic.

COMMENTS: The stout stature; mottled grayish, lilac-gray, and/or beige-tan cap with a silvery virgate center; and clustered growth under hardwoods help distinguish this species. A number of *Tricholoma* grow in the same habitat and can look very similar, but never develop pinkish tones on the gills and rarely if ever grow in fused clusters. *E. ferruginans* has a strong bleachy odor, and members of the *E. lividoalbum* complex are larger and more brown capped.

Entoloma medianox

C. F. Schwarz

MIDNIGHT ENTOLOMA

CAP: 3–15 cm across, dome shaped when young, becoming broadly convex to plane or irregularly uplifted. Very young fruitbodies buried in duff pale whitish beige to cream colored, soon grayish blue, sky blue, or powder blue, dark grayish blue to navy blue or midnight blue at maturity, often with a pale streaky area over center of cap. Older specimens often lose pigmentation irregularly, showing grayish or beige splotches and streaks. Surface smooth to strongly wrinkled-rivulose, slightly viscid to dry. **GILLS:** Attached and notched, fairly close, edges often irregular or eroded-serrate. Whitish to creamy at first, often with a pale blue wash near margin (sometimes more extensive), becoming pinkish as spores mature. **STIPE:** 6–10 cm long, 1.5–3 cm thick, club shaped or cylindrical, sometimes tapered at base. Whitish, developing grayish blue to sky blue or navy blue areas, sometimes yellowish near base. Fairly robust, surface often with a metallic sheen and a marbled appearance. Surface nearly smooth and usually with thin, appressed chevrons of silvery or whitish tissue near apex. **FLESH:** Whitish to creamy or pale bluish (especially near cap surface). **ODOR:** Indistinct to farinaceous. **TASTE:** Farinaceous. **KOH:** No reaction. **SPORE DEPOSIT:** Cinnamon pinkish to salmon. **MICROSCOPY:** Spores 6–11 x 5.5–9.5 µm, rather weakly angled, with 4 or 5 sides. Pileipellis a cutis of entangled, clamped hyphae embedded in a gelatinous layer. Cystidia absent.

ECOLOGY: Solitary, scattered, or in loose clusters, forming arcs or large troops in duff under redwood and California Bay Laurel, but also under in Madrone-Tanoak woods and others in forest settings. Not found in grassland or meadows, unlike its European counterparts. Very common throughout our range. Fruiting from fall through midwinter.

EDIBILITY: Edible. Firm and flavorful, and often abundant. However, caution is recommended due to a lack of historical experience with this species.

COMMENTS: This species is easily recognized by its large size, robust stature, dark blue cap color, and gills that become pink when the spores mature. Occasionally specimens are encountered with entirely light pink to rosy caps; DNA data shows that these are pigment aberrations rather than a different species. A species with smaller spores and perhaps darker grayish blue gills has been collected in Mendocino County; it is genetically similar to *E. caesiolamellatum*. Most other *Entoloma* are duller, while the blue *Leptonia* are much more slender.

MISAPPLIED NAMES: *Entoloma madidum* and *E. bloxamii*.

NOMENCLATURE NOTE: *E. medianocte* is an illegitimate, superfluous name. *Mea gran culpa.*

Entocybe trachyospora

(Largent) Largent, T. J. Baroni & V. Hofstetter

CAP: 1–6 cm across, bell shaped or domed when very young, often rounded-conical with a distinct umbo, expanding to flat, but often retaining an umbo. Color quite variable, from dark blackish brown to dull reddish brown, light grayish beige or grayish tan (see comments); darkest at disc, paler outward and margin strongly translucent-striate. Surface thinly viscid to moist or dry in age, smooth or very finely roughened near center. **GILLS:** Obscurely attached, closely spaced, margins even to eroded. Whitish to pale gray, often dingy grayish tan in age with salmon tones as spores mature. **STIPE:** 4–10 cm long, 0.3–1 cm thick, cylindrical with a slightly enlarged base, usually somewhat twisted and irregular, wavy and bumpy. Color variable, whitish gray to pale bluish, sky blue or grayish blue, whitish to dingy yellowish toward base. Surface nearly smooth, but distinctly vertically striate. **FLESH:** Thin, fragile in cap, whitish. Stipe fibrous, becoming hollow in age. **ODOR:** Indistinct. **TASTE:** Not sampled. **SPORE DEPOSIT:** Dull pinkish tan to salmon. **MICROSCOPY:** Spores 6–8 x 6–7 μm, isodiametric, distinctly bumpy with 6 to 9 angles, clamps present.

ECOLOGY: Solitary or scattered in small troops, occasionally in larger swarms, in leaf litter and mossy humus under various conifers from midfall into winter. Can be locally common on the Far North Coast, but is uncommon over most of our range, with one record from the Santa Cruz Mountains.

EDIBILITY: Unknown.

COMMENTS: The cap is smoother and the stipe usually thicker and more distinctly silvery striate than those of similarly colored *Leptonia*. The often wavy surface of the stipe and less upright stature are more subtle clues. Microscopically, the subglobose spores with many obtuse angles are very characteristic. *E. nitidum* is a Pacific Northwest species that is quite similar, but is larger and more evenly sky blue to royal blue overall; there are no confirmed records of it from California, but it might turn up on the Far North Coast. A number of published varieties reflect the wide range of colors shown by this species: Northern specimens with whitish gray stipes, pale gills, and grayish to hazel brown caps are the nominate as variety *trachyospora*. *E. t.* var. *griseoviolacea* has grayish brown tones in the cap, while *E. t.* var. *purpureoviolacea* is more purplish blue overall, with bluish gray gills (the latter variety likely represents a distinct species).

Leptonia "Pewter Blue"

CAP: 2–5 cm across, convex to flat, margin slightly inrolled or downcurved when young, becoming wavy, uplifted, and often splitting in age. Sky blue to grayish blue or pearly pewter blue. Surface dry, with a granular or scurfy appearance due to the presence of many fine, slightly uplifted, fibrillose scales, sometimes breaking into large plaques near center of cap. In age scales may disappear over much of cap, making it appear smooth and rather metallic. **GILLS:** Variably attached but usually sinuate, close to moderately spaced. Whitish to light grayish white, soon washed with sky blue or more extensively bluish gray in age, developing pink tones as spores mature, drying to yellowish. **STIPE:** 3–6 cm long, 0.75–1.5 cm thick, usually tapered slightly upward, cylindrical or compressed and grooved. Off-white or washed with grayish blue. Surface dry, slightly felty or woolly. **FLESH:** Whitish to creamy or yellowish, distinctly stringy-fibrous in stipe. Stipe stuffed when young, usually becoming hollow in age. **ODOR:** Indistinct when young, somewhat sour or cheesy in age. **TASTE:** Indistinct. **SPORE DEPOSIT:** Dull pinkish to pinkish tan. **MICROSCOPY:** Spores 8–10 x 6–7.5 μm, with 5 or 6 distinct angles, heterodiametric. Cheilocystidia cylindrical, many with a narrow nipple at tip, others septate or knobbed. Pileipellis a short, entangled trichoderm with brownish intracellular pigment.

ECOLOGY: Scattered in groups or small clusters and arcs in needle duff under redwood. Locally common but so far only known from Santa Cruz County. Fruiting from early winter into spring.

EDIBILITY: Unknown.

COMMENTS: The sky blue to pewter blue cap and stipe; whitish to bluish gray gills; and finely scaly, felty, or woolly cap and stipe textures identify this beautiful and distinctive *Leptonia*. Other bluish *Leptonia* are usually much darker in color, and most have smoother cap and stipe surfaces.

Leptonia decolorans group

CAP: 1–4 cm across, domed at first, becoming broadly convex, occasionally flattening out or becoming wavy, center sometimes sunken (when cap tissue collapses into hollow stipe). Dark bluish gray to navy blue, margin usually opaque. Surface dry, finely scurfy at center, appressed-fibrillose outward to nearly smooth in age. **GILLS:** Attached, whitish (but see comments) becoming pinkish as spores mature. **STIPE:** 3–8 cm long, 0.2–0.5 cm thick, cylindrical with a slightly enlarged base. Light grayish blue to steel blue, occasionally darker, nearly navy blue. Base often with extensive whitish mycelium. **FLESH:** Thin, fibrous, whitish. Hollow in stipe. **ODOR:** Indistinct or somewhat fragrant or cheesy in age. **TASTE:** Not sampled. **SPORE DEPOSIT:** Dull pinkish. **MICROSCOPY:** Spores 7.5–11.5 x 5–8 μm, on average greater than 10 μm. Pileipellis a palisade-trichoderm. Cheilocystidia present. Clamps absent.

ECOLOGY: Solitary or scattered in small groups in leaf litter or mossy duff, or on bare soil of trail cuts and roadsides. Most often under redwood, but also in mixed live oak and California Bay Laurel groves. Common throughout our range. Fruiting from late fall through spring, earlier in north.

EDIBILITY: Unknown

COMMENTS: This is right in the middle of the "Little Blue *Leptonia*" swarm; members of this group are very difficult to identify with certainty, even with microscopic data and technical references. The bluish gray to navy blue cap and stipe (lacking brown tones even in age), nonstriate cap margin, whitish gills, abundant cheilocystidia; and spores averaging greater than 10 μm long are important (but not definitive) characters. The name *L. decolorans* was based on mushrooms from New Zealand and is probably misapplied to mushrooms on the Pacific Coast. Largent described *L. d.* forma *atropruinosipes* based on specimens with dark blackish blue scurfs near the stipe apex, as well as *L. d.* forma *cystidiosa* for those with pale scurfiness near the upper stipe; these may be elevated to species rank. *L. yatesii* has distinctly decurrent gills and grows singly, while *L. viridiflavipes* has weakly decurrent gills, fruits in clusters, and has yellowish tones near the stipe base.

Leptonia subviduense group

CAP: 1–6.5 cm across, convex with a narrow incurved margin, becoming broadly convex, only occasionally flattening out, margin often wavy and splitting in age. Deep royal blue to navy blue or very dark bluish gray. Surface dry, finely fuzzy-tomentose to granular-felty at center, mostly remaining so as it expands, but outer areas becoming more flattened-scaly. Margin opaque or obscurely translucent-striate. **GILLS:** Obscurely attached, fairly close, edge even to slightly eroded (but see comments). Bluish white to sky blue or light gray with a bluish gray wash (often blotchy), in some collections evenly bright bluish. **STIPE:** 3–8 cm long, 0.2–0.7 cm thick, cylindrical with an enlarged base or more often flattened with a longitudinal groove. Light grayish with a sky blue wash or slightly darker grayish blue. Surface dry, smooth, base often with extensive covering of cottony white mycelium. **FLESH:** Thin, fibrous, hollow in stipe, light grayish blue to whitish with a blue wash. **ODOR:** Indistinct. **TASTE:** Not sampled. **SPORE DEPOSIT:** Pinkish to dingy salmon. **MICROSCOPY:** Spores 7.5–12 x 5.5–8.5 µm, 5 or 6 sided, moderately angled. Cheilocystidia usually abundant, colorless (see comments). Pileipellis a trichoderm. Clamps absent.

ECOLOGY: Solitary, in clumps of up to five fruitbodies, or scattered in small troops in humus and needle litter of redwood, but can be found in disturbed forest settings of other types. Fairly common and widespread. Fruiting in late fall into early winter, occasionally well into spring.

EDIBILITY: Unknown.

COMMENTS: This species is difficult to reliably distinguish from the multitudes of other "Little Blue *Leptonia*." The bluish colors of the gills, fuzzy-tomentose cap without a translucent-striate margin, average spore length less than 10 µm, and abundant cheilocystidia are helpful but not definitive characters. This species is similar to *L. chalybaea* (sensu Largent, 1994), but Largent's reports regarding the presence of cheilocystidia for that species are contradictory. He considers *L. chalybea* synonymous with *L. nigra* Murrill, which was collected in San Mateo County; that name may be more appropriate for our area since *L. chalybaea* is based on a European type. Members of the *L. decolorans* group have larger spores (on average greater than 10 µm long) and whiter gills when young (but may show bluish gills when rain soaked or in age).

Leptonia serrulata group

CAP: 1.5–6 cm across, hemispheric to broadly convex when young, more or less plane or uplifted and slightly wavy in age. Often with a broad, round umbo that is slightly depressed at the center (umbilicate). Most often deep navy blue but ranging from royal blue in bright forms to grayish steel blue in dull forms. Surface dry, finely tomentose-scruffy when young, becoming appressed-tomentose with fine radially arranged scales that become uplifted as cap expands. Translucent-striate when mature, sometimes more pronounced when wet. **GILLS:** Variably attached, most often notched with a slight decurrent tooth, often seceding. Grayish white to bright bluish when young, becoming bluer especially near margin, sometimes vivid blue overall, edges marginate, beaded with contrasting navy blue to royal blue clusters of cheilocystidia. **STIPE:** 2–8 cm long, 0.3–0.5 cm thick, cylindrical. Grayish to navy blue, steel blue to gray in dull forms, brighter blue in others. Surface finely punctate with dark clusters of cystidia at least at apex, often finely so overall, otherwise smooth to obscurely vertically striate. **FLESH:** Very thin, pale grayish to dark bluish. Stipe hollow. **ODOR:** Indistinct. **TASTE:** Not sampled. **SPORE DEPOSIT:** Dull salmon to pale brownish pink. **MICROSCOPY:** Spores 8–12 x 5.5–8.5 µm (average greater than 10 µm long), moderately to strongly angled. Clamps absent. Cheilocystidia abundant, grayish brown.

ECOLOGY: Solitary or scattered in small groups in duff, moss, or bare soil of road cuts, most often under redwood, also in other forest types. Fruiting in fall in the north, well into winter or early spring in the southern part of our area. Infrequent overall but can be locally common.

EDIBILITY: Unknown.

COMMENTS: The dark royal blue to navy blue colors, striate cap, and dark blue marginate gills make a distinctive suite of features. However, this taxon is a good example of a bad habit: assigning an existing name based on a single highly distinctive feature (in this case the dark blue marginate gills). Preliminary genetic data suggests that the species illustrated above ranges north at least to British Columbia and is not conspecific with the European or eastern North American collections going by this name.

Leptonia carnea

Largent

CAP: 3–7 cm across, round with an inrolled margin when young, becoming convex to plane, margin occasionally wavy and splitting in age. Deep navy blue to indigo blue or with splotches of iridescent bright blue, metallic when dry. Surface dry, covered in small, fine tufts or scales. **GILLS:** Attached, often notched, subdistant, with many short gills. Dull grayish to lilac-gray or dingy bluish to purplish when young, becoming dusted with pinkish to salmon color as spores mature and often developing dark grayish purple or navy blue streaks and mottling. **STIPE:** 4–10 cm long, 0.6–1.5 (2) cm thick, equal, or with a slightly enlarged base. Navy blue, indigo to deep purplish blue, with bands or chevrons of fine fibrillose tufts; base with a large area of cottony whitish mycelium. Sometimes adorned with bright greenish yellow droplets when young. **FLESH:** Whitish with traces of blue near surface, solid, fibrous. **ODOR:** Farinaceous. **TASTE:** Farinaceous. **SPORE DEPOSIT:** Pinkish tan to dull pinkish brown. **MICROSCOPY:** Spores 9–13 x 6–10 μm, 5 or 6 angled, heterodiametric. Clamps abundant in pileipellis.

ECOLOGY: Solitary or in small groups of scattered individuals, in humus and duff, almost always under redwood, but we have seen it once in a pure stand of Monterey Cypress. Known from Humboldt County south to Big Sur, especially common in Santa Cruz County. Nearly endemic to California (one record from Washington). Fruiting late fall through early spring.

EDIBILITY: Unknown.

COMMENTS: Although superficially similar to many bluish *Leptonia*, the large size (for a *Leptonia*), solid white flesh, and shaggy surface of the stipe make this gorgeous species easy to recognize. *L. pigmentosipes* is often confused for this species, but has much paler (whitish) gills that occasionally become markedly purplish in age, a more fragile, hollow stipe, and a tendency to become strongly umbilicate in age, with the cap flesh caving into the top of the hollow stipe. *Entoloma medianox* is usually much larger, with a smoother, often slightly viscid cap and a thicker, paler stipe.

SYNONYM: *Entoloma subcarneum (largent)* Blanco-Dios.

Leptonia pigmentosipes

Largent

CAP: 3–8 cm across, broadly convex to round-conical with an incurved margin, eventually nearly flat. Indistinctly umbonate when young, soon umbilicate to depressed or with a crater into top of stipe; margin slightly inrolled, lobed and wavy in age. Deep purple-blue to bluish black but usually showing some true purple or violet tones, sometimes reddish near margin. Surface dry, densely tomentose-scaly at first, then shaggy-fibrillose overall with uplifted tufts and scales, scales becoming flattened and radially splitting to reveal paler whitish gray flesh. **GILLS:** Attached, notched or seceding, with cross veins. White, becoming pinkish, in age sometimes with a light purple or bluish wash. **STIPE:** 4.5–13 cm long, 0.4–0.8 cm thick, cylindrical, but often compressed and grooved. Mostly metallic navy blue, but usually with purple tones near apex and at base; extreme apex almost always with a distinct whitish zone. Surface dry, with chevrons of shaggy-fibrillose scales, scarcer towards apex, sometimes absent in age. **FLESH:** Thin, whitish gray to yellowish, stipe hollow. **ODOR:** Indistinct. **TASTE:** Indistinct. **SPORE DEPOSIT:** Dull pinkish to salmon. **MICROSCOPY:** Spores 9–12 x 5.5–9 μm (on average greater than 10 μm), heterodiametric, distinctly angled. Clamps absent.

ECOLOGY: Solitary or in small groups in humus and needle duff, mostly under redwood and Monterey Cypress. Locally common from Santa Cruz north to Mendocino County. Fruiting from midfall through early spring.

EDIBILITY: Unknown.

COMMENTS: The purplish cap, pallid gills, and scruffy stipe with a white zone at the apex are good features by which to identify this beautiful *Leptonia*. It is sometimes confused with *L. carnea* due to its deeply colored, iridescent fruitbodies, but that species has a solid stipe that is pigmented all the way to the apex, gills that are never so white, and a more robust stature. *L. occidentalis* and *L. cyanea* var. *occidentalis* are similarly colored and have scaly stipe surfaces, but both have clamps in the pileipellis. The latter species grows on wood and is similar to Largent's concept of *L. tjallingiorum*.

Leptonia "Brown and Blue" group

CAP: 1.5–5 cm across, convex or rounded-conical with a wavy and slightly incurved margin, expanding to broadly convex, in age slightly uplifted and wavy, sometimes with a low, broad umbo or slightly umbilicate. Tan-brown, dingy beige-brown to reddish brown, often with a dark gray or bluish gray center; margin translucent-striate when wet. Surface extensively covered with fine scales composed of flattened fibrils; can be slightly tomentose at center and smooth at margin. **GILLS:** Obscurely to broadly attached, close to subdistant, margin even or slightly irregular. Whitish to pale cream when young, becoming pinkish to dingy salmon in age. **STIPE:** 4–9 cm long, 0.3–0.5 cm thick, cylindrical, fairly straight, base slightly enlarged. Color somewhat variable but always with a distinct bluish hue: sky blue to gunmetal blue, grayish blue, or navy blue. Surface dry, smooth, or with fine, pallid scurfiness near apex. **FLESH:** Thin, pale, hollow in stipe. **ODOR:** Indistinct. **TASTE:** Not sampled. **SPORE DEPOSIT:** Pinkish to dull salmon tan. **MICROSCOPY:** Variable, differing among the 4 or 5 species in complex. Spores distinctly angled, with 5 or 6 sides, 8–12 μm. Cheilocystidia present in some members, absent in others.

ECOLOGY: Solitary or scattered in small troops, occasionally weakly clustered, in leaf litter and mossy humus under redwood, also in other forest types. Uncommon from midfall into midwinter.

EDIBILITY: Unknown.

COMMENTS: Mushrooms in this this complex have brownish caps that contrast noticeably with the bluish stipes. Largent applied the European names *L. lividocyanula*, *L. asprella*, *L. gracilipes*, and *L. sodalis* to various members of this complex in California. However, there are no species in this complex described from California, and Largent stated that he could not reliably tell them apart—hence our use of this nickname. The caps of *L. parva* start out dark bluish gray and become browner in age, whereas *L. foliocontusa* has a reddish gray to brown cap and gills that slowly turn orangey tan.

Leptonia formosa group
GOLDEN LEPTONIA

CAP: 1–7 cm across, domed when young with a finely incurved margin, becoming broadly convex to slightly boxy (with a broader, flat center), in age nearly flat or shallowly uplifted with a slightly wavy margin. Warm ocher to orangey brown, sometimes darker brown at center, becoming golden yellow to yellowish beige but covered with fine brownish scales. Margin usually bright to pale golden yellow, distinctly translucent-striate with brownish tan lines. Surface dry, covered in many tiny scales overall when young, but soon appressed-fibrillose outward, with a streaky appearance, sometimes smooth except at very center. **GILLS:** Obscurely attached, fairly closely spaced, edge even to strongly serrate or eroded; sometimes with brown marginate areas. White to pale cream, soon with dingy pinkish tones. **STIPE:** 3–10 cm long, 0.3–0.7 cm thick, cylindrical with a slightly enlarged base or compressed with a central groove. Creamy yellow to ocher, dingy pale tan, or light brownish. Base usually with a large cottony area of whitish mycelium. **FLESH:** Thin, fibrillose, hollow in stipe, whitish to beige. **ODOR:** Indistinct. **TASTE:** Not sampled. **SPORE DEPOSIT:** Pinkish salmon. **MICROSCOPY:** Spores 9–12 x 6–10.5 μm, 5 or 6 sided, distinctly angular, clamps absent. Cheilocystidia absent.

ECOLOGY: Solitary or in small groups or clusters of two to four fruitbodies in moss or leaf humus, or on bare soil of road cuts and trailsides. Common in forested habitat throughout our area, especially under redwood. Fruiting from late fall through early spring.

EDIBILITY: Unknown.

COMMENTS: The yellowish ocher cap with minute brownish scales, smooth yellow stipe, and whitish gills help distinguish this species. A very similar species (or perhaps just a richly pigmented form?) with a carrot orange stipe and orange-brown cap occurs under Monterey Cypress in our area. *Leptonia exalbida* is very similar but has a whitish stipe. Largent mentions a number of other species in this complex: *L. exilis* supposedly differs by its yellow-green stipe when very young; *L. badissima* and *L. turci* both have darker chocolate brown caps and brownish stipes; and *L. xanthochroa* has brownish marginate gills. We use many of these names following Largent (1994), but most are based on specimens collected far from the Redwood Coast; it may be that none of these species are present in our area. Largent's *L. formosa* var. *microspora* has spores averaging less than 10 μm long and has mostly 2-spored basidia. *Rhodophana nitellina* has a smooth, carrot orange cap and an orange stipe, and smells like cucumber. Similar species of *Nolanea* have darker brown, hygrophanous, and often more conical caps.

Leptonia exalbida

Largent

WHITE-STEMMED LEPTONIA

CAP: 1–4 cm across, domed to broadly convex with a slightly incurved margin when young, expanding to nearly plane, sometimes with a slightly wavy margin. Warm tan to light brown, usually dark brown at center, margin becoming translucent-striate. Surface dry, covered in small scales made of flattened clumps of fibrils, smoother toward margin. **GILLS:** Obscurely attached or seceding, edges mostly even. Whitish when young, light grayish cream, becoming pinkish with spores. **STIPE:** 2–6.5 cm long, 0.2–0.4 cm thick, cylindrical with a slightly enlarged base. Whitish to very light grayish cream, apex sometimes with an obscure light grayish wash. Surface completely smooth, with cottony white mycelium at base. **FLESH:** Thin, fragile to slightly cartilaginous, hollow in stipe. Whitish. **ODOR:** Indistinct. **TASTE:** Not sampled. **SPORE DEPOSIT:** Light pinkish to salmon. **MICROSCOPY:** Spores 8–12 x 6–8 µm, with 5 or 6 sides, heterodiametric. Cheilocystidia abundant. Clamps absent. Pileipellis a palisade-trichoderm.

ECOLOGY: Solitary, in pairs, or scattered in small groups in mossy humus or twig and needle litter, most often under cedar and redwood from late fall through early spring. Generally infrequent throughout our area, but can be locally common in some years.

EDIBILITY: Unknown.

COMMENTS: The brownish tan cap and white stipe in combination with the small size, attached gills, growth on the ground, and pinkish spore deposit identify this species. Species in the *L. formosa* group can look quite similar, but have yellower stipes, more conical, ocher-colored caps, and usually lack cheilocystidia. Both *L. formosa* and *L. exalbida* could be mistaken for *Pluteus romellii*, which has free gills, thicker flesh, yellowish gills when young, and grows on wood.

Leptonia "Ruby Grapefruit"

CAP: 1.5–5 cm across, dome shaped to round-conical, expanding to broadly conical or irregularly wavy to nearly plane, smooth or with small scales and clumps of fibrils. Magenta-pink, ruby red or coral pink, scales slightly darker when present. **GILLS:** White to creamy, becoming pinkish even before spores mature. **STIPE:** 1–5 cm long, 0.3–0.5 cm thick, equal, slender and cylindrical in some collections, squat and grooved others. Creamy beige to pinkish, usually whitish toward base. **FLESH:** Thin, whitish pale pinkish, hollow in stipe. **ODOR:** Indistinct. **TASTE:** Not sampled. **SPORE DEPOSIT:** Pinkish. **MICROSCOPY:** Spores 9–13 x 6–10.5 µm, with 5 to 7 angles.

ECOLOGY: Solitary or in small groups under redwood and California Bay Laurel. Rare, known from only a few sites in the southern part of our range. Fruiting in late fall and winter.

EDIBILITY: Unknown.

COMMENTS: This lovely mushroom is recognizable by its ruby red to coral pink cap, pinkish beige stipe, and pinkish spore print. There are at least three pink-capped *Leptonia* in our area. *L. subrubinea* is very similar (and may even be the same) but appears to be a bit paler pink with a slightly more matted-scaly cap texture; it is known from the Far North Coast. *Leptonia rosea* var. *marginata* has a paler whitish beige stipe and contrasting pinkish red gill edges. Another, undescribed species has a rosy pinkish cap with a narrow bluish gray margin and a smooth gray stipe. All of these appear to be quite rare. Some *Laccaria* species are pink and have a finely scurfy cap, but have widely spaced thick-edged gills, a tougher texture, white spore deposits, and have spiny (not angled) spores.

Inocephalus cystomarginatus

Largent

CAP: 1.5–4 cm across, convex to broadly convex, sometimes nearly flat or weakly uplifted, center of cap even or slightly depressed (but not really umbilicate), occasionally with a small umbo. Dingy pinkish tan to brick red, usually darkest brownish red near center, tan to pale beige near margin, often metallic looking when dry. Surface dry, radially fibrillose to smooth, sometimes with darker uplifted scales, margin often splitting, opaque. **GILLS:** Attached, often notched, close to fairly widely spaced, relatively thick. Dull creamy to pallid beige, soon tan to pinkish brown, edges darker (often blotchy) with clusters of pinkish brown cheilocystidia. **STIPE:** 3–8 cm long, 0.2–0.4 cm thick, straight, cylindrical. Creamy, beige or tan, becoming light pinkish brown to brick red, base usually remaining whitish. Surface somewhat fibrillose and dark streaked, apex densely coated with small, pale scurfy scales. **FLESH:** Thin, fibrous, pale. **ODOR:** Indistinct. **TASTE:** Not sampled. **SPORE DEPOSIT:** Dull pinkish tan. **MICROSCOPY:** Spores 8–13 x 6.5–10 μm, heterodiametric, with 5 or 6 angles. Cheilocystidia club shaped, with internal brown pigment. Clamps present. Cells of cap cuticle incrusted.

ECOLOGY: Solitary, in pairs or small troops in duff or on soil under redwood. Widespread, but uncommon, although it can be locally abundant in some areas. Fruiting in winter and spring

EDIBILITY: Unknown.

COMMENTS: The tall, thin stipe, pinkish beige to brick red cap, dark reddish gill edges, and pink spore print are fairly distinctive. Although the radially fibrillose cap resembles an *Inocybe*, the gills do not turn dingy brown, the stipe is never bulbous at the base, and the odor is never spermatic. Other *Inocephalus* in our area have smoother caps and evenly colored gills. The poorly known *I. minimus* has a darker brown, translucent-striate cap, isodiametric spores, and septate cheilocystidia, and lacks clamps; it is known from the Far North Coast.

Rhodophana nitellina

T. J. Baroni & Bergemann

CUCUMBER ENTOLOMA

CAP: 2–6 cm across, broadly convex to plane, often with a depressed center, margin becoming wavy and sometimes slightly uplifted in age. Carrot orange to dull orange to dark yellowish orange outward, often with irregularly distributed reddish orange blotches and translucent-striate when wet. Hygrophanous; fading to pale beige (especially where cracked) when dry. Surface dry, smooth, sometimes with small uplifted scales. **GILLS:** Broadly attached, fairly widely spaced, partial gills numerous, rarely forked. Dingy pallid white, sometimes cream-orange or dull beige. **STIPE:** 4–12 cm long, 0.3–0.8 cm thick, cylindrical. Usually long, slender, straight or often sinuous or curved. Dull, often somewhat translucent, bright orange to orangish beige, apex with pale beige or whitish scurfy scales. Base with streaks of white mycelium. Surface smooth or sometimes bumpy. **FLESH:** Thin, very fragile. Dingy pale orange or yellowish orange. **ODOR:** Strongly cucumber-farinaceous. **TASTE:** Farinaceous. **KOH:** No reaction. **SPORE DEPOSIT:** Dull pinkish tan. **MICROSCOPY:** 6.5–7.5 x 4.5–5 μm, teardrop shaped or elongated; distinctly roughened.

ECOLOGY: Solitary or scattered in small groups in duff under redwood, sometimes in other dark forest habitats. Common throughout our range. Fruiting from late fall into spring.

EDIBILITY: Unknown.

COMMENTS: Although this mushroom is fairly nondescript in shape, the bright orange color, striate margin, cucumber-farinaceous odor, and pinkish spores are disctinctive. Look-alikes include *Lactarius rubidus*, with a sweetish odor, milky-watery latex, and white spores, and *Tubaria* species, with dull brown spore deposits and a less brittle texture.

Rhodocybe "Blue-Gray"

CAP: 1–4 cm across, convex with an inrolled and sometimes finely ribbed margin, expanding to broadly convex, often weakly lobed. Dull grayish tan to bluish gray, often pale whitish gray toward margin with darker "water spots." Surface dry, smooth but very finely pruinose, especially dense at margin and appearing very finely matted. **GILLS:** Broadly attached or slightly decurrent, widely spaced, thick edged, a few often forked near stipe. Grayish to gray-tan, edges paler but not strongly contrasting. **STIPE:** 2–5 cm long, 0.4–0.7 cm thick, usually widest at apex, tapered toward base. Cylindrical or often slightly compressed. Dingy beige-tan to grayish tan but often with a bluish ring around apex; most of surface covered in a whitish bloom, smoother in age, except for persistent pale scurfs near apex. **FLESH:** Fairly thin, slightly cartilaginous, solid or becoming hollow in stipe, dingy pale grayish. **ODOR:** Usually indistinct or slightly farinaceous. **TASTE:** Indistinct or weakly bitter-acrid. **SPORE DEPOSIT:** Dull pinkish tan. **MICROSCOPY:** Data is lacking from western North American specimens.

ECOLOGY: Solitary, in pairs, in duff or humus or on mossy soil, usually near redwood in dark, moist forests. Widespread, generally infrequent, but can be locally common, especially in the northern part of our range. Fruiting in fall and winter.

EDIBILITY: Unknown.

COMMENTS: The grayish blue colors, fairly thick and well-spaced subdecurrent gills, and somewhat stout stipes help distinguish this species. This taxon belongs to a poorly known group of species, most of which remain undescribed. A somewhat common *Rhodocybe* that occurs in the same habitat is slightly larger and taller, has a dark brown cap (without any bluish tones), and has contrasting pale scurfs on the dark upper stipe. *Rhodocybe hondensis* has a pinkish tan or light brownish cap, broadly attached pale gills, and a pale creamy stipe; it is not uncommon under redwoods. *Rhodocybe nuciolens* is very similar to the latter species, but may be more robust on average and has cheilocystidia (absent in *R. hondensis*).

Nolanea bicoloripes

Largent & Thiers

CAP: 1–4 cm across, bell shaped at first, expanding to broadly conical and then nearly flat, almost always with a small but distinct nipple-like grayish umbo. Creamy whitish to silvery grayish white; margin opaque when young, translucent-striate when mature, opaque again as it dries. Surface dry, silky-fibrillose when young, nearly smooth and often with raised and depressed concentric zones at maturity. **GILLS:** Narrowly attached or seceding, moderately widely spaced, edges even. Creamy whitish when young, becoming pinkish as spores mature. **STIPE:** 4–12 cm long, 0.2–0.4 cm thick, cylindrical, usually tall, thin, and rather straight. At first pale grayish, soon darkening upward and becoming extensively dark yellowish brown or dark grayish tan, contrasting strongly with gills and cap at maturity. Entirely smooth. **FLESH:** Thin, whitish grayish, rather cartilaginous and rigid in stipe. **ODOR:** Indistinct. **TASTE:** Not sampled. **SPORE DEPOSIT:** Pinkish to salmon tan. **MICROSCOPY:** Spores 6.5–8.5 x 5–6 μm, 5 or 6 sided, distinctly angular. Gills lacking cystidia, clamps rare to absent.

ECOLOGY: Solitary or scattered in troops in moss and needle humus of conifers, especially redwood and Western Redcedar (*Thuja plicata*). Locally common on the Far North Coast, absent south of Sonoma County. Fruiting from early fall into winter.

EDIBILITY: Unknown.

COMMENTS: The tall, thin, straight gray stipe, often with a darker base, contrasting with the pale cap; pinkish gills; and overall small size of the fruitbodies help identify this species. *Mycena galericulata* can look similar when it develops pinkish gills in age, but is usually larger, with a tougher, more curving stipe, and is usually clearly attached to wood. Most similar species in the family Entolomataceae have a different stature and/or different coloration (less contrasting caps and stipes). *Inocephalus rigidipus* also has a tall, cartilaginous stipe, but has a browner, slightly umbilicate or round cap (not umbonate) and lacks the silkiness of this species.

Nolanea edulis

(Peck) Largent

CAP: 2–5 cm across, convex, rarely flat, occasionally bell shaped and with a rounded umbo, but usually slightly umbilicate or depressed at center, margin sometimes slightly wavy in age. Dull, dark brown at first, soon grayish brown, fading to pale grayish or brownish beige, usually with a distinctly mottled or radially streaked to virgate, slightly metallic appearance. Often concentrically zonate, with narrow bands of dark and pale colors. Margin not translucent striate or only obscurely so. Surface dry, slightly velvety or tomentose when very young (especially near margin), soon smooth with innate, pale fibrils. **GILLS:** Usually notched or broadly attached with a decurrent tooth, somewhat distant, edges smooth to irregularly serrate. Dingy pale grayish beige, becoming brownish tan and then pinkish as spores mature. **STIPE:** 2–5 cm long, 0.25–0.5 cm thick, cylindrical, often tapered toward base, sometimes weakly club shaped. Dingy beige-tan to light grayish tan or silvery. Silky-fibrillose or finely scurfy, soon becoming smoother. **FLESH:** Thin, whitish gray, fibrous or slightly rubbery. Stipe solid or hollow in age. **ODOR:** Mild to slightly farinaceous. **TASTE:** Mild to slightly nutty or weakly farinaceous. **SPORE DEPOSIT:** Pinkish tan to dull pinkish brown. **MICROSCOPY:** Spores (7.5) 8–8.5 (10) x 7–8 (9) μm, 5 to 6 sided, distinctly angled. Cheilocystidia absent. Cells of cap surface coarsely incrusted.

ECOLOGY: Solitary or more often scattered in troops on bare soil amid small herbaceous plants and areas with short grass, often in exposed settings along trails or roads and in urban areas. Fairly common throughout our range in appropriate habitats. Fruiting from late fall into spring.

EDIBILITY: Likely nontoxic.

COMMENTS: The small size, usually umbilicate brownish cap, broadly attached gills with a pinkish cast at maturity, and growth in grassy, open areas help identify this mushroom. It may not be closely related to other *Nolanea*, but for now remains in that genus. *N. sericea* is a catchall name for many small grass-dwelling *Nolanea* with brown, distinctly hygrophanous, conical or umbonate caps. However, Largent described many species from northern California that match this description and can be differentiated only by microscopic examination and consultation with his monograph *Entolomatoid Fungi of the Western United States and Alaska*. Genetic work from Tom Bruns and Else Vellinga at University of California, Berkeley, indicates that there are at least five macroscopically similar species in the *N. sericea* group at Point Reyes National Seashore alone!

Nolanea holoconiota group

CAP: 3–8 cm across, rounded-conical to bell shaped with a distinct umbo when young (3–5 cm tall), becoming broadly conical, sometimes with a fairly pointed umbo, expanding to nearly flat, margin lobed or wavy. Light brown to yellowish brown or grayish tan, often with ocher tones, umbo and center of cap often paler yellowish tan to warm beige; hygrophanous, pale in age and along cracks, translucent-striate when moist (becoming opaque when dry). Surface moist to dry, quite smooth. **GILLS:** Attached, fairly closely spaced, broad and soft, edges of gills often slightly irregular. Pale grayish beige to tan, becoming dingy salmon as spores mature. **STIPE:** 4–10 cm long, 0.4–1 (1.5) cm thick, cylindrical or often compressed with a longitudinal groove, base slightly enlarged to club shaped. Light brownish to tan, grayish tan, often with a pale sheen over upper half. Surface nearly smooth, but vertically striate, often twisted and silvery looking. **FLESH:** Fairly thin, cap often very fragile, pale grayish tan to whitish. Stipe fibrillose-stringy, becoming hollow in age. **ODOR:** Indistinct. **TASTE:** Indistinct. **SPORE DEPOSIT:** Dull pinkish to salmon tan. **MICROSCOPY:** 8.5–11 x 8–10 μm, 5 or 6 sided, distinctly angled. Basidia typically 4-spored, but can be predominantly 2-spored in some collections. Pileipellis with internal globules of brown pigment. Stipe surface with distinctive caulocystidia: cylindrical to capitate (with an enlarged, rounded head); often emerging from center of hyphae (rather than from end).

ECOLOGY: Solitary or in clumps (not fused) in conifer duff, often in disturbed areas, mostly from late winter well into spring. Fairly common throughout our range.

EDIBILITY: Unknown.

COMMENTS: The brownish gray to beige-tan colors, conical to umbonate, hygrophanous cap, stipe covered in a whitish bloom, and habit of fruiting in spring help distinguish this species group. Many other *Nolanea* generally match the description above: *N. proxima* is very similar, but has slightly smaller spores (7.5–10 x 7–10 μm), and lacks capitate cystidia on the upper stipe. *N. pseudostrictia* is also large and often fruits in spring, but the cap cuticle cells are densely and coarsely incrusted. *N. minutostriata* has a very small cap (less than 4 cm wide), a thin, cylindrical stipe, and white gills; it has relatively small spores (7.5–9 x 7–8 μm) and weakly incrusted cap cuticle cells. *N. pseudopapillata* has an acutely umbonate, translucent-striate cap, spores averaging 7.5–9 x 6–7 μm, and cap cuticle cells that are strongly incrusted; it is known from coastal forests throughout our range, mostly fruiting under redwood in fall and early winter. Keep in mind that identifying *Nolanea* is difficult and often impossible given our current state of knowledge. Although your mushroom may match one of the above descriptions fairly well, proving its identity in any sort of definitive way will take much more work.

Pouzarella alissae

Largent & Bergemann

ALISSA'S POUZ

CAP: 0.8–3 cm across, convex, soon expanding to nearly flat, margin sometimes slightly wavy in age; caps of all stages can show a small or broad umbo. Silvery when young, soon grayish tan to gray, occasionally with an olive cast, streaked brownish to blackish in age. Dry, silky-fibrillose when young, soon shaggy looking with small, uplifted scales made of clumped fibrils. Edge of cap fringed with small tufts of pale hairs.
GILLS: Broadly attached, rather widely spaced with fairly thick edges. Pale grayish tan when young, becoming dingy grayish brown with smoky grayish blotches, and a pinkish cast in age.
STIPE: 4–8 cm long, 0.2–0.3 cm thick, cylindrical with a slightly enlarged base. Silvery whitish when young, becoming dingy tan to grayish brown, darker brown where handled. Surface entirely covered with whitish fibrils and scurfs when young, only slightly smoother in age but still noticeably silky-striate. **FLESH:** Thin, pale grayish. **ODOR:** Indistinct. **TASTE:** Not sampled. **SPORE DEPOSIT:** Pinkish to dingy salmon tan. **MICROSCOPY:** Spores 8.5–11.5 x 6–8.5 µm, heterodiametric, with 5 to 6 distinct angles or slightly nodulose (bumpy or warty). Cheilocystidia scattered, hymenial cystidia occasionally present on gill faces as well. Cap cuticle coarsely incrusted. Clamps absent.

ECOLOGY: Solitary, in clumps or loose clusters in duff under Bigleaf Maple and ferns. Rare; so far only known from the type location made (in fall) in Prairie Creek Redwoods State Park in northern Humboldt County.

EDIBILITY: Unknown.

COMMENTS: *Pouzarella* are distinguished from other members of Entolomataceae by their silky-fibrillose or hairy-scaly cap and stipe surfaces, which give them a shaggy look; the colors are often dingy or dark with a silvery sheen, but not usually as brown as many *Nolanea* and never blue or purple as in many *Leptonia*. Microscopically, they tend to have incrusting pigment in the cap cuticle cells, but lack clamps in all tissues. *Inocybe* can be told apart by their brown spore deposits and smooth or knobby (but not angular) spores. *Pouzarella* are rare on the California coast and are known from fewer collections. *P. alissae* is distinctive due to its olive tones, moderate size, fairly widely spaced gills, and tendency to develop grayish or sooty tones. Another similar *Pouzarella* in our area appears to be undescribed. It has obvious olive to jade greenish tones in its basal mycelium, a warmer brownish cap, and more closely spaced gills. Other species include *P. fulvostrigosa*, which has reddish hairs around the stipe base, and *P. araneosa*, which lacks the greenish olive hues and has spores that average slightly larger. These latter two are based on type specimens collected well outside our biogeographic area, suggesting that some of the *Pouzarella* in our area are undescribed. This species was named after its discoverer, Alissa Allen.

10 • Pluteus and Allies

This is an easy-to-recognize group of mushrooms: the distinctly free gills and pinkish spore deposit separate them from the other groups in this book. Most *Pluteus* species grow on wood or woody debris. The rare *Volvariella* are found in a variety of habitats: decaying wood, rich humus, or, in one case, on old mushrooms of the genus *Clitocybe*. *Volvopluteus* are often found in rich, cultivated soil or in rich humus, especially in disturbed areas. The last two genera are easily distinguished from *Pluteus* by the universal veil that leaves a volva sheathing the base of the stipe. Interestingly, recent phylogenetic research suggests that *Pluteus*, *Volvopluteus*, and *Volvariella* are quite closely related to *Amanita*!

Important features to note when identifying members of this group include texture of the cap, overall coloration, and identity of the rotting wood substrate. Reliable identification often requires microscopy. Pay particular attention to the structure of the cap surface; the presence or absence of clamp connections; the presence or absence of pleuro- and/or cheilocystidia, as well as their size and shape; and the size of the spores.

Pluteus exilis

Singer

WESTERN DEER MUSHROOM

CAP: 4–12 (15+) cm across, egg shaped or domed when young, often wrinkled, with concentric humped levels or a flat-topped appearance, soon smoothing out and expanding to broadly convex and then flat, margin sometimes slightly wavy. Dark blackish brown to dark brown, soon medium brown or gray to light brown, often more metallic gray in dry weather. Sometimes with a distinct white band around margin. Occasional "albino" forms nearly pure white. Surface greasy to moist, soon dry. Mostly distinctly smooth, often slightly tomentose at center. **GILLS:** Distinctly free, close to crowded, broad, soft to mushy. White at first, becoming light pink and then salmon to dingy pinkish tan in age. **STIPE:** 5–15 cm long, 0.7–1.5 cm thick at apex, cylindrical, straight to curved, larger toward base (2+ cm thick). White to creamy but covered with a sparse (sometimes dense) layer of grayish to dark blackish brown vertically arranged fibrillose patches. These scales can wear off or become flattened in age, then stipe appears extensively creamy or with dingy light tan tones. **FLESH:** Thick, white, moderately firm when young, soon soft. **ODOR:** Radishy. **TASTE:** Mild to radishy. **SPORE DEPOSIT:** Salmon. **MICROSCOPY:** Spores 7–8 x 5–5.5 µm, broadly ellipsoid to subglobose, smooth. Pleurocystidia abundant, thick walled, bottle shaped, with a crown of 2 to 4 hooks or "antlers." Pileipellis a cutis without clamps. Cheilocystidia club shaped.

ECOLOGY: Solitary, in pairs, or in small groups on stumps and logs of hardwoods and conifers. Most common on very decayed oak, Tanoak, or California Bay Laurel wood, often fruiting in large numbers early in fall, but can be found throughout wet season. Very common throughout our area, but exact distribution uncertain due to confusion with similar species (see comments).

EDIBILITY: Edible, but soft fleshed and often unpleasant tasting.

COMMENTS: This common species can be recognized by its fairly large size, dark brown to grayish brown cap, free white gills that turn pink, brownish scaly fibrils on the stipe, and growth on decaying wood. For a long time this species was identified as *P. cervinus*. That species is also present in California, but usually fruits on wood chips, primarily in the spring. It usually has a paler brown to tan cap and fewer fibrils on the stipe. However, these characters overlap, and some collections may not be identifiable without a genetic sequence. "Albino" forms of *P. exilis* are found regularly and could be confused with many other species (see comments under *P. petasatus*). *P. atromarginatus* also has a dark brown cap and antlered cystidia, but has distinctly blackish brown gill edges; it's occasionally found on wood chips in our area. *P. primus* can be told apart its long, cylindrical cheilocystidia (greater than 120 µm) and abundant clamps in the cap cuticle; it has been recorded on wood chips in our area. *P. brunneidiscus* also has clamps in the pileipellis, but has much shorter cheilocystidia than *P. primus*; it is quite rare in our area.

Pluteus petasatus

(Fr.) Gillet

CAP: 5–20 cm across, convex at first, becoming broadly convex, often with an inrolled and lobed margin, becoming nearly flat in age. Extensively whitish to light gray or pallid tan, center of cap often darker, with broad, flattened brown scales. Surface smooth to bumpy, often metallic when dry. **GILLS:** Distinctly free, close to crowded, broad. White, becoming pinkish and then rich salmon tan in age. **STIPE:** 5–10 (15) cm long, 0.7–1.5 cm thick at apex, cylindrical and upright, slightly enlarged toward base (up to 2.5 cm thick). White, slightly dingier in age. Surface smooth or vertically silky, sometimes with brown fibrils near base. **FLESH:** Fibrous, white, solid but becoming quite soft. **ODOR:** Indistinct or radishy. **TASTE:** Mild. **SPORE DEPOSIT:** Salmon. **MICROSCOPY:** Spores 6–7 × 4–5 μm, broadly ellipsoid, smooth. Pleurocystidia with 2 or 3 points at top, or often lance shaped with a single point. Pileipellis a cutis without clamps.

ECOLOGY: Solitary or more often in large clusters in wood chips or on dead wood of exotic trees in parks and landscaped urban areas. Most abundant in fall and spring, but fruiting throughout season. Locally common from Sonoma County southward.

EDIBILITY: Edible. Mediocre like *P. exilis*, but a bit firmer.

COMMENTS: The large size and often stout stature, usually relatively pale cap, free, white gills that become pinkish tan, and habit of growing in disturbed and urban settings help identify this species. *P. cervinus* grows in the same habitats (in our area), but is usually not as pale and often a bit smaller, with slightly larger spores on average (7–8 x 5–6 μm). *P. exilis* is darker, with a scaly stipe, and grows in forest habitats. Pale gray to whitish-capped *Pluteus* found in forest settings should be compared to *P. leucoborealis*, which has slightly larger spores on average (6.5–8 x 5–6 μm). It has been recorded from Oregon and might be found on alder and maple logs on the Far North Coast. Unlike all the species mentioned above, *P. tomentosulus* has bottle-shaped gill cystidia without "horns" at the tip; it is usually smaller and has a finely fuzzy, white cap.

Pluteus umbrosus group

CAP: 3–6 cm across, bell shaped to convex, expanding to broadly convex, sometimes flat. Surface dry, covered in ocher-brown to dark golden brown tufts of fibrils that are usually granular looking and arranged in radiating veins, but are densest and darkest at center, more widely spaced and often forming flattened scales toward margin, with white to cream surface showing through. **GILLS:** Distinctly free, close, broad. Cream with light golden tan edges when young (soon eroded), becoming pinkish to salmon tan. **STIPE:** 4–9 cm long, 0.5–1 cm thick, cylindrical, straight to curved, slightly enlarged toward base (up to 1.5 cm thick). Creamy to light tan, covered in bands of golden brown to dark brown fibrils and chevrons. **FLESH:** Fibrous, firm at first, soon rather soft, white to pale cream. **ODOR:** Indistinct or slightly earthy or radishy. **TASTE:** Mild. **SPORE DEPOSIT:** Pinkish tan. **MICROSCOPY:** Spores 5–6.5 x 4–5 μm, broadly ellipsoid, smooth. Cap cuticle made of upright, clumped, long-cylindrical, brownish cells without clamps.

ECOLOGY: Solitary or in small groups on well-decayed hardwoods, most common on live oak and Tanoak from midfall into winter. Widespread but rarely encountered in our area.

EDIBILITY: Unknown, likely nontoxic.

COMMENTS: The granular-scaly, golden brown cap, fibrillose-granular bands on the stipe, free gills that turn pinkish, and growth on wood help set this mushroom apart. Specimens of *P. umbrosus* from Europe have more strongly dark-marginate gills, but are genetically quite similar. *P. leoninus* has a more evenly velvety cap and is usually taller and more slender. Also see comments under *P. plautus*. A few *Pholiota* are superficially similar, but usually have more pointed cap scales and rings or large scales on the stipe, and produce a brown spore deposit.

Pluteus leoninus

(Schaeff.) P. Kumm.

LION PLUTEUS

CAP: 2–6.5 (8) cm across, rounded-conical to bell shaped, becoming broadly convex, flattening in age. Dark golden brown to ocher-brown or warm orange brown, in some forms bright yellow when young, duller yellowish brown in age. Surface dry, evenly velvety when young with a pale sheen, smoother in age. **GILLS:** Free (sometimes obscurely), close, broad. Pinkish white when young, sometimes very light yellow, salmon as spores mature. **STIPE:** 3–8 (10) cm long, 0.5–1.5 cm thick, cylindrical, usually quite tall and straight, base slightly enlarged. Creamy whitish to light creamy yellow at first, yellowish to light tan with twisting, silvery vertical striations in age, often orangey pink near very base. Surface dry, smooth. **FLESH:** Fibrous, whitish, soft. **ODOR:** Indistinct. **TASTE:** Mild. **SPORE DEPOSIT:** Dull pink. **MICROSCOPY:** Spores 7–8.5 x 6–7 μm, broadly ellipsoid to subglobose, smooth. Cystidia on gills large, bottle shaped, thin walled, tips rounded or with odd warts and nipples. Cells of cap surface long, upright, cylindrical.

ECOLOGY: Solitary or in small groups on branches, logs, and woody debris of hardwoods and conifers. Fruiting from fall into early winter. Uncommon throughout our range.

EDIBILITY: Unknown, likely nontoxic.

COMMENTS: The beautiful yellow-gold to warm tan-brown colors, velvety cap surface, free gills that tend to be slightly pinkish white even when young, and growth on wood are good identifying features for this species. *P. romellii* can be very similar in color and habitat, but has a smooth to wrinkled-veiny cap surface (made up of round to broadly club-shaped cells); it is often duller, with an olive to tan-brown cap. *P. umbrosus* has a more granular-scaly cap and a finely dark granular stipe surface; it is often more stout in stature.

SYNONYM: *Pluteus flavofuligineus* G. F. Atk.

Pluteus romellii

(Britzelm.) Sacc.

CAP: 1–5 cm across, domed or oval, expanding to broadly convex, occasionally flat or slightly wavy-uplifted. At least two color variants occur in our area: one starts out dark chestnut brown and becomes lighter orange-brown or golden fawn as it ages; the other has a beautiful deep green to forest green cap when young (often with a narrow, bright yellow edge), becoming duller yellowish green to fawn or olive-tan in age. Both variants can be slightly translucent-striate when fresh and moist. Surface moist to dry, often extensively rugose-wrinkly when young (sometimes even reticulate at center of cap), smoother in age (not velvety or with upright hairs). **GILLS:** Free, close, broad. Lemon to pale yellow when young, soon pale creamy pink, dingy salmon tan in age. **STIPE:** 1.5–5 (6) cm long, 0.2–0.5 cm thick, cylindrical, upright, slightly enlarged toward base. Bright lemon yellow to pale yellow, with twisting silvery white striations, some collections with orangey tones near base. **FLESH:** Thin, soft, pale yellowish. **ODOR:** Indistinct. **TASTE:** Mild. **SPORE DEPOSIT:** Dull pinkish. **MICROSCOPY:** Spores 5–6 x 6–7 μm, oval, smooth. Cystidia on gill surfaces broadly bottle shaped or swollen, with a tapered base and round tip, smooth. Pileipellis a dense layer of brown club-shaped to globose cells.

ECOLOGY: Solitary or in small scattered groups on well-decayed woody debris, logs, branches, and stumps of hardwoods, especially Tanoak and live oak. Infrequently encountered to locally common. Found throughout our range. The spread of Sudden Oak Death has created a surplus of habitat for this species and other wood rotters that prefer Tanoak. Fruiting from fall into spring.

EDIBILITY: Unknown, likely nontoxic.

COMMENTS: This species can be recognized by its yellowish gills (when young), yellow stipe, distinctly wrinkled brownish cap, small size, and growth on woody debris. *P. leoninus* has a velvety-plush cap surface (at least when young) and is often brighter yellow capped. The two can be extremely similar, so be sure to look closely at the cap surface. *P. phaeocyanopus* sometimes has a yellowish green stipe, but usually has a darker gray-brown cap and develops a bluish gray tone to the stipe in age.

Pluteus phaeocyanopus

Minnis & Sundb.

CAP: 1.5–5 cm across, domed at first, expanding to broadly convex. Blackish brown to dark grayish brown at center, paler to fawn brown or tan outward, margin whitish tan and often obscurely translucent-striate and/or sulcate. Surface dry, finely or sometimes coarsely wrinkled. **GILLS:** Free, fairly close to somewhat widely spaced. Whitish to dingy cream, slightly pinkish in age, but often weakly so. **STIPE:** 3.5–6 cm long, 0.2–0.5 cm thick, cylindrical, straight to slightly curved, base slightly enlarged. Pallid whitish gray, developing yellowish green tones, then extensively bluish gray in age, with many twisting, vertical, silvery striations. Surface smooth or with small, pale hairs and fibrils. **FLESH:** Thin, fibrous, pale. **ODOR:** Indistinct. **TASTE:** Not sampled. **SPORE DEPOSIT:** Dull pinkish. **MICROSCOPY:** Spores 6–8.5 x 6–8 µm, globose to subglobose, smooth. Pleurocystidia and cheilocystidia broadly bottled shaped to club shaped with a narrowed base. Pileipellis a layer of short, fat, club-shaped or round cells.

ECOLOGY: Solitary, in small groups, or more often scattered in troops on woody debris of Tanoak and live oak. Fruiting in late fall and winter. Locally common in southern part of our range (probably occurs farther north).

EDIBILITY: Unknown. The blue tones in the stipe may indicate the presence of psilocybin.

COMMENTS: The small size, dark brown cap with a finely wrinkled texture, blue-gray tones on the stipe, and growth on hardwood debris in forest settings distinguish this species. *P. romellii* is similar in cap texture and habitat, but usually has yellowish gills when young, a yellow stipe without blue-green tones, and often a warmer brown cap. Other *Pluteus* in our area with blue-gray discoloration on the stipe occur primarily on wood chips in urban areas; these include a member of the *P. nanus* group (smaller fruitbodies, with light brown to gray-brown caps), as well as a species in the *P. salicinus* group, which can be told apart by its pileipellis structure (a cutis with clamps) and pleurocystidia with small hooks at the tips. A species in the *P. longistriatus* group is sometimes found on wet, rotten wood in bathrooms and kitchens in our area; it produces small fruitbodies with strongly radially fibrillose, often pleated grayish caps, but lacks any blue-gray tones on the stipe. A member of the *P. thomsonii* group occurs on oak and Tanoak debris in our area; it can look similar but lacks blue-gray tones on the stipe, has a brownish tan cap that is veined and reticulate at the center, and has a distinctly translucent-striate margin.

Pluteus plautus group

CAP: 3–7 (10) cm across, domed at first, expanding to broadly convex and then nearly flat, sometimes a bit wavy in age. Dark reddish brown to brown, or ocher-brown to tan, distinctly hygrophanous and fading to beige, margin translucent-striate (especially obvious in age). Surface dry to moist, often slightly bumpy or wrinkly, densely tomentose at first, soon breaking up into small granular-hairy scales (often flattening or wearing off), then extensively smooth near margin. **GILLS:** Free, close, broad. Light creamy, becoming pinkish beige and then dingy salmon tan. **STIPE:** 3–6 cm long, 0.2–0.5 cm thick, cylindrical, slightly enlarged near base. Whitish gray to tan, with twisting, silvery striations, surface mostly smooth but usually covered with fine brown hairs near base. **FLESH:** Fibrous, pale, often quite soft and waterlogged (especially in cap). Bruising slowly olive brownish where injured. **ODOR:** Indistinct or radishy. **TASTE:** Mild. **SPORE DEPOSIT:** Dingy salmon. **MICROSCOPY:** Spores 6–7 x 5–6 μm, broadly ellipsoid to subglobose. Pileipellis made of upright, long, cylindrical brown cells. Pleurocystidia smooth, pear shaped or club shaped.

ECOLOGY: Solitary or in small groups on well-decayed hardwoods, with some variants also on conifers. Fruiting throughout wet season. Collectively, members of this group are fairly common throughout our range.

EDIBILITY: Unknown, likely nontoxic.

COMMENTS: This is the least well-understood group of *Pluteus* in California. A handful of different species (at least) are present in our area, and some are quite distinct from the others. One common variant matches the description above but is larger (cap to 10 cm across); another has a less distinctly tomentose cap surface and is shorter. Features that unify and distinguish the group are the finely fuzzy to distinctly granular-hairy cap surface (not usually strongly netted and wrinkled), the lack of distinct yellow tones to the cap and stipe, the hygrophanous nature of the cap (usually obviously fading to beige), the striate margin, and the tendency for the base of the stipe to be covered in dark fibrils. Species in the *P. leoninus* group can be very similar, but are usually taller or have larger caps and usually show more yellow to golden brown coloration. *P. umbrosus* is also similar but usually larger, with darker bands of fibrils over the entire length of the stipe.

Volvariella surrecta

(Knapp) Singer

CAP: 2–5 cm across, round-conical to convex, expanding to broadly convex, margin often splitting in age. White to pale grayish white, center developing pale yellowish brown colors. Surface dry, silky-fibrillose. **GILLS:** Free, close to crowded. Whitish to pale pinkish at first, becoming pinkish buff. **STIPE:** 3–8 cm long, 0.4–1 cm thick, equal or enlarged toward base. White to pale grayish white. Surface dry, silky-fibrillose, slightly scruffy at apex. **VOLVA:** Forming a white, membranous, lobed cup at stipe base. **FLESH:** Thin, fibrous, whitish. **ODOR:** Indistinct. **TASTE:** Not sampled. **SPORE DEPOSIT:** Pinkish brown. **MICROSCOPY:** Spores 5.5–7.5 x 3.5–7 μm, ovoid, smooth.

ECOLOGY: Solitary or in troops on misshapen, slightly decayed *Clitocybe nebularis*. Extremely rare, only known from a couple of collections on the Far North Coast. Fruiting in fall.

EDIBILITY: Unknown.

COMMENTS: The silky white cap, pinkish gills and spore deposit, white saclike volva, and growth on deformed *Clitocybe* make this species unmistakable. Similar *Volvariella* grow on the ground or in rich humus: *V. smithii* is a small, stocky species with a whitish to pinkish buff cap and small spores (4.5–7 x 3–4 μm). *V. hypopithys* also has a small, whitish cap, but is taller and more slender, with larger spores (6–10 x 3.5–6 μm). *V. taylori*, a grass-dwelling species, has dark gray to pale grayish fibrils covering the cap, and a dark grayish to brown volva. *V. bombycina* is a rather large, wood-dwelling species (often on standing trees) with a large, sheathing, saclike volva. All these species have been found in California but are quite rare, and are a real treat to find.

Volvopluteus gloiocephalus

(DC.) Vizzini, Contu & Justo

CAP: 5–15 cm across, egg shaped to round-conical or domed when very young, expanding to broadly convex and often with a low, round umbo, occasionally uplifted and wavy. Extensive white to cool pale to medium gray with a brownish or olive wash, darkest at center (margin often whitish at all ages), often radially streaked with appressed dark grayish fibrils, overall paler and metallic in age. Margin opaque, sometimes sulcate in age when tissue collapses. Surface very smooth, viscid to moist when young, remaining so in wet weather, otherwise soon becoming dry and shiny. **GILLS:** Distinctly free, rather soft, very close to crowded. White or creamy when young, soon pinkish cream and then distinctly salmon pink to tan-pink as spores mature. **STIPE:** 5–20 cm long, 1–4 cm thick, narrowest at apex, base enlarged and club shaped, silky or smooth overall. White to creamy, becoming dingy tan in age, or covered partially with pinkish spores. **VOLVA:** Fairly thick and membranous (but fragile), sheathing stipe base when young, occasionally breaking up and becoming indistinct in age. White to grayish. Never leaving a patch on cap. **FLESH:** Moderately thick, firm at first, becoming rather soft, white or marbled with grayish. **ODOR:** Mild to slightly musty-mealy. **TASTE:** Mild. **SPORE DEPOSIT:** Often heavy, salmon to pinkish tan. **MICROSCOPY:** Spores 13–15 (20) x 8–9.5 (12) μm, ellipsoid, smooth. Cystidia absent from gill tissue. Pleurocystidia and cheilocystidia present, swollen, club shaped or tapered at both ends, smooth.

ECOLOGY: Solitary, scattered in groups or clusters, occasionally occurring in large troops in lignin-rich mulch, wood chips, nutrient-rich gardens, and even dark agricultural soil. Mostly found in disturbed and landscaped urban areas; like many introduced species quite common in some areas and seemingly absent in others. Fruiting throughout year, most commonly in spring and fall.

EDIBILITY: Edible but soft and not particularly tasty. Familiarize yourself with the Death Cap (*Amanita phalloides*; p. 44) before eating.

COMMENTS: The fairly large size, pinkish gills, and presence of a volva distinguish this mushroom. If the volva is destroyed or overlooked, the fruitbodies are very similar to some larger species of *Pluteus*, but those are usually found on more intact wood and usually have cystidia with hooks at the tip. Some *Amanita* in section *Vaginatae* (the Grisettes) are similar, but have a less bulbous stipe base, often show rusty stains on the outer surface of the volva, and have prominently grooved (sulcate) cap margins as well as white spores. Greenish gray specimens with a silvery appearance have been confused with the deadly poisonous *Amanita phalloides*, but that species has a ring on the stipe and whitish creamy gills at maturity, and produces a white spore deposit. See comments under *Volvariella surrecta* for other similar species. *Volvopluteus michiganensis* is quite similar but poorly known; it has spores less than 12.5 μm in length on average.

SYNONYMS: *Volvariella gloiocephala* (DC.) Boekhout & Enderle, *V. speciosa* (Fr.) Singer.

11 • *Russula and Lactarius*

Russula and *Lactarius* are some of the most familiar and abundant of our woodland mushrooms. Fruitbodies of both genera are recognizable by their generally stout statures (with a stipe that is shorter than the width of the cap), often uplifted caps, and chalky-granular or hard-brittle textures. Those that exude milk or latex when broken are *Lactarius*; those that lack any latex are *Russula*. A few relatives of *Lactarius* and *Russula* produce hypogeous, trufflelike fruitbodies, but these are much less common in our area. Both *Lactarius* and *Russula* are mycorrhizal and especially important in the ecology of temperate forests around the world.

Although *Lactarius* are relatively diverse in California, and some can be a bit tricky to identify, our concepts for the local species are fairly strong. Be sure to note the color of the latex (and whether or not it changes color after exposure to air), taste (mild or peppery?), any distinctive odor, and the coloration of the cap (especially the presence of concentric zones). Microscopically, the structure of the amyloid spore ornamentation (which requires the use of Melzer's reagent) can be an important identifying feature.

Identification of *Russula* is more difficult (or frequently impossible). There are at least a hundred species in California (possibly many more), but most remain poorly characterized and/or undescribed. Take note of overall coloration (which can be extremely variable even within a single species), size and stature, presence or absence of a sulcate cap margin, taste (peppery or mild?), presence or absence of short gills, and any bruising reactions. Microscopically, the structure of the amyloid spore ornamentation (which requires the use of Melzer's reagent) and the structure of the cap tissues are just two of the many important features to examine. Also keep in mind that many of our *Russula* are using borrowed European names, and will likely change as we gain a better understanding of the western North American *Russula* flora.

Russula "Ocher Oaks"

CAP: 3–7 cm across, convex at first, becoming irregular broadly convex, often with a depressed center and wavy margin. Yellowish to yellow-white, mottled with pink when young, to mostly pinkish mottled with yellow or pinkish orange blotches. Surface slightly viscid when wet, to dry. **GILLS:** Broadly attached at first, becoming notched or narrowly attached, close, narrow, very fragile, forking or fused near stipe, partial gills occasional. Cream to pale yellow when young, ocher-orange as spores mature. **STIPE:** 4–7 cm long, 1–2.5 cm thick, equal or enlarged slightly toward base. White, becoming dingy where handled. Surface smooth, dry. **FLESH:** Thin, firm, but fragile. White. **ODOR:** Indistinct. **TASTE:** Mild. **SPORE DEPOSIT:** Ocher. **MICROSCOPY:** Spores 8–11 x 7.5–8.5 μm, subglobose to ovoid, ornamented with sparse scattered warts 0.5 μm high, with connecting lines or ridges.

ECOLOGY: Solitary or scattered on ground under live oak. Common in drier oak forests from Mendocino County southward. Fruiting from fall into early winter.

EDIBILITY: Edible.

COMMENTS: This description matches two undescribed, oak-associated, yellowish-capped *Russula* with ocher-orange mature gills, white stipes, and a mild taste; we have yet to find any macro-morphological differences between them. The dark ocher spores alone help distinguish them from other similar *Russula*. *R. basifurcata* has a whitish to creamy buff cap that becomes yellow and often develops pink tones around the margin. It also has cream to pale yellow gills that are fused and/or forked near the stipe, pale yellow spores, and a mild to slightly bitter taste. *R. chamaeleontina* has a pink to rose-colored cap when young (often with a yellowish disc), becoming salmon pink with a yellowish orange disc in age. It has creamy yellow to ocher gills and a mild taste. *R. maculata* is very similar to the previous species, but has a red cap with a paler disc. It is very common during wet winters in southern California.

Russula bicolor

Burl.

CAP: 3–7 cm across, convex when young, becoming broadly convex to plane, often with a depressed center, margin often grooved in age. Pink to pale red, often with yellow to peachy orange center, with pinkish spots, rarely mostly yellowish with pale pinkish margin. Surface smooth, thinly viscid and shiny when wet, dull if dry. **GILLS:** Broadly attached, but sometimes becoming free in age, close to subdistant, narrow, fragile, partial gills rare or absent. White. **STIPE:** 3.5–6 cm long, 1–1.8 cm thick, equal or enlarged slightly toward base. White. Surface smooth, dry. **FLESH:** Thin, fragile, white. **ODOR:** Indistinct. **TASTE:** Quickly and strongly acrid, leaving a lingering peppery, burning sensation in mouth. **SPORE DEPOSIT:** White. **MICROSCOPY:** Spores 9–11 x 8–10 μm, subglobose to subelliptical, ornamented with isolated warts 0.5–1 μm high, occasionally connected with ridges, forming a partial reticulum.

ECOLOGY: Solitary or scattered on ground, well-decayed wood, or moss-covered stumps and logs in coniferous forests. Locally common on the Far North Coast, almost always near Sitka Spruce. Fruiting from fall into early winter.

EDIBILITY: Unknown, not recommended.

COMMENTS: This beautiful *Russula* can be recognized by the pink cap with a yellow to peachy orange center; white gills, stipe, and spores; strongly acrid taste; and growth with northern conifers. *R. ellenae* has a similar rose pink to coral pink cap with a yellow disc when young, but becomes more yellow in age. It has white to pale yellow gills, pale yellow spores, a slightly acrid taste, and more southern distribution, occurring south to Santa Cruz County. *R. veternosa* has a variable cap color, ranging from pink to red to violet, often with yellow blotches. It also has yellow gills, yellow-ocher spores, an acrid taste, and often a honeylike odor. *R. simillima* has a yellowish to yellowish brown disc and a salmon pink to reddish brown margin when young, becoming dingy yellow overall in age. It has a slowly acrid taste and white spores, and often grows in oak-pine woods. *R. basifurcata* has a yellow to whitish cap with a pinkish to red margin, but has yellow mature gills, pale ocher spores, and a mild taste, and grows under oaks.

Russula cremoricolor

Earle

WINTER RUSSULA

CAP: 3–10 cm across, rounded to convex when young, becoming broadly convex to plane, often slightly depressed at center, and sometimes wavy in age. Two color forms occur. Pale form: yellowish cream to creamy white, often with a paler margin in age. Red form: bright cherry red, sometimes with whitish creamy blotches, or irregularly mottled with pale colors. Some collections may have intermediate-colored caps. Surface slightly greasy or viscid when wet, soon dry, smooth. **GILLS:** Narrowly attached, close to subdistant, narrow, very fragile, leaving shards when handled, partial gills absent. White when young, creamy white in age. **STIPE:** 3–10 cm long, 0.6–2 cm thick, equal. White. Surface smooth, dry. **FLESH:** Thin, firm when young, granular and brittle in age. **ODOR:** Indistinct. **TASTE:** Extremely acrid, quickly leaving a hot-pepper burning sensation in mouth. **SPORE DEPOSIT:** White. **MICROSCOPY:** Spores 7.5–9.5 x 5.5–8 μm, subglobose to subellipsoid, ornamented with scattered warts 0.2–0.5 μm high, with ridges and scattered lines, not forming a reticulum.

ECOLOGY: Solitary, in pairs, or scattered in troops and arcs on ground in mixed forests, especially under Tanoak. Widespread and common throughout our range. Fruiting from late fall through winter.

EDIBILITY: Slightly toxic, causing mild gastrointestinal distress.

COMMENTS: Although there are many red- and cream-capped *Russula* species in California, the occurrence of red and creamy capped fruitbodies in close proximity is a good clue indicating this species in our area. The very acrid taste, rather round cap that lacks grooves around the margin, and white gills and stipe are important identification features. Like many *Russula*, it fruits abundantly in some years and is nearly absent in others. *R. californiensis* also has an evenly red cap, but grows under pine and has a weakly acrid taste and a white stipe that turns gray when bruised or in age.

MISAPPLIED NAME: The names *R. emetica* and *R. silvicola* are often incorrectly applied to the red-capped forms before the two color forms were determined to be genetically identical.

Russula sanguinea

Fr.

ROSY RUSSULA

CAP: 4–9 (12) cm across, rounded to convex when young, becoming broadly convex to plane, often with a depressed center, margin occasionally uplifted in age. Bright scarlet red, red to rosy red, often slightly darker on disc. Can fade to pink to dingy white after heavy rain or prolonged exposure to sun. Surface smooth, thinly viscid to tacky, often shiny when wet, dull when dry. **GILLS:** Broadly attached to slightly decurrent, close to subdistant, narrow, fragile, partial gills rare, occasional forking near stipe. White to creamy white at first, becoming cream to pale yellow in age. **STIPE:** 2.5–7 (10) cm long, 1–2.5 cm thick, equal or enlarged slightly toward base. Rosy pink to pale pink, often with a whitish bloom, darkening when handled. Surface smooth, dry. **FLESH:** thin to moderately thick, firm, fragile. White. **ODOR:** Indistinct. **TASTE:** Quickly and strongly acrid, leaving a lingering peppery taste in mouth. **SPORE DEPOSIT:** Cream to pale yellow. **MICROSCOPY:** Spores 7.5–9.5 x 6.5–8.5 μm, subovoid to subellipsoid, ornamented with fine isolated warts 0.4–0.6 μm high.

ECOLOGY: Scattered or in troops, rarely solitary, on ground or in duff under pines. Very common throughout our range. Fruiting from early fall in north through winter southward.

EDIBILITY: Extremely acrid; to be avoided.

COMMENTS: The scarlet red cap, pink to rosy stipe, white to creamy gills, acrid taste, and growth with pine make the Rosy Russula easy to recognize. *R. americana* is nearly identical, but has larger spores and grows with Grand Fir on the Far North Coast. *R. rhodopus* grows with live oak, has a red cap that develops yellowish colors on the disc, and has a mild to slightly acrid taste. Members of the *R. queletii* group have dark wine red to reddish purple caps and rosy or purple-blushed stipes. Members of the *R. xerampelina* group can have red caps and pinkish blushed stipes, but have a mild taste, fishy odor, and slow yellowish brown staining on the stipe when scratched or handled. There are a large number of poorly understood red-capped *Russula* with white stipes; see *R. cremoricolor* (p. 229), *R. bicolor* (p. 228), and *R. californiensis* (p. 231).

Russula californiensis

Burl.

CALIFORNIAN RED RUSSULA

CAP: 5–15 cm across, rounded to convex when young, becoming broadly convex to plane, often with a depressed center and an uplifted or wavy margin in age. Dark rosy red, scarlet red to red, usually slightly darker on disc, and often developing whitish to yellowish spots in age. Surface smooth, thinly viscid to tacky when wet, dull when dry. **GILLS:** Narrowly attached, notched, or pulling away and becoming free in age, close to subdistant, broad, very fragile, partial gills rare, occasional forking near stipe. Creamy white at first, becoming creamy yellow to pale ocher-yellow in age. **STIPE:** 4–10 cm long, 1–3 cm thick, more or less equal. White, or with a faint pinkish blush at base, slowly bruising grayish when handled, to grayish in age. Surface smooth, dry. **FLESH:** Thin to moderately thick, firm when young, becoming fragile. White, stipe often pale grayish in age. **ODOR:** Indistinct. **TASTE:** Mildly to strongly acrid when young (typically gills are acrid, flesh mild), becoming more mild in age. **SPORE DEPOSIT:** Yellow. **MICROSCOPY:** Spores 8.5–11.5 x 6–9 μm, subglobose, subovoid to subellipsoid, ornamented with fine warts 0.1–0.5 μm long, with connecting fine lines and ridges, forming a partial to complete reticulum.

ECOLOGY: Solitary, scattered, or in troops on ground or in duff under pine. Locally common under Monterey Pine in the southern part of our range, occasional to rare under Bishop Pine north of the San Francisco Bay Area. Fruiting from midfall into early spring.

EDIBILITY: Edible, at least when well cooked.

COMMENTS: The rosy red cap, creamy to yellow gills, growth with pine, and white stipe that bruises grayish when handled are distinctive. Although there are a number of red-capped *Russula* in our area, the graying stipe sets it apart from all others. *R. decolorans* is another species with a graying stipe, but has a copper orange cap and grows with northern conifers. Red-capped members of the *R. xerampelina* group have yellowish ocher to brown-staining stipes and fishy odors.

Russula queletii group

CAP: 3–10 cm across, rounded to convex when young, becoming broadly convex to plane, occasionally with a depressed center in age. Dark wine red to purplish when young, staying dark wine red on disc, but becoming paler, purple-red toward margin. Can fade to pale purple to almost white after heavy rain or prolonged exposure to sun. Surface smooth, thinly viscid to tacky, often shiny when wet, dull when dry. **GILLS:** Broadly attached to slightly decurrent, close, narrow, fragile, partial gills rare, occasional forking near stipe. White at first, becoming cream in age. **STIPE:** 3–7 cm long, 0.8–2 cm thick, equal or enlarged slightly toward base. Blushed rosy purple to pale pinkish purple, darkening when handled. Surface smooth, dry. **FLESH:** Thin to moderately thick, firm, fragile. White. **ODOR:** Indistinct to mildly fruity. **TASTE:** Quickly and strongly acrid, at least when young, leaving a lingering peppery taste in mouth. **SPORE DEPOSIT:** Cream to pale yellow. **MICROSCOPY:** Spores 6.5–9.5 x 6–8 μm, subglobose to subelliptical, ornamented with well-developed isolated warts.

ECOLOGY: Solitary, scattered, or in troops on ground under conifers, especially pines. Common in fall and winter throughout our range.

EDIBILITY: Strongly acrid; to be avoided.

COMMENTS: The dark wine red to purple cap, white to creamy gills, rosy purple stipe, and acrid taste are distinctive. *R. sanguinea* is similar but has a scarlet to red cap and a pink-to red-blushed stipe. *R. pelargonia* is a small species with a purplish red to purplish gray cap, a pinkish-blushed stipe, an acrid taste, and a sweet odor of geraniums (at least when the stipe base is crushed). Members of the *R. xerampelina* group can have wine red caps and blushed stipes, but have a fishy odor, a mild taste, and yellowish to orangey brown staining on the stipes when handled. There are also a number of purple-capped *Russula* with white stipes: see *R. mendocinensis* (p. 233) and *R. punicea* (p. 232) for acrid-tasting species matching this description and *R. murrillii* for those that are mild.

Russula punicea

Thiers

CAP: 3–7 cm across, convex when young, becoming broadly convex to plane, often with a depressed center. Deep vinaceous purple to dark reddish purple, evenly colored when young, slightly paler on margin in age. Surface smooth, viscid and shiny when wet. **GILLS:** Broadly attached, close to subdistant, narrow, fragile, partial gills absent. Creamy white at first, becoming yellowish to pale ocher. **STIPE:** 3–7 cm long, 1–2 cm thick, equal, enlarged slightly, or pinched toward base. White, discoloring orangish at base when handled. Surface smooth, dry. **FLESH:** Thin, firm, but fragile, white. **ODOR:** Indistinct. **TASTE:** Quickly and strongly acrid, leaving a lingering peppery, burning sensation in mouth. **SPORE DEPOSIT:** Yellow. **MICROSCOPY:** Spores 8–9.5 x 6.5–7.5 μm, subglobose to subelliptical, ornamented with coarse isolated warts 0.3–0.7 μm high, with heavy ridges, not forming a reticulum.

ECOLOGY: Solitary or scattered on ground, in duff or moss in coniferous forests, especially under Sitka Spruce on the Far North Coast. Locally common, but apparently quite range restricted in our area. Fruiting in fall.

EDIBILITY: Unknown; not recommended.

COMMENTS: This *Russula* can be recognized by the deep vinaceous purple, evenly colored cap, creamy yellow to pale ocher gills, white stipe, yellow spores, and strongly acrid taste. *R. mendocinensis* is larger and often has some dingy yellow-orange blotches mixed into the dark purple cap. *R. murrillii* has a dark purple to rosy purple cap, white to pale yellow gills, pale yellow spores, and a mild taste. Members of the *R. queletii* group can have deep vinaceous purple caps, but have white to creamy gills, a rosy-blushed stipe, and creamy to pale yellow spores.

Russula murrillii

Burl.

CAP: 2–7 cm across, convex when young, becoming broadly convex to plane, often with a depressed center. Dark purple to vinaceous purple, occasionally purple-brown when young, or rosy purple with paler patches in age. Surface viscid when wet, dull and pruinose if dry. **GILLS:** Broadly attached to notched, close to subdistant, narrow, fragile, partial gills absent. White at first, becoming cream colored to pale yellow as spores mature. **STIPE:** 3–8 cm long, 1–3 cm thick, equal or enlarged slightly toward base. White, occasionally bruising yellowish when handled. Surface smooth, dry. **FLESH:** Thin, fragile, white. **ODOR:** Indistinct. **TASTE:** Mild. **SPORE DEPOSIT:** Cream to pale yellow. **MICROSCOPY:** Spores 7.5–10 x 5.5–8.5 μm, subglobose to subovoid, ornamented with isolated warts 0.2–0.5 μm high, connected with ridges, forming a partial to complete reticulum.

ECOLOGY: Solitary, scattered, or in troops on ground in mixed coniferous-hardwood forests. Widespread in our range, but especially common north of the San Francisco Bay Area. Fruiting from fall into early winter.

EDIBILITY: Edible.

COMMENTS: Distinctive features include the purple cap that is viscid when wet, but pruinose when dry (covered in a powdery bloom), mild taste, creamy to pale yellow gills and spores, and white stipe that may stain yellowish when handled. Members of *R. amethystina* group have vinaceous purple to violet caps (often with yellowish spots in age) and dark yellow to ocher gills and spores. *R. turci* has a dark purple to purple-black cap with brownish tones on the disc, dark yellow mature gills and spores, and often an iodine odor in the flesh of the stipe base. *R. lilacea* has a rosy purple to violet-gray cap, white to pale creamy gills, and a mild to slightly acrid taste. *R. abietina* can have a purplish cap when young, but it often becomes variegated with olive tones in age.

Russula mendocinensis

Thiers

CAP: 6–14 cm across, convex when young, becoming broadly convex to plane, with a depressed center. Margin downcurved at first, becoming irregular and occasionally uplifted in age. Vinaceous purple to dark rosy purple, occasionally with brownish tones, or darker or paler patches, and often developing dingy yellowish orange blotches or spots. Surface smooth, viscid and shiny when wet, dull and often with debris stuck to it when dry. **GILLS:** Attached, and sometimes notched, close to subdistant, broad, fragile, with occasional partial gills. Creamy white at first, becoming yellowish to pale ocher as spores mature. **STIPE:** 5–9 cm long, 1–3 cm thick, equal or enlarged slightly toward base. White, rarely with a faint rosy blush, discoloring orangish when handled. Surface smooth, dry. **FLESH:** Moderately thick, firm when young, soft and fragile in age. White. **ODOR:** Indistinct. **TASTE:** Quickly and strongly acrid, leaving a lingering peppery, burning sensation in mouth. **SPORE DEPOSIT:** Yellow. **MICROSCOPY:** Spores 8–9 x 6–8 μm, ovoid, subglobose to broadly ellipsoid, ornamented with scattered heavy warts 0.5–1 μm high, with heavy ridges, forming a partial to complete reticulum.

ECOLOGY: Solitary, scattered, or in troops on ground, in moss or duff of mixed conifer-hardwood forests on the North Coast. Locally common, known from north of the San Francisco Bay Area, but probably in the mountains of the San Francisco Peninsula as well. Fruiting in fall and early winter.

EDIBILITY: Not recommended; potentially slightly toxic due to acrid taste.

COMMENTS: The medium to large fruitbodies, vinaceous purple caps that often develop dingy yellowish orange blotches, creamy white to pale ocher gills, white stipes that discolor orangish, yellow spores, and strongly acrid taste help identify this species. Care should be taken not to confuse this species for the Shrimp Russula (*R. xerampelina* group); although the two can have identical cap colors, the latter has darker ocher-yellow gills at maturity, usually a distinctly rosy-blushed stipe that stains yellowish to orangish brown when handled, a mild taste, and a fishy odor. *R. punicea* is smaller, has an evenly colored, dark vinaceous purple cap, and grows under Sitka Spruce. The widespread *R. atroviolacea* has a dark purple, reddish purple, or blackish purple cap with a paler rosy purple, striate margin; it also differs by its white to creamy gills, whitish to pale yellow spores, and slightly acrid taste. *R. placita* has a dark purple-brown to reddish brown cap with a reddish purple disc, pale yellow spores, and white stipe that sometimes becomes rosy blushed in age. Members of the *R. queletii* group have deep vinaceous purple caps that fade to rosy purple at the disc, white to creamy gills, rosy-blushed stipes, and creamy to pale yellow spores. See *R. murrillii* (p. 232) for similar species with purple caps and a mild taste.

Russula cyanoxantha (sensu CA)
MAUVE RUSSULA

CAP: 5–15 cm across, convex when young, becoming broadly convex, plane, or uplifted and irregular; disc often depressed at any stage. Color extremely variable, from violet-purple to rosy or purplish gray, usually variegated with paler areas and/or orangish spots; rarely olive overall. Generally fading as it ages, becoming more muted, grayish lilac with more olive tones and an orangish to yellowish disc. Surface dull, greasy to dry or moist, or thinly viscid when wet. Smooth, or finely cracked into a "quilted" or areolate pattern, especially in age and/or near margin. **GILLS:** Broadly attached to slightly decurrent, close, narrow, less fragile than other *Russula* and often greasy feeling. Often fused near stipe, then forking, sometime repeatedly, especially near margin. White to creamy in age, at times with pale orangish brown spots or stains. **STIPE:** 5–12 cm long, 1.5–5 cm thick, equal, enlarged toward base or irregular. White, or occasionally blushed pale rosy pink, and sometimes developing orangish brown stains in age. Surface smooth, dry. **FLESH:** Thick, very firm when young, soon mealy and often infested with insects in age. White. **ODOR:** Indistinct. **TASTE:** Mild to sweet. **SPORE DEPOSIT:** White. **MICROSCOPY:** Spores 6.5–10 x 5.5–7.5 μm, elliptical to subovoid, ornamented with short isolated warts 0.1–0.3 μm high, occasionally with fine lines, but not forming a reticulum.

ECOLOGY: Solitary or scattered on ground or in duff, usually under Tanoak or live oak. Common throughout our range.

Abundant in some years, very scarce in others. Fruiting from late fall through winter.

EDIBILITY: Edible and excellent when fresh; the flesh is thick and firm, with a sweet flavor. However, even young specimens can become soft and insect infested; these should be avoided.

COMMENTS: The medium to large fruitbodies with variably magenta to olive-variegated caps; mild taste; growth with hardwoods; and, most importantly, the white, forking gills make this an easy *Russula* to identify. *R. olivacea* is a somewhat similar large species with variably colored caps ranging from rosy purple to olive, but has a rosy-blushed stipe, yellow spores, and nonforking gills. *R. occidentalis* has an olive, greenish yellow to grayish green cap with purple tones and yellowish spores, and slowly stains pinkish and then gray on the stipe. Some members of the *R. xerampelina* group are similarly colored, but grow with conifers, stain yellowish and then brownish when scratched, develop a fishy odor, and have creamy to yellowish gills in age.

MISAPPLIED NAME: *Russula cyanoxantha* is a European name, and although ours is known to be a different species, it remains unnamed. We will continue to use the well-known European placeholder name until ours is formally described.

Russula olivacea (sensu CA)
PURPLE-OLIVE RUSSULA

CAP: 5–20 (30) cm across, convex when young, becoming broadly convex to plane with a depressed center, occasionally slightly uplifted and irregular-wavy in age. Color extremely variable, generally some shade of rosy purple variegated with olive, but can be any combination of olive-brown, lilac, gray, mauve, green, yellow, or even red. Typically variegated with a few of these colors, with olive tones near disc, and a smoky lilac margin. Surface often with a pale bloom when young, finely textured, often with a roughened or wrinkled band around margin, occasionally concentrically cracking. Dry to thinly viscid when wet. **GILLS:** Broadly attached, close to subdistant, broad, fragile, occasionally forking near stipe. White when young, becoming cream to pale yellow in age. **STIPE:** 4–15 (25) cm long, 1.5–4 cm thick, equal or club shaped, with a slightly enlarged base. All white to rosy pink blushed at first, blush becomes more pronounced in age, base occasionally with orangish spots or stains. Not staining, or only slightly brownish when handled. Surface smooth, dry. **FLESH:** Thick, very firm when young, often brittle and granular in age. White. **ODOR:** Indistinct. **TASTE:** Mild to sweet. **SPORE DEPOSIT:** Yellow. **MICROSCOPY:** Spores 8–12 x 6.5–9.5 μm, ovoid to subglobose, ornamented with isolated warts 1–2 μm high, connecting lines or ridges lacking or scarce.

ECOLOGY: Solitary, scattered, or in troops on ground, in duff or moss in mixed coniferous forest. Common and widespread, occurring throughout our range. First fruiting in fall, but continuing well into winter.

EDIBILITY: Edible and excellent; however, a small percentage of people appear to be allergic and experience gastrointestinal distress after eating it (see comments).

COMMENTS: This large, variably colored *Russula* is one of the better edible mushrooms in California; it has firm flesh when young and a very sweet flavor, and is abundant most years. Eat only a small amount (well cooked) the first time you try it, since a small number of people experience gastrointestinal discomfort after eating it. The size alone helps separate it from most other *Russula*; other distinguishing features include the olive to purplish cap with a band of concentric wrinkles near margin, creamy gills, rosy-blushed stipe, and mild taste. The Shrimp Russula (*R. xerampelina*) often has a darker purple or red-colored cap, white to creamy gills that darken to yellowish ocher in age, a fishy odor, and stains yellowish to ocher-brown on the stipe when scratched and handled. *R. occidentalis* also has a variegated olive to purple cap, but has a white stipe that stains gray when scratched and in age.

Russula xerampelina group
SHRIMP RUSSULA

CAP: 5–15 (20) cm across, convex when young, becoming broadly convex to plane, with a depressed center. Color extremely variable, even within the same patch. Generally some shade of deep purple, reddish purple, reddish brown to rosy purple. But can have any shade of these colors, as well as brown, gray, greenish olive, olive-brown, or yellowish orange. Surface smooth, thinly viscid when wet, finely tomentose and often with debris stuck to it when dry. **GILLS:** Broadly attached to notched, close to subdistant, broad, fragile, rarely forking and with occasional partial gills. Creamy white when young, becoming cream to ocher-yellow in age. **STIPE:** 4–10 (15) cm long, 1–4 cm thick, equal or club shaped, with a slightly enlarged base, or narrower in middle. White, generally with a rosy pink flush, which is more pronounced in age, occasionally entirely white or rosy pink, staining pale ocher-yellow, then orange-brown when handled. Surface smooth, dry. **FLESH:** Thick, very firm when young, often brittle in age. White, discoloring pale ocher-yellow to pale orangish brown, or light brown overall in age. Stipe with softer pithlike core. **ODOR:** Mild when young, soon developing a fishy odor, often described as shrimp- or crablike. **TASTE:** Mild to sweet. **SPORE DEPOSIT:** Yellow. **MICROSCOPY:** Spores 7.5–9.5 x 6.5–7.5 μm, ovoid to elliptical, ornamented with isolated warts 0.5–1 μm high, with a few or no connecting lines or ridges.

ECOLOGY: Scattered, in small troops, or occasionally solitary on ground, in duff or moss in mixed conifer forests; two common forms are especially fond of Douglas-fir and pines,

while a deep purple form thrives in Sitka Spruce forest on the Far North Coast. Common and widespread throughout our range, extending north into Pacific Northwest. Generally fruiting in one large flush in fall, but scattered fruitings can occur throughout fall and winter.

EDIBILITY: Edible and excellent. The firm flesh, sweet-savory flavor, and reliable field marks make this a good species to seek out for the table.

COMMENTS: Shrimp Russula refers to a complex of at least five species in California. A number of European names have been applied to them, but our taxa undoubtedly represent different species from those in Europe. Members of the group can be recognized by the variably colored caps, with dark purple to reddish purple colors being the most common; the white to creamy gills that darken to ocher-yellow in age; the partially rosy-blushed stipes (rarely entirely pinkish); and especially the yellow to light ocher-brown staining when scratched. This staining reaction is most pronounced when the stipes are scraped, but often develops after normal handling as well; it is not a quick reaction, but after a few minutes (or by the time you get home with a full basket), the stipe will often be covered in ocher-brown fingerprints. The odor can be mild in young fruitbodies or cold weather, but is usually quite strong and sometimes even repulsive in mature fruitbodies. Some of the names used for species in this complex include *R. elaeodes* (olive-brown-capped forms); *R. graveolens* (deep purple to vinaceous brown caps and slightly larger spores); *R. semirubra* (deep red, purplish

red to brownish red caps, slightly larger spores with longer spines); and *R. grundii* (smaller, pale yellow to brownish yellow caps, described from southern California oak savannah). Also see *Russula* "Green Madrone". Look-alikes outside this group include the edible *R. olivacea* and the acrid-tasting *R. mendocinensis*, which should not be eaten.

Shrimp Russula Identification Checklist

- Cap color quite variable, most commonly wine purple to reddish brown
- Gills creamy when young, yellowish to ocher in age
- Stipe creamy, often blushed with pink tones
- Stipe staining slowly yellow and then brown when scratched or handled
- Stipe with a firm exterior and a pithy core
- Taste mild
- Odor slightly fishy or like shrimp (especially at maturity)

Russula "Green Madrone"

CAP: 5–15 cm across, convex when young, becoming broadly convex to plane, often with a depressed center. Green, yellowish green to yellowish olive at first, often with a whitish bloom, developing brownish green to orangish tones, and becoming orangish olive to olive-yellow in age. Surface smooth to slightly wrinkled, dry to moist, sometimes cracking in age. **GILLS:** Broadly attached to notched, close to subdistant, broad, fragile, often forking at stipe, partial gills rare. Creamy white when young, becoming cream to pale yellow in age. **STIPE:** 3–7 cm long, 2–3.5 cm thick, equal or enlarged toward base. White when young, staining ocher-orange in age or when handled. Surface smooth, dry. **FLESH:** Thick, firm when young, often brittle in age. White. **ODOR:** Indistinct. **TASTE:** Mild to sweet. **SPORE DEPOSIT:** Pale yellow. **MICROSCOPY:** Spores 8–10 x 8–9 μm, ovoid to nearly globose, coarsely ornamented with scattered warts, with occasional connecting lines or ridges.

ECOLOGY: Scattered, in small troops or clusters; occasionally solitary, on ground in duff under Madrone and manzanita. Locally common and widespread, occurring throughout our range. Fruiting from midfall into early winter.

EDIBILITY: Edible and excellent. The firm flesh and sweet taste are similar to *R. olivacea*.

COMMENTS: The green to yellow-olive cap, white to creamy yellow gills, ocher staining, mild taste, lack of a distinctive odor, and growth with Madrone or manzanita make this an easy *Russula* to recognize. Although distinctive and in some years quite common, it appears to be undescribed. *R. olivacea* has a highly variable cap (occasionally similar to "Green Madrone"), but typically develops purplish, reddish, or brownish tones in age. It also has slightly darker spores and a rosy-blushed stipe. *R. grundii* has a paler yellow to brownish yellow cap, darker yellow to ocher-yellow gills at maturity, and a fishy odor. It is known only from open oak woodlands near Santa Barbara. See *R. aeruginea* (p. 238) for additional green-capped species.

Russula aeruginea
Lindbl.

GREEN RUSSULA

CAP: 4–10 cm across, convex when young, becoming broadly convex to plane, but with a depressed center, and/or an uplifted margin in age. Deep green, olive-green to grayish green at first, fading to grayish green or yellowish green and usually developing yellowish brown stains on disc in age. Surface smooth, dull and often with a powdery bloom when young and dry, moist to thinly viscid when wet, margin often grooved in age. **GILLS:** Broadly attached to notched, close to subdistant, broad, fragile, often forking near stipe, partial gills rare. White when young, cream to pale yellow in age. Occasionally discoloring brownish when old. **STIPE:** 3–8 cm long, 1.5–3 cm thick, equal. White when young, often developing yellowish to dingy yellowish orange stains in age. Surface smooth, dry. **FLESH:** Moderately thick, firm when young, brittle in age. White. **ODOR:** Indistinct. **TASTE:** Mild. **SPORE DEPOSIT:** Pale yellow. **MICROSCOPY:** Spores 6.5–9 (11) x 5.5–7 (9) μm, subovoid to subglobose, ornamented with isolated warts 0.2–0.5 μm high, with short ridges or lines, but not forming a reticulum.

ECOLOGY: Solitary, scattered, or in small troops on ground, in moss or duff in a wide variety of forest types. Common and widespread, occurring throughout our range. Fruiting from fall into winter.

EDIBILITY: Edible and good, especially the young, firm buttons.

COMMENTS: The dark to light greenish gray to yellowish green cap, pale yellow gills that often fork near the stipe, pale yellow spores, and mild taste help identify this common *Russula*. *R. urens* has a dark green cap that can fade to yellowish green, but has dark yellow to ocher spores and a strongly acrid taste; it occurs under Sitka Spruce on the Far North Coast. *Russula* "Green Madrone" is a common undescribed species found under Madrone and manzanita; it differs by its bright yellow-green caps that develop orangish colors in age and by orangish staining on the stipe when handled. *R. heterophylla* has a dark olive-green, greenish brown, or greenish purple cap that often ages more purplish brown. It develops distinctive lightning-bolt-like cracks and areolate patches on the cap and has whitish gills that develop orangish stains, a mild taste, and white spores. It appears to be restricted to the Far North Coast, where it is quite common. *R. smithii* is very similar, but has a dark green to greenish brown cap, a pink- to rosy-blushed stipe, and larger spores. Both *R. heterophylla* and *R. smithii* are edible.

Russula occidentalis

(Singer) Singer

CAP: 5–15 cm across, rounded to convex when young, becoming broadly convex, plane, with a depressed center in age. Color variable, olive, olive-gray, olive-purple to all purple at first, often variegated with a combination of these colors as it expands, usually darker on disc, becoming paler, developing yellowish olive to grayish lilac colors in age. Surface smooth, dull and often with a powdery bloom when young; dry to moist, thinly viscid when wet, margin often grooved in age. **GILLS:** Broadly attached to notched, close to subdistant, broad, fragile, often with short, irregular partial gills near margin. White when young, cream to pale creamy yellow in age. Occasionally blackening slightly around margin, or on edges. **STIPE:** 4–8 cm long, 1.5–3 cm thick, equal or enlarged slightly toward base. White when young, discoloring to grayish in age. Slowly staining pinkish red, then gray when handled or scratched. Surface smooth, dry. **FLESH:** Thick, firm when young, brittle in age. White to gray, slowly staining pale pinkish red, then gray when young. **ODOR:** Indistinct. **TASTE:** Mild. **SPORE DEPOSIT:** Pale yellow to cream colored. **MICROSCOPY:** Spores 8.5–11.5 x 6.5–8 μm, subovoid to subglobose, ornamented with pointed isolated spines or warts up to 2 μm high, occasionally with short ridges or lines, but not forming a reticulum.

ECOLOGY: Solitary, scattered, or in troops in moss or duff under Sitka Spruce. Very common on the Far North Coast, but restricted to spruce forests. Fruiting in fall into early winter, occasionally with a smaller flush in spring and early summer.

EDIBILITY: Reported to be edible.

COMMENTS: This species can be separated from other large, variably olive- to purple-capped *Russula* by the fleeting pinkish red staining reaction (slowly turning gray) on the stipe and flesh. The staining is best observed by scratching the stipes of young fruitbodies. Other distinguishing features are the grayish stipe and flesh of older specimens, mild taste, pale yellow spores, and growth with Sitka Spruce. *R. pacifica* supposedly differs by the lack of red staining before turning gray and the slightly smaller spores, but we believe it to be a synonymous variant of *R. occidentalis*. *R. olivacea* can share similar colors, but is generally larger in size, has a rosy-blushed stipe and yellow spores, and does not stain. *R. decolorans* has an orange-red to copper orange cap that fades in age, has whitish gills and stipe, and also stains extensively grayish; it appears to be restricted to the extreme northwest portion of California. *R. californiensis* has a red to pink cap, a white stipe that stains gray (especially at the base), and a slightly acrid taste; it grows under pines farther south. See *R. albonigra* (p. 244) and *R. nigricans* (p. 245) for hard-fleshed, compact, dingy or dull-colored black- and gray-staining species.

Russula versicolor

Jul. Schäff

VARIABLE BIRCH RUSSULA

CAP: 2–6 cm across, convex when young, becoming broadly convex to plane, often with a depressed center in age, margin grooved. Color extremely variable, often some shade of, or a variegated mix of, purple, wine red, red, pink, brown, gray, olive, green, yellow, or whitish. Generally with a darker disc and paler margin. Surface thinly viscid and shiny when wet, dull if dry. **GILLS:** Broadly attached, but sometimes becoming free in age, close to subdistant, narrow, fragile, occasional forking near stipe. White at first, becoming cream colored to pale yellow in age. **STIPE:** 2–5 cm long, 0.5–1.5 cm thick, equal or enlarged slightly toward base. White at first, occasionally developing yellowish stains. Surface smooth, dry. **FLESH:** Thin, very fragile, white, slowly discoloring pale yellow when cut or in age. **ODOR:** Indistinct. **TASTE:** Quickly and strongly acrid, leaving a lingering peppery, burning sensation in mouth. **SPORE DEPOSIT:** Yellow to ocher-yellow. **MICROSCOPY:** Spores 5.5–8.5 x 4.5–6.5 μm, subelliptical to subovoid to nearly globose, ornamented with coarse isolated warts with scattered ridges, forming a partial to complete reticulum.

ECOLOGY: Solitary or scattered on ground or in grass under planted ornamental birch trees. Like many of the birch associates often first fruiting in late summer around irrigated trees, continuing into fall or early winter. Second flush sometimes appears during wet spring seasons.

EDIBILITY: Not recommended. Small acrid-tasting *Russula* are not recommended as edibles, as some are toxic.

COMMENTS: Even though this mushroom is highly variable in color, it can be easily recognized by the growth under non-native birch trees, small size, yellowish spores, and acrid taste. Aside from its habitat, it resembles a number of native species. *R. gracilis* (sensu CA) has a pink, rosy purple, or reddish cap with an olive- or yellow-tinged disc, and a pinkish-tinged stipe. It often occurs in wet areas in mixed hardwood–Douglas-fir forests. Also see *R. fragilis* (at right), an acrid-tasting species with white spores that grows under conifers.

Russula fragilis

Fr.

FRAGILE RUSSULA

CAP: 2–6 cm across, convex when young, becoming broadly convex to plane, often with a depressed center, margin occasionally translucent-striate or grooved in age. Color extremely variable, generally with a dark purple, gray, reddish brown to pinkish center, transitioning to a zone that can be any of these shades, but often with olive tones, to a paler margin, rarely unicolor. Surface smooth, thinly viscid and shiny when wet, dull if dry. **GILLS:** Broadly attached, but sometimes becoming free in age, close to subdistant, narrow, fragile, partial gills rare or absent. White to pale creamy white. **STIPE:** 2–5 cm long, 0.5–1.5 cm thick, equal. White at first, occasionally developing yellowish stains near base. Surface smooth, dry. **FLESH:** Thin, very fragile, white. **ODOR:** Indistinct to sweet, fruity. **TASTE:** Quickly and strongly acrid, leaving a lingering peppery, burning sensation in mouth. **SPORE DEPOSIT:** White. **MICROSCOPY:** Spores 6–9 x 5.5–7.5 μm, subglobose to subovoid, ornamented with coarse isolated warts 0.8–1.2 μm high, connected with ridges, forming a partial to complete reticulum.

ECOLOGY: Solitary or scattered on well-decayed wood, moss-covered logs, or stumps, occasionally on ground. Generally fruiting in wet areas under conifers, rarely under hardwoods (see comments). Widespread and common throughout our range. Fruiting from fall into winter.

EDIBILITY: As with most acrid *Russula*, not recommended.

COMMENTS: This species can exhibit a wide range of cap colors, sometimes all on the same fruitbody! The small size; variable cap color (almost always with a dark disc) with some olive or purple tones; white gills, stipe, and spores; and acrid taste help identify it. *R. versicolor* is similar but grows with introduced birch trees and has yellow spores. *R. gracilis* (sensu CA) also has a variably colored cap—usually some shade of pink, rosy purple to reddish with an olive- or yellow-tinged disc—and a pinkish-tinged stipe. It often grows in wet areas in mixed forests of Douglas-fir and hardwoods. It is our belief that the reports of *R. fragilis* with hardwoods represent this and possibly other species.

Russula puellaris

Fr.

CAP: 2–6 cm across, convex when young, becoming broadly convex to plane, often with a depressed center. Margin downcurved at first, occasionally uplifted and often grooved in age. Deep reddish purple, rosy purple to brick red when young, developing paler colors toward the margin as it expands; light purple, pinkish to dingy brownish pink, then developing a yellow-orange base color in age. Surface thinly viscid and shiny when wet, dull if dry. **GILLS:** Broadly attached, close, narrow, very fragile. White at first, becoming creamy to yellow, staining dark yellow to pale orange when damaged, to yellow-orange overall in age. **STIPE:** 2–6 cm long, 0.8–1.5 cm thick, equal or enlarged slightly toward base. White at first, staining yellow to pale orange when handled, becoming dingy yellow to pale orange overall in age. Surface smooth, dry to moist. **FLESH:** Thin, very fragile. White, staining yellow to pale orange when cut or in age. **ODOR:** Indistinct. **TASTE:** Mild. **KOH:** Orange on cap. **SPORE DEPOSIT:** Yellow to ocher. **MICROSCOPY:** Spores 6.5–9 x 6–7.5 µm, subellipsoid to subovoid to nearly globose, ornamented with scattered isolated warts 0.5–1 µm high, connecting ridges sparse or absent.

ECOLOGY: Occasionally solitary, more often scattered in troops on ground, in duff or moss under conifers, especially Western Hemlock and Sitka Spruce. Uncommon from Mendocino County northward. Fruiting in fall and early winter.

EDIBILITY: Nontoxic, but small and fragile.

COMMENTS: This small *Russula* can easily be identified by the dark purple to brick red cap that fades and develops yellow to orange colors in age, white gills that darken as the spores mature, white stipe that stains yellow to pale orange when handled, and mild taste. *R. abietina* is similarly colored, but has olive tones on the disc (and less of a tendency to develop yellow-orange colors in age), white to yellow gills, and a white stipe.

Russula stuntzii

Grund

STUNTZ'S RUSSULA

CAP: 3.5–8 cm across, rounded to convex when young, becoming broadly convex to plane, and often with a depressed center in age. Whitish, grayish to pale lilac-gray, often darker on disc, paler toward margin, or in age. Surface smooth, viscid to tacky when wet, shiny when dry. **GILLS:** Broadly to narrowly attached, close, narrow, fragile, partial gills rare. White to pale creamy white. **STIPE:** 3–8 cm long, 1–2 cm thick, equal or enlarged slightly toward base. White to pale grayish white, often discoloring pale grayish when handled. Surface smooth, dry. **FLESH:** Thin to moderately thick, firm, fragile, white. **ODOR:** Indistinct. **TASTE:** Quickly and strongly acrid, leaving a lingering peppery taste in mouth. **SPORE DEPOSIT:** White to pale creamy white. **MICROSCOPY:** Spores 6–9 x 6–7.5 µm, ovoid, subglobose to broadly ellipsoid, ornamented with scattered warts 0.5–1 µm high, with rare lines or ridges.

ECOLOGY: Solitarily or scattered, usually on well-decayed moss-covered logs or stumps, occasionally on ground. Locally common in coniferous forests on the Far North Coast, especially under Grand Fir. Occurs as far south as Sonoma County, but rare south of central Mendocino County. Fruiting in fall and winter.

EDIBILITY: Unknown; the strongly acrid taste is a deterrent.

COMMENTS: The whitish cap with grayish or lilac tones, white gills, whitish stipe that becomes grayish, acrid taste, and habit of growing on well-decayed logs are distinctive. There are a few other pale-colored *Russula*, but they all have "warmer" tones to the cap: *R. raoultii* has a whitish to creamy cap, white gills and stipe, and a strongly acrid taste. *R. albidula* has a creamy white to creamy yellow cap (but can become dingy in age), cream-colored spores, a white stipe that stains tan when handled, and a mild to slightly acrid taste. Also compare with *R. crassotunicata* and the pale-capped form of *R. cremoricolor*.

Russula cerolens

Shaffer

POCKET-STALKED RUSSULA

CAP: 3–10 cm across, rounded when young, becoming broadly convex to plane, with a depressed center at all times. Margin often becoming wavy and often uplifted in age. Yellowish brown, tan to dark brown, often darker on disc, often developing orangish stains and occasionally becoming dingy grayish brown, darkening overall in age. Surface smooth, with groves around margin, viscid and shiny when wet, dull and often cracking when dry. **GILLS:** Broadly attached to notched, close to subdistant, broad, fragile, partial gills rare or absent. White to creamy at first, developing orangish brown to olive-brown spots and stains, becoming dingy cream colored with brownish stains in age. **STIPE:** 3–8 cm long, 1–2.5 cm thick, equal or enlarged slightly toward base. White but often with an orange base when young, developing dingy brownish stains in age. Surface smooth, dry. **FLESH:** Thin, firm, fragile when young, mealy and brittle in age; stipe often with 2 or 3 hollow, diamond-shaped pockets or chambers. White, occasionally with brownish stains. **ODOR:** Mild at first, soon unpleasant, like burnt hair, or more chemical, like burnt plastic or rubber. **TASTE:** Disagreeable, bitter at first, soon acrid. **SPORE DEPOSIT:** White. **MICROSCOPY:** Spores 6.5–9 x 5.5–6.5 μm, subglobose to broadly ellipsoid, ornamented with scattered warts 0.5–0.7 μm high, occasionally with ridges, but not forming a reticulum.

ECOLOGY: Solitary, scattered, or in troops on ground in a wide variety of forest types, especially with live oak and pines. Very common throughout our range. Fruiting from fall through winter.

EDIBILITY: Unknown; likely toxic.

COMMENTS: This ubiquitous *Russula* can fruit by the hundreds under live oak or Monterey Pine in the southern part of our range and with Shore Pine on the North Coast. However, it is not limited to these forest types and can be found in many kinds of hardwood and conifer forests throughout our range. It is easily recognizably by the yellow-brown to brown cap with a distinctly grooved margin, white gills, and white stipe that develops reddish orange stains. The unpleasant odor, similar to burnt hair or burnt rubber, is also distinctive. The Almond-scented Russula (*R. fragrantissima* group) looks similar, but as the name suggests, it has an almond or maraschino cherry odor when young and also is quite a bit larger. *R. crassotunicata* lacks an odor, has a paler cap (when young) with a thick rubbery skin, and lacks the ground margin.

MISAPPLIED NAMES: *Russula amoenolens* and *R. pectinatoides*.

Russula fragrantissima (sensu CA)
ALMOND-SCENTED RUSSULA

CAP: 5–15 (25) cm across, round-convex, convex to plane, often with a depressed center in age. Honey brown to beige-brown, often with red-brown to orange-brown blotches and stains. Surface smooth to bumpy, with a sulcate margin in age, viscid when wet, to dry, skin fairly thick and rubbery, easily peeled and often cracking if dry. **GILLS:** Broadly attached to notched, close to subdistant, broad, partial gills present. Pale tan to yellowish beige or clay colored, often with ocher-brown stains. Often decorated with clear droplets, especially when young. **STIPE:** 6–20 cm long, 3–6 cm thick, club shaped or swollen in middle. Whitish to clay tan, with dingy brownish stains in age. **FLESH:** Moderately thick to thin, firm when young, soon softer and brittle or mealy in age. Stipe chambered at first, soon hollow throughout. **ODOR:** Very strong, sweet like almond extract or maraschino cherries, fouler in age. **TASTE:** Bitter or unpleasant at first, soon very acrid. **SPORE DEPOSIT:** Cream to pale yellow. **MICROSCOPY:** Spores 8.5–10.5 x 6.5–9.5 µm, globose to subglobose, ornamented with scattered warts 0.8–1.2 µm high, with heavy ridges or lines, forming a partial to full reticulum.

ECOLOGY: Solitary or in scattered patches on ground in mixed hardwood-conifer forest. Often with Tanoak in the southern part of our range, with Sitka Spruce in the north. Locally common on the North Coast, uncommon south of the San Francisco Bay. Fruiting from early fall into winter.

EDIBILITY: Likely toxic.

COMMENTS: The pungent sweet odor, yellow-brown cap, and chalklike or brittle flesh distinguish this mushroom from all others. A few *Hygrophorus* have a similar odor, but are more fibrous-fleshy, taste mild, and have thick, waxy gills. *R. cerolens* looks very similar, but is smaller and has a distinctly different odor (more sour). *R. laurocerasi* (sensu CA) is slightly smaller and has a similar, persistently sweet odor and smaller spores with longer warts (1.5–2.5 µm high). These names used for our Almond-scented Russula are European and will likely change.

Russula crassotunicata
Singer

CAP: 3–8 (10) cm across, rounded to convex when young, becoming broadly convex to plane, often with a depressed center and occasionally a uplifted margin in age. Pale creamy yellow when young, soon developing yellowish to ocher-brown stains and discoloration, occasionally becoming completely light brown in age. Surface viscid, covered with a thick slime layer when wet, often dull and cracked if dry. Cap skin thick, gelatinous and rubbery when wet, almost leathery when dry. **GILLS:** Broadly attached to notched, close to subdistant, narrow, fragile, partial gills rare, occasional forking near stipe. White to creamy white at first, developing brownish spots and blotches, occasionally becoming all brown in age. **STIPE:** 3–5 cm long, 0.8–2 cm thick, often short, stout, equal or enlarged slightly toward base. White at first, developing brownish spots and stains. Surface smooth, dry. **FLESH:** Thin to moderately thick, rubbery-gelatinous near cap skin, very firm when dry. White, slowly staining brownish when cut. **ODOR:** Indistinct. **TASTE:** Usually quickly and strongly acrid and bitter, but can be mildly acrid to mostly bitter. **SPORE DEPOSIT:** White. **MICROSCOPY:** Spores 9–11 x 7–9 µm, subglobose to broadly ellipsoid, ornamented with fine isolated warts 0.8–1.2 µm high.

ECOLOGY: Solitary, scattered, or in small troops on ground or in duff under conifers. Locally common on the Far North Coast in fir and hemlock forest. Not recorded south of Sonoma County. Fruiting in fall and winter.

EDIBILITY: Unknown; not recommended.

COMMENTS: The thick, gelatinous, and rubbery cap skin (when wet); creamy yellow colors with brown stains; and growth with northern conifers help identify this *Russula*. *R. albidula* has a whitish cap that becomes pale yellow (especially on the disc), an acrid taste, and slightly smaller, cream-colored spores; it usually grows with oaks. *R. albida* also has a whitish to pale yellow cap with a rubbery cuticle, but has yellow spores and a mild to slightly bitter taste. *R. raoultii* has relatively small, fragile fruitbodies with white to pale yellowish caps, white spores, and a strongly acrid taste. The pale form of *R. cremoricolor* has a creamy white cap, white spores, and lacks a rubbery cap skin. *R. stuntzii* has a whitish to pale grayish cap and a white stipe that discolors grayish.

Russula cantharellicola

D. Arora & N. H. Nguyen

CAP: 8–25 (30) cm across, convex with an incurved margin and a depressed center when young, to vase shaped with a wavy uplifted margin in age. Dingy whitish to cream-beige, soon mottled with tan-beige to dull brown, sometimes extensively dingy brown, usually with a distinctive spotted and streaked appearance, developing blackish gray areas in age. Surface weakly viscid to dry, smooth, sometimes metallic. **GILLS:** Broadly attached to slightly decurrent, close, broad, fragile. Creamy whitish to beige-tan, becoming irregularly dingy brick red to pinkish or dull grayish wine colored in age, dusted with whitish spores. **STIPE:** 5–15 cm long, 2.5–8 cm thick, equal or enlarged toward base. Whitish to creamy, developing tan or brownish blotches, often mottled with dingy reddish in age. Bruising slowly pink to dingy reddish, sometimes more reddish brown or occasionally brown without pinkish red phase. **FLESH:** Thick, firm to very hard in stipe, breaking cleanly when broken (but can be difficult to do!). White, slowly bruising liked the stipe. **ODOR:** Often fairly strong, musty, unpleasant. **TASTE:** Mild to slightly bitter. **SPORE DEPOSIT:** White to off-white. **MICROSCOPY:** Spores 7–10 x 6–8 μm, subglobose, broadly ellipsoid to ovoid, ornamented with dense low warts 0.1–0.5 μm high, occasionally connected with fine lines, forming a partial reticulum.

ECOLOGY: Solitary or in small arcs under Coast Live Oak where *Cantharellus californicus* mycelium is present, fruiting from late summer into early fall (but fruitbodies are very slow to decay and can be found into early winter). Locally common south of Sonoma County. The association with chanterelles is quite intimate, but the exact nature of the relationship is unclear (probably parasitic).

EDIBILITY: Closely related species are extremely toxic. Avoid!

COMMENTS: This species is quite distinctive. Other similar species of *Russula* are darker and dingier, or less extensively red staining in age (see *R. nigricans*, p. 245), or lack red tones entirely (*R. albonigra*). The tendency of this species to fruit near *Cantharellus californicus* (p. 24) under Coast Live Oak early in the fall is an excellent clue to its identity.

Russula albonigra

(Krombh.) Fr.

WHITE AND BLACK RUSSULA

CAP: 5–15 (20) cm across, convex when young, becoming broadly convex to plane, often with a depressed center, occasionally funnel shaped and often irregular, with a wavy or uplifted margin in age. Dingy white at first, darkening in patches, gray (occasionally grayish brown) to black, becoming completely black in age. Surface dull, smooth, slightly viscid or moist at first, soon dry. **GILLS:** Broadly attached to subdecurrent, close to subdistant, thick, broad, fragile, occasionally forking or inter-veined, partial gills interspersed around margin. White when young, becoming creamy white to pale grayish yellow, staining black when bruised, becoming dark gray to all black in age. **STIPE:** 3–9 cm long, 2– 4 cm thick, equal, swollen in middle or enlarged toward base. White to pale gray, staining black when handled, to dark gray to black in age. Surface smooth, dry. **FLESH:** Thick, very firm. White when young, often grayish in age, staining black. **ODOR:** Indistinct. **TASTE:** Mild. **SPORE DEPOSIT:** White. **MICROSCOPY:** Spores 7–10.5 x 5.5–6.5 μm, broadly ellipsoid to subovoid, finely ornamented with short isolated warts and spines 0.1–0.2 μm high, with occasional short ridges or lines, forming a partial reticulum. Cap cuticle 100–150 μm, made up with horizontal, parallel hyphae, with scattered free hyphal tips.

ECOLOGY: Solitary, scattered, or in troops on ground in mixed forest, often with hardwoods. It especially prefers oak-Madrone forests in the greater San Francisco Bay Area, Tanoak-Madrone forests on the North Coast. Common and widespread throughout our range (but fruitings can be very scarce in some years). Fruiting in fall, but fruitbodies often persist well into winter.

EDIBILITY: To be avoided; some blackening *Russula* are poisonous.

COMMENTS: This large *Russula* can be identified by the white fruitbodies that discolor patchily grayish and then black (with no intermediate reddish phase). *R. atrata* is nearly identical macroscopically, but has a thicker, gelatinous cap cuticle (200–450 μm thick). Other blackening *Russula* stain pinkish or red before they blacken (see *R. cantharellicola* and *R. nigricans* and the species mentioned in their comments sections).

Russula nigricans group
BLOODY BLACKENING RUSSULA

CAP: 5–15 (20) cm across, convex when young, becoming broadly convex with a depressed center, to funnel shaped and often irregular, with a wavy or uplifted margin in age. Dingy white at first, darkening in patches, gray to grayish brown, to eventually black. Surface dull, finely velvety to smooth, often cracking into areolate patches in age, slightly viscid if wet when young, soon dry. **GILLS:** Broadly attached, subdistant to distant, thick, broad, fragile, partial gills numerous. Creamy to pale grayish yellow, slowly staining red, then black when damaged, becoming dark gray to all black in age. **STIPE:** 2–7 cm long, 1.5–5 cm thick, stout, more or less equal. Whitish at first, soon pale gray to grayish brown, to dark gray to black in age, slowly staining red, then black when handled. Surface smooth, dry. **FLESH:** Thick, very firm. Whitish when young, staining red, with an intermediate reddish brown phase, then dark gray to black. Or often grayish to all black in older fruitbodies. **ODOR:** Mild, to musty or earthy when young, generally unpleasant in age. **TASTE:** Mild to slightly acrid. **SPORE DEPOSIT:** White. **MICROSCOPY:** Spores 6.5–8 x 5–7 μm, broadly ellipsoid to subglobose, ornamented with short isolated warts and spines 0.1–0.3 μm high, with short ridges or lines, forming a partial to full reticulum. Cap cuticle 100–150 μm thick, interwoven epicutis, (uppermost cells in cap surface) embedded in a gelatinous matrix, with free hyphal tips up to 5 μm wide.

ECOLOGY: Solitary, scattered, or in troops on ground in coniferous forests, especially under pine and Sitka Spruce.

Very common north of the San Francisco Bay Area. Fruiting from early fall into winter, or in summer in areas with heavy fog drip.

EDIBILITY: Some blackening *Russula* are poisonous; to be avoided.

COMMENTS: This *Russula* is one of the more common members of the blackening *Russula* group in California, often found fruiting by the hundreds in coastal pine and spruce forests on the North Coast. It can be identified by the finely velvety, whitish cap that becomes smooth, grayish brown and then black in age, grayish yellow gills, and slow red staining when damaged. The red staining reaction usually takes five or more minutes, then fades to reddish brown, grayish, and then black after about twenty minutes. *R. dissimulans* is slightly smaller and has browner colors, more crowded gills, and larger spores *R. densifolia* has a thinly viscid or shiny, whitish cap that becomes brown; crowded, creamy to pale yellowish gray gills; and a staining reaction progressing from pinkish red to dark gray. *R. adusta* has a thinly viscid to shiny, whitish cap that becomes brown to brownish gray, and stains dingy grayish brown to dark gray (sometimes with a fleeting pinkish stage). *R. albonigra* slowly stains black, with no red phase. The whole group is in need of modern taxonomic work to establish solid species concepts.

Russula brevipes group
SHORT-STALKED RUSSULA

CAP: (4) 7–30 (35) cm across, convex with an incurved margin and a depressed center when young, to vase shaped with a wavy uplifted margin in age. White at first, soon dingy white to whitish buff with yellowish brown spots or stains. Surface dry, finely tomentose to smooth. **GILLS:** Broadly attached to slightly decurrent, close, broad, fragile, occasionally forking, or fusing near the stipe, partial gills numerous. White when young, sometimes with a bluish cast (see comments), often developing brownish spots and stains in age. Edges often have clear droplets when young. **STIPE:** 2–8 cm long, 2–5 cm thick, short, stout, more or less equal. White, with brownish spots or stains. Occasionally with a bluish band at apex. **FLESH:** Thick, very firm when young, granular in age. White, but often with brownish stains around larva tunnels. **ODOR:** Mild or occasionally unpleasant in age. **TASTE:** Mild to slightly acrid. **SPORE DEPOSIT:** White to cream. **MICROSCOPY:** Spores 8–10.5 x 6.5–9 μm, subglobose, broadly ellipsoid to ovoid, ornamented with scattered warts 0.3–1.3 μm high, and heavy ridges forming a partial reticulum.

ECOLOGY: Solitary, in small clusters, or scattered in large troops in a wide variety of forest types. Widespread and very common throughout our range. Fruiting from early fall well into winter.

EDIBILITY: Some forms edible and very good, others downright awful (it is a species complex). Firm, young buttons are best, especially those growing with live oak and manzanita.

COMMENTS: This *Russula* is capable of producing extremely large fruitbodies; caps measuring 25–30 cm across are regularly encountered. The white colors overall (often with yellowish to brown stains), cap that becomes uplifted and funnel shaped in age, lack of latex, and absence of gray to black staining are helpful identifying features. Genetic data indicate that there are at least five species in the *R. brevipes* complex in California, some of which have only slight macroscopic differences. The name *R. brevipes* var. *acrior* has been applied to forms with a bluish green band at the stipe apex. *R. brevipes* var. *megaspora* has larger spores (9–14 x 8–12 μm). Those growing with hardwoods are generally smaller and brighter white in color. *R. cascadensis* has a white cap and gills, but is more likely to develop orangish stains as it ages, has a strongly acrid taste, and has smaller spores (6.5–8 x 5.5–6.5 μm); it is locally common on the Far North Coast. Members of the blackening *Russula* group all show grayish to black staining. Some large *Lactarius* species can look similar, but bleed white or colored latex when the gills are broken.

Hypomyces lactifluorum

(Schwein.) Tul.

LOBSTER MUSHROOM

FRUITBODY: 4–20 cm across, often highly irregular and distorted, occasionally resembling the shape of the host (*Russula brevipes*). Cap surface dry to moist, often cracked or pitted. Underside dry, covered with small pimples (perithecial). Gills completely covered, appearing as nondescript shallow ridges. Generally bright orange, can be whitish with pale orange blotches when young and buried in duff, developing reddish purple stains and blotches as it starts to decay. **STIPE:** Often short, stout, and distorted, but can be fully formed. **FLESH:** Very firm and white with a thin orange crust at first, becoming mottled with orange and softer as it ages. **ODOR:** Indistinct. **TASTE:** Mild. **KOH:** Bright magenta-purple. **SPORE DEPOSIT:** White. **MICROSCOPY:** 35–40 (45) x 4.5–7 μm, spindle shaped, distinctly warted.

ECOLOGY: Solitary, scattered, or in large troops, often buried in duff, before erupting at maturity. Growing on fruitbodies of *Russula brevipes*, nonparasitized *Russula* usually present within patches. Common on the Far North Coast, especially under Sitka Spruce, occasional in Mendocino County, not known from south of Sonoma County. Fruiting from late summer into early winter.

EDIBILITY: Edible and good. Only firm young specimens should be eaten.

COMMENTS: A fungus on a fungus—what could be better? The Lobster Mushroom is a parasitic fungus that forms a bright orange crust engulfing its host mushroom (*Russula brevipes* in most, if not all cases). There are a number of other *Hypomyces* species, many of which are specific to a single species or genus of host mushrooms. *H. lateritius* grows on *Lactarius* (most often those in the *L. deliciosus* group) and covers only the gills of its host with a whitish layer; it is uncommon in California. *H. luteovirens* covers the gills and stipe of *Russula* with a pimpled greenish yellow crust; it is rare in California. *H. aurantius* forms a bright orange layer on the undersides of polypores, especially older fruitbodies of *Fomitopsis pinicola*; it is common on the Far North Coast. *H. rosellus* is a rare, bright pink to rosy, pimpled crust that grows on the decaying fruitbodies of many different kinds of mushrooms.

Macowanites species

The following description is an overview of many different species, many of which are rarely found and poorly known.

CAP: 1–4 (6) cm across, rounded to convex, margin pinched to or attached to the stipe in some, and not opening, others downcurved. Creamy, beige, pink, red, or purple. Surface dry to viscid. **GILLS:** Completely deformed into a convoluted, chambered mass, to moderately deformed with wrinkled cross veins. **STIPE:** 0.5–7 cm long, 0.3–1 cm thick, short and indistinct to "normal"; equal to club shaped. White, beige to pinkish blushed. **FLESH:** Thin (fruitbodies typically mostly made of contorted "gills"). **ODOR:** Mild or in other species fruity, almond scented, bleachlike, or disagreeable. **TASTE:** Mild to acrid. **SPORE DEPOSIT:** Not obtainable. **MICROSCOPY:** Spores globose to ellipsoid, amyloid, variably ornamented with warts, and/or spines and ridges.

ECOLOGY: Solitary or scattered in small groups, rarely abundant. Often at least partially buried in duff or moss. Commonly encountered on the Far North Coast under Sitka Spruce, but other species occur throughout our range with conifers as well as hardwoods. Typically fruiting in the fall and early winter.

EDIBILITY: Unknown.

COMMENTS: *Macowanites* is an artificial genus comprising many *Russula* that are not all closely related, but which were nevertheless grouped because of their secotioid fruitbodies.

These mushrooms have lost the ability to forcibly discharge their spores and instead rely on animals to eat the fruitbodies and disperse the spores in their droppings. *M. luteolus* (upper left) is probably the most commonly encountered species in our region; it has a creamy white to pale creamy yellow cap and a short stipe. The gills are mostly enclosed by the cap, highly convoluted, and creamy buff to pale ocher-yellow. *M. luteolus* lacks a distinct odor and has an acrid taste. *M. chlorinosmus* is very similar and has a unpleasant (but not acrid) taste and a strong bleachy odor (especially in age). Both appear to be restricted to the spruce-hemlock forests on the Far North Coast. *M. olidus* has a brownish cap that often has ocher-red stains (similar to *Russula cerolens*), highly convoluted gills, and a burnt hair odor; it typically grows buried in duff under live oak. A similar species with an almond odor also grows under live oak; it is often buried in the soil and has a poorly formed stipe (sometimes very rudimentary) and an irregular cap with highly deformed gills. This species appears to be undescribed. *M. nauseosus* has a purplish to reddish cap (often with brown tones), a mild odor, and a strongly disagreeable taste. Another species (lower right) found under pine on the North Coast has a rosy pink cap (often variegated with brownish tones in age), pale ocher orangish, slightly deformed gills, a normal-looking white stipe, and a mild odor and taste.

Lactarius rufus

(Scop.) Fr.

RED HOT MILK CAP

CAP: 3–10 (12) cm across, convex with an incurved margin when young, becoming broadly convex to plane or slightly uplifted, often with a depressed center, with or without an umbo. Dull brick-red, reddish brown, pinkish brown to orange-brown. Surface dull, smooth to slightly wrinkled, dry to moist (never viscid). **GILLS:** Broadly attached to subdecurrent, close to crowded, often forking near stipe, narrow, fragile. Pale orangish white, buff to pale orange, sometimes developing reddish spots or discoloration in age. **LATEX:** Copious, white, usually rather thick, unchanging and not staining tissue. **STIPE:** 3–12 cm long, 0.5–2 cm thick, equal or nearly so, but often pinched or misshapen at base. Reddish brown, orange-brown to reddish tan, often paler toward base. Surface dry, often covered with a whitish sheen of very fine powder. **FLESH:** Thin, fragile, off-white, tan to pale orange. Stipe stuffed at first, becoming hollow. **ODOR:** Indistinct. **TASTE:** Slowly very acrid, sometimes taking up to 30 seconds, can be mild to slightly bitter at first, but will have a lingering hot peppery taste. **KOH:** Olive-green on cap and stipe. **SPORE DEPOSIT:** White to pale creamy white. **MICROSCOPY:** Spores 7–9.5 x 6–7 μm, broadly ellipsoid, ornamented with scattered warts and irregular ridges and lines, forming a partial reticulum, up to 0.5 μm high.

ECOLOGY: Solitary, scattered, or in small clusters, often forming large troops in duff and moss under pines. Very common on the Far North Coast, becoming less common southward to the San Francisco Bay Area, where it is rare. Fruiting in fall and winter.

EDIBILITY: Poisonous.

COMMENTS: The drab brick red to pinkish brown cap, pale buff to orangish white gills, unchanging white latex, growth with pines, and very acrid taste (sometimes slow to develop) help distinguish this species from many other small, reddish orange *Lactarius*. *L. rufulus* which grows with Coast Live Oak, is brick red or orange-red overall and has a solid stipe and comparatively mild taste. *L. rubidus* has an orange-brown, roughened cap surface, mild taste, and sweet odor. Other small reddish orange *Lactarius* have moist to viscid caps.

Lactarius rufulus

Peck

SOUTHERN CANDY CAP

CAP: 3–8 (11) cm across, convex to plane, becoming wavy, disc sunken or umbonate at times, margin incurved at first, becoming ruffled and wavy in age. Brick red, reddish brown to reddish orange. Surface smooth to slightly wrinkled, dry to greasy. **GILLS:** Broadly attached to slightly decurrent, fairly close to subdistant, narrow. Pale orangish buff to pale orange, developing reddish brown spots and stains in age. **LATEX:** Very scant, watery white. **STIPE:** 3–10 cm long, 0.5–2.5 cm thick, equal or tapered slightly toward base, solid, stocky. Reddish orange to reddish brown with a whitish to buff sheen, base paler. Surface smooth, dry, base covered with whitish to bright orange hairs that often bind to duff and soil. **FLESH:** Moderately thick to thin, firm, brittle. Ivory-buff to pale brown. Stipe flesh firm, solid but becoming soft in age, rarely hollow. **ODOR:** Faint to moderately strong, sweet or slightly unpleasant. **TASTE:** Mild to slightly sweet. **KOH:** Pale olive-orange to olive-yellow on all parts. **SPORE DEPOSIT:** Pale yellowish buff. **MICROSCOPY:** Spores 7–9 x 7–9 μm, globose to subglobose, ornamented with scattered warts and irregular ridges and lines, forming a partial reticulum, up to 0.5 μm high.

ECOLOGY: In clusters, small groups, or solitary on ground under Coast Live Oak from Mendocino County south. Uncommon north of the San Francisco Bay Area, common to the south. Often fruiting from late fall though spring, most abundant in late winter or early spring.

EDIBILITY: Edible, not quite as good as *L. rubidius*, but often used like candy caps. It's more likely to develop musty tones when dried.

COMMENTS: The Southern Candy Cap frequently fruits alongside the Candy Cap (*L. rubidus*) in Coast Live Oak woodlands in late winter and spring. It can be told apart by the darker brick red to reddish orange cap, the stouter stature, and the solid stipe with lots of spiky orange hairs around the base. *L. rufus* is similar, but grows under pines, has a dull pinkish brown to brick red cap and white latex, and is very acrid tasting (and poisonous). *L. luculentus* var. *laetus*, *L. subviscidus*, *L. substriatus*, and *L. subflammeus* all have viscid to moist caps.

Lactarius rubidus

(Hesler & A. H. Sm.) Methven

CANDY CAP

CAP: 1.5–8 (12) cm across, broadly convex to plane, becoming uplifted, wavy, disc often slightly sunken. Margin downcurved at first, becoming wavy and finely ruffled in age. Orange-brown, warm orange to pale buff-orange, occasionally reddish orange in age. Surface dry to moist, wrinkled, slightly roughened (somewhat like the skin of a citrus fruit). **GILLS:** Broadly attached to slightly decurrent, close to subdistant, narrow. Buff to pale orange. **LATEX:** Scant to variably absent, watery white to clear. **STIPE:** 2–10 cm long, 0.3–1.5 cm thick, equal or tapered slightly toward base, slender. Pale orangish buff to pale orange, with a whitish sheen that is easily rubbed off when picked. Surface smooth, dry, base covered with small whitish hairs. **FLESH:** Thin, brittle. Buff to pale orange. Stipe very fragile, hollow (sometimes stuffed when young). **ODOR:** Distinctive, but difficult to describe; pleasantly fresh, sweet, and sharp, like fresh fenugreek, becoming much more pronounced, sweeter, like brown sugar or maple syrup when dried. **TASTE:** Mild. **KOH:** Pale olive-orange on cap and stipe. **SPORE DEPOSIT:** White to pale cream. **MICROSCOPY:** Spores 6–8 x 6–7.5 μm, globose to subglobose, ornamented with scattered warts and irregular ridges and lines, forming a partial reticulum, 0.5–1 μm high.

ECOLOGY: Solitary, in small groups or clusters on ground, or on well-decayed stumps and logs in a wide range of forest types. Most abundantly fruiting with live oak and pine over much of our range, also found with Tanoak and Douglas-fir (the latter especially on the Far North Coast). Very common on the coast from Santa Cruz County northward into southern Oregon, uncommon outside this range. Fruiting from late fall through spring, but in most years reaching peak of fruiting abundance in early winter (slightly later, into early spring southward).

EDIBILITY: A real treat, a sweet-flavored little morsel of a mushroom. It's best after gentle (but thorough) drying. It develops a strong maple sugar taste and odor that can been used to flavor ice cream, cookies, cheesecake, and even vodka. Although it can also be used fresh (some people favor it as an addition to meats), many find the flavor slightly bitter or disagreeable.

COMMENTS: This species makes it possible to have mushroom ice cream that actually tastes good! The dry, warm orange, wrinkled-roughened cap with an often scalloped or ruffly margin, watery-white latex, hollow stipe, and sweet fenugreek odor (when fresh) help identify this mushroom. Although the odor when fresh is distinctive, it is different from the maple sugar odor that develops when the mushroom is dried (the potent odor of dried specimens can persist for decades!). It has many look-alikes; a number of other orange- or orangish-capped *Lactarius* are similar but lack one or more of the features mentioned above. *L. rufulus* has a darker brick red to reddish orange, frequently solid stipe, and stouter stature, and often has a pronounced coating of spiky orange hairs around the stipe base. The poisonous *L. rufus* grows only

under pines and has a dull pinkish to reddish brown cap, creamy white latex, and a very acrid taste (which can be slow to develop). The oak-loving *L. xanthogalactus* has a pinkish to peachy buff cap and white latex that quickly turns yellow; it is poisonous. *L. thiersii* is a tiny, rare species with a dry, orangish brown cap and lacks an odor. Also compare *L. substriatus*, *L. subviscidus*, and *L. luculentus* var. *laetus*. From above, *Paralepista flaccida* looks very similar to the Candy Cap, but has fibrous flesh, no latex, and a slight odor of black pepper. Be especially cautious of collection specimens growing on or near decayed stumps; the wood-dwelling Deadly Galerina (*G. marginata*) can have a vaguely similar warm color, but has a smooth, hygrophanous, honey brown cap, pale tan to brown gills, a partial veil that leaves a faint ring on the stipe, and brown spores. Finally, *Rhodophana nitellina* has a similarly colored cap, but is usually translucent-striate around the margin and has a strong cucumber odor and pinkish buff spores.

SYNONYM: *Lactarius fragilis* var. *rubidus* Hesler & A. H. Sm.

Candy Cap
Identification Checklist

- Cwap warm orange to bright orange-brown
- Cap surface dry, distinctly roughened or wrinkled
- Cap margin often ruffly
- Latex cloudy white to nearly clear, like skim milk
- Stipe fragile and hollow
- Spore deposit white (not yellowish)
- Odor of fresh mushrooms sweet, like fresh fenugreek
- Odor of dried mushrooms sweet, like maple sugar

Lactarius luculentus var. *laetus*
Hesler & A. H. Sm.

CAP: 2–7 cm across, convex to plane, disc slightly depressed, margin downcurved when young, slightly uplifted in age. Orange to orange-brown, slightly paler toward margin and in age. Surface smooth, viscid to subviscid, rarely dry, but greasy when so. **GILLS:** Broadly attached to slightly decurrent, narrow, fragile, close to crowded, partial gills numerous. Pale buff to orangish buff, with orangish brown spots or stains in age. **LATEX:** Copious, thick, white, unchanging, staining gills pale brownish orange. **STIPE:** 2–7 cm long, 0.5–1 cm thick, equal or enlarged toward base. Pale buff-orange to pale orange. Surface smooth, moist to dry, with fine whitish hairs at base. **FLESH:** Thin, fragile, pale orange to pale orangish tan. Stipe firm to fragile, stuffed with tan pith when young, becoming hollow. **ODOR:** Indistinct. **TASTE:** Mild to slightly bitter. **KOH:** Unknown. **SPORE DEPOSIT:** White. **MICROSCOPY:** Spores 8–10 x 7–8.5 μm, ellipsoid to broadly ellipsoid, ornamented with scattered warts and irregular ridges and lines, forming a nearly complete reticulum, 0.5 μm high.

ECOLOGY: Solitary or scattered on ground or in moss in northern coniferous forest, often under Sitka Spruce. Uncommon in California. Fruiting in fall or early winter on the Far North Coast.

EDIBILITY: Unknown, probably nontoxic.

COMMENTS: The orange to orange-brown cap, white, unchanging latex, mild to slightly bitter taste, and growth under Sitka Spruce help identify this small, uncommon milk cap. *L. substriatus* is more reddish orange, has white latex that slowly dries to pale yellow, and tastes slightly acrid. *L. subviscidus* and *L. subflammeus* have moist to subviscid caps and taste acrid. Other mild-tasting species include *L. rubidus*, *L. rufulus*, and *L. thiersii*, all of which have dry to moist (but not viscid) caps.

Lactarius substriatus

A. H. Sm.

CAP: 2–5 (7) cm across, broadly convex to plane, often with a slightly depressed center. Margin downcurved when young, often ruffled and slightly uplifted in age, translucent-striate when wet. Reddish orange, orange to brownish orange, paler toward margin. Surface smooth to slightly roughened, viscid to moist. **GILLS:** Broadly attached, close, narrow, fragile. Pale orangish buff to pale orange, sometimes with orange-brown spots in age. **LATEX:** Copious to scant, white, thin, wheylike, slowly and erratically turning pale yellowish or creamy. **STIPE:** 2–8 cm long, 0.4–1 cm thick, equal or enlarged slightly toward base. Pale orangish buff to orange, often with a whitish bloom when young. Surface dry to moist, smooth, often with whitish to orangish hairs at base. **FLESH:** Thin, very fragile, pale orange. Stipe stuffed when young, hollow in age. **ODOR:** Indistinct. **TASTE:** Slightly acrid. **SPORE DEPOSIT:** White to creamy. **MICROSCOPY:** Spores 7.5–9 x 5.5–7 µm, subglobose to broadly ellipsoid, ornamented with scattered warts and irregular ridges and lines, forming a partial reticulum, 0.5–1 µm high.

ECOLOGY: Solitary or scattered on ground or in moss in northern coniferous forests, usually under Sitka Spruce. Uncommon to locally common, at least from central Mendocino County northward (confusion with similar species clouds an understanding of its range). Fruiting from fall into winter.

EDIBILITY: Unknown, probably nontoxic.

COMMENTS: The small fruitbodies, deep orangish red to orange-brown caps that are translucent-striate at the margin, white latex that slowly goes yellow (an often overlooked feature), and slightly acrid taste help identify this species. *L. subviscidus* is similar but usually slightly larger and has an opaque orange cap and thick, white latex that slowly turns pale yellow. *L. subflammeus* has a deep reddish orange opaque cap and unchanging latex; it is common under conifers, especially pines. *L. luculentus* var. *laetus* has a viscid orange to orange-brown cap, white unchanging latex, and a mild taste, and grows under Sitka Spruce.

Lactarius subviscidus

Hesler & A. H. Sm.

CAP: 2–6 (8) cm across, convex to plane, disc slightly depressed. Margin downcurved when young, wavy, ruffled and slightly uplifted in age. Dark orange, orange-red to orange-brown, slightly paler toward margin and in age. Surface wrinkly or slightly roughened to smooth, slightly viscid when wet, greasy when dry. **GILLS:** Broadly attached, close to crowded, narrow, fragile. Pale buff to orangish buff, with orange spots or stains in age. **LATEX:** Copious, thick, white at first, slowly discoloring creamy yellow, drying pale yellow, staining gills orangish. **STIPE:** 3–10 cm long, 0.5–1.5 cm thick, equal or enlarged toward base. Pale orange to orange-brown with a whitish sheen or streaks. Surface smooth, moist to dry, with fine whitish hairs at base. **FLESH:** Thin, fragile, pale tan. Stipe stuffed with tan pith when young, becoming hollow. **ODOR:** Indistinct. **TASTE:** Slowly acrid, usually taking 10–15 seconds. **KOH:** Pale olive-orange to olive-yellow on cap and stipe, pale yellow-olive on gills. **SPORE DEPOSIT:** White. **MICROSCOPY:** Spores 8–9.5 x 7–8 µm, broadly ellipsoid, ornamented with scattered warts and irregular ridges and lines, forming a partial reticulum, 0.5–1 µm high.

ECOLOGY: Solitary or scattered in small groups on ground or in mossy areas under conifers, especially Sitka Spruce. Occurring from Mendocino County northward, becoming very common in Humboldt and Del Norte Counties. Fruiting from late fall into spring.

EDIBILITY: Nontoxic; frequently mistaken for Candy Cap.

COMMENTS: The orange to orange-brown, opaque caps, pale buff gills, white latex that slowly dries yellowish, slowly acrid taste, and growth under Sitka Spruce help distinguish this species from other small orange *Lactarius*. *L. subflammeus* has a slightly more reddish orange cap and white latex that doesn't change to yellow, and often grows under pine; it is otherwise nearly identical (and in fact may be the same species based on preliminary DNA work). *L. substriatus* also has white latex that slowly turns yellowish, but is more reddish orange and has a translucent-striate margin when wet. *L. luculentus* var. *laetus* has an orange to orange-brown cap, white unchanging latex, and a mild taste; it also occurs in northern spruce forests. *L. rubidus* has a dry, wrinkled, orange-brown cap, scant watery-white latex, and a mild taste.

Lactarius atrobadius

Hesler & A. H. Sm.

DARK ALDER MILK CAP

CAP: 2.5–6 cm across, broadly convex, plane or uplifted, often with a depressed center and sometimes an umbo. Margin downcurved when young, becoming uplifted and often ruffled. Deep dark purple-red, liver red to dark orange-red, margin slightly lighter in age, center remaining dark. Surface dry to slightly greasy, smooth to wrinkled or roughen. **GILLS:** Broadly attached to slightly decurrent, narrow, close to crowded, forking occasionally. Light pinkish buff to tan, becoming orange-buff, often with dark red-brown stains. **LATEX:** Scant, skim-milk white. **STIPE:** 4–9 cm long, 0.7–1.3 cm thick, enlarged slightly toward base, creased or pinched at times, brittle, base covered with whitish orange mycelial hairs. Reddish orange, paler toward apex. Surface smooth, dry to moist. **FLESH:** Thin, fragile, pale tan. Stipe stuffed with pith when young, hollow in age. **ODOR:** Indistinct. **TASTE:** Mild. **KOH:** Slight olive-orange on all parts. **SPORE DEPOSIT:** Creamy white. **MICROSCOPY:** Spores 8–10 x 6–8 μm, broadly ellipsoid to ellipsoid, ornamented with scattered warts and a broken or partial reticulum, up to 1 μm high.

ECOLOGY: Solitary or scattered on ground in wet areas or along streams near alder. Very common in fall and early winter in Humboldt and Del Norte Counties, rare south of Mendocino County.

EDIBILITY: Unknown, probably nontoxic.

COMMENTS: The dark liver red to dark orange-red cap, scant white latex, mild taste, and growth under alder help distinguish this species from an otherwise frustrating group of small orange- to red-capped *Lactarius* species. *L. occidentalis* is a similar small species found under alder, but has an olive-brown cap that becomes dark vinaceous brown to orange-brown in age, scant white latex, and a mild taste; it is also common on the Far North Coast. *L. alpinus* var. *mitus* has a brownish orange to brown, finely textured cap that fades to beige when it dries. *L. riparis* is a montane species with a brownish orange, reddish brown, or brown cap that often has a darker center and paler margin; it often grows with alder, willow, and fir. *L. rufulus* can have a similar color but is much stockier, and grows with live oak.

Lactarius "Acrid Orange"

CAP: 2 7 cm across, broadly convex to plane, often with a slightly depressed center, margin downcurved when young, slightly uplifted in age. Dull pale orange, peachy orange, orange-buff to orange-brown. Surface smooth, viscid to moist. **GILLS:** Broadly attached, close, narrow, fragile, partial gills numerous. Pale orangish buff to pale orange. **LATEX:** Copious, white, unchanging. **STIPE:** 2–5 cm long, 0.5–1 cm thick, more or less equal, or pinched slightly toward the base. Pale orangish buff with a whitish bloom. Surface dry to moist, smooth, often with whitish hairs at base. **FLESH:** Thin, fragile, stipe stuffed when young, hollow in age. Whitish to pale buff-orange. **ODOR:** Indistinct. **TASTE:** Slightly to strongly acrid, latex stronger than flesh. **SPORE DEPOSIT:** Creamy white. **MICROSCOPY:** Spores 5.5–7 x 5.5–7 μm, globose to subglobose, ornamented with irregular ridges and lines, forming a partial to complete reticulum, up to 0.5 μm high.

ECOLOGY: Solitary or scattered on ground or in moss in mixed Tanoak–Douglas-fir forest. Uncommon, known from Santa Cruz County to Mendocino County but probably more widespread. Fruiting in fall and early winter.

EDIBILITY: Unknown.

COMMENTS: The small to medium-size fruitbodies, dull orangish colors, unchanging white latex, and acrid taste help identify this uncommon milk cap. *L. manzanitae* has a beige to tan disc, unchanging white latex, and a slowly acrid taste, and grows under Madrone and manzanita. *L. xanthogalactus* can be similar in color, but usually has a pinker brown cap and latex that quickly turns from white to bright yellow. *L. rufus* has a dry, dull pinkish to reddish brown cap and an extremely acrid (but slow to develop) taste. *L. subviscidus, L. substriatus,* and *L. subflammeus* are all more brightly colored than "Acrid Orange". The Candy Cap (*L. rubidus*) is slightly paler and dry capped, have watery latex, and taste mild.

NOMENCLATURE NOTE: This species was included as *L. desjardinii* in Andrew Methven's volume on Lactarius in *Agaricales of California,* but that name was never officially published. The secotoid *Arcangeliella desjardinii* has since been transferred to *Lactarius,* making *L. desjardinii* the legitimate name for that species, leaving *L.* "Acrid Orange" without a name.

Lactarius deliciosus group

CAP: 3–12 (14) cm across, convex with an inrolled margin when young, becoming plane with a depressed center, margin often wavy and uplifted in age. Color variable, often mottled, zonate to obscurely zonate or water spotted. Orange, brownish orange to orangish buff, greenish orange to greenish orange-brown or with patchy greenish turquoise discoloration to all turquoise-green in age or when very young. Surface smooth, moist to dry. **GILLS:** Broadly attached to slightly decurrent, close to crowded, narrow, fragile. Bright orange, to orangish buff, staining deep red to wine red and then slowly turquoise-green when damaged or with a greenish cast, to all green in age. **LATEX:** Scant, bright neon orange, discoloring and staining tissue deep red, then slowly turquoise-green to deep green. **STIPE:** 2–7 cm long, 1–3 cm thick, equal, slightly enlarged or tapered toward base. Orange to orangish buff, with a whitish sheen when young, often paler at apex, developing greenish to turquoise-green stains, especially when damaged or in age. Surface smooth, dry to moist, not scrobiculate, or rarely so. **FLESH:** Thin to thick, firm, fragile and very grainy. Orange to buff, discoloring deep orangish red to wine red and then slowly green, paler orangish white in stipe center up into cap. Stipe can become hollow in age. **ODOR:** Indistinct. **TASTE:** Slowly slightly acrid. **SPORE DEPOSIT:** Creamy to pale yellow. **MICROSCOPY:** Spores 7–9 x 5.5–6.5 µm, ellipsoid, ornamented with isolated warts and ridges, forming a partial reticulum, up to 0.5 µm high.

ECOLOGY: Solitary, scattered, in small clusters or troops in duff under pine (especially Monterey and Bishop Pine).

Widespread, but most common in the San Francisco Bay Area. Fruiting in fall and winter.

EDIBILITY: Edible, but most people don't find it particularly delicious. The grainy texture is off-putting, and it lacks in the flavor department. Slow cooking helps a bit, and firm, younger fruitbodies are less grainy.

COMMENTS: This common *Lactarius* (upper left) can be distinguished from other species in this group by the growth with pines, orange latex that discolors deep red, strongly zonate (and often spotted) orangish cap, and extensive dark green staining. *Lactarius deliciosus* is a species complex in California, none of which are the true European *L. deliciosus*. Members of the group can be differentiated by their tree associations, latex color and oxidation reactions, and degree of green staining. *L. deliciosus* var. *areolatus* (lower left) grows with Shore Pine on the Far North Coast and differs by its scant, unchanging orange latex and light turquoise-green stains. *L. deliciosus* var. *piceus* (sensu CA; upper right) grows with Sitka Spruce and has an orange cap that readily discolors green and orange latex that slowly goes vinaceous purple (lower right). *L. aestivus* has a faintly zonate, orange cap and unchanging, bright orange latex, and grows with firs. *L. aurantiosordidus* has an orange-olive cap and dingy orange, unchanging latex. *L. rubrilacteus* has wine red to brownish red latex from the start. Be sure not to confuse them with the poisonous *L. xanthogalactus*; although it can have a vaguely similar cap color, it has white latex that turns bright yellow.

Lactarius aestivus

Nuytinck & Ammirati

CAP: 3–12 cm across, convex with an incurved margin when young, becoming plane with a depressed center or broadly vase shaped in age. Bright orange to peachy orange, paler in age or when dry, often zonate with alternating light and dark rings. Can develop slight greenish stains or discolorations in age. Surface moist to dry, smooth. **GILLS:** Broadly attached to slightly decurrent or even decurrent in age, close to crowded, some forking near stipe, partial gills numerous, broad, fragile. Bright orange to peachy orange or orangish buff. **LATEX:** Scant, bright neon orange, not discoloring, slowly staining tissue pale greenish to greenish gray. **STIPE:** 3–6 cm long, 1–2.5 cm thick, more or less equal or tapered slightly toward the base. Marbled orange, peachy orange to pale orangish buff, often scrobiculate, with bright orange spots. **FLESH:** Thin to moderately thick, firm, fragile. Bright orange to peachy orange, sometimes slowly staining pale greenish to greenish gray when damaged. Stipe stuffed with pale orange pith, becoming hollow in age. **ODOR:** Indistinct. **TASTE:** Mild or slowly slightly acrid. **KOH:** Orange on stipe, little change elsewhere. **SPORE DEPOSIT:** Creamy to pale yellow. **MICROSCOPY:** Spores 8.5–10 x 6.5–8 µm, ellipsoid, ornamented with irregular ridges, forming a partial reticulum, up to 0.5 µm high.

ECOLOGY: Solitary or scattered in moss, duff, or soil under true firs. Common under Grand Fir from coastal Sonoma County northward. Fruiting in fall and early winter.

EDIBILITY: Edible. In our opinion the best tasting of the "Delicious Milk Cap" group in California. The flesh is firmer and less grainy than the related species found under pine. Slow baking brings out a sweeter flavor and firms up the texture.

COMMENTS: The bright orange zonate cap, orange gills, scrobiculate stipe, neon orange latex, sparse greenish gray staining, and growth under Grand Fir help set this species apart from the other *Lactarius* with orange latex. It is most likely to be confused with *L. deliciosus* var. *areolatus*, which grows with Shore Pine and stains more bluish green. Other species in this group either have orange latex that changes to deep red or vinaceous, and discolor greenish more extensively. The poisonous *L. xanthogalactus* has a pinkish brown cap and white latex that becomes bright yellow.

MISAPPLIED NAME: *Lactarius deliciosus* var. *deliciosus.*

Lactarius aurantiosordidus

Nuytinck & S. L. Miller

CAP: 2.5–5 (7) cm across, convex with an incurved margin when young, becoming plane with a depressed center or broadly vase shaped. Color variable, orange to dingy orangish brown to olive-orange, often with dark turquoise to olive-green discoloration, especially in center of cap and in age. Zonate at times with darker concentric bands and/or mottled with darker areas. Surface moist to dry, smooth. **GILLS:** Broadly attached to slightly decurrent, close to crowded, narrow, fragile. Orange, ocher to orangish buff, developing turquoise to greenish olive discolorations, especially when damaged or in age. **LATEX:** Scant, dark dingy orange, not discoloring, slowly staining tissue turquoise-green. **STIPE:** 2–6 cm long, 0.7–2 cm thick, equal, enlarged slightly or tapering slightly at base. Orange to mottled orangish brown, discoloring turquoise-green to olive green. **FLESH:** Thin, firm, fragile. Orangish to orange-buff or greenish, slowly discoloring turquoise-green. Stipe fragile, stuffed when young, becoming hollow in age. **ODOR:** Indistinct. **TASTE:** Mild or slowly slightly acrid. **KOH:** Dingy orange on stipe, no reaction elsewhere. **SPORE DEPOSIT:** Creamy to pale yellow. **MICROSCOPY:** Spores 8.5–10 x 6.5–7.5 μm, broadly ellipsoid to ellipsoid, ornamented with isolated warts and ridges, forming a partial reticulum.

ECOLOGY: Solitary, scattered, or in groups on ground, in moss or duff under Sitka Spruce. Locally very common, but limited to coastal spruce forest from central Mendocino County northward. Fruiting abundantly in fall and early winter with scattered fruitings into spring.

EDIBILITY: Edible.

COMMENTS: This species was recently described from Redwood National Park; it can be told apart from the other *Lactarius* with orange latex by the small size, dingy orange, green-stained cap, unchanging orange latex, and growth with Sitka Spruce. It often grows alongside the similar-looking *L. deliciosus* var. *piceus*, which is usually larger, with brighter orange colors and bright orange latex that slowly turns deep red to vinaceous.

SYNONYM: *Lactarius deliciosus* var. *olivaceosordidus* (sensu CA).

Lactarius rubrilacteus

Hesler & A. H. Sm.

BLEEDING MILK CAP

CAP: 4–12 (15) cm across, convex with an inrolled margin when young, becoming broadly convex to plane with a depressed center, or uplifted in age. Color variable, orange-brown, pinkish orange, orangish buff to buff-brown, often mottled, zonate to obscurely zonate or water spotted, staining or discoloring greenish turquoise when very young or in age. Surface smooth, slightly viscid to dry. **GILLS:** Broadly attached to slightly decurrent, close to crowded, narrow, fragile, often forking near stipe. Pinkish orange, orangish buff to buff-brown, often with greenish stains, staining reddish brown and then slowly turquoise-green when damaged. **LATEX:** Scant, vinaceous red to reddish brown, discoloring and staining tissue reddish brown and then slowly turquoise-green, orange in stipe base only. **STIPE:** 2–8 cm long, 1–3 cm thick, equal or slightly tapered toward base. Pinkish orange to orange to buff, often with a pale bloom when young, developing greenish stains in age. Surface dry to moist, smooth, occasionally scrobiculate, often with orangish hairs at base. **FLESH:** Thin to thick, firm and very grainy. Orangish buff to reddish, discoloring reddish brown and then slowly greenish, paler buff in stipe center up into cap, stipe becoming hollow in age. **ODOR:** Indistinct. **TASTE:** Mild. **SPORE DEPOSIT:** Pale yellow. **MICROSCOPY:** Spores 7.5–9 x 6–7 μm, broadly ellipsoid to ellipsoid.

ECOLOGY: Solitary, scattered, or in small clusters on ground or in duff under Douglas-fir. Common throughout our range. Fruiting from late fall into early spring, most abundant in early winter.

EDIBILITY: The young fruitbodies are good edibles, but mature specimens have a disagreeable grainy texture. Slow baking or roasting helps keep the flesh texture palatable.

COMMENTS: This common *Lactarius* of Douglas-fir forest can be identified reliably by the dingy salmon to pinkish orange cap, greenish staining, and vinaceous red to reddish brown latex. Some members of the *L. deliciosus* group have orange latex that slowly turns to the same color as the fresh latex of *L. rubrilacteus*, but most are associated with other conifer hosts. The poisonous *L. xanthogalactus* has white latex that quickly changes to bright yellow.

Lactarius xanthogalactus

Peck

YELLOW-STAINING MILK CAP

CAP: 3–10 (12) cm across, convex when young, becoming broadly convex to plane with a depressed center, margin downcurved when young, slightly uplifted in age. Color variable, pinkish orange, peachy orange, orange-red to orangish brown, often zonate to obscurely zonate, mottled or water spotted, discoloring orange-brown in age. Surface smooth, slightly viscid to dry. **GILLS:** Broadly attached to slightly decurrent, close to crowded, narrow, fragile. Creamy, pinkish yellow to orangish buff, occasionally with orangish brown spots or stains in age. **LATEX:** Copious to scant, creamy white at first, quickly to slowly becoming neon yellow to pale yellow. **STIPE:** 2–7 (10) cm long, 0.6–2.5 cm thick, equal, slightly tapered or enlarged toward base. Creamy tan to pale orange, darkening in age. Surface dry, smooth, often with whitish hairs at base. **FLESH:** Thin to thick, firm, stipe stuffed when young, hollow in age. Creamy to creamy orange, discoloring yellow from latex. **ODOR:** Indistinct. **TASTE:** Slightly bitter-acrid. **SPORE DEPOSIT:** Pale yellow. **MICROSCOPY:** Spores 7–9 x 5.5–7 μm, broadly ellipsoid to ellipsoid, ornamented with scattered warts and irregular ridges and lines, forming a partial reticulum, 0.5 μm high.

ECOLOGY: Solitary, scattered, or in small clusters on ground, in moss or duff in mixed forests. Grows with hardwoods and conifers alike, and is especially common with Douglas-fir and live oak. Very common from southern California into Oregon. Fruiting from midfall through winter.

EDIBILITY: Poisonous.

COMMENTS: The pinkish orange, zonate, and often spotted cap, creamy to pale orange gills, and white latex that turns bright yellow are a distinctive set of features. The color change of the latex usually takes 5–10 seconds but can be much quicker or take longer depending on the age and dryness of the fruitbody. *Lactarius* "Acrid Orange" is similarly colored, but has an evenly colored, viscid to moist cap, unchanging white latex, and a more strongly acrid taste. *L. rubrilacteus* can look similar but has vinaceous red to reddish brown latex and develops greenish stains. Members of the *L. deliciosus* group have scant orange latex when first damaged and stain greenish. *L. resimus* and *L. scrobiculatus* both have white latex that changes to yellow, but have matted-tomentose to appressed-fibrillose caps. *L. alnicola* has a creamy to yellowish cap, a strongly acrid taste, and white latex that rarely changes to pale yellow.

MISAPPLIED NAME: Identified both as *Lactarius vinaceorufescens* and *L. chrysorrheus* in the past, but these are now known to be distinct, eastern North American species.

Lactarius alnicola

A. H. Sm.

GOLDEN MILK CAP

CAP: 5–15 (22) cm across, broadly convex and centrally depressed when young, becoming uplifted or funnel shaped. Margin inrolled at first, becoming uplifted and wavy. Creamy yellow, whitish, yellowish to yellowish orange, often with conspicuous, but sometimes inconspicuous concentric zonations. Surface slightly viscid to dry, margin finely pubescent when young, quickly becoming smooth overall. **GILLS:** Broadly attached to decurrent, narrow, fragile, close, often forking near stipe. Creamy, yellowish to pale yellowish orange, or pale pinkish. **LATEX:** Copious, thick, white to creamy, erratically turning or drying yellowish (see comments). **STIPE:** 2–6 cm long, 1–3.5 cm thick, stout, equal or tapering toward base. Creamy white to yellowish, often developing orangish stains, slightly to heavily scrobiculate, with darker grayish orange spots and pockmarks. Surface smooth to slightly pitted, dry to moist. **FLESH:** Firm, but brittle, white to creamy, discoloring pale yellowish, stipe hollow. **ODOR:** Indistinct. **TASTE:** Quickly (more rarely slowly) acrid. **KOH:** Slight orangish yellow on cap, no reaction elsewhere. **SPORE DEPOSIT:** White to creamy white. **MICROSCOPY:** Spores 5.5–7 x 7–9 µm, ellipsoid, ornamented with isolated warts and ridges, forming a partial reticulum, 0.5–1 µm high.

ECOLOGY: Solitary, scattered, or in large groups or clusters on ground under Coast Live Oak. Very common from Mendocino County southward, uncommon outside this range. Long fruiting period from midfall through spring.

EDIBILITY: Very acrid, possibly poisonous.

COMMENTS: This milk cap is one of our most common mushrooms, often fruiting by the hundreds in Coast Live Oak forest in early winter. It can be identified by the creamy to yellowish cap, pale gills, white latex, and strongly acrid taste. The concentric zonations on the cap can form multiple complete rings or are occasionally lacking entirely (entire cap then appearing evenly colored). Likewise, the pits on the stipe (scrobiculations) are usually present, but occasionally are absent or in other cases are hyperabundant, covering the entire stipe and even some of the cap! The color change of the latex is also quite variable; it has been reported to dry to yellow or slowly stain the flesh pale yellow, but we have rarely observed such reactions. *L. subvillosus* is a similar-looking species with a zonate, orangish buff to peachy orange cap that has a bearded margin when young, and pale pinkish buff gills; it grows in Tanoak-Madrone forests. *L. scrobiculatus* var. *pubescens* is a montane conifer associate with a pale yellow, appressed-fibrillose cap and white latex that quickly turns yellow.

Lactarius subvillosus

Hesler & A. H. Sm.

CAP: 5–11 cm across, convex, broadly convex to plane, centrally depressed, margin inrolled when young and covered in a band of dense fuzz composed of fine whitish to pale pinkish hairs. Fringed margin becomes less conspicuous as cap expands and sometimes disappears in age. Orangish buff to peachy orange, pinkish buff to yellowish buff, often with conspicuous concentric zonations of alternating dark and light rings when young, which blend together in age. Surface moist to slightly viscid, consisting of appressed fibrils, especially around margin, giving it a slightly scruffy look, becoming smooth or nearly so in age. **GILLS:** Slightly decurrent to decurrent, close to crowded, narrow, fragile. Pale pinkish to pinkish buff, discoloring slightly yellowish buff when damaged. **LATEX:** Scant, thin, creamy white, discoloring pale yellowish as it dries. **STIPE:** 3–7 cm long, 1.5–3.5 cm thick, equal or tapering slightly toward base. Off-white to pale pinkish or pinkish orange, slightly to heavily scrobiculate, with darker orangish spots and pockmarks. Surface smooth, moist to dry. **FLESH:** Firm, pale pinkish orange, not staining, stipe hollow. **ODOR:** Indistinct to slightly sweet. **TASTE:** Quickly and strongly acrid. **KOH:** Slight yellowing on all parts. **SPORE DEPOSIT:** Creamy yellow. **MICROSCOPY:** Spores 6–8 x 5–6 μm, ellipsoid, ornamented with scattered warts and irregular ridges and lines, forming a partial reticulum, up to 0.5 μm high.

ECOLOGY: Scattered or in groups on ground in duff of mixed Madrone, Tanoak, and Douglas-fir forest. Locally common from the Santa Cruz Mountains northward. Fruiting in fall and early winter.

EDIBILITY: Possibly poisonous.

COMMENTS: The zonate, orangish buff cap with a bearded margin when young, pale pinkish buff gills, creamy white latex, scrobiculate stipe, and acrid taste help identify this pretty *Lactarius*. *L. alnicola* looks similar, but has a paler cap that is only finely fringed (not bearded) around the margin, and mostly grows under live oak. *L. pubescens* has pinkish white cap with a heavily bearded margin and only grows under birch.

Lactarius resimus (sensu CA)

CAP: 5–20 (25) cm across, broadly convex and centrally deeply depressed when young, becoming broadly funnel shaped, margin tightly inrolled at first, finely bearded. Milk white to creamy when young, becoming pale dingy yellow, developing orange-brown stains in age. Surface finely pubescent around margin when young, becoming matted to appressed-fibrillose in a slime layer, disc smooth. **GILLS:** Broadly attached to decurrent, narrow, fragile, close to crowded, often forking near stipe. White to creamy when young, becoming pinkish yellow to pinkish buff, often with sordid yellowish brown stains in age. **LATEX:** Scant, white at first, quickly turning yellow. **STIPE:** 2.5–6 cm long, 1–3.5 cm thick, stout, equal or tapering toward the base. White to pale yellowish, developing dingy orange-brown stains in age, sometimes scrobiculate, with small grayish orange spots and pockmarks. Surface smooth to slightly roughened, moist to dry. **FLESH:** Firm, thick, white, becoming pale yellow from latex change, stipe stuffed with pith when young, becoming hollow. **ODOR:** Indistinct. **TASTE:** Mild to slowly, slightly acrid. **SPORE DEPOSIT:** Creamy white. **MICROSCOPY:** Spores 6–8 x 5–6 μm, broadly ellipsoid, ornamented with scattered warts and irregular ridges and lines, forming a partial reticulum, up to 0.5 μm high.

ECOLOGY: Solitary, scattered, or in small groups on ground in mixed coastal forests, usually around oaks or Chinquapin. Locally common throughout our range. Fruiting from early fall into winter.

EDIBILITY: Unknown.

COMMENTS: The white to pale yellow cap with a viscid-bearded margin, white to pinkish gills, white latex that quickly turns yellow, and preference for Chinquapin help identify this species. Older specimens can easily be confused for *L. scrobiculatus* var. *canadensis*, which also has white latex that quickly turns yellow, but has a yellowish cap even when young and a coarsely scrobiculate stipe. *L. controversus* has a whitish cap (often with purplish stains), pinkish gills, unchanging white latex, and a slowly acrid taste, and grows under aspens or cottonwoods (*Populus*) trees. *L. pubescens* var. *betulae* has a pinkish white cap with a bearded margin and strongly acrid taste, but only grows with birch trees.

Lactarius scrobiculatus group

CAP: 6–15 (20) cm across, broadly convex and centrally depressed when young, becoming plane to funnel shaped. Margin tightly inrolled at first, finely bearded, often covered with clear droplets when young. Yellowish, yellowish orange to brownish orange, sometimes with inconspicuous concentric zonations. Surface viscid when wet, especially around margin, matted-tomentose to appressed-fibrillose, giving it a roughened feel. **GILLS:** Attached to decurrent, close, narrow, fragile. Pale yellow to pale yellowish orange, discoloring dingy orangish brown. **LATEX:** Creamy white, drying yellowish. **STIPE:** 3–8 cm long, 1–4 cm thick, stout, equal. Pale yellowish to yellowish orange, heavily scrobiculate, with darker orangish spots and pockmarks. Surface rough, pitted, moist to dry. **FLESH:** Firm, off-white to pale yellowish, stipe hollow, usually with orangish discolorations. **ODOR:** Indistinct. **TASTE:** Quickly and strongly acrid. **KOH:** No reaction. **SPORE DEPOSIT:** Creamy white. **MICROSCOPY:** Spores 7–9.5 x 5.5–7.5 µm, broadly ellipsoid, ornamented with scattered amyloid warts and irregular ridges and lines forming a partial reticulum, up to 0.5 µm high.

ECOLOGY: Scattered in groups or loose clusters on ground under Sitka Spruce on the Far North Coast. Common to locally very common from Mendocino County northward. Fruiting from early fall into winter.

EDIBILITY: Not edible, possibly poisonous.

COMMENTS: This is a confusing species complex; the description above pertains to the common form of L. "scrobiculatus" found on the Far North Coast. It can be told apart from similar milk caps by the yellowish, funnel-shaped cap, coarsely scrobiculate stipe, white latex, and strongly acrid taste. L. scrobiculatus var. canadensis is slightly smaller and paler capped and has white latex that quickly changes to yellow and a mild to slightly acrid taste; it is commonly found in coastal Mendocino County under Grand Fir and Western Hemlock. L. resimus is similar, but has a white to pale yellow cap with a finely bearded margin, white latex that turns yellow, and a mild to slightly peppery taste. L. repraesentaneus can look almost identical to L. scrobiculatus, but has latex that slowly stains lilac-purple; rare in California, it is known from only a few locations on the Far North Coast under Sitka Spruce.

Lactarius controversus

(Pers.) Pers.

POPLAR MILK CAP

CAP: 6–15 (20) cm across, broadly convex, centrally depressed, becoming broadly funnel shaped in age, margin inrolled and finely pubescent when young, becoming uplifted in age. White, dingy white to pale pinkish tan, often with vinaceous to lavender-brown stains. Surface slightly viscid to dry, matted-fibrillose, especially around margin, becoming smooth or nearly so. **GILLS:** Broadly attached to slightly decurrent, narrow, fragile, crowded, partial gills numerous. Pale pinkish, creamy pink to pale pinkish buff, slightly darker in age. **LATEX:** Copious, white, unchanging and not staining. **STIPE:** 2.5–7 cm long, 1.5–3 cm thick, equal or tapering slightly toward base, fragile, hollow. Dingy white to pale pinkish white. Surface smooth, moist to dry. **FLESH:** Firm, whitish, not staining. **ODOR:** Indistinct. **TASTE:** Slowly, strongly acrid. **SPORE DEPOSIT:** Pale yellow. **MICROSCOPY:** Spores 6–7.5 x 4.5–5.5 µm, ellipsoid, ornamented with isolated warts and ridges, forming a partial reticulum, up to 0.5 µm high.

ECOLOGY: Scattered or in small clusters on ground, often in riparian zones under aspens or cottonwoods (Populus spp.). Rare on the coast, mostly found along major rivers in northern part of our area and inland areas of California. Fruiting in fall and early winter.

EDIBILITY: Not edible, possibly poisonous.

COMMENTS: The whitish cap, pinkish gills, white latex, slowly acrid taste, and growth under cottonwoods help identify this milk cap. L. resimus has a white to pale yellow cap with a bearded margin, creamy white gills that can become pinkish, white latex that quickly turns yellow, and a mild to slightly peppery taste. L. alnicola grows under oaks and has a yellowish cap. L. scrobiculatus has a yellowish cap and heavily scrobiculate stipe; it is found under northern conifers.

Lactarius pubescens var. betulae

(A. H. Sm.) Hesler & A. H. Sm.

BEARDED BIRCH MILK CAP

CAP: 3–10 (13) cm across, convex, broadly convex to plane, disc centrally depressed. Margin finely to coarsely bearded and tightly inrolled to stipe when young, staying inrolled and coarsely bearded until cap fully expands, sometimes becoming ruffled or wavy in age. Pale pinkish buff, pinkish cinnamon to tan on disc, lighter toward margin when young, sometimes with inconspicuous concentric zonations. Surface smooth to slightly hairy on disc, coarsely hairy around margin, becoming matted-tomentose to nearly smooth, slightly viscid to dry. **GILLS:** Broadly attached to slightly decurrent, narrow, fragile, close to crowded. Creamy white, pale pinkish to pale pinkish orange. **LATEX:** White, often scant, sometimes drying yellowish. **STIPE:** 2–7 cm long, 1–2.5 cm thick, stout, equal. Creamy white, pinkish to pinkish yellow, often pinker at apex. Surface smooth, moist to dry. **FLESH:** Firm, fragile, moderately thin. Pale pinkish white to creamy, stipe stuffed when really young, becoming hollow. **ODOR:** Indistinct. **TASTE:** Slowly, strongly acrid, usually taking 15–30 seconds. **SPORE DEPOSIT:** Creamy yellow. **MICROSCOPY:** Spores 6–7.5 x 4.5–6 μm, broadly ellipsoid to ellipsoid, ornamented with scattered warts and irregular ridges and lines, forming a partial reticulum, 0.5–1 μm high.

ECOLOGY: Scattered, in groups or small clusters on lawns under introduced birch trees. Fruiting from late summer through fall, but can appear in spring as well. Locally common in urban areas with lots of planted birch.

EDIBILITY: Not edible, possibly poisonous.

COMMENTS: The pale pinkish white cap with a coarsely bearded margin, strongly acrid taste, and growth under birch help identify this species. It is often found growing in large troops mingling with other common birch associates including *Paxillus cuprinus* and *Leccinum* species. None of the following look-alikes grow with birch: *Lactarius torminosus* var. *nordmanensis* is a bearded milk cap found in the California mountains; it has a pinker cap, grows under willow, aspen, or cottonwood, and has larger spores (7.5–9.5 x 6–7.5 μm). *L. subvillosus* has a peachy orange cap and scrobiculate stipe, and grows in mixed Tanoak-Madrone and Douglas-fir woods. *L. resimus* has a bearded, white to pale yellow cap, white latex that turns yellow, and a mild to slightly peppery taste. Members of the *L. scrobiculatus* group have yellowish, centrally depressed to funnel-shaped caps, strongly scrobiculate stipes, white to pale yellow latex, and a strongly acrid taste; they are found under northern conifers. *L. alnicola* grows under live oak and has a yellowish cap that is not strongly bearded around the margin.

Lactarius argillaceifolius var. *megacarpus*
Hesler & A. H. Sm.

GIANT GRAY MILK CAP

CAP: 5–15 (30) cm across, convex to broadly convex when young, becoming plane to undulating, and centrally depressed. Margin even, inrolled when young, becoming wavy in age. Color variably, lilac-gray, pinkish gray to creamy gray when young, becoming grayish orange to grayish brown, often with orangish stains or blotches. Surface slightly viscid to moist, smooth to wrinkled or slightly roughened. **GILLS:** Broadly attached to slightly decurrent, broad, fragile, close to subdistant. Creamy to pale buff when young, becoming pale orangish buff to orange-brown, staining orange-brown. **LATEX:** Copious when fresh, creamy white to pale yellow, staining tissue brownish. **STIPE:** 2.5–12 cm long, 2–5 cm thick, equal or enlarged at base. Creamy white to buff, developing orange-brown discolorations. Surface smooth, moist to dry. **FLESH:** Firm, thick, whitish to ivory, stipe stuffed to hollow. **ODOR:** Indistinct. **TASTE:** Slowly, strongly acrid. **KOH:** Unknown. **SPORE DEPOSIT:** Creamy yellow. **MICROSCOPY:** Spores 7.5–9.5 x 6–7.5 μm, broadly ellipsoid to ellipsoid, ornamented with scattered warts and a broken or partial reticulum.

ECOLOGY: Solitary or in groups and clusters on ground under live oak and occasionally Tanoak. Common from Mendocino County southward; rarer to the north. Fruiting in late fall and winter.

EDIBILITY: Unknown.

COMMENTS: Few other milk caps are as large and stocky as this species; the combination of the lilac grayish cap, creamy gills that become orangish brown, thick whitish latex, growth under oaks, and acrid taste make this a very distinctive species. Similar species include *L. kauffmanii*, which has a viscid, dark grayish brown to grayish tan cap and grows under Western Hemlock and Grand Fir. *L. californiensis* is generally smaller and produces white latex that stains the tissues lilac-purple. *L. circellatus* var. *borealis* has a zonate and often spotted, lilac-gray cap, buff to pale ocher gills, and white, unstaining latex. *L. pallidiolivaceus* is smaller and has a dry, finely velvety, beige to light gray cap and white latex that stains the tissues pale orange.

NOMENCLATURE NOTE: *Lactarius argillaceifolius* is a distinct species found in eastern North America; this variety should be elevated to species rank.

Lactarius californiensis

Hesler & A. H. Sm.

CAP: 4–12 (14) cm across, convex with an incurved margin when young, becoming broadly convex to plane, often with a depressed center. Margin lobed, uplifted or wavy in age. Color variable, dark gray, gray to grayish white when young and patchy at times if covered with leaves, becoming gray to dark brownish gray, pale with orange-brown discolorations or patchy brownish gray stains in age. Occasionally with dark concentric zones or "water spots" when young. Slowly staining lilac-purple when damaged. Surface smooth, viscid and slimy when wet, moist to shiny when dry. **GILLS:** Broadly attached to slightly decurrent, close to subdistant, narrow, fragile, partial gills numerous. Creamy white, ivory to buff at first, becoming dingy buff to yellowish tan, often with brownish stains in age. Staining lilac-purple when damaged, then slowly going purplish brown to brownish orange. **LATEX:** Copious, bright white to creamy white, staining tissue lilac-purple. **STIPE:** 4–12 cm long, 1.5–4 cm thick, equal, pinched or slightly misshapen toward base. Creamy white, ivory to pale buff, often developing orange stains in age, staining dingy lilac-purple when handled or bruised. Surface smooth, moist to slippery or sticky, but not viscid. **FLESH:** Thick, firm, ivory to pale buff, stipe stuffed, solid when young, becoming soft but rarely hollow in age. All parts can slowly stain pale dingy lilac-purple. **ODOR:** Indistinct. **TASTE:** Mild to slowly slightly acrid. Previous reports say it tastes acrid, but all the collections we have tasted are mild to just slightly acrid. **KOH:** Pale orange to orange-yellow on cap and stipe. **SPORE DEPOSIT:** Creamy to yellowish. **MICROSCOPY:** Spores 7–9 x 6–7 µm, ellipsoid, ornamented with isolated warts and ridges forming a partial reticulum, up to 0.5 µm high. Cap cuticle an ixocutis.

ECOLOGY: Solitary or scattered in troops in moss or duff in mixed forests, usually under Tanoak. Very common from Sonoma County to southern Humboldt County, uncommon south to Santa Cruz County. Fruiting in early fall and winter.

EDIBILITY: Unknown.

COMMENTS: The variably grayish, viscid cap, creamy white gills that discolor ocher in age, white latex that stains lilac-purple, and growth with Tanoak help identify this milk cap. *L. pallescens* is very similar but paler (milk white to light gray) when young, and grows mostly with conifers. However, discolored older specimens of these two species can be tough to tell apart; see comments under *L. pallescens* for distinguishing features. *L. montanus* (=*L. uvidus* var. *montanus*) has a pinkish gray cap that turns green in KOH and whitish gills that stain purple, and grows in the western mountains. *L. aspideoides* is another purple-staining montane species with a viscid yellow cap and stipe. Other similar gray milk caps (*L. kauffmanii*, *L. argillaceifolius* var. *megacarpus*, and *L. pseudomucidus*) do not stain violet-purple.

MISAPPLIED NAMES: *L. pallescens* and *L. uvidus* have been indiscriminately applied to any violet-staining, grayish-capped *Lactarius* in California.

Lactarius pallescens

Hesler & A. H. Sm.

CAP: 3–10 cm across, convex with an incurved margin when young, becoming broadly convex to plane, often with a depressed center. Bright white when young, becoming dingy white to pale grayish white or pale gray in age, occasionally with faint concentric zones or "water spots." Staining lilac-purple when damaged and often with orange to orange-brown stains in age. Surface smooth, viscid, slimy when wet, moist to shiny when dry. **GILLS:** Broadly attached to slightly decurrent, close to subdistant, narrow, fragile, partial gills numerous. White to off-white, becoming creamy white, often with orangish spots in age, slowly staining lilac-purple when damaged and then slowly going purplish brown to brownish orange. **LATEX:** Copious, bright white, staining the tissue lilac-purple. **STIPE:** 3–10 cm long, 0.8–2 cm thick, equal or enlarged slightly toward base. White to off-white, often developing orange stains in age, slowly staining lilac-purple when handled or bruised. Surface smooth, viscid or slimy. **FLESH:** Thick to thin, firm, fragile, white to dingy white, stipe stuffed when young, can becoming hollow in age, all parts slowly staining lilac-purple. **ODOR:** Indistinct. **TASTE:** Mild to slowly slightly acrid. **KOH:** Pale yellow on cap. **SPORE DEPOSIT:** White to creamy. **MICROSCOPY:** Spores 9–10.5 x 7–9 µm, ellipsoid, ornamented with scattered warts and irregular ridges and lines, forming a partial reticulum, 0.5–1 µm high. Cap cuticle an ixotrichoderm (erect cells in a gelatinized matrix) when young and an ixolattice in age (collapsing and becoming more spread out).

ECOLOGY: Solitary or scattered in moss or duff in mixed forests, usually with Western Hemlock. Uncommon in California, more frequent north. Fruiting in fall and early winter.

EDIBILITY: Unknown; may be toxic.

COMMENTS: The white viscid cap and stipe and pale gills that all slowly stain lilac-purple when damaged and growth with conifers are helpful field marks in identifying this mushroom. There has been past confusion between L. pallescens and the similar L. californiensis, which has a darker gray to grayish white cap, grows with Tanoak, and often has orangish stains on the stipe. Microscopically, L. pallescens has larger spores, and the cap cuticle is an ixotrichoderm when young.

Lactarius circellatus var. borealis

Hesler & A. H. Sm.

CAP: 3–8 cm across, convex to broadly convex, becoming plane and centrally depressed, margin inrolled when young, often undulating in age. Gray, lilac-gray, pale vinaceous gray to grayish brown, often marbled, and with watery spots and paler concentric zonations. Surface smooth, slightly viscid to dry. **GILLS:** Broadly attached to slightly decurrent, narrow, fragile, close to subdistant, often forking near stipe, partial gills numerous. Pale pinkish buff, pale buff-ocher to orangish, unstaining. **LATEX:** Copious, white, unchanging. **STIPE:** 2–6 cm long, 0.7–2 cm thick, often tapered toward base and swollen in middle. Creamy white to pale orangish buff, discoloring orangish when handled. Surface smooth, moist to dry. **FLESH:** Firm, thick to thin, whitish, unchanging. Stipe stuffed to hollow. **ODOR:** Indistinct. **TASTE:** Slowly but strongly acrid. **KOH:** Unknown. **SPORE DEPOSIT:** Creamy to pale yellow **MICROSCOPY:** Spores 7–9 x 5–6.5 µm, broadly ellipsoid, ornamented with isolated warts and ridges, forming a partial reticulum, up to 0.5 µm high.

ECOLOGY: Solitary or scattered on ground in mixed coniferous-hardwood forests. Rare; known from Marin County north into southern Oregon. Fruiting in fall.

EDIBILITY: Unknown.

COMMENTS: The small to medium-size fruitbodies, violet-gray to gray-brown cap with zonations and spots, pinkish buff to pale ocher gills, unchanging white latex, and acrid taste help identify this rare milk cap. L. argillaceifolius var. megacarpus is much larger and has a dingy lilac-gray to grayish cap that becomes blotched with ocher stains, and creamy white latex that stains the gills brownish. L. pallidiolivaceus has a dry, finely velvety, blotchy beige-gray cap and has white latex that slowly stains the tissues orange.

Lactarius kauffmanii

Hesler & A. H. Sm.

CAP: 4–15 cm across, convex to plane, occasionally with a slightly depressed center, margin inrolled when young. Dark grayish brown to grayish tan, often fading and becoming more brownish as it ages. Surface smooth, viscid when wet, becoming shiny and often with debris stuck to it if dry. **GILLS:** Broadly attached to slightly decurrent, narrow when young, becoming broad in age, fragile, close to crowded, often forking near stipe. Creamy white to pale ocher-white, developing dingy tan, orangish or ocher discoloration as it ages. **LATEX:** Copious, white to dingy white, discoloring grayish as it dries, staining tissue orangish tan. **STIPE:** 4–12 cm long, 1–3 cm thick equal, tapering toward base or with a swollen middle. Dingy white to creamy white, becoming orangish tan and discoloring dingy tan, orangish or ocher in age. Surface smooth, moist to tacky but not viscid. **FLESH:** Firm, brittle, creamy white, slowly staining orange to brownish orange, stipe hollow in age. **ODOR:** Indistinct. **TASTE:** Slowly acrid. **KOH:** Bleaching cap color to pale yellowish, no change elsewhere. **SPORE DEPOSIT:** White to creamy. **MICROSCOPY:** Spores 8–9.5 x 6–7.5 μm, broadly ellipsoid to ellipsoid, ornamented with irregular ridges and lines forming a partial reticulum, 0.5–1 μm high.

ECOLOGY: Solitary, scattered, or in troops on soil, in moss or duff under Western Hemlock and Grand Fir. Very common in the northern part of our range, occurring south to Sonoma County. Fruiting in fall and early winter.

EDIBILITY: Unknown.

COMMENTS: The viscid, grayish brown cap, creamy white gills, and stipe that become stained dingy ocher-tan in age, and the growth under Western Hemlock or Grand Fir are the best features to identify this mushroom. *L. pseudomucidus* is generally smaller and has a viscid, dark to pale gray or grayish tan cap and stipe, and white gills. *L. argillaceifolius* var. *megacarpus* has large fruitbodies with violet-gray to dingy gray caps, and grows under live oak. *L. caespitosus* is a montane species with a grayish brown to brown cap, pale yellow to grayish yellow gills that stain light brown, white latex that slowly turns yellowish, and a strongly acrid taste. Also compare with the purple-staining species *L. californiensis*.

Lactarius pseudomucidus

Hesler & A. H. Sm.

SLIMY MILK CAP

CAP: 2–8 (10) cm across, convex with an incurved margin when young, becoming broadly convex to plane, often with a depressed center and an uplifted margin in age. Dark gray to dark grayish brown, often paler around margin in age. Surface smooth, viscid when wet, moist to shiny when dry. **GILLS:** Broadly attached to slightly decurrent, close to crowded, narrow, fragile. White to off-white, becoming creamy white to pale yellowish buff, staining brownish when damaged. **LATEX:** Copious to scant, bright white to watery white, drying pale yellow and staining gills brownish. **STIPE:** 3–10 cm long, 0.5–1.5 cm thick, equal or enlarged slightly toward base, at times with a crease or fold. Dark gray to brownish gray, often concolorous with or slightly lighter than cap, paler towards base and often off-white at extreme base, usually a distinct pale zone at apex. Surface smooth, viscid or tacky. **FLESH:** Thin, fragile, off-white to gray, stipe stuffed when young, becoming hollow in age. **ODOR:** Indistinct. **TASTE:** Acrid. **SPORE DEPOSIT:** Creamy white to pale yellow. **MICROSCOPY:** Spores 7–9 x 6–7 μm, broadly ellipsoid to ellipsoid, ornamented with scattered warts and irregular ridges
and lines forming a partial to nearly complete reticulum, 0.5–1 μm high.

ECOLOGY: Solitary or scattered in moss or duff under conifers, especially under Western Hemlock and Grand Fir. Very common in northern part of our range, rare south of Mendocino County. Fruiting in fall in north, later into winter southward.

EDIBILITY: Unknown.

COMMENTS: The viscid, dark gray to light gray cap and stipe, contrasting white to creamy white gills, white latex, acrid taste, and growth under Western Hemlock and Grand Fir help distinguish this species from among other gray-capped *Lactarius*. Older specimens can look quite similar to *L. kauffmanii*, which is usually found growing in the same habitat. However, *L. kauffmanii* has a dry to moist (never viscid) creamy white to pale buff stipe that is concolorous with the gills. *L. glutigriseus* is a very similar montane species that is a bit more orangish brown. *L. californiensis* has a viscid gray cap and white latex that stains its tissues lilac-purple.

Lactarius pallidiolivaceus

Hesler & A. H. Sm.

CAP: 2–10 (12) cm across, convex, plane, wavy, lobed or misshapen. Color variable, often patchy, can be or have a mix of dark gray to grayish olive, pale gray, grayish orange to creamy white. Typically patchy gray to gray-olive with paler spots and/or blotches with orange discoloration or stains. Surface dry, finely velvety, smooth to slightly wrinkled. **GILLS:** Attached, subdistant to close, at times crowded, often with cross veins, wrinkled or distorted. Creamy to pinkish white when young, becoming pale orange or peachy orange, developing pale orange stains when damaged. **LATEX:** Copious to scant, white, staining tissue pale orange to orangish brown. **STIPE:** 2–7 cm long, 0.5–2 cm across, equal or tapered toward base. Creamy white to pale orangish, staining orange to orange-brown when handled. **FLESH:** Thin to thick, firm, white to creamy white, staining orange to reddish brown when cut. Stipe stuffed when young, at times becoming hollow in age. **ODOR:** Indistinct. **TASTE:** Mild. **SPORE DEPOSIT:** Pale yellow. **MICROSCOPY:** Spores 7.5–9 x 7.5–9 μm, subglobose to globose, ornamented with scattered warts and irregular ridges and lines, forming a partial to complete reticulum, 0.5–1.5 μm high.

ECOLOGY: Solitary or scattered on ground in mixed forest, probably growing with Tanoak. Uncommon to rare in most of our area. Fruiting from early to midfall on the North Coast, into early winter in the San Francisco Bay Area.

EDIBILITY: Unknown.

COMMENTS: The dry, often misshapen, beige to gray cap, white latex that stains the tissue pale orange, and cream to pale ocher gills are distinctive identifying features of this milk cap. Specimens of *L. pallidiolaviceus* with distorted gills can be nearly indistinguishable from the secotioid species *Lacatarius (Arcangeliella) desjardinii*. Contorted fruitbodies are occasionally even found in the same patch as "normal" gilled specimens, which raises the question of whether the two should be considered distinct species. *L. californiensis* has a smooth, viscid, grayish cap, gills and stipe that discolor dingy ocher in age, and latex that stains its tissues violet. *L. circellatus* var. *borealis* has a darker, zonate, lilac-gray cap.

Lactarius fallax var. concolor

A. H. Sm. & Hesler

CHOCOLATE MILK CAP

CAP: 2.5–6 cm across, convex to plane, at times with a depressed center and/or a sharp umbo. Margin incurved when young, can become ruffled, wavy, or uplifted in age. Dark brown to brown, or deep reddish brown. Surface velvety, dry to slightly moist, finely to coarsely wrinkled. **GILLS:** Broadly attached to slightly decurrent, close to crowded, narrow, fragile. White, creamy, to creamy yellow or pale buff-yellow, discoloring pale orange-brown when damaged. **LATEX:** Copious to scant, white, slowly staining tissue orange-brown. **STIPE:** 3–10 cm long, 0.5–1.5 cm thick, equal or enlarged slightly and often curved toward base. Dark brown to brown or grayish brown, paler toward base where it can be off-white to pale buff. **FLESH:** Thin, fragile, off-white to buff-brown, stipe stuffed with buff pith, can become hollow in age. **ODOR:** Indistinct. **TASTE:** Mild. **KOH:** No reaction. **SPORE DEPOSIT:** Creamy white to pale yellowish. **MICROSCOPY:** Spores 7–9 x 7–9 μm, subglobose to globose, ornamented with irregular ridges and lines, forming a partial reticulum, 1–2 μm high.

ECOLOGY: Solitary or scattered on ground or often on well-rotted logs and stumps under conifers, especially Sitka Spruce. Locally common from central Mendocino County northward. Fruiting from fall into winter.

EDIBILITY: Edible and good, but often thin fleshed and rarely found in large numbers.

COMMENTS: The velvety, wrinkled, dark brown cap and the creamy white gills that contrast strongly with the brown stipe make this a distinctive and easily recognized species. *L. fallax* var. *fallax* is identical in all respects except that the edge of each gill is dark brown (marginate). Darker forms of *L. pallidiolivaceus* can look vaguely similar but are usually more grayish or olive-gray in color and thicker fleshed, and grow under Tanoak.

Lactarius turpis

(Weinm.) Fr.

CAP: 4–10 (15) cm across, convex with an inrolled margin when young, often staying incurved well into maturity, becoming plane with a depressed center or slightly uplifted in age. Dark dingy olive-green, olive-brown to olive-orange, often mottled, spotted or streaked. Surface slightly viscid to dry, wrinkled to pitted at times, scaly-fibrillose to radially fibrillose, margin tomentose when young. **GILLS:** Broadly attached to slightly decurrent, close to crowded, often forking near stipe. Creamy white to pale orangish when young, becoming dingy, dirty cream to greenish gray, developing olive-brown spots and stains in age. **LATEX:** Copious, thick, white, drying pale greenish, staining gills olive-brown. **STIPE:** 2–6 cm long, 1–4 cm thick, equal or tapered and pinched toward base. Olive, olive-brown to dingy olive-yellow, often paler toward apex; sometimes scrobiculate, with darker spots. Surface smooth, slightly viscid to dry. **FLESH:** Thin to thick, firm, fragile. Whitish, ivory-gray to pale dingy yellow, slowly darkening olive-tan. Stipe flesh stuffed with whitish pith when young, becoming hollow and dingy olive-brown. **ODOR:** Indistinct. **TASTE:** Slowly acrid. **KOH:** Violet-purple on cap. **SPORE DEPOSIT:** Creamy to pale yellow. **MICROSCOPY:** Spores 7–9 x 4.5–6 µm, broadly ellipsoid to ellipsoid, ornamented with scattered warts and irregular ridges and lines, forming a partial reticulum, 0.5 µm high.

ECOLOGY: Solitary, scattered, in troops or clusters on ground under introduced birch trees. Uncommon in California, but very common in the Pacific Northwest. Like most introduced birch associates, it often fruits in late summer or early fall, even before rains arrive.

EDIBILITY: Not edible, likely toxic.

COMMENTS: The dark, dingy olive-green to olive-brown cap and stipe, stout stature, and growth under introduced birch trees help identify this milk cap. *L. pubescens* var. *betulae* (also introduced) is far more common under birches in California, but as you go north, it is largely replaced by *L. turpis*. A very similar native species is *L. olivaceoumbrinus*. It occurs on the North Coast under conifers, especailly Sitka Spruce.

SYNONYMS: *Lactarius necator, Lactarius plumbeus.*

12 • Waxy Caps

There are two major groups in this section: the medium-size to large, often dull-colored, mycorrhizal *Hygrophorus* and the smaller, slender, often brightly colored *Hygrocybe*-like taxa. Many familiar waxy caps from this latter group are now placed in smaller genera including *Cuphophyllus*, *Neohygrocybe*, *Porpolomopsis*, and *Gliophorus*.

Hygrocybe, for the most part, are brightly colored, with lots of red, orange, and yellow species. In addition, they often have dry to thinly viscid caps and dry to slightly greasy stipes (but not truly viscid). *Gliophorus*, on the other hand, have extremely viscid caps and stipes, which often makes them tough to handle because they are so slippery. Due to a nomenclature issue, *Cuphophyllus* is the correct generic name for the mushrooms formerly known as *Camarophyllus*. They can be recognized by their decurrent, widely spaced gills and frequently drab colors. *Neohygrocybe* are also quite drab and have fragile beige to light brown fruitbodies that often stain pale pinkish and sometimes dark brown to blackish after being damaged. *Porpolomopsis* is represented by a single species in California, easily recognizable by its splitting, umbonate cap and pink color. In addition to these genera, the wood-dwelling, omphalinoid mushrooms *Chrysomphalina* and *Chromosera* are actually closely related to *Gliophorus*. The bryophyte-associated *Arrhenia* and the basido-lichen *Lichenomphalia* are also in *Hygrophoraceae* as is the drab gray *Ampulloclitocybe*.

Many of the names we are using for waxy caps in California are of European origin. In many cases, genetic evidence suggests that our species are distinct, however, much more work is needed to clarify this group.

The ecological roles of these genera are somewhat unclear. Research by Seitzman et al. (2011) suggests that many members of the Hygrophoraceae that were formerly thought to be saprobic are likely in biotrophic relationships with understory plants and/or bryophytes.

Important identification characters to note include the colors of all surfaces (subtle differences in color can be very important), stature, and texture of the cap and stipe (especially degree of viscidity). For the mycorrhizal *Hygrophorus*, be sure also to note the identity of nearby trees.

Hygrophorus camarophyllus

(Alb. & Schwein.) Dumée

CAP: 3–10 cm across, rounded to convex, often broadly umbonate, with an inrolled margin when young, becoming broadly convex to plane and then uplifted, rarely funnel shaped in age. Dark gray to dark grayish brown, usually with dark brown or blackish radial streaking. Surface appressed-fibrillose, dry or slightly tacky if wet. **GILLS:** Broadly attached to decurrent, well spaced, thick, broad, with short partial gills, greasy feeling. Contrastingly white at first, becoming slightly creamy or grayish with a pale edge. **STIPE:** 2–10 cm long, 1–3 cm thick, more or less equal or tapering toward base. Dark to light beige or grayish, usually with longitudinal brownish streaks, often with darker blotches at gill attachment. Surface appressed-fibrillose to smooth, dry. **FLESH:** Moderately thick, firm, fibrous. White. **ODOR:** Indistinct. **TASTE:** Indistinct. **SPORE DEPOSIT:** White. **MICROSCOPY:** Spores 7–10 x 4–6 μm, ellipsoid, smooth, inamyloid, colorless.

ECOLOGY: Solitary or scattered on ground, in duff or moss in mixed conifer and Tanoak forests on the North Coast. Rather uncommon, fruiting in late fall and winter.

EDIBILITY: Edible.

COMMENTS: The robust stature; dry, streaked grayish brown cap; widely spaced, decurrent, white gills; and dry, gray stipe are distinctive. *H. marzuolus* (sensu CA) is a common spring mushroom in the mountains, often fruiting on the edges of melting snow banks; it is slightly paler and often has a disagreeable odor. *H. atramentosus* is quite similar, but has a viscid, dark gray cap with a blackish disc, creamy white gills, and a gray, longitudinally streaked stipe; it is rare in our area, known only from a few locations on the Far North Coast. *H. calophyllus* is also similar, but is much rarer in our area; it has a viscid, dark grayish brown cap and stipe and pinkish gills when young. *Clitocybula atrialba* looks quite similar, but has a distinctly scurfy, slender stipe; it grows on or around decaying wood.

Hygrophorus olivaceoalbus (sensu CA)
SHEATHED WAXY CAP

CAP: 2–7 cm across, rounded-conical with an incurved margin at first, becoming broadly convex to plane, occasionally uplifted in age. Brown with a darker center and darker streaks, lighter brown outward with a slightly darker disc in age. Surface viscid, covered with glutinous slime, over appressed fibrils. **GILLS:** Broadly attached to slightly decurrent, subdistant to distant, narrow, waxy to greasy. White. **STIPE:** 2–8 cm long, 0.7–1.5 cm thick, tall, slender, often thicker in middle, with a tapered base. White base color with irregular pattern of sheathing brown fibrils or chevrons below a glutinous-fibrous, upward-flaring ring, white above ring. Surface viscid, silky-fibrillose, covered with glutinous slime. **PARTIAL VEIL:** Glutinous-covered fibrils, leaving a dry, smooth annular zone at apex. **FLESH:** Thin, soft, fibrous in stipe. White. **ODOR:** Indistinct. **TASTE:** Indistinct. **SPORE DEPOSIT:** White. **MICROSCOPY:** Spores 10–12 x 6–7 μm, ellipsoid, smooth, inamyloid, colorless.

ECOLOGY: Solitary, scattered, or in small clusters on ground, in moss or thick duff in northern coniferous forest, always with Sitka Spruce. Locally common on the Far North Coast. Fruiting in winter.

EDIBILITY: Edible.

COMMENTS: This beautiful waxy cap can be recognized by the viscid brown cap, pale gills, viscid white stipe ornamented with brown fibrils, and growth with Sitka Spruce. It doesn't fit the European concept (as per usual) and probably needs a new name. Based mostly on spore size, Largent used the name *Hygrophorus persoonii* for this taxon in his monograph on Hygrophoraceae of California. However, that name refers to a European species that grows with oaks. *H. hypothejus* has a similar stature and can be brownish, with a slimy cap and stipe, but has (or soon develops) yellow, orange, or reddish tones; it grows with pines. *H. inocybiformis* has a dry cap covered with gray scaly fibrils and a gray scaly stipe with a silky partial veil; it is rare in California, restricted to the Far North Coast. *H. pustulatus* has a viscid, grayish cap and a dry white stipe with grayish scurfs.

Hygrophorus hypothejus (sensu CA)

CAP: 2–8 cm across, broadly conical to convex with a inrolled margin at first, becoming broadly convex to plane, sometimes umbonate, often uplifted and wavy in age. Color extremely variable, from dark olive-brown, yellow-brown, yellow-orange to orange-red, most often some combination, with orange-red colors developing with age. Surface smooth, viscid to tacky when wet, often with debris stuck to it and shiny when dry. **GILLS:** Broadly attached to decurrent, close to subdistant, broad, greasy. Whitish to pale yellowish buff at first, developing yellowish to ocher color as they age. **STIPE:** 3–12 cm long, 0.5–3 cm thick, tapering toward base. Whitish, creamy yellow, dark yellow, or pale orange. Surface silky-fibrillose on upper portions, more or less smooth below, viscid to moist. **PARTIAL VEIL:** Thin, consisting of fine glutinous fibers that leave a vague ring when young, disappearing in age. **FLESH:** Thin to thick, soft, fibrous in stipe. Yellow to whitish. **ODOR:** Indistinct. **TASTE:** Indistinct. **SPORE DEPOSIT:** White. **MICROSCOPY:** Spores 8-10 x 4–5 µm, ellipsoid, smooth, inamyloid, colorless.

ECOLOGY: Solitary, scattered, or clustered on ground or in duff under pines. Locally common throughout our range. Fruiting from late fall through winter.

EDIBILITY: Edible.

COMMENTS: The viscid, variably colored cap with olive-brown to bright orange-red or yellow tones, whitish to yellowish ocher gills, viscid stipe, and growth with pines are helpful identifying characters. *H. olivaceoalbus* has a dark to light brown cap, white gills, and a stipe with brownish bands; it grows with spruce.

Hygrophorus gliocyclus (sensu CA)

CAP: 4–10 cm across, convex with an inrolled margin at first, soon broadly convex to plane. Whitish to cream, with a pale yellow to yellowish buff center or entirely dull beige-yellow. Surface smooth, viscid, covered with glutinous slime. **GILLS:** Broadly attached to decurrent, close to distant, thick edged, greasy. Off-white to pale creamy yellow. **STIPE:** 2–5 cm long, 0.8–2.5 cm thick, stout, equal, enlarged in middle or toward base. Dingy creamy white, with a distinct dry white zone above veil. Surface viscid, covered with glutinous slime. **PARTIAL VEIL:** Glutinous, leaving an annular zone near apex of stipe. **FLESH:** Thick, firm, fibrous in stipe. White. **ODOR:** Indistinct. **TASTE:** Indistinct. **SPORE DEPOSIT:** White. **MICROSCOPY:** Spores 8–10 (12) x 4.5–6 µm, ellipsoid, smooth, inamyloid, colorless.

ECOLOGY: Solitary, scattered, or in groups, on ground in thick duff under pines. Fruiting in late fall and winter. Locally common under Bishop Pine and Monterey Pine along coast and scattered under Ponderosa Pine in Santa Cruz Sandhills.

EDIBILITY: Edible and good if the slime is removed. Alternatively, the mushroom can be rolled in bread crumbs (they stick to the slime) and then fried.

COMMENTS: The pallid cap with a yellowish buff center, soft, greasy-feeling decurrent gills, slimy stipe, and growth with pines are a distinctive combination. *H. flavodiscus* has a creamy cap with a golden yellow disc and whitish gills that are tinged pinkish; it is rare on the coast (reportedly occurring under Ponderosa Pine), but we have only seen it in the mountains with Sugar Pine (*Pinus lambertiana*).*H. sordidus* is a robust species with a slightly viscid to dry, white or creamy cap and a dry, white stipe; it is associated with live oak. *H. ponderatus* is also pallid and has a dry stipe, but is associated with conifers and possibly Tanoak.

Hygrophorus ponderatus

Britzelm.

CAP: 4–14 cm across, convex with an inrolled margin at first, becoming broadly convex to plane, wavy in age. White or with slight yellowish discoloration at the center. Surface smooth, viscid to tacky when wet, often with adhering debris when dry. **GILLS:** Broadly attached to decurrent, close to subdistant, broad, greasy. White, occasionally with a hint of pinkish. **STIPE:** 3–7 cm long, 1–3 cm thick, thickest near base, sometimes slightly bulbous. White. Surface smooth and dry at the apex, fibrous, viscid, often with debris stuck to lower portion. **PARTIAL VEIL:** Thin, consisting of fine fibers that leave a faint ring relatively low on stipe, often disappearing. **FLESH:** Thick, firm, fibrous in stipe. White. **ODOR:** Indistinct.. **TASTE:** Indistinct. **SPORE DEPOSIT:** White. **MICROSCOPY:** Spores 6.5–10 x 5–6 µm, ellipsoid, smooth, inamyloid, colorless.

ECOLOGY: Solitary or scattered on ground under conifers and possibly Tanoak. Uncommon, on coast from Sonoma County northward, but range poorly known. Fruiting in late fall and winter.

EDIBILITY: Edible.

COMMENTS: The stocky stature, decurrent, waxy gills, pallid overall colors, and viscid cap are a few of the not particularly distinctive features of this nondescript species. The partial veil can be easy to miss if only mature specimens are found, but in those cases, the dry stipe apex and viscid base should be distinctive. *H. subalpinus* is very similar, but fruits in spring in the mountains, often near melting snowbanks. *H. subellenae* has a medium-size, whitish waxy cap that lacks a partial veil and often develops a yellowish cap center (especially as it dries out). *H. sordidus* has a thinly viscid to dry, creamy whitish cap, broadly attached to slightly decurrent gills, and a dingy whitish stipe that often tapers downward; it can be locally common under live oak.

Hygrophorus chrysodon (sensu CA)
GOLD DUST WAXY CAP

CAP: 2.5–8 cm across, convex with an inrolled and often scalloped margin at first, soon broadly convex to plane. Whitish to creamy, but covered with golden yellow fibrils and patches when young, especially near margin, often brighter whitish with bright yellow splotches in wet weather. Surface slightly fibrillose to smooth, viscid to dry. **GILLS:** Subdecurrent to decurrent, close to distant, thick edged, often fairly narrow, greasy. Whitish to creamy or slightly pinkish. **STIPE:** 3–8 cm long, 0.5–1.5 cm thick, equal or tapered slightly toward base. Whitish when young but with yellow flakes, sometimes nearly entirely white in age. Surface granular to smooth, dry to slightly viscid. **FLESH:** Soft, fibrous, white. **ODOR:** Indistinct. **TASTE:** Indistinct. **KOH:** Yellow to orange. **SPORE DEPOSIT:** Whitish. **MICROSCOPY:** Spores 7–10 x 3–4 µm, ellipsoid, smooth, inamyloid, colorless.

ECOLOGY: Solitary, clustered, or in troops on ground, in duff, or in moss in a wide variety of coniferous and hardwood forests. Widespread, but uncommon throughout our range. Fruiting from late fall through winter.

EDIBILITY: Edible.

COMMENTS: The gold flakes on the cap and stipe and the white to creamy gills make this one of the more lovely and distinctive waxy caps. Although the yellow flakes can wash off in wet weather, the cap usually will have flakes adhering on the disc or at the margin, likewise with the stipe apex. The western North American species is distinct from the true European species and needs a new name. *H. sitchensis* has a viscid white cap, a stipe that turns yellow from the base up as it ages, and an almond extract odor. *H. eburneus* and *H. piceae* are pure white when young, but may discolor very slightly yellowish as they age. The former is dramatically slimy, unlike *H. chrysodon*.

Hygrophorus eburneus (sensu CA)
COWBOY'S HANDKERCHIEF

CAP: 2–8 (12) cm across, conical to convex with an inrolled margin at first, becoming broadly convex to plane, often uplifted, wavy and irregular in age. Bright white, occasionally with slight yellowing in age. Surface smooth, very viscid, covered with glutinous slime. **GILLS:** Broadly attached to decurrent, subdistant to distant, thick, broad, waxy or greasy feeling. White, occasionally tinged very pale yellowish or pinkish in age. **STIPE:** 2–10 (15) cm long, 0.5–1.5 cm thick, more or less equal or tapering toward base. White, discoloring pale dingy yellowish white in age. Surface covered with white punctations on apex when young, to smooth overall in age, viscid, covered with glutinous slime. **FLESH:** Thin, soft, stipe fibrous. White. **ODOR:** Indistinct. **TASTE:** Indistinct. **SPORE DEPOSIT:** White. **MICROSCOPY:** Spores 6.5–10 x 4–6 μm, ellipsoid, smooth, inamyloid, colorless.

ECOLOGY: In small clusters, scattered troops, or solitary on ground under hardwood trees and shrubs from late fall through winter. Common under oak, Madrone, and manzanita south of Mendocino County, rare on coast farther north, where it is more common inland in drier oak woodlands.

EDIBILITY: Edible but very slimy.

COMMENTS: The extremely glutinous, slimy white cap and stipe, decurrent gills, and growth under hardwoods are distinctive. This taxon is known to be distinct from the European *H. eburneus*, but has yet to be given a distinct name. *H. piceae* differs in having a dry to moist (never viscid) stipe and is smaller and more slender; it occurs under Sitka Spruce on the Far North Coast. Another Sitka Spruce associate, *H. sitchensis*, has a viscid white cap and stipe, but develops yellow color from the base up, and has an almond extract odor. *H. gliocyclus* is stockier, has an off-white to yellowish beige cap with a darker center, and grows with pines.

Hygrophorus agathosmus
(Fr.) Fr.
GRAY ALMOND-SCENTED WAXY CAP

CAP: 4–12 (15) cm across, convex to plane, margin inrolled and often tomentose at first, becoming plane, to slightly uplifted in age. Gray, grayish buff to grayish tan, sometimes slightly browner. Surface smooth, viscid and shiny when wet but often drying out with stuck-on bits of duff. **GILLS:** Broadly attached, close to well spaced, with short gills, narrow to broad, waxy feeling. Bright white when young, staying white or becoming pale grayish white. **STIPE:** 4–15 cm long, 0.7–2 cm thick, equal or tapering toward base. White to off-white. Surface very finely scurfy or pruinose at first, soon smooth, dry. **FLESH:** Thin to moderately thick, soft, fibrous, white to pale gray. **ODOR:** Mildly to strongly sweet, like almond extract. **TASTE:** Indistinct. **SPORE DEPOSIT:** White. **MICROSCOPY:** Spores 8–10 x 4–5.5 μm, ellipsoid, smooth, inamyloid, colorless.

ECOLOGY: Solitary or scattered, in small clusters or sometimes large troops on ground or in thick duff of coniferous forest, especially under Douglas-fir. Uncommon to locally common throughout our range. Fruiting in late fall and winter.

EDIBILITY: Edible, but the sweet odor doesn't persist when cooked.

COMMENTS: The gray cap, white gills, dry, white stipe, and almond extract odor are distinctive. *H. pustulatus* is similar, but is usually more slender, has grayish dots on the stipe, and lacks any distinctive odor; it is common in winter on the Far North Coast under Sitka Spruce. *H. bakerensis* has a similar almond extract odor, but has a yellowish brown cap with a whitish margin. *H. occidentalis* has a darker gray cap, white gills, a viscid white stipe, and no odor.

Hygrophorus bakerensis

A. H. Sm. & Hesler

BROWN ALMOND-SCENTED WAXY CAP

CAP: 4–12 (15) cm across, conical to convex with an inrolled margin at first, becoming broadly convex to plane, or wavy and irregular in age. Yellowish brown to orangey brown at center, margin contrastingly whitish. Surface very viscid and covered with glutinous slime when wet, tacky in dry weather, smooth, but with appressed fibrils under the gluten. **GILLS:** Broadly attached to decurrent, subdistant to distant, thick, broad, with a greasy feel. White when young, often with creamy yellow or pinkish tones in age. **STIPE:** 4–10 (15) cm long, 0.7–2.5 cm thick, more or less equal, but often tapering or rooting toward base. White, discoloring pale yellowish in age. Surface dry to moist, covered with white punctations on apex when young, smooth overall in age. **FLESH:** Moderately thick, soft, stipe fibrous. White. **ODOR:** Strongly of almond extract when young, weaker in age. **TASTE:** Mild to slightly disagreeable. **SPORE DEPOSIT:** White. **MICROSCOPY:** Spores 7–9 x 4–6 μm, ellipsoid, smooth, inamyloid, colorless.

ECOLOGY: Often in small clusters or in scattered troops, rarely solitary, on ground in duff under conifers, fruiting from late fall into spring. Locally common on the North Coast under Western Hemlock and Grand Fir, rare south of Mendocino County.

EDIBILITY: Edible, but slimy and bland (the odor doesn't remain after cooking).

COMMENTS: This attractive waxy cap can be recognized by the slimy, warm brownish to yellowish brown cap with a paler edge, white gills and stipe, and odor of almond extract. The cap, gills, and stipe of young fruitbodies are often adorned with clear water droplets. *Rhodocollybia oregonensis* also has a strong almond extract odor, but usually has a darker reddish brown cap with a warm beige margin and a taller, more slender, usually rooting stipe. Its gills are usually clearly serrate along the margin at maturity, and the stipe often develops reddish brown blotches. *H. agathosmus* also has an almond extract odor, but has a gray to grayish buff cap; it is more common overall and has a much more extensive southern range. *H. variicolor* is a poorly known species with a yellowish brown cap, an almond extract odor, and a partial veil (according to the description). *H. roseibrunneus* is vaguely similar, but lacks a strong odor, grows with hardwoods, and has a variably pallid to pinkish brown cap.

Hygrophorus roseibrunneus

Murrill

ROSY BROWN WAXY CAP

CAP: 2–8 cm across, convex with an inrolled margin at first, becoming broadly convex to plane, often with a low, broad umbo, occasionally uplifted and wavy in age. Color variable, generally pinkish brown to orangish brown with a darker center, but can be dark brown at center and yellowish brown to pale beige outward; paler toward margin at all stages. Surface smooth, viscid to tacky when wet, with debris stuck to it and shiny when dry. **GILLS:** Broadly attached to decurrent, close to subdistant, broad, greasy. Bright white to cream. **STIPE:** 4–12 cm long, 0.8–2.5 cm thick, equal or tapering toward base. White or blushed with pale brown. Surface at first finely scurfy, especially upper portions, becoming more or less smooth in age, moist to dry. **FLESH:** Thin, soft, fibrous in stipe. White. **ODOR:** Indistinct. **TASTE:** Indistinct. **SPORE DEPOSIT:** White. **MICROSCOPY:** Spores 6–10 x 4–6 µm, ellipsoid, smooth, inamyloid, colorless.

ECOLOGY: In small clusters, scattered in troops, or occasionally solitary in duff under live oak and Tanoak. Fruiting from late fall into spring. Common under Coast Live Oak in the southern part of our range, occurring north at least to Mendocino County.

EDIBILITY: Edible.

COMMENTS: The variation in cap colors shown by this species can be confusing, but the range is actually quite distinctive: usually some shade of pinkish brown, but can vary from dark brown to beige. The white gills, finely scurfy upper stipe, and growth with oak are also helpful identifying features.

NOMENCLATURE NOTE: There has been some confusion with regards to the correct name for this species; *H. albicastaneus* might refer to the same species, in which case that name would take precedence. Largent's concept of *H. roseibrunneus* in *The Agaricales (Gilled Fungi) of California* (vol. 5 *Hygrophoraceae*) corresponds to the species included here as *H. erubescens*. He also described *H. brunneus*, which we believe to be a browner-capped form of *H. roseibrunneus*.

Hygrophorus capreolarius

sensu Hesler & Smith

CAP: 3–8 cm across, convex with an inrolled margin at first, soon broadly convex to plane, margin becoming wavy and occasionally uplifted in age. Reddish purple with a grayish hue and darker wine red scales at first, soon extensively patchy with dark, sordid purplish colors. Often with darker concentric rings or a ring of dark spots around margin. Surface appressed-fibrillose to finely scaly, thinly viscid when wet, often with stuck debris when dry. **GILLS:** Broadly attached to decurrent, subdistant to distant, thick edged, greasy. Pinkish gray to vinaceous buff at first, soon developing darker wine-colored stains, becoming dark, dingy reddish purple in age. **STIPE:** 3–8 cm long, 0.5–1.2 cm thick, equal or enlarged toward base. Whitish, but streaked with vinaceous pink, darkening to wine red from the base up as it ages. Surface smooth to finely scurfy, dry. **FLESH:** Thin to moderately thick, firm, fibrous in stipe. Pinkish. **ODOR:** Indistinct. **TASTE:** Indistinct. **SPORE DEPOSIT:** Whitish. **MICROSCOPY:** Spores 6–9 x 4–5 μm, smooth, inamyloid, colorless.

ECOLOGY: Solitary or scattered on ground or in moss, almost always with Grand Fir or Western Hemlock present. Locally common from Sonoma County north. Fruiting from late fall through winter.

EDIBILITY: Edible.

COMMENTS: This smallish waxy cap can be identified by the vinaceous gray cap and gills that darken as it ages, darker ring or spots around the cap margin, and growth with northern conifers. One could easily mistake it for *H. russula*, but that species is usually larger, has a paler color scheme, and grows with hardwoods. *H. erubescens* has taller fruitbodies with pale pinkish caps that develop vinaceous pink blushes and yellowish stains in age; it grows primarily with spruce. *H. purpurascens* is also similar, but has a cortina-like partial veil, is usually larger, and has a rosy purple cap.

Hygrophorus russula

(Schaeff.) Kauffman

RUSSULA WAXY CAP

CAP: 4–12 (18) cm across, convex with an inrolled margin at first, soon broadly convex to plane, margin becoming wavy and occasionally uplifted in age. Pinkish with a whitish margin, irregularly streaked and spotted with vinaceous pink fibrillose patches, often extensively wine red overall when weathered and in age. Surface more or less smooth when young, soon extensively, but finely fibrillose-scaly, viscid when wet, often with adhering debris when dry. **GILLS:** Broadly attached to slightly decurrent, close to subdistant, broad, greasy. Off-white to pale pinkish at first, soon developing pinkish to vinaceous edges and blotches, occasionally evenly vinaceous pink in age. **STIPE:** 3–10 cm long, 1.5–4 cm thick, more or less equal, often with a pinched base. Whitish at first, soon extensively pinkish streaked, darkening to wine red from base up as it ages. Surface smooth to finely fibrous, dry. **FLESH:** Thick, firm, fibrous. White to pinkish with vinaceous stains around larva tunnels. **ODOR:** Indistinct. **TASTE:** Mild or slightly pine-like. **SPORE DEPOSIT:** White. **MICROSCOPY:** Spores 5–7 x 3.5–5 μm, ellipsoid, smooth, inamyloid, colorless.

ECOLOGY: Solitary, in small clusters, or scattered in troops or arcs under hardwoods in late fall and winter. Widespread but generally uncommon, occurring in mixed forests, especially under Chinquapin and Tanoak, from Santa Cruz County north, at least to inland Humboldt County.

EDIBILITY: Edible.

COMMENTS: The stature of this mushroom resembles that of a *Russula*, but the waxy gills, streaked pinkish to wine-colored cap, and fibrous flesh rule out that genus. The pinkish cap with dark wine red spots around the margin, paler gills and stipe, and growth with hardwoods make this a fairly distinctive species. It could be mistaken for *H. purpurascens*, which has an obscure cortina-like partial veil when young and grows with conifers in the mountains (it is very rare on the coast). *H. capreolarius* is a smaller, darker vinaceous purple species that grows with Grand Fir and Western Hemlock. *H. erubescens* has a paler pinkish-streaked cap, often develops yellowish stains as it ages, has a taller stature, and grows with Sitka Spruce. *H. fragrans* has an evenly colored rich pinkish to peachy orange cap and white stipe.

Hygrophorus erubescens

(Fr.) Fr.

CAP: 4–12 cm across, convex, broadly convex to plane, often wavy and uplifted in age, margin inrolled when young and covered with water droplets. Rosy pink to dark pink on disc, paler pink or whitish toward margin, often streaked and splashed with pink, overall paler and often developing yellowish discolorations in age. Surface smooth, viscid to tacky when wet, shiny and often with adhering debris when dry. **GILLS:** Broadly attached to subdecurrent, close to subdistant, broad, greasy. White or pale creamy white to pinkish white, developing pinkish stains and spots in age. **STIPE:** 6–17 cm long, 0.8–2.5 cm thick, more or less equal, but often tapering to a curved and pointed base. White to pinkish white, sometimes with yellow tones. Surface finely scurfy or granular, becoming smooth, moist to dry, upper stipe often covered with clear droplets when young. **FLESH:** Moderately thick to thin, firm, fibrous in stipe, at times becoming hollow in age. Creamy to pale pinkish white, discoloring rosy pink around larva tunnels. **ODOR:** Indistinct. **TASTE:** Slightly bitter at first, becoming acrid. **KOH:** Slight yellowing on all parts. **SPORE DEPOSIT:** Whitish to cream colored. **MICROSCOPY:** Spores 6.5–10 x 4–6 μm, ellipsoid, smooth, inamyloid, colorless.

ECOLOGY: Scattered or in large groups in duff under conifers, fruiting from late fall into winter. Can be locally common under Sitka Spruce and Grand Fir, but restricted to northernmost parts of the Redwood Coast.

EDIBILITY: Unknown; the bitter, acrid taste is unpleasant.

COMMENTS: This large waxy cap can be recognized by the viscid, rosy pink-splashed cap, yellow discolorations, and creamy white gills. The droplets on the upper stipe are often a helpful clue as well. *H. purpurascens* is similarly colored, but has a faint cortina when young; it is common in the California mountains, but very rare on the coast. *H. fragrans* is very similar; it has a viscid peachy orange to peachy pink cap and a creamy white stipe that turns bright orange at the apex with the application of KOH. Also compare with *H. russula* (hardwood associated) and *H. capreolarius* (smaller and much darker).

NOMENCLATURE NOTE: Largent considered this to be *Hygrophorus roseibrunneus* in *The Agaricales (Gilled Fungi) of California* (vol. 5 *Hygrophoraceae*), and described what we believe to be *H. roseibrunneus* as *H. brunneus*. See comments under *H. roseibrunneus* for more information.

Hygrophorus fragrans

(Murrill) Hesler & A. H. Sm.

FRAGRANT PINK WAXY CAP

CAP: 3–15 (20) cm across, rounded to convex with an inrolled margin at first, becoming broadly convex to plane, occasionally wavy and uplifted in age. Pink, pinkish orange or salmon at first, fading slightly in age. Surface smooth, viscid to tacky when wet, shiny and often with adhering debris when dry. **GILLS:** Broadly attached to subdecurrent, close to subdistant, broad, waxy to greasy. Whitish, creamy to pinkish cream. **STIPE:** 4–12 cm long, 1–2.5 cm thick, more or less equal, often with a slightly rooting base. Whitish, pinkish white to orangish white, staining orange at base when handled. Surface punctate at apex, fibrous to smooth over lower portion, moist to dry, upper stipe covered with clear droplets when young. **FLESH:** Moderately thick, firm, fibrous in stipe. Whitish to pale pinkish white. **ODOR:** Mild to fragrant; chemical sweet. **TASTE:** Indistinct. **KOH:** Bright orange on stipe apex. **SPORE DEPOSIT:** White. **MICROSCOPY:** Spores 7–9.5 x 4.5–5.5 μm, ellipsoid, smooth, inamyloid, colorless.

ECOLOGY: Scattered or in small clusters in duff under northern conifers, fruiting from fall into early winter. Uncommon to locally common, especially with Grand Fir and Western Hemlock on the North Coast, absent (or very rare) south of Sonoma County.

EDIBILITY: Edible.

COMMENTS: This beautiful waxy cap is easily recognizable by the pink to pinkish orange cap, whitish to pinkish white stipe, and bright orange KOH reaction of the stipe apex. *H. erubescens* has a rosy-pink-splashed cap and creamy white gills that stain pinkish, and lacks a KOH reaction on the stipe. A pale form called *H. pudorinus* var. *fragrans* forma *pallidus* differs solely by its creamy cap color; it is rare on the coast and occurs more often in the mountains under true firs.

SYNONYM: Long known as *Hygrophorus pudorinus* var. *fragrans*. It appears that William Murrill's *H. fragrans* (described from Oregon in 1912) is the correct name for this species.

Hygrophorus discoideus (sensu CA)

CAP: 1.5–5 cm across, broadly convex to plane, often with a slight umbo, occasionally slightly uplifted in age. Whitish to creamy buff near margin with an orangey brown to reddish brown spot of color at center, fading in age. Surface smooth, viscid when wet, shiny and often with debris stuck to it when dry. **GILLS:** Broadly attached to subdecurrent, close to subdistant, narrow, greasy. Off-white to cream colored. **STIPE:** 3–6 (9) cm long, 0.3–0.7 cm thick, more or less equal. Off-white, with pale orangish stains on lower portion in age. Surface smooth to finely fibrillose, viscid to tacky. **PARTIAL VEIL:** Glutinous and inconspicuous, disappearing in age. **FLESH:** Very thin, fibrous in stipe. Creamy white, discoloring orangey brown in lower stipe. **ODOR:** Indistinct. **TASTE:** Indistinct. **SPORE DEPOSIT:** White. **MICROSCOPY:** Spores 5.5–7 x 3.5–4.5 μm, ellipsoid, smooth, inamyloid, colorless.

ECOLOGY: Uncommon in our area, fruiting in duff in coniferous forests on the Far North Coast. Large troops sometimes found in late fall and winter under Sitka Spruce.

EDIBILITY: Unknown.

COMMENTS: This small waxy cap is easily identified by the viscid cap with an orangey to reddish brown disc and paler margin, whitish gills, and viscid stipe. Like many of our *Hygrophorus*, it is going by a European name (despite being a genetically distinct taxon in need of its own name). *H. pusillus* is quite similar, but has a whitish cap with pinkish red blotches and a nonviscid stipe; it is more widespread and in particular occurs much farther south.

Hygrophorus pusillus

Peck

CAP: 1.5–3 cm across, broadly convex to plane or centrally depressed, margin incurved and often scalloped at first, becoming wavy and occasionally uplifted in age. White to pale pinkish white, with irregular reddish pink splashes of color near center when young, fading and becoming pinkish blushed in age. Surface smooth, viscid when wet, shiny to dull when dry. **GILLS:** Broadly attached to subdecurrent, subdistant to distant, broad, thick edged. White at first, becoming creamy white to pale pinkish yellow. **STIPE:** 2–7 cm long, 0.3–0.5 cm thick, more or less equal. White, pale creamy white to pale pinkish white. Surface finely granular or powdered at apex, smooth below, dry to moist. **FLESH:** Thin, fibrous in stipe. White to pale pinkish cream. **ODOR:** Faintly fruity-spicy or indistinct. **TASTE:** Indistinct. **SPORE DEPOSIT:** Whitish to cream colored. **MICROSCOPY:** Spores 8–10 x 4–5 µm, narrowly ellipsoid, smooth, inamyloid, colorless.

ECOLOGY: Solitary or scattered on ground, in duff or moss in mixed coniferous forest. Fruiting from late fall through winter. Generally uncommon to rare in our area. We see it most often in slightly disturbed soil under Douglas-fir, also occurring under Western Hemlock and true firs.

EDIBILITY: Unknown.

COMMENTS: The small size, white to pale pinkish cap with darker splashes of pinkish red, widely spaced pale gills, and dry white stipe are helpful distinguishing features of this species. *H. piceae* is a similar all-white viscid-capped species under spruce. *H. discoideus* has a viscid cap with an orange-brown to reddish brown center and is associated with spruce. *H. saxatilis* has a similarly pale to peachy cap, but is larger and has dark pinkish cinnamon to pinkish buff gills. Also compare with the creamy white *Cuphophyllus* species, which differ primarily in their decurrent gills and more gracile stature.

Cuphophyllus virgineus

(Wulfen) Kovalenko

CAP: 1.5–6 cm across, conical to convex when young, becoming broadly convex to plane, often with a low, broad umbo, margin incurved and often ribbed at first, translucent-striate when wet. White to creamy white, sometimes tinged pale yellowish at center in age. Surface smooth, moist to greasy or dry. **GILLS:** Subdecurrent to decurrent, well spaced, often with cross veins, thick, soft and greasy. White to creamy white. **STIPE:** 3–7 cm long, 0.3–1 cm thick, equal or tapered slightly toward base. White to creamy white, often tinged yellowish in age, occasionally with a pinkish base. Surface smooth to slightly fibrillose, moist to dry. **FLESH:** Thin, white. **ODOR:** Indistinct. **TASTE:** Indistinct. **SPORE DEPOSIT:** White. **MICROSCOPY:** Spores 9–12 x 5.5–7 µm, ellipsoid or teardrop shaped, smooth, inamyloid, colorless.

ECOLOGY: Solitary, scattered, in small clusters or large troops in duff, moss, or soil. Generally under redwood, California Bay Laurel, or Monterey Cypress, but can occur in a variety of different habitats. Widespread in our area, but uncommon. Fruiting from late fall into early spring.

EDIBILITY: Nontoxic, rarely eaten.

COMMENTS: This species is recognizable by the overall pale coloration, thick, decurrent gills, and lack of any distinctive odor or taste. Fruitbodies of *C. niveus* are slightly smaller and have viscid to moist caps and smaller spores (averaging 8.5 x 5.3 µm). The very similar *C. russocoriaceus* has a pronounced cedarwood odor and medicinal-bitter taste.

SYNONYM: *Camarophyllus virgineus* (Wulfen) P. Kumm.

Cuphophyllus russocoriaceus

(Berk. & T. K. Mill.) Bon

CEDAR-SCENTED WAXY CAP

CAP: 0.7–4.5 cm across, convex when young, becoming plane in age, margin incurved, translucent-striate and often ribbed when young, occasionally uplifted in age. White to creamy white, sometimes tinged pale yellowish, hygrophanous, brighter white when dry. Surface smooth, moist to dry. **GILLS:** Decurrent, relatively widely spaced, often veined with narrow to broad cross ridges, soft and greasy. White to creamy white. **STIPE:** 4–10 cm long, 0.2–1 cm thick, equal or enlarged slightly toward base. White to cream at first, often pinkish at base in age. Surface smooth to vertically fibrillose, moist to dry. **FLESH:** Thin, white. **ODOR:** Strongly fragrant, like cedarwood. **TASTE:** Usually disagreeable, metallic or medicinal, sometimes mild. **KOH:** No reaction. **SPORE DEPOSIT:** White. **MICROSCOPY:** Spores 7–10 x 4.5–6 µm, averaging 8.8 x 5.4 µm, ellipsoid, smooth, inamyloid, colorless.

ECOLOGY: Solitary or scattered in small to large troops, especially in redwood or California Bay Laurel duff, also in mixed forests. Common throughout our range. Fruiting from late fall through winter.

EDIBILITY: Unknown.

COMMENTS: One sniff of this mushroom is practically enough to identify it; the strong odor of cedarwood is shared by only a few other mushrooms, none of which really resemble this species. Occasionally the odor will be faint or absent in older specimens; in these cases, the white to cream colors, decurrent gills, and disagreeable taste help distinguish it from other white *Cuphophyllus*. *C. virgineus* is very similar, but lacks any distinctive odor, has a mild taste, and is typically slightly larger. *Marasmius calhouniae* has more widely spaced, wavier gills and lacks any odor.

NOMENCLATURE NOTE: If the "true" European *C. russocoriaceus* is shown to be distinct from the western North American taxon, the name *C. lawrencei* (described by Hesler and Smith as a *Hygrophorus*) is the correct name for this taxon.

SYNONYMS: *Hygrocybe russocoriacea* (Berk. & T. K. Mill.) P. D. Orton & Watling, *Camarophyllus russocoriaceus* (Berk. & T. K. Mill.) J. E. Lange.

"Cuphophyllus" graveolens

(A. H. Sm. & Hesler) N. Siegel & C. F. Schwarz comb. prov.

CAP: 2.5–6 cm across, convex to plane with a depressed center in age, margin inrolled and often ribbed at first, often ragged in age. Pale peachy orange, pinkish cinnamon to creamy buff base color, with a whitish silky sheen, paler overall in age. Surface silky-fibrillose to smooth, moist to dry. **GILLS:** Broadly attached to decurrent, fairly widely spaced, occasionally with cross veins, fairly thick, soft and greasy. Pale peachy orange to creamy buff. **STIPE:** 4–10 cm long, 0.7–1.5 cm thick, equal, club shaped, with an enlarged base, or enlarged in middle. Pale creamy peach to pale creamy buff. Surface smooth, longitudinally striate, moist to dry. **FLESH:** Thin, pale peachy orange to orangish white. **ODOR:** Strong, fragrant, almost sickening sweet with a fruity-chemical component (weaker in age or cold weather, or even in some young collections). **TASTE:** Indistinct. **SPORE DEPOSIT:** White. **MICROSCOPY:** Spores 6–9 x 4–6 µm, ellipsoid, smooth, inamyloid, colorless.

ECOLOGY: Solitary or scattered in small troops in duff or moss, or on ground in mixed coniferous-hardwood forests, often in mixed Douglas-fir and Tanoak forests as well as under redwood. Widespread in our area, but only locally common, more frequent north of the San Francisco Bay Area. Fruiting from late fall through winter.

EDIBILITY: Unknown, likely nontoxic.

COMMENTS: The combination of a silky-looking pale peachy cap, thick, widely spaced, decurrent gills, and fragrant odor is distinctive. Often the first few sniffs of the odor are pleasant, after which it often becomes unpleasantly overpowering. *C. pratensis* is very similar, but usually is larger and stockier, with a darker orange to peachy orange cap lacking the silky sheen, and has little odor.

SYNONYM: *Hygrophorus graveolens* A. H. Sm. & Hesler.

Cuphophyllus pratensis

(Fr.) Bon

SALMON WAXY CAP

CAP: 2.5–8 (10) cm across, convex to plane, often with a broad umbo, margin inrolled and sometimes ribbed at first, becoming uplifted in age. Ochraceous orange to salmon-orange or paler buff-apricot in age. Surface smooth, moist to dry. **GILLS:** Subdecurrent to deeply decurrent, well spaced, often with cross veins, thick, soft and greasy. Ochraceous orange to salmon orange at first, then pale pinkish beige. **STIPE:** 4–10 cm long, 0.7–2 cm thick, equal, club shaped, or with a tapered base. Pale ochraceous orange to whitish. Surface smooth, moist to dry. **FLESH:** Moderately thick to thin, pale orangish buff to whitish. **ODOR:** Indistinct to faintly fragrant. **TASTE:** Mild to sweet. **SPORE DEPOSIT:** White. **MICROSCOPY:** Spores 5–8 x 4–5.5 µm, ellipsoid, smooth, inamyloid, colorless.

ECOLOGY: Solitary, scattered, or in small clusters in duff or moss, or on ground in mixed forest, especially fond of mixed California Bay Laurel and oak woodland and open redwood forests. Common throughout our area. Fruiting from late fall through spring.

EDIBILITY: Edible and good. Some people find it flavorless.

COMMENTS: The salmon orange to pinkish buff cap, thick, widely spaced, decurrent gills, and lack of a strong odor distinguish this species. *C. graveolens* is very similar, but has a paler cap and often has a strong, pungent-fragrant odor. *Hygrophorus nemoreus* var. *raphaneus* looks quite similar, but is larger, has broadly attached to slightly decurrent gills, and often has a slightly darker ochraceous orange cap; it is rare, found under live oak in the southern part of our range. *Chroogomphus tomentosus* is similar in color, but has a slightly matted-shaggy cap and black spores.

SYNONYMS: *Camarophyllus pratensis* (Fr.) P. Kumm., *Hygrocybe pratensis* (Fr.) Murrill.

Cuphophyllus fornicatus

(Fr.) Lodge, Padamsee & Vizzini

CAP: 2.5–6 cm across, broadly conical to convex, often umbonate, margin incurved at first, sometimes splitting in age. Light buff-brown to grayish buff or gray on disc, gradually paler gray to white toward margin, sometimes with obscure radial streaks. Margin sometimes translucent-striate. Surface smooth, thinly viscid to moist or dry. **GILLS:** Narrowly attached with a distinct notch, subdistant, broad, soft and greasy. White to off-white or very pale buff. **STIPE:** 3–8 cm long, 0.5–1.2 cm thick, equal or enlarged slightly toward base. White, watery translucent-white to pale grayish white. Surface smooth, moist to slightly viscid. **FLESH:** Thin, white. **ODOR:** Indistinct. **TASTE:** Indistinct. **SPORE DEPOSIT:** White. **MICROSCOPY:** Spores 7–9 x 4.5–6 µm, ellipsoid, smooth, inamyloid, colorless.

ECOLOGY: Solitary or scattered on ground, in duff or moss in older redwood forests or under California Bay Laurel. Rather uncommon to rare; occurring throughout our range. Fruiting from late fall through winter in north, continuing well into spring in Santa Cruz County.

EDIBILITY: Unknown.

COMMENTS: This rare and rather undistinguished waxy cap can be recognized by the grayish cap, decurrent and widely spaced, waxy-feeling whitish gills, smooth, whitish stipe, and relatively stocky stature (for a waxy cap). *Gliophorus irrigatus* has a viscid cap that is sometimes light gray when faded, but is more often dark gray to brownish gray. It has a viscid, gray to brown stipe that is more slender than that of *C. fornicatus*. *Hygrophorus agathosmus* also has a solid gray cap, but is more robust and smells like almond extract. *H. pustulatus* has more closely spaced gills and fine gray scurfs at the stipe apex; it is more northern in distribution. *Tricholoma inamoenum* has a white cap, less waxy gills, and a strong odor of mothballs or coal tar.

SYNONYM: *Hygrocybe fornicata* (Fr.) Singer.

Cuphophyllus colemannianus

(A. Bloxam) Bon

CAP: 1.5–5 (7) cm across, rounded to convex, broadly convex to plane, at times with a slight umbo. Dull brown to reddish brown or light vinaceous brown, usually darker in center, paler toward margin. Surface dull, smooth, moist to slightly viscid, margin translucent-striate when wet. **GILLS:** Decurrent, very widely spaced, thick, waxy, sometimes forking and with many inter-gill veins. Pinkish tan, pale tan, or creamy white. **STIPE:** 3–8 cm long, 0.3–1 cm thick, more or less equal, often curved at base. Pale pinkish tan to whitish, often paler toward base. Surface dry to slightly moist, smooth to appressed-fibrillose when dry. **FLESH:** Thin, dull brown to reddish brown in cap, whitish and fibrous in stipe. **ODOR:** Indistinct. **TASTE:** Indistinct. **SPORE DEPOSIT:** White. **MICROSCOPY:** Spores 8–10 x 5–6 µm, ellipsoid, smooth, inamyloid, colorless.

ECOLOGY: Solitary or scattered in duff, mostly found in mixed redwood forest, also in dark microclimates in mixed Bay Laurel and live oak woodlands. Uncommon to rare, occurring at least as far south as the Santa Cruz Mountains, north into the Pacific Northwest. Fruiting in winter and early spring.

EDIBILITY: Unknown.

COMMENTS: The smooth, dull brown to ruddy brown cap, thick, widely spaced decurrent gills, and dry stipe help set this waxy cap apart from others. *Camarophyllus (Cuphophyllus) recurvatus* is similar, but has a dark brown to olive-brown cap and white gills, and is often smaller (caps 1–3 cm across). *Cuphophyllus subviolaceus* has a lilac to smoky purple cap, lilac-gray gills and a whitish stipe; it is also rare in California, found in the duff of redwood forests from winter well into spring.

SYNONYMS: *Camarophyllus colemannianus* (A. Bloxam) Ricken, *Hygrocybe colemanniana* (A. Bloxam) P. D. Orton & Watling.

Neohygrocybe nitrata group

CAP: 2–6 cm across, convex with a downcurved margin at first, then wavy and irregular, often uplifted and splitting in age. Yellowish beige to light brown when young, becoming more dingy in age. Surface smooth to finely scaly, moist to dry. **GILLS:** Attached, distinctly notched, subdistant to distant, often inter-veined. White to creamy. **STIPE:** 4–9 cm long, 0.5–1.2 cm thick, equal, enlarged or tapered at base. Whitish beige to light brown, sometimes staining pinkish, then light brown when damaged. Surface smooth, moist to dry. **FLESH:** Thin, very fragile, stipe fibrous, whitish in color, occasionally slowly staining pinkish. **ODOR:** Faint to strong, like bleach to chemicallike. **TASTE:** Unpleasant, chemical. **SPORE DEPOSIT:** White. **MICROSCOPY:** Spores 7.5–9 x 5–6 μm, ellipsoid, smooth, inamyloid, colorless.

ECOLOGY: Solitary or scattered in small groups in duff or on bare soil under redwood, fruiting from late fall into winter. Occurs throughout our range, but uncommon in northern reaches and very rare southward.

EDIBILITY: Unknown.

COMMENTS: The pallid dingy yellow to tan cap, widely spaced and thick-edged gills, and chemical or bleach odor are helpful distinguishing characters. *Neohygrocybe* are taxonomically troubled in California, as there appear to be a few undescribed species. *N. ovina* is a very distinctive (but rare) species found on the Far North Coast; it has a dark gray-brown to brown cap, reddish staining, and, often, a fruity odor.

Porpolomopsis calyptriformis

(Berk.) Bresinsky

PINK WAXY CAP

CAP: 2–6 cm across, acutely conical to plane or uplifted in age, with distinct (and often sharp) umbo at all times. Margin often splitting, giving it a star-like shape. Bright pink to pale pink, in age sometimes more peachy or with dull reddish tones. Surface smooth, slightly viscid when wet, moist to dry in drier conditions. **GILLS:** Attached, distinctly notched, subdistant to distant. Pale pink to pinkish white, sometimes peachy in age. **STIPE:** 5–15 cm long, 0.5–1 cm thick, equal. White to creamy white. Surface smooth, moist to dry. **FLESH:** Thin, fragile, white to pale pink in cap, white, fibrillose, hollow in stipe. **ODOR:** Indistinct. **TASTE:** Indistinct. **SPORE DEPOSIT:** White. **MICROSCOPY:** Spores 7.5–9 x 5–7 μm, ellipsoid to slightly almond shaped, smooth, inamyloid, colorless. Pleurocystidia spindle shaped with tapering ends, or cylindrical; cheilocystidia similar, but smaller.

ECOLOGY: Solitary or scattered in small groups in duff or moss or on bare soil in a variety of forests, especially under live oak, redwood, and California Bay Laurel (and at edges of meadows near such forests). Uncommon to rare throughout our range. Fruiting from winter into late spring.

EDIBILITY: Unknown.

COMMENTS: The pastel pink, narrowly conical, pointed cap and tall white stipe make this beautiful mushroom practically unmistakable. It is probably more common than we think, but seems to fruit after other coastal mushrooms have stopped fruiting and most mushroomers have gone into hibernation. This stunning mushroom alone should serve as reason enough to keep poking about in redwood and bay laurel forests well into spring.

SYNONYM: *Hygrocybe calyptriformis* (Berk.) Fayod.

Gliophorus psittacinus (sensu CA)
PARROT MUSHROOM

CAP: 1–4 cm across, rounded-conical to convex when young, becoming broadly convex to plane, often with an umbo. Bright or dark green to bluish green, olive-green, olive-brown, or orange-brown, fading to light yellowish green, peachy orange, pinkish, or lilac, paler toward margin. Strongly hygrophanous, translucent striate when wet, colors duller when dry. Surface smooth, extremely viscid, with a thick layer of slime. **GILLS:** Attached, with distinct notch and decurrent tooth, thick, greasy, subdistant to distant. Colors variable, creamy white to pale yellow, or brilliant yellow, yellowish green, or greenish white, sometimes pinkish cream to peachy. **STIPE:** 2–6 cm long, 0.3–0. 7 cm thick, equal, often curved. Yellow to greenish yellow to pale yellow or creamy white, often brighter yellow toward base, at times with a bluish green or lilac band at apex. Surface smooth to bumpy, viscid and shiny. **FLESH:** Thin, fragile, pallid yellowish. **ODOR:** Indistinct. **TASTE:** Indistinct. **KOH:** Bleaches out cap pigment. **SPORE DEPOSIT:** White. **MICROSCOPY:** Spores 8–10 x 5–5.5 µm, averaging 9 x 5.3 µm, ellipsoid, smooth, inamyloid, colorless. Basidia 4-spored, 35–45 x 7–8 µm.

ECOLOGY: Solitary or scattered, sometimes clustered, often in redwood and California Bay Laurel duff, also in other kinds of forested habitats. Very common throughout our range, but often overlooked because of small size of fruitbodies and colors that often blend in with surroundings. One of the first waxy caps to fruit as fall transitions to winter; continues fruiting well into spring.

EDIBILITY: Nontoxic.

COMMENTS: Chameleon Waxy Cap would be a fitting name for this species, as the range of color variants, even in a single fruitbody as it ages, is extraordinary. Although the typical sequence is from greenish to orangish or pinkish yellow, there are also "backward Parrots" that start off orange and become green! Most have at least some green color, although sometimes only when very young or only at the stipe apex. The rare variety *Gliophorus psittacinus* var. *californicus* (facing page) has a bluish jade to teal-blue cap and bluish stipe, as well as larger basidia and smaller spores. Other similar species include *G. irrigatus*, which has a slimy grayish brown to brown cap and stipe and white gills. The colors of *G. laetus* are also bright and variable (yellowish green, orange, pink, lilac), but it can be distinguished by its decurrent gills and roughened to finely bumpy cap texture.

SYNONYM: *Hygrocybe psittacina* (Schaeff.) P. Kumm.

The Many Genera of Waxy Caps

Hygrophorus—A highly varied group; all are ectomycorrhizal. Fruitbodies range from small to quite large, most have solid flesh. Caps and stipes range from viscid to dry. Colors usually fairly muted, with a few more strongly colored species.

Hygrocybe—Fruitbodies brightly colored, (often red, orange, yellow etc.) small to medium sized, caps dry to viscid, and with dry to slightly viscid stipes.

Gliophorus—Fruitbodies viscid on both cap and stipe. Colors often very bright (some muted), including greens, pinks, yellows, reds, grays, and blues.

Cuphophyllus—Colors often muted, ranging from white to cream, gray, salmon-orange to reddish brown. Caps dry to slightly lubricious caps, dry stipes, often slightly to deeply decurrent gills.

Neohygrocybe—Fruitbodies drab, usually yellowish beige to light brown, sometimes dark brown to blackish; many staining pinkish. Surfaces usually dry to slightly greasy. Odors often pungent and bleach-like.

Porpolomopsis—Caps umbonate, splitting radially, slightly viscid to dry. Only one species in California.

Chrysomphalina*—Growing on wood, cap and stipe dry, gills brightly colored, slightly to deeply decurrent.

Chromosera*—Growing on wood, very viscid, color variable, but with lilac gills when young. Closely related to *Gliophorus*. One species in California.

Arrhenia*—Omphalinoid or pleurotoid stature, dry caps and stipes. Mostly moss-associated species.

Lichenomphalia*—A dull pinkish-colored lichenized species, usually growing out of green algal mats on decaying logs or nutrient-poor soil.

Ampulloclitocybe*—Dry gray cap, often club shaped gray stipe, decurrent and often contrasting creamy gills. Growing in moss, well decayed wood and duff.

* In the family *Hygrophoraceae*, but included in other chapters in this book

Gliophorus laetus

(Pers.) Herink

CAP: 1–3 (4) cm across, convex to plane, often depressed or sunken at center, margin translucent striate, incurved at first, becoming wavy and sometimes uplifted in age. Colors highly variable, often darker in center, greenish, purplish green, purplish pink, pinkish orange to peachy orange when young, losing green and purple colors quickly as it ages, becoming peachy orange to pinkish orange to orangish buff. Surface bumpy to smooth, viscid when wet, but often drying out. **GILLS:** Subdecurrent to decurrent, close to subdistant, thin, soft, greasy. Off-white to pale pink, pale pinkish orange to pale peachy orange. **STIPE:** 2–8 cm long, 0.3–0.7 cm thick, enlarged slightly toward base. Colors variable, greenish, purplish green, purplish pink, pinkish orange to peachy orange, often brighter yellow-orange toward base, and retaining greenish colors longer than cap. Surface smooth, viscid to moist. **FLESH:** Thin, concolorous. **ODOR:** Indistinct. **TASTE:** Indistinct. **KOH:** No reaction. **SPORE DEPOSIT:** White. **MICROSCOPY:** Spores 5–8 x 3–4.5 μm, averaging 6.7 x 3.5 μm, broadly ellipsoid, smooth, inamyloid, colorless.

ECOLOGY: Solitary or scattered on ground, in duff or moss, especially under redwood. Fruiting from late fall through winter. Common in northern coastal forests, rare south of Sonoma County.

EDIBILITY: Unknown.

COMMENTS: This waxy cap competes with *G. psittacinus* for most variably colored mushroom! It can be told by the bright, often multicolored fruitbodies that have viscid caps and stipes, and decurrent gills. The fruitbodies of *G. psittacinus* can be similarly colored, but don't have decurrent gills and tend to have bell-shaped or umbonate-convex caps.

SYNONYM: *Hygrocybe laeta* (Pers.) P. Kumm.

"*Gliophorus*" *flavifolius*

(A. H. Sm. & Hesler) N. Siegel & C. F. Schwarz comb. prov.

GOLDEN-GILLED WAXY CAP

CAP: 1–5 cm across, rounded to convex when young, becoming broadly convex to nearly flat, occasionally wavy, irregular or slightly uplifted in age; margin often with overhanging gelatinous rim when young. Lemon yellow, pale yellow to golden, with a creamy white or clear disc, young caps occasionally all creamy white; margin distinctly translucent-striate. Surface smooth, viscid to glutinous. **GILLS:** Broadly attached to slightly decurrent, thick, moderately well spaced. Bright golden yellow, orange-yellow to chrome, staying bright yellow in age. **STIPE:** 2–5 cm long, 0.4–0.8 cm thick, equal or tapered slightly toward base. Porcelain white, opaque or slightly translucent. Surface smooth, slimy-viscid. **FLESH:** Thin, fragile. Pale yellow in cap, white in stipe. **ODOR:** Indistinct. **TASTE:** Indistinct. **KOH:** No reaction. **SPORE DEPOSIT:** White. **MICROSCOPY:** Spores 7–9 x 4–5 µm, ellipsoid, smooth, inamyloid, colorless.

ECOLOGY: Solitary or scattered in small groups (rarely in large troops) in humus or duff, usually in dark, damp redwood or California Bay Laurel forests; also found under Douglas-fir. Occurs from Santa Cruz County to Oregon but is vary rare (only known from a handful of locations). Fruiting from winter into early spring.

EDIBILITY: Likely nontoxic.

COMMENTS: This beautiful mushroom is a treat to find. The bright golden gills, viscid white stipe, and yellow cap with a white disc set this species apart from all others. There is a rare, undescribed waxy cap in the San Francisco Bay Area that has a very similar cap, but differs in having creamy white gills and an opaque, dry stipe.

SYNONYM: *Hygrocybe flavifolia* (A. H. Sm. & Hesler) Singer.

Gliophorus irrigatus

(Pers.) A. M. Ainsw. & P. M. Kirk

SLIMY SOOTY WAXY CAP

CAP: 1–3.5 cm across, rounded to convex, becoming flat, often with an umbo. Color variable, dark brown to blackish, grayish brown or grayish white, paler toward margin. Surface smooth, slimy-viscid, margin translucent-striate. **GILLS:** Attached, often notched with a decurrent tooth, subdistant to distant, thick, greasy. White to off-white or pale gray. **STIPE:** 2–6 cm long, 0.2–0.5 cm thick, equal or slightly tapered at base, hollow or stuffed with white pith. Dark brown to grayish brown or grayish white, often concolorous with cap, and with translucent zones. Surface smooth, viscid. **FLESH:** Thin, fragile, white. **ODOR:** Indistinct. **TASTE:** Indistinct. **SPORE DEPOSIT:** White. **MICROSCOPY:** Spores 8–10 x 5.5–6 µm, ellipsoid, smooth, inamyloid, colorless.

ECOLOGY: Solitary or scattered in small patches on ground, in moss, or in humus under redwood and other conifers. Fruiting from late fall into winter, but rarely in large numbers. Uncommon throughout our range, occurring as far south as Santa Cruz County.

EDIBILITY: Likely nontoxic.

COMMENTS: This uncommon waxy cap is easily distinguished from its brethren by the slimy, dark brown to grayish cap and stipe and the white gills. *Mycena quinaultensis* looks quite similar, but grows in troops and has a tall, slender, fragile stipe; it is more northern in distribution. *Hygrophorus (Gliophorus) subaromaticus* has a creamy white to pale gray cap, whitish stipe, and slightly chemical odor (like new plastic or burnt rubber). *Cuphophyllus fornicatus* has a grayish brown, opaque cap and white, nonviscid stipe.

SYNONYMS: *Hygrocybe irrigata* (Pers.) Bon, *H. unguinosa* (Fr.) P. Karst.

"Gliophorus fenestratus"

C. F. Schwarz nom. prov.

WINDOWED WAXY CAP

CAP: 1–2.5 cm across, rounded to convex when young, becoming broadly convex to plane, often with a small umbo, margin inrolled at first, expanding to plane or slightly uplifted. Bright red and often with an orangish margin at first, soon fading to orange and then entirely yellow in age. Strikingly translucent-striate, and with a distinctive milky to translucent disc or umbo. Surface smooth, very viscid. **GILLS:** Attached, often notched with a decurrent tooth, close to subdistant, thick, waxy. Orange with a yellowish edge when young, pale yellow to orange in age. **STIPE:** 2–6 cm long, 0.2–0.5 cm thick, equal or slightly enlarged at base. Somewhat translucent, yellow to yellow-orange at the apex, paler yellowish white toward base. Surface smooth to slightly bumpy, viscid. **FLESH:** Very thin, fragile, pallid yellowish. **ODOR:** Indistinct. **TASTE:** Indistinct. **KOH:** No reaction. **SPORE DEPOSIT:** White. **MICROSCOPY:** Spores 7–9 x 4–4.5 (5) µm, elongate-ellipsoid to subcylindric, with some misshapen or triangular, smooth, inamyloid, colorless.

ECOLOGY: Solitary or scattered in small patches in duff in redwood and Monterey Cypress forest, or in moss in mixed forest. Uncommon to locally common from midfall through early spring. Known from Santa Cruz County to Mendocino County, but probably occurs farther south, as well as on the Far North Coast.

EDIBILITY: Likely nontoxic.

COMMENTS: This colorful waxy cap can be recognized by the viscid, translucent-striate cap with a milky or translucent center, rapidly fading colors, and slimy stipe. Although it starts off bright red, it will quickly fade to orange and then yellow, seemingly changing before your eyes. *G. minutulus* is another small, viscid, red-capped species found in the same habitats. However, it retains a red disc even in age and lacks the milky translucent umbo. See *Hygrocybe marchii* for other little red waxy caps without distinctly viscid stipes.

Hygrocybe virescens

(Hesler & A. H. Sm.) Montoya & Bandala

LIME-GREEN WAXY CAP

CAP: 2–8 cm across, convex or irregularly broadly conical when young, becoming plane with an umbo, margin often uplifted, becoming ragged and splitting in age. Bright lime green to neon green, often with orange near center, especially in age. Fading to yellow with light green hues, rarely entirely orangish when young or old. Surface smooth to finely appressed-fibrillose, dry to slightly moist. **GILLS:** Attached with a distinct notch, but can appear free at times. Close to subdistant, thick, waxy. Whitish to pale greenish white, sometimes pallid greenish yellow. **STIPE:** 2–10 (12) cm long, 0.4–1 cm thick, equal or tapering toward base, often grooved or split. Lime green to yellowish green, often paler, off-white to greenish white at base and with a whitish sheen at apex. Surface smooth to fibrous, dry to moist. **FLESH:** Fibrous, thin and fragile in cap, stipe hollow. Off-white to yellowish green. **ODOR:** Indistinct. **TASTE:** Indistinct. **KOH:** Little change to slight bleaching on cap. **SPORE DEPOSIT:** White. **MICROSCOPY:** Spores 7–10 x 5–6.5 μm, ellipsoid, smooth, inamyloid, colorless.

ECOLOGY: Solitary, scattered, or in troops in duff or disturbed areas under redwood or occasionally in mixed coniferous woods. Occurring throughout our range, but fruiting irregularly (see comments) from midfall into winter.

EDIBILITY: Unknown.

COMMENTS: The lime green tones of the cap (and to a lesser extent the gills and stipe) make this a distinctive waxy cap. Older, yellowish orange specimens can be very similar to *H. flavescens*, but fruitbodies of that species are usually pallid yellow, with smooth, moist to viscid caps. Alexander H. Smith described it from a single collection from Trinidad, California, in 1956. David Largent, who wrote the waxy cap volume of *The Agaricales of California*, never saw it, despite collecting extensively near the type locality. David Arora found a large fruiting near Mendocino, California in 1971 and went 40 years without seeing it again! Then, in late fall of 2011, both authors and a handful of other collectors turned up a number of *H. virescens* fruitings (some very profuse) at scattered locations along the Redwood Coast. This illustrates the difficulty of determining the true rarity of macrofungi. Even dedicated searchers can go years without seeing a single fruitbody, despite the presence of (presumably persistent) mycelia. This fruiting also raised questions about the phenomenon of apparently synchronized fruitings: after years of obscurity, *H. virescens* underwent simultaneous, profuse flushes in both central Mendocino County and southern Santa Cruz County, more than 300 km distant from one another and at very different points in the progression of the mushroom season. Similar stories involving other species have played out in other parts of the United States as well as Europe. Although the idiosyncrasies of detection and reporting bias toward rare species may explain much of this pattern, the difficulty of studying such a phenomenon will likely preclude convincing explanations for a long time to come.

Hygrocybe flavescens

(Kauffman) Singer

YELLOW WAXY CAP

CAP: 2–8 cm across, convex to plane with a small umbo, margin becoming uneven, mature specimens often wavy. Pale lemon yellow to bright yellow or yellowish orange, fading in age. Surface moist to viscid when wet, often dry in older specimens or in dry weather, and often with debris stuck to the cap. **GILLS:** Notched to narrowly attached, subdistant, narrow to broad, soft, greasy. Off-white to pale yellow when young, staying pale yellow or becoming darker yellow to yellowish orange in age. **STIPE:** 3–5 cm long, 0.5–1.5 cm thick, more or less equal, usually developing a longitudinal groove in age. Yellow, pale lemon yellow or yellowish orange. Surface fibrous to smooth, moist to slippery when wet, quickly drying out. **FLESH:** Creamy white to pale yellow, stipe stuffed when young, becoming hollow. **ODOR:** Indistinct. **TASTE:** Indistinct. **KOH:** Bleaching cap pigment. **SPORE DEPOSIT:** White. **MICROSCOPY:** Spores 7.5–9 x 4–5 μm, averaging 8.4 x 4.3 μm, ellipsoid, smooth, inamyloid, colorless.

ECOLOGY: Solitary, scattered, or in small clusters in humus and duff, most often under redwood, but can be found in many forest types. Common throughout our range. Fruiting from late fall into spring.

EDIBILITY: Reportedly edible.

COMMENTS: This common waxy cap is best identified by the viscid, convex to plane, yellow to yellow-orange cap without any red tones and the whitish to yellow gills and stipe. Some authors consider *H. chlorophana* to be a separate species, distinguished by a viscid stipe and more orange cap. We consider these more orange and slightly viscid-stalked specimens to be within the range of variation for *H. flavescens*. *H. acutoconica* is similar, but has a more peaked-conical, yellow-orange to reddish cap and a fibrous stipe. Faded fruitbodies of the much rarer *H. virescens* can be similar, but have dry caps and usually at least faint traces of greenish hues.

Hygrocybe acutoconica

(Clem.) Singer

ACUTE CONIC WAXY CAP

CAP: 2.5–9 (11) cm across, conical with a distinct pointed umbo when young, becoming broadly conical to convex, sometimes wavy and uplifted, margin often splitting. Whitish when very young, very soon pale to lemon or golden yellow, often extensively orange or red-orange, usually darker at disc, fading in age. Surface smooth, very viscid when wet, but can become dry and shiny with adhering debris. **GILLS:** Narrowly attached with a notch, close to subdistant, thin, greasy. White to pale yellowish white when young, becoming pale yellow. **STIPE:** 2–15 cm long, 0.5–1.5 cm thick, more or less equal. Yellow to yellowish orange, whitish toward base, sometimes slowly discoloring blackish gray at base. Surface smooth to fibrous, moist (but not viscid) to dry. **FLESH:** Thin, pale yellow. Fragile in cap, fibrous in stipe, stipe hollow in age. **ODOR:** Indistinct. **TASTE:** Indistinct. **KOH:** No reaction. **SPORE DEPOSIT:** White. **MICROSCOPY:** Spores 11–15 x 5.5–8 μm, averaging 12.4 x 6.9 μm, broadly oblong-ellipsoid, occasionally irregular in outline, or constricted, inamyloid, colorless. *H. acutoconica* var. *microspora* has smaller, broadly ovoid to ellipsoid spores averaging 8.8 x 6.1 μm.

ECOLOGY: Solitary or in troops in duff. Very common throughout our range, primarily under redwood and in California Bay Laurel forest, but occurring in a wide variety of habitats. Usually the first waxy cap to fruit in fall, often disappearing when weather gets cold and most other waxy caps are fruiting.

EDIBILITY: Likely nontoxic.

COMMENTS: The conical or sharply umbonate, viscid, reddish orange to yellow cap; whitish to pale yellow gills; and yellow stipe and lack of black staining (except occasionally on the stipe base) are distinctive. Members of the *H. singeri* complex can look very similar, but the gills are soon gray and often extensively black overall in age. The fruitbodies of *H. flavescens* can show orange-yellow caps, but are never acutely umbonate.

Hygrocybe singeri

(A. H. Sm. & Hesler) Singer

WESTERN WITCH'S HAT

CAP: 3–10 cm across, bell shaped to narrowly conical, becoming irregularly broadly conical, umbonate, often pointed. Yellow-orange to reddish orange, rarely extensively scarlet, sometimes yellowish black to olive without orange-red tones, developing extensive black streaking, and often entirely black in age. Surface smooth, viscid and shiny when wet, dull when dry. **GILLS:** Attached to nearly free, close to subdistant. Off-white when young, becoming pallid gray, irregularly developing black edges and blotches, sometimes entirely black in age. **STIPE:** 5–20 cm long, 0.7–2 cm thick, equal or enlarged toward base, hollow, fibrous, longitudinally striate. Yellow, yellow-orange, or olive-grayish, often with a whitish base, slowly developing black stains. Surface smooth, viscid to dry. **FLESH:** Thin, fragile, fibrous, off-white, slowly staining black where damaged. **ODOR:** Indistinct. **TASTE:** Indistinct. **KOH:** No reaction. **SPORE DEPOSIT:** White. **MICROSCOPY:** Spores 9–13 x 5–7 μm, averaging 10.2 x 5.8 μm, ellipsoid, smooth, inamyloid, colorless.

ECOLOGY: Solitary or scattered, often in large troops under redwood and Monterey Cypress, but can occur in many forest types. Very common throughout our range. Fruiting from late fall through winter.

EDIBILITY: Not recommended. There are (unfounded?) reports of this species and its relatives being poisonous.

COMMENTS: The bright colors, narrowly conical cap, and blackening on all parts make this an easy mushroom to identify, at least in a general sense. Other blackening species in California include *H. olivaceoniger*, which is smaller and has a yellow to greenish yellow cap without orange-red tones (it may end up being a synonym of *H. singeri*). Perhaps more distinct is *H. nigrescens*, with larger, consistently redder caps and occurrence under hardwoods. Another similar species is *H. acutoconica*, which has a slimy, orange-yellow, conical cap and does not discolor gray and black. We have found a pigmentless form of *H. singeri*, with watery whitish fruitbodies that stained black in age. Some authors have used the name *H. conica* for redder-capped specimens with dry stipes, but we consider most of these to fall within the extensive range of variation shown by *H. singeri*. The fruitbody size and colors of *H. singeri* are different from the eastern North American and European taxa referred to as *H. conica*. Recent molecular work has shown that even in Europe *H. conica* has been applied to a complex of ten or more species, many of which remain unnamed.

Hygrocybe marchii

(Bres.) Singer

CAP: 1–2 cm across, rounded-conical, convex to plane, margin inrolled and often scalloped when young, plane in age. Bright red at first with a distinctly paler orange edge, becoming orange-red with an orange to yellow-orange margin. Surface slightly viscid to dry, smooth. **GILLS:** Broadly attached, thick, waxy, close to subdistant. Pale yellow to yellowish orange. **STIPE:** 2–5 cm long, 0.2–0.4 cm thick, equal, or tapered toward a curved base. yellow-orange to orange, often paler toward base. Surface smooth, moist (but never viscid) to dry. **FLESH:** Very thin, fragile, concolorous. **ODOR:** Indistinct. **TASTE:** Indistinct. **KOH:** No reaction. **SPORE DEPOSIT:** White. **MICROSCOPY:** Spores 6.5–8 x 3–4.5 μm, averaging 7.1 x 3.8 μm, narrowly ellipsoid to ellipsoid, sometimes centrally constricted (like peanut shells), smooth, inamyloid, colorless.

ECOLOGY: Solitary or scattered in duff and humus, especially under redwood and Monterey Cypress. Common throughout our range. Fruiting from late fall through winter.

EDIBILITY: Unknown.

COMMENTS: The small size, slightly viscid to dry red caps, yellow to orange gills, and moist to dry stipes are helpful characters for identifying this mushroom. Look-alikes include the very similar *Gliophorus minutulus*, which has a decidedly viscid stipe. Less similar is *G. fenestratus*, with a deeply translucent-striate reddish orange to yellow cap with a translucent umbo and a viscid stipe. The following species are less well-understood: *Hygrophorus (Gliophorus) mamillatus* has an orange to orange-yellow, acutely umbonate cap and a viscid stipe. *Hygrocybe reae* (=*H. mucronella*?) has an orange to yellowish orange, viscid cap, whitish gills, a dry, yellowish orange stipe, and a bitter taste. *H. moseri* is slightly larger and has an orange cap and pale orangish white decurrent gills. *H. miniata* has a finely scruffy orange cap and often more decurrent gills. Larger fruitbodies of any of these (caps broader than 2 cm or so) can cause confusion and should be compared with *H. coccinea* and similar species.

Hygrocybe miniata

(Fr.) P. Kumm.

CAP: 1–3 cm across, convex to plane, often with a slightly depressed center, margin downcurved and often ruffled when young. Bright red to orange-red, fading to reddish orange or yellow-orange. Surface dry to moist, never viscid, center with small tuffs or scales, more noticeable when it gets older, giving it a felty appearance. **GILLS:** Broadly attached, thick, waxy, widely spaced. Creamy yellow to pale yellow when young, becoming yellow to yellow orange in age. **STIPE:** 2–7 cm long, 0.2–0.4 cm thick, equal. Reddish orange to orange, becoming yellow-orange to yellow, often paler toward base. Surface smooth, moist to dry, often with lighter longitudinal striations as it dries. **FLESH:** Thin, fragile, yellow-orange. **ODOR:** Indistinct. **TASTE:** Indistinct. **KOH:** No reaction. **SPORE DEPOSIT:** White. **MICROSCOPY:** Spores 7–9 x 4–6 μm, averaging 7.8 x 5.1 μm, variably shaped, generally broadly ellipsoid to ellipsoid, pear shaped or cylindrical and constricted (like peanuts in a shell), or like corn kernels, smooth, inamyloid, colorless.

ECOLOGY: Scattered, in small clusters, or solitary, often in duff or on ground in mixed forest, or in mossy areas. Common from late fall through winter in the northern part of our range, uncommon south of Sonoma County.

EDIBILITY: Edible.

COMMENTS: The small size, the reddish orange colors, and especially the minutely scruffy cap help distinguish this waxy cap. *H. moseri* has a nearly smooth orange cap and pallid decurrent gills. See *H. marchii* (and the species mentioned in the comments) for other small, smooth-capped species.

Hygrocybe ceracea

(Sowerby) P. Kumm.

CAP: 1–4 cm across, convex to broadly convex, at times with a slight umbo, sometimes uplifted and splitting in age. Bright golden orange when young, margin paler and translucent-striate, hygrophanous, fading, often in a patchy manner, to pale orange or yellowish orange to yellow. Surface smooth, viscid when young. **GILLS:** Broadly attached to subdecurrent, close to subdistant, thick, waxy. Golden yellow to orange with a paler edge when young, soon fading to yellowish in age. **STIPE:** 2–4 cm long, 0.2–0.5 cm thick, equal or tapering slightly toward base. Orange to gold, fading to pale yellow. Surface smooth, moist to dry. **FLESH:** Thin, fragile, concolorous with cap and stipe, stipe hollow in age. **ODOR:** Indistinct. **TASTE:** Indistinct. **KOH:** Bleaching pigment on all parts. **SPORE DEPOSIT:** White. **MICROSCOPY:** Spores 7–9 x 5–6.5 µm, ellipsoid, smooth, inamyloid, colorless.

ECOLOGY: Solitary, scattered, or in troops on grassy lawns or moss in late fall and winter. Uncommon, restricted to northern part of our range.

EDIBILITY: Unknown.

COMMENTS: The translucent-striate, viscid, orange cap that fades to yellow, subdecurrent gills, small to medium-size fruitbodies, and growth in grassy and mossy areas help set this species apart from other small, colorful waxy caps. *Gliophorus fenestratus* usually has a red cap that fades to orange then yellow as it ages, with a translucent to milky umbo and a viscid stipe. Compare with the larger *H. aurantiosplendens* and the smaller, red-capped *H. marchii*, both of which typically grow in duff in the woods (not on lawns).

Hygrocybe aurantiosplendens

R. Haller Aar.

CAP: 2.5–7 cm across, convex to plane, sometimes umbonate, margin incurved at first, becoming wavy and splitting, sometimes uplifted in age. Color variable, red to orange-red at first, fading irregularly as it ages, becoming orange to yellowish orange with reddish patches, eventually entirely orangey yellow. Surface smooth, viscid to moist. **GILLS:** Narrowly attached with a distinct notch, subdistant, narrow to broad, soft, greasy. Yellow at first, becoming yellowish orange to peachy; often more reddish in age. **STIPE:** 3–10 cm long, 0.4–1.5 cm thick, equal or enlarged slightly downward, but tapering at base or narrowed at center. Yellow to orange, paler toward base. Surface smooth, moist to dry, never viscid. **FLESH:** Thin, yellowish to yellow-orange in cap, stipe hollow. **ODOR:** Indistinct. **TASTE:** Indistinct. **SPORE DEPOSIT:** White. **MICROSCOPY:** Spores 7–10.5 x 4–6 µm, ellipsoid, smooth, inamyloid, colorless.

ECOLOGY: Solitary, scattered, or in small clusters in duff, moss, or soil, generally under redwood or California Bay Laurel. Common from late fall into winter along the North Coast. Fruiting in late winter and in spring southward.

EDIBILITY: Unknown.

COMMENTS: The smooth, quickly fading red to orangey yellow cap, yellow-orange to peachy gills, and smooth stipe are some of the ambiguous identifying features of this waxy cap. *H. splendidissima* has a more persistently scarlet red cap, pale pinkish red gills with a pallid edge, and a stipe that becomes longitudinally striate in age. Other similar species include *H. laetissima*, which has a bright red cap that can fade to orangish red, creamy to pale yellow gills that become pinkish red in age, and a much more fibrillose stipe. *H. coccinea* has and retains a deeper red to blood red coloration to the cap, gills, and stipe. *H. flavescens* has a yellow cap, pallid gills, and often slightly slippery stipe. *H. acutoconica* has a conical, viscid, yellow-orange to reddish orange cap. Also compare with the group of species mentioned under *H. marchii*, most of which are distinctly smaller.

Hygrocybe coccinea

(Schaeff.) P. Kumm.

SCARLET WAXY CAP

CAP: 2–5 (7) cm across, rounded-conical to convex, becoming broadly convex to plane, occasionally with a low umbo, margin incurved at first, becoming wavy and irregular, sometimes uplifted and splitting in age. Deep scarlet red to blood red, sometimes becoming patchy orangish red in age. Surface smooth to very slightly roughened, thinly viscid to greasy or dry. **GILLS:** Attached, notched with a decurrent tooth, close to well spaced, broad, soft, greasy. Pinkish red to red, with a paler whitish yellow to pinkish edge at first, becoming orange-red in age. **STIPE:** 2–7 cm long, 0.5–1.5 cm thick, equal or enlarged slightly toward base, occasionally with a longitudinal groove in age. Red, often with a whitish to orangish base. Surface smooth, only faintly longitudinally striate when old and dry. **FLESH:** Thin, fragile, red in cap, orangish red in stipe. **ODOR:** Indistinct. **TASTE:** Indistinct. **KOH:** No reaction. **SPORE DEPOSIT:** White. **MICROSCOPY:** Spores 7–10 x 4–5.5 µm, ellipsoid to broadly ellipsoid, smooth, inamyloid, colorless.

ECOLOGY: Solitary, in small clusters, or scattered in small to large troops in duff, moss, or soil, usually under redwood. Very common throughout our range. Fruiting in late fall or winter in north, later in winter and spring in south.

EDIBILITY: Reportedly edible, but not recommended. Closely related species (*H. punicea* group) have caused problems for some people.

COMMENTS: The medium-size fruitbodies, deep scarlet red to blood red cap, red gills, and smooth stipe are helpful in separating this species from the multitude of other red waxy caps. *H. splendidissima* has a bright scarlet red cap; pale pinkish red gills with a whitish edge that darkens to orangish red in age; and a whitish to yellow stipe that darkens and becomes longitudinally striate. *H. laetissima* and *H punicea* are larger, and both have paler, longitudinally striate, fibrillose stipes. *H. aurantiosplendens* has a red cap when young, but quickly fades patchily to orange and yellow. Members of the *H. marchii* group are quite a bit smaller and don't have extensively red gills.

Hygrocybe laetissima

(A. H. Sm. & Hesler) Singer

CHERRY-RED WAXY CAP

CAP: 2–8 cm across, rounded to convex with an inrolled margin when young, becoming plane, uplifted or wavy in age. Bright scarlet to cherry red at first, becoming red to reddish orange, lighter toward margin, often with orange blotches or spots in age. Surface smooth, moist to tacky, never viscid. **GILLS:** Attached with a distinct notch, widely spaced, thick, waxy. White to creamy yellow when young, becoming yellowish orange, pinkish red or red in age. **STIPE:** 2–8 cm long, 0.5–3 cm across, equal or tapering toward base. Whitish to pale yellow at first, soon yellowish orange with a reddish blush on lower half, becoming redder at maturity. Surface dry, longitudinally striate with whitish fibers, often finely scruffy in age. **FLESH:** Thin, somewhat fragile in cap, fibrous in stipe. White to reddish blushed in cap, paler to whitish in stipe. Stipe stuffed at first, often hollow in age. **ODOR:** Indistinct. **TASTE:** Indistinct. **KOH:** Bleaching cap pigment. **SPORE DEPOSIT:** White. **MICROSCOPY:** Spores 7–11 x 4–6 µm, ellipsoid to narrowly ellipsoid, smooth, inamyloid, colorless.

ECOLOGY: Scattered in small to large patches, occasionally solitary, on ground, in moss, or in well-decayed duff under California Bay Laurel and redwood. Widespread, occurring throughout our range; very common from Santa Cruz County to Sonoma County, less common northward, rare on the Far North Coast. Fruiting in late fall in northern part of range, winter and spring southward.

EDIBILITY: Not recommended. Although some people apparently eat it with no problems, there are a few reports of minor gastrointestinal distress.

COMMENTS: This mushroom shines like a bright light in the understory of dark, damp redwood forests. The combination of robust stature, scarlet young caps, pale young gills that develop peachy tones, and longitudinally fibrillose stipe that is often extensively white when young (developing yellow to orange or reddish tones) separates this species from other *Hygrocybe*. It belongs to a confusing group of Pacific Coast species that are desperately in need of modern taxonomic work. This species has most often been called *H. punicea* by Californians, but Hesler and Smith applied that European name to an as-yet-undescribed species that differs by its darker blood red to deep red cap and more orange tones in the gills and stipe. Another poorly known similar species, *H. splendidissima*, has medium-size fruitbodies with slightly viscid red caps and smoother stipes. *H. aurantiosplendens* has a reddish cap that quickly and blotchily fades to orange and a smooth stipe. *H. coccinea* is more slender and has a darker red cap, reddish gills, and a smooth reddish stipe.

Hygrocybe punicea (sensu CA)
CRIMSON WAXY CAP

CAP: 2–10 cm across, convex with an inrolled margin when young, becoming plane, uplifted or wavy in age, sometimes with a very broad, round umbo. Blood red to deep red at first, fading, often brick reddish with orangish blotches in age. Surface smooth, slightly viscid to greasy or dry. **GILLS:** Broadly attached or notched, widely spaced, thick, greasy feeling. Pale yellow to yellowish orange when young, soon yellowish orange, orange to dull reddish orange in age. **STIPE:** 2–10 cm long, 0.5–3 cm across, equal or tapering toward base. Pale yellowish to orange at first, soon orange to dark reddish. Surface dry, longitudinally striate with paler fibrils, often scruffy in age. **FLESH:** Thin, somewhat fragile, reddish blushed in cap. Stipe fibrous, stuffed at first, often hollow in age; whitish to pale yellow or orange. **ODOR:** Indistinct. **TASTE:** Indistinct. **KOH:** Bleaching cap pigment. **SPORE DEPOSIT:** White. **MICROSCOPY:** Spores 8–10 x 5–6 μm, ellipsoid to narrowly ellipsoid, smooth, inamyloid, colorless.

ECOLOGY: Solitary or scattered on ground, in moss or well-decayed duff under redwood. Common north of Mendocino County, less common southward, not recorded south of San Francisco Bay Area. Fruiting from midfall into winter in northern part of range, later into spring southward.

EDIBILITY: Not recommended. Although some people eat it with no problems, a few have reported minor gastrointestinal distress.

COMMENTS: This waxy cap can be recognized by the relatively large size (for a *Hygrocybe*), dark red cap, gills that become reddish in age, and longitudinally striate stipe. *H. laetissima* is very similar, but the cap is scarlet red, and the stipe is brighter yellow; although it occurs throughout our range, it is more common south of Mendocino County, largely replacing *H. punicea*. See comments under *H. laetissima* for more distinguishing features and other similar species.

13 • The White-Spored Multitude

Although this group includes a number of interesting mushrooms, it also contains many of the plainest-looking, most featureless fungi in our area. The list of genera included here is quite long, and most were at one time placed in the *Tricholomataceae*, which was long used as a "taxonomic waste bin" for white-spored mushrooms whose evolutionary relationships were difficult to guess. Little by little, many of the genera represented here are being sorted into named lineages of their own; likewise, individual species are being critically examined to determine which genus they "truly" belong in. You can expect a significant amount of name changes pertaining to these species in the near future—*Clitocybe* and the genera centered around *Gymnopus* are especially likely to be divided and/or redefined.

Ecologically, most of the genera represented here are generalist saprobes and litter decayers, although a few have rather odd and/or specific substrates (for example, only on spruce cones or on rotting fruitbodies of certain mushrooms). However, *Laccaria*, *Tricholoma*, and *Catathelasma* are ectomycorrhizal; the first two genera are especially important as symbiotic partners of trees in temperate forests around the globe.

We've developed a number of subgroups (mostly based on size and stature type) to help you navigate the White-Spored Multitude. Microscopic examination paired with good technical references will be necessary to identify some of the little white-spored mystery mushrooms you come across.

The White-Spored Multitude Groups

Group A (p. 299–301)—Tiny mushrooms (caps usually less than 1 cm apart) with a slender stature, slightly to deeply decurrent gills, a fragile, dry stipe, and conical, convex, or slightly umbilicate caps: *Hemimycena*, *Phloeomana*, *Mycena picta*, *Rickenella*.

Group B (p. 302–315)—Stipe fragile, cap broadly conical, peaked, or convex, usually translucent-striate when fresh and moist, fruitbodies dull to brightly colored: *Atheniella*, *Mycena*, *Roridomyces*, *Mycopan*.

Group C (p. 316–323)—Stipe thin, tough, and often wiry, many with fruitbodies that can dry out and revive when wet: *Xeromphalina*, *Gymnopus*, *Mycetinis*, *Micromphale*, *Crinipellis*, *Cryptomarasmius*, *Marasmius*. *Rhizomarasmius*, *Marasmiellus*.

Group D (p. 324–332)—Stipe usually cartilaginous or tough but not wiry, sometimes tapering and rooting, frequently hollow or stuffed with a stringy pith, fruitbodies small to medium size (a few species occasionally have caps 8+ cm wide): *Marasmius*, *Connopus*, *Gymnopus*, *Macrocystidia*, *Rhodocollybia*, *Caulorhiza*.

Group E (p. 333–335)—Mushrooms small, caps typically less than 2.5 cm across. Cap margin finely ribbed or "crimped," cap surface opaque, often with a fine pale bloom, stipe often finely dandruffy, base of stipe usually with long spiky or shaggy mycelium and rhizomorphs: *Pseudobaeospora*, *Baeospora myriadophylla*.

Group F (p. 335–336)—Small mushrooms growing on cones of various conifers, cap convex to plane: *Baeospora myosura*, *Strobilurus*. (Many of the groups above will occasionally fruit from cones, whereas species in this group are *restricted* to this substrate.)

Group G (p. 337–340)—Growing on mushrooms or decaying remains of mushrooms, gills usually present (absent or rudimentary in two species), most with a sclerotium (small hard mass of tissue) at base of stipe (but often embedded in substrate of decaying mushroom tissue), two species with branches coming off stipe: *Collybia*, *Dendrocollybia*, *Asterophora*.

Group H (p. 341–348)—Small (but not tiny) mushrooms with convex caps often greater than 1 cm across that have a distinctly depressed center and broadly attached to decurrent gills (this distinctive stature type is called omphalinoid): *Chromosera*, *Chrysomphalina*, *Xeromphalina*, *Aphroditeola*, *Contumyces*, *Lichenomphalia*, *Pseudoomphalina*, *Camarophyllopsis*, *Myxomphalia*, *Arrhenia*.

Group I (p. 349–355)—Small to medium-size mushrooms with broadly attached to decurrent gills, stipe fleshy-fibrous (not cartilaginous or tough): *Ampulloclitocybe, Pseudoclitocybe, Paralepista, Clitocybe.*

Group J (p. 356–358)—Fruitbodies tough, with fibrous stipes and thick, often widely spaced gills. Fruitbody coloration usually in warm pastels, from pinkish buff to orangey, or in one species with intense purple colors: *Laccaria.*

Group K (p. 358–365)—Medium-size to large mushrooms with notched, broadly attached, or decurrent gills. Often in rings, arcs, or dense clusters: *Clitocybe, Leucopaxillus, Melanoleuca, Lyophyllum, Lepista.*

Group L (p. 366–383)—Medium-size to large mushrooms, white or pale yellowish, typically close to crowded, notched (sinuate) gills, and fibrous flesh, often with a farinaceous odor. Cap textures range from smooth to (more often) fibrillose or scaly: *Tricholoma.*

Group M (p. 384–390)—Contains two *Tricholoma,* the unrelated *Floccularia* and *Catathelasma,* and the often clustered *Armillaria.* All have a membranous to flocculent partial veil (at least when young). Also see *T. cingulatum* (p. 377), a small gray species that has a distinct partial veil when young.

Group N (p. 391–396)—Miscellaneous medium-size to large mushrooms growing from wood (can be quite decayed). If you can't find your species here, check other white-spored sections, especially Groups A, B, C, and D, above: *Tricholomopsis, Callistosporium, Clitocybula, Flammulina, Omphalotus, Panus, Neolentinus, Pleurotus dryinus.*

Group A

Hemimycena delectabilis

(Peck) Singer

CAP: 0.2–1 cm across, conical to convex at first, expanding to broadly convex or plane, often with a small rounded umbo. Watery white to pale creamy white with a translucent-striate margin when wet; dull, opaque, chalk white when dry. Surface smooth, moist to dry. **GILLS:** Broadly attached to decurrent, moderately spaced, often with short inter-gill veins, edges fringed when young. White. **STIPE:** 2–4 cm long, 0.05–0.2 cm thin, fragile. Watery white to white. Surface smooth, dry to moist, base with sparse whitish hairs. **FLESH:** Concolorous, very thin, fragile. **ODOR:** Bleachlike, especially when crushed. **TASTE:** Indistinct. **SPORE DEPOSIT:** White. **MICROSCOPY:** Spores 5–7 x 3.5–4 µm, ellipsoid, smooth, inamyloid. Pleurocystidia scattered to abundant, 30–60 x 7–12 µm, spindle shaped to elongated, cheilocystidia abundant, and similar.

ECOLOGY: Scattered in troops, rarely solitary, in duff and needles, especially that of Douglas-fir. Common and widespread, occurring throughout northern California. Fruiting soon after first rain in fall into winter.

EDIBILITY: Inedible.

COMMENTS: *Hemimycena* are relative easy to identify to genus, by basis of the small to tiny size, white to translucent white colors, and often decurrent gills. They are among our more common litter decayers, often occurring in large troops soon after the first rains in the fall, especially on the Far North Coast. Most need microscopic examination to differentiate species; if you are happy with knowing *Hemimycena* and don't need a species name, stop here; it will save you some frustration. *H. delectabilis* stands out by the large size (relatively speaking) and the odor of bleach, especially when crushed. *H. delicatella* is very similar, but lacks an odor, and has cylindrical spores 7–12 x 2.5–3.5 µm, from both 2- and 4-spored basidia. The pleurocystidia and cheilocystidia are similar and abundant, 18–26 x 4–6 (9) µm, and nearly cylindrical or ventricose, with the top rounded into a small head. *H. tortuosa*, another commonly encountered species, grows on the bark of both hardwoods and conifers. It is much smaller, as it rarely gets more than 0.5 cm across, and is often adorned with dew droplets. Microscopically, it has very distinct, corkscrewlike pileocystidia. *H. gracilis* generally develops yellowish tones in age, has much larger, cylindrical spores, lacks pleurocystidia, and has short, cylindrical cheilocystidia. *H. hirsuta* has a minutely fuzzy cap and poorly formed gills. *H. albidula*, which grows on hardwood leaves or bark, often has a pointed, umbonate cap, ellipsoid spores, 7–9 x (3) 4–5 µm, and pleurocystidia and cheilocystidia that look like the basidia. *Resinomycena saccharifera* is a tiny species that grows on sedge and grass stems, has a powdery or resinous cap, sparse, broadly attached, widely spaced gills, and a short, thin stipe.

Phloeomana speirea
(Fr.) Redhead

CAP: 0.3–1 cm across, rounded, convex to omphalinoid, often ribbed, or with broad sulcate grooves. Disc light brown, grayish brown to grayish yellow at first, fading, buff, yellowish to grayish, margin paler, grayish buff, pale gray to whitish, and often with darker lines in groves. Surface very finely powdery at first, soon smooth, moist to dry, translucent-striate when wet. **GILLS:** Broadly attached to decurrent, close to subdistant, narrow. White to pale grayish white. **STIPE:** 1–3.5 (5) cm long, 0.05–0.1 cm thick, very fragile, slender, equal or slightly enlarged toward base. Water whitish to pale grayish white, occasionally with pale yellowish to brownish coloration at base in age. Surface finely powdery at first, to smooth, dry to slightly moist, base with fine whitish hairs. **FLESH:** Very thin, fragile, whitish. **ODOR:** Indistinct. **TASTE:** Indistinct. **SPORE DEPOSIT:** White. **MICROSCOPY:** Spores 8.5–11.5 x 4.5–6.5 µm, ellipsoid to almond shaped, with a prominent hilar appendage, smooth, thin walled, inamyloid. Basidia narrowly club shaped, 2-spored. Cheilocystidia elongate to narrowly club shaped, often forking at tip with one to several knoblike projections. Pleurocystidia absent.

ECOLOGY: Solitary, scattered, or in small clusters, on moss-covered bark or small woody debris. Mostly found on Tanoak, but on other hardwoods and conifers. Fruiting soon after first rains in fall, sporadically occurring through winter into spring. Locally common, but easily overlooked.

EDIBILITY: Unknown.

COMMENTS: The small size, grayish to grayish brown convex cap with a paler margin, broadly attached to decurrent gills, thin stipe, and growth on tree bark help identify this dainty mushroom. Microscopically, the inamyloid spores and 2-spored basidia are useful diagnostic characters. The rare *Mycena clavata* is similar, but has a darker cap and grows on the bark of Western Redcedar (*Thuja plicata*). There are many small bark-dwelling species of *Mycena* in California, but most have conical to bell-shaped caps and amyloid spores.

SYNONYM: *Mycena speirea*. Even though this species resembles a *Mycena* macroscopically, the inamyloid spores are unlike true *Mycena*. Phylogenetic research has shown that it belongs in the family *Porotheleaceae*.

Mycena picta
(Fr.) Harmaja

CAP: 0.1–0.7 cm across, cylindrical to bell shaped, with an sunken disc and distinct groves, margin occasionally wavy in age. Dark brown to ocher-brown on disc, paler, yellowish toward margin. Surface smooth, moist to dry, hygrophanous, translucent-striate when wet, dull when dry. **GILLS:** Broadly attached to decurrent, moderately spaced, extending horizontally from stipe to cap margin. Creamy white to pale gray at first, to grayish buff in age. **STIPE:** 1–6 cm long, 0.1 cm thick, more or less equal, or slightly wider toward apex. Yellowish brown to darker, reddish brown at base. Surface dry to moist, smooth above, base covered in fine yellowish brown hairs. **FLESH:** Very thin, fragile, concolorous. **ODOR:** Indistinct. **TASTE:** Mild. **SPORE DEPOSIT:** White. **MICROSCOPY:** Spores 8–10 x 4–5 µm, elliptical, smooth, thin walled, amyloid.

ECOLOGY: Solitary or scattered in moss or coniferous duff. Rare, only known from the Far North Coast in California, continuing into the Pacific Northwest. Fruiting in fall.

EDIBILITY: Unknown.

COMMENTS: The brown, boxy, or bell-shaped and often umbilicate cap, unusual gills (forming a straight edge from the stipe to the cap margin), and tiny size are distinctive. So far it is only known from a couple of collections in California, but the color and size make it very easy to overlook. It could be mistaken for *Rickenella swartzii*, which has a purplish to gray convex cap and a dusky purple or gray stipe apex.

Rickenella swartzii

(Fr.) Kuyper

CAP: 0.3–0.8 (1.2) cm across, broadly convex to plane with a sunken center, margin downcurved when young, becoming slightly uplifted or wavy. Dark grayish purple to purplish black on young disc, fading to orangish purple, paler toward margin. Surface dry, appearing slightly powdered (pruinose) when young to smooth. **GILLS:** Slightly to deeply decurrent, close to subdistant. Whitish to pale vinaceous gray. **STIPE:** 2–4 (6) cm long, 0.05–0.2 cm thick, equal, thin. Dark purple to purplish gray at apex, transitioning to orangish at base. Surface dry, powdered (pruinose) to smooth. **FLESH:** Very thin, pallid. **ODOR:** Indistinct. **TASTE:** Mild. **SPORE DEPOSIT:** White. **MICROSCOPY:** Spores 4–5 (7) x 2–3 µm, narrowly ellipsoid, smooth, inamyloid. Pleurocystidia and cheilocystidia abundant, cap surface with numerous pileocystidia, caulocystidia abundant on stipe apex.

ECOLOGY: Scattered in mossy woods or in mossy lawns, fruiting in fall and winter. Common on the Far North Coast, rare south of Mendocino County.

EDIBILITY: Unknown.

COMMENTS: This small, pretty, and often overlooked mushroom is recognizable by the plane, umbilicate cap with a purplish center that fades to purplish orange, decurrent gills, and contrasting dusky purplish stipe apex (orangish toward the base). *R. fibula* often grows near this species in moss beds and has the same stature, but is orange overall. There are lots of small, grayish, moss-dwelling *Mycena*, but they rarely have the same stature or the dark purple colors.

MISAPPLIED NAME: *Rickenella setipes.*

Rickenella fibula

(Bull.) Raithelh.

ORANGE MOSS AGARIC

CAP: 0.3–1.2 cm across, broadly convex to plane with a sunken center, margin downcurved and often ribbed when young, becoming slightly uplifted or wavy in age. Orange at first, soon fading to pale orange to orangish buff in age, but often retaining orange color on disc. Surface dry, appearing slightly powdered (pruinose) when young to smooth. Translucent-striate when wet. **GILLS:** Slightly to deeply decurrent, close to subdistant. Whitish to pale creamy orange. **STIPE:** 1–4 (6) cm long, 0.05–0.2 cm thick, equal. Bright to pale orange, often paler at the base. Surface dry, pruinose to smooth. **FLESH:** Very thin, fragile, orangish, stipe often hollow. **ODOR:** Indistinct. **TASTE:** Mild. **KOH:** No reaction. **SPORE DEPOSIT:** White. **MICROSCOPY:** Spores 4.5–6 x 2–3 µm, narrowly ellipsoid, smooth, inamyloid. Pleurocystidia and cheilocystidia abundant, subcylindric to spindle shaped. Cap surface covered with elongated pileocystidia, caulocystidia abundant on stipe apex.

ECOLOGY: Solitary, scattered in troops or small clusters in moss, with a preference for open mossy areas such as mossy yards or roadsides. Common in the northern part of our range, rarer south of the San Francisco Bay Area. Fruiting from late fall into spring.

EDIBILITY: Unknown.

COMMENTS: This small moss inhabitant can be recognized by the orange convex to plane cap with a depressed disc, pale, decurrent gills, and long, thin, orangish stipe. Many species of orangish brown to buff *Galerina* often grow alongside this species in moss beds; however, they have conical to convex caps (lacking the "omphalinoid" stature) and have light rusty brown spores. *Xeromphalina campanella* is slightly larger, has an orange-brown cap, and grows in clusters and swarms on wood. *Chrysomphalina aurantiaca* is also orange, but has a stockier stature and grows on wood. Also compare the orange *Mycena* and *Atheniella* species.

Group B

Atheniella aurantiidisca

(Murrill) Redhead, Moncalvo, Vilgalys, Desjardin & B. A. Perry

CAP: 0.3–1.5 cm across, conical at first, becoming broadly conical to bell shaped. Color variable, bright orange over disc, pale orangish white margin when young, soon orange on disc, transitioning to a yellow midportion and creamy white margin, to all yellow with a whitish margin or even pale yellowish to creamy white overall in age. Surface smooth, moist to dry, translucent-striate when wet. **GILLS:** Narrowly attached, notched, occasionally with a decurrent tooth, close to subdistant, narrow. Whitish, pale yellowish to yellowish orange. **STIPE:** 1.5–5 cm long, 0.1–0.2 cm thick, fragile, equal, or enlarged slightly toward base. Pale yellowish, translucent yellow to whitish. Surface finely powdered at first, becoming smooth, moist to dry. Base with coarse whitish hairs. **FLESH:** Very thin, fragile, concolorous with cap, stipe hollow, translucent yellow to whitish. **ODOR:** Indistinct. **TASTE:** Mild. **SPORE DEPOSIT:** White. **MICROSCOPY:** Spores 6–9 x 3–4.5 µm, ellipsoid, smooth, thin walled, inamyloid.

ECOLOGY: Solitary, scattered, or in small to large troops in moss, duff, or other coniferous debris. Seems restricted to northern coniferous forest, especially under Western Hemlock. Infrequent in Sonoma and Mendocino Counties, common on the Far North Coast. Fruiting in fall or early winter.

EDIBILITY: Unknown.

COMMENTS: The three-toned (orange, yellow, whitish), conical cap, pallid to orangish gills, thin, translucent yellow to whitish stipe, and growth in duff make this an easy mushroom to recognize. *Mycena strobilinoidea* is similar but has a more rounded to bell-shaped cap that is bright reddish orange to orange, and has orangish gills with a dark reddish orange edge. *M. oregonensis* is bright yellow overall and has a bell-shaped to convex cap and a very thin stipe. *M. acicula* is a tiny species with a reddish orange or coral-colored cap that fades to light orange, and has a long, thin, yellowish stipe. *Atheniella adonis* has a bright pink cap when young that fades to pale pink or pinkish white in age.

SYNONYM: *Mycena aurantiidisca* (Murrill) Murrill.

Atheniella adonis

(Bull.) Redhead, Moncalvo, Vilgalys, Desjardin & B. A. Perry

PINK MYCENA

CAP: 0.4–2 cm across, conical, convex to uneven-convex, usually with an umbo, margin often splitting in age. Bright pink to salmon pink when young, fading as it ages to paler pink, pinkish yellow to pinkish white. Surface smooth, moist to dry, translucent-striate when wet. **GILLS:** Attached, close to subdistant, often inter-veined. Pale pink, off-white to creamy or pale yellowish. **STIPE:** 2–4 cm long, 0.05–0.3 cm thick, very fragile, equal or slightly thinner and curved toward base. Pale pink to pinkish white, creamy white or translucent, often lighter, to white toward base. Surface often dusted with whitish powder when young, becoming smooth, moist to dry, base with whitish fibrils. **FLESH:** Very thin, fragile, pinkish. **ODOR:** Indistinct. **TASTE:** Mild. **SPORE DEPOSIT:** White. **MICROSCOPY:** Spores 7–8 x 3–5 μm, ellipsoid to broadly ellipsoid, with a hilar appendage, smooth, inamyloid.

ECOLOGY: Solitary, scattered, or in small groups or clusters in humus and duff, especially Douglas-fir needle litter, but occurs in a wide variety of forest types. Uncommon occurring throughout our range. Fruiting from midfall through winter.

EDIBILITY: Unknown.

COMMENTS: The small size, and pink conical-convex caps of young specimens are usually enough to separate this mushroom from all but the similarly colored *Mycena rosella*, which differs in having pink gills with darker pink edges. Faded older specimens of *A. adonis* can be quite pale and then resemble a large *Hemimycena* species, which are white to cream colored and have subdecurrent gills. Also compare the tiny *Mycena acicula*, which has a bright orange to orange-red cap when young that can fade to pinkish in age; it usually has a yellowish stipe.

SYNONYM: *Mycena adonis* (Bull.) Gray.

Mycena rosella

(Fr.) P. Kumm.

ROSY PINK MYCENA

CAP: 0.3–1.5 (2) cm across, rounded to bell shaped at first, becoming convex, broadly convex to plane, occasionally with an small umbo, or depressed on disc. Deep pink to rosy red at first, soon fading pinkish to pinkish gray, and often with yellowish over disc in age. Surface smooth, dry to moist, translucent-striate when wet. **GILLS:** Broadly attached, often with a decurrent tooth, close to subdistant, narrow. Pinkish at first, becoming pale pinkish peach to pale pinkish white. Marginate; edges darker pink to rosy pink. **STIPE:** 1–5 (7) cm long, 0.05–0.2 cm thick, fragile, equal or slightly enlarged toward base. Watery pink, pinkish gray to light pinkish brown at first, becoming dingy pinkish gray to light grayish brown. Surface smooth, dry, base with sparse whitish fibrils. **FLESH:** Very thin, fragile, pinkish to pinkish gray. **ODOR:** Indistinct to faintly radishlike. **TASTE:** Mild. **SPORE DEPOSIT:** White. **MICROSCOPY:** Spores 8–9.5 (10.5) x 4–5 μm, ellipsoid, smooth, thin walled, amyloid.

ECOLOGY: Solitary, scattered, or in small troops on ground, in conifer needles and duff, or in moss. Locally common in coniferous forest on the Far North Coast, occurring southward to Sonoma County, but rather rare south of Humboldt County. Fruiting in fall and early winter.

EDIBILITY: Unknown.

COMMENTS: The rosy cap that fades to pinkish gray, broadly attached pinkish gills with darker rosy pink edges, and pinkish gray stipe help identify this interesting *Mycena*. *Atheniella adonis* shares the same pink colors, but generally has a more bell-shaped cap and lacks the darker pink gill edges; it also has inamyloid spores. Older specimens could be mistaken for *M. capillaripes*, which has a dark to light gray cap that sometimes shows pink tones and pinkish gill edges, but has an odor of chlorine or bean sprouts when crushed, and is much more common. *M. sanguinolenta* is generally smaller and more reddish, reddish brown, or vinaceous, has reddish gill edges, and bleeds a reddish brown juice from the stipe when broken.

Mycena acicula

(Schaeff.) P. Kumm.

CORAL PINK MYCENA

CAP: 0.2–0.8 cm across, rounded to bell shaped at first, becoming convex, occasionally with an umbo. Bright orange-red to orange, often with a paler margin, fading to pinkish orange to yellowish orange. Surface dusted with fine whitish granules when young, becoming smooth, dry to slightly viscid. **GILLS:** Attached, close to subdistant. Pale yellow, yellowish pink, pale pink to off-white. **STIPE:** 1–5 cm long, 0.05–0.15 cm thick, very fragile, equal or slightly thinner and often rooting toward base. Pale yellow to yellow-orange, becoming pale yellow to creamy white. Surface smooth, moist, base with whitish fibrils. **FLESH:** Very thin, fragile, yellowish. **ODOR:** Indistinct. **TASTE:** Mild. **SPORE DEPOSIT:** White. **MICROSCOPY:** Spores 9–12 x 3–4 μm, elongate-ellipsoid to spindle shaped, smooth, inamyloid.

ECOLOGY: Solitary or in small groups on woody debris (particularly fond of undersides of Tanoak logs and bits of bark), as well as on moss-covered wood and twig or leaf litter. Quite common but tiny and often under layers of brambles and easily overlooked. Fruiting from first rains in fall through spring.

EDIBILITY: Unknown. It would take hundreds to make a bite.

COMMENTS: The tiny size, coral-colored cap, and yellow stipe, and tendency to fruit from well-rotted and/or moss-covered wood help distinguish this species from other brightly colored small *Mycena*. Similar species include *M. oregonensis*, which has a bright yellow color and bell-shaped cap, and typically grows in conifer duff. Another species common in conifer duff is *Atheniella aurantiidisca*, which differs by having a conical orange cap when young that fades to yellow, and a translucent white to pale yellowish stipe. *A. adonis* is stockier (but still small) and has a bright pink cap that fades to pale pink or pinkish white.

Mycena aurantiomarginata

(Fr.) Quél

ORANGE-GILLED MYCENA

CAP: 0.5–2 cm across, conical to bell shaped, becoming nearly plane, occasionally with an umbo. Dark olive-brown to orange-brown, becoming gray-brown at center, margin paler; light orange-brown at first, becoming yellow-brown to gray-brown, often with an olive tinge and dark gray to gray-brown striations. Strongly hygrophanous, translucent-striate to center when wet. Surface smooth, dull, moist to dry. **GILLS:** Attached, close. Bright yellow-orange when young, becoming paler yellowish orange and finally yellowish gray. Marginate, gill edge remaining bright yellow-orange throughout. **STIPE:** 2–6 cm long, 0.1–0.3 cm thick, fragile, equal or slightly enlarged toward base. Yellow to yellow-orange, becoming grayish yellow to yellow-brown in age. Surface smooth, moist to dry, base covered in yellowish orange to yellowish gray hairs. **FLESH:** Very thin, fragile, yellowish to brownish gray in cap, yellowish to almost translucent in stipe. **ODOR:** Indistinct. **TASTE:** Mild. **SPORE DEPOSIT:** White. **MICROSCOPY:** Spores 9–11 x 5–6 μm, ellipsoid, smooth, amyloid. Cheilocystidia club shaped, often with a long, narrowed base, or with a rounded head and spiny tips, filled with a bright orange pigment.

ECOLOGY: Solitary, scattered, or in small groups on ground or in well-decayed duff of many forest types. Widespread and very common, on the Far North Coast especially abundant under Sitka Spruce early in fall. Generally fruits under Douglas-fir and Tanoak in late fall or winter in southern part of our range, but will occur in live oak duff as well.

EDIBILITY: Unknown.

COMMENTS: The combination of orange gills with dark edges and translucent-striate olive-brown to yellowish brown caps make this beautiful *Mycena* easy to recognize. *M. strobilinoidea* is similar, but has a bright red to reddish orange cap that fades to pale yellow, and has yellow gills with a dark red to reddish orange edge; it seems to prefer higher-elevation coniferous forests in the far northern part of our range.

Mycena oregonensis

A. H. Sm.

WESTERN YELLOW MYCENA

CAP: 0.2–1 cm across, egg shaped to bell shaped when young, expanding to convex or broadly convex, occasionally with small umbo. Bright yellow to bright yellow-orange, fading to yellow or pale yellow, but often remaining darker on disc. Surface dull, dry, at times dusted with fine yellowish granules. **GILLS:** Attached, close to subdistant. Pale yellow, yellowish orange to yellowish white or creamy white. **STIPE:** 1.5–5 cm long, 0.05–0.1 cm thick, very fragile, equal, often curved, sometimes rooting. Bright yellow to yellow-orange, becoming paler yellow to orangish yellow. Surface finely powdered, covered with tiny orange granules, base covered with yellowish orange hairs. **FLESH:** Very thin, fragile, yellowish. **ODOR:** Indistinct. **TASTE:** Mild. **SPORE DEPOSIT:** White. **MICROSCOPY:** Spores 7–10 x 2.5–4 µm, broadly cylindrical with a pointed end, smooth, inamyloid. Cheilocystidia abundant, club to spindle shaped, with a rounded head, and with yellow contents. Pleurocystidia similar, infrequent.

ECOLOGY: In small groups, scattered troops, or solitary in duff and debris of all forest types, especially common under Douglas-fir and redwood. Fruiting from first rains in fall until spring, but fruitbodies dry out fast and thus are mostly found during peak of wet spells.

EDIBILITY: Unknown. It would take hundreds to make a bite.

COMMENTS: The tiny size, bright yellow to yellow-orange coloration, and thin, brightly colored stipe help distinguish this species from other small, brightly colored *Mycena*. *M. acicula* has a reddish orange cap when young and a pale yellowish stipe. *M. aurantiidisca* differs by having a conical orange cap when young that fades to yellow; it is stockier overall.

Mycena epipterygia

(Scop.) Gray

YELLOW-STEMMED MYCENA

CAP: 0.7–2 cm across, egg shaped at first, soon conical to bell shaped, becoming broadly conical to convex in age. Grayish olive to grayish green, yellowish gray to yellowish white. Surface smooth, or dusted with fine whitish powder, sticky to viscid. **GILLS:** Attached with a decurrent tooth, close to subdistant. Whitish to pale gray, sometimes developing pale pinkish or yellowish colors in age. **STIPE:** 3–7 cm long, 0.1–0.2 cm thick, equal or slightly enlarged toward base. Bright yellowish green when young, becoming pale yellowish green to pale yellowish or yellowish white. Surface smooth, viscid to sticky, base with fine off-white to grayish hairs. **FLESH:** Very thin, fragile, grayish white. **ODOR:** Faint, pleasant to unpleasant, often like cucumbers or sometimes rancid-farinaceous in age. **TASTE:** Farinaceous. **SPORE DEPOSIT:** White. **MICROSCOPY:** Spores 9–12 (13) x 5–7µm, ellipsoid, smooth, thin walled, amyloid.

ECOLOGY: Solitary or in small groups on duff and in moss, or on rotting coniferous wood in a wide variety of forest types. Uncommon in the southern part of our range, becoming locally common northward. Fruiting from early fall through winter.

EDIBILITY: Unknown.

COMMENTS: The viscid yellow stipe and some combination of grayish olive to grayish yellow cap colors make this *Mycena* easy to identify. The many described varieties of *M. epipterygia* include var. *viscosa*, which has a strong farinaceous odor and taste, and a tendency to develop dirty brownish stains. *M. nivicola* has a dark brown to yellow-brown cap and a viscid to dry, yellow stipe; however, it is restricted to the mountains and fruits on woody debris or duff at the edges of melting snowbanks.

Mycena californiensis

(Berk. & M. A. Curtis) Sacc.

CALIFORNIAN BLEEDING MYCENA

CAP: 0.7–2.5 cm across, conical to bell shaped at first, becoming broadly conical, often with an umbo and a pleated margin. Brownish orange, peachy orange to buff-brown, darker toward center, margins paler buff-brown to orange brown. Surface often with a minutely velvety sheen, smooth as it ages, sulcate and translucent-striate on margin. **GILLS:** Attached, widely spaced. Pinkish white to pale peach-white, often developing brownish orange blotches. Marginate, with an orange-buff to peachy orange edge that is often concolorous with stipe. **STIPE:** 2–8 cm long, 0.1–0.4 cm thick, fragile, equal or slightly enlarged toward base. Peachy orange, orange-buff to brownish orange. Surface smooth, moist to dry, base covered with white to pale orange hairs. Hollow, exuding reddish orange juice when fresh specimens are broken. **FLESH:** Very thin, fragile, almost translucent in stipe. **ODOR:** Indistinct. **TASTE:** Mild. **SPORE DEPOSIT:** White. **MICROSCOPY:** Spores 8–12 x 4–6 μm, ellipsoid to almond shaped, smooth, thin walled, amyloid.

ECOLOGY: Solitary, scattered, or in small clusters or troops under oak. Very common in duff under Coast Live Oak in the southern half of our range. Fruiting from midfall into winter.

EDIBILITY: Unknown.

COMMENTS: This common "bleeding" *Mycena* can be told from similar species by the orange-brown colors, orange gill edges, and preference for live oak litter. Similar species include *M. sanguinolenta*, which is generally smaller and more reddish brown or vinaceous overall, bleeds dark red juice, and appears to be restricted to conifer debris (at least in California). *M. haematopus* grows on wood and bleeds reddish juice, but does not have marginate gills. *M. purpureofusca* is purple to purplish gray, has vinaceous marginate gills, and usually grows on pinecones or pine branches. *M. bulliformis* also has a pinkish purple to purplish gray cap, but has whitish gills and grows on wood, often in clusters. *M. aurantiomarginata* has a duller olive brownish cap (although often with a brilliant orange margin) and bright orange to peachy or yellow-orange gills, and more often grows under redwood, Douglas-fir or in mixed Tanoak forest.

Mycena haematopus

(Pers.) P. Kumm

BLEEDING MYCENA

CAP: 0.8–3.5 cm across, rounded to bell shaped at first, becoming broadly convex. Pinkish brown, reddish brown to vinaceous brown at center, lighter reddish brown to pinkish buff toward margin, often developing reddish brown to purplish brown stains. Surface very finely powdered, giving it a sheen at first, becoming smooth, dry to moist, translucent-striate in age. Margin with overhanging flaps of sterile tissue, crimped at first, becoming ragged in age. **GILLS:** Attached, close to subdistant. Pale pinkish white to pale pink when young, becoming pinkish and developing dark reddish to purplish brown stains, spots, and blotches. **STIPE:** 2–10 cm long, 0.15–0.5 cm thick, fragile, equal or slightly enlarged toward base. Reddish, reddish brown, reddish gray to violet-gray. Surface smooth, dry to moist, base covered with off-white to grayish hairs. Hollow, exuding dark red, reddish brown to wine-colored juice when fresh specimens are broken. **FLESH:** Thin, fragile, reddish brown, bleeding red, like stipe. **ODOR:** Indistinct. **TASTE:** Mild. **SPORE DEPOSIT:** White. **MICROSCOPY:** Spores 9–11 x 5.5–7 µm, ellipsoid, broadly ellipsoid to almond shaped, smooth, thin walled, amyloid.

ECOLOGY: Fruiting in clusters, occasionally in scattered troops or solitary on rotting wood of both hardwoods and conifers. Very common in fall and winter throughout our range.

EDIBILITY: Unknown.

COMMENTS: This bleeding *Mycena* is easily identified by the reddish brown cap with ragged overhanging tissue at margin, growth on wood, and dark red juice it bleeds when broken (especially from the stipe). *M. sanguinolenta* is quite similar, but is smaller, generally grows on conifer duff, and has colored gill edges. *M. californiensis* is more orange-brown, has orange gill edges, grows on oak duff, and bleeds reddish orange juice. *M. purpureofusca* has a purplish to violet-gray cap, grows on pine debris, does not bleed, and has purplish gill edges. *M. bulliformis* also grows in clusters on rotting wood, but does not bleed when broken; nor does it have the overhanging tissue on the margin. *M. maculata* has a gray cap that develops reddish stains and blotches in age, grows on decaying conifer wood, and does not bleed when broken.

Mycena purpureofusca

(Peck) Sacc.

CAP: 0.8–2.5 cm across, conical to bell shaped at first, becoming broadly bell shaped, convex to nearly flat. Dark purple to reddish purple when young, becoming purple to purplish pink over disc, paler toward margin, becoming purplish brown to purplish gray in age. Distinctly translucent-striate when young. Surface dull, dry to moist. **GILLS:** Attached, close to subdistant. White to off-white at first, becoming pale pinkish white to pale purplish white. Marginate, edges purple to purplish red. **STIPE:** 2–8 cm long, 0.1–0.3 cm thick, fragile, slightly enlarged toward base. Dark purplish gray to purple at first, becoming purplish, reddish purple, purplish brown to purplish gray. Surface finely pubescent to smooth, dry to moist, base covered with purplish, off-white to grayish hairs, hollow, not exuding juice when broken. **FLESH:** Very thin, fragile, purplish. **ODOR:** Indistinct. **TASTE:** Mild. **SPORE DEPOSIT:** White. **MICROSCOPY:** Spores 9–12 x 6–8 µm, broadly ellipsoid, smooth, thin walled, amyloid.

ECOLOGY: Solitary, scattered, or in small clusters on cones, twigs, branches, bark, or other debris of pines or occasionally other conifers. Very common in fall and winter in pine woods throughout our range.

EDIBILITY: Unknown.

COMMENTS: This *Mycena* is readily identified by the purple coloration, white to pale pinkish purple gills with dark purple edges, and growth on cones and woody debris of pines. *M. bulliformis* is very similar, but generally grows in clusters on decaying logs; it has white gills when young that are only faintly marginate, with violet-brown edges. It has a cap cuticle made of doorknob-shaped (bulliform) cells. *M. sanguinolenta* usually grows on conifer duff, has reddish brown gill edges, and bleeds a dark reddish juice from the stipe when broken. *M. californiensis* has an orange-brown cap and orange gill edges, grows on oak duff, and bleeds reddish orange juice from the stipe when broken. *M. maculata* has a much grayer cap that develops reddish stains, grows on decaying conifer wood, and has nonmarginate gill edges.

Mycena pura group
LILAC MYCENA

CAP: 1–4 cm across, bell shaped to convex at first, becoming broadly convex to plane, usually with a broad umbo. Margin incurved when young, often wavy, ragged and occasionally slightly uplifted in age. Color extremely variable, generally some shade of purple, purple-gray, lilac, pinkish lilac to pinkish gray at first, becoming more gray as it ages, or becoming dull grayish lilac, gray to whitish when dry. Strongly hygrophanous, translucent-striate and shiny when wet, dull, opaque when dry. Surface smooth, moist to dry.
GILLS: Broadly attached or notched with a slightly decurrent tooth, close to subdistant, broad, partial gills numerous, occasionally strongly inter-veined. Lilac-gray, pinkish lilac to purplish when young, becoming grayish, whitish to brownish gray in age. **STIPE:** 3–10 cm long, 0.2–0.8 cm thick, equal or enlarged toward base. Lilac-gray, purplish, gray to grayish brown, often developing yellowish to brownish discoloration at base in age. Surface dry, slightly scurfy at apex when young, becoming smooth, base with whitish to yellowish hairs. **FLESH:** Thin, fragile, stipe hollow, cartilaginous. Purplish, grayish to whitish. **ODOR:** Cucumber- or radishlike.
TASTE: Indistinct, mildly earthy, or radishlike. **SPORE DEPOSIT:** Whitish. **MICROSCOPY:** Spores (7) 7.5–9.5 x 3–5 μm, narrowly ellipsoid, cylindrical to oblong, smooth, thin walled, weakly amyloid (or occasionally inamyloid).

ECOLOGY: Solitary, scattered, or in small clusters in duff or occasionally well-decayed wood. Mostly under conifers, occasionally under hardwoods. In California it prefers redwood and Sitka Spruce duff, but can be found in a wide variety of habitats. Very common throughout our range. Fruiting from early fall in north, well into spring in central and southern part of range.

EDIBILITY: Possibly poisonous.

COMMENTS: The strongly hygrophanous, translucent-striate cap with purple, lilac-gray, or pinkish colors (when young), lilac-gray to pinkish lilac gills and stipe, and radishy odor are distinctive. This is a highly variable species that changes drastically when wet versus dry, or young versus old. There are many described varieties and forms, and recent genetic work has shown that it is part of a species complex. As for most groups, more work is needed before the spores settle. *M. rutilantiformis* shares a similar stature, but has a dark purple to violet-gray cap and purple marginate gills; it is very rare in California and is only known from a couple of collections. *Inocybe lilacina* is often violet-gray when young, but has a cobwebby partial veil, a silky, opaque cap, and brown spores. *Laccaria amethysteo-occidentalis* is larger and has a tough, fibrous stipe and thick, widely spaced gills. *Clitocybe tarda* grows in grass and has pinkish buff spores. *C. nuda* is a large, stocky mushroom unlikely to be mistaken for a *Mycena*.

SYNONYM: *Prunulus purus* (Pers.) Murrill.

Mycena capillaripes

Peck

BEAN SPROUT MYCENA

CAP: 0.5–2.5 cm across, conical at first, becoming broadly conical to bell shaped; occasionally umbonate, margin sometimes uplifted in age. When wet, dark gray to dark grayish brown, fading to pale gray to pale grayish brown; grayish buff to pale pinkish gray or grayish white when dry. Strongly hygrophanous, translucent-striate, with darker lines when wet, dull, much paler when dry. Surface smooth, dry to moist. **GILLS:** Attached, notched, often with a decurrent tooth, close to subdistant, narrow. Whitish, pale grayish white to pale pinkish buff. Marginate, edges darker, although sometimes only faintly so, pinkish brown to pale reddish brown. **STIPE:** 2–8 cm long, 0.05–0.25 cm thick, very fragile, equal or slightly enlarged toward base. Light gray to grayish brown and often developing pinkish tones on lower portions, paler, whitish to pale gray toward apex. Surface smooth, dry, base with whitish to grayish fibrils or hairs. **FLESH:** Very thin, fragile, pallid, grayish. **ODOR:** Like chlorine or bean sprouts, especially when crushed. **TASTE:** Radishy to unpleasant. **SPORE DEPOSIT:** White. **MICROSCOPY:** Spores 8.5–11 (12.5) x 4.5–6.5 (7) μm, ellipsoid, smooth, thin walled, amyloid.

ECOLOGY: Scattered or in small clusters forming large troops in duff and needle litter under pines and other conifers. Very common throughout our range, frequently fruiting in huge numbers soon after first fall rains, but scattered fruitings often persist into spring.

EDIBILITY: Unknown.

COMMENTS: The gray to grayish brown, translucent-striate cap that fades in age or when dry; whitish to pale gray gills that have a darker pinkish brown edge (use a hand lens to check); chlorine or bean sprout–like odor; and growth in conifer duff are distinctive. Older specimens could be mistaken for faded *M. rosella,* which has pinkish gills with darker rosy pink edges. *M. olivaceomarginata* has a brown to yellowish brown cap and grayish gills with yellowish pink to reddish brown edges, and seems to be restricted to grassy lawns; it is common in urban areas on the Far North Coast. *M. citrinomarginata* has an extremely variable cap color (sometimes overlapping with the range shown by *M. capillaripes*), but is usually more yellowish brown and has gills that are whitish to pale grayish and have a darker yellow, yellowish orange, or yellowish brown edge. It is uncommon on the Pacific Coast, but occurs under Sitka Spruce in the northern part of our range. *M. leptocephala* has a darker cap, strong chlorine odor, and often grows on woody debris.

Mycena amicta

(Fr.) Quél.

BLUE-FOOT MYCENA

CAP: 0.4–1.5 cm across, egg shaped to conical at first, becoming broadly conical to bell shaped, often with a rounded translucent umbo, margin occasionally upturned in age. Color variable, extensively bluish, turquoise-blue to bluish gray, or occasionally lacking blue tones; more dark gray to gray-brown at first, soon fading to gray or grayish brown, retaining bluish on margin, to pale gray or whitish in age. Surface slightly powder at first, becoming smooth, moist to slightly viscid. Skin often separable as a gelatinous layer. **GILLS:** Narrowly attached, notched, occasionally with a decurrent tooth, often breaking away from stipe in age, close to subdistant, narrow. Whitish, pale grayish to pale grayish brown. **STIPE:** 3–10 cm long, 0.1–0.2 cm thick, equal, often curved and rooting at base. Pale gray to grayish brown, base blue to blue-gray and often covered with fine to coarse whitish to bluish hairs. Surface often covered with fine, very short whitish hairs, giving it a powdered appearance, becoming smooth, dry. **FLESH:** Very thin, fragile, concolorous with cap, stipe hollow, grayish. **ODOR:** Indistinct. **TASTE:** Mild. **SPORE DEPOSIT:** White. **MICROSCOPY:** Spores 7–9 x 4–5 µm, ellipsoid, smooth, thin walled, amyloid.

ECOLOGY: Solitary or scattered in small groups in duff or on well-rotted branches or other woody debris of both conifers and hardwoods. Widespread, but only locally common in California. Easily overlooked or confused for any of the other myriad gray *Mycena* after bluish colors fade. Fruiting in fall in north, well into winter southward.

EDIBILITY: Unknown.

COMMENTS: When this *Mycena* is young, the blue coloration renders it practically unmistakable; however, it fades quickly as it matures, and the blue often remains only at the very base of the stipe in age. Other distinguishing features include the easily separable, gelatinous cap cuticle, often translucent umbo, and fine whitish fibrils and scurfs covering the young stipes.

Mycena galopus

(Pers.) P. Kumm.

MILKY MYCENA

CAP: 0.7–2 (2.5) cm across, conical to bell shaped at first, becoming broadly bell shaped to convex. Dark grayish brown to gray at center; lighter gray, pale buff-gray to grayish white toward margin, often with darker lines of striations. Surface smooth, or dusted with fine whitish powder, dry to moist, translucent-striate. **GILLS:** Attached, subdistant. Bright white, whitish to pale gray, occasionally developing dirty stains. **STIPE:** 3.5–9 (12) cm long, 0.1–0.3 cm thick, equal, but often with a slightly enlarged and curved base. Gray to pale gray or grayish brown, becoming grayish white in age. Hollow, exuding a white to watery whitish juice when broken. Surface dry to moist, smooth or with fine fibers above, base covered in long white hairs, occasionally exuding whitish droplets near apex. **FLESH:** Thin, fragile, grayish white, exuding white juice when broken, at least in fresh, young specimens. **ODOR:** Indistinct. **TASTE:** Mild. **SPORE DEPOSIT:** White. **MICROSCOPY:** Spores 9–12 x 5–7 µm, ellipsoid to almond shaped, smooth, thin walled, amyloid.

ECOLOGY: Scattered or in small clusters, often in large troops in duff. Typically under redwood, but occurs under other conifers and hardwoods. Very common in northern part of our range, fruiting soon after first fall rains. Locally common southward, fruiting well into winter.

EDIBILITY: Unknown.

COMMENTS: The Milky Mycena is easily dismissed as one of myriad gray *Mycena*, but is distinguished from all the others by the opaque white "milk" that it exudes when the stipe is broken. Although this feature can be fleeting, it is not shown by any other *Mycena* in California.

Mycena leptocephala

(Pers.) Gillet

DARK MYCENA

CAP: 0.5–3.5 cm across, egg shaped to conical at first, becoming broadly conical to bell shaped. Dark gray, dark grayish brown to nearly black at first, fading slightly, gray to light grayish brown but retaining the dark striations. Extreme margin often paler, whitish to pale gray. Surface with a powdery bloom when young and dry, becoming smooth, moist to dry, translucent-striate. **GILLS:** Narrowly attached, notched, occasionally with a decurrent tooth, close to subdistant. Gray, grayish to whitish, edges often paler. **STIPE:** 2–8 cm long, 0.1–0.3 cm thick, fragile, equal, or enlarged slightly toward base. Dark gray to dark grayish brown, fading slightly in age. Surface finely powdery at first, becoming smooth, moist to dry. Base tomentose, covered with fine to coarse whitish hairs. **FLESH:** Very thin, fragile, concolorous with cap, stipe hollow, dark gray. **ODOR:** Usually strongly chlorinelike when crushed, becoming fainter in age. **TASTE:** Indistinct. **SPORE DEPOSIT:** White. **MICROSCOPY:** Spores 7.5–12 x 5–6 μm, ellipsoid to nearly teardrop shaped, smooth, thin walled, amyloid.

ECOLOGY: Scattered or clustered in small to large troops in wood chips, soil, duff, and woody debris, or on bark of living and dead trees. Common throughout our range. Fruiting from late fall through winter.

EDIBILITY: Unknown.

COMMENTS: This species is distinguished by the dark cap, ranging from blackish to dark grayish brown (retaining dark striations even after the cap fades), grayish gills, and strong chlorine odor. No other *Mycena* in California is consistently as dark as *M. leptocephala*. We have seen large fruitings in urban areas landscaped with wood chips and in garden soil. Gray *Mycena* are notorious for being tough to identify to species, and most require microscopic examination to distinguish them. *M. abramsii* has a gray to grayish brown cap that fades to pale gray in age, white to pale grayish gills, and a mild to slightly chlorinelike odor; it grows on wood, often in tufts of two or three fruitbodies. It is very common, occurring throughout our range. *M. alnicola* is nearly identical to *M. abramsii*, but generally has a pale bluish to bluish gray powdery bloom when young; it is common on alder and maple on the Far North Coast. *M. atroalboides* has dark brown to grayish brown cap with a powdery bloom, grayish gills that often develop dingy spots or stains, a gray stipe, and a chlorine odor; it grows in conifer duff. The growth in duff, brownish cap tones, and smaller spores (7–10 x 3–5 μm) distinguish it from *M. leptocephala*. Also compare with the paler duff-dwelling *M. capillaripes*.

Mycena maculata

P. Karst

RED-SPOTTED MYCENA

CAP: 1–3 (4) cm across, conical to bell shaped at first, becoming broadly bell shaped to convex, usually with a distinct, low broad umbo. Dark gray, dark grayish brown to blackish brown at first, remaining so on disc, but fading outward to light gray to grayish buff. Developing reddish to vinaceous spots and stains in age. Surface smooth, moist to dry, hygrophanous, translucent-striate when wet, dull when dry. **GILLS:** Narrowly attached, with a broad notch, close to subdistant. Off-white to pale gray, developing dingy reddish brown to vinaceous spots and stains in age. **STIPE:** 3–10 (13) cm long, 0.1–0.4 cm thick, more or less equal, occasionally with a rooting base. Gray to whitish gray, paler toward base, developing reddish brown to vinaceous spots and stains from base up in age. Surface dry to moist, smooth above, base covered in coarse white hairs. Hollow, not bleeding when broken. **FLESH:** Thin, fragile, watery gray to grayish white, slowly staining reddish brown. **ODOR:** Indistinct to faintly farinaceous. **TASTE:** Mild to mildly farinaceous. **SPORE DEPOSIT:** White. **MICROSCOPY:** Spores 9–10.5 x 5–6.5 μm, ellipsoid, smooth, thin walled, amyloid.

ECOLOGY: Often scattered in small to large clusters, rarely solitary, on well-decayed logs, stumps, or woody debris. Prefers coniferous debris, but occurs on that of hardwoods as well. Very common throughout our range. Fruiting from early fall in the north, well into winter in the south.

EDIBILITY: Unknown.

COMMENTS: The gray cap, the often broadly notched gray gills, the growth in clusters on decaying wood, and especially the development of reddish brown to vinaceous spots and stains in age make this *Mycena* easy to identify. Young specimens often lack the reddish staining or show it only at the stipe base. *M. haematopus* has more reddish brown colors, has bits of sterile tissue hanging from the cap margin, and bleeds a reddish juice when broken. *M. galericulata* is very similar, but generally has a taller stature, a tough stipe, and white to pale grayish or pinkish white gills. It lacks reddish stains and generally grows on hardwood stumps and logs. See *M. leptocephala* (p. 311) and *M.capillaripes* (p. 309) for grayish duff-dwelling species.

Mycena filopes

(Bull.) P. Kumm.

CAP: 0.5–2 cm across, egg shaped to rounded at first, becoming conical to bell shaped. Margin with overlapping sterile tissue, often resembling a partial veil when young, becoming jagged as cap expands or disappearing altogether in age. Grayish brown, brown to gray at center, margin paler when young, fading to light gray, pale grayish brown to grayish white. Surface with a powdery whitish to pale grayish bloom when young, becoming smooth, moist to dry, translucent-striate around margin when old and wet. **GILLS:** Narrowly attached, notched, occasionally with a decurrent tooth, close, narrow. Whitish to pale grayish white. **STIPE:** 2–12 (15) cm long, 0.1–0.2 cm thick, equal or enlarged slightly toward base, often tall and slender. Whitish to pale gray at apex, darker, gray, dark gray to grayish brown lower. Surface finely powdery at first, becoming smooth, moist to dry. Base covered with fine to coarse, long whitish hairs. **FLESH:** Very thin, fragile, concolorous with cap, stipe hollow, gray. **ODOR:** Indistinct, or occasionally iodine-like. **TASTE:** Mild. **SPORE DEPOSIT:** White. **MICROSCOPY:** Spores 10.5–11 (12) x 5.5–6.5 μm from 2-spored basidia, 8–10.5 x 4.5–6 μm from 4-spored basidia, ellipsoid, smooth, thin walled, amyloid.

ECOLOGY: Scattered or in loose clusters or troops on branches, twigs, woody debris, and thick duff of both conifer needles and hardwood leaves. Occurs throughout our range, more common northward. Fruiting in fall in the north, well into winter in the south.

EDIBILITY: Unknown.

COMMENTS: The gray to grayish brown cap that often fades to whitish, overhanging tissue at the cap margin, and gray stipe with coarse whitish hairs at the base are distinctive. *M. metata* is nearly indistinguishable from *M. filopes* macroscopically, except it lacks the sterile tissue around the margin of the cap, and its gills often go pinkish in age.

Mycena galericulata

(Scop.) Gray

CAP: 1.5–6 cm across, conical to bell shaped at first, becoming broadly convex to plane, usually with a distinct, broad umbo at all times. Margin incurved when young, often wavy, splitting and occasionally slightly uplifted in age. Dark grayish brown, brown to gray at center, paler toward margin in age, fading to lighter grayish brown, light brown to whitish. Surface often wrinkled or finely sulcate to smooth, slightly viscid to dry, translucent-striate when wet. **GILLS:** Broadly attached to decurrent, sometimes with a broad notch and decurrent tooth, close to subdistant, broad, partial gills numerous, occasionally strongly inter-veined. White to pale whitish gray, developing pinkish tones in age. **STIPE:** 3–15 cm long, 0.2–0.6 cm thick, equal or enlarged lower, slightly compressed or with a twisting grove down middle, and often with a rooting base. Dark to pale grayish brown, paler toward apex, often developing yellowish discoloration in age. Surface dry, smooth, base tomentose, covered in dense white hairs. **FLESH:** Thin, fragile, watery brown to grayish. Stipe hollow, somewhat tough, cartilaginous. **ODOR:** Mild to farinaceous or radishlike. **TASTE:** Indistinct, mildly farinaceous or earthy. **SPORE DEPOSIT:** Whitish to creamy. **MICROSCOPY:** Spores (8) 9–11 (12) x 6–8 μm, ellipsoid to broadly ellipsoid, smooth, thin walled, containing oil droplets, amyloid.

ECOLOGY: Scattered or solitary, rarely in small clusters on stumps, logs, or other coarse woody debris. Seems to prefer hardwoods, especially Tanoak and alder, but occurs on conifers as well. Very common throughout our range. Fruiting from early fall in the north, well into winter in the south.

EDIBILITY: Unknown.

COMMENTS: This *Mycena* can be recognized by the brownish gray cap, whitish gills that often develop pinkish tones in age, tall and often slightly rooting, cartilaginous stipe, and growth on wood. The large size and growth on wood alone help separate it from most other *Mycena*. *M. maculata* is a similar large species on wood, but generally grows in clusters, has grayish gills, and develops reddish brown to vinaceous spots and stains in age. See *M. leptocephala* (p. 311) and its look-alikes for smaller, gray *Mycena*. *Mycopan scabripes* has a slightly darker, brownish to grayish brown, often wrinkled cap, whitish gills that become dingy pinkish white to grayish white in age, and a shorter stipe with fine fibrils. *Inocephalus rigidipus* can look very similar, but grows on the ground, develops darker pink gills in age, and has pinkish buff spores.

Mycena tenax

A. H. Sm.

CAP: 1–3.5 cm across, conical at first, becoming convex, broadly convex to plane, margin sometimes uplifted in age. Dark brown to grayish brown at center, paler, grayish brown to grayish on margin, fading slightly in age. Surface smooth, viscid to moist, translucent-striate when wet. **GILLS:** Broadly attached, with a decurrent tooth, close to subdistant, narrow to moderately broad. White to pale grayish, often developing brownish, reddish brown to dingy pinkish brown spots and stains in age. **STIPE:** 3–7 cm long, 0.1–0.3 (0.4) cm thick, equal or enlarged toward base. Grayish brown to brown, apex paler, grayish white. Surface with sparse whitish powdery specks at first, soon smooth, glutinous to viscid. Base finely tomentose, with fine to whitish to grayish hairs. **FLESH:** Very thin, fragile, grayish, stipe hollow, dark grayish brown. **ODOR:** Farinaceous to rancid-farinaceous. **TASTE:** Farinaceous to disagreeable. **SPORE DEPOSIT:** White. **MICROSCOPY:** Spores 7.5–9.5 x 4–5 μm, ellipsoid, smooth, thin walled, amyloid. Cheilocystidia club shaped or more irregular, covered with cylindrical to irregular, often branching projections on tip or along side. Pleurocystidia spindle shaped to nearly cylindrical.

ECOLOGY: Solitary, scattered, or in small clusters, often forming large troops carpeting ground in duff and needle litter. Very common under Sitka Spruce on the Far North Coast, occasional under redwood as far south as the San Francisco Bay Area. Fruiting from early fall in the north into early winter in the southern part of its range.

EDIBILITY: Unknown.

COMMENTS: The viscid to moist, dark brown to grayish brown cap (often with a paler margin); whitish gills that often develop brownish to reddish brown spots in age; viscid stipe; and farinaceous odor and taste are distinctive. In most years, it is by far the most common *Mycena* in our spruce forests, often carpeting the duff with thousands of fruitbodies. *M. vulgaris* is similar, but often has more decurrent gills; it also differs in its lack of pleurocystidia. *M. clavicularis* also has a slimy stipe, but has a dry to slightly moist cap, subdecurrent gills, more grayish brown to yellowish brown cap colors, and lacks any distinctive odor or taste. It can also occur in large troops under spruce in the north; widespread, but less common in duff under other conifers. *Roridomyces roridus* is generally smaller, has a rounded-convex grayish to whitish cap, white, decurrent gills, and a white stipe coated with a thick layer of slime. *Mycena quinaultensis* is typically larger and has a conical to bell-shaped cap with a darker center and a viscid gray stipe. Stockier specimens of *M. tenax* could be confused with *Gliophorus irrigatus*, but that species is generally larger and has bright white, thicker, waxy gills and a translucent gray to brownish gray stipe.

Roridomyces roridus

(Fr.) Rexer

CAP: 0.4–1.2 cm across, convex to broadly convex, with a sunken center and often pleated margin. Pale beige-white to grayish white, fading to off-white. Surface smooth, moist. **GILLS:** Broadly attached to subdecurrent, close to subdistant. White, off-white to pale grayish cream. **STIPE:** 1–4 cm long, 0.05–0.15 cm thick, equal or with a slightly enlarged base. Whitish, pale gray to translucent white. Surface extremely viscid, covered with a clear gel, quite thick at base. **FLESH:** Very thin, fragile, whitish. **ODOR:** Indistinct. **TASTE:** Mild. **SPORE DEPOSIT:** White. **MICROSCOPY:** Spores 8–10 x 4–5 μm, narrowly ellipsoid, smooth, thin walled, amyloid. Cheilocystidia spindle shaped to nearly cylindrical, often quite irregular. Cap cuticle with a dense layer of inflated cells with narrow stalks and a dingy brownish content.

ECOLOGY: Solitary or in small, scattered clusters on twigs, small branches, and other small woody debris, especially huckleberry and Tanoak twigs. Less often in duff or humus. Locally common throughout our range. Fruiting throughout the wet season.

EDIBILITY: Unknown.

COMMENTS: The small size, slightly pleated beige-white to grayish white cap, and whitish to pale gray stipe with a sheathing coat of slime make this mushroom easy to identify. *Mycena clavicularis* also has viscid stipe, but lacks the thick sheath of slime and has a yellowish to grayish cap. See *M. tenax* (p. 314) for darker, slimy *Mycena*.

SYNONYM: *Mycena rorida* (Fr.) Quél.

Mycopan scabripes

(Murrill) Redhead, Moncalvo & Vilgalys

CAP: 1.5–4 cm across, conical to bell shaped when young, becoming broadly conical to convex, often umbonate at all stages. Margin downcurved when young, occasionally uplifted and splitting in age. Dark gray to grayish brown at center, with a paler margin, hygrophanous, fading slightly in age or when dry. Surface dry, with a powdery whitish bloom when young, smooth to wrinkled. **GILLS:** Narrowly attached, often notched with a decurrent tooth, close to subdistant, narrow to broad, with short partial gills around margin and inter-gill veins. Off-white to pale grayish, occasionally with a pinkish cast, at times developing reddish brown stains in age. **STIPE:** 3–10 cm long, 0.3–0.5 cm thick, equal or enlarged slightly toward base. Pale gray with a whitish, silky sheen, occasionally with darker specks near apex. Surface covered with longitudinally striate fibrils or small scales when young, to nearly smooth in age. **ODOR:** Indistinct. **TASTE:** Mild. **SPORE DEPOSIT:** White. **MICROSCOPY:** Spores 7–10 x 4–6 μm, broadly ellipsoid, smooth, amyloid. Pleurocystidia and cheilocystidia abundant, spindle shaped to cylindrical.

ECOLOGY: Solitary or scattered in small troops in duff or soil under redwood. Uncommon, but widespread; occurring throughout our range. Fruiting from fall into spring.

EDIBILITY: Unknown.

COMMENTS: This mushroom is often mistaken for an *Entoloma* or *Nolanea* at first glance, due to its longitudinally striate fibrils of the stipe, and sometimes pinkish tinged gills, but the wrinkled cap and ellipsoid white spores help differentiate it. Similar gray *Mycena* species either grow on wood (*M. galericulata*) or have translucent-striate caps and thin, fragile stipes. Although *Mycopan scabripes* had been placed in both *Mycena* and *Hydropus*, genetic data suggest a closer relationship to *Baeospora*, hence the new genus. The name invokes the Greek god Pan (and his furry legs) as a reference to the scurfy or silky fibrils of the stipe.

SYNONYMS: *Hydropus scabripes* (Murrill) Singer, *Mycena scabripes* (Murrill) Murrill.

Group C

Xeromphalina cauticinalis

(Fr.) Kühner & Maire

CAP: 1–5 cm across, convex, occasionally with a small umbo and downcurved margin at first, soon plane with a broadly depressed center, margin slightly or strongly uplifted, and usually wavy and/or ruffled when expanded. Variably dull or bright yellowish orange with a brown center; edge of young caps often with a very narrow white edge. Darker specimens more reddish brown near center and yellow toward margin, often with orange-brown blotches. Translucent-striate when wet. Surface slightly viscid to shiny when wet, otherwise moist to dry, rough or slightly wrinkled. **GILLS:** Slightly decurrent, fairly widely spaced, short gills and many cross veins present. Dull yellowish cream when young, soon yellow with occasional orange blotches, edge usually whitish. **STIPE:** 3–8 cm long, 0.2–0.3 cm thick, cylindrical or broader at apex, straight, curved or slightly sinuous. Yellowish beige at first, darkening upward from orangey brown to dark reddish brown to black, in age extensively black with only narrow reddish and yellow bands near apex. Surface covered in fine, pale fuzz, often most conspicuous as pale scurfs at apex. Base with ocher mycelial fuzz, extreme base with fine, wiry black, branching rhizomorphs. **FLESH:** Very thin, concolorous; tough in stipes, cap somewhat fragile. **ODOR:** Indistinct. **TASTE:** Bitter, sometimes slowly. **KOH:** Orangey red to cherry red on cap,

gills, pale part of stipe. **SPORE DEPOSIT:** White. **MICROSCOPY:** Spores 4.5–7 x 3–3.5 µm, ellipsoid to egg shaped, smooth, amyloid. Hyphae near cap margin not encrusted.

ECOLOGY: Large, often contiguous troops of evenly spaced single and clustered fruitbodies. Common throughout our area on coniferous needle duff, especially under redwood. Fruiting from fall into spring, most abundant during wet periods in winter.

EDIBILITY: Unknown.

COMMENTS: The yellow to orangey brown caps, tough, extensively dark stipes, and habit of fruiting in troops or clusters help distinguish this species from most others. Substrate (conifer needle litter or intact logs?) and taste (bitter, unpleasant, or mild?) are important features to note in this genus. A few other *Xeromphalina* are difficult to distinguish, including *X. cornui*, which generally is smaller in stature, has a mild taste, and often grows solitarily or scattered in small troops, not in gregarious clusters like *X. cauticinalis*. Both have longer spores and incrusted hyphae near the cap margin, distinguishing them from *X. cauticinalis*. Another bitter-tasting species, *X. fulvipes*, generally grows on small branches or twigs, but occasionally in thick duff; it differs by having broadly attached gills that are never decurrent and very narrow spores.

Gymnopus confluens

(Pers.) Antonin, Halling & Noordel.

CLUSTERED GYMNOPUS

CAP: 1–5 (6.5) cm across, convex to plane, often umbonate, margin becoming upturned and wavy in age. Light ocher-brown when young and moist, to cinnamon buff in age, strongly hygrophanous; pinkish brown to grayish buff when dry. Surface smooth, moist to dry. **GILLS:** Narrowly attached, with a distinct notch, to free, close to very crowded, narrow, partial gills numerous. Pinkish buff to whitish. **STIPE:** 3–12 (15) cm long, 0.2–0.6 cm thick, equal. Pinkish buff to light brown, paler, whitish toward apex, darkening overall in age. Surface dry, minutely fuzzy over entire length, with a whitish basal tomentum. **FLESH:** Concolorous, thin, tough, leathery, stipe fibrous. **ODOR:** Mild to slightly garlicky. **TASTE:** Indistinct. **SPORE DEPOSIT:** White. **MICROSCOPY:** Spores 7–9.5 (11) x 3.5–4.5 μm, ellipsoid, smooth, inamyloid.

ECOLOGY: Often in arcs or troops of small to large fusing clusters, rarely solitary, in duff in coniferous forest. Locally common on the Far North Coast. Fruiting in fall.

EDIBILITY: Inedible.

COMMENTS: The thin fleshed, ocher-brown to buff, hygrophanous caps, crowded gills, tall fuzzy stipes, and clustered growth help distinguish this *Gymnopus*. *G. villosipes* is similar, but is generally smaller and darker vinaceous brown when young; although it is often clustered, its stipes are rarely fused. *Connopus acervatus* grows in fused clusters, but has reddish brown to lilac-gray caps that fade in age, and smooth stipes.

SYNONYM: *Collybia confluens* (Pers.) P. Kumm.

Gymnopus villosipes

(Cleland) Desjardin, Halling & B. A. Perry

CAP: 1–3.5 cm across, convex to plane with a sunken center, often wrinkled to irregular in age. Dark reddish brown, vinaceous brown to dark brown when young and wet, dull brown when dry, becoming dingy brown to beige in age. Strongly hygrophanous, translucent-striate when wet, dull, much paler when dry. Surface dry to moist, smooth to wrinkled, often leathery in age. **GILLS:** Attached to a gelatinous collar, which can dissolve, making them appear free in age, close to subdistant, narrow, partial gills numerous. Beige to light brown. **STIPE:** 1.5–8 cm long, 0.1–0.3 cm thick, more or less equal, tough, wiry. Reddish brown to dark brown, often paler toward apex. Surface dry, covered with fine, stiff hairs, base with binding mycelium and rhizomorphs. **FLESH:** Very thin, leathery, stipe tough, fibrous, hollow. **ODOR:** Indistinct. **TASTE:** Mild. **KOH:** No reaction. **SPORE DEPOSIT:** White. **MICROSCOPY:** Spores 6.5–10.5 x 3.5–4.5 μm, ellipsoid, smooth, thin walled, inamyloid.

ECOLOGY: Scattered in troops or clusters, occasionally solitary, in duff under conifers or in wood chips. Often in disturbed areas, such as trailsides or urban landscaping. Very common throughout our range, fruiting in large numbers soon after first rains and persisting into spring. Occasionally fruiting in summer in irrigated areas.

EDIBILITY: Unknown.

COMMENTS: This common and widespread little brown mushroom can be recognized by its small size; hygrophanous cap that can be anywhere from reddish to vinaceous brown, light brown to beige; and thin, wiry stipe covered in a fine pale coating of tiny hairs. *G. fuscopurpureus* is quite similar, but is often larger, with a darker vinaceous cap, pale vinaceous gills, a smooth stipe, and a bright green KOH reaction. *G. confluens* has pale beige to yellowish beige colors and grows in fused clusters. Another common wood chip species, *G. subpruinosus*, has a light brown to beige, translucent-striate to sulcate cap and a pruinose stipe.

Mycetinis copelandii

(Peck) A. W. Wilson & Desjardin

GARLIC MUSHROOM

CAP: 0.8–2.5 cm across, convex to plane, occasionally wavy and uplifted in age. Light brown to pale ocher-brown at first, fading to pinkish buff, often paler toward margin. Hygrophanous, translucent-striate when moist, fading in a patchy manner, opaque, dull when dry. Surface smooth to wrinkled, dry. **GILLS:** Broadly attached, close to subdistant. Buff to whitish beige. **STIPE:** 2–7 cm long, 0.1–0.4 cm thick, thin, tough, equal, or occasionally irregularly flattened and grooved. Reddish brown with a beige apex when young, darkening, dark brown to almost blackish at base, light brown at apex in age. Surface finely pubescent, covered with fine whitish hairs. **FLESH:** Concolorous, very thin, tough and leathery in cap, fibrous in stipe. **ODOR:** Very strong, pleasant, like that of garlic, or disagreeable, of rotting garlic. **TASTE:** Garliclike. **SPORE DEPOSIT:** White. **MICROSCOPY:** Spores 13–17 x 3–4 µm, teardrop to club shaped, smooth, inamyloid.

ECOLOGY: Solitary, scattered, or in small clusters and troops on fallen Tanoak leaves, occasionally live oak and Chinquapin as well. Very common throughout our range, often fruiting in early fall, with scattered flushes through early winter.

EDIBILITY: Edible, with caution (see comments) and in moderation. It can be used as a seasoning, although be sure to use specimens that don't smell like rotting garlic (they taste that way, too).

COMMENTS: The light brown cap, pallid gills, minutely fuzzy stipe, growth on Tanoak leaves, and, most importantly, the strong garlic odor are distinctive. You can often smell this species long before you see it! *M. salalis* is nearly indistinguishable macroscopically, except for its habitat. It only grows on the leaves of Salal (*Gaultheria shallon*) and Oregon Grape (*Berberis* species); it is occasionally found in slightly drier areas on the Far North Coast. Another similar species with a strong garlic odor is *M. scorodonius*, which is rare in California; it differs by its paler cap and a smooth stipe. This species has caused gastrointestinal problems when eaten. *Gymnopus brassicolens* has a strongly hygrophanous, reddish brown to dark brown cap with a paler margin, and a twisted, flattened, smooth, and often shiny stipe with a black lower portion. It has a strong, disagreeable odor like rotting cabbage.

SYNONYMS: *Marasmius copelandii* Peck. Recent genetic work has suggested that *Mycetinis* belongs in *Gymnopus*, but it has not been formerly transferred yet.

Micromphale sequoiae

Desjardin

CAP: 0.6–1.2 cm across, convex to plane, center often depressed in age. Light brown to pallid light orangish brown when young, fading to buff to beige in age. Surface smooth to wrinkled, moist to dry. **GILLS:** Broadly attached, close to subdistant, narrow, partial gills numerous. Buff to light brownish, occasionally with orangish tones. **STIPE:** 2–5 cm long, 0.75–1.5 cm thick, tough, thin. Dark reddish brown to black at the base, with a pale brownish to buff apex, darkening in age. Surface pubescent; covered with minute fuzz, dry. **FLESH:** Concolorous, very thin, tough. **ODOR:** Indistinct when young, occasionally slightly disagreeable in age. **TASTE:** Mild to slightly garlicky. **KOH:** No reaction. **SPORE DEPOSIT:** White. **MICROSCOPY:** Spores 6.5–7.5 x 3–4 µm, ellipsoid, smooth, inamyloid.

ECOLOGY: Scattered in troops or in small clusters on twigs and needle litter of redwood. Common throughout our range. Fruiting in fall and early winter.

EDIBILITY: Inedible.

COMMENTS: This species can be identified by the light brown to buff cap, thin stipe with a fine, pale scurfy coating and dark base, lack of a distinct odor, and obligate growth on redwood twigs. *Gymnopus androsaceus* is probably most similar, but has a darker reddish brown cap and a thin, wiry, entirely black stipe with black rhizomorphic threads growing into the surrounding needles; although it often grows in redwood duff, it is not restricted to that habitat and can grow in mixed conifer or hardwood duff. *Mycetinis copelandii* and *M. salalis* are similar in color and stature, but can easily be separated by the pungent garlic odor and the habitat. *M. copelandii* grows on Tanoak or oak leaves, and *M. salalis* on Salal (*Gaultheria shallon*) leaves. A common but undescribed *Gymnopus* species is similar, but differs by its dark brown cap and a smooth, twisted or slightly flattened black stipe. It occurs on redwood duff in the late fall and winter throughout our range.

Gymnopus androsaceus

(L.) Della Maggiora & Trassinelli

HORSEHAIR MUSHROOM

CAP: 0.3–1 cm across, convex to plane, center often depressed in age. Becoming wavy, wrinkled to irregular when dry, then reconstructing after rains. Dark reddish brown when young, margin fading to brown, drying to buff-brown with a darker disc. Surface smooth to wrinkled, moist to dry. **GILLS:** Broadly attached, close to subdistant, narrow, partial gills numerous. Buff to light brownish, occasionally with orangish tones. **STIPE:** 2–6 cm long, 0.05–0.1 cm thick, tough, thin, wiry. Dark brown to black with a light brown to reddish brown apex. Surface smooth, dry, base with thin, hairlike rhizomorphs. **FLESH:** Very thin, tough. **ODOR:** Indistinct. **TASTE:** Indistinct. **KOH:** No reaction. **SPORE DEPOSIT:** White. **MICROSCOPY:** Spores (6) 7–9 x 3–4.5 µm, ellipsoid, to almond shaped, smooth, inamyloid.

ECOLOGY: Scattered in small to large troops on conifer needles and duff, especially pine and redwood. Surrounding needles will often be coated with hairlike black rhizomorphic threads. Very common throughout our range, fruiting in fall into winter.

EDIBILITY: Inedible.

COMMENTS: This tiny species can be readily identified by the reddish brown cap, thin, wiry black stipe, and presence of black rhizomorphic threads covering the surrounding needles. *Micromphale sequoiae* is similar and often grows intermixed with *G. androsaceus*. It differs by having a pallid cap, a minutely fuzzy, reddish brown stipe with a buff apex (and lacking the rhizomorphic threads), and a slightly stockier stature. Another (undescribed) *Gymnopus* is very similar, but typically has a dark brown cap and a slightly thicker, twisted or slightly flattened black stipe. It also occurs on redwood duff, but lacks the surrounding black rhizomorphic threads.

SYNONYM: *Marasmius androsaceus* (L.) Fr.

Micromphale arbuticola

Desjardin

MADRONE MICRO

CAP: 0.3–1.5 (2) cm across, convex when young, becoming broadly convex to plane, occasionally umbonate, more often with a depressed center, margin often ruffled and occasionally slightly uplifted in age. Light to dark brown, becoming pinkish brown to orangey tan in age, paler, to whitish near margin. Surface moist to dry, finely silky with whitish to pale grayish fibrils especially when young, soon nearly smooth. **GILLS:** Broadly to narrowly attached, broad, quite distant, sometimes forked, short gills present but scarce. Pinkish tan to dingy tan-brown, paler to creamy white toward edge. Edge very finely serrate (use hand lens!). **STIPE:** 0.2–3 cm long, 0.1–0.3 cm thick, cylindrical, curved, often slightly off-center. Dark blackish brown at first, becoming dingy pinkish brown or orangish tan. Surface nearly smooth or with a few pale scurfs at apex. **FLESH:** Very thin, quite tough, dark in stipe, fragile and pallid in cap. **ODOR:** Often noticeably unpleasant, like rotten green onions or rotten garlic, or even slightly fecal. **TASTE:** Mild to unpleasant (like odor). **KOH:** No reaction. **SPORE DEPOSIT:** Off-white. **MICROSCOPY:** Spores 6.5–8 x 3–4.5 µm, ellipsoid, smooth, inamyloid.

ECOLOGY: Scattered groups (often in fairly large numbers) on bark plaques of mature Madrone, usually on the darker, sheltered side of trunk. Fruiting whenever ambient humidity is high and bark is soaked, most frequently in winter through spring. Extremely common in appropriate habitat throughout much of our range, not yet reported south of Santa Cruz County.

EDIBILITY: Unknown.

COMMENTS: Although many species of small, trooping mushrooms make their homes on Madrone bark, this species stands out among them due to the very short, dark, curved stipes, widely spaced pinkish tan gills, brownish caps, and unpleasant odor. A few similar-looking *Gymnopus*, *Micromphale*, and *Marasmius* grow on duff and have much taller stipes. The fruitbodies of *Phaeomarasmius* species are hairier, darker, and lack an odor; in our area, they are less common and prefer oak.

Rhizomarasmius undatus

(Berk.) R. H. Petersen

CAP: 0.8–2 cm across, convex to plane, margin occasionally scalloped. Whitish, gray to buff-brown, soon developing ocher to reddish brown spots and stains, or evenly light brown with darker stains in age. Surface smooth, often covered with tiny dew droplets in humid weather, to dry. **GILLS:** Broadly attached to slightly decurrent, well spaced, often with inter-gill veils and short partial gills around margin. White when young, developing brownish spots and stains. **STIPE:** 5–12 cm long, 0.1–0.2 cm thick, tall, thin, often rooting. Reddish brown with a white apex when young, darkening in age. Surface covered with whitish, silky-tomentose fibrils; often adorned with tiny dew droplets in humid weather. **FLESH:** Very thin, whitish, tough in cap, fibrous, brownish in stipe. **ODOR:** Indistinct. **TASTE:** Indistinct. **SPORE DEPOSIT:** White. **MICROSCOPY:** Spores 8–12 x 4–6.5 µm, almond shaped, smooth.

ECOLOGY: Solitary or scattered in duff in mixed redwood forest, fruiting from fall into early winter. Widespread, but quite rare. We have seen it as far south as Big Sur and north to Oregon.

EDIBILITY: Unknown.

COMMENTS: This rare mushroom can be readily identified by its whitish to light brown cap, white gills that stain brownish, and long, thin, often rooting stipe covered with whitish scurfs. Perhaps most striking is its tendency to become entirely covered with clear droplets of moisture in foggy weather.

SYNONYMS: *Marasmius undatus* (Berk.) Fr., *M. chordalis* Fr.

Gymnopus quercophilus

(Pouzar) Antonín & Noordel.

OAK-LEAF PINWHEEL

CAP: 0.2 –0.7 cm across, convex to broadly convex, with an umbo at times, becoming wrinkled or misshapen when it dries out. Buff-brown to pale reddish brown disc with a paler off-white margin when young, fading to buff, pinkish buff to grayish buff on disc, off-white to pale whitish buff on margin. Surface dry, finely powdery to smooth, often slightly sulcate or grooved and often slightly ruffled around margin. **GILLS:** Broadly attached, subdistant to distant. White, off-white to pale pinkish white. **STIPE:** 1–2.5 cm long, 0.05–0.1 cm thick, thin, tough, wiry. Orangish brown to reddish brown with an off-white to pinkish white apex when young, becoming dark reddish brown to reddish black with a paler apex in age. **FLESH:** Very thin, tough. **ODOR:** Indistinct. **TASTE:** Indistinct. **KOH:** No reaction. **SPORE DEPOSIT:** White. **MICROSCOPY:** Spores 8.5–10 (11.5) x 3–4.5 µm, ellipsoid to almond shaped, smooth, inamyloid.

ECOLOGY: Small troops (5–20+ caps) directly on individual, relatively intact (not much decayed) leaves of Tanoak and live oak. Each leaf can host multiple genetically distinct individuals of this species, with zigzagging black lines indicating boundaries where two individuals meet. Very common under Tanoak throughout our range, fruiting from early fall into early winter. Generally less common on live oak leaves, but in some areas abundant on that host, especially where Tanoak is absent.

EDIBILITY: Tiny, tasteless, probably nontoxic, but not certainly known.

COMMENTS: This tiny mushroom sometimes fruits in carpets of thousands on Tanoak and oak leaves; this growth habit in combination with the tiny fruitbodies, pale caps, and dark wiry stipes helps identify it. *Marasmius epiphyllus* is a tiny mushroom with a flimsy whitish cap and pinkish brown to reddish brown stipe; it grows on wet, rotting leaves of Pacific Dogwood and cottonwoods. *G. androsaceus* is a small reddish brown *Gymnopus* with black rhizomorphs that grows on conifer debris. *Micromphale sequoiae* is slightly larger and grows on redwood duff and needles.

SYNONYM: *Marasmius quercophilus* Pouzar.

Crinipellis piceae

Singer

SPRUCE NEEDLE CRINIPELLIS

CAP: 0.2–0.8 cm across, convex to plane. Whitish base color, with brown radiating, matted fibrils, and a reddish brown disc. Surface dry, covered with radiating, matted fibrils that often overlap margin. **GILLS:** Narrowly attached with a deep notch to free, close to subdistant, with short partial gills. White. **STIPE:** 1.5–3.5 cm long, 0.05–0.1 cm thick, thin, tough, wiry. Dark brown to blackish with a paler apex and a darker base, with a whitish bloom when young. Surface dry, covered with fine fibrils. **FLESH:** Very thin, tough. **ODOR:** Indistinct. **TASTE:** Indistinct. **KOH:** No reaction **SPORE DEPOSIT:** White. **MICROSCOPY:** Spores 7–9 x 3–4.5 µm, narrowly ellipsoid to cylindrical, smooth, inamyloid.

ECOLOGY: Solitary or scattered in troops on needles of Sitka Spruce. Very common on the Far North Coast, locally common southward to central Mendocino County. Fruiting in flushes soon after first rains in fall, but can be found throughout season in areas with heavy fog drip.

EDIBILITY: Unknown; minuscule, tough.

COMMENTS: This tiny mushroom can easily be identified by the brown, radiating, matted fibrils on the cap, tough, wiry stipe, and growth on spruce needles. It is one of the most common fungi of the coastal forests of Humboldt and Del Norte Counties, but because of its small size, it is easily overlooked. *Gymnopus androsaceus* grows in a host of different coniferous duff and has a smooth, reddish brown cap. *Micromphale sequoiae* is slightly larger, grows on redwood needles, and has a smooth cap.

Cryptomarasmius corbariensis

(Roum.) T. S. Jenkinson & Desjardin

OLIVE-LEAF PINWHEEL

CAP: 0.1–0.7 cm across, convex to plane, with pleated groves, and a depressed center in age. Dark brown overall at first, soon with a dark brown center and a yellow-brown to orange-brown margin. Surface dry, smooth to wrinkled, with pleated groves. **GILLS:** Broadly attached sparse, very well spaced, occasionally inter-veined. White when young, cream to pale creamy yellow in age. **STIPE:** 0.5–3 cm long, 0.04–0.1 cm thick, very thin, wiry, tough. Dark brown to black with a paler apex. Surface smooth, shiny, dry. **FLESH:** Concolorous, very thin, tough. **ODOR:** Indistinct. **TASTE:** Indistinct. **KOH:** No reaction. **SPORE DEPOSIT:** White. **MICROSCOPY:** Spores 8–10 x 3.5–5 µm, ellipsoid to cylindrical, smooth, inamyloid.

ECOLOGY: Scattered in troops on dead leaves and duff of olive trees, fruiting after soaking rains from fall into spring. Range in California is poorly known. Else Vellinga and Nhu Nguyen have found it to be common under olive trees around Berkeley, and there are reports from Santa Cruz, California and the Sierra Foothills.

EDIBILITY: Better stick to the olives.

COMMENTS: This tiny mushroom can be readily identified (but not found) by its tiny size, pleated orange-brown cap, widely spaced white gills, and growth on olive leaves. Interestingly, it is not closely related to *Marasmius* despite its superficial resemblance to many members of that genus. *Marasmius curreyi* is similar, but a bit more orange overall and grows in grass.

SYNONYM: *Marasmius corbariensis* (Roum.) Sacc.

Marasmiellus candidus

(Bolton) Singer

CAP: 0.5–3.5 (6) cm across, round to convex when young, becoming broadly convex to nearly plane, center often depressed in age. Margin translucent-striate when wet, grooved or pleated; often wavy when dry. White to translucent white at first, staying white, or developing creamy yellowish, pinkish to brick red blotches in age (young caps rarely entirely bubblegum pink). Surface finely powdery to smooth, moist to dry. **GILLS:** Broadly attached to decurrent, sparse, very widely spaced, with few to many inter-gill veins. White, or developing creamy yellowish or pink to reddish spots, splotches, or overall color in age. **STIPE:** 0.5–2.5 (4) cm long, 0.1–0.4 cm thick, central to off-center, curved away from center of cap into substrate; more or less equal, often with a circular pad at attachment. Whitish to pale grayish, darkening upward from base, from brown to ashy black, in age more than half of stipe may be dark. Surface dry, with fine, silky appressed fibrils to smooth **FLESH:** Concolorous, very thin, tough-leathery. **ODOR:** Indistinct. **TASTE:** Mild. **KOH:** No reaction. **SPORE DEPOSIT:** White. **MICROSCOPY:** Spores 11.5–16 x 3.5–5.5 μm, elongated teardrop shaped to subfusiform (an elongated spore tapered toward one end and rounded on the other), smooth, inamyloid.

ECOLOGY: Scattered in small to large troops consisting of small clusters of fruitbodies, occasionally solitary, on twigs, branches, and mossy logs of both conifers and hardwoods, shrubs and herbaceous plants, especially *Rubus* spp. Very common throughout our range. Generally fruiting with a large flush in early to midfall in north, late fall in south, then scattered fruitings in wet periods through winter and spring.

EDIBILITY: Unknown.

COMMENTS: The white cap; sparse, widely spaced gills with radiating veins between them; short, curved stipe with a dark base; and growth on woody debris make this common mushroom easy to identify. Occasionally it will develop creamy yellow, pink or reddish colors, especially in age. *Campanella* "Veiny" differs by the more veinlike "pseudogills" and tiny or absent stipe. Although whitish when young, it is never bright white like *M. candidus* and often develops bluish gray colors overall. *Tetrapyrgos subdendrophora* is much smaller and has a white cap that develops blackish spots and stains in age, a very short bluish black stipe, and veinlike pseudogills; it grows primarily on the stems of dune grass (*Ammophila* spp.). *M. calhouniae* has a taller stature, more fragile flesh, and a white stipe; it grows in duff (especially that of redwood and pine).

Group D
Marasmius calhouniae
Singer
CALHOUN'S PINWHEEL

CAP: 1–4.5 cm across, convex to plane with a depressed center, margin often uplifted and wavy with a finely scalloped edge. Dull grayish white to bright white, often with a pale yellow wash in age, translucent-striate when wet. One rarely encountered form has a greenish blue to blue-gray cap, especially when young or dry. Surface smooth, moist to dry. **GILLS:** Broadly attached to decurrent, quite widely spaced, often wavy and wrinkly, short gills present as well as many cross veins. White. **STIPE:** 3–7 cm long, 0.2–0.6 cm thick, equal or tapered toward base, cylindrical or irregularly flattened or grooved. White to translucent whitish gray, base with orange-tan to yellowish tones. Surface mostly smooth, to obscurely finely punctate or slightly hairy near base. **FLESH:** Concolorous, thin, somewhat fragile, stipe usually hollow, fragile, base often breaking off when picked. **ODOR:** Indistinct. **TASTE:** Mild. **SPORE DEPOSIT:** White. **MICROSCOPY:** Spores 9–11.5 x 3.5–5 μm, ellipsoid to almond shaped, smooth, inamyloid.

ECOLOGY: In small or large arcs or in huge scattered troops in thick conifer-needle duff, especially that of pines and redwood (rarely under hardwoods). Fairly common throughout much of our range. Fruiting from midfall through early winter.

EDIBILITY: Unknown.

COMMENTS: This species is recognized by its small, whitish, thin-fleshed fruitbodies, broadly attached or slightly decurrent gills that are often distinctly wavy and inter-veined, and gregarious growth in conifer duff. Various *Cuphophyllus* can be similar, but have more distinctly decurrent gills, less wavy gills, and solid stipes. *Marasmiellus candidus* can be similar, but has a small, curved stipe with a dark base and grows on woody debris. Some similar-looking *Mycena* form huge troops, but usually have fragile conical caps and longer, more slender stipes.

SYNONYM: Called *Marasmius sp.* (unidentified) in *Mushrooms Demystified*.

Marasmius albogriseus
(Peck) Singer

CAP: 1.5–4 cm across, hemispheric when very young, soon convex, broadly convex to nearly plane with a broadly depressed center and a ruffled and wavy margin. Tan to grayish beige when young, soon grayish tan to whitish beige, hygrophanous; fading when dry. Usually deeply translucent-striate (sometimes only narrowly at margin) when wet; opaque when dry. Surface smooth, greasy when young, soon dry. **GILLS:** Narrowly attached, often obscured by broadness of gills, moderately to widely spaced, occasionally with cross veins. Creamy to beige when young, soon tan, often with grayish pink or vinaceous stains in age. **STIPE:** 2.5–7 (10) cm long, 0.3–0.7 cm thick, equal or occasionally with a rooting base. Ivory white at apex, to pale creamy yellow downward with a yellowish beige to ocher-yellow base. Surface dry, smooth at apex, matted hair over lower stipe; base of stipe usually embedded in a thick, white, corky mycelium binding together large chunks of substrate. **FLESH:** Thin, tough, whitish to pale gray. **ODOR:** Indistinct. **TASTE:** Indistinct. **KOH:** No reaction. **SPORE DEPOSIT:** Whitish. **MICROSCOPY:** Spores 6.5–8 x 3.5–4.5 (5) μm, broadly ellipsoid to almond shaped, smooth, inamyloid.

ECOLOGY: In small clusters, troops, arcs, or rings under trees with heavy accumulations of leaf litter and humus. Especially in Monterey Cypress groves, as well as *Ceanothus* spp. thickets and live oak woodlands. Common in southern California, uncommon in our area. Fruiting from winter into spring.

EDIBILITY: Unknown.

COMMENTS: This mushroom could be mistaken for many others, but to a trained eye, it is quite distinctive: the whitish beige to grayish cap and tough stipe with a yellowish, hairy lower portion help distinguish it. Many *Clitocybe* are also grayish tan, but generally have close, decurrent gills and squatter stipes and are less tough fleshed. *Marasmius oreades* is very similar in all respects except color and habitat; it has a warmer beige-ocher cap without grayish tones and grows in grassy lawns. *M. calhouniae* is flimsier, with thin flesh, more decurrent gills, and a fragile stipe.

Marasmius oreades

(Bolton) Fr.

FAIRY RING MUSHROOM

CAP: 1–6 cm across, rounded to bell shaped when young, becoming broadly convex to plane, margin often wavy in age, edge sometimes ruffled. Pinkish tan to ocher-beige with a darker edge at first, soon paler to beige, strongly hygrophanous; fading when dry, or in age, often to nearly white. Surface smooth, greasy when young, soon dry. **GILLS:** Narrowly attached or free, moderately close to well spaced, broad. Creamy whitish to pale beige, becoming ocher blotched or overall warm tan in age. **STIPE:** 3–7 cm long, 0.3–0.5 cm thick, cylindrical and often quite straight. Ivory to creamy or beige-tan near apex, ocher-tan downward. Smooth to finely pruinose near apex, finely velvety to obscurely matted-tomentose over lower half when young, or smooth in age. Base attached to a sometimes noticeable corky mycelial mat. **FLESH:** Thin, somewhat tough, buff to whitish in cap; tough, fibrous, whitish in stipe. **ODOR:** Mild to slightly almondy or sweet, chemical-like. **TASTE:** Mild. **KOH:** No reaction. **SPORE DEPOSIT:** Whitish. **MICROSCOPY:** Spores 6.5–9 x 4–6 μm, broadly ellipsoid, almond shaped to lemon shaped, smooth, inamyloid.

ECOLOGY: In clumps, arcs, and rings, often fruiting in early fall and spring, or summer in watered lawns. Occurring almost exclusively on grass lawns. Irregularly distributed over our entire area, can be locally abundant.

EDIBILITY: Edible and good, but not recommended, as it often grows in areas treated with herbicides or polluted by urban emissions.

COMMENTS: Few other mushrooms that regularly occur on grass lawns could be confused for this species. The warm buff-beige caps, tough stipes with a fuzzy lower surface, and habit of forming rings or arcs on lawns are a distinctive combination of characters. As with many mushrooms, care should be taken when eating it from treated lawns or roadsides, two places where it commonly grows, as the fruitbodies of some mushrooms concentrate heavy metals. *M. albogriseus* is similar, but generally has slightly grayer tones and grows in duff in forested habitats. *M. oreades* shares the same habitat as the poisonous *Clitocybe rivulosa*, but is unlikely to be mistaken for that species.

Marasmius plicatulus

Peck

RED PINWHEEL

CAP: 1–4 (5.5) cm across, conical to bell shaped when young, becoming broadly conical to convex, often umbonate in age. Deep wine red, burgundy to reddish brown, but more orange-brown to reddish tan colors are possible. Often paler toward margin, at least in age. Surface dull, dry, finely velvety and often with a whitish sheen. Water droplets often ball up on cap when wet. **GILLS:** Narrowly attached, subdistant to distant, broad, often with 1 or 2 partial gills in between each full gill. Pale pinkish white to creamy white when young, becoming pale buff to pale yellowish buff. **STIPE:** 6–12 cm long, 0.2–0.4 cm thick, more or less equal, tough, wiry, fibrous. Colors quite variable, white at extreme apex when young, then neon or pastel magenta, to reddish or purplish, transitioning through dark reddish brown or purplish brown to black toward base; often extensively polished reddish brown or blackish in age. Surface smooth, shiny, base covered with a "pom-pom" of spiky whitish mycelial hairs. **ODOR:** Indistinct. **TASTE:** Indistinct. **SPORE DEPOSIT:** White. **MICROSCOPY:** Spores 12–16.5 x 4.5–6.5 µm, almond shaped to spindle shaped, smooth, inamyloid.

ECOLOGY: Solitary, scattered, or in small groups on duff or leaves. Very common throughout our range in wide variety of habitats. Particularly fond of cypress, eucalyptus, redwood, and oak duff. Fruiting in large numbers in fall and winter over much of our range, commonly from early to midfall in northern Sitka Spruce forests.

EDIBILITY: Unknown.

COMMENTS: This beautiful and unmistakable mushroom can be found in large troops in many different habitats; it is a notable habitat generalist among gilled mushrooms in our area. The deep wine red to purplish red cap is typical of collections from central Mendocino County southward. An orange-brown to reddish brown or brown-capped form becomes more common northward. A pastel pink form (lacking any black tones on the stipe) is rarely encountered and may be a different species.

Connopus acervatus

(Fr.) K. W. Hughes, Mather & R. H. Petersen
CLUSTERED CONNOPUS

CAP: (0.7) 1.5–4 cm across, convex with a thinly inrolled margin at first, becoming broadly convex, plane or centrally depressed in age. Reddish purple when young, soon lilac brown, fading to brownish in age, strongly hygrophanous; pinkish brown to buff when dry. Surface smooth, moist to dry. **GILLS:** Narrowly attached, with a curved notch, to free in age, close to crowded, narrow, partial gills numerous. Whitish to pale pinkish buff. **STIPE:** 3–10 cm long, 0.2–0.5 cm thick, equal, fragile. Reddish purple when young, soon reddish brown, darkening in age. Surface smooth, base covered with a whitish tomentum when young. **FLESH:** Concolorous, thin, fragile in cap, brittle, but fibrous in stipe. **ODOR:** Indistinct. **TASTE:** Indistinct. **KOH:** No reaction. **SPORE DEPOSIT:** White. **MICROSCOPY:** Spores 5.5–7 x 2.5–3 μm, cylindrical to ellipsoid, smooth, inamyloid.

ECOLOGY: Small or large fused clusters of overlapping fruitbodies, often growing from well-decayed stumps, logs or buried roots, or thick duff. Common under Sitka Spruce on the Far North Coast, occasional in mixed coniferous forest south to Mendocino County. Fruiting in fall or early winter.

EDIBILITY: Not edible, possibly mildly poisonous.

COMMENTS: The clustered growth, reddish brown to lilac-gray, hygrophanous caps, and smooth stipes are distinctive. Faded specimens could be mistaken for *Gymnopus confluens*, but that species is usually taller and less tightly clustered, and has a minutely fuzzy stipe. *Baeospora myriadophylla* is never as strongly clustered and is more purple when fresh. *G. fuscopurpureus* is much darker, is rarely clustered, and turns green in KOH. *G. brassicolens* has a brown cap that fades to ocher-buff or tan when dry, but has a black lower stipe and a strong disagreeable odor.

SYNONYMS: *Gymnopus acervatus* (Fr.) Murrill, *Collybia acervata* (Fr.) P. Kumm.

Gymnopus subpruinosus

(Murrill) Desjardin, Hailing & Hemmes

CAP: 1–7 cm across, broadly convex to plane, often quite wavy in age. Dull beige to tan with brownish grooves, disc sometimes darker brown, usually paler toward margin. Surface dry, with pallid fibrils or a powdery bloom when young; smooth when young, soon bumpy and extensively grooved nearly all the way to disc. **GILLS:** Attached, well spaced, occasionally inter-veined, partial gills numerous. Distinctively clay tan to grayish beige. **STIPE:** 3–10 cm long, 0.2–0.4 cm thick, more or less equal, thin, tall. Whitish to tan. Surface dry, floccose-pruinose over entire length. Base often with white, binding mycelial rhizomorphs. **FLESH:** Very thin in cap, tough and slightly wiry in stipe. **ODOR:** Indistinct. **TASTE:** Indistinct. **KOH:** Darkening slightly. **SPORE DEPOSIT:** White. **MICROSCOPY:** Spores 7.5–9 x 3.5–4 μm, ellipsoid to teardrop shaped, smooth, thin walled, inamyloid, colorless in KOH.

ECOLOGY: Often clustered, sometimes in large arcs or rings in lignin-rich humus, wood chips, or thick duff, usually in urban areas. Fruiting in early fall through spring.

EDIBILITY: Unknown.

COMMENTS: The deeply striate, pallid beige cap and clay-colored gills in combination with the tough stipe and urban milieu distinguish this species. It is most likely to be confused with *G. villosipes*, which has a much darker brown to vinaceous brown color overall, and a more distinctly umbilicate cap.

Gymnopus brassicolens

(Romagn.) Antonin & Noordel.

STINKING GYMNOPUS

CAP: 1.5–7 cm across, rounded at first, soon broadly convex to plane, with a low umbo, becoming wavy and uplifted, and often with a scalloped margin in age. Color variable, strongly hygrophanous; dark mahogany at first, soon paler reddish brown, and becoming blotchy pinkish beige to tan around center of cap and at margin, often pallid to nearly white at maturity if dry. Margin slightly to deeply translucent-striate. Surface greasy to dry, smooth or bumpy, sometimes with very small patches of silky fibrils when very young. **GILLS:** Narrowly attached to free, close, broad, edges finely serrated. Whitish to cream when young, becoming clay colored or brownish tan, usually with a paler whitish edge. **STIPE:** 4–10 cm long, 0.3–0.6 cm thick, equal or rooting, cylindrical when young but most often compressed and flattened with a central groove or vertically splitting. Dark reddish brown at first, darkening from base up, soon very dark brown to black with a paler apex, covered with a fine whitish dusting over most of stipe. Base with fairly extensive contrasting white mycelium. **FLESH:** Thin, tough and dark in stipe, caps more fragile. **ODOR:** Like rotting cabbage or kale, sometimes reported as garlicky or like sewer gas. **TASTE:** Unpleasant, like odor. **KOH:** Indistinct. **SPORE DEPOSIT:** Whitish. **MICROSCOPY:** Spores 5.5–7.5 x 2.5–4 μm, ellipsoid to teardrop shaped, smooth, inamyloid.

ECOLOGY: Clumps or troops, sometimes in large contiguous patches on wood chips or in decaying leaf litter in urban habitats and other disturbed areas, occasionally in redwood duff or other forested habitats. Locally common throughout our range (tied to appropriate habitat). Fruiting primarily in fall and winter, but can occur any time of year.

EDIBILITY: Unknown.

COMMENTS: This *Gymnopus* is readily distinguished by its tough, extensively black stipe that tends to be grooved and split, unpleasant odor, brownish to pale, deeply striate cap, and growth in disturbed habitats. *G. erythropus* also grows in clusters in wood chips, but lacks an odor and has an orange to reddish orange stipe. *Macrocystidia cucumis* has a dark mahogany cap with a narrow pale margin and a cucumber odor; it tends to have a taller stature, and microscopically is distinguished by very large gill cystidia. *Gymnopus* "Dark Redwood Duff" is a smaller, undescribed species that has a dark brown cap, smooth, twisted or slightly flattened black stipe, and mildly disagreeable odor; it is common in redwood duff throughout our range.

Gymnopus dryophilus

(Bull.) Murrill

OAK-LOVING GYMNOPUS

CAP: 2–8 cm across, convex with a narrow inrolled margin and often with a low, round umbo at first, becoming plane, to uplifted and wavy and occasionally centrally depressed in age. Warm brick to ocher-brown at first, then honey colored to beige, finally pale tan, hygrophanous, fading to pale beige or whitish when dry. Surface smooth, greasy when young, soon dry. **GILLS:** Quite variably attached, ranging from broadly attached with a decurrent tooth to nearly free or seceding, close to crowded, soft, greasy feeling, partial gills numerous. Pale beige-tan, pinkish white to bright white. **STIPE:** 3–10 cm long, 0.3–1 cm thick, usually curved, cylindrical at first but often becoming irregularly flattened and grooved. Dull pale beige to tan, in some collections brighter honey tan to ocher, especially in age; usually darker ocher-brown near base. Smooth to fibrillose-wrinkled. Base covered in blotches of hairy-spiky, pale contrasting whitish tomentum, sometimes with stringy white rhizomorphs. **FLESH:** Soft, somewhat tough in cap, cartilaginous in stipe. **ODOR:** Pleasant. **TASTE:** Indistinct. **KOH:** No reaction. **SPORE DEPOSIT:** White. **MICROSCOPY:** Spores 5–7.5 x 3–4 μm, ellipsoid, smooth, inamyloid.

ECOLOGY: In arcing troops or clusters, occasionally solitary. Found in many forest types, but has a particular fondness for live oak humus. Fruiting in early fall, with scattered fruitings into spring.

EDIBILITY: Edible, but not widely collected for food.

COMMENTS: The cheery ocher cap with a smooth greasy surface, smallish (but not tiny) size, and growth in troops help distinguish this relatively featureless fungus. The very similar *Rhodocollybia butyracea* has ragged-edged gills and is larger, while *R. badiialba* is both larger and more darkly colored (reddish brown). *Gymnopus subsulphureus*, which is only known from the redwood forest on the Far North Coast, differs by having a yellow-ocher cap and yellowish gills. *Melanoleuca* can be similarly colored, but have taller, straighter stipes, more crowded gills, and amyloid ornamentation on the spores. *Clitocybe* generally have broadly attached or decurrent gills and paler or "cooler" colors.

Gymnopus fuscopurpureus

(Pers.) Antonin, Halling & Noordel.

CAP: 2–4 cm across, convex with a narrow inrolled margin at first, becoming plane, to uplifted and wavy and occasionally centrally depressed in age. Deep purplish brown to reddish brown, hygrophanous, fading to dull purplish brown to reddish brown when dry. Surface smooth, greasy when young, soon dry. **GILLS:** Broadly attached to nearly free or seceding in age, close to subdistant, soft, greasy feeling, partial gills numerous. Pale purplish brown, with a whitish sheen when young, to light reddish brown in age. **STIPE:** 3–7 cm long, 0.2–0.4 cm thick, more or less equal, or becoming irregularly flattened and grooved in age. Dark purplish brown to reddish brown. Surface finely pruinose, or matted-tomentose over lower portion when young, to smooth or finely fibrillose in age. **FLESH:** Concolorous, soft, somewhat tough in cap, cartilaginous in stipe. **ODOR:** Indistinct. **TASTE:** Indistinct. **KOH:** Green to forest green on all parts. **SPORE DEPOSIT:** White. **MICROSCOPY:** Spores 6.5–8.5 x 3–5 μm ellipsoid, teardrop to spindle shaped, smooth, inamyloid, colorless to greenish in KOH.

ECOLOGY: Solitary or scattered in small groups or clusters in duff or well-rotted woody debris under conifers. Uncommon, fruiting in fall on the Far North Coast, rare south of Mendocino County.

EDIBILITY: Unknown.

COMMENTS: The dark purple-brown colors, hygrophanous cap, and green KOH reaction are distinctive. There are a number of similar *Gymnopus* (most of which are undescribed), but all lack the green KOH reaction. One very similar *Gymnopus* found in coniferous duff on the North Coast is a dead ringer for this species when young, but develops more orangish brown colors in age and has a slight yellowish orange KOH reaction. An undescribed species we have been calling *Gymnopus* "Bay Cap" has a dark bay-brown cap and lacks a distinct KOH reaction; it is common in forest duff throughout our range. *G. villosipes* has a tough, thin-fleshed fruitbody, with a dark reddish brown cap and a thin, finely pubescent stipe. *Macrocystidia cucumis* has a reddish brown cap with a pale margin and a strong odor of cucumbers.

Macrocystidia cucumis

(Pers.) Joss.

CUCUMBER-SCENTED MUSHROOM

CAP: 1–4 cm across, hemispherical when young, becoming broadly convex in age, occasionally with a wavy margin. Blackish brown to dark mahogany with a whitish sheen, paler, reddish brown outward, to an abruptly pale, nearly white extreme margin when wet. Hygrophanous, generally much paler; pale mahogany buff when dry, surface moist or dry, very finely pruinose to smooth. Margin with a narrow sterile band exceeding gills, translucent-striate when wet. **GILLS:** Finely attached to free, close, broad. Pale creamy to beige, becoming cinnamon to pinkish brown in age. **STIPE:** 3–8 cm long, 0.3–0.5 cm thick, equal and rather straight, surface irregularly wavy. Dark reddish brown to nearly black over lower portion, variably pale pinkish tan to brownish over upper portion, to only near extreme apex in age; covered with a whitish pruinose sheen, darkening from base up in age. Surface dry, covered with a dusting of pruinose cystidia, sometimes like crushed velvet, becoming nearly smooth in age. **FLESH:** Thin, tough, springy. Dark in stipe, paler in cap. **ODOR:** Usually distinctive, like cucumber or sometimes unpleasant. Described, rather aptly, by one person as smelling like sushi. **TASTE:** Slightly farinaceous to mild. **KOH:** No reaction. **SPORE DEPOSIT:** Pinkish to dingy salmon tan. Microscopy Spores 7.5–10 x 4–5 μm, ellipsoid, narrowed toward hilar appendage. Cystidia quite large with pointed tips, abundantly scattered over faces and edges of gills, clamps abundant.

ECOLOGY: Often in clumps, small clusters, or scattered troops in wood chips and other disturbed landscaping areas from early fall into winter. Uncommon in our area (recorded as far south as Santa Cruz County), generally more common northward, regular in Pacific Northwest wood chip beds.

EDIBILITY: Unknown.

COMMENTS: The dark cap with a narrow pale edge, dark, tough stipe, and cucumber odor make this a very distinctive species with only one real look-alike: *Gymnopus brassicolens* is similar in many respects, but is dark capped only when young, soon fading and becoming much paler. Its cap also becomes plane or uplifted in age, and the stipe is typically grooved and less upright; its odor is very disagreeable, like rotting cabbage.

Rhodocollybia badiialba

(Murrill) Lennox

CAP: 2.5–7 (10) cm across, convex to plane, often slightly uplifted and wavy in age. Dark reddish brown to vinaceous brown when young, often with a paler margin and disc, fading slightly overall in age. Surface smooth, dry to greasy. **GILLS:** Narrowly attached, with a distinct notch, very close to crowded, soft, edges ragged or serrated, like a saw blade. White to pale pinkish white, occasionally with reddish spots. **STIPE:** 4–12 cm long, 0.5–2 cm thick, equal, or with a tapered, rooting base. Pinkish buff to pale orange-buff, often paler toward apex, and developing reddish brown stains in age. Smooth, dry, striate or grooved, at times with a powdery apex, base with white rhizomorphs. **FLESH:** Thin, soft, somewhat tough in cap, fibrous in stipe. **ODOR:** Mild to pungent. **TASTE:** Slowly bitter. **KOH:** No reaction. **SPORE DEPOSIT:** Creamy pink to pale pinkish orange. **MICROSCOPY:** Spores 3.5–5.5 x 3–4.5 µm, globose to broadly ovoid, smooth, dextrinoid.

ECOLOGY: In small clusters, troops, or occasionally solitary in duff and needles under pine and Sitka Spruce. Locally common on north coast, occasional south of San Francisco Bay Area. Fruiting in the fall in north, well into winter southward.

EDIBILITY: Inedible.

COMMENTS: This species can be readily identified by its dark reddish brown cap, crowded gills with ragged edges, relatively long, fibrous stipe, and pinkish buff spores (globose under a microscope). *R. butyracea* is quite similar, but generally has a greasy, dark brown cap (ellipsoid spores). *R. maculata* has the same stature as *R. badiialba*, but has a creamy beige cap, stipe, and gills, all of which develop reddish spots and stains in age. *R. oregonensis* has a dark reddish brown cap with a paler margin and has a strong almond extract odor.

Rhodocollybia oregonensis

(A. H. Sm.) Lennox

CAP: 3–10 cm across, convex to plane, often with a low, broad umbo, often slightly uplifted and wavy in age. Dark reddish brown to dark brown overall when young, soon fading slightly, to more orange-brown and becoming distantly two toned, with a darker center, and a paler, buff to orange-brown margin. Surface smooth, moist to dry. **GILLS:** Broadly to narrowly attached, often with a distinct notch, close to subdistant, edges finely serrated when young, becoming more ragged in age. Creamy white to pale pinkish buff. **STIPE:** 6–20 (30) cm long, 0.5–2 cm thick, equal, enlarged lower, then tapering to a rooting base. Whitish when young, developing reddish brown stains in age. Smooth, dry, longitudinally striate or fibrous, at times with a powdery apex. **FLESH:** Thin, tough, fibrous in stipe. **ODOR:** Strongly like almond extract. **TASTE:** Mild to slightly bitter. **SPORE DEPOSIT:** Creamy pink to pale pinkish orange. **MICROSCOPY:** Spores 6–8 x 4–5 µm, ovoid, smooth, dextrinoid.

ECOLOGY: Solitary or scattered in thick duff, or around decaying stumps and logs in mixed hardwood-conifer forest. Scattered throughout our range, but uncommon. Fruiting in fall in north, well into winter in the south.

EDIBILITY: Inedible.

COMMENTS: The reddish brown, often two-toned cap, serrated to ragged gill edge, long rooting stipe that develops reddish spots and stains, and, most importantly, strong almond extract odor are distinctive. *R. badiialba* generally has a dark reddish brown cap that evenly fades and does not become two toned; it also lacks a distinctive odor. Probably most similar is *Hygrophorus bakerensis*, which also has a distinct almondy odor, but differs by having a two-toned cap that is more of a yellow-brown with a pale, whitish margin, and rather thick, waxy white gills.

Rhodocollybia butyracea
(Bull.) Lennox

CAP: 2–7 cm across, convex to plane, with a low, broad umbo, often slightly uplifted and wavy in age. Dark brown, reddish brown to dark cinnamon brown at first, quickly fading to light brown or cinnamon brown, to buff if dry. Surface smooth, lubricious and slippery when wet, to dry. **GILLS:** Narrowly attached to free, close to crowded, edges finely serrated when young, becoming more ragged in age. White to pale pinkish. **STIPE:** 2–7 cm long, 0.4–1 cm thick, equal, to club shaped with a swollen base. Pale pinkish buff when young, becoming cinnamon brown in age. Smooth, dry to greasy, often translucent and longitudinally striate, upper portion smooth, lower portion covered with whitish hairs. **FLESH:** Concolorous or slightly paler, thin, stipe fibrous, often hollow in age. **ODOR:** Indistinct. **TASTE:** Mild. **SPORE DEPOSIT:** Creamy buff to pale pinkish buff. **MICROSCOPY:** Spores 5.5–7.5 x 2.5–4 μm, ellipsoid, smooth, some dextrinoid.

ECOLOGY: Solitary, in small clusters, or in scattered troops in duff and needles of both hardwoods and conifers. Occurring throughout our range, but not particularly common. Fruiting in fall in north, well into winter or even spring southward.

EDIBILITY: Edible.

COMMENTS: The dark brown to cinnamon brown, greasy cap, serrated to ragged gill edges, and creamy buff to pinkish buff spores are the best identifying features. The most similar species is *Gymnopus dryophilus,* which generally has a warmer, ocher-brown to orangish cap, white gills with an even edge, and white spores. *R. badiialba* is generally larger and has a dark reddish brown cap and a stockier, often rooting stipe.

Caulorhiza umbonata
(Peck) Lennox
REDWOOD ROOTER

CAP: (1) 3–10 (15) cm across, conical when young, often with a lobed and irregular margin, expanding to bell shaped, and then flat or uplifted and slightly wavy, usually with a small umbo, often quite distinct and pointed. Dark chestnut to ocher-brown when young, becoming ocher-beige to cream, paler outward, whitish and sometimes translucent-striate near margin. Surface greasy to dry, smooth, often strongly radially wrinkled. **GILLS:** Free or obscurely attached, close, broad. White to cream. **STIPE:** 7–15 (30) cm long, 0.5–3 cm thick, cylindrical or often strongly grooved, compressed or splitting in age; usually tall and straight, strongly tapered and very elongated rooting base, often extending 6–15 cm below surface of soil. Creamy, beige, or pinkish tan, with twisting white vertical striations, or finely scurfy or powdery near apex, base covered in spiky to felty white or creamy mycelium. **FLESH:** Thin, stringy-fibrous to cartilaginous, hollow in stipe. **ODOR:** Indistinct. **TASTE:** Indistinct. **SPORE DEPOSIT:** White. **MICROSCOPY:** Spores 5–8 x 3–5 μm, ellipsoid, smooth, amyloid.

ECOLOGY: Solitary or scattered in loose groups, almost exclusively under redwood. Often in thick duff, but also on logs and stumps on the Far North Coast. Very rarely encountered in wood chips, or in mixed forests of California Bay Laurel or Monterey Cypress without redwood nearby. Fruiting from late fall into early spring.

EDIBILITY: Unknown, likely nontoxic.

COMMENTS: The conical or umbonate, honey-colored cap, pale gills, long rooting stipe, and growth under redwood combine to make this species quite distinctive. The wide variation in size of the fruitbodies and the occasional specimens found in habitats other than redwood forest raise the question of whether additional species are involved.

Group E
Pseudobaeospora aphana
Vellinga
BORING PSEUDO

CAP: 0.5–1.5 cm across, hemispheric to conical or bell shaped with an inrolled margin when young, expanding to broadly convex or flat, margin often remaining finely inrolled. Dingy cream to beige, tan, or light brown, margin often whitish. Surface dry, very finely velvety to nearly smooth. **GILLS:** Free or obscurely attached, moderately spaced, thick edged. Dull creamy when young, becoming dingy tan to yellowish or light brown in age. **STIPE:** 1.5–4 cm long, 0.2–0.3 cm thick, cylindrical, often curvy, base only slightly enlarged. Light grayish beige, soon becoming tan. Base conspicuously covered in spiky whitish rhizomorphs. Surface dry, fibrous, finely scaly-scurfy. **FLESH:** Very thin, fibrous, pale. **ODOR:** Unpleasant ("like iodine and smelt"). **TASTE:** Not sampled. **KOH:** No reaction. **SPORE DEPOSIT:** Whitish, often hard to obtain. **MICROSCOPY:** Spores 4–5.5 x 3–4 µm, broadly ellipsoid to subglobose. Gill cystidia absent.

ECOLOGY: In scattered groups in rich humus. We have found it under Monterey Cypress in winter, and it was described from "mixed forest," but it is very poorly known. Currently known from only a few places around San Francisco Bay Area.

EDIBILITY: Unknown.

COMMENTS: This mushroom is so featureless and easy to overlook that it is even named for these qualities (*aphana*, meaning "inconspicuous"). With some practice, it can be recognized by its small, dingy cream to tan-brown fruitbodies with an inrolled cap margin, thick-edged and well-spaced gills, and stipe base with distinct white rhizomorphs. A very similar and equally drab undescribed species differs by having a pinkish to purplish KOH reaction. It, too, is only known from a couple of locations, in cypress duff in the San Francisco Bay Area. *P. stevensii* has a slightly darker brown cap that turns green in KOH and often has more distinct fine chevrons covering the stipe. It can easily be confused for some *Clitocybe*, which are usually larger and stockier and have broadly attached to decurrent gills.

Pseudobaeospora stevensii
Desjardin
FRED'S PSEUDO

CAP: 0.7–3.5 cm across, buttons tightly rounded with a strongly inrolled and often ribbed margin, becoming broadly convex to plane, often with an umbo, or slightly uplifted and wavy in age. Dull brown to reddish brown, darker over center, tan-brown to grayish or dingy buff outward, sometimes vaguely zonate. Surface usually appearing covered in pale powder, which is easily rubbed off when cap is handled or where in contact with sticks in duff. **GILLS:** Finely attached or notched, but sometimes appearing free, moderately thick, edges somewhat thick. Pale dingy cream to cream-tan, tan brown to grayish brown in age. **STIPE:** 3–8 cm long, 0.3–0.5 cm thick, cylindrical, slightly enlarged toward base. Dull tan to light brown, often with a yellowish cast, evenly covered in small whitish or creamy chevrons or scabers. Lower portion of stipe and base covered in long, fairly stiff, white to creamy hairs. **FLESH:** Dull brownish, somewhat cartilaginous, stipe hollow in age. **ODOR:** Indistinct. **TASTE:** Mild. **KOH:** Distinctly teal-green. **SPORE DEPOSIT:** Sparse, white. **MICROSCOPY:** Spores 3–4 x 2.8–3.2 µm, globose to subglobose, smooth. Cheilocystidia abundant, irregular-cylindrical.

ECOLOGY: Solitary, scattered, or in small clusters in soil, humus, or duff under Monterey Cypress and redwood. Generally infrequent, but can be locally common. Known from Santa Cruz to Mendocino Counties. Fruiting from fall into spring.

EDIBILITY: Unknown.

COMMENTS: The dull brown to reddish brown colors, slender stipe covered in prominent pale chevrons, teal-green KOH reaction, and spiky mycelium at the base of the stipe distinguish this species. A similar, undescribed species we have been calling *P. "pseudostevensii"* has a stouter stature, a stipe with a finely pruinose (but not scaly-chevroned) appearance, and a dull olive KOH reaction, and lacks spiky whitish rhizomorphs at the base.

Pseudobaeospora deckeri

C. F. Schwarz

DECKER'S PSEUDO

CAP: 0.7–4 cm across, buttons rounded with a finely but distinctly inrolled and often ribbed margin, expanding to broadly convex or even uplifted and wavy in age. Royal purple to amethyst or pinkish purple, to dull tan-purple in age. Covered with a fine pale pinkish or lilac powder, which is easily rubbed off when cap handled or where in contact with twigs and leaves. In age and dry weather, becoming quite brownish and almost entirely lacking lilac tones. **GILLS:** Narrowly attached or deeply notched, rarely broadly attached or appearing free, moderately spaced, thick edged. Pale tan to lilac-tan, becoming more purple when wet, often developing orange-brown blotches in age. **STIPE:** 3–7 cm long, 0.2–0.5 cm thick, equal, sometimes with a slightly enlarged base. Lilac to amethyst pinkish or purplish, covered with very fine, whitish or pale lilac scurfy scales, often disappearing over lower part of stipe. Base covered with coarse lilac hairs. **FLESH:** Cartilaginous-fibrillose, amethyst purple or grayish purple, darkest in cap, lighter toward base of stipe. **ODOR:** Indistinct. **TASTE:** Slightly peppery-acrid. **KOH:** Dark blue-green, but appearing dark purple against color of darker caps. **SPORE DEPOSIT:** Whitish, often scant. **MICROSCOPY:** Spores 3.5–4 x 3–3.5 µm, subglobose, smooth, some dextrinoid. Cheilocystidia absent. Clamps present.

ECOLOGY: Solitary, scattered, or in small to large, close clusters on ground or in humus under Monterey Cypress or redwood. Generally uncommon, but can be locally abundant. Known from throughout our range, fruiting from late fall into spring.

EDIBILITY: Unknown.

COMMENTS: *Pseudobaeospora* as a genus can be recognized by their opaque caps with finely inrolled margins; rather thick, moderately spaced gills; and slender stipes, often with a scurfy apex and variable amount of coarse hairs at the base. KOH reactions are usually important for identification. Until 2004, there were no described species from California; since then two additional species have been described, and we have found eleven additional undescribed taxa. The evenly lilac to amethyst or royal purple colors of *P. deckeri* combined with the small stature help set it apart from almost all other species. Most similar are two undescribed species: a rosy pink to pinkish beige species (under redwoods, in Big Sur) and a slender, dark purplish black-capped species with a yellowish KOH reaction (in Humboldt County). *P. deckeri* often fades to brownish and closely resembles *P. stevensii* (with which it often grows); however, the latter lacks all traces of purple, has more spiky-shaggy mycelium at the stipe base, and usually has a scalier appearance to the stipe.

Baeospora myriadophylla

(Peck) Singer

LAVENDER BAEOSPORA

CAP: 1–2.5 cm across, convex when young, becoming broadly convex to plane, margin incurved at first, often becoming wavy in age. Lavender to lilac-gray when young, becoming lilac-brown to grayish brown in age, fading to grayish tan to buff if dry. Surface moist to dry, covered with a whitish bloom when young, finely floccose to nearly smooth. **GILLS:** Attached, sometimes with a small notch, very crowded, narrow, partial gills numerous. Purplish, lilac to lilac-gray when young, becoming purplish brown to purplish gray in age. **STIPE:** 1.5–6 cm long, 0.2–0.4 cm thick, equal, often rather long and thin. Purplish, lilac to lilac-gray and covered with a powdery white bloom when young, becoming purplish brown to grayish. Surface dry, powdery when young, becoming smooth, base covered with coarse to tomentose, whitish to lilac hairs. **FLESH:** Very thin, cartilaginous. **ODOR:** Strongly fungal. **TASTE:** Mild. **KOH:** No reaction. **SPORE DEPOSIT:** Whitish. **MICROSCOPY:** Spores 2.5–4.5 x 2–3 µm, rounded-ellipsoid, smooth, weakly amyloid. Cystidia large and encrusted.

ECOLOGY: In small, scattered groups on rotting wood of both conifers and hardwoods. We have seen it on fir, alder, and cottonwood. Rare in California, only known from a few locations on the Far North Coast and northern mountains. More common to north in Oregon and Washington. Fruiting from fall through spring.

EDIBILITY: Unknown.

COMMENTS: Though infrequently encountered, it is readily identified by its crowded purple gills, lavender to lilac-gray to grayish brown cap, and growth on wood. *Chromosera cyanophylla* is viscid and has distinctly decurrent, moderately spaced, purplish gills and a yellowish cap. Other small purplish mushrooms include *Mycena pura* and *Pseudobaeospora deckeri*, neither of which grow on wood. *Connopus acervatus* grows in much larger clusters and has pallid gills.

Group F

Baeospora myosura

(Fries) Singer

CONIFER-CONE BAEOSPORA

CAP: 0.5–3 cm across, broadly convex, soon becoming flat or weakly wavy, usually with a small umbo at center. Light brown, brownish tan to butterscotch tan or yellowish, usually paler around margin. Surface opaque, smooth, moist to dry. **GILLS:** Finely attached to free, narrow, very crowded. Off-white to pale grayish white. **STIPE:** 2.5–6 cm long, 0.05–0.3 cm thick, equal, often rooting and curved. Pinkish beige to tan-brown with fine, contrasting pale scruffiness over most of length, especially near apex. Base covered with fuzzy whitish hairs. **FLESH:** Very thin, stipe fibrous. **ODOR:** Indistinct. **TASTE:** Indistinct. **KOH:** No reaction. **SPORE DEPOSIT:** Dull whitish to off-white. **MICROSCOPY:** Spores 3–4 x 2–2.5 µm, ellipsoid, smooth, amyloid.

ECOLOGY: Solitary, scattered, or in small troops fruiting directly out of conifer cones, most often those of Sitka Spruce or Douglas-Fir, occasionally pine. Very common in Sitka Spruce forest on the Far North Coast, uncommon elsewhere. Fruiting from right after the first rains in fall into spring, most abundant in early fall.

EDIBILITY: Unknown.

COMMENTS: The broadly convex and often slightly umbonate, light caramel-colored cap, very crowded whitish gills, small size, and growth on conifer cones distinguish this species. It is more common northward, where it co-occurs with other cone-dwelling species—*Strobilurus trullisatus* and *S. occidentalis*. Both of these have whitish yellow to grayish caps, more widely spaced white gills, and inamyloid spores.

Strobilurus occidentalis

V. L. Wells & Kempton

SPRUCE-CONE MUSHROOM

CAP: 0.5–1.5 cm across, broadly convex to plane, or wavy in age. Creamy to yellowish gray at first, darkening to gray to grayish brown, often with a yellowish gray center in age. Margin translucent-striate when wet. Surface smooth, moist to dry. **GILLS:** Attached, close to subdistant, moderately broad, partial gills numerous. White. **STIPE:** 1.5–6 cm long, 0.1–0.2 cm thick, thin, tough, often with a rotting base. Orange-brown over lower portions with a white apex, darkening in age. Surface dry, appearing powdery, with a whitish sheen, often with whitish to pale orange hairs at base. **FLESH:** Concolorous, very thin, tough. **ODOR:** Indistinct. **TASTE:** Mild. **KOH:** No reaction. **SPORE DEPOSIT:** White. **MICROSCOPY:** Spores 4–6 x 2–3 μm, ellipsoid to cylindrical, smooth, inamyloid.

ECOLOGY: Solitary or scattered in small troops, growing directly out of Sitka Spruce cones (sometimes buried in duff). Locally common on Far North Coast, absent south of Mendocino County. Fruiting soon after first fall rains and continuing into early winter.

EDIBILITY: Unknown.

COMMENTS: The creamy cap that develops grayish brown as it ages; white gills; thin stipe with a white apex and an orange-brown lower portion; and, most importantly, growth on Sitka Spruce cones are helpful in identification. *S. trullisatus* is very similar, but has a paler cap and a yellowish to yellow-brown stipe with a white apex; although it can occur on spruce cones, it is far more common on Douglas-fir cones. *S. albipilatus* is a common montane species, usually on pine cones or fir cone scales, but can also occur in thick duff or woody debris. It is generally larger and has a light to dark gray or grayish brown cap and yellowish stipe. A montane species, *S. diminutivus*, is (as the name suggests) tiny, with caps less than 0.5 cm across; it generally grows on Sugar Pine (*Pinus lambertiana*) or Ponderosa Pine cones. *Baeospora myosura* has a light brown cap and very crowded gills; it also occurs on cones, mostly those of spruce and Douglas-fir.

Strobilurus trullisatus

(Murrill) Lennox

DOUGLAS-FIR CONE MUSHROOM

CAP: 0.5–2 cm across, convex to plane, occasionally wavy and uplifted in age. Whitish with a pinkish orange to buff cast to center, occasionally pale grayish white when young, to all white. Lightly hygrophanous, faintly translucent-striate when moist, opaque, dull when dry. Surface smooth, moist to dry. **GILLS:** Broadly attached, close to subdistant. White or with a pale pinkish buff cast. **STIPE:** 2–6 cm long, 0.1–0.2 cm thick, thin, tough, with a deep rotting base if cone is buried. Yellowish to yellow-brown over lower portions with a white apex, darkening slightly, to orangish brown overall in age. Surface dry, appearing powdery, with a whitish sheen, often with whitish to pale orange hairs at base. **FLESH:** Concolorous, very thin, tough. **ODOR:** Indistinct. **TASTE:** Mild. **KOH:** No reaction. **SPORE DEPOSIT:** White. **MICROSCOPY:** Spores 3.5–6 x 2–3.5 μm, ellipsoid to narrowly almond shaped, smooth, inamyloid.

ECOLOGY: Solitary or scattered in small troops, growing directly out of Douglas-fir cones, rarely those of other conifers. Very common throughout our range. Fruiting soon after first rains in fall, occasionally into winter.

EDIBILITY: Inedible.

COMMENTS: The whitish cap (often with a yellowish or buff blush at the disc), white gills, thin stipe with a yellowish to yellow-brown, fuzzy base, and growth on Douglas-fir cones are distinctive. *S. occidentalis* is very similar, but is restricted to growth on Sitka Spruce; it has a whitish or often grayish cap and a more orangey lower stipe. *Baeospora myosura* has a light brown cap and very crowded gills; it is most common on spruce cones (but does occur on Douglas-fir cones over most of our range).

Group G
Collybia cirrhata
(Schumach.) Quél.
COMMON COLLYBIA

CAP: 0.2–1 cm across, convex to plane, occasionally with a small rounded umbo, margin incurved when young, occasionally slightly uplifted or wavy in age. White to creamy buff, often with a slightly darker, pinkish to yellowish buff center. Surface dry, smooth. **GILLS:** Broadly attached to slightly decurrent in age, close to crowded. narrow. Whitish to pinkish buff. **STIPE:** 1.5–5 cm long, 0.05–0.2 cm thick, equal, often with a rooting base. Whitish to buff, or pale orangish buff over lower portion. Surface dry, minutely hairy or fuzzy, to smooth, base often with white hairs. **FLESH:** Concolorous, thin, fragile in cap, fibrous in stipe. **ODOR:** Indistinct. **TASTE:** Mild. **KOH:** No reaction. **SPORE DEPOSIT:** White. **MICROSCOPY:** Spores 5–6.5 x 2–3.5 μm, ovoid to ellipsoid, smooth, inamyloid.

ECOLOGY: Scattered or clustered in troops on blackish remains of decaying mushrooms, especially *Russula* and *Lactarius*. Although this *Collybia* can show many white rhizomorphs in the rotting mushroom substrate, it lacks sclerotia. Common in northern part of our range, occasional southward at least as far as Santa Cruz County. Fruiting from fall through early winter.

EDIBILITY: Unknown.

COMMENTS: The description here more or less matches the fruitbodies of three different species, which can only be reliably distinguished by pulling apart the rotting host mushroom and looking for sclerotia. If no sclerotia are present, you have *Collybia cirrhata* (upper left photo). If you find a swollen or elongated, reddish brown sclerotium (resembling a large apple seed), you have *C. tuberosa* (two right photos). If you find a small, round sclerotium (sometimes resembling a shriveled corn kernel), you have *C. cookei* (bottom left photo). As a group, they are easily recognized by the small size, whitish to creamy buff caps, and growth on decaying mushrooms. *C. cookei* is generally slightly smaller than the other two and seems to prefer Honey Mushrooms (*Armillaria* spp.) and especially old fruitbodies of *Bondarzewia mesenterica*; it is widespread (occurring south to Santa Cruz County), but is uncommon throughout our range. The Tuberous Collybia (*C. tuberosa*) is very similar to *C. cirrhata*, save for the distinctive reddish brown sclerotium (bottom left photo); it is rare south of Humboldt County, but common on the Far North Coast on old *Russula* (especially those in the *R. nigricans* group). *Dendrocollybia racemosa* and *D. pycnoramella* also grow on decaying mushrooms (sometimes alongside one or more of these species), but have opaque gray caps and multiple side branches on their stipes.

Dendrocollybia racemosa

(Pers.) R. H. Petersen & Redhead

LONG-BRANCHED DENDRO

CAP: 0.5–1.5 cm across, bell shaped to convex, often with a low umbo, remaining convex or becoming slightly more flat in age. Opaque, light gray to grayish beige, sometimes dark gray or tan to brownish. Surface moist to dry, smooth. **GILLS:** Obscurely attached, close to subdistant. Dull, pale gray to dingy yellowish tan. **STIPE:** 1–6.5 (9) cm long, 0.1–0.2 cm thick, cylindrical for much of length, but rather sinuous and tapering downward into a very long lower portion mostly below level of duff. Surface irregularly covered in straight to curved branches 0.3–0.5 cm long with enlarged, rounded tips. These branches, called conidial pegs, can be lost in age, and stipe can appear mostly smooth. Base of stipe emerging directly from small, black, oval or egg-shaped sclerotia. **SCLEROTIUM:** 0.2–0.7 cm wide, overall ellipsoid-rounded, surface smooth. Matte black, hard. **FLESH:** Very thin, grayish to tan. **ODOR:** Indistinct. **TASTE:** Indistinct. **KOH:** No reaction. **SPORE DEPOSIT:** Whitish, scant. **MICROSCOPY:** Spores 4–5 x 2–3 µm, narrowly ellipsoid, smooth. Asexual spores from stipe 7–12 x 3–5 µm, highly irregular, cylindrical, peanut shaped, or bent.

ECOLOGY: Often in clusters, rarely solitary, on remains of rotting mushrooms, particularly *Lactarius* and *Russula*. In our area, some of the favored "hosts" for *D. racemosa* are *Lactarius alnicola* and various blackening *Russula*. Very common, but often overlooked, fruiting fall into early winter.

The sclerotia are generally seen on rotting mushrooms in winter into early spring, the year before the fruitbodies.

EDIBILITY: Unknown.

COMMENTS: With the opaque gray to grayish beige cap, grayish gills, small size, and long, sinuous, tapering stipe with irregular branches, this is an easy mushroom to identify. Specimens that have lost their side branches are harder to identify, but the smooth, black sclerotia set them apart. However, these are easy to overlook unless carefully excavated. The only truly similar species is *D. pycnoramella*, with much shorter, closely spaced conidial branches and shorter fruitbodies arising directly from more knobby-warted (rather than smooth) sclerotia. Older specimens can lose their conidial branches, especially on the upper part of the stipe, and could be mistaken for a *Pseudobaeospora*. Although this species was perhaps once thought to be rare, our experience suggests that it is mostly just inconspicuous. It often fruits in areas rich with other small-capped trooping mushrooms (*Mycena* and others), making it very difficult to initially detect. Once you become very familiar with the mushrooms in your area, you can key in on the unique gestalt impression made by the small, tightly clustered, grayish brown, nonstriate caps.

SYNONYM: *Collybia racemosa* (Pers.) Quél.

Dendrocollybia "pycnoramella"

N. Siegel & C. F. Schwarz nom. prov.

SHORT-BRANCHED DENDRO

CAP: 0.2–0.6 (0.8) cm across, bell shaped to convex, with a low, round umbo, margin uneven to weakly ribbed. Opaque, beige-tan to grayish tan, usually darkest or most richly colored at center of cap, sometimes appearing weakly zonate. Surface smooth, moist to dry. **GILLS:** Obscurely attached, close to subdistant, somewhat thick edged. Dull, pale gray when young, becoming buff-gray to dingy tan. **STIPE:** 0.7–2 cm long, 0.1–0.2 cm thick, equal, base of stipe directly emerging from a sclerotium. Pale gray to grayish tan. Surface densely and evenly covered in small branches with slightly enlarged, rounded tips. **SCLEROTIUM:** 0.3–1 cm wide, irregularly ellipsoid and often lobed, with a bumpy-warted surface. Black, hard. **FLESH:** Thin, grayish to tan. **ODOR:** Indistinct. **TASTE:** Indistinct. **KOH:** No reaction. **SPORE DEPOSIT:** Whitish to pale buff, scant. **MICROSCOPY:** Spores 4–5 x 2–2.5 µm, ellipsoid, smooth. Asexual spores from stipe 5–10 x 2.5–3 µm, highly irregular, cylindrical, peanut shaped, or bent. Surface of sclerotium made of brown, mazelike or branched thick-walled cells.

ECOLOGY: In small clusters on remains of rotting mushrooms in late fall and winter. Very rare, currently only known from the type locality in the Santa Cruz Mountains.

EDIBILITY: Unknown.

COMMENTS: With its tiny size, beige to grayish colors, small and densely packed conidial branches, and stipe emerging directly from a black, bumpy sclerotium, this is a very distinctive mushroom. We have encountered this species only once, in the late fall on the University of California, Santa Cruz, campus, in a mossy area in a dark gulch near Tanoak, Douglas-fir, and a very rotten live oak snag. It was emerging from a pile of rotting mushrooms, fruiting next to *Collybia cookei*. Our best guesses at the identity of the host mushrooms are *Armillaria mellea* or possibly *Hypholoma fasciculare*, but we are far from certain. If you think you've found *D. pycnoramella*, please photograph and/or voucher it and let us know (see Resources, p. 576); it appears to be the rarest species in this book! *D. racemosa* is similar, but has much longer, less closely spaced conidial branches and a very long, sinuous stipe that emerges from clusters of round, smooth sclerotia.

Asterophora parasitica

(Bull.) Singer

CAP: 0.3–2 cm across, rounded, bell shaped to conical with an inrolled and crimped margin when young, becoming broadly conical, convex, to plane with a low, broad umbo. Pale gray to grayish beige base color, covered with silvery white fibrils; becoming gray to gray-brown in age. Surface dry, silky when young, to smooth in age. **GILLS:** Attached, distant, very thick, irregular, often with a fuzzy appearance. Whitish, light gray to grayish brown. **STIPE:** 0.5–3 cm long, 0.1–0.4 cm thick, equal or tapered toward base. Pale gray to grayish brown base color, covered with silvery white, silky fibrils, becoming darker, gray to grayish brown when handled or in age. Surface dry, silky-fibrous to smooth. **FLESH:** Very thin, soft, fragile in cap, fibrous in stipe. Grayish brown to watery brown. **ODOR:** Unpleasant, rancid farinaceous. **TASTE:** Strongly farinaceous. **SPORE DEPOSIT:** Whitish. **MICROSCOPY:** Basidiospores 5–7 x 3–4 µm, ellipsoid, smooth. Asexual chlamydospores 12–17 (20) x 9–11 µm, spindle shaped to oval, thick walled, with large droplets.

ECOLOGY: In small to large clusters, rarely solitary on decaying *Russula* (especially black-staining species and *R. brevipes* group). Usually growing from the cap disc or near the stipe base of the host. Rare in our area, known only from the Far North Coast. Fruiting throughout year, whenever there are decaying fruitbodies of its hosts.

EDIBILITY: Unknown. Not recommended.

COMMENTS: The silky, whitish to gray cap, widely spaced, irregular gills, and growth on decaying *Russula* are distinctive. It is most likely to be mistaken for *A. lycoperdoides*, which has a white to light brown, rounded cap that becomes powdery as it ages and usually lacks gills. *Volvariella surrecta* is a silky white species that grows on large *Clitocybe* species; it is larger, with more crowded, white to light pink gills, and a volva forming a saclike cup at the stipe base. It is extremely rare, known in our area from only a handful of records. Also see the *Collybia* species, all of which colonize and grow from rotting mushrooms.

Asterophora lycoperdoides

(Bull.) Ditmar

POWDER CAP

CAP: 0.5–2 cm across, rounded to convex with an inrolled margin. White at first, soon becoming a light brown powdery mass. **GILLS:** Often absent, or only present as rudimentary veins or ridges. Grayish to watery brown. **STIPE:** 3–5 cm long, 0.3–0.6 cm thick, equal or tapered toward base. White when young, becoming grayish to dingy gray in age. Surface dry, pubescent to silky. **FLESH:** Thin, soft and white at first in cap, soon light brown and powdery. Stipe firm, fibrous, becoming hollow in age, gray to watery brown. **ODOR:** Unpleasant, rancid farinaceous (and like rotten mushrooms when still on host). **TASTE:** Strongly farinaceous. **SPORE DEPOSIT:** Basidiospores rare, whitish, chlamydospores light brown. **MICROSCOPY:** Basidiospores 3–6 x 2–4 µm, ellipsoid, smooth. Asexual chlamydospores 13–20 x 10–20 µm, ovoid to nearly round, covered with blunt spines or warts.

ECOLOGY: In small groups or clusters on decaying *Russula*, especially the Blackening Russula (*R. nigricans*, *R. albonigra*, etc.) and the *R. brevipes* group, more rarely on *Lactarius* species. Locally common on the Far North Coast, unrecorded south of Humboldt County. Often fruits in late summer or early fall on *Russula* that have fruited following fog drip or during dry spells in fall. In wet years, excess moisture rots the *Russula* faster than the *Asterophora* can establish itself and fruit.

EDIBILITY: Unknown. Not recommended.

COMMENTS: This oddball is one of a handful of mushrooms that grow on decaying mushrooms. It is easily identified by the white to light brown cap that becomes powdery as it ages, lack of defined gills, and growth on rotting mushrooms. *A. parasitica* shares this habitat, but has a silky cap and thick, irregular gills. The powdery cap of this *A. lycoperdoides* is mostly made up of asexual chlamydospores, which are dispersed by wind and splashing raindrops. Upon reaching a suitable host, these spores can grow much faster then the sexual basidiospores, allowing them colonize and fruit from the host before it is fully decayed. The genus *Dendrocollybia* also produces asexual spores, but they occur on conidial pegs (short side-branches on the stipe).

Group H

Chromosera cyanophylla

(Fr.) Redhead, Ammirati & Norvell

CAP: 0.4–2.5 cm across, bell shaped to convex when young, becoming broadly convex to plane, with a depressed center. Bright to dull yellow to peachy yellow, fading in age to yellowish beige, translucent-striate with darker lines most of the way to disc. Surface viscid to tacky, smooth. **GILLS:** Broadly attached to decurrent, moderately spaced, broad, fragile, partial gills short and numerous. Lilac-purple when young, fading to grayish lilac, pinkish beige and occasionally becoming pale yellowish in age. **STIPE:** 1–4 cm long, 0.1–0.3 cm thick, equal or club shaped with an enlarged base. Lilac-purple at apex and yellowish lower when young, fading and becoming more yellow in age. Base with bright lilac-purple disc when young. Surface viscid, smooth, often slightly fuzzy at base. **FLESH:** Very thin, fragile, yellowish, stipe stuffed with watery pith or hollow in age. **ODOR:** Indistinct. **TASTE:** Mild. **SPORE DEPOSIT:** White. **MICROSCOPY:** Spores 6–7 x 3–3.5 μm, broadly ellipsoid to almond shaped, smooth. Pleurocystidia and cheilocystidia absent.

ECOLOGY: Solitary, scattered, or in small clusters on decaying conifer logs and branches, especially those of Grand Fir and Western Hemlock. Fairly common north of Mendocino County, very scarce southward on the coast at least as far south as Point Lobos in Monterey County (on cypress). Generally fruiting from midwinter into spring.

EDIBILITY: Unknown.

COMMENTS: The translucent-striate, yellow, viscid cap, lilac-purple color of the young gills, viscid stipe, and growth on conifers are distinctive. Older specimens lose their vivid colors, but most look-alikes lack the distinctly viscid cap and stipe. Other brightly colored omphalinoid species on wood include *Chrysomphalina aurantiaca*, which has a bright orange, dry cap and stipe. *C. chrysophylla* has a yellow-brown, scaly cap when young (becoming orange-yellow in age), and has bright golden gills. *C. grossula* has pale yellowish green colors when young, fading to yellowish beige; it is quite rare on the Redwood Coast. *Contumyces rosellus* is a rosy pink species that grows in nutrient-poor or sandy soil. *Gliophorus* species have thick "waxy" gills and grow on the ground. *Chromosera cyanophylla* has bounced from genus to genus, never quite fitting into any of them. The genus *Chromosera* was erected to accommodate this species; in the process, the eastern North American species (*Mycena*) *lilacifolia* was synonymized with the European species *C. cyanophylla*. Recent genetic work shows that the western species is distinct from the European species, but has yet to determine its relationship to the eastern North American species.

SYNONYMS: *Mycena lilacifolia, Clitocybe lilacifolia, Omphalina lilacifolia.*

Chrysomphalina aurantiaca

(Peck) Redhead

CAP: 0.5–3 cm across, convex to plane, often with a depressed center in age. Bright orange at first, fading to pale orange to yellowish orange or sometime creamy orange in age. Surface dry, covered with fine stiff hairs when young, becoming smooth or nearly so, often translucent-striate in age. **GILLS:** Broadly attached to slightly decurrent, moderately spaced, narrow, fragile, partial gills short and numerous. Creamy orange to pale orange at first, becoming yellowish orange in age. **STIPE:** 1–3 cm long, 0.1–0.3 cm thick, equal or enlarged slightly at base. Orange to yellowish orange. Surface dry, smooth, base often with fuzzy white hairs. **FLESH:** Very thin, fragile, orange, stipe stuffed with orangish pith, occasionally hollow in age. **ODOR:** Indistinct. **TASTE:** Mild. **KOH:** No reaction. **SPORE DEPOSIT:** Whitish. **MICROSCOPY:** Spores 6.5–9 (11.5) x 4–5.5 µm, ellipsoid, smooth. Basidia usually 4-spored, rarely 1-, 2-, or 3-spored. Pleurocystidia, cheilocystidia, and clamps absent.

ECOLOGY: Scattered, in small clusters or troops on decaying conifer stumps and logs, most commonly Grand Fir. Generally fruiting in winter. Very common in Pacific Northwest and California mountains, rather uncommon on the coast. We have seen it as far south as the San Francisco Bay Area, but it is rarely found south of Mendocino County.

EDIBILITY: Unknown.

COMMENTS: The depressed cap center, orange cap with fine hairs around the margin (obscure in age), pale orange gills, and growth on wood help identify this colorful mushroom. Brightly colored, dry-capped *Hygrocybe* species (such as *H. miniata*) are similar but don't usually grow on wood. *C. chrysophylla* is often mistaken for *C. aurantiaca*, but is usually larger and has golden gills and a yellow-brown, scaly cap that becomes orange-yellow in age. *C. grossula* occurs in the same habitat and has the same stature, but has pale yellowish green colors when young, fading to yellowish beige.

SYNONYM: *Omphalia aurantiaca* Peck.

Chrysomphalina chrysophylla

(Fr.) Clémençon

CAP: 1–5 cm across, convex to plane, with a depressed center. Color variable, often warm brown, ocher-brown to grayish brown to olive-gold at first, with an golden base color, becoming paler and more golden yellow, but often with some brownish scales on disc. Surface dry to moist, covered with fine tufts or scales when young, becoming mostly smooth, but retaining some scales on disc. **GILLS:** Broadly attached to decurrent, moderately spaced, narrow, partial gills numerous. Bright chrome yellow to golden when young, fading slightly, but staying golden yellow in age. **STIPE:** 1–5 cm long, 0.1–0.5 cm thick, more or less equal, or enlarged at base. Bright chrome yellow to golden yellow, fading slightly in age. Surface dry, smooth, base often with fuzzy white hairs. **FLESH:** Very thin, somewhat fragile to rubbery, orangish yellow, stipe stuffed with yellowish pith when young, hollow in age. **ODOR:** Indistinct. **TASTE:** Mild. **KOH:** No reaction. **SPORE DEPOSIT:** Off-white, buff to pale pinkish buff. **MICROSCOPY:** Spores 8.5–15.5 x 4.5–6 (7) µm, ellipsoid, elongate-ellipsoid to subcylindric, inamyloid, smooth. Basidia mostly 4-spored, rarely 1- or 2-spored. Pleurocystidia, cheilocystidia, and clamps absent.

ECOLOGY: Scattered, in small clusters or troops on or around decaying conifer stumps and logs. Uncommon in coniferous forests of North Coast, as well as higher-elevation fir forests in the mountains. Fruiting from fall through early spring.

EDIBILITY: Unknown.

COMMENTS: The grayish brown to bright gold cap (often with darker scales), bright chrome yellow gills, and growth on wood are distinctive. *C. aurantiaca* is generally smaller, has a bright orange cap that is finely fringed when young, and has pale orange to yellowish orange gills. *Hygrophoropsis aurantiaca* is similar, but often has an orange-brown cap and orange gills that are wavy and repeatedly forked. *Craterellus tubaeformis* is superficially similar, but has pale "gills" with cross veins, and a central hollow running from the cap through the stipe.

SYNONYM: *Omphalina chrysophylla* (Fr.) Murrill.

Xeromphalina campanella
(Fr.) Kuehner & Maire
FUZZY FOOT

CAP: 0.3–2.5 (3) cm across, convex to plane, with a depressed center, margin downcurved, occasionally splitting in age. Dull orange, orange-brown to yellowish brown. Surface dry to moist, smooth, translucent-striate when wet. **GILLS:** Subdecurrent to decurrent, moderately spaced, broad, partial gills numerous, strongly inter-veined. Yellow-beige to pale orange. **STIPE:** 1–3 (5) cm long, 0.1–0.3 cm thick, equal or enlarged slightly toward base, thin, somewhat wiry, fibrous. Yellow to orangish at apex, transitioning to orange, to a dark brown or reddish brown at lower portion and often with a fuzzy orange base. Surface smooth, dry, with fine hairs at base. **FLESH:** Very thin, concolorous. **ODOR:** Indistinct. **TASTE:** Indistinct. **KOH:** Orangish brown. **SPORE DEPOSIT:** White. **MICROSCOPY:** Spores 5–8 x 3–3.5 μm, ellipsoid to cylindrical, smooth, amyloid.

ECOLOGY: Scattered or clustered in small to larger troops on decaying conifer logs or stumps, or rarely on small branches. Widespread and very common throughout our range. Fruiting from early fall well into spring, but most abundant in midfall.

EDIBILITY: Unknown, probably nontoxic.

COMMENTS: The slightly decurrent gills, orange cap, dark lower stipe, and growth on wood make this a distinctive species. *X. brunneola* is very similar, but has a brown to orange-brown cap and an unpleasant to bitter taste; it is generally uncommon, but occasional large fruitings can be found, usually on the North Coast. See *X. cauticinalis* (p. 316) for similar species growing in duff or moss. *Hygrophoropsis aurantiaca* is larger and has crowded orange gills that fork repeatedly. Some of the small wood-dwelling *Gymnopilus* and *Galerina* can superficially resemble *Xeromphalina*, but have brown spores.

Aphroditeola olida
(Quel.) Redhead & Manfr. Binder
PINK BUBBLEGUM MUSHROOM

CAP: 1–3 (4.5) cm across, broadly convex to nearly plane with a inrolled margin and a depressed center at first, becoming irregular, more broadly funnel shaped, with a wavy or lobed margin. Pink to pale coral pink, occasionally with dark pink splotches and stains, fading to pinkish white to pinkish buff in age. Surface dry to slightly tacky, finely tomentose to smooth. **GILLS:** Decurrent, quite shallow, forking repeatedly, often wrinkled and with cross veins. Whitish to pale pinkish **STIPE:** 1–3.5 (5) cm long, 0.2–0.5 cm thick, equal or tapered downward, occasionally rooting. Pale pinkish to whitish, often spotted with darker pink stains. **FLESH:** Thin, fibrous, white. **ODOR:** Very pleasant, fruity, like bubblegum and/or root beer. **TASTE:** Mild. **SPORE DEPOSIT:** White. **MICROSCOPY:** Spores 3–6 x 2.5–4 μm, ellipsoid to spindle shaped, smooth, inamyloid.

ECOLOGY: Solitary or scattered in small groups in duff or moss under conifers. Rare in our range, primarily known from the Far North Coast. Fruiting in the fall.

EDIBILITY: Unknown.

COMMENTS: The pink cap, forking, decurrent gills, and, most importantly, fruity bubblegum or root beer–like odor make this rare mushroom distinctive. The much commoner *Contumyces rosellus* has more widely spaced gills and thin, almost translucent flesh, and lacks an odor. Members of the *Hygrophoropsis aurantiaca* group also have branched gills, but typically have orange colors rather than pink (although some have creamy white gills), and most lack any distinctive odor.

SYNONYMS: *Hygrophoropsis olida* (Quel.) Metrod, *H. morganii* (Peck) H. E. Bigelow.

Contumyces rosellus
(M. M. Moser) Redhead, Moncalvo, Vilgalys & Lutzoni

CAP: 0.5–3 cm across, convex when young, soon uplifted and depressed at center, margin often wavy or scalloped. Rosy pink to light pastel pink, paler variants creamy buff or pale pinkish beige, hygrophanous to nearly white but edges often with dark wine red discoloration; obscurely translucent-striate when wet. Surface slightly rough, moist to dry. **GILLS:** Broadly attached to decurrent, widely spaced, sometimes forked or branched, edge quite thick. Often strongly rosy pink when very young, soon ghostly whitish to pale cream or pale pink, margins often darker, to wine red. **STIPE:** 0.5–1.5 cm long, 0.1–0.3 cm thick, cylindrical, base variably slightly enlarged or tapered. Pinkish to beige-pink, often darkest at apex (contrasting with gills). **FLESH:** Thin, pale, translucent white. **ODOR:** Indistinct. **TASTE:** Mild. **SPORE DEPOSIT:** Hard to obtain, white. **MICROSCOPY:** Spores 11–14 x 5–6 μm, elongated teardrop shape, smooth. Clamps present.

ECOLOGY: Solitary or in loose clusters or scattered in troops, usually on bare soil, gravel, or sand with short grass or weeds, often on road cuts or along trails. Fairly common in our area, quite common in southern California but often overlooked. Fruiting from late fall through spring.

EDIBILITY: Unknown.

COMMENTS: Spotting the fruitbodies can be tough, but identification is straightforward: the tiny size, widely spaced gills, pastel pink to rosy coloration, and growth on nutrient-poor soil distinguish it. *Lichenomphalia umbellifera* is quite similar in shape, but is rarely extensively pink, and grows on mossy or algae-covered wood or soil and is usually taller. *Rickenella fibula* is orange and grows in beds of moss. Confusion with *Omphalina pyxidata* seems to have been common in California for some time; we suspect that only *C. rosellus* occurs in our area.

Lichenomphalia umbellifera
(L.) Redhead, Lutzoni, Moncalvo & Vilgalys
LICHEN AGARIC

CAP: 0.5–2.5 (3) cm across, convex to plane, with a depressed center, margin downcurved, ribbed and often scalloped, becoming plane to uplifted in age. Light buff, cinnamon buff to light brown when young, quickly fading to creamy beige, with darker lines or groves. Surface dry to moist, pleated or grooved, otherwise smooth. **GILLS:** Slightly to deeply decurrent, well spaced, thick, broad, partial gills often short, inter-veined and numerous. Buff to beige at first, becoming creamy beige to dingy whitish beige, edges usually concolorous, but occasionally darker. **STIPE:** 1–3 (4) cm long, 0.1–0.3 cm thick, equal or enlarged slightly toward base. Buff, beige to creamy beige, often darker, with a vinaceous tint at apex. Surface dry, smooth. **FLESH:** Very thin, off-white to beige. **ODOR:** Indistinct. **TASTE:** Mild. **KOH:** No reaction. **SPORE DEPOSIT:** White. **MICROSCOPY:** Spores 7–9 x 4–6 µm (but larger, 12 x 8 µm from 1-spored basidia), ellipsoid, smooth, inamyloid. Basidia often 2-spored but can be 1-, 3-, or 4-spored, clamps absent.

ECOLOGY: Solitary or scattered in varied habitats. Often on logs, stumps, or bark of old redwood trees, but also in areas with nutrient-poor sandy or clay soil. Associated with a *Coccomyxa* algae, which forms the green mats on logs or soil from which the fruitbody grows. Common and widespread, known from Santa Cruz County northward. Fruiting from late fall into spring.

EDIBILITY: Unknown.

COMMENTS: The omphalinoid stature; buff to light brown cap that fades to creamy beige; thick, widely spaced, creamy gills; and growth from a green algal mat help identify this common lichenized agaric. It is one of a handful of basidio-lichens (nearly all other lichens are ascomycetes). *Lichenomphalia* have specialized modified hyphae that loosely encase the cells of *Coccomyxa* algae and siphon off some of their photosynthetic products. Older *Contumyces rosellus* can look similar, but generally have some pinkish coloration and grow in sandy areas or on bare soil.

SYNONYMS: *Omphalia ericetorum* (Pers.) S. Lundell, *O. umbellifera* (L.) Quél. The genus *Lichenomphalia* was described to accommodate these lichenized agarics. They are closely related to the mostly bryophilous (moss-associated) *Arrhenia*. They were also called *Clitocybe* in the not-so-distant past.

Pseudolaccaria pachyphylla

Vizzini & Contu

CAP: 0.5–2.5 cm across, convex to plane, with a depressed center, margin downcurved and often ribbed when young, becoming ragged in age. Buff to pale straw colored, with paler appressed fibrils. Surface dry, covered with appressed fibrils, which often break up into small uplifted scales when old. **GILLS:** Broadly attached, close to subdistant, broad, thick, and often slightly irregular. Off-white to pale buff. **STIPE:** 1–3 (5) cm long, 0.1–0.3 cm thick, more or less equal, often curved at base. Off-white to pale buff. Surface dry, covered with fine silken fibrils to smooth. **FLESH:** Very thin, fibrous, off-white to pale buff. **ODOR:** Indistinct. **TASTE:** Mild. **KOH:** No reaction. **SPORE DEPOSIT:** White. **MICROSCOPY:** Spores 4.5–6.5 x 3.5–4.5 μm, nearly round to broadly ellipsoid, smooth, amyloid. Basidia 4-spored. Pleurocystidia and cheilocystidia abundant, cylindrical to spindle shaped, clamps present.

ECOLOGY: Solitary or scattered on nutrient-poor, moss-covered soil, clay, or sandy areas, especially old logging roads or disturbed areas. Locally common, occurring throughout our range, but often overlooked because of habitat and small size. Fruiting in late fall or winter.

EDIBILITY: Unknown.

COMMENTS: The omphalinoid stature, buff to pale straw colors, slightly scurfy cap and stipe, and growth in nutrient-poor soil are distinctive. *Lichenomphalia umbellifera* has a ribbed cap and widely spaced gills, and grows on green algal mats. Small *Clitocybe* generally have smooth caps and stipes and often grow in duff or moss.

Camarophyllopsis paupertina

(A. H. Sm. & Hesler) Boertm.

MOTHBALL MUSHROOM

CAP: 0.5–2 cm across, bell shaped, convex to plane, often with a depressed center. Yellowish tan to pale grayish tan or very light brown when young, darkening to gray brown or dingy brown in age, fading when old or dry. Surface dry to moist, smooth or very finely scurfy. **GILLS:** Broadly attached to decurrent, distantly spaced, quite narrow, often splitting in age. Beige to pallid brown. **STIPE:** 1–5 cm long, 0.3–0.6 cm thick, often tapered and curved toward base. Beige to light brown, darkening from base up in age. Surface dry, smooth. **FLESH:** Very thin, fragile. **ODOR:** Very strong, distinctive odor of mothballs. **TASTE:** Disagreeable. **KOH:** No reaction. **SPORE DEPOSIT:** Whitish, often scant. **MICROSCOPY:** Spores 5–6 x 4–5 μm, subglobose to broadly ellipsoid, inamyloid, smooth. Cap cuticle a trichoderm.

ECOLOGY: Scattered, in small clusters or troops in duff or on bare soil under redwood and other conifers, typically in disturbed areas such as trailsides or along old logging roads. Found throughout our area and can be locally common, but is easily overlooked due to small size and muted colors. Fruiting from late fall into spring, often near earth tongues (*Trichoglossum* spp.) and club fungi (*Clavaria* spp.).

EDIBILITY: Unknown.

COMMENTS: The small size, drab colors, and overpowering mothball odor make this mushroom easy to identify. The very similar *C. subfuscescens* lacks an odor (or smells sweet and fruity) and often has a yellowish stipe; it is much rarer. Genetic data have shown that *C. paupertina* is more closely related to coral fungi of the genus *Clavulina* than to other gilled mushrooms.

MISAPPLIED NAME: Based on the mothball odor and similar stature, the name *Camarophyllopsis foetens* has been applied broadly to California collections. We think that such records probably refer to *C. paupertina*.

Myxomphalia maura

(Fr.) Hora

CAP: 0.5–3 (5) cm across, convex to plane, with a depressed center, margin incurved when young, becoming plane to uplifted in age. Dark blackish brown, dark gray to grayish brown when wet or young. Strongly hygrophanous, fading to pale gray when dry, translucent-striate when wet. Surface slightly viscid when wet, to dry, smooth. **GILLS:** Broadly attached to subdecurrent, close to somewhat crowded, narrow, partial gills numerous. Bright white at first, contrasting with cap and stipe color, becoming off-white to pale grayish in age. **STIPE:** 1–5 (6) cm long, 0.2–0.5 cm thick, equal or enlarged slightly toward base. Dark blackish brown, dark gray to grayish brown. Surface dry, apex with whitish powder when young, soon smooth overall. **FLESH:** Very thin, grayish, stipe stuffed with whitish pith, occasionally hollow in age. **ODOR:** Mild to slightly farinaceous. **TASTE:** Mild to slightly farinaceous. **KOH:** No reaction. **SPORE DEPOSIT:** White. **MICROSCOPY:** Spores 4.5–6.5 x 3.5–4.5 μm, nearly round to broadly ellipsoid, smooth, amyloid. Basidia 4-spored. Pleurocystidia and cheilocystidia abundant, cylindrical or to spindle shaped, clamps present.

ECOLOGY: Solitary, scattered, or in small clusters on burnt ground, around ash or charcoal. Found throughout our range, but generally restricted to campfire pits or brush pile burns on coast. Very common after forest fires in the mountains. Fruiting in fall, often again in spring.

EDIBILITY: Unknown.

COMMENTS: The umbilicate cap, decurrent whitish gills, dark blackish brown to grayish brown, hygrophanous cap, and growth in burnt areas are distinctive. It is most likely to be mistaken for *Tephrocybe anthracophila*, another small dark gray-brown mushroom found at burn sites. The latter species differs by its more conical-convex or plane cap (often umbonate) and grayer gills. *Arrhenia obscurata* is similar in cap color, but has darker, thicker, more widely spaced gills; it often grows on bare soil or in moss. *A. epichysium* grows on decaying wood and has darker gills. *Pseudoclitocybe cyathiformis* is larger, has grayish gills, and doesn't usually grow in burnt areas.

SYNONYMS: *Mycena maura* (Fr.) Kühner, *Omphalina maura* (Fr.) Quél.

Arrhenia epichysium

(Pers.) Redhead, Lutzoni, Moncalvo & Vilgalys

CAP: 1.5–6 cm across, convex to plane, with a distinct depression at center, margin, becoming slightly uplifted in age; sometimes appearing like a small goblet or vase. Dark, dull brownish black when young, often becoming paler, gray, brownish gray to tan-brown. Surface smooth, dry to moist, translucent-striate when wet. **GILLS:** Broadly attached to decurrent, close to subdistant, even, narrow. Dull gray, pale whitish gray to dingy grayish tan. **STIPE:** 2.5–6 cm long, 0.1–0.4 cm thick, equal or with a slightly enlarged base, straight or curved. Gray, pale tan, sometimes much darker, to brownish black. Surface smooth or slightly bumpy, dry, base often with whitish to pale gray fuzzy mycelium. **FLESH:** Very thin, cartilaginous, stipe hollow. **ODOR:** Indistinct. **TASTE:** Mild. **KOH:** No reaction. **SPORE DEPOSIT:** White, but often difficult to obtain. **MICROSCOPY:** Spores 7–8.5 x 4–4.5 μm, ellipsoid, smooth, colorless, inamyloid. Pileipellis clamped and strongly encrusted.

ECOLOGY: Solitary or in small groups on well-decayed wood. Widespread but uncommon, often fruiting in cool, wet winter months.

EDIBILITY: Nontoxic.

COMMENTS: This species confuses beginners and experienced mushroomers alike. With a bit of experience, it can be recognized its drab, rather dark gray, brown, or blackish cap; pale gray gills; growth on wood; and small size. *A. obscurata* is similarly colored but grows in moss or on bare soil. *Myxomphalia maura* only grows in recently burnt areas (such as in campfire pits or burnt forest); it has a dark gray to blackish brown cap (fading to pale gray) and strongly contrasting white gills when young. *Pseudoclitocybe cyathiformis* looks very similar, but is generally larger and has a strongly hygrophanous cap that is usually only translucent-striate near the margin; although it can grow on rotting wood, it is more often found on the ground.

SYNONYMS: *Omphalina epichysium* (Pers.) Quél., *Clitocybe epichysium* (Pers.) H. E. Bigelow.

Arrhenia chlorocyanea

(Pat.) Redhead, Lutzoni, Moncalvo & Vilgalys

CAP: 0.3–2 cm across, convex to plane, with a depressed center, margin downcurved and often ribbed when young, becoming plane to slightly uplifted in age. Dark navy blue to turquoise-green to dark bluish green when young, fading to grayish blue to grayish green with yellowish undertones in age. Surface dry to moist, slightly roughened when young, becoming smooth. **GILLS:** Subdecurrent to decurrent, well spaced, fairly thick, broad, partial gills often irregular, inter-veined and numerous. Pale bluish at first, becoming bluish white to dingy white in age. **STIPE:** 1–4 cm long, 0.1–0.3 cm thick, equal or enlarged slightly toward base. Dark turquoise-green to dark bluish green when young, fading slightly in age. Surface dry to moist, smooth. **FLESH:** Very thin, turquoise-green to grayish green, stipe stuffed when young, often hollow in age. **ODOR:** Indistinct. **TASTE:** Mild. **KOH:** No reaction. **SPORE DEPOSIT:** White. **MICROSCOPY:** Spores (7.5) 8.5–11 x 4.5–6 μm, broadly ellipsoid to oval, smooth, inamyloid.

ECOLOGY: Solitary, scattered, or occasionally in small clusters in moss, often in areas with nutrient-poor or disturbed soil, such as old logging roads or areas of sandy clay. Associated in some way with moss, probably as a parasite. Found throughout our range, but rare, or at least rarely spotted due to its small size. Fruiting from late winter well into spring.

EDIBILITY: Unknown.

COMMENTS: The umbilicate cap, decurrent, bluish white gills, dark turquoise-green to bluish cap and stipe, and small size are distinctive. It is unlikely to be mistaken for anything else. *Lichenomphalia umbellifera* often occurs in the same habitat and has the same stature, but has a light brownish to pinkish beige cap.

SYNONYMS: *Clitocybe atroviridis* H. E. Bigelow, *Omphalina chlorocyanea* (Pat.) Singer, *O. viridis* (Hornem.) Kuyper.

Group I

Ampulloclitocybe clavipes

(Pers.) Redhead, Lutzoni, Moncalvo & Vilgalys

CLUB-FOOT

CAP: 2–8 cm across, broadly convex with an inrolled and often ribbed or crimped margin at first, expanding to flat, occasionally with a small umbo, often becoming slightly uplifted in age. Gray, grayish brown at center or grayish white to creamy outward. Often with scattered dark "water spots" and a whitish bloom. Surface smooth, moist to dry. **GILLS:** Decurrent, close to subdistant, clean looking. Whitish to cream. **STIPE:** 2–7 cm long, 0.3–1 cm thick at apex, base often swollen, up to 2.5 cm across. Grayish to pale grayish brown, often marbled looking with whitish streaks or specks. Surface moist to dry, smooth. **FLESH:** Thin to moderately thick, whitish to watery white. **ODOR:** Indistinct or sweet, like grape soda. **TASTE:** Mild. **SPORE DEPOSIT:** White. **MICROSCOPY:** Spores 4.5–7 (9) x 3–4.5 µm, ellipsoid, smooth, inamyloid.

ECOLOGY: Solitary or scattered in small troops in moss and needle duff under Sitka Spruce. Locally common, but restricted to the Far North Coast. Fruiting from early fall into spring.

EDIBILITY: Edible, but negative interactions with alcohol have been reported. Although we can't find any concrete information about such incidents in California, caution is warranted.

COMMENTS: The grayish cap, creamy, decurrent gills, and grayish, club-shaped stipe with a swollen base are helpful identifying features; most collections also have a distinct grape soda odor. *A. avellaneoalba* has a browner cap with a more distinctly ribbed margin, is generally larger and taller, and often grows from well-decayed wood. *Pseudoclitocybe cyathiformis* has a dark gray, distinctly hygrophanous cap, and grayish, broadly attached (not decurrent) gills. Although similarly colored, *Clitocybe nebularis* is much larger and has a skunky odor.

SYNONYMS: *Clitocybe clavipes* (Pers.) P. Kummel. *Ampulloclitocybe* belongs in the family *Hygrophoraceae*, well removed from *Clitocybe*, more closely related to *Cuphophyllus*.

Pseudoclitocybe cyathiformis

(Bull.) Singer

GOBLET CLITOCYBE

CAP: 2–8 cm across, broadly convex with an inrolled margin and a low, round umbo when young, soon becoming nearly plane with a low umbo, eventually uplifted with a strongly depressed center, sometimes with a wavy margin. Dark grayish to gray-brown, covered with a fine pale bloom when young, hygrophanous; fading to pale gray when dry. Surface slightly greasy or dry, smooth. **GILLS:** Broadly attached to slightly decurrent, close to subdistant. Dingy grayish or pale grayish white when young, then grayish to brownish tan in age. **STIPE:** 4–10 cm long, 0.5–1 cm thick, cylindrical with a club-shaped base. Grayish tan, slightly vertically striate and roughened. **FLESH:** Often soft, slightly rubbery or flabby in cap, fibrous in stipe. **ODOR:** Indistinct. **TASTE:** Mild. **SPORE DEPOSIT:** Whitish or off-white. **MICROSCOPY:** Spores 6.5–10.5 (13) x (4) 5–6.5 µm, ellipsoid to broadly ellipsoid, smooth, inamyloid to weakly amyloid.

ECOLOGY: Solitary or scattered in small groups, usually on or near decayed logs, stumps, and bark of hardwoods, occasionally in humus or soil. Locally common from greater San Francisco Bay Area south to Santa Cruz County, rare on the North Coast. Fruiting from fall into winter.

EDIBILITY: Nontoxic.

COMMENTS: This species' dark colors, tall stature, and depressed cap at maturity set it apart (a little, anyway) from hordes of similar *Clitocybe*. *Clitocybula atrialba* has a similar stature, but has more widely spaced contrasting gills and a scurfy stipe. *Arrhenia epichysium* is quite a bit smaller, always grows on wood, and has a distinctively translucent-striate cap when wet. *Ampulloclitocybe* have deeply decurrent, creamy white gills and grow in moss or duff. The smaller *Myxomphalia maura* has a strongly umbilicate cap even when young and more contrasting pale whitish gills, and grows in burnt areas.

Paralepista flaccida

(Sowerby) Vizzini

CAP: 3–9 cm across, broadly convex with a distinctly inrolled margin at first, center of cap soon slightly depressed, becoming strongly uplifted with a wavy margin. Rich ocher to orange with reddish brown spots and streaks, weakly hygrophanous, fading to dingy tan. Lighter variants dull buff to pinkish beige and creamy at margin. Surface greasy or dry, smooth, but often with a pale sheen when fresh and with small moldy-looking patches on cap in age. **GILLS:** Decurrent, very close to crowded, often narrowly forked near attachment to stipe. Whitish to cream, becoming beige, often dingy pinkish tan in age. **STIPE:** 5–12 cm long, 0.4–1 cm thick, cylindrical, upright but often curvy, base slightly enlarged. Creamy to pinkish tan or ocher brown (darker in age). Smooth to fibrillose, base often covered in conspicuous pale mycelium. **FLESH:** Fibrous, thin, pale. **ODOR:** Usually fairly distinct, like black pepper. **TASTE:** Indistinct. **SPORE DEPOSIT:** Off-white. **MICROSCOPY:** Spores 4–5 x 3.5–4 µm, globose to broadly ellipsoid, finely warted, inamyloid.

ECOLOGY: Scattered in small clumps and clusters, or in denser troops forming arcs and rings. Found in rich humus and needle duff of a wide range of forest types, especially abundant under live oak and redwood in late winter and early spring. Common throughout our range.

EDIBILITY: Unknown.

COMMENTS: This species is characterized by its ocher-colored, uplifted caps, close, pale, decurrent gills, and distinctive odor of black pepper. This can be a tough species to recognize. It resembles various *Clitocybe* and particularly the paler species of *Rhodocybe*; the latter can be told apart by their usually smaller, less depressed caps, lack of distinctive smells, pinkish spore prints, and elongate, roughened-bumpy spores. *P. flaccida* is sometimes mistaken for the Candy Cap (*Lactarius rubidus*), but that species has brittle flesh, exudes a clear latex, smells sweetish, and is usually smaller.

SYNONYMS: *Clitocybe inversa* (Scop.) Quél., *C. flaccida* (Sowerby) P. Kumm., *Lepista flaccida* (Sowerby) Pat.

Clitocybe trulliformis

(Fr.) P. Karst.

CAP: 0.7–4 cm across, convex with a narrow inrolled margin and often a small umbo when young, expanding to broadly convex, often depressed at center with a wavy or crinkly margin. Dark grayish at center (blackish in some specimens), to gray overall with a whitish bloom, or grayish tan to grayish beige outward with scattered gray-tan spots, extreme margin often white. Surface dry, slightly fuzzy when young, smooth in age. **GILLS:** Broadly attached to slightly decurrent, quite widely spaced. White to creamy. **STIPE:** 1–5 cm long, 0.2–0.5 cm thick, cylindrical, usually upright, base slightly enlarged. Grayish to beige or whitish gray in age, usually markedly contrasting with bright whitish gills. Smooth to fibrillose. **FLESH:** Fibrous, thin, white. **ODOR:** Indistinct. **TASTE:** Not sampled. **SPORE DEPOSIT:** White. **MICROSCOPY:** Spores 4.5–6.5 x 2.5–4 µm, ellipsoid, smooth, inamyloid.

ECOLOGY: Growing in small clumps or scattered in troops in duff of many forest types, especially common under Douglas-fir and redwood, fruiting in late fall and winter. Widespread in our area, can be locally common but generally infrequent. Probably underreported, as it is tough to recognize and easily overlooked.

EDIBILITY: Unknown.

COMMENTS: The small size, dark grayish, finely textured cap, and widely spaced pale gills that contrast with the gray to tan stipe are the feebly distinguishing characters of this species. Many other *Clitocybe* are superficially similar, but are slightly larger and paler, and/or have more closely spaced gills. *C. trulliformis* could possibly be confused for a *Melanoleuca*; they also tend to have pale gills contrasting with a darker cap and stipe, but have notched (not decurrent) crowded gills.

Clitocybe californiensis

H. E. Bigelow

CALIFORNIA CLITOCYBE

CAP: 1–6 cm across, broadly convex with an inrolled margin, becoming plane, in age sometimes wavy, but often retaining a downcurved margin. Dull pinkish tan or pinkish beige, sometimes orangey pink at center, to pale beige outward, becoming much paler overall in age or dry weather. Surface smooth, slightly greasy, becoming dry and shiny in dry weather. **GILLS:** Broadly attached, close to crowded. Pale, whitish to beige, sometimes light tan in age. **STIPE:** 3–10 cm long, 0.5–1 cm thick, cylindrical or slightly flattened, straight or often curved, slightly enlarged toward base. Pale tan-beige or pinkish tan, relatively smooth but covered in fine whitish vertical lines. **FLESH:** Thin, somewhat rubbery. **ODOR:** Indistinct. **TASTE:** Mild. **SPORE DEPOSIT:** White. **MICROSCOPY:** Spores 5–6 x 3–3.5 (4) µm, ellipsoid, smooth, inamyloid.

ECOLOGY: In small scattered clumps or clusters on lignin-rich oak leaf humus or well-decayed oak wood. Fairly common throughout the southern half of our range. Fruiting from fall into early spring.

EDIBILITY: Unknown.

COMMENTS: *C. californiensis* is a rather boring (that is, typical) *Clitocybe* that manages to set itself apart (barely) by its pinkish buff cap and preference for decaying hardwood substrate. *C. salmonilamella* also grows on small woody debris, but has a belly button at the center of the cap, pinkish gills at maturity, and larger spores.

Clitocybe salmonilamella

H. E. Bigelow

PINK-GILLED CLITOCYBE

CAP: 1.5–6 cm across, convex with a slightly depressed center and weakly inrolled margin when young, expanding to flat and then uplifted with a wavy margin in age. Brownish tan with a whitish bloom, becoming light beige-tan to creamy beige, margin translucent-striate in age. Hygrophanous, fading outward to nearly white when dry. Surface moist to greasy or dry, smooth. **GILLS:** Broadly attached or occasionally weakly decurrent, close to somewhat widely spaced. Creamy whitish to pale grayish beige, becoming light pastel pink in age or occasionally dingy, light salmon when old. **STIPE:** 4–10 cm long, 0.4–1 cm thick, cylindrical, or more often flattened and strongly grooved, base slightly enlarged. White to dingy light grayish beige, darkening slightly to yellowish tan upward from base in age. Smooth to vertically fibrillose, covered in a fine white bloom over most of length, base with conspicuous white mycelial fuzz and rhizomorphs. **FLESH:** Fibrous, thin, rubbery-cartilaginous, hollow in stipe. **ODOR:** Indistinct. **TASTE:** Indistinct to slightly peppery. **SPORE DEPOSIT:** Whitish to light buff. **MICROSCOPY:** Spores 7–9 x 3.5–4.5 µm, ellipsoid, smooth, inamyloid. Some hyphae of cap cutis encrusted.

ECOLOGY: Solitary or in small groups of single or paired fruitbodies on small-diameter woody debris, especially on decaying live oak branches buried in duff (also in surrounding humus). Fruiting in fall and early winter. Fairly common in the southern half of our range, probably more widespread but probably underreported.

EDIBILITY: Unknown.

COMMENTS: The umbilicate, hygrophanous cap, often grooved stipe, pinkish mature gills, and growth on small bits of wood help distinguish this species. The gills often turn pinkish slowly and/or weakly, making this species quite difficult to recognize for beginners and seasoned identifiers alike. In such cases, the relatively large spores can be a helpful feature. There are many very similar *Clitocybe* that usually have less umbilicate caps or more cylindrical stipes, or grow primarily on humus rather than small woody debris.

Clitocybe vermicularis subsp. americana

H. E. Bigelow

CAP: 2–6 cm across, broadly convex with an inrolled margin at first, expanding to flat with a depressed center, eventually uplifted with a strongly depressed center and a wavy, sometimes crimped margin in age. Pale cinnamon buff to beige, with a whitish bloom if dry. Surface smooth, moist to dry. **GILLS:** Decurrent, close to crowded. Whitish to cream or pale buff. **STIPE:** 2–5 cm long, 0.3–0.7 cm thick, cylindrical or tapered downward. Surface pale buff with a whitish pruinose bloom. Base with many short, but conspicuous white rhizomorphs. **FLESH:** Thin, slightly fibrous, whitish. **ODOR:** Indistinct. **TASTE:** Mild. **SPORE DEPOSIT:** White. **MICROSCOPY:** Spores 4–6 x 2.5–3.5 µm, ellipsoid, smooth, inamyloid.

ECOLOGY: Solitary, scattered, or in small clusters in needle duff under Sitka Spruce. Very common on the Far North Coast. Fruiting in winter and spring.

EDIBILITY: Unknown.

COMMENTS: This species superficially resembles many different *Clitocybe*; the pale cinnamon buff cap and white rhizomorphs are the best clues. Another species with conspicuous rhizomorphs is *C. albirhiza*, a very common species that fruits in the spring with snowmelt in higher-elevation coniferous forests.

Clitocybe spp. "The Gray Group"

CAP: 2–5.5 (7) cm across, convex with an inrolled margin when young, becoming broadly with or without a distinctly umbilicate center. Gray to brown at first, hygrophanous, fading to beige-gray, or beige-tan in age, often with a "haloed" appearance. Margin opaque when dry, slightly translucent-striate when wet. Surface greasy to dry, smooth. **GILLS:** Broadly attached, fairly close. Dingy pale beige or creamy tan at first, often becoming extensively grayish with a paler margin in age. **STIPE:** 2–5 cm long, 0.4–0.7 cm thick, cylindrical with a slightly enlarged base or compressed and grooved. Dingy beige to tan or grayish. **FLESH:** Thin, fibrous, pale dingy, stipe usually hollow at maturity. **ODOR:** Indistinct to farinaceous. **TASTE:** Indistinct. **SPORE DEPOSIT:** White to off-white. **MICROSCOPY:** Spores 5.5–6 x 3–4 µm, ellipsoid, smooth, inamyloid. Clamps present.

ECOLOGY: Small clusters or scattered troops in duff, often in disturbed areas in redwood forest, but found in other forest types as well. Fruiting from midfall through winter, with occasional spring flushes.

EDIBILITY: Unknown.

COMMENTS: Many *Clitocybe* in our area will approximately match the description above. Given the current state of knowledge, they are very difficult to identify: original concepts for many species are weak, and the few modern attempts to clarify the situation have mostly failed due to a complete lack of photography and excessive reliance on secondhand information. A great deal of careful work using multiple approaches will be needed to shed light on the diversity and identification of grayish *Clitocybe* on the Redwood Coast. These are a few names commonly applied to members of this group in our area: *C. subditopoda*, growing on conifer needle duff, has distinctly grayish gills and a farinaceous odor and taste. *C. vibecina*, also farinaceous and under conifers, has larger spores (6–8 x 4 µm). *C. metachroa* (something like the photographed species above) has light gray or pallid gills that are sometimes forked and shorter spores (5.6–6 x 3–4 µm). *C. pseudodicolor* has a grayish cap that fades to nearly white, is often only shallowly depressed, and has pallid gills and spores measuring 5.5–7 x 3–4 µm.

Clitocybe sclerotoidea

(Morse) Bigelow
PARASITIC CLITOCYBE

CAP: 0.5–4 (5) cm across, broadly convex with a distinctly inrolled margin, sometimes with a weak umbo, eventually plane, to irregular, wavy, and uplifted in age. Whitish to dingy grayish white, grayish tan, or brownish, often mottled with brown or dark gray "water spots" or rings and radial streaks. Surface dry and finely felty when very young, soon smooth, becoming slightly greasy. **GILLS:** Broadly attached to slightly decurrent, widely spaced, quite thick. Pallid grayish at first, soon dark dingy grayish brown with pinkish brown hues, contrasting strongly with paler stipe where gills terminate at apex of usually much paler stipe. **STIPE:** 2–8 cm long, 0.3–1 cm thick, cylindrical or irregular, often flattened and grooved, usually strongly curving at least near base. Whitish to grayish, with brown tones in age or where handled. Apex sometimes strongly scurfy, entire surface covered in a fine white bloom (destroyed when handled and then stipe appearing brownish). **FLESH:** Fairly firm and fibrillose-fleshy, grayish. **ODOR:** Indistinct. **TASTE:** Not sampled. **SPORE DEPOSIT:** Whitish. **MICROSCOPY:** Spores 9–10.5 x 3.5–4.5 μm, elongate-ellipsoid to spindle shaped, smooth, inamyloid.

ECOLOGY: In small or large clusters of fused stipes, attached to an irregular, lumpy, rubbery-fleshy mass of marbled white and yellowish orange tissue (see comments), usually near *Helvella vespertina*. Widely distributed in our area and can be locally common in some years, but generally infrequently encountered. Fruiting from late fall into early spring.

EDIBILITY: Unknown.

COMMENTS: The marbled mass of tissue invariably found at the stipe bases of the fruitbodies was at one time thought to be a sclerotium (hence the scientific name of this species). Closer study revealed that this is actually a mass of the *Clitocybe* mycelium and tissue from a *Helvella* (confirmed only from *H. vespertina* and not the closely related *H. dryophila*). This distinctive structure goes a long way to distinguish an otherwise rather drab mushroom. However, even specimens that come in from the field separated from the ball of their host's tissue are identifiable. The dark, thick, widely spaced gills and drab colors of the fruitbodies are fairly distinctive. *C. ditopa* is somewhat similar, but the cap is browner, and the gills are not as thick and are paler (thus not contrasting strongly with the stipe).

Clitocybe rivulosa

(Pers.) P. Kumm.
SWEAT-PRODUCING CLITOCYBE

CAP: 1–5 cm across, broadly convex with a finely inrolled margin when young, becoming nearly plane and slightly depressed at center, and often uplifted and ruffled in age. White to whitish gray with grayish tan "water spots," concentric rings, and radial streaks, often becoming pinkish brown in blotches or in cracks. Surface dry, more or less smooth, often wrinkled and cracking in age. **GILLS:** Broadly attached to decurrent. Dingy pale gray to creamy gray or gray. Often with irregular, contrasting lighter orange zones or bands. **STIPE:** 2–6 cm long, 0.5–2 cm thick, cylindrical or more often flattened and grooved and irregularly curved or twisted, usually slightly or distinctly enlarged near base. Pale whitish gray to creamy grayish, often vertically lined and roughened, becoming dingy grayish brown or ocher-brown where handled or in age. **FLESH:** Thin to moderately thick, rather firm-stringy or rubbery, pale grayish to grayish or dingy brownish. **ODOR:** Indistinct. **TASTE:** Mild. **SPORE DEPOSIT:** Whitish. **MICROSCOPY:** Spores 5–5.5 x 3–3.5 μm, ellipsoid, smooth, inamyloid.

ECOLOGY: Scattered in tight clusters or solitary in large troops, often forming arcs or rings in grassy areas such as pastures and lawns. Fruiting throughout wet season, more common in fall and spring.

EDIBILITY: Very poisonous, containing muscarine. Sweating, salivation, and gastrointestinal distress are typical symptoms. Although small, the fruitbodies are frequently abundant in urban areas, and if many are eaten, they are potentially deadly for small children or pets.

COMMENTS: Definitely a *Clitocybe* in the pejorative sense, this rather bland-looking mushroom is told apart by its growth in clusters and arcs in grassy areas, drab whitish gray colors accented with pinkish tan hues, often flattened and grooved stipe, and frequently concentrically zoned and "water-spotted" caps. *Clitopilus prunulus* is similar but has pinkish spores and a farinaceous odor and taste.

Clitocybe fragrans

(With.) P. Kumm.

ANISE CLITOCYBE

CAP: 3–7 cm across, broadly convex when young, becoming plane, usually with a shallow depression at center of cap, margin ruffled or wavy in age. Yellowish beige, to tan, creamy white to nearly white, margin usually translucent-striate, hygrophanous. Surface smooth, slightly greasy, but not viscid, becoming dry. **GILLS:** Broadly attached to weakly decurrent, or with a decurrent tooth, close to moderately spaced. Pale at all ages, from whitish to watery beige. **STIPE:** 2.5–7.5 cm long, 0.2–0.8 cm thick, cylindrical, sometimes slightly curved. Pallid whitish to beige-tan. Surface smooth to slightly fibrillose, dry. **FLESH:** Very thin and watery. **ODOR:** Quite strong, sweet or aniselike when fresh. **TASTE:** Slightly sweet or mild. **SPORE DEPOSIT:** Whitish. **MICROSCOPY:** Spores 6.5–8.5 x 3.5–4.5 μm, ellipsoid to teardrop shaped, smooth, inamyloid.

ECOLOGY: Solitary or scattered in troops, most often in rich redwood duff throughout our range, also in other mixed evergreen forests. Fruiting from late fall through early spring.

EDIBILITY: Edible, but easy to confuse with other species of unknown edibility. The flavor doesn't seem to be retained after cooking.

COMMENTS: This species would be one of the most featureless of local gilled mushrooms were it not for its strong fragrance. Although not everyone detects the anise character, almost everyone can sense the strength and sweetness of the odor. Similar species with a distinct anise odor include *C. deceptiva*, which is extremely similar, but generally slightly smaller with a darker cap when young and a light pinkish white spore print. *C. oramophila* has a pale reddish orange to pinkish cap and grows in spruce duff on the Far North Coast. *C. odora var. pacifica* also shares the sweet-anise odor, but is entirely bluish to blue-green or olive bluish in age. Many other *Clitocybe* are structurally similar and drab colored, but smell unremarkable or at least not sweet. *Rhodocybe hondensis* and *R. nuciolens* are not as pale (darker pinkish tan), are thicker fleshed, and have pinkish spore prints and unremarkable odors.

Clitocybe odora var. *pacifica*

Kauffman

BLUE-GREEN ANISE CLITOCYBE

CAP: 2–8 cm across, broadly convex with an inrolled margin at first, becoming flat or centrally depressed, often uplifted and wavy in age. Pastel turquoise-grayish to sky blue with teal hues with fine, dense innate silvery pale streaks, sometimes with "water spots" and blotches of darker blue-green, disc often beige-tan. Becoming dingy olive-gray with bluish hues. Surface dry or greasy, smooth. **GILLS:** Broadly attached, edge of gills often eroded. Pallid grayish sky blue, then more dingy turquoise-blue and then sordid olive grayish; often with whitish gray blotches as spores mature. **STIPE:** 2–6 cm long, 0.5–2 cm thick, shape quite variable, cylindrical, often enlarged toward base, sometimes tapering (or bulbous and tapering), usually with adhering leaves and substrate, sometimes with elastic whitish rhizomorphs. Pallid whitish with sky blue and grayish tones, beige-tan toward base, irregularly developing dingy olive and orange blotches in age or where handled. **FLESH:** Thin to moderately thick, white, dingy tan to variably pallid bluish gray. **ODOR:** Strong and sweet like aniseed or fennel. **TASTE:** Mild. **SPORE DEPOSIT:** Dull whitish or off-white. **MICROSCOPY:** Spores 6–8 x 4–5 μm, ellipsoid, smooth, inamyloid.

ECOLOGY: Often in small clusters or scattered in troops or arcs, often in deep, moist live oak duff, also under Tanoak and conifers. Can be locally common in southern portion of our range, but generally infrequent, rare on the Far North Coast. Fruiting from fall into winter.

EDIBILITY: Edible, but not commonly eaten. The odor is not reflected in the flavor when cooked, but some people have tried infusing the mushrooms into vodka, with varying degrees of success.

COMMENTS: No other species in our area really approaches this oddly lovely species in appearance. The muted blue-green hues and strong, sweet anise odor handily set it apart. *Marasmius calhouniae* very rarely appears to age bluish green (possibly from a viral infection), but is much smaller, has very widely spaced gills, and lacks a strong odor. The pallid whitish tan *Clitocybe fragrans* shares the strong aniseed odor.

Group J

Laccaria amethysteo-occidentalis
G. M. Muell.

WESTERN AMETHYST LACCARIA

CAP: 1.5–10 cm across, convex to plane, often with a depressed center, or wavy to highly irregular. Color extremely variable, depending on age and moisture content. Often distinctly royal purple when young, but can be purplish, lilac, pinkish buff to gray. Typically developing tan tones in age, or fading to beige or grayish buff when dry. Surface dry to moist, smooth to finely fibrillose-scaly. **GILLS:** Broadly attached, notched with a decurrent tooth or slightly decurrent, fairly well spaced, often cross veined. Royal purple to lilac when young, and retaining purple tones longer than cap, before fading to pinkish buff. **STIPE:** 3–12 (16) cm long, 0.4–2 cm thick, equal, enlarged toward base, or quite twisted and distorted. Purple, purplish gray, purple-brown, pinkish buff to grayish beige, typically with numerous whitish or pale buff loose fibrils and scurfs. Basal mycelium distinctly violet, often well after rest of fruitbody has faded. **FLESH:** Thin to moderately thick, rubbery-tough, fibrous. Violet, grayish lilac to whitish beige. **ODOR:** Indistinct. **TASTE:** Mild. **SPORE DEPOSIT:** White. **MICROSCOPY:** Spores (6.5) 7.5–10.5 x 6.5–9 μm, subglobose, ellipsoid to almond shaped, ornamented with sparse to crowded spines, 0.5–1.5 μm long. Basidia 4-spored.

ECOLOGY: Solitary, scattered, or in small to large clusters on ground under both conifers and hardwoods. Especially fond of Douglas-fir, live oak, and pine. Very common throughout much of our range, more local and patchy on the Far North Coast. Fruiting from fall into spring, most abundant in early winter.

EDIBILITY: Edible and good (at least the caps; the stipes are rather tough). Avoid eating collections from urban areas, along roads, or potentially polluted areas, as *Laccaria* concentrate heavy metals in their fruitbodies.

COMMENTS: When fresh, the vibrant purple colors, widely spaced gills, fibrous stipe, and white spores make this species practically unmistakable. When faded, it can easily be mistaken for other *Laccaria*. The larger size, markedly scurfy, fibrous stipe, violet basal mycelium, and lack of warm pinkish buff tones help distinguish it. *L. bicolor* is generally smaller, and has a warmer pinkish brown to buff cap and paler gills (but often with a violet cast when young). *Clitocybe nuda* has a smooth cap, softer flesh, and buff spores. Some *Cortinarius* and *Inocybe* can look similar, but have brown spores.

Laccaria fraterna

(Sacc.) Pegler

CAP: 1–4 cm across, convex to plane, occasionally wavy in age. Reddish brown to pinkish buff; hygrophanous, translucent-striate when wet, fading and becoming opaque when dry. Surface dry to moist, smooth to very finely scaly. **GILLS:** Broadly attached, at times with a fine decurrent tooth, fairly well spaced, edges even, becoming ragged in age. Pinkish to pinkish buff. **STIPE:** 2–5 (7) cm long, 0.2–0.5 cm thick, equal, reddish brown to pinkish brown, often slightly darker than cap. Surface dry, striate, often with paler twisting fibrils. Basal mycelium white. **FLESH:** Thin, tough, fibrous, pinkish buff. **ODOR:** Indistinct. **TASTE:** Mild. **SPORE DEPOSIT:** White. **MICROSCOPY:** Spores 8.5–11 x 8–10.5 μm, globose to subglobose, ornamented with sparse spines 1–2 μm long. Basidia 2-spored.

ECOLOGY: Solitary, scattered, or in small clusters on ground under eucalyptus, *Melaleuca*, and *Acacia*. Especially common in urban areas, fruiting from late fall into spring.

EDIBILITY: Not recommended, as it often grows in urban areas that are often polluted.

COMMENTS: The small size, translucent-striate, reddish to pinkish caps, fibrous flesh, and growth under Australian trees and shrubs help distinguish this species. *L. laccata* is generally slightly larger, paler, and also grows with native trees.

Laccaria proxima (sensu CA)

CAP: (1.5) 3–8 cm across, convex to plane, becoming irregular and wavy in age. Dull ocher-buff, pinkish brown to pale buff; hygrophanous, fading when dry. Surface moist to dry, smooth to finely scaly at first, soon distinctly and extensively fibrillose-scaly. **GILLS:** Broadly attached, or with a slight notch, well spaced, edges even, occasionally becoming ragged in age. Pinkish, pinkish buff to pale buff. **STIPE:** 3–8 (12) cm long, 0.3–1 cm thick, equal, club shaped or twisted. Pinkish buff, ocher-buff to pale buff. Surface dry, often with paler twisting fibrils. Basal mycelium white. **FLESH:** Thin, tough, fibrous, pale pinkish buff to pale buff. **ODOR:** Indistinct. **TASTE:** Mild. **SPORE DEPOSIT:** White. **MICROSCOPY:** Spores 8–11 (12.5) x 7–9 μm, broadly ellipsoid to ellipsoid, ornamented with sparse to crowded spines, 0.5–1 μm long. Basidia 4-spored.

ECOLOGY: Solitary, scattered, or in small clusters on ground under pine and occasionally other conifers. Common on the North Coast, distribution patchier elsewhere. Fruiting from late fall through winter.

EDIBILITY: Edible and good, but see the edibility comments for *Laccaria amethysteo-occidentalis* (p. 356).

COMMENTS: The larger size and stockier stature, scaly cap, and growth with pine are helpful in distinguishing this species from *L. laccata*. The similar conifer-loving *L. bicolor* has violet basal mycelium and often a violet cast to the young gills. *L. nobilis* is usually much taller, grows with Sitka Spruce on the Far North Coast, and has slightly smaller spores.

Laccaria laccata group
LACKLUSTER LACCARIA

CAP: 1.5–5 cm across, convex to plane, often rather irregular and wavy in age. Dull orange-brown, pinkish brown to pinkish buff; hygrophanous, very weakly translucent-striate when wet, fading and becoming opaque when dry. Surface dry to moist, smooth to slightly fibrillose-scaly. **GILLS:** Broadly attached, fairly well spaced, edges even, becoming ragged in age. Pinkish, pinkish buff to pale buff. **STIPE:** 2–7 cm long, 0.2–0.7 cm thick, equal or twisted. Pinkish buff to pale buff. Surface dry, often with twisting fibrils. Basal mycelium white. **FLESH:** Thin, tough, fibrous, pale pinkish buff to pale buff. **ODOR:** Indistinct. **TASTE:** Mild. **SPORE DEPOSIT:** White **MICROSCOPY:** Spores (7) 9–11 (13) x 7–10 (11.5) µm, globose to subglobose, ornamented with sparse to crowded spines, 1–2 µm long. Basidia 4-spored.

ECOLOGY: Often in small clusters, scattered, or solitary on ground under live oak and Tanoak, occasionally with pine. Especially common in southern part of our range. Fruiting in late winter and spring, occasionally in fall.

EDIBILITY: Edible, but see the edibility comments for *Laccaria amethysteo-occidentalis* (p. 356).

COMMENTS: This *Laccaria* can be recognized by its small to medium-size orangey cap; pallid to orange or dull pinkish gills; and fibrous stipe. The name *L. laccata* is applied to a few difficult-to-distinguish, poorly understood, and probably undescribed species in California. The most common is a clustered species with live oak. Both *L. bicolor* and *L. proxima* look very similar, but are slightly larger; in addition, *L. bicolor* has violet basal mycelium, and *L. proxima* has a distinctly scaly cap and often grows with conifers. *L. fraterna* is more pinkish and grows with introduced Australian trees. *L. nobilis* is larger and much taller, and grows with conifers (especially Sitka Spruce) on the Far North Coast.

Group K
Clitocybe tarda
Peck
GRASS BLEWIT

CAP: 1–4 (6) cm across, convex with an inrolled margin when young, expanding to flat or slightly uplifted with a wavy margin, sometimes also with a weak umbo. Lavender or pinkish buff when young, soon becoming tan or brown; hygrophanous, fading to beige or grayish white as it dries out. Surface smooth, greasy to dry, often shiny in dry weather. **GILLS:** Broadly attached, slightly notched, close to subdistant. Light lilac at first, soon buff to beige or light gray, dingy tan in age. **STIPE:** 2–7 cm long, 0.3–0.7 cm thick, cylindrical, upright, base slightly enlarged. Lilac to gray, fading to whitish, grayish, or tan in age. Smooth or more often vertically fibrillose, dry. **FLESH:** Fibrous, thin, pale to lilac or grayish. **ODOR:** Indistinct. **TASTE:** Mild. **SPORE DEPOSIT:** Pinkish buff. **MICROSCOPY:** Spores (4.5) 6 –8 x 3–5 µm, ellipsoid, smooth to finely warty.

ECOLOGY: In small clumps or scattered in troops in lawns with dead grass debris, occasionally in compost or other nutrient-rich disturbed areas. Uncommon to rare in California, perhaps overlooked. Many patches are undoubtedly mowed into tiny pieces or trampled by soccer players as soon as they fruit. Typically fruiting from late fall into spring, but possible at any time of year in watered lawns.

EDIBILITY: Edible but tiny and bland.

COMMENTS: The small size, faint lilac-lavender tones, growth in groups in lawns, and buff spore deposit help distinguish this species, which resembles a small, slender, drab version of *Lepista nuda*. It could be confused with *Laccaria amethysteo-occidentalis*, but that species is usually quite a bit larger, with more intense purple colors (especially the gills and mycelium at the stipe base), and has a shaggy-hairy fibrous stipe surface.

SYNONYM: *Lepista tarda* (Peck) Murrill.

Clitocybe nuda

(Bull.) H. E. Bigelow & A. H. Sm.

BLEWIT

CAP: 3–15 cm across, rounded with a strongly inrolled margin, soon convex, becoming broadly convex to nearly flat, often wavy or uplifted in age. Color of cap varies substantially. Typical fruitbodies are amethyst purple or pinkish purple when young with a paler margin, but many are light lavender, pale pinkish beige with a lilac hue, buffy beige, or even entirely brownish in age. Those growing in exposed areas or in dry weather sometimes show striking silvery gray. Surface smooth, soapy or slightly greasy when fresh and moist, often dry and often shiny in age. **GILLS:** Attached, often notched, close. Beautiful deep lavender or amethyst when young, becoming duller in age, developing ocher-brownish hues, sometimes extensively pale. **STIPE:** 4.5–10 cm long, 2–4 cm thick, cylindrical or strongly swollen at base. Pale silvery lilac to evenly light purple, sometimes with ocher or brownish tones, especially in age. Covered in pale scurfs (especially near apex), otherwise smooth to fibrillose. Base usually covered in a layer of fluffy lilac-white or purple mycelium, often binding to humus, and with pale lilac rhizomorphs as well. **FLESH:** Thick, firm, marbled amethyst or pallid, when fresh often strikingly full of water. **ODOR:** Often slightly sweet or floral. David Arora describes it as "frozen orange juice." **TASTE:** Mild to slightly spicy. **SPORE DEPOSIT:** Pinkish buff. **MICROSCOPY:** Spores 5.5–8 x 3.5–5 μm, ellipsoid, finely warted, inamyloid.

ECOLOGY: Solitary, scattered, or clustered in small loose troops, sometimes in large rings or arcs. Deep live oak, cypress, or eucalyptus duff is a favored habitat, but can be found almost anywhere, so long as some rotting organic matter is present. Common throughout our area. Fruiting from late fall into spring.

EDIBILITY: Edible and popular, but not consistently good. The habitat in which the fruitbodies grow seems to affect the taste, and those from eucalyptus and cypress duff in particular are reported as often being musty or bitter tasting. It is best used in cream-based dishes, where it enhances the flavor.

COMMENTS: This species is quite variable in color within a consistent range: purple to brown, with varying shades of beige, pink, and silvery gray. The gills and stipe base tend to stay purple after the cap fades, and this can be a good clue when examining browned-out older fruitbodies. The stout stipe and smooth cap with an inrolled margin also give it a distinctive gestalt. The fragrant odor when fresh and pale pinkish buff spore print further help separate this species from all others, except *Clitocybe brunneocephala*, which is similar in all respects but never shows purple tones and is almost always very squat. It is uncommon in our area, but quite common in grassy areas in Southern California. Some *Cortinarius* can look strikingly similar to the Blewit, but have cobwebby fibers on the stipe, rusty brown spores, and often a finely fibrillose cap. Although they do not appear to have caused any poisonings (since they have undoubtedly been confused for the Blewit), they should be avoided due to their poorly known edibility status.

SYNONYM: *Lepista nuda* (Bull.) Cooke.

Clitocybe nebularis

(Batsch) P. Kumm.

CLOUDY CLITOCYBE

CAP: 4–16 cm across, convex with a strongly inrolled margin, becoming plane, uplifted, and often wavy in age. Grayish white to darker gray with a fine whitish bloom, in age mottled grayish tan with a white margin. Appearing virgate, and frequently with dark grayish "water spots" around margin. Surface dry, smooth and slightly metallic, or when moist with scattered patches of moldy-looking whitish fuzz. **GILLS:** Broadly attached or slightly decurrent, fairly close to moderately spaced. Whitish cream to dingy beige or grayish tan in age. **STIPE:** 6–15 cm long, 1.5–4 cm thick, equal to slightly enlarged toward base. Whitish to grayish or tan-gray. Smooth or with darker scales near apex of stipe, base quite often with irregularly scattered moldy-looking, fuzzy whitish patches. **FLESH:** Fibrous, fleshy, whitish. **ODOR:** Strong and distinctive, a musty-musky mixture often compared to skunk, swamp gas, rancid flour, or marijuana. **TASTE:** Difficult to assess separately from odor. **SPORE DEPOSIT:** Creamy. **MICROSCOPY:** Spores 5.5–8 x 3–4 µm, ellipsoid, smooth, inamyloid.

ECOLOGY: In small groups or large rings and arcs, occasionally solitary in forest duff. Most abundant in dense redwood forest from late fall into early spring, but can be found in almost any forest type. Common throughout our range.

EDIBILITY: Although some people eat this species, there are reports of it causing mild gastrointestinal upsets. The off-putting odor is enough of a deterrent for many people.

COMMENTS: The grayish cap, large size, slightly decurrent white gills, and distinctive musky odor distinguish this species. *C. harperi* is a poorly known species that is quite similar in all respects except for its darker grayish brown gills, lack of a skunky odor, and lack of clamp connections in all tissues; it is uncommon to rare in the coastal mountains south of San Francisco. Similar species include *Leucopaxillus gentianeus*, which has a brown cap and strong bitter taste, and *L. albissimus*, which has whitish fruitbodies (becoming yellowish) lacking grayish tones and does not have the strong unpleasant odor. The less common *Ampulloclitocybe clavipes* produces smaller fruitbodies with more decurrent gills and strongly swollen stipe bases, and often has a pleasant sweet odor. Very rarely parasitized by *Volvariella surrecta*, a tiny mushroom that emerges from the caps of this species.

Leucopaxillus albissimus

(Peck) Singer

LARGE WHITE LEUCOPAX

CAP: 6–30 cm across, domed with an inrolled margin when young, becoming broadly convex to flat, often wavy and lobed in age. White to cream or dingy yellowish tan at center in age. Surface dry, finely velvety to smooth, cracking into small uplifted scales or more deeply cracked in age or dry weather. **GILLS:** Subdecurrent to decurrent, close to crowded. White to creamy, becoming light yellowish in age. **STIPE:** 9–30 cm long, 2–6 (9!) cm thick, usually swollen downward but then tapered at base, sometimes cylindrical or irregularly swollen above. White to creamy, developing yellowish beige tones in age. Smooth to vertically fibrillose, often rough or with small, pointed scales in age. Base usually with adhering pad of white mycelium and rhizomorphs bound to duff and needle litter. **FLESH:** Very thick and firm, solid-rubbery, whitish, solid; tough and slow to decay. **ODOR:** Distinctive, odd, usually unpleasant. **TASTE:** Mild to bitter. **SPORE DEPOSIT:** White. **MICROSCOPY:** Spores 5.5–8.5 x 4–6 μm, ellipsoid, with amyloid warts and spines.

ECOLOGY: Solitary or more often forming arcs and rings in duff. Prefers growing under redwood in our area, but can be found in almost any kind of forest. One patch on Upper Campus at the University of California, Santa Cruz, forms giant (20 m across), nearly complete rings every winter. Fruiting from midfall through winter, fruitbodies persistent, sometimes lasting for months.

EDIBILITY: Inedible; tough, musty smelling, and often bitter.

COMMENTS: The large to huge fruitbodies, white to dingy creamy colors, decurrent gills, white mycelial mat, and habit of growing in arcs and troops make this species fairly easy to recognize. *Clitocybe nebularis* can be similar in colors, shape, and fruiting habits, but has a grayish cap, and usually doesn't have a mat of mycelium and litter at the base.

Leucopaxillus gentianeus

(Quél.) Kotl.

BROWN LEUCOPAX

CAP: 7–15 cm across, domed to broadly convex with a strongly inrolled, lobed, and usually strongly ribbed margin when young, expanding to broadly convex or flat, sometimes uplifted and slightly wavy. Reddish brown to medium brown, ocher-brown or tan, fading to dull beige, margin often nearly whitish. Surface finely velvety, becoming smoother and often finely cracked into plaques. **GILLS:** Broadly attached to slightly decurrent, close to crowded, white to cream. **STIPE:** 5–10 (15) cm long, 1–3 cm thick, cylindrical with an enlarged, club-shaped base. White to cream, discoloring beige to yellowish. Base usually attached to a thick grayish white mat of mycelium holding bits of twigs and humus. **FLESH:** Fibrous, white, solid, firm and rubbery at first but often riddled with maggot larvae and becoming crumbly. **ODOR:** A bit musty, distinct, slightly unpleasant. **TASTE:** Extremely bitter. **SPORE DEPOSIT:** White. **MICROSCOPY:** Spores 4–6 x 3–5 μm, broadly ellipsoid to subglobose, ornamented with amyloid warts and ridges.

ECOLOGY: Solitary, scattered, or clustered, often in large troops, rings, or arcs in duff of many different forest types, with a preference for redwood. Very common, occurring throughout our range. Fruiting from late fall through winter.

EDIBILITY: Inedible, extremely bitter.

COMMENTS: The finely velvety, brownish cap often ribbed around the margin, white gills, and stout, white stipe attached to a mat of pale mycelium make this species fairly easy to recognize. Mature caps look remarkably like pancakes. A few *Tricholoma* in our area are vaguely similar, but have notched gills that are often not as white (beige to tan, or blotchy orange brown) and usually have less robust stipes that are never attached to extensive mats of mycelium.

SYNONYM: *Leucopaxillus amarus* (Alb. & Schwein.) Kühner.

Melanoleuca subpulverulenta

(Pers.) Singer

CAP: 3–7.5 cm across, rounded with an inrolled margin at first, expanding to broadly convex or becoming flat with a slightly wavy margin, sometimes with a low, round umbo. Dark gray to brownish gray at center, lighter grayish beige to nearly whitish, and often with dark grayish or brown "water spots" near margin and developing ocher streaks in age. Surface dry, smooth, but usually entirely covered in a fine, pale bloom. **GILLS:** Broadly attached, notched, with a slight decurrent tooth, edges serrate and eroded in age. Whitish cream to dingy light beige. **STIPE:** 3–8 cm long, 0.8–2 cm thick, cylindrical, base slightly enlarged. Whitish cream to beige, becoming dingy in age and developing reddish orange spots and streaks. Finely velvety or nearly smooth but upper part finely and densely scurfy. **FLESH:** Fibrous, whitish, solid. **ODOR:** Indistinct. **TASTE:** Indistinct. **SPORE DEPOSIT:** White. **MICROSCOPY:** Spores 7–7.5 x 4–5 μm, ellipsoid to almond shaped. Covered in amyloid warts and ridges (appearing smooth when mounted in water). Cheilocystidia long and pointed, some with a swollen base, most with clusters of crystals at tip and appearing barbed.

ECOLOGY: In small troops on cellulose-rich debris (decaying grass thatch in lawns, manure piles with straw mixed in) in disturbed areas. Widespread, but uncommon. Mostly found in winter and spring.

EDIBILITY: Unknown.

COMMENTS: This odd species can be recognized by its grayish cap with dark spots and a pale bloom, finely notched gills, ocher spots and streaks, and growth on grassy or compost substrates. The amyloid-ornamented spores and harpoonlike cheilocystidia help confirm the genus. *M. lewisii* was described from California, but has a creamy white cap and stipe; most other *Melanoleuca* in our area are darker (see comments under *M. melaleuca*).

Melanoleuca melaleuca group

CAP: 3–8 cm across, convex when young, becoming broadly convex to plane, usually with a low, round umbo. Dark blackish brown, dark brown, to grayish brown. Surface moist to greasy, very finely velvety to completely smooth. Sometimes metallic in dry weather. **GILLS:** Distinctly notched, close to crowded. Bright white, sometimes a bit creamy or light grayish in age, usually contrasting strongly with stipe. **STIPE:** 4–10 cm long, 0.6–1 cm thick, cylindrical, straight or slightly curved, often wider at apex and with a club-shaped base, narrower at midpoint. **FLESH:** Fibrous, white, soft, stipe becoming slightly stringy and hollow. **ODOR:** Indistinct. **TASTE:** Indistinct. **SPORE DEPOSIT:** White. **MICROSCOPY:** Spores 6–8 x 4–5.5 μm, ellipsoid, ornamented with amyloid warts and ridges (appearing smooth in water). Cystidia long and often pointed, tipped with barblike crystals.

ECOLOGY: Solitary or scattered in small groups in winter and spring. Prefers rich duff along trails and other disturbed areas in forests, also grows in wood chips and in rich soil in parks.

EDIBILITY: Edible, but not recommended; known to concentrate heavy metals.

COMMENTS: The smooth cap and sinuate, bright white gills that contrast strongly with the brown cap and stipe make this species fairly easy to recognize. A number of *Melanoleuca* in our area occur in more forested settings. One common species has a grayish gold cap and yellowish ocher stipe, and is often found under redwoods. Another species has a squat stature and dark grayish brown, heavily spotted cap (the "Frogskin" *Melanoleuca*); it is sometimes found in thick duff under Monterey Cypress. This genus remains very poorly known in our area.

Lyophyllum semitale group
BLACKENING LYOPHYLLUM

CAP: 3–10 cm across, convex to plane, often becoming uplifted and wavy in age, to flat or slightly irregular and wavy in age. Beige to light grayish brown, soon developing irregular blackish streaks and stains, to all black in age. Surface moist, greasy, smooth. **GILLS:** Broadly attached, close to subdistant. Warm gray to creamy beige, staining slowly yellowish to gray, then black. **STIPE:** 5–12 cm long, 0.5–1.5 cm thick, enlarged toward base, but often with a tapering root. Grayish beige to gray, staining like gills, becoming blackish in age. **FLESH:** Thin, rubbery in cap, cartilaginous-fibrous in stipe. **ODOR:** Rancid-farinaceous. **TASTE:** Farinaceous. **SPORE DEPOSIT:** White. **MICROSCOPY:** Spores 6–9 x 3–5 μm, ellipsoid, smooth, inamyloid.

ECOLOGY: Solitary or in small groups in duff and moss, usually in mixed conifer forests in fall and winter. Uncommon, but found throughout most of our range.

EDIBILITY: Unknown.

COMMENTS: Species in this group of *Lyophyllum* are recognizable by their brownish gray caps, pallid gills, stipe that becomes bluish gray to sooty blackish when bruised, white spore deposit, and medium-size fruitbodies with a cartilaginous-fibrous texture. There are a number of poorly known species in this group; little work has been done on this group in California.

Lyophyllum decastes group
FRIED CHICKEN MUSHROOM

CAP: 3–10 cm across, rounded with an inrolled margin when young, expanding to broadly convex and then flat, often wavy in age. Color quite variable, dark brown, gray-brown, mottled grayish beige, ocher-tan, or honey colored. Often with spots, blotches, and streaks of darker color. Young caps of darker variants often have a pale grayish bloom overall. Surface greasy to dry, smooth or with fine fibrils near center. **GILLS:** Narrowly attached, fairly close, soft and waxy. White to creamy, dingy in age. **STIPE:** 4–12 cm long, 0.7–2 cm thick, cylindrical, usually curved, fused in large clusters. Whitish cream at first, developing pale yellowish tones, then becoming tan. Smooth, moist to dry. **FLESH:** Fibrous-rubbery, pale, solid or with a small hollow in stipe. **ODOR:** Indistinct. **TASTE:** Indistinct. **SPORE DEPOSIT:** White, often copious. **MICROSCOPY:** Spores 4–6 μm, round, smooth, inamyloid.

ECOLOGY: In dense clumps with 5–50+ fruitbodies, often on nutrient-poor soil or in thin humus along trails in forests, parks, and other disturbed areas throughout our range. Mostly found in early fall and winter, but can fruit any time of year.

EDIBILITY: Edible, fairly popular due to its firm texture and abundance (a single clump is often enough for a couple of meals), but the flavor is fairly bland and often grows in "unsavory" habitats.

COMMENTS: The greasy or dry, smooth caps fruiting in large to huge clusters, lack of any veil tissue, and rubbery white flesh help identify this mushroom. Many of the differences between described *Lyophyllum* are subtle, and the genus as a whole is poorly known in our area. A similar species reported from our area is *L. loricatum*, which supposedly differs by its dark blackish brown cap with a thicker cuticle. *Lepista subconnexa* also grows in clusters (but never as dense or prolific), but is paler grayish white overall and has a pinkish buff spore print. *Armillaria mellea* also grows in clusters, but has a flaring ring near the apex of the stipe and looks quite different in may other respects.

Lepista subconnexa

(Murrill) Harmaja

CAP: 4–10 (12) cm across, broadly convex with an inrolled and often ribbed margin, becoming flat or slightly uplifted and wavy, sometimes with a depressed center. Dingy whitish to grayish beige, or light dingy buff, often with darker "water spots" and fine whitish streaks outward. Surface smooth, greasy to dry. **GILLS:** Broadly attached, close to crowded, some forked near stipe. Light creamy to beige or grayish buff in age. **STIPE:** 4–12 cm long, 0.7–2 cm thick, cylindrical, upright, straight to slightly curved, base enlarged or club shaped, but usually with extreme bases of multiple stipes tapered into bundles. Whitish to dingy beige, developing dull tan tones in age. Surface smooth to vertically fibrillose. Base usually with cylindrical white rhizomorphs. **FLESH:** Fibrous, solid, whitish to tan. **ODOR:** Fragrant or musty. **TASTE:** Mild to bitter. **SPORE DEPOSIT:** Buff to light pinkish tan. **MICROSCOPY:** Spores 4.5–6 x 3–4 µm, ellipsoid, warted-spiny, inamyloid.

ECOLOGY: In dense clumps in rich duff (rarely solitary), especially in disturbed, open areas under redwood and oak, but can be found in most any mixed forest. Fruiting in winter and early spring. Widely distributed in our area, but generally uncommon.

EDIBILITY: Nontoxic.

COMMENTS: The dingy whitish to grayish buff colors overall, growth in clusters, and whitish rhizomorphs are good clues; the pinkish buff spore deposit helps clinch the identification. *Lyophyllum* in our area are quite similar in stature and clustered growth, but have darker caps and white spores. Fruitbodies of *L. subconnexa* found fruiting singly can be very confusing; the spore deposit color is a good clue for such specimens.

SYNONYM: *Clitocybe subconnexa* Murrill.

Group L

Tricholoma saponaceum group
SOAPY TRICH

Note: This description pertains to many different forms of this "species."

CAP: 1.5–15 cm across, conical to convex when young, becoming broadly convex, plane, with or without an umbo, margin inrolled when young, often slightly uplifted, lobed and wavy in age. Greenish gray, gray, bluish gray, grayish brown, brown, yellowish brown to yellow. Surface dry to slightly moist, smooth or with scattered appressed scales on disc, radially splitting and cracking into large scales in dry weather. **GILLS:** Narrowly to broadly attached, often with a distinct notch and/ or a decurrent tooth, or even appearing free at times, close to subdistant, broad, partial gills numerous. Whitish, creamy white to pale yellowish tan, sometimes staining pinkish. **STIPE:** 3–15 (20) cm long, 0.5–3 cm thick, more or less equal, bulbous, tapering or rooting at base. Whitish, creamy white to pale yellowish tan, staining or discoloring pinkish at base. Surface smooth to slightly fibrous, moist to dry. **FLESH:** Moderately thick to very thin, firm to fragile, whitish to pale gray, stipe solid, fibrous, whitish to pale gray with a pinkish base. **ODOR:** Distinctive, like old-fashioned unscented soap. **TASTE:** Mild to disagreeable, often described as "soapy." **SPORE DEPOSIT:** White. **MICROSCOPY:** Spores 5–7 x 3.5–5 μm, ellipsoid, smooth.

ECOLOGY: Solitary, scattered, or in small clusters on ground in a wide variety of forests. Common throughout fall and winter.

EDIBILITY: Unknown, possibly poisonous.

COMMENTS: In coastal California, we have at least five different "forms" of *T. saponaceum*; the above description covers four of them. A slender form found under Tanoak (see lower left photo) has a gray to grayish buff convex cap and a long, rooting stipe with a pinkish blushed base. It could be mistaken for small gray *Hygrophorus*, but it lacks a viscid cap. Another form has a broadly umbonate, two-toned cap (brown on the disc, gray to grayish yellow at the margin) and a long, tapering stipe with a bright pink base; it is found mostly on the Far North Coast under Shore Pine (see upper left photo). Occurring under pines is a stocky, pale to bright yellow or yellowish green-capped form with a short, often bulbous stipe (see upper right photo). A large gray, bluish gray, or grayish brown-capped form with a pointed, pinkish orange stipe base is found in mixed forests with Douglas-fir and oaks. Finally, *T. saponaceum* var. *squamosum* has a dark grayish green, convex to plane cap with black scales on the disc, and has matted dark gray fibrils on the stipe; it is mostly found on the North Coast under Grand Fir. All these forms share the pinkish flesh and slow staining in the stipe base in age and the soapy odor. However, the pink is often slow to develop, and not everybody detects the soapy odor. There appear to be intergrades between some of the forms, suggesting that this is a highly variable species.

Tricholoma saponaceum var. *squamosum*

(Cooke) Rea

SCALY SOAPY TRICH

CAP: 3–10 cm across, convex when young, becoming broadly convex to plane, margin inrolled when young, occasionally slightly uplifted and wavy in age. Dark olive-green to olive-gray with fine blackish scales over disc and often with "water spots" around margin when young, fading to dingy greenish gray with a darker disk, often developing pinkish spots or stains. Surface dry to slightly moist, disc covered with small appressed scales, radially splitting and cracking into larger scales in dry weather. **GILLS:** Narrowly to broadly attached, often with a distinct notch, or even close to subdistant, broad, partial gills numerous. Whitish when young, becoming creamy white, occasionally discoloring yellowish in age and slowly staining pinkish when damaged. **STIPE:** 3–8 cm long, 1–2.5 cm thick, enlarged toward the lower portion or slightly bulbous, but tapering at extreme base. Whitish to pale gray with dark gray to blackish fibrils or scales, base pinkish, staining pink to pale yellow when handled. Surface moist to dry, covered with appressed fibrils or small scales. **FLESH:** Moderately thick, firm, whitish to pale gray, staining pinkish, stipe solid, fibrous, whitish to pale gray, pinkish at base, staining like cap flesh. **ODOR:** Distinctive, like old-fashioned unscented soap. **TASTE:** Disagreeable, often described as "soapy." **SPORE DEPOSIT:** White. **MICROSCOPY:** Spores 5–7 x 3.5–5 μm, ellipsoid, smooth.

ECOLOGY: Solitary, scattered, or in small clusters on ground or in mossy areas, usually under Grand Fir. Locally common from Sonoma County north. Fruiting from midfall into winter.

EDIBILITY: Unknown, possibly poisonous.

COMMENTS: The olive to greenish gray cap with blackish scales on the disc, whitish gills that often stain pinkish when damaged, and pale stipe covered in blackish fibrils are a striking combination. The other common large form of *T. saponaceum* generally grows with Douglas-fir and oaks, and usually has a gray to gray-brown or bluish gray cap, whitish gills, and a pointed to slightly rooting stipe with a pinkish orange base. *T. subsejunctum* can look similar, but has a yellowish green viscid cap with appressed, radiating black fibrils that give it a streaked (rather than scaly) appearance.

Tricholoma inamoenum

(Fr.) Gillet

STINKY TRICH

CAP: 1.5–5 cm across, convex, becoming broadly convex to plane or wavy in age. White, creamy, pale pinkish white to pale tan, slightly darker on disc. Surface dry, smooth. **GILLS:** Narrowly attached with a distinct notch, or appearing free at times, widely spaced, broad, partial gills numerous, often becoming ragged in age. White, creamy white to pale tan in age. **STIPE:** 4–12 cm long, 0.3–0.8 cm thick, tall, spindly, equal, enlarged toward, or pointed at base. White, creamy white to pinkish white, with pale orangish tan stains and often with a rosy pink base. Surface dry, silky-fibrillose to smooth. **FLESH:** White, very thin, fragile in cap, stipe solid, fibrous. **ODOR:** Weirdly musty-floral and chemical. Nauseatingly strong, of mothballs mixed with lilacs. **TASTE:** Repulsive, burnt-tar-like. **SPORE DEPOSIT:** White. **MICROSCOPY:** Spores 9.5–14.5 x 5–8 μm, almond shaped, spindle shaped, or broadly ellipsoid, smooth.

ECOLOGY: Solitary or scattered on ground or in moss under conifers, especially Sitka Spruce. Locally very common in Del Norte County and northern Humboldt County, occurring as far south as Mendocino County, rare or absent in much of southern part of our range. Fruiting mostly in fall and early winter.

EDIBILITY: Possibly poisonous; in any case, the odor and taste are major deterrents.

COMMENTS: The slender stature, whitish colors, widely spaced gills, and peculiar odor make this a distinctive mushroom. The strong odor is reminiscent of coal tar or mothballs, although some people get a heavy floral scent (like lilacs) at first, while others get nauseated at the faintest whiff. *T. sulphureum* has a similar odor and stature, but is pale yellow to yellowish buff overall, and generally slightly stockier; it is rare in California (recorded as far south as Marin County) and more commonly found in the Pacific Northwest. *Camarophyllopsis paupertina* is a small, brownish buff to gray mushroom with a similar mothball odor. Some *Cuphophyllus* can look similar, but have decurrent gills and lack this particular odor.

Tricholoma arvernense

Bon

CAP: 5–12 (15) cm across, conical to convex when young, becoming broadly convex to plane or uplifted with a low broad to pronounced umbo, margin often crimped when young, becoming wavy, uplifted and often splitting in age. Color extremely variable, generally olive-yellow to olive-brown with darker blackish olive scales when in button stage, fading as it expands, to brownish yellow, golden yellow to grayish yellow and eventually pale yellow-tan to yellowish white. Surface dry, covered with fine scale or tufts, to appressed-fibrillose, often radially cracking in age. **GILLS:** Narrowly attached with a distinct notch, close to crowded, broad, partial gills numerous. Orangish yellow to creamy yellow when very young, fading to whitish to creamy with yellow around margin. **STIPE:** 5–15 cm long, 0.8–3 cm thick, often twisted and curved, more or less equal but often pointed toward base. Whitish to pale yellowish white, pinkish white to pale buff. Surface dry, pruinose at apex, appressed-fibrillose, silky to smooth. **FLESH:** Thin to moderately thick, fibrous to fragile, whitish to pale grayish yellow in cap, stipe tough, fibrous, whitish to pale grayish white. **ODOR:** Farinaceous. **TASTE:** Farinaceous. **SPORE DEPOSIT:** White. **MICROSCOPY:** Spores 4.5–7 x 3.5–5.5 μm, ellipsoid, smooth.

ECOLOGY: Solitary, scattered, or in small to large groups or clusters in duff under conifers. Occurs under Knobcone Pine at least as far south as the Santa Cruz Mountains, but slightly more common northward to coastal Mendocino County with Bishop Pine. Fruiting in late fall and winter.

EDIBILITY: Unknown.

COMMENTS: The olive-yellow, brownish yellow, or yellowish white cap with darker scales, yellowish to whitish gills, and growth with conifers help distinguish this uncommon species. The highly variable cap can resemble a number of other *Tricholoma*, depending on the stage of development. Young specimens could be mistaken for *T. subsejunctum*, which differs by having a greenish yellow cap with blackish streaks. *T. intermedium* has a yellow to yellow-brown cap with small scales over the disc, but has white gills and a stockier stature. *T. equestre* also has a yellow to yellow-brown cap, but has bright yellow gills. Some forms of the *T. saponaceum* group can look similar, but are generally more dull in color and have smoother, more evenly colored caps and pinkish stipe bases. *T. davisiae* is similar, but has an acutely umbonate cap that is more greenish yellow when young, fading to pinkish yellow; it also tends to discolor pinkish in the stipe base. To our knowledge, it hasn't been found in California, but has been found in Oregon and Washington. A *Tricholoma* with a dark-streaked, light greenish gray cap and dingy, pallid greenish gray gills grows under Shore Pine in sandy soil on the Far North Coast; it has been identified as *T. aestuans*, but is almost certainly different from the true European concept of that species and needs its own name.

SYNONYMS: *Tricholoma tumidum* (sensu Shanks). *T. arvernense* was described from France, but it very closely resembles California collections.

Tricholoma equestre group
CANARY TRICH

CAP: 3.5–12 cm across, rounded to convex when young, becoming broadly convex to nearly plane, margin downcurved when young, often wavy or lobed and sometimes slightly uplifted in age. Golden yellow to pale yellow with brownish scales on disc when young, becoming dingy yellow to yellowish brown or even all brown. Surface thinly viscid at first, becoming dry and often with adhering debris. Mostly smooth, but often appressed-fibrillose over disc, breaking up into fine scales. **GILLS:** Narrowly attached with a distinct notch, or appearing free at times, close to crowded, broad, partial gills numerous. Pale to bright yellow to greenish yellow when young, becoming slightly paler in age. **STIPE:** 2–10 cm long, 0.7–2 cm thick, equal or enlarged toward or even bulbous at base. Whitish to pale yellow, darkening slightly in age. Surface dry, fibrillose. **FLESH:** Thin to moderately thick, firm, white to pale yellowish near cap surface, stipe fibrous, solid to hollow in age, whitish. **ODOR:** Farinaceous. **TASTE:** Mild to sweet-farinaceous. **SPORE DEPOSIT:** White. **MICROSCOPY:** Spores 5–8.5 x 3.5–6 μm, broadly ellipsoid, smooth.

ECOLOGY: Solitary or scattered on ground under conifers, usually buried in duff. Other forms grow with oak and Madrone. Especially common along coast in sandy soil with pine. Fruiting from late fall well into winter.

EDIBILITY: Considered by many to be a great edible. However, it apparently contains a rhabdomyolitic compound implicated in several deaths in Europe (large amounts of *T. equestre* consumed daily for a week). There haven't been any poisonings reported from North America, but if you choose to eat this mushroom, you should be cautious.

COMMENTS: The golden yellow cap with brownish scales, yellow gills, and whitish to pale yellow stipe make this an easy mushroom to identify. *T. intermedium* is quite similar, but has white gills without any yellow tones. *T. subsejunctum* has a greenish yellow cap with olive and blackish streaks or fibrils and whitish gills that become blotchy with cream and yellow tones. *T. sulphureum* also has a yellow cap and gills, but has smaller fruitbodies with an awful coal tar odor; it is much rarer in California.

SYNONYMS: Although better known as *T. flavovirens*, *T. equestre* is an older name and thus takes precedent. However, we know that the species in California is not the same as the true European *T. equestre* and needs a new name. To add to the confusion, more than one species goes by the name *T. equestre* in California. Those growing under live oak are often taller and have brighter yellow gills and slightly greenish yellow caps appear to be *T. yatesii*, a species described from Berkeley. Other common names for members of this species complex include Man On Horseback and Yellow-Gilled Trich.

Tricholoma subsejunctum

Peck

GREEN-STREAKED TRICH

CAP: 3–8 cm across, conical to convex when young, becoming broadly convex to plane or uplifted with a low broad to pronounced umbo, margin often crimped when young, becoming wavy, slightly uplifted, and often splitting in age. Dark greenish black disc over a yellow to yellowish green base color, with dark streaks or appressed virgate fibers, becoming dark greenish yellow with blackish streaks overall. Surface slightly fibrous or finely scaly to smooth, viscid when wet, but often dries out. **GILLS:** Narrowly attached with a distinct notch or appearing free at times, close to subdistant, broad, partial gills numerous. Off-white, pale yellowish white to pale yellowish green, often darker toward margin. **STIPE:** 3–10 cm long, 1–2 cm thick, often twisted and curved, enlarged or slightly bulbous toward base. Color variable and often patchy, whitish, yellowish to greenish yellow. Surface dry, slightly fibrous to smooth. **FLESH:** Fragile to fibrous, thin. Off-white to pale gray in cap, fibrous, whitish to yellowish near surface in stipe. **ODOR:** Faint, mildly farinaceous or indistinct. **TASTE:** Mildly bitter-farinaceous to strongly bitter or unpleasant. **SPORE DEPOSIT:** White. **MICROSCOPY:** Spores 5–8 x 4–6 μm, broadly ellipsoid to subglobose, smooth.

ECOLOGY: Solitary or scattered on ground or in moss in northern coastal coniferous forest. It seems to prefer Grand Fir and Western Hemlock, but also grows in Sitka Spruce forest. Occurs from northern coastal Sonoma County up into the Pacific Northwest, but is uncommon south of Humboldt County. Fruiting from late in fall into winter.

EDIBILITY: Unknown, possibly poisonous.

COMMENTS: The smooth yellow to yellow-green cap with radial blackish streaks or fibrils, whitish to yellowish gills and stipe, and growth under conifers help identify this species. *T. arvernense* can look similar, but lacks the dark streaks and is generally more yellow-brown when young, fading to whitish yellow as it ages. The edible *T. portentosum* has a more evenly colored dark gray to gray-brown, streaked, viscid cap, whitish gills that become yellowish, and a mild to slightly farinaceous taste. *T. equestre* has a golden yellow to yellow-brown cap that can have small scales (but is never streaked), bright yellow gills, and a farinaceous taste and odor. *T. intermedium* has a cap like *T. equestre*, but has white gills. Members of the *T. saponaceum* group can have greenish caps, but are dry, dull colored, and not streaked.

NOMENCLATURE NOTE: There is still some confusion as to which species we have in California. It has been called *T. sejunctum,* which is a hardwood-associated species described from Europe; it is doubtful that it is the same as ours. *T. viridilutescens* is a spruce-associated European species that may be genetically similar enough to supplant *T. subsejunctum* as the "correct" name for the West Coast species. We have chosen to use the North American name until current work on *Tricholoma* clarifies the situation.

Tricholoma portentosum

(Fr.) Quél.

STREAKED TRICH

CAP: 4–12 (15) cm across, broadly convex to plane, margin downcurved when young, becoming uplifted, wavy and often splitting in age. Dark gray to grayish brown over a paler gray base, with dark streaks or appressed virgate fibers, sometimes developing yellowish gray colors in age. Surface smooth to appressed fibrous, viscid to moist, or shiny if dry; skin often peels easily. **GILLS:** Narrowly attached with a distinct notch or appearing free at times, close to subdistant, broad and often ragged in appearance, creamy white when young, becoming creamy to pale yellowish, especially around margin. **STIPE:** 5–15 cm long, 1–3 cm thick, often curved or twisted, equal, tapering downward or abruptly bulbous at base. White to creamy when young, becoming dingy white to pale yellow. Surface dry, smooth to slightly fibrous. **FLESH:** Fibrous, relatively thin, off-white to pale gray or yellowish gray, stipe fibrous, solid, rarely becoming hollow, off-white to pale gray or yellowish near surface. **ODOR:** Faint, mildly farinaceous to indistinct. **TASTE:** Mildly farinaceous. **SPORE DEPOSIT:** White. **MICROSCOPY:** Spores 5–8.5 x 3–6 μm, broadly ellipsoid, smooth.

ECOLOGY: Solitary, scattered, or in small clusters on ground, in moss or duff in northern coastal coniferous forests. Especially fond of Grand Fir and Western Hemlock, occasionally with pine. Found from coastal Sonoma County northward, but uncommon south of Humboldt County.

EDIBILITY: Edible and very good, but caution is warranted because it has a number of toxic look-alikes.

COMMENTS: The dark gray to gray-brown, streaked viscid cap, whitish gills that become yellowish, mild to slightly farinaceous taste, and growth under conifers help identify this species. *T. griseoviolaceum* is quite similar, but can be told apart by its growth under hardwoods (oak and Tanoak), violet-gray cap, and stronger farinaceous odor. *T. mutabile* is a somewhat similar hardwood associate with a viscid, pale gray to dark gray cap (paler near the margin), white gills and stipe, and strong farinaceous odor; it is reportedly toxic. *T. pardinum* has a dry, finely scaly, light grayish cap; it is very toxic. *T. virgatum* is a similar northern conifer associate, but has a peaked or umbonate streaked cap that is dry to moist, dingy whitish gills, and a strongly acrid taste. Members of the *T. saponaceum* group can look similar, but have dry, dull-colored, unstreaked caps and often stain pink (especially in the lower stipe). Similar *Entoloma* species (some of which are poisonous) usually have pinkish mature gills and pinkish spores.

NOMENCLATURE NOTE: *Tricholoma portentosum* is a European name, and our species may end up being called *T. avellaneifolium*, which was described from the Cascade foothills of Oregon. *T. griseoviolaceum* and *T. mutabile* were once lumped under the species concept of *T. portentosum* in California.

Tricholoma griseoviolaceum

Shanks

CAP: 4–10 (12) cm across, convex when young, becoming broadly convex to plane, sometimes with a low broad umbo, margin often wavy, slightly uplifted and splitting in age. Dark violet-gray to gray, often with dark gray streaks or appressed virgate fibers, but can be whitish to pale lilac-gray when buried under duff. Surface smooth, viscid to moist, or shiny if dry. **GILLS:** Narrowly attached with a distinct notch or appearing free at times, close, broad. Bright white when young, becoming off-white to pale pinkish white. **STIPE:** 4–15 cm long, 1–2.5 cm thick, equal or tapering and often curving and/or slightly rooting toward base. Bright white when young, becoming dingy white, something discoloring yellowish orange on lower portions. Surface dry, smooth to slightly fibrous. **FLESH:** Fibrous, relatively thin. Off-white to pale gray, stipe stuffed with white pith when young becoming hollow and pale grayish in age. **ODOR:** Sweet-farinaceous or like cucumbers. We've also gotten a slight wintergreen odor from a few collections. **TASTE:** Farinaceous, but pleasant. **SPORE DEPOSIT:** White. **MICROSCOPY:** Spores 3.5–5 x 5–7 μm, broadly ellipsoid, smooth.

ECOLOGY: Solitary, scattered, or in small clusters on ground under live oak or occasionally Tanoak, often buried in thick duff. Generally uncommon, but abundant in some years. More common in southern California, uncommon to rare north of the San Francisco Bay Area (at least on coast). Fruiting from late fall through winter.

EDIBILITY: Edible and good, but caution is warranted, as there are a number of poisonous look-alikes.

COMMENTS: This species can be told from other similar *Tricholoma* by the violet-gray, viscid cap, white gills and stipe, farinaceous odor and taste, and growth under oaks. *T. portentosum* has a dark gray to gray-brown, streaked, viscid cap, whitish gills that become yellowish, and a mild to slightly farinaceous taste, and grows with northern conifers. *T. virgatum* also grows with northern conifers and differs by its umbonate, streaked cap that is dry to moist (not viscid), dingy whitish gills, and acrid taste. *T. mutabile* has a viscid cap with a paler grayish disc and broader pale margin, white gills and stipe, and a strong farinaceous odor; it is a rare species, known only from a few locations on the North Coast and Sierra Foothills under Tanoak. Both *T. virgatum* and *T. mutabile* are probably poisonous. Some similar *Entoloma* are also poisonous and can be told apart by their pinkish buff spores.

SYNONYMS: *Tricholoma griseoviolaceum* was split out from the Californian concept of *T. portentosum* by Kris Shanks, who described this violet-gray oak associate as a new species in her thesis on Californian *Tricholoma*.

Tricholoma virgatum

(Fr.) P. Kumm.

VIRGATE TRICH

CAP: 4–10 (12) cm across, conical when young, becoming convex to plane or with a slightly uplifted margin when mature, with a well-defined umbo at all stages. Dark gray to gray or brownish gray with black or dark gray streaks or appressed virgate fibers. Surface dry to moist, smooth when young, but made up of appressed streaked fibers that can form little uplifted scales when old or dry. **GILLS:** Narrowly attached with a distinct notch or appearing free at times, close, broad, often ragged in appearance. Off-white, dingy white to pale gray, edge often darker gray. **STIPE:** 7–15 cm long, 1–2 cm thick, equal or slightly swollen in middle, base with an abrupt bulb at times. White to dingy white, sometimes discoloring yellowish orange on lower portions. Surface dry, smooth. **FLESH:** Fibrous, relatively thin. Off-white to pale gray in cap, stipe stuffed with white pith when young, becoming hollow in age. **ODOR:** Faint, farinaceous. **TASTE:** Slightly bitter-farinaceous or disagreeable at first, slowly (20–30 seconds) becoming strongly acrid. **SPORE DEPOSIT:** White. **MICROSCOPY:** Spores 6–8.5 x 4.5–6.5 µm, broadly ellipsoid, smooth.

ECOLOGY: Solitary or scattered on ground in coastal coniferous forest, especially under Grand Fir. Common from late fall into winter on the Far North Coast, uncommon south of Humboldt County.

EDIBILITY: Probably poisonous.

COMMENTS: The umbonate, gray, streaked cap that is never truly viscid, dingy whitish gills, and acrid taste help identify this species. *T. portentosum* often grows in the same habitats, but differs by its dark gray, viscid cap that is never as distinctly umbonate or peaked, whitish gills that become yellowish in age, and mild to slightly farinaceous taste. *T. griseoviolaceum* has a dark violet-gray, viscid cap and white gills, a mild taste, and grows under Tanoak and live oak. *T. nigrum* has a moist to tacky cap that is dark gray on the disc with a paler margin, whitish gills, a white stipe that often has small blackish scales on the upper portions, and a very strong farinaceous odor and taste; rare in California, it is more common in Oregon and Washington.

Tricholoma atroviolaceum

A. H. Sm.

CAP: 4–12 (15) cm across, rounded to conical in button stage, quickly broadly convex to plane, margin downcurved and pinched when young, often uplifted, wavy or lobed in age. Dark grayish black to dark violet-gray on disc with a paler margin that often has darker spots or water stains when young, becoming grayish violet to grayish brown, Surface dry, covered with fine scales. **GILLS:** Narrowly attached with a distinct broad notch, or appearing free at times, close to subdistant, thick, broad, becoming ragged in age, partial gills numerous. Pale gray to pinkish gray, becoming dingy grayish buff, edges darker, often with dark grayish black to purplish black spots or stains. **STIPE:** 5–12 cm long, 1–3.5 cm thick, equal or enlarged at base. Whitish, grayish to pale violet-gray when young, becoming grayish buff and often developing pale orangish tan discolorations at base when handled. Surface dry, fibrillose, and often with fine dark scales on upper portions. **FLESH:** Thin, fibrous to fragile in cap, whitish to pale gray, staining reddish gray and becoming dingy grayish brown. Stipe tough, fibrous, solid, colored and staining like cap, although may be orangish tan in base. **ODOR:** Farinaceous. **TASTE:** Strongly farinaceous. **SPORE DEPOSIT:** White. **MICROSCOPY:** Spores 7–10 x 5–7 µm, broadly ellipsoid, smooth.

ECOLOGY: Solitary or scattered in duff of mixed coniferous forests. Widespread, occurring throughout our range, but rather uncommon and local. Fruiting from late fall into winter.

EDIBILITY: Unknown.

COMMENTS: The dry, dark grayish black to violet-gray, finely scaly cap, dingy pinkish gray to grayish buff gills (often with darker edges), and reddish gray staining of the flesh are the distinguishing features of this species. A number of other *Tricholoma* can be difficult to tell apart: *T. luteomaculosum* also has a grayish cap and often blackish-edged gills, but the cap surface is less distinctly tufted-scaly, and the overall coloration tends to be quite a bit paler (especially the gills); it grows with oak and Tanoak. *T. pardinum* has a whitish gray cap with small dark scales, whitish gills, and usually a club-shaped stipe. *T. atrosquamosum* has gray to blackish tufts and scales covering the cap (contrasting with the white background flesh) and a pale gray stipe with contrasting blackish fibrils and scales. A few *Entoloma* species can look somewhat similar, but have smoother caps and pink spores.

Tricholoma pardinum

(Pers.) Quél.

LEOPARD TRICH

CAP: 4–12 (15) cm across, convex, broadly convex to plane; margin downcurved when young, becoming wavy and sometimes splitting in age. Whitish to pale gray with darker gray to blackish tufted scales on the disc or appressed fibers that break up into dark gray scales. Surface dry, appressed-fibrillose when young, becoming scaly. **GILLS:** Narrowly attached, with a distinct notch, close to crowded, broad, partial gills numerous, becoming ragged in age. Whitish, creamy to pale creamy buff in age. **STIPE:** 7–11 cm long, 1–3 cm thick, equal to club shaped with an enlarged base. White to dingy white, often bruising pale orangish brown near base. Surface dry, smooth to appressed-fibrillose. **FLESH:** Thick, firm whitish to pale gray in cap; stipe fibrous, firm, whitish, pale gray with orangish stains at base. **ODOR:** Farinaceous. **TASTE:** Farinaceous. **SPORE DEPOSIT:** White. **MICROSCOPY:** Spores 6–10.5 x 4–7 μm, ellipsoid, smooth.

ECOLOGY: Solitary or scattered on ground in mixed forests. Prefers Tanoak-Madrone woods in southern part of our range, but seems to grow with conifers in north. Uncommon in most years. Fruiting from late fall into winter.

EDIBILITY: Poisonous, causing severe gastrointestinal distress.

COMMENTS: The dry, whitish to gray cap with gray scales, white gills and stipe, and large, stocky stature are the best features for identifying this mushroom. Younger or wetter caps sometimes lack the gray color and can be whitish and without the gray scales. *T. venenatum* is very similar, but has a whitish to tan cap with slightly larger grayish brown scales; it is so far known only from the mountains, where it is rare. *T. vernaticum* is a large, gray, montane species that fruits in the spring or summer; it has a smooth, whitish to gray cap that often has a whitish patch on the disc, a faint partial veil that leaves a slight ring on the stipe, and a strong cucumber odor.

Tricholoma terreum var. *cystidiotum*

(Shanks) Blanco-Dios

MOUSE TRICH

CAP: 1.5–6 cm across, conical to convex when young, becoming broadly convex to nearly plane and often with a low broad umbo, margin tomentose, inrolled at first, often slightly uplifted and splitting in age. Dark grayish black to mouse gray with a paler margin, becoming pale gray to grayish tan in age or when dry. Surface dry, matted-tomentose to felty when young, becoming more appressed-fibrillose to finely scaly in age.
GILLS: Narrowly attached with a distinct notch, or appearing free at times, close to subdistant, broad, partial gills numerous. Pale gray when young, becoming grayish white to whitish, occasionally with darker spots or stains. **STIPE:** 2–6 cm long, 0.4–1 cm thick, equal or slightly bulbous at base. Whitish to pale gray, becoming yellowish gray, often discoloring yellowish when handled. Surface dry, silky-fibrillose. **PARTIAL VEIL:** Faint, silky cortina leaving gray hairs on stipe, usually only visible when young. **FLESH:** Very thin, fragile in cap, whitish to pale gray, stipe fibrous, solid to hollow, whitish. **ODOR:** Indistinct. **TASTE:** Indistinct. **SPORE DEPOSIT:** White. **MICROSCOPY:** Spores 6–8.5 x 3.5–5.5 µm, broadly ellipsoid, smooth.

ECOLOGY: Solitary, scattered, or in small clusters on ground or duff under conifers, especially Douglas-fir or pine. Widespread and common along coast. Fruiting from midfall into winter.

EDIBILITY: Edible but not reccomended. Be careful with your identification, as the edibility of many gray *Tricholoma* is still unknown, and a few are poisonous.

COMMENTS: The dark gray to mouse gray, felty cap, pale gray to whitish gills, small size. and growth with conifers help identify this dainty Trich. *T. moseri* is a very similar species that fruits in the mountains shortly after snowmelt. It has an appressed-fibrillose, gray cap (often paler than that of the Mouse Trich), whitish to pale gray gills, a white to pale grayish stipe, and no partial veil. *T. atrosquamosum* has a gray cap with blackish fibrils or scales, white gills, and a pale gray stipe ornamented with appressed blackish fibrils or scales; it is rare in California. *Tricholoma* "Stanford Gray" has slightly larger fruitbodies, with a strong farinaceous odor and taste; it grows with live oak.

SYNONYMS: Many consider the European *Tricholoma terreum* and *T. myomyces* to be the same, with *T. terreum* being the older and preferred name. However, we believe that the Californian variety, *cystidiotum*, is a distinct species

Tricholoma "Stanford Gray"

CAP: 3–8 (10) cm across, convex when young, becoming broadly convex to plane, margin inrolled or downcurved when young, becoming wavy and occasionally slightly uplifted in age. Color variable, dark gray or whitish if buried in duff when young, becoming lighter gray and often blotchy, with a brownish disc. Surface dry with appressed dark gray fibrils and smaller appressed scales at disc. **GILLS:** Attached with a distinct notch, close to somewhat crowded, broad, edged eroding and often becoming ragged in age. White to dingy white when young, becoming grayish white and often developing dingy yellowish stains in age. **STIPE:** 2.5–7 cm long, 0.7–2 cm thick, equal, tapering or enlarged slightly at base. Off-white to pale grayish white often with grayish veil fibers, and sometimes with orangish stains at base in age. Surface dry, fibrous. **PARTIAL VEIL:** Faint, grayish white cortina-like silky fibers, often leaving a faint ring on upper stipe and fine hairs on cap margin when young, or disappearing altogether in age. **FLESH:** Relatively thin and fragile in cap, off-white to pale gray; stipe fibrous, solid, rarely becoming hollow, off-white. **ODOR:** Farinaceous. **TASTE:** Strongly farinaceous. **SPORE DEPOSIT:** White. **MICROSCOPY:** Spores 6.4 x 3.8 μm on average, broadly ellipsoid, smooth.

ECOLOGY: Solitary, scattered, or in large groups or clusters on ground or in duff under Coast Live Oak. Rare to locally common, often found fruiting in large patches. Uncommon and local on the San Francisco Peninsula, more common southward. Fruiting throughout winter and well into spring in wet years.

EDIBILITY: Unknown.

COMMENTS: This gray *Tricholoma* can be identified by its variable gray to grayish brown cap with appressed fibrils and scales, whitish gills, strong farinaceous odor and taste, and growth under live oak. *T. terreum* var. *cystidiotum* has a dark gray or "mouse gray" cap that is densely tomentose when young, giving it a soft felty feeling; however, in age it becomes more appressed-fibrillose or scaly and can be mistaken for "Stanford Gray", but lacks the farinaceous odor and taste, and grows with conifers. *T. atrosquamosum* has a gray cap with blackish fibrils or scales, white gills, and a pale gray stipe with appressed blackish fibrils or scales; it grows with northern conifers. *T. pardinum* is a large, stocky, grayish white species that is more common to the north under conifers; it is very toxic. *T. cingulatum* has a dry pale gray cap, a cottony partial veil leaving a distinct ring around the stipe, and grows with willow. It is rare on the coast, mostly occurring along rivers in the central valley.

NOMENCLATURE NOTE: Although *Tricholoma* "Stanford Gray" is similar to the European *T. scalpturatum* (Kris Shanks determined it as such), it is a distinct species and in need of a name.

Tricholoma cingulatum

(Almfelt) Jacobasch

CAP: 2–6 cm across, conical to convex when young, becoming broadly convex to plane with a low broad umbo, margin incurved and cottony when young, becoming wavy in age. Pale gray or with a bluish gray cast when young, to grayish white with darker gray scales or fibrils, or grayish tan in age. Surface dry, appressed-fibrillose to finely scaly. **GILLS:** Broadly attached and often with a decurrent tooth, close to crowded, narrow, partial gills numerous. White to pale grayish white, discoloring to pale yellowish, especially around margin in age. **STIPE:** 3–8 cm long, 0.4–1 cm thick, more or less equal or enlarged slightly toward base. Whitish to pale grayish white, often with gray specks or small scales at apex. Surface dry, pruinose above the ring, appressed-fibrillose with bands of white veil remnants below ring. **PARTIAL VEIL:** Cottony, white, ragged and flaring upward, becoming appressed or matted in age. **FLESH:** Thin, fragile, whitish to grayish, stipe fibrous, whitish to pale grayish. **ODOR:** Farinaceous. **TASTE:** Farinaceous. **SPORE DEPOSIT:** White. **MICROSCOPY:** Spores 5–7 x 3–4.5 μm, ellipsoid, smooth.

ECOLOGY: Solitary, scattered, or in small clusters in swampy areas under willow (*Salix* spp.), fruiting from fall into spring. Seemingly rare, but willow thickets are not often surveyed for fungi. Widespread throughout our range, slightly more frequent in the Central Valley, on the North Coast dunes, or in areas along rivers.

EDIBILITY: Unknown.

COMMENTS: This species is easy to recognize by the dry, gray, slightly scaly cap, whitish gills, partial veil that leaves a ring on the stipe, and association with willows. *Tricholoma* "Stanford Gray" is larger, grows with live oak, and has a faint, cortina-like partial veil leaving only a slight ring on the stipe that can disappear in age. *T. terreum* var. *cystidiotum* is a small, dark gray species that grows under conifers; it has a faint, silky partial veil that rarely leaves a ring on the stipe.

Tricholoma imbricatum (sensu CA)

(Fr.) P. Kumm.

CAP: 3.5–12 (15) cm across, rounded to convex when young, becoming broadly convex, margin inrolled when young, staying downcurved well into maturity, often splitting in age. Dull dark brown but can be pinkish buff to light brown when young and buried in duff. Surface dry, matted-fibrillose to appressed-fibrillose, often with small upturned tufts or scales, usually radially cracking in age, revealing buff flesh. **GILLS:** Narrowly to broadly attached, often with a distinct notch and a decurrent tooth, close, broad, partial gills numerous. Whitish to pale buff, often with reddish brown spots or stains in age. **STIPE:** 5–12 cm long, 1–3 cm thick, equal or club shaped with an enlarged lower portion, but with a pointed base. Dingy buff to light brown, darkening in age. Surface dry, matted-fibrillose, slightly scaly to smooth. **FLESH:** Thick, firm, fibrous, whitish to buff, especially under cap cuticle, stipe solid, very fibrous, whitish, slowly staining brownish. **ODOR:** Indistinct. **TASTE:** Indistinct to bitter. **SPORE DEPOSIT:** White. **MICROSCOPY:** Spores 5–7 x 4–6 μm, ellipsoid, smooth.

ECOLOGY: In small to large clusters, scattered, or solitary on ground, often buried in duff. Very common under pine throughout our range. Generally fruiting from fall through winter.

EDIBILITY: Bitter, tough, inedible.

COMMENTS: The dry, dark brown, finely scaly cap, whitish to buff gills, light brown stipe, and growth with pine are the best features by which to identify this common Trich. *T. vaccinum* is smaller; has a reddish brown to orangish brown, finely scaly cap with a cottony margin and a partial veil (when young); and grows with Sitka Spruce. See comments under *T. vaccinum* for another similar species (*Tricholoma* "Fuzzy Margin"). Most other brown *Tricholoma* in California have viscid caps when wet.

NOMENCLATURE NOTE: *Tricholoma imbricatum* is a European name for a species that is distinct from western North American taxon, but no new name has been proposed for the species here.

Tricholoma vaccinum

(Schaeff.) P. Kumm.

CAP: 2–8 cm across, conical to rounded when young, becoming convex to plane, margin tightly inrolled and cottony in button stage, often ribbed and tufted with veil remnants when young and occasionally splitting in age. Reddish brown to rusty brown, lighter toward margin. Surface dry, matted-fibrillose to appressed-fibrillose, covered with small upturned tufts or scales and often radially cracking in age. **GILLS:** Narrowly to broadly attached, often with a distinct notch or even appearing free at times, close to subdistant, moderately broad, partial gills numerous. Creamy to pale buff, with orangish brown spots or stains in age. **STIPE:** 3–9 cm long, 0.5–2 cm thick, more or less equal or tapered toward base. Whitish to dingy buff base color with darker orangish buff scales or fibers, darkening to orangey brown in age. Surface dry, matted-fibrillose, covered with tufts and scales and pubescent veil remnants when young. **PARTIAL VEIL:** Cottony, often clinging to cap margin and not leaving a ring on the stipe. **FLESH:** Thin, fibrous, whitish to buff, stipe stuffed when young, becoming hollow, all parts slowly staining orangish brown. **ODOR:** Farinaceous. **TASTE:** Farinaceous-bitter, although mild at times. **SPORE DEPOSIT:** White. **MICROSCOPY:** Spores 6–7 x 4–5.5 µm, ellipsoid, smooth.

ECOLOGY: Solitary, scattered, or in small clusters on ground, in moss, or on well-decayed moss-covered logs or stumps in coastal spruce forest. Locally common on the coast from Mendocino County northward. Fruiting from late fall into winter.

EDIBILITY: There are conflicting reports on the edibility of this species, but the precautionary principle (as well as the tough flesh and bitter-farinaceous taste) rule this species out as an edible.

COMMENTS: The dry, scaly, reddish brown to orange-brown cap, whitish to buff gills that discolor orangish, cottony partial veil when young, finely scaly stipe, and growth with spruce help identify this pretty little Trich. There is a similar undescribed species under Shore Pine on the North Coast dunes we have nicknamed *Tricholoma* "Fuzzy Margin". It has a creamy orange to pale orange-brown cap that is soft and matted-tomentose with a cottony margin when young (becoming finely scaly at the disc, but not nearly as scaly as *T. vaccinum*); its gills are creamy white when young, becoming creamy orange, and the stipe is whitish with orange streaks and stains (more orangish in age). *T. imbricatum* is larger; has a dark brown cap with small scales, whitish to buff gills, and a light brown stipe; and grows with pine. It is generally more common south of Humboldt County.
T. aurantio-olivaceum is a small, slender species that has an ocher-beige to buff-brown cap, and grows in mixed forests of Tanoak and Douglas-fir.

Tricholoma dryophilum

(Murrill) Murrill

OAK-LOVING TRICH

CAP: 4–12 (15) cm across, convex when young, becoming broadly convex to nearly plane, margin downcurved to slightly inrolled and often ribbed when young, becoming wavy or lobed and slightly uplifted in age. Color variable, creamy orange to pinkish buff with darker streaks and patches and an orangish brown disc when young, becoming orange, orange-brown to brown. Surface viscid to moist or with debris stuck to it if dry, smooth or sometimes with appressed radiating fibers. **GILLS:** Narrowly attached with a distinct notch, close to crowded, narrow to relatively broad, partial gills numerous. White to creamy white, developing orange-brown spots or stains in age and slowly bruising rusty orange-brown when damaged. **STIPE:** 5–15 cm long, 1–3 cm thick, more or less equal, or with a swollen lower portion and slightly tapered at base. White to creamy white, with orangish to orange-brown streaks, becoming darker, orange-brown to brown from base up in age, staining orange-brown when damaged. Surface dry, slightly fibrous or finely scaly or scruffy at apex. **FLESH:** Firm, thick, tough, whitish, slowly staining orange-brown in cap, stipe tough, fibrous, solid or sometimes hollow in age, whitish, slowly staining orange-brown. **ODOR:** Strongly farinaceous. **TASTE:** Cap surface very bitter, flesh farinaceous. **SPORE DEPOSIT:** White. **MICROSCOPY:** Spores 5–8 x 4–6 μm, broadly ellipsoid, smooth.

ECOLOGY: Scattered in large troops or clusters on ground under oaks, often buried in duff. Very common throughout the range of Coast Live Oak, also grows with deciduous oaks in the Central Valley. Fruiting from midfall through winter.

EDIBILITY: Poisonous.

COMMENTS: This is probably our most common of the confusing group of *Tricholoma* with viscid, orange-brown caps. The cap color varies from creamy with orangish streaks when young to entirely orange or brown; the combination of the whitish gills, whitish stipe that develops orangish streaks and stains, strongly farinaceous odor and taste (viscid cap layer is bitter), and growth under oaks help identify it. A similar species also associated with oaks is *T. ustale*, which differs by its grayish orange to orangish brown cap and strongly bitter taste. *T. manzanitae* has a creamy white to pinkish buff cap when young, but darkens as it ages and could be mistaken for *T. dryophilum*; however, it has a creamy stipe with yellowish granules or fine scales at the apex and grows with manzanita or Madrone. *T. populinum* has a viscid, pinkish brown to reddish brown cap, whitish gills that stain reddish brown, a strong farinaceous odor and taste, and grows with aspen or cottonwood. *T. muricatum* has a pinkish brown to reddish brown cap, buff to orangish brown gills, and a bitter-farinaceous taste, and grows with pine. *T. fracticum* is a pine associate with a red-brown to orange-brown cap and a stipe that is often bicolored, with a sharp distinction between the white upper and orange lower parts.

Tricholoma manzanitae

T. J. Baroni & Ovrebo

MANZANITA TRICH

CAP: 5–12 cm across, convex when young, becoming broadly convex to nearly plane, margin often ribbed, inrolled and soft and cottony when young. creamy white when young, becoming creamy buff to pinkish buff, darkening to orangish buff on disc and eventually orangish brown to buff-brown and staining orange-brown in age. Surface with small appressed scales or matted fibrils, viscid to sticky when wet to dry, often with debris stuck to it, especially when dry. **GILLS:** Narrowly attached with a distinct notch, close to crowded, relatively narrow, partial gills numerous, creamy white to pale pinkish white, developing orange-brown spots or stains in age and slowly bruising orange-brown when damaged. **STIPE:** 3–9 cm long, 1–3 cm thick, more or less equal or enlarged slightly at base. Creamy white to pale yellow, with yellowish granules or fine scales, especially near apex, bruising and developing orange-brown stains, especially near base. Surface dry, often with granules or fine scales on upper portions, especially when young, base often with whitish binding mycelium. **FLESH:** Firm, thick, tough, whitish, slowly staining orange-brown in cap, stipe tough, fibrous, solid, whitish, slowly staining orange to orange-brown. **ODOR:** Indistinct. **TASTE:** Indistinct to slightly bitter-farinaceous. **SPORE DEPOSIT:** White. **MICROSCOPY:** Spores 5–7 x 3.5–5 μm, broadly ellipsoid, smooth.

ECOLOGY: Solitary, scattered, or in small clusters on ground, often buried in duff under manzanita and Madrone. So far known only from California and is fairly common in appropriate habitat. Widespread in our area in mature Madrone forest on coast and in manzanita thickets on nutrient-poor soil inland. Fruiting from midfall into late winter.

EDIBILITY: Unknown, not recommended.

COMMENTS: The creamy white to pinkish buff cap that darkens as it ages, creamy white to pinkish white gills, stipe with yellowish granules or fine scales near the apex, and growth under manzanita or Madrone identify this species Trich. It has many look-alikes: *T. dryophilum* has a creamy cap that develops orange-brown tones, a whitish stipe that develops orangish streaks and stains, and a bitter-farinaceous taste; it grows with live oak. *T. populinum* has a viscid pinkish brown to reddish brown cap, whitish gills that stain reddish brown, and a strong farinaceous odor and taste, and grows with Quaking Aspen and cottonwood; in California, it appears to be restricted to the Sierras. *Leucopaxillus* can look similar, but have broadly attached to decurrent gills and lots of white mycelium at the base of the stipe.

Tricholoma nictitans (sensu CA)

CAP: 3.5–12 (15) cm across, conical to convex when young, becoming broadly convex to plane with a low broad umbo, margin inrolled and ribbed when young, staying downcurved well into maturity. Rusty brown, orange-brown to brown, often with darker reddish brown appressed fibrils or fine scales at disc. Surface viscid, appressed-fibrillose, finely scaly on disc to smooth. **GILLS:** Narrowly to broadly attached, with a distinct notch, close to subdistant, broad, partial gills numerous. Creamy white to pale buff, developing reddish brown spots or stains in age. **STIPE:** 5–15 cm long, 1–2 cm thick, tall, often twisted, equal, enlarged lower, but tapering or slightly rooting at base. Dingy buff to orange-brown when young, darkening to brown from base up in age. Surface dry, matted-fibrillose, often with fine silky hairs, base often with matted whitish mycelium. **FLESH:** Relatively thin, firm, fibrous, whitish to creamy buff, stipe stuffed when young, quickly becoming hollow, very fibrous, buff to brownish. **ODOR:** Farinaceous. **TASTE:** Farinaceous. **SPORE DEPOSIT:** White. **MICROSCOPY:** Spores 5–8 x 3.5–6 μm, ellipsoid, smooth.

ECOLOGY: Solitary, scattered, or in small clusters on ground or in thick duff under conifers on the Far North Coast. Common under Shore Pine on coastal dunes in California and Oregon, occasionally found in coastal spruce forest. Fruiting from early fall into winter.

EDIBILITY: Unknown, probably poisonous.

COMMENTS: This Trich can be recognized by its tall stature, rusty brown, viscid cap, creamy to pale buff gills that are often spotted, and orangish brown, hollow stipe. *T. muricatum* is a similar species that also grows under pine, but has a streaked, dull reddish brown to pinkish brown cap and creamy orange gills. *T. fracticum* also has an orange-brown to red-brown cap, but differs by having a viscid, cortina-like partial veil when young that leaves a line on the stipe (which is often distinctly white above and orange below this line). *T. imbricatum* also grows under pines, but has a dark brown, dry cap. Also compare *T. dryophilum* and *T. ustale*, both oak associates.

NOMENCLATURE NOTE: There's a lot of confusion about the correct name for this species. *Tricholoma nictitans* is a European species that many mycologists consider to be a synonym of *T. fulvum*, and they consider ours to be *T. transmutans* (a species originally described from New York). However, we believe that the western North American species is distinct from all of these and that it requires a new name. For the time being, we continue using the name applied by Kris Shanks in her monograph on Californian *Tricholoma*.

Tricholoma aurantio-olivaceum

A. H. Sm.

CAP: 1.5–6.5 cm across, conical to convex when young, becoming broadly convex to plane, often with a low broad to pronounced umbo, margin downcurved when young, slightly uplifted and often splitting in age. Color variable, generally creamy orange when young, becoming orange to orange-brown on disc, creamy orange to orangish tan on margin, with dark orange-brown streaks or stains, darkening to rusty orange to orange-brown. Surface dry, with appressed fibrils on disc, becoming finely scaly. **GILLS:** Narrowly attached with a distinct notch, close to subdistant, broad, partial gills numerous. White to pale buff at first, soon developing dingy rusty orange spots or stains, or all rusty orange to orange-brown in age. **STIPE:** 3–10 cm long, 0.3–0.7 cm thick, tall, slender, often twisted and curved, more or less equal, but often tapered and rooting toward base. Whitish to creamy base color with orangish to brownish fibrils or scales, darkening orange-brown from the base up in age. Surface dry, silky-fibrillose to finely scaly. **FLESH:** Thin, fibrous, off-white to pale buff, becoming grayish buff, stipe fibrous, stuffed to hollow, whitish at first, grayish brown in age. **ODOR:** Indistinct. **TASTE:** Indistinct to slightly acrid. **SPORE DEPOSIT:** White. **MICROSCOPY:** Spores 5.5–7.5 x 3.5–5.5 μm, broadly ellipsoid, smooth.

ECOLOGY: Solitary, scattered, or occasionally in small clusters on ground in mixed coastal forest, probably associated with Tanoak or Douglas-fir. Fruiting in late winter and early spring. Generally rare, but can be locally common, and widespread throughout our area, recorded south to the Santa Cruz Mountains.

EDIBILITY: Unknown.

COMMENTS: The creamy orange cap that darkens to orange-brown as it ages, whitish buff gills that tend to become completely rusty brown and blotchy, and silky, slender stipe help identify this rather atypical *Tricholoma*. Because of the tendency of the gills to become dingy, older specimens are easily mistaken for *Inocybe* (which have brown spores and tend to have more radially fibrillose caps). *T. vaccinum* looks similar, but has a reddish brown to orange-brown, scaly cap and a partial veil when young; it grows under spruce. *Tricholoma* "Fuzzy Margin" (see comments under *T. vaccinum*) has a paler creamy orange to pale orange-brown cap that is soft and matted-tomentose with a cottony margin when young; it grows with pine. *T. aurantium* is a stocky species with an orange cap that is often stained jade green to olive even when young, white gills that slowly stain orange-brown, and a dense pattern of orange chevrons on the stipe.

Tricholoma aurantium

(Schaeff.) Ricken

ORANGE TRICH

CAP: 3–10 cm across, convex to broadly convex, margin inrolled when young, staying downcurved well into maturity. Color variable, often deep olive-green to forest green with some orange color and orange droplets around margin when young, becoming orange, reddish orange to orange-brown with darker, reddish brown appressed scales. Surface viscid to moist, smooth or with small appressed scales at disc to finely scaly in dry weather. **GILLS:** Narrowly attached with a distinct notch, close to crowded, relatively narrow, partial gills numerous. White to dingy white, staining and becoming orange-brown. **STIPE:** 4–9 cm long, 0.7–2 cm thick, more or less equal, but tapering or pointed at base. Orange to greenish orange chevrons or belts of scales covering lower three-quarters of stipe, but terminating abruptly below a smooth white band near apex. Surface dry, but often covered in orangish droplets when young, finely scaly with chevrons, as noted above. **FLESH:** Firm, fibrous, whitish to watery gray, staining pale orange-brown, stipe fibrous, solid, whitish, staining orange-brown. **ODOR:** Farinaceous. **TASTE:** Unpleasant, strongly bitter-farinaceous. **KOH:** Bright red on cap, reddish on stipe, little reaction elsewhere. **SPORE DEPOSIT:** White. **MICROSCOPY:** Spores 5–6 x 3.5–4 µm, broadly ellipsoid, smooth.

ECOLOGY: Solitary or scattered on ground, in moss, or in duff of conifer and mixed-evergreen forests. Fruiting in late fall and winter. Generally uncommon in California, rare south of Mendocino County, occurring south to Santa Cruz County.

EDIBILITY: Inedible due to unpleasant taste.

COMMENTS: This beautiful *Tricholoma* is easily identified by its orange (or variably mottled with jade green to olive) cap, white gills that slowly stain orange-brown, and orange chevrons on the stipe. *T. focale* has similar colors, but has a membranous partial veil and mostly grows with Tanoak, pine, and manzanita in our area. *T. aurantio-olivaceum* is a smaller, slender species that has a dry, creamy orange to orange-brown cap and white gills that become dingy ocher-brown in age. *T. fracticum* has a viscid, reddish brown to orange-brown cap and a bicolored stipe (white on the upper half, orange on the lower half); it grows with pine.

Tricholoma fracticum

(Britzelm.) Kreisel

CAP: 2.5–12 cm across, convex when young, becoming broadly convex to plane or lobed and wavy in age, margin inrolled and ribbed when young, staying downcurved well into maturity. Dark reddish brown to orange-brown, often with darker appressed fibrils at disc. Surface viscid to dry, but shiny if so, smooth or with scattered appressed fibrils. **GILLS:** Narrowly to broadly attached, with a distinct notch and often a decurrent tooth, close to crowded, broad, partial gills numerous. Creamy white to orangish white, or pale orangish buff, developing orangish brown spots or stains in age. **STIPE:** 2–8 cm long, 0.7–2.5 cm thick, more or less equal but tapering to a pinched base. With a well-defined annular zone when young, whitish above ring, orange below, becoming more blended, orangish overall in age. Surface slightly viscid to dry, pruinose above ring, appressed-fibrillose below. **PARTIAL VEIL:** Forming a faint viscid cortina, leaving an orangish band on stipe, but becoming obscure in age. **FLESH:** Moderately thick, firm, whitish buff, stipe hollow, fibrous, whitish buff to pale orangish buff. **ODOR:** Indistinct to slightly farinaceous. **TASTE:** Strongly bitter. **SPORE DEPOSIT:** White. **MICROSCOPY:** Spores 5–7 x 3.5–5.5 µm, ellipsoid, smooth.

ECOLOGY: Solitary, scattered, in clusters, rows, or arcs on ground or duff under pine. Locally common in greater San Francisco Bay Area, generally uncommon elsewhere. Fruiting in late fall through winter.

EDIBILITY: Unknown; the bitter taste and closely related poisonous species are major deterrents.

COMMENTS: In an otherwise confusing group of *Tricholoma*, this species may be the easiest to recognize by the partial veil and bicolored stipe. However, older specimens lose this feature to varying degrees and can be tough to identify. Similar pine associates include *T. muricatum*, which lacks the distinct partial veil and bicolored stipe, and *T. nictitans*, which has a tall stature, rusty brown, viscid cap, and orangish brown, hollow stipe. Also compare the oak-associated *T. dryophilum* and *T. ustale*.

Group M

Tricholoma magnivelare

(Peck) Redhead

MATSUTAKE

CAP: 6–20 cm across, peg-shaped to convex buttons expanding to broadly convex and eventually plane, margin inrolled and covered with veil when young, to downcurved with veil remnants, eventually flattening or flaring slightly upward. White to creamy buff at first, with fibrils becoming orangish to pale brown. Surface dry to slightly viscid when wet, often with appressed cottony patches or fibrils. **GILLS:** Narrowly to broadly attached, close to really crowded, narrow, partial gills numerous. Whitish, creamy white to pale creamy pink when young, developing orangish spots or stains in age. **STIPE:** 5–15 cm long, 1.5–4.5 cm thick, equal or tapering to a pointed base. Whitish to creamy with orangish to brownish scales or fibrils below ring, whitish above. Surface dry to slightly viscid, fine chevrons of powdery scales above veil, appressed flocculent scales below veil that can gelatinize slightly in wet weather. **PARTIAL VEIL:** Thick, cottony, flaring upward at first, becoming appressed in age. **FLESH:** Thick, very firm to rubbery, whitish to creamy white at first, slowly discoloring orangish or with orangish stains around worm holes. **ODOR:** A mix of sweet cinnamon spiciness and odd mustiness. Some people only smell the "good" spicy odor, while others recoil with disgust. David Arora's famous description, "a provocative compromise between Red Hots and dirty gym socks," is spot-on. **TASTE:**

Indistinct. **SPORE DEPOSIT:** White. **MICROSCOPY:** Spores 6–8 x 4.5–6 µm, broadly ellipsoid, smooth.

ECOLOGY: Solitary, in small "nests," or scattered on ground, often buried in duff. Found in a wide variety of forest types, preferring Tanoak over most of our range, especially abundant in inland Tanoak-Madrone forests from Sonoma County northward. Also frequently found in coastal areas with sandy soil under pine or Douglas-fir, as well as in open-understory forests of true firs at higher elevations. One of the better habitats in the southern part of our range is in manzanita thickets on sandy or clay-rich soil, where it will often fruit in rings around sprawling manzanita branches, buried in duff. Generally fruiting after cold weather in late fall or early winter in northern forests, later into January farther south.

EDIBILITY: Edible and highly prized. The strong flavor goes a long way, and a small amount can season a whole dish. Most people recommend thinly slicing the mushrooms into a stir-fry or using them to flavor rice or other grains. They also make a great addition to clear soups.

COMMENTS: Other common names for this mushroom include Pine Mushroom, White Matsutake, or simply "Tanoaks" (especially in areas where their conifer hosts are less common). The whitish cap; crowded, creamy white gills; thick, cottony veil; firm, rubbery flesh; and distinctive cinnamon odor make the Matsutake quite distinctive. Other similar mushrooms in our area include a member of the *T. caligatum* group (a species complex), which is usually smaller and has brown scales on

the cap and stipe; although it can have the same odor, it has a bitter taste. The often massive *Catathelasma ventricosum* has a whitish to gray cap, decurrent gills, a thick partial veil that forms a double ring, and a strongly farinaceous odor and taste. Care should be taken not to mistake Matsutake for the potentially deadly poisonous *Amanita smithiana* and *A. silvicola*; both have relatively fragile flesh and an unpleasant, wet dog–chlorine odor, but occasional young specimens of *A. smithiana* can smell like Matsutake.

Tricholoma murrillianum: A confused past, and an uncertain future

Although often called *Tricholoma magnivelare*, the correct name for our Matsutake is a little complicated.

Charles Peck described a Matsutake collected in upstate New York in 1873 as *Agaricus ponderosus* (most gilled fungi were still called *Agaricus* back then). This was an illegitimate name because it was previously used for a different species. In 1878 described it as *Agaricus magnivelaris*. In 1887 Saccardo described *Armillaria ponderosa*, which was a valid name and one that was commonly used into the 1990s. Even though Rolf Singer placed it in *Tricholoma* (=*T. ponderosum*)

in 1949, this name wasn't used. When Scott Redhead studied these species in the early 1980s, he synonymized *T. ponderosum* with the older species name *magnivelaris* and transferred it into *Tricholoma*, which gave us the name most often used today: *Tricholoma magnivelare*.

But…recent genetic work has shown that the western North American species is different from the eastern species! Further investigation reveals that Murrill described a Matsutake from Newport, Oregon, as *Armillaria arenicola* in 1912. He also described a different species as *Tricholoma arenicola* at the same time, which only matters because when Singer examined *Armillaria arenicola* in 1942, he correctly recognized that it belonged in *Tricholoma*, but had to give it a new species name because Murrill had already used the name *T. arenicola* for a different fungus. So Singer gave our western Matsutake the name *T. murrillianum*, which is currently the earliest and most technically correct name for the species profiled here.

However, all of this may be a moot point! Because of cultural and economic importance of this mushroom, a case has been made for preserving the name *Tricholoma magnivelare* for the western Matsutake, and a proposal is in the works to formalize this; the final resolution remains to be seen.

Tricholoma focale

(Fr.) Ricken

VEILED ORANGE TRICH

CAP: 3–12 (15) cm across, convex when young, becoming broadly convex to nearly plane, margin inrolled when young. Color variable, orange but often with deep olive-green to forest green colorations in button stage, with darker reddish brown appressed scales or fibers, becoming orange-brown to reddish orange in age. Surface viscid to moist, with small appressed scales or matted fibrils, to nearly smooth. **GILLS:** Narrowly attached with a distinct notch, or even appearing free at times, close to crowded, relatively narrow, partial gills numerous. Dingy white to pale buff-white, developing orange-brown stains in age and slowly bruising orange-brown when damaged. **STIPE:** 4–11 cm long, 1–3 cm thick, more or less equal, but tapering or pointed at base. Orangish belts of scales over a paler base color, whitish above veil, discoloring orange-brown in age below veil. Surface dry, covered with scaly belts or fibrous tufts below veil, smooth above. **PARTIAL VEIL:** Thick, membranous, somewhat sheathing, leaving a flaring ring and often orangish scaly belts. **FLESH:** Moderately thick to thin, firm, fibrous, whitish, slowly staining orange-brown to dingy brown in cap, stipe fibrous, solid, whitish, slowly staining orange to orange-brown. **ODOR:** Strongly farinaceous. **TASTE:** Unpleasant, strongly bitter-farinaceous. **KOH:** Bright red on cap, reddish on stipe, little reaction elsewhere. **SPORE DEPOSIT:** White. **MICROSCOPY:** Spores 5–6 x 3–4 µm, broadly ellipsoid, smooth.

ECOLOGY: Solitary or scattered on ground, in moss or duff in a wide variety of forest types. We commonly find it in Tanoak-Madrone and pine forests along the coast, and often in manzanita thickets as well. Often occurs in the same habitat as Matsutake, but generally fruits a little earlier (making it a good indicator species). Fruiting from fall into winter. Uncommon south of Mendocino County, locally common as far south as the Santa Cruz Mountains.

EDIBILITY: Inedible due to unpleasant taste.

COMMENTS: The orangey to greenish streaked cap can look very similar to *T. aurantium*, but this species has a membranous partial veil that leaves a ring on the stipe. *Armillaria badicephala* (which is actually a *Tricholoma*) is like a brown-colored version of *T. focale*; it is rare in California found only on the Far North Coast dunes under spruce and pine. *Floccularia albolanaripes* has a yellow to yellow-brown cap, creamy white to pale yellow gills, a flocculent white partial veil, and a similarly shaggy-scaly stipe that is white to creamy yellow with yellow belts or bands.

NOMENCLATURE NOTE: This species may turn out to be distinct from the true (European) *T. focale*, and for this reason, some mycologists use the name *T. zelleri* (based on type specimens from the Pacific Northwest) for this species.

Floccularia albolanaripes

(G. F. Atk.) Redhead

CAP: 4–12 (16) cm across, egg shaped to convex with a rounded-conical or somewhat pointed umbo, becoming broadly convex, sometimes slightly uplifted and wavy in age. Dark brown at center, ocher-brown to golden brownish outward, bright yellow outward, often with appressed-fibrillose brown scales and streaks, creamy yellow to whitish at extreme margin. In age fading overall to pale grayish beige. Surface dry, often retaining remnants of partial veil as a cottony band around entire margin, or as thin, triangular flaps of tissue. **GILLS:** Notched, close. Whitish to creamy or buttery yellow, in age often with a dull apricot wash. **STIPE:** 5–10 cm long, 1.2–4 cm thick, straight or slightly curved, base slightly enlarged. Upper part smooth and creamy or yellowish to light apricot, most of lower stipe pale creamy and abundantly covered in shaggy, cottony white veil with scattered bright lemon yellow to golden brown plaques and chevrons. **FLESH:** Fairly firm, fibrous-fleshy. Creamy white. **ODOR:** Indistinct to slightly farinaceous. **TASTE:** Mild. **SPORE DEPOSIT:** White. **MICROSCOPY:** Spores 6–9 x 4–5 μm, ellipsoid, smooth, amyloid.

ECOLOGY: Singly, in fused pairs or scattered in troops in duff. In our area most often under live oak, Douglas-fir, and redwood. Fruiting from late fall into early spring. Widely distributed throughout coastal and montane California, but generally uncommon. Patches tend to fruit reliably year after year.

EDIBILITY: Edible. Very good when young and fresh, older collections often slightly musty.

COMMENTS: The yellow to golden brown cap, cream-white gills, and creamy stipe with yellowish to brownish shaggy scales make this attractive species easy to recognize. It is uncommon throughout our area and always a treat to find. Similar brightly colored species include *Amanita augusta*, which has a more membranous partial veil (not as shaggy) and yellowish warts on the cap surface; it is usually taller and larger as well. *Tricholoma equestre* has a similar cap color, but has a smooth yellow stipe and brilliant yellow gills.

Catathelasma ventricosum (sensu CA)
LARGE CAT

CAP: 8–25 (35) cm across, tightly rounded and peglike when young, with an inrolled margin, becoming convex to broadly convex, eventually becoming plane to slightly wavy and uplifted in age. Color variable within a range; pale to dark gray or grayish brown, often with whitish blotches when young, becoming paler, to dingy gray, pale gray to almost white, with darker and paler areas, and often with whitish, flocculent patches, then typically becoming grayish brown in age. Occasionally with orangish stains. Surface dry to slightly tacky, smooth or with flocculent patches. **GILLS:** Decurrent, often in a curving manner, close to crowded. Creamy white to pale creamy buff. **STIPE:** 10–30 (40) cm long, 2–7 (10) cm thick, equal aboveground, then with a long tapering taproot. Occasionally swollen at ground level, then tapering below. Grayish to whitish, staining or discoloring orangish. **PARTIAL VEIL:** Layered, a grayish membranous outer covering leaving bands on stipe and a thick white fibrillose-flocculent inner layer forming a persistent flaring to ragged ring rather high on stipe and often leaving patches on cap margin. **FLESH:** Very thick, firm, rubbery, stipe hard. White. **ODOR:** Strongly farinaceous. Older specimens can be more rancid. **TASTE:** Farinaceous. **SPORE DEPOSIT:** White. **MICROSCOPY:** Spores 8–12 (14) x 4–5.5 μm, ellipsoid, smooth, amyloid.

ECOLOGY: Solitary or scattered, more rarely in small clusters of 2–4 fruitbodies in thick humus, needle duff, or mossy coniferous forest on the North Coast. Common under Sitka Spruce on Far North Coast, rarer south to Mendocino County. Fruiting from late summer through winter, most abundant in the midfall.

EDIBILITY: Edible, but very tough.

COMMENTS: This mushroom forms massive fruitbodies that might very well be the heaviest gilled mushrooms in California. The cap diameter is only rivaled by occasional specimens of *Gymnopilus ventricosus* and *Russula brevipes*. Besides the large size and white colors, the decurrent gills, thick double veil, tapered, rooting stipe, and farinaceous odor help clinch the identification. Although there are reports of both *Catathelasma ventricosum* and *C. imperiale* occurring in California, it is our belief that we have a single, somewhat variable species on the coast that doesn't quite fit either species. The name *C. ventricosum* is problematic, as it was originally described from Alabama, referring to a mushroom with a "shining white" cap growing with hardwoods; it is highly doubtful that it is the same as ours. *C. imperiale* is a European species (with brown to golden brown caps), but is applied in California to darker, larger, and/or weathered specimens. *Tricholoma magnivelare* has a musty cinnamon-like odor and notched (not decurrent) gills. *Amanita smithiana* has a powdery partial veil that breaks up and leaves hanging remnants around the cap margin.

Armillaria nabsnona

T. J. Volk & Burds.

CAP: 4–11 cm across, convex at first, soon flat, sometimes with a low, broad umbo. Warm brown, ocher-brown, to more ocher-orange toward the margin. Distinctly translucent-striate when wet. Surface thinly viscid when wet, finely matted-tomentose at very center and with a few small tufts of hairs outward, but extensively smooth (usually entirely smooth in age). **GILLS:** Broadly attached with a decurrent tooth, close to subdistant. Creamy white to creamy buff, occasionally with a pinkish cast, discoloring yellowish when damaged, often with brownish stains in age. **STIPE:** 7–14 cm long, 0.3–1 cm thick, enlarged downward to a swollen base. Buff to light brown, covered with whitish fibers at first, darkening in age. **PARTIAL VEIL:** Thin, cottony, forming a flaring ring at first, often disappearing, or leaving a weak ring zone. **FLESH:** Thin, fibrous, white. **ODOR:** Indistinct. **TASTE:** Mild. **SPORE DEPOSIT:** White. **MICROSCOPY:** Spores (6) 8–10 x 5.5–6.5 μm, ellipsoid, inamyloid, moderately thick walled, smooth.

ECOLOGY: In small clusters, scattered in troops, or occasionally solitary in swampy areas with alder and maple. Either growing directly from decaying wood or emerging from ground near base of trees. Common on the North Coast, scarcer southward at least to the San Francisco Bay Area. Fruiting in fall and early winter.

EDIBILITY: Edible with caution (see *A. mellea*, p. 390).

COMMENTS: The rather smooth, translucent-striate, ocher-brown cap, thin partial veil, white spores, and growth near alder help identify this species. Other *Armillaria* have less richly orange-brown caps and are often more densely fibrillose-scaly. Probably the most similar-looking mushroom is *Kuehneromyces mutabilis*, which has a more strongly hygrophanous cap, extensive shaggy rings of tissue on the stipe, and brown spores. *A. mellea* is somewhat similar, but has a thick, membranous partial veil and tapered stipe base, and often grows in very large clusters.

Armillaria solidipes

Peck

HONEY MUSHROOM

CAP: 5–14 cm across, rounded or domed with an inrolled margin at first, soon broadly convex to flat or slightly uplifted and wavy in age. Dark brown at center, pinkish to pinkish beige or creamy tan outward, fading as it loses moisture. Margin slightly translucent-striate, often with adhering white "teeth" of partial veil tissue. Surface slightly viscid to dry, covered in dark hairs (most dense at center). **GILLS:** Broadly attached, slightly notched and often with a fine decurrent tooth, close. Whitish to light creamy, aging to dingy beige. **STIPE:** 5–15 cm long, 0.7–1.7 cm thick at apex, fairly upright, base enlarged and often club shaped (up to 3 cm thick). White to creamy, pinkish tan near apex, beige-tan near base. Surface dry, smooth to vertically striate but covered in whitish to grayish brown felty patches and fibrillose scales. Older specimens become extensively dark grayish behind coating of pale, cottony tissue. **PARTIAL VEIL:** Ragged, felty band or cottony white skirt set high on stipe, edge often turning beige or brown. **FLESH:** Fibrous, white, solid. Stipe with a rind and stringy interior. **ODOR:** Indistinct. **TASTE:** Mild. **SPORE DEPOSIT:** White, often quite heavy. **MICROSCOPY:** 7.5–9.5 x 5–6.5 μm, ellipsoid, smooth. Clamp connections present, scattered.

ECOLOGY: In clusters or large troops on trunks of standing dead trees, as well as on large, decaying conifer logs. Common in the northern part of our range, quite uncommon south of Mendocino County, largely replaced by *A. sinapina* (see comments). Fruiting from fall into winter.

EDIBILITY: Edible with precautions. See comments on edibility for *A. mellea* (p. 390).

COMMENTS: This species (like most *Armillaria*) is extremely variable, but typical collections can be identified by their pinkish brown caps; pale, broadly attached gills, stipes covered in scales and with club-shaped bases; and growth on conifers (in our area). *A. sinapina* often has a darker pinkish brown to orange-brown cap, a more cobwebby partial veil, and extensive yellow colors (especially the fuzz around the stipe base); it frequently grows solitary or in small clumps on the ground (from buried roots), and is quite common in mixed forests in our area. *A. mellea* has longer stipes that are pointed at the base and lacks clamps.

Armillaria mellea

(Vahl) P. Kumm.

HONEY MUSHROOM

CAP: 4–15 (25) cm across, domed at first with an inrolled margin, expanding to broadly convex and then flat, often uplifted or wavy in age. Occasionally with a broad, round umbo. Color variable, dark brown at center, pinkish brown outward, fading to beige from center to margin as it loses moisture, or extensively golden brown, fading to yellowish brown to creamy tan. Edge of cap pale creamy beige, translucent-striate, often with small bits of white veil tissue hanging off edge. Surface thinly viscid at first, becoming moist to dry, covered with small dark hairs and/or yellow fibrillose tufts. **GILLS:** Broadly attached, often with a small decurrent tooth, close, white to creamy, then yellowish beige, becoming pinkish tan to dingy ocher-tan in age. **STIPE:** 9–20 (30) cm long, (0.6) 1–2 (6) cm thick, cylindrical, straight to curved or sinuous, often tapered toward base and pointed (especially when in fused clusters). Pale cream, beige to tan, then pinkish brown, in some collections strongly darkening to smoky gray upward from base, sometimes with yellowish fuzz near base as well. Surface dry, innately fibrillose to smooth, but with many scattered bands and chevrons of white tissue, especially just below partial veil. **PARTIAL VEIL:** White, thick, membranous, flaring skirt set high on stipe, margin often fringed, sometimes with small brown or yellow scales around edge. **FLESH:** Thick, white in cap. Stipe with a darker cartilaginous rind and a distinctly stringy fibrous, white interior. **ODOR:** Indistinct. **TASTE:** Mild. **SPORE DEPOSIT:** White. Often very heavy. Older clusters often leave the lower caps and surroundings covered in a thick layer of white spore powder. **MICROSCOPY:** Spores 7–9 x 5.5–6.5 µm, ellipsoid, smooth. Clamps absent from all tissues.

ECOLOGY: Often in large, dense clusters at base of dead or dying hardwoods, especially oaks, Tanoak, and Madrone, or on stumps and standing trees and shrubs. Very common throughout our range, but rare on immediate coast in the north. Often fruiting in fall and winter, but can fruit almost any time of year.

EDIBILITY: Edible and quite good. However, it is a frequent cause of stomach upsets, so only use the caps of young specimens (the stipes are tough) and always cook them thoroughly.

COMMENTS: The white, flaring partial veil, long stipes with tapered bases, and growth in clusters (sometimes with 100+ caps) help distinguish this species. *A. solidipes* can be similar, but usually has a shaggier-looking stipe that is club shaped at the base. This species is an aggressive parasite and is ecologically very versatile; it has been recorded on conifers as well as hardwoods and has a rather cosmopolitan distribution. *Lyophyllum decastes* often fruits in large clumps of overlapping caps, but grows in disturbed soil and lacks a veil. Some *Gymnopilus*, *Pholiota*, and *Kuehneromyces* are similar, but produce brown spore prints.

Group N

Tricholomopsis rutilans

(Schaeff.) Singer
PLUMS AND CUSTARD

CAP: 5–11 cm across, conical-convex to rounded when young, soon convex to broadly convex and then plane with a low, broad umbo. Rosy pink, magenta to wine red, fading in age to pale pinkish over a pale creamy yellow background, edge often very pale to creamy or nearly white (yellowish in age). Surface densely hairy-velvety, in age fine fibrils becoming more appressed. **GILLS:** Attached, notched, moderately close. Creamy when young, soon buttery yellow. **STIPE:** 5–16 cm long, 1–2.5 cm thick, cylindrical, slightly curved. Fibrils purplish red to pinkish red over whitish to creamy yellow surface. Extreme apex sometimes contrastingly lilac to royal purple. Covered with chevrons of fibrils, very dense when young, becoming flattened or disappearing in age. **FLESH:** Thin, fleshy-fibrous. Creamy yellow, slowly bruising ocher-brown where cut. **KOH:** Reddish orange on gills and stipe. **ODOR:** Slightly farinaceous or pleasantly nutty. **TASTE:** Slightly nutty or indistinct. **SPORE DEPOSIT:** Whitish. **MICROSCOPY:** Spores 5–7 (9) x 3.5–5 (6.5) µm, ellipsoid to teardrop shaped, inamyloid, smooth.

ECOLOGY: Solitary or scattered in troops on rotting coniferous wood, wood chips, or lignin-rich humus. Widespread and fairly common. Fruiting from fall through late winter.

EDIBILITY: Edible, but not to everyone's liking.

COMMENTS: The reddish pink to purplish fibrillose cap, yellowish gills, and contrasting reddish chevrons on the pale creamy stipe help distinguish this species. Although the medium- or large-size fruitbodies have the stature of a *Tricholoma*, the brighter colors and lignicolous habit distinguish it from that genus (but note that this species often grows in lignin-rich forest soil rather than clearly on rotting wood, and this can cause confusion). *Gymnopilus luteofolius* can look very similar, but has a partial veil and rusty brown spores. *Rugosomyces onychina* has a purple cap that lacks the fibrils and also has yellowish gills and a purplish stipe. It is rare on the coast and more common in the mountains in the spring.

Tricholomopsis decora

(Fr.) Singer
DECORATED MOP

CAP: 2–6 cm across, rounded at first, expanding to broadly convex to flat, margin often finely inrolled. Bright lemon yellow to dingy yellow, covered in dark gray, brown to ocher-brown fibrils and scales. Hygrophanous, fading to yellowish creamy as it loses moisture, edge of cap sometimes translucent-striate. Surface moist to greasy or dry, densely tufted-scaly when young, flattened-scaly or somewhat smoother in age. **GILLS:** Broadly attached to notched, broad. Bright canary yellow to duller yellow, often with peachy tones in age or after drying out. **STIPE:** 3–7 cm long, 0.4–1.2 cm thick, cylindrical, usually curved, and often deeply inserted into wood substrate, usually broken when collected. Canary yellow to dingy light yellow or pale creamy beige. Surface smooth or slightly scurfy. **FLESH:** Thin, fibrous, hollow in stipe, yellow. **KOH:** Orange on cap. **ODOR:** Indistinct. **TASTE:** Indistinct. **SPORE DEPOSIT:** White. **MICROSCOPY:** Spores 6–7.5 x 4.5–5 µm, ellipsoid, smooth.

ECOLOGY: Solitary or in small groups on decaying conifer wood in late fall and winter. Rather uncommon over most of our area, more regularly found from Sonoma County northward.

EDIBILITY: Unknown.

COMMENTS: The small size, bright yellow gills and stipe, yellow cap with dark fibrils and scales on the cap, and growth on wood help distinguish this species. *T. sulfureoides* is similarly bright yellow (lacking the dark cap fibrils), but has a sparse, silky partial veil when young and has smaller spores (5.5–6.5 x 4.5–5 µm). *T. fulvescens* has a paler orange-yellow to creamy tan cap with paler fibrils, a yellowish stipe that stains rusty orange, and larger spores (8–10 x 6–7 µm). Both are restricted to the Far North Coast. *Chrysomphalina chrysophylla* has a slender stature, a golden orange cap with dark scales that becomes brighter and smoother in age, and bright golden gills. *Callistosporium luteo-olivaceum* is fairly similar, but has a smooth cap and is usually duller, with olive to brown tones on the cap.

Callistosporium luteo-olivaceum
(Berk. & M. A. Curtis) Singer

CAP: 2–6 cm across, convex with an incurved margin when young, becoming broadly convex to nearly plane. Dingy olive-green to olive-yellow at first, center often darker, becoming more golden brown in age. Weakly hygrophanous, fading to dull olive-yellow to yellow-buff. Surface dry, slightly wrinkly, opaque. **GILLS:** Narrowly attached, often notched, close, narrow. Dingy cream or pale yellowish. **STIPE:** 3–8 cm long, 0.3–0.7 cm thick, equal, round to compressed and slightly grooved. Olive-yellow at first, losing olive tones and darkening from base up in age. Surface appressed-fibrillose, often pruinose toward apex. **FLESH:** Concolorous, thin, fibrous. **KOH:** Red on cap, orangey red to reddish brown on gills and stipe. **ODOR:** Faint, sweet-fragrant like some *Clitocybe*. **TASTE:** Not sampled. **SPORE DEPOSIT:** Whitish. **MICROSCOPY:** Spores 6–7 x 4–5 µm, broadly ellipsoid to ovoid, inamyloid, smooth, reddish in KOH.

ECOLOGY: Solitary or scattered in small troops on well-decayed logs and stumps, with a preference for pine. Common throughout our range, but often overlooked. Fruiting from early fall into spring.

EDIBILITY: Unknown.

COMMENTS: No other species in our area shares the combination of growth on wood, olive-yellow coloration, white spores, and red KOH reaction. Perhaps most likely to be confused are weathered specimens of *Tricholomopsis decora* and some of its relatives, but they usually have dark hairs on the cap and brighter gills, and do not turn distinctly red in KOH. *Simocybe centunculus* is similarly colored and grows on wood, but is much smaller, has a finely velvety cap, and has brown spores.

Clitocybula atrialba
(Murrill) Singer

CAP: 2.5–7 (10) cm across, convex when young, becoming plane with a depressed center, to broadly funnel shaped in age. Dark gray to dark gray-brown at first, becoming pale grayish brown to gray. Surface finely appressed-fibrillose, to smooth, moist to dry. **GILLS:** Subdecurrent to decurrent, fairly widely spaced, broad. White to pale gray. **STIPE:** 5–15 cm long, 0.4–1.5 cm thick, very tall and slender, equal or enlarged slightly toward base. Gray to grayish brown, paler toward base and in age. Surface covered with small scurfs and scales, especially pronounced near apex. **FLESH:** Thin, fibrous in stipe, whitish to pale gray. **ODOR:** Indistinct. **TASTE:** Indistinct. **SPORE DEPOSIT:** White. **MICROSCOPY:** Spores 6–9 x 7–8 µm, subglobose to globose from 4-spored basidia, up to 13 x 9 µm and broadly ellipsoid to ovoid from 1- and 2-spored basidia, smooth, amyloid.

ECOLOGY: Solitary or scattered on well-decayed branches and logs of hardwoods (often buried). Has a preference for deciduous oaks, alder, and Bigleaf Maple, but also grows on live oak, and probably other species. Widespread but uncommon. Fruiting from late fall into early spring.

EDIBILITY: Unknown.

COMMENTS: The grayish brown cap, whitish to gray, subdecurrent gills, long, slender scurfy stipe, and growth on wood are a distinctive set of features. Probably most similar are *Hygrophorus* species like *H. camarophyllus* and *H. pustulatus*, which don't grow on wood and which have thick gills with a waxy feel. *Ampulloclitocybe avellaneoalba* and *A. clavipes* have smooth stipes and creamy white gills that are typically more crowded.

Flammulina velutipes var. *lupinicola*

Redhead & R. H. Petersen

LUPINE VELVET-FOOT

CAP: 2–5 cm across, convex to plane, occasionally with a low umbo. Ocher-brown to orange-brown with a slightly paler margin when young, or wet; hygrophanous, fading to light orange-brown to orange-buff when dry or in age. Surface smooth, thinly viscid and translucent-striate when wet, to dry and opaque. **GILLS:** Broadly attached or narrowly attached with a distinct notch, close to subdistant, greasy feeling, partial gills numerous. Whitish to pale ocher-buff. **STIPE:** 2–5 cm long, 0.2–0.7 cm thick, more or less equal, often curved. Pale whitish at apex, gradually darker, from orange to orange-brown, with a dark brown base. Surface dry, distinctly velvety, especially over lower portion. **FLESH:** Thin, concolorous, cartilaginous in stipe. **ODOR:** Indistinct. **TASTE:** Mild. **KOH:** No reaction. **SPORE DEPOSIT:** White. **MICROSCOPY:** Spores 7–15 x 4–6.5 μm, elliptic, cylindrical to apple seed shaped, smooth, inamyloid, colorless.

ECOLOGY: In small to medium-size clusters on dead bush lupine (*Lupinus* spp), most common on the Far North Coast on Yellow Bush Lupine (*Lupinus arboreus*). Common, but restricted to coastal dunes, riparian zones, and sandy areas near ocean where bush lupine grows. Fruiting from fall into spring.

EDIBILITY: Edible, with good flavor. However, it is rather thin fleshed and the stipes are tough. The form of *Flammulina velutipes* that grows on hardwood trees is usually larger and a better choice for the table.

COMMENTS: This small, pale variety (species?) can be recognized by its hygrophanous, ocher-brown to orange-brown cap, whitish to pale ocher-buff gills, velvety stipe, and growth on bush lupine. *F. velutipes* is generally larger, grows in larger clusters, and has a darker reddish brown cap that fades in age and much smaller spores (7–8.5 x 3.5–4.5 μm); it is uncommon in California, occurring on hardwoods (especially oak and Tanoak). *F. populicola* is larger still and stockier, with a yellow-orange cap; it grows off the roots and lower trunks of cottonwood and aspen (*Populus* spp.) All these species are edible.

Omphalotus olivascens

H. E. Bigelow, O. K. Miller, & Thiers

WESTERN JACK O'LANTERN

CAP: 4–25 cm across, broadly convex to plane, becoming uplifted and wavy in age. Olivaceous orange, yellow-orange, dingy orange to olive, occasionally more olive-brown, often with water marks, splotches, and stains, or completely olive-green in age in wet weather. Becoming more reddish in dry weather. Surface moist to dry, smooth and innately fibrillose, often becoming coarsely cracked in dry weather. **GILLS:** Decurrent, fairly widely spaced, gills often forked near stipe. Olivaceous orange, dull orange to grayish olive, often appearing paler when dusted with white spores. Often with irregular, contrasting lighter zones or bands. **STIPE:** 3–15 cm long, 1.5–7 cm thick, cylindrical to spindle shaped, base curved and often tapered, rooting deeply into wood substrate and thus often broken when collected. Central to off-center, sometimes truly lateral or very reduced. Olivaceous orange, yellowish orange to dull olive. Surface smooth or roughened and fibrillose. **FLESH:** Thick, fibrous, firm or rubbery, pale grayish white to vinaceous gray, often with more orange tones in age, often marbled with olive tones throughout. **ODOR:** Indistinct. **TASTE:** Indistinct. **SPORE DEPOSIT:** Whitish (though mushroom often stains paper orange when attempting a print). **MICROSCOPY:** Spores 6.5–8 x 6–6.5 μm, globose to ovoid.

ECOLOGY: Usually in overlapping clusters, sometimes solitary or in small, scattered clusters. These clusters can sometimes be quite massive! Fruiting from stumps and dead roots, but also from logs, rarely on trunks of standing trees. Typically on oak, Chinquapin, or eucalyptus, also on a wide variety of other hardwoods (also rarely on conifers). Common from Mendocino County south. Fruiting from fall into spring, most abundant in late fall or early winter.

EDIBILITY: Poisonous, causing severe vomiting and diarrhea.

COMMENTS: This distinctive species can easily be recognized by its olive-orange to yellow-orange colors, clustered growth on wood, decurrent gills, and pale spores. Beginners sometimes confuse this species for chanterelles because of the yellow-orange color and decurrent gills, resulting in serious poisonings. *O. olivascens* can be easily told apart from chanterelles by their true gills, growth on wood, flesh that is not pure white, and darker orange-olive colors. *O. olivascens* is a great dye fungus that yields colorfast violet, gray, and green dyes on wool. Interestingly, it is bioluminescent, hence the common name! The gills glow faintly green, a feature that can observed by staring at fresh, mature specimens in a very dark room. Although prime fruitbodies should produce a visible glow within 10 minutes, some specimens glow only faintly, and others do not produce a visible glow at all. At first glance, it might be mistaken for *Gymnopilus ventricosus*, another large, wood-dwelling species, but on closer inspection, the brown spores and a partial veil easily distinguish that species.

Panus conchatus

(Bull.) Fr.

CAP: 3–15 cm across, buttonlike when very young, soon flattened with a tightly inrolled margin, becoming broadly convex, often asymmetrical or lobed. Becoming uplifted and wavy, but with margin usually remaining inrolled in age. Royal purple to amethyst at first, then lavender to reddish brown with darker bands, becoming tan in blotches or evenly beige overall in age. Surface dry, velvety (like suede) or matted-fibrillose, margin often strikingly fuzzy when young, smoother when rain beaten or in age. **GILLS:** Decurrent, often forking near stipe and wavy or wrinkled. Lilac to lavender or dingy purplish tan at first, paler and slightly ocher or creamy blotched in age. **STIPE:** 3–8 cm long, 1.5–4 cm thick, off-center, sometimes central, often much wider than cap and disproportionately stocky at first (young fruitbodies appearing pinheaded). Quite stout and cylindrical, often embedded in substrate. Lilac, purplish tan to grayish beige, base often with whitish fuzz. Surface dry, suedelike when young, becoming smooth. **FLESH:** Whitish, tough, firm and rubbery at first, leathery in age. **ODOR:** Indistinct. **TASTE:** Indistinct. **SPORE DEPOSIT:** White. **MICROSCOPY:** Spores 5–7.5 x 2.5–3.5 μm, ellipsoid, inamyloid, smooth.

ECOLOGY: Solitary, scattered, or in small clusters on dead hardwood logs and larger branches. Very common wherever oaks and Tanoaks are abundant, also occurring on other hardwoods. Fruiting from fall through spring. Fruitbodies resist rot and can persist for many weeks.

EDIBILITY: Inedible, too tough to eat.

COMMENTS: No other species in our area produces such fuzzy and strikingly purple fruitbodies. Faded specimens can resemble *Sarcomyxa serotina*, but that species is much fleshier (not as tough) and has a much smaller, lateral stipe and a smooth cap.

Neolentinus kauffmanii

(A. H. Sm.) Redhead & Ginns

CAP: 1.5–5 (8) cm across, broadly convex to plane, often with a depressed center. Whitish beige to pale pinkish beige at first, darkening to pinkish tan in age. Surface dry, finely pruinose or hairy, becoming smooth. **GILLS:** Broadly attached to decurrent, close to crowded, partial gills numerous. Edges very finely serrate when young, becoming more ragged in age. Whitish to pale pinkish white at first, darkening slightly, pale creamy buff to pale pinkish buff in age. **STIPE:** 2–6 cm long, 0.3–1 cm thick, central to off-center, rarely lateral, equal or slightly rooting, often curved. Whitish, pale pinkish buff to pale tan. Surface dry, finely pruinose, covered with a white sheen when young. **FLESH:** Thin, fibrous and tough, especially in stipe, whitish to pale buff. **ODOR:** Indistinct. **TASTE:** Slightly acrid. **SPORE DEPOSIT:** White. **MICROSCOPY:** Spores 5–6 x 2–2.5 μm, oblong to ellipsoid, inamyloid, smooth.

ECOLOGY: Solitary or scattered in troops on well-decayed logs, stumps, and branches of Sitka Spruce. Common on the Far North Coast, occurring south to Mendocino County. Fruiting throughout year, but most abundant in winter.

EDIBILITY: Unknown.

COMMENTS: The whitish to pinkish buff colors, serrate gills, small size, and growth on decaying spruce wood are helpful identifying features. *Lentinellus flabelliformis* is smaller and has a short, lateral stipe and more widely spaced serrate gills. *Ossicaulis lignatilis* has a white cap and very crowded white gills that discolor pale yellowish in age, and usually grows inside the hollows of hardwood stumps and logs. Some *Clitocybe* are similar, but generally have more widely spaced gills with even edges.

Pleurotus dryinus

(Pers.) P. Kumm.

VEILED OYSTER

CAP: 5–13 cm across, spatulate at first, convex with an inrolled edge, becoming broadly convex to plane, and often slightly wavy and uplifted in age. Gray to grayish tan, becoming grayish white to creamy white overall. Surface dry, covered with pale, broad appressed-fibrillose scales when young, nearly smooth in age, margin often with thicker whitish remnants of partial veil. **GILLS:** Deeply decurrent, pale creamy to white. **STIPE:** 3–8 cm long, 1.5–4 cm thick, attached asymmetrically to cap, tapered toward base and often deeply embedded in substrate. **PARTIAL VEIL:** White, tissue ragged, membranous, forming a weak zone on stipe or more often remaining attached to margin of cap. **FLESH:** Thick, firm, rubbery, white. **ODOR:** Slightly fragrant (aniselike) to musty. **TASTE:** Mild. **SPORE DEPOSIT:** White. **MICROSCOPY:** Spores 9.5–14 x 3.5–4.5 μm, cylindrical to elongate-ellipsoid, inamyloid, smooth.

ECOLOGY: Solitary, in fused pairs or in small groups on dead or partially dead hardwoods, more often Tanoak or live oak. Rare, appearing early in season, primarily from late summer to fall.

EDIBILITY: Edible, but uncommon and not nearly as prolific as other *Pleurotus*.

COMMENTS: The growth on wood, off-center stipe, decurrent white gills, and partial veil (often as remnants at cap edge or in a weak zone on stipe) are distinctive. Although confusion with other *Pleurotus* species is possible when the partial veil is not apparent, the distinctively textured cap (with appressed-fibrillose patches) sets this species apart.

14 • Pleurotoid Mushrooms

Species in this group are all similar in stature: gilled mushrooms growing sideways from decaying wood (or in a few cases on moss or soil). They tend to have short, off-center stipes or no stipe at all; most are saprobes, but some (including *Pleurotus*) are known to prey on nematodes! Aside from these features, they vary widely in size, texture, and spore color. The genera with white spore deposits are *Pleurotus*, *Hohenbuehelia*, *Cheimonophyllum*, *Tetrapyrgos*, and *Campanella*. As expected for members of the family Entolomataceae, *Claudopus* and *Clitopilus* have pinkish spore deposits, but the gills of their fruitbodies can remain pale for a long time, leading to confusion with the white-spored genera mentioned above. *Crepidotus* species have dingy brown spore deposits and usually have brownish gills to match.

Identification features to note include structure and attachment of the gills, presence or absence of a stipe, gill color, texture of the fruitbody, and substrate (what kind of wood and what state of decay). The color of the spore deposit goes a long way in this group!

Pleurotus ostreatus group
OYSTER MUSHROOM

CAP: 3–15 (25) cm across, semicircular, fan shaped (or oyster shell shaped), with an inrolled margin when young, expanding to plane or uplifted in age. Color variable, generally gray to grayish white when fresh, although buff to grayish brown colors are common, becoming paler in age. Often off-white to chalky white when dry and with yellow discoloration in age. Surface moist to dry, smooth. **GILLS:** Decurrent, close to subdistant, broad. White, cream, to pinkish buff, darkening as it matures, yellowing when old. **STIPE:** 0.5–3 (6) cm long, 0.3–2 (4) cm thick, laterally attached (or nearly central in some forms), often stout, more or less cylindrical and equal. White to creamy, yellowing in age, surface dry, smooth (sometimes riged or tripe textured), with a pubescent or hairy base. **FLESH:** Thin to thick, rubbery, often tough, especially in stipe, fibrous, whitish. **ODOR:** Distinct, but tough to describe, other than like a *Pleurotus*. Mix of anise-like sweetness and mushroomy-earthy tones. **TASTE:** Mild. **KOH:** No reaction. **SPORE DEPOSIT:** White, buff to lilac-gray. **MICROSCOPY:** Spores 7.5–9.5 x 3–4 μm, narrowly elliptical, smooth, inamyloid, colorless in KOH.

ECOLOGY: Often in overlapping clusters or clumps on deadwood, rarely solitary. Most often on hardwoods, especially Tanoak, live oak, and alder, but also on conifers on the North Coast. Very common, occurring throughout forested parts of California. Often fruiting in a large flush soon after first rains in fall, with scattered fruitings into spring.

EDIBILITY: Edible and excellent. A good species for beginners, as it has few look-alikes and is often abundant.

COMMENTS: The fan-shaped fruitbodies, grayish buff to white caps, whitish decurrent gills, lateral stipes, and growth on wood are distinctive. The *P. ostreatus* complex consists of a few known species. *P. ostreatus* and *P. pulmonarius* are nearly indistinguishable; however, those found on conifers are more likely to represent *P. pulmonarius*. *P. populinus*, as the name suggests, often grows on cottonwood and aspen, and has whitish to buff spores (with no lilac tones). *Hohenbuehelia petaloides* has thinner flesh, close to crowded gills, and ocher-brown caps. *Pleurocybella porrigens*, the Angel Wing, has thin, rubbery flesh, pure white oblong or wavy caps, and grows on decaying conifers. It is not known from California, but is very common from central Oregon north into Alaska. *Lentinellus ursinus* has a hairy cap and ragged, sawtooth gill edges. *Crepidotus* species are smaller and thinner fleshed and have brown spores. Compare also to *Pleurotus dryinus*, which looks similar to *P. ostreatus*, but has felty-silky patches on the cap, and a cottony or membranous (but often disappearing) partial veil on the stipe. It is uncommon to rare on the Redwood Coast, primarily found on hardwoods.

Hohenbuehelia petaloides

(Bull.) Schulzer

SHOEHORN OYSTER

CAP: 2–7 (10) cm across, often in fused clusters up to 50 cm across, shoehorn to funnel shaped, becoming more fan shaped in age, margin tightly inrolled when young, often staying downcurved in age. Brown, ocher-brown, grayish brown to gray at first, fading to whitish gray or whitish buff in age. Surface smooth, slightly viscid when wet, otherwise lubricious or moist, often with a white bloom when young. **GILLS:** Decurrent, very narrow near attachment, close to crowded, becoming wavy. White, pale grayish white to creamy, at times with a yellowish cast in age. **STIPE:** 1–4 cm long, 0.5–3 cm thick, off-center to lateral, stout, often flat or compressed rather than cylindrical. Whitish to grayish buff. Surface dry, covered with whitish hairs. **FLESH:** Moderately thin to thick, rather tough, rubbery and with a gelatinized layer when wet. Watery whitish to gray. **ODOR:** Indistinct to farinaceous. **TASTE:** Mild to slightly farinaceous. **KOH:** No reaction. **SPORE DEPOSIT:** Whitish. **MICROSCOPY:** Spores 5–6.5 x 3.5–5.5 µm, elliptical, smooth, inamyloid, colorless in KOH. Pleurocystidia spindle shaped, thick walled, with incrusted crystals over tip. Cheilocystidia similar, or more cylindrical and thin walled.

ECOLOGY: In fused and overlapping clusters in wood chips, or solitary or in small clusters on decaying hardwood logs and stumps. Generally scarce, but locally common in urban areas, uncommon in natural habitats. Fruiting opportunistically whenever there is enough moisture to fruit, but mostly in fall and winter.

EDIBILITY: Edible, but rather chewy and bland.

COMMENTS: This mushroom can be recognized by its rubbery texture, brownish to gray cap, and narrow, crowded, decurrent gills. When growing in wood chips, it can form large fused clusters, but rarely does so in the "wild." *Pleurotus ostreatus* is larger and often paler, lacks the gelatinized flesh, and has broader, more widely spaced gills.

Hohenbuehelia grisea

(Peck) Singer

CAP: 0.5–2.5 cm across, generally bell shaped with an inrolled margin when very young, expanding to fan shaped with an even margin in age. Dark gray to dark bluish gray at first, more grayish brown and often fading, especially around margin in age. Surface finely tomentose, velvety near attachment point, more or less smooth in age. Dry to wet, or slightly gelatinous. **GILLS:** Radiating out from attachment point, close to subdistant. Grayish at first, becoming white, then creamy in age. **STIPE:** Absent, or very rudimentary and lateral. **FLESH:** Thin, often gelatinous when wet or rubbery when dry. Dark gray near cap surface, whitish under that layer. **ODOR:** Slightly farinaceous. **TASTE:** Slightly farinaceous. **KOH:** No reaction. **SPORE DEPOSIT:** White. **MICROSCOPY:** Spores 6–9 x 3–4.5 µm, elliptical to cylindrical, smooth, inamyloid. Cheilocystidia often club shaped or like a bowling pin. Pleurocystidia lance to spindle shaped, tips heavily incrusted with crystals.

ECOLOGY: Solitary, scattered, or forming overlapping shelves on deadwood or woody shrubs. Particularly fond of bush lupine, Scotch Broom (*Cytisus scoparius*), and live oak branches. Widespread but uncommon. Fruiting in fall and winter.

EDIBILITY: Unknown.

COMMENTS: The small, pleurotoid fruitbodies with dark gray caps, grayish to creamy gills, and slightly rubbery-gelatinous texture are helpful distinguishing features. *H. nigra* is similar, but has a black cap when young (fading to grayish in age) and blackish to grayish gills. Microscopically, *H. nigra* lacks the incrusted cystidia and has round spores. Like *H. grisea*, it grows on hardwood branches; it is widespread but rather rare. *Resupinatus applicatus* is very similar to *H. nigra*, but generally has thinner flesh (without a gelatinized layer) and typically has thinner and more crowded gills (however, without side-by-side comparisons, this can be a difficult distinction).

SYNONYM: *Hohenbuehelia atrocoerulea* var. *grisea* (Peck) Thorn & G. L. Barron.

Hohenbuehelia mastrucata

(Fr.) Singer

CAP: 2.5–6 cm across, fan shaped, kidney shaped to nearly circular, margin inrolled into maturity, cap often becoming wavy and lobed in age. Gray to buff, frosted with whitish hairs and scales, especially toward center, becoming more watery buff when wet or grayish buff when dry in age. Surface covered with blunt spines or tufted scales and matted hairs at first, losing most in age, dry to wet, with a thin skin covering a gelatinous layer, up to 0.7 cm thick if wet, more rubbery when dry. **GILLS:** Radiating out from attachment point, close to subdistant. White to beige to pale whitish buff in age. **STIPE:** Laterally attached to wood and often lacking, or occasionally with an indistinct stipe. **FLESH:** Thin, gelatinous or rubbery when dry. Watery gray to watery buff on cap surface, whitish under that layer. **ODOR:** Indistinct. **TASTE:** Mild. **KOH:** No reaction. **SPORE DEPOSIT:** White. **MICROSCOPY:** Spores 7–9 x 4–5.5 µm, elliptical, smooth, inamyloid. Cheilocystidia often club to spindle shaped. Pleurocystidia with swollen bases and narrow necks, tips incrusted with crystals.

ECOLOGY: Solitary or scattered, occasionally as overlapping shelves on well-decayed hardwoods. We have only seen it on Tanoak in California. Very rare in our region, known from a handful of records in late fall and winter.

EDIBILITY: Unknown.

COMMENTS: This species can easily be recognized by its pleurotoid shape and gelatinized cap covered with soft, white tufts. *H. grisea* is a smaller, dark gray species that has a finely tomentose cap (except near the attachment to the substrate, where it is distinctly velvety). *H. petaloides* is larger, has a smooth, gray to brownish cap and rubbery flesh, and at times shows a thin gelatinous layer; it is much more common.

Panellus stipticus

(Bull.) P. Karst.

CAP: 2–6 cm across, convex, soon fan shaped, then plane to irregularly lobed and wavy in age, margin tightly inrolled at first, remaining so well into maturely. Beige-tan to dull, pale ocher-tan, dusty whitish near margin, sometimes zonate with a grayish area near attachment. Surface moist to dry, suedelike to smooth, soon finely areolate with fibrillose patches. **GILLS:** Broadly attached to shallowly decurrent, often crisped and wrinkled, forking and sometimes anastomosing, with many cross veins. Beige-tan to brown. **STIPE:** 0.5–2 cm long, 0.05–0.2 cm thick, very short and rudimentary, laterally attached to cap, thickest near gills and tapered to attachment to wood. Orange-tan to beige or brownish. Surface dry, covered with a fine whitish tomentum, breaking into pale chevrons near gills. **FLESH:** Thin, rubbery-tough, pliant. **ODOR:** Indistinct. **TASTE:** Mild to bitter. **KOH:** No reaction. **SPORE DEPOSIT:** Whitish. **MICROSCOPY:** Spores 3.5–5.5 x 2–3.5 µm, narrowly elliptic, smooth, amyloid, colorless in KOH.

ECOLOGY: Often in clumps (with fused stipe bases) or scattered in troops or rows on rotting branches and logs, especially Tanoak. Occurring throughout our range, but uncommon. Fruitbodies quite persistent and can be encountered throughout the season, usually most conspicuous after prolonged wet spells.

EDIBILITY: Inedible.

COMMENTS: This species is distinguished by its ocher-tan to beige colors, small off-center stipe, and growth in troops on wood. Most similar species are found in *Crepidotus* (with a flabbier, moister texture and brown spore deposits). *Panellus mitis* is a similar whitish species that grows on conifer branches in the mountains. *Phyllotopsis nidulans* is larger and brighter orange, and often stinks. Eastern North American specimens of *P. stipticus* glow green in the dark, while ours appears to lack the bioluminescent chemistry; this may be indicative of species-level difference.

Rimbachia bryophila

(Pers.) Redhead

CAP: (0.1) 0.3–1 (1.5) cm across, at first downward-facing and bell-shaped with an incurved margin, becoming convex, semicircular to fan shaped, wavy and occasionally lobed in age. Bright white, chalky white, or pale cream. Surface dry, very finely hairy to nearly smooth. **GILLS:** Irregular, shallow, radiating out from attachment, often with numerous cross veins. White. **STIPE:** Absent. **FLESH:** Very thin, concolorous. **ODOR:** Indistinct. **TASTE:** Mild. **KOH:** No reaction. **SPORE DEPOSIT:** White. **MICROSCOPY:** Spores 5–7 x 4.5–7 µm, subglobose or shaped like apple seeds with a pointed apiculus, smooth, inamyloid, colorless in KOH.

ECOLOGY: Solitary or in scattered troops on moss, often laterally attached to the fronds. Very common in mossy habitats throughout our range. Fruiting from late fall into spring.

EDIBILITY: Unknown.

COMMENTS: This common moss parasite can be identified by its small size, white colors, and shallow gills (often with cross veins). *Arrhenia retiruga* is similar in appearance and habitat, but can be told apart by its smooth to veined underside and slightly more grayish colors. *Muscinupta laevis* also has a smooth underside, but is bright white in color. *Cheimonophyllum candidissimum* grows on deadwood.

Arrhenia retiruga

(Bull.) Redhead

CAP: (0.1) 0.3–1 cm across, at first a downward-facing cup to bowl shape, becoming convex, semicircular to fan shaped, wavy and occasionally lobed in age. Grayish white to pale grayish brown, paler when dry. Surface dry, very finely hairy to nearly smooth. **GILLS:** Absent. Underside smooth at first, becoming wrinkled or with shallow veins radiating out from attachment point in age. Grayish white, to pale grayish brown, paler when dry. **STIPE:** Lacking, generally laterally attached directly to moss. **FLESH:** Very thin, concolorous. **ODOR:** Indistinct. **TASTE:** Mild. **KOH:** No reaction. **SPORE DEPOSIT:** White. **MICROSCOPY:** Spores 6–9 (11) x 3–5 µm, shortly cylindrical to elliptical, smooth, inamyloid, colorless in KOH.

ECOLOGY: Solitary or in scattered troops on stems and leaves of moss. Common, but easily overlooked. Fruiting from fall into spring from Santa Cruz Mountains north.

EDIBILITY: Unknown.

COMMENTS: The pallid grayish colors, lack of gills, and growth on moss make this dainty mushroom distinctive. *Muscinupta laevis* (=*Cyphellostereum laeve*) is very similar, but is bright white in color; it has not yet been reported from California, but we have seen it on the southern Oregon coast. *Rimbachia bryophila* has more distinct gills with numerous cross veins and is white in color. *Calyptella capula* has hanging bell-shaped or irregular, downward-facing pie-shaped fruitbodies and grows on small rotting branches or herbaceous stems.

Cheimonophyllum candidissimum

(Berk. & M. A. Curtis) Singer

CAP: 0.3–2 cm across, fan shaped with an inrolled margin at first, becoming plane, or shriveled upon drying. White, at times chalky, rarely developing cream tones. Surface dry, finely tomentose to smooth. **GILLS:** Radiating out from stipe or attachment point, close to subdistant, often ragged in age. White at first, becoming cream colored. **STIPE:** Usually absent or indistinct, usually only visible from underside. **FLESH:** Very thin, soft, whitish. **ODOR:** Indistinct. **TASTE:** Indistinct. **KOH:** No reaction. **SPORE DEPOSIT:** White. **MICROSCOPY:** Spores 5–6.5 x 4.5–5.5 µm, broadly elliptical to nearly round, smooth, inamyloid. Cheilocystidia thin, sometimes branching hairs.

ECOLOGY: Often scattered in troops, or solitary on decaying logs and branches, generally those of hardwoods, especially Madrone and Tanoak, more rarely on conifers. Very common throughout our range, fruiting through the wet season.

EDIBILITY: Unknown.

COMMENTS: This dainty pleurotoid mushroom can be recognized by its white cap, white to creamy gills, and lack of a distinct stipe. Although very common, it is often overlooked. *Clitopilus hobsonii* is very similar, but has pinkish spores and often develops pinkish tones to the gills in age. It also has a fuzzy attachment point and often white rhizomorphs as well. A number of *Crepidotus* are similar, but soon develop buff to brown colors on the gills and have yellowish brown to brown spores: *C. versutus* often grows on deciduous bark, *C. herbarum* grows on grass stems or leaves, *C. variabilis* grows on stems and small branches (and has more pinkish buff spores), and *C. subverrucisporus* has more clay-colored gills.

Tetrapyrgos subdendrophora

(Redhead) E. Horak

CAP: 0.1–1 cm across, convex to semicircular. White to creamy white, developing bluish black spots in age. Surface dry, wrinkled, very finely hairy to smooth. **GILLS:** 3–5 primary anastomosing veins radiating out from stipe, with shallow irregular cross veils. Whitish to cream colored, developing bluish black spots in age. **STIPE:** 0.1–0.3 cm long, 0.03–0.07 cm thick, off-center, appearing lateral. Grayish with a white sheen when young, to bluish black in age. **FLESH:** Extremely thin, somewhat tough. **ODOR:** Indistinct. **TASTE:** Mild. **KOH:** No reaction. **SPORE DEPOSIT:** White. **MICROSCOPY:** Spores 8–10 x 6–7 µm, triangular to pyramidal, smooth, inamyloid.

ECOLOGY: Solitary or scattered on stems of dune grass (*Ammophila* spp.), less often on other grasses, sedges, or *Rubus* canes. Locally common on North Coast dunes, rare elsewhere. Fruiting from fall into early winter.

EDIBILITY: Inedible.

COMMENTS: This tiny mushroom is distinguished by the whitish cap that develops bluish black spots, irregularly veined gills, short, off-center stipe, and growth on dune grass (*Ammophila* spp.). *Campanella* "Veiny" also develops bluish gray colors in age and has highly irregular anastomosing pseudogills, but is usually larger and often lacks a stipe completely; it grows on small branches, especially those of Scotch Broom (*Cytisus scoparius*). *Calyptella capula* has hanging, downward-facing bell-shaped or pipe-shaped fruitbodies with smooth or faintly veined fertile surfaces; it grows on small branches or herbaceous stems (we have seen it on debris of *Rubus*, Grand Fir, and Western Hemlock). Also see *Arrhenia retiruga* (at left) and *Rimbachia bryophila* (p. 401), two small pleurotoid fungi with pseudogills on moss. *Marasmiellus candidus* is generally larger and has a central stipe and widely spaced gills. A similar species that shares the same habitat is *Resinomycena saccharifera*. It has widely spaced gills without pronounced cross veins and a short but distinct central to off-center stipe.

Campanella "Veiny"

CAP: 0.5–4 cm across, semicircular, ear shaped to scallop shaped, often pleated, ruffled or wavy. Off-white, creamy white to pale grayish at first, staying so, or darkening irregularly, to bluish gray in age. Surface dry, wrinkled or ruffled. **GILLS:** Irregular anastomosing veins radiating out from 3–5 primary veins; occasionally looking irregular-poroid. Whitish to pale grayish, rarely cream colored, occasionally developing dark grayish stains in age. **STIPE:** Usually absent, attaching directly to wood from cap; if present, up to 0.2 cm long, 0.1 cm thick, off-centered to lateral, often coming from top of cap. **FLESH:** Concolorous, very thin, tough. **ODOR:** Indistinct. **TASTE:** Mild. **KOH:** No reaction. **SPORE DEPOSIT:** White. **MICROSCOPY:** Spores 7–8 x 4–5.5 µm, irregular ellipsoid, smooth, inamyloid.

ECOLOGY: Scattered or solitary, often on undersides of small branches and twigs, occasionally growing on the sides (in which case it usually has a stipe). Most commonly on Scotch Broom (*Cytisus scoparius*) or *Rubus* species, also seen on alder and Salal (*Gaultheria shallon*) branches. Uncommon to rare, occurring along coast north of San Francisco Bay Area. Fruiting in fall into early winter.

EDIBILITY: Inedible.

COMMENTS: The whitish to grayish blue colors, "pseudogills" made up of irregular veins, tiny or absent stipe, and growth on small woody branches are distinctive. This fruitbody morphology is rather rare in temperate areas, but much more common in the tropics. Some have suggested that this species is a form of *Tetrapyrgos subdendrophora*, which we believe to be different. *T. subdendrophora* is generally much smaller and has a white to creamy cap that develops bluish black spots and stains in age, an off-center, short bluish black stipe, and more pronounced, veinlike "pseudogills."

Lentinellus flabelliformis

(Bolton) S. Ito

CAP: 0.5–3 cm across, fan shaped, semicircular, to kidney shaped, convex, sunken around stipe. Margin downcurved when young, becoming wavy and occasionally lobed in age. Pinkish buff, light cinnamon brown to beige; hygrophanous, fading when dry. Surface moist to dry, smooth. **GILLS:** Broadly attached to decurrent, close to well spaced, edges sawtoothed to ragged, often more noticeable in age. Whitish to pale pinkish buff. **STIPE:** 0.2–1.5 (2.5) cm long, 0.1–0.3 cm thick, lateral, or rarely off-centered, short and nearly indistinct, to elongated Whitish to pale pinkish buff, darkening to cinnamon in age. **FLESH:** Concolorous, thin, moderately tough, and often fibrous. **ODOR:** Mild to slightly farinaceous. **TASTE:** Slightly acrid-farinaceous. **ODOR:** Mild to slightly farinaceous. **KOH:** No reaction. **SPORE DEPOSIT:** White. **MICROSCOPY:** Spores 4.5–6.5 (7) x 4–5 μm, subglobose to broadly ovoid, minutely spiny, amyloid, colorless in KOH. Pleurocystidia and cheilocystidia spindle shaped, often quite long and slender. Cap cuticle with pileocystidia.

ECOLOGY: Scattered in small troops or clusters, or solitary on small branches and twigs. Seems to prefer Scotch Broom (*Cytisus scoparius*) and willow (*Salix* spp.), but can be found on a number of hardwoods and conifers. Locally common on North Coast, scattered elsewhere throughout our range. Fruiting in fall and early winter.

EDIBILITY: Unknown.

COMMENTS: The pleurotoid shape, serrate gills, and white spores help distinguish *Lentinellus*; the small size, pale colors, and growth on small branches are indicative of *L. flabelliformis*. *L. occidentalis* supposedly differs by having slightly shorter spores. Also see *L. ursinus* for similar, larger *Lentinellus*. *Neolentinus kauffmanii* has a central to slightly off-centered stipe and grows on spruce logs and stumps.

Lentinellus ursinus group

CAP: 2–10 cm across, fan shaped, semicircular, to kidney shaped, soon irregular lobed. Often brown to reddish brown over disc, paler light brown to cinnamon buff on margin, darkening in age; hygrophanous, fading overall when dry. Surface moist to dry, coarsely hairy to velvety over much of cap with a smooth margin. **GILLS:** Decurrent, close to crowded, edges sawtoothed and often ragged. Whitish to buff, developing cinnamon color in age. **STIPE:** Absent or indistinct indistinct and lateral. **FLESH:** Concolorous, thin, and moderately tough. **ODOR:** Mild, like black pepper, or slightly fragrant. **TASTE:** Acrid and often bitter. **KOH:** No reaction. **SPORE DEPOSIT:** White. **MICROSCOPY:** Spores 4–4.5 x 3–3.5 μm, subglobose to broadly ovoid, roughened with small warts, amyloid.

ECOLOGY: Clustered in overlapping shelves, scattered, or rarely solitary on decaying logs or dead trees. Often on hardwoods, also occurring on conifers. Common throughout our range, fruiting from fall into spring.

EDIBILITY: Inedible.

COMMENTS: The pleurotoid shape, velvety to coarsely hairy cap, strongly serrate gills, white spores, and growth on wood are distinctive. *L. castoreus* is generally slightly smaller, paler, and less acrid tasting, and has more crowded gills. (Note that both are European names and may be misapplied here; it is also likely that we have additional undescribed taxa.) *L. vulpinus* forms fused clusters from a common base, grows on hardwoods, and has smaller spores. *L. montanus* grows on conifer branches near melting snowbanks in the mountains and has a light brown to pinkish buff cap with a felty disc. *L. flabelliformis* is smaller and has a smooth cap. Other mushrooms with a similar shape and growth on wood lack the jagged gill edges.

Sarcomyxa serotina
(Pers.) P. Karst.
GREEN OYSTER

CAP: 3–12 (15) cm across, fan shaped, semicircular, to kidney shaped, margin inrolled, becoming wavy and often lobed in age. Olive-green to bluish gray or olivaceous gray, often developing yellow, ocher or brownish tones as it ages. Rarely ocher even when young. Surface viscid to moist, shiny if dry, more or less smooth, but often with a velvety patch near attachment point, occasionally finely hairy and with scales near disc when dry. **GILLS:** Radiating out from a broad, curb-like stipe, close to crowded, occasionally forking, partial gills numerous. Orange-buff to orange at first, fading; yellowish buff to beige in age. **STIPE:** Laterally attached to wood, indistinct from above, appearing as a short, curblike attachment on underside. **FLESH:** Moderately thick to thin near margin, firm, rubbery. Watery whitish. **ODOR:** Often indistinct, or slightly earthy. **TASTE:** Mild to bitter-metallic. **SPORE DEPOSIT:** Creamy yellowish. **MICROSCOPY:** Spores 3.5–6.5 x 1–2 μm, cylindrical to sausage shaped, smooth, amyloid. Cheilocystidia abundant, club to spindle shaped, pleurocystidia similar, sparse.

ECOLOGY: Scattered in overlapping or solitary shelves, occasionally in clusters, on decaying logs or dead trees, almost always on Red Alder (*Alnus rubra*) or Bigleaf Maple. Known from Santa Cruz Mountains northward, rare south of Humboldt County. Scattered fruitings occur in the fall, but often needs cold weather to fruit well.

EDIBILITY: Edible, but of poor quality.

COMMENTS: The pleurotoid-shape, olive cap, and crowded creamy or orangish gills with a very rudimentary stipe are distinctive. *Phyllotopsis nidulans* has a fuzzy cap, brighter orange gills and often a disagreeable odor; it is much more common in our area, growing on hardwoods. *Tapinella panuoides* has a dry brown cap, wavy, often cross-veined gills, and brown spores. *Hohenbuehelia petaloides* is smaller and thinner, and has a brown to gray cap and white to gray gills. Also known as Late Oyster or Fall Oyster.

SYNONYM: *Panellus serotinus* (Pers.) Kühner. It is more closely related to the Hygrophoraceae than to *Panellus* (*P. stipticus* and allies), which are related to *Mycena*.

Phyllotopsis nidulans
(Pers.) Singer
ORANGE MOCK OYSTER

CAP: 2–8 cm across, fan shaped, semicircular, to kidney shaped, margin inrolled, often into maturity, often wavy and irregular lobed. Orange, but covered with whitish hairs, and appearing paler, occasionally all orange, or orangish buff in age. Surface moist to dry, covered with finely pubescent hairs, which can become matted or occasionally disappear in age. **GILLS:** Radiating out from attachment point, close to crowded, partial gills numerous. Orange-buff to orange at first, fading slightly, to yellow-orange. **STIPE:** Absent. Laterally attached to wood with white velvety fuzz at attachment point. **FLESH:** Concolorous, thin, moderately tough and rubbery. **ODOR:** From mild and indistinct to strong, rancid and disagreeable; like rotting cabbage, swamp gas, or mouse urine. **TASTE:** Pungent, unpleasant, occasionally bitter. **SPORE DEPOSIT:** Pinkish orange to creamy peach. **MICROSCOPY:** Spores 5–7 x 2–3.5 μm, cylindrical to bead shaped, smooth, inamyloid.

ECOLOGY: Often clustered in overlapping shelves, scattered, or solitary on decaying logs or dead trees, generally Tanoak and live oak. Common throughout our range, fruiting throughout the wet season, more abundant in late winter or spring.

EDIBILITY: Inedible.

COMMENTS: The pleurotoid shape, velvety cap (at least when young), and orange colors are distinctive of this common "mock oyster." The odor is often downright repulsive (but can be nearly absent). *Crepidotus crocophyllus* also has a hairy cap and orangey gills (from light orangish beige to rather bright orange), but differs by its lack of a distinct odor, smaller fruitbodies, and brown spores. *Sarcomyxa serotina* has a mostly smooth cap with olive tones and yellowish to orange gills with a short, wide, wedgelike stipe. *Tapinella panuoides* has crimped, often cross-veined gills and brown spores. *Lentinellus* has jagged, sawtoothed gill edges and generally lacks the orange colors.

Schizophyllum commune

Fr.

SPLIT GILL

CAP: 1–4 (5) cm across, fan shaped, margin inrolled when young, becoming lobed and wavy and often ragged in age. Pinkish buff, light grayish brown, grayish to whitish. Surface coarsely hairy, dry. **GILLS:** Radiating out from attachment point, consisting of closely spaced, "split" gills that open up in wet weather. **STIPE:** Absent or rudimentary, laterally attached to wood. **FLESH:** Thin, tough, leathery. **ODOR:** Indistinct. **TASTE:** Mild. **SPORE DEPOSIT:** Whitish. **MICROSCOPY:** Spores 3–4 x 1–1.5 μm, cylindrical, smooth, colorless in KOH.

ECOLOGY: In overlapping, shelving clusters, in small groups, or solitary on dead hardwoods, especially live oak and California Bay Laurel. Common throughout our range. Fruiting in wet weather, but fruitbodies persist throughout year.

EDIBILITY: Edible, but small and rather tough. Not recommended.

COMMENTS: The small size, pale color, and fuzzy margin are similar to many pleurotoid wood rotters, but the blunt-edged, "split" gills are diagnostic. These grooved gills close during dry weather and open back up to produce spores when conditions are favorable.

Clitopilus hobsonii

(Berk.) P. D. Orton

CAP: 0.5–3.5 cm across, fan shaped, often with an inrolled margin, expanding in age. Very young emerging caps appearing completely round. Whitish to beige or pale tan near margin, usually distinctly white nearer to attachment to wood. Surface dry, silky, finely fuzzy, smoother toward margin in age, covered in fluffy or hairy white mycelial tomentum near attachment to substrate (tomentum often continues to grow in humid tackle boxes, and after storage can be noticeably moldy looking). **GILLS:** Radiating out from central attachment to wood, many short gills present. Often asymmetrical, with shorter gills on edge of cap bordering wood, forming a funnel shape. Whitish to dull, pale beige-tan, often with a pinkish cast in age. **STIPE:** Absent or extremely reduced. Attachment to wood often with radiating sheets of mycelium, sometimes also with poorly formed white rhizomorphic threads. **FLESH:** Very thin, soft, fragile. **ODOR:** Farinaceous. **TASTE:** Farinaceous. **KOH:** No reaction. **SPORE DEPOSIT:** Pinkish buff to dull pinkish brown (paler and less brown than in similar *Crepidotus*). **MICROSCOPY:** Spores 6–9 x 3.8–5 μm, elongated-ellipsoid to almond shaped, with longitudinal ridges (often difficult to see), appearing ribbed in face view.

ECOLOGY: In groups of solitary fruitbodies or small overlapping clusters, often on inner side of peeling bark or on well-decayed heartwood of oak and Tanoak. Sometimes on rather long stretches (1+ m) of substrate. Fairly common throughout our range, but easily overlooked. Fruiting in fall and spring,

EDIBILITY: Unknown.

COMMENTS: This species is difficult to separate from pale *Crepidotus* species in the field and also somewhat resembles a few white-spored mushroom genera (*Cheimonophyllum*, *Panellus*). The pallid pinkish brown spore deposit distinguishes this species from those taxa (although a light deposit from a *Crepidotus* could be interpreted as pinkish). The spores appear clear and ridged under the microscope (unlike the brown, smooth spores of *Crepidotus*). *Cheimonophyllum candidissimum* can be told apart by the lack of rhizomorphic or sheet-forming mycelium and white spores. *Claudopus byssisedus* has similar mycelial sheets and rhizomorphs, but has a darker grayish brown cap.

Claudopus byssisedus

(Pers.) Gillet

CAP: 1.5–4.5 cm across, inrolled when very young but soon fan shaped or plane, margin often quite wavy in age. Grayish brown to mouse gray or pallid grayish tan, often with a sheen of silvery fibrils, usually opaque, often covered with cobwebby white mycelium near attachment to substrate. **GILLS:** Radiating out from attachment point, widely spaced, fairly thick edged, partial gills present. Creamy or grayish white at first, becoming pinkish to salmon in age. **STIPE:** Absent or rudimentary, whitish to gray with a pale bloom, laterally attached to substrate. Base usually with prominent with sheets and cords of white mycelium. **FLESH:** Thin, fragile. **ODOR:** Indistinct to farinaceous. **TASTE:** Indistinct. **SPORE DEPOSIT:** Pinkish to salmon. **MICROSCOPY:** Spores 7.5–11 x 5–8.5 µm, elongate, multifaceted with 5 or 6 sides. Cheilocystidia rare to frequent, capitate.

ECOLOGY: Solitary or in small to large troops. Uncommon in much of our area, but can be locally abundant, especially on soil in hollows or under woody debris on the Far North Coast. We have also seen it on the surface of muddy walls deep inside the Empire Cave in Santa Cruz (pictured).

EDIBILITY: Unknown.

COMMENTS: The small, fan-shaped grayish caps, practically absent stipe, and pinkish gills help identify this species. William Murrill described *C. avlleaneus* from Oregon; it may be a more appropriate name for this species, but use of the European name persists. The sheeting, rhizomorphic mycelium is a great help in distinguishing it from *Clitopilus hobsonii*, *Crepidotus* spp., and *Cheimonophyllum candidissimum*. The last two have brownish and white spores, respectively. Other *Claudopus* reported from California include *C. dulcisasporus*, with a spermatic odor and sweet flavor; *C. graveolens*, with tiny caps, gray gills, and a skunky odor; and *C. parasiticus*, which grows on the gills of chanterelles. The latter species is extremely rare in our area.

Crepidotus epibryus

(Fr.) Quél.

SNOWY CREP

CAP: 0.5-3.5 cm across. Half-dome or shell-shaped to fan-shaped, margin often inrolled when young, often becoming wavy in age. Bright white when young, becoming duller white to pale beige, usually whiter nearer attachment to wood, hygrophanous to chalky white when dry. Surface dry, silky to finely fuzzy, smoother in age, covered in fluffy or hairy white mycelial tomentum near attachment to substrate. **GILLS:** Radiating out from central attachment to wood, many short gills present. Whitish-gray to dull, pale beige tan when young, to yellowish brown or dull brown as spores mature. **STIPE:** Absent. **FLESH:** Very thin, soft. **SPORE DEPOSIT:** Dull yellowish-brown to pale brown. **MICROSCOPY:** Spores 6-8 x 2.5-3.5 (4) µm; shaped like apple seeds; elongate-ellipsoid and tapered or pointed at one end, smooth. Cheilocystidia thin, long and curvy or slightly coiled. Clamps rare or absent.

ECOLOGY: In small groups on rotting wood. Usually twigs or small branches among moss; also on wood chips, or herbaceous stems. Common, occurring throughout our range, fruiting from late fall into early spring.

EDIBILITY: Unknown.

COMMENTS: This species can be separated from other Crepidotus by its bright white colors, relatively light spore deposit, apple seed-shaped spores, long, thin cheilocystidia, and lack of abundant clamp connections. There are many other small Crepidotus in our area, most of which require microscopic examination to differentiate. The common C. applanatus is larger, has a strongly hygrophanous, smooth, dull whitish-beige cap, and small, spiny, round spores. Clitopilus hobsonii is very similar, but has a pinkish spore deposit, a less fuzzy cap texture, and longitudinally-ridged spores under the microscope. Cheimonophyllum candidissimum has a smoother cap, creamy white gills, a tiny rudimentary stipe, and white spores.

Crepidotus crocophyllus

Berk.

SAFFRON CREP

CAP: 1–7 cm across, convex with an inrolled margin, soon fan shaped, expanding to wavy in age. Brown to orange-brown, fading to ocher, creamy or nearly white. Slightly velvety when young, evenly covered with fine, pointed tufts of hairs. **GILLS:** Radiating out from attachment point, partial gills present, sometimes forked. Creamy to ocher or pale orange. **STIPE:** Absent, laterally attached to substrate, but often with a large distinct pale to ocher-tan patch of mycelium near attachment. Base usually with prominent sheets and cords of white mycelium. **FLESH:** Thin, cream to ocher. **ODOR:** Indistinct. **TASTE:** Indistinct. **KOH:** Dull orange on cap, yellow, then bright orange, finally nearly cherry red on gills. **SPORE DEPOSIT:** Dull brownish. **MICROSCOPY:** Spores 5–8 x 5.5–7 µm, subglobose, finely punctate-spiny. Cheilocystidia highly irregular, with elongate and sometimes branched appendages.

ECOLOGY: Solitary or scattered in troops with overlapping caps on well-decayed wood, especially of Tanoak and live oak in early fall, with smaller numbers in winter and spring. Fairly common in the southern part of our range, rarer northward.

EDIBILITY: Unknown.

COMMENTS: The semicircular, orangey brown, densely velvety cap, orange-hued gills, and lack of a stipe distinguish this species from all others in our area. Faded older specimens could be confused with other species of *Crepidotus*, but can still be separated by their bright KOH reaction. *C. mollis* has a less densely velvety cap, with scattered brown fibrils and larger (7–10 x 4.5–6 µm), smooth, ellipsoid spores. *Phyllotopsis nidulans* has brighter orange colors and pale pinkish orange to creamy peach spores; it often smells foul.

Crepidotus mollis

(Fr.) Staude

HAIRY CREP

CAP: 2–6 cm across, fan shaped to semicircular, margin slightly inrolled when very young but soon expanding to plane or slightly scalloped or wavy at margin. Dull tan brown to beige-brown, strongly hygrophanous and becoming watery grayish buff to light beige as it dries out. Surface dry to moist to waterlogged, in age slightly viscid or flabby. Finely hairy when very young, becoming nearly smooth with scattered brownish fibrils in age. Sometimes obscurely striate. **GILLS:** Radiating out from attachment point, many partial gills present. Grayish white to dingy cream, browner in age. **STIPE:** Absent, but with distinct tuft of whitish mycelium at attachment to substrate. **FLESH:** Moderately thin, soft, often waterlogged, dingy whitish gray to yellowish tan. **ODOR:** Indistinct. **TASTE:** Indistinct. **KOH:** No reaction. **SPORE DEPOSIT:** Dull brownish. **MICROSCOPY:** Spores 7–10 x 4.5–6 µm, smooth, ellipsoid. Pleurocystidia absent (except in var. *cystidiosus*).

ECOLOGY: Solitary or in small groups on deadwood, especially hardwoods. Often on fairly intact but decorticated (barkless) Madrone logs in our area in early fall, but can fruit any time throughout the wet season.

EDIBILITY: Unknown.

COMMENTS: The semicircular, hygrophanous, dingy grayish beige to light brownish cap helps identify this species. *C. applanatus* is similar, but has a nearly smooth or weakly fibrillose whitish or dingy grayish white cap (which can become covered with brownish spores in age) and small (4–5.5 µm), globose spores. Although many identifiers use *C. mollis* and *C. applanatus* for two common mushrooms in our area, it is unlikely that these names are appropriate for species on the Pacific Coast. We encourage readers to consult Hesler and Smith's monograph on this genus, which is freely available at MykoWeb (see Resources p. 576).

Deconica horizontalis

(Bull.) Noordel.

CAP: 0.2–2.5 cm across, convex, plane to fan shaped, margin finely inrolled and often ribbed when young, sometimes wavy in age. Dark purplish brownish to reddish brown or ocher-brown, fading to beige or tan as it loses moisture. Surface moist to dry, finely velvety or silky. **GILLS:** Broadly attached, often with a decurrent tooth. Cinnamon brown when young, mottled with lilac-gray, and dark gray-brown to brown in age, edges fringed whitish. **STIPE:** 0.2–0.7 cm long, 0.1–0.2 cm thick, off-center (but not lateral), short, curved. Light to dark brown, but often covered with a white tomentose when young, becoming darker, to blackish brown in age. Surface dry, covered with a fine pale bloom to smooth. **FLESH:** Very thin, dark. **ODOR:** Indistinct. **TASTE:** Indistinct. **SPORE DEPOSIT:** Dark purplish gray. **MICROSCOPY:** Spores 6–7.5 x 4–5 μm, ellipsoid to almond shaped, with a germ pore, thick walled, smooth.

ECOLOGY: Solitary, scattered, or in troops on small branches and sticks, canes of blackberry (*Rubus* spp.) and Scotch Broom (*Cytisus scoparius*), rotting wood, almost any cellulose-rich debris (even discarded jeans!). Common, but often overlooked because of small size and inconspicuous nature. Fruiting in fall and spring.

EDIBILITY: Unknown, probably nontoxic.

COMMENTS: The small size, off-center stipe, growth on wood, and dark purplish brown spore print distinguish this species. *Crepidotus* and *Panellus* can look quite similar, but have dull brown and white spore deposits, respectively. *Phaeomarasmius* have finely scaly caps and stipes and have ocher-brown spores. Some *Simocybe* can look similar when stunted, but usually are more olive-brown and have a central stipe.

SYNONYM: *Melanotus horizontalis* (Bull.) P. D. Orton.

Tapinella panuoides

(Batsch) E.-J. Gilbert

CAP: 2–12 cm across, oblong to fan shaped, with an inrolled margin when young, expanding to plane or uplifted, and often lobed in age. Light brown, yellow-brown to olive-brown, often paler toward margin. Surface moist to dry, finely velvety to nearly smooth in age. **GILLS:** Radiating out from attachment point, often wrinkled and sometimes anastomosing near attachment, with many cross veins, narrow. Creamy buff, yellowish beige to tan, at times with an olive cast. **STIPE:** Absent or rudimentary, laterally attached to substrate. **FLESH:** Moderately thin, soft, pale whitish to pale yellow-brown. **ODOR:** Indistinct. **TASTE:** Mild to slightly bitter. **KOH:** Olive-yellow on all parts. **SPORE DEPOSIT:** Pale brown to yellowish brown. **MICROSCOPY:** Spores 4.5–6 x 3–4 μm, elliptical to ovoid, smooth, yellowish in KOH.

ECOLOGY: Solitary or in overlapping clusters on rotting coniferous logs, or in clumps or clusters on wood chips or woody debris. Generally uncommon, but large fruitings can occur in urban areas in wood chips and along skid roads following logging. Fruiting throughout wet season, more common in late winter and spring.

EDIBILITY: Unknown.

COMMENTS: The pleurotoid shape, brownish to ocher colors, wavy-wrinkled gills, brown spores, and growth on wood are distinctive. *Crepidotus crocophyllus* and *C. mollis* have fragile flesh, more regular gills, and darker brown spores. *Tapinella atrotomentosa* has a fuzzy, dark brown, often central stipe. *Hygrophoropsis* is usually larger, and has forking gills, central stipes, and white spores.

15 • Gilled Bolete Relatives

You might be surprised to find out these gilled mushrooms are related to boletes, but a quick look under the microscope will reveal that these mushrooms share the same elongated spore shape commonly found in members of the Boletales.

The *Gomphidius* have decurrent, somewhat thick-edged gills that start pale but become smoky gray or blackish as the spores mature, and often have bright yellow stipe bases. They are found near *Suillus* in various nonpine conifer forests. *Chroogomphus* are very similar in stature to *Gomphidius*, but have orange flesh and are mostly found in pine forests (one species is found with Western Hemlock). Both genera have partial veils, but those of *Gomphidius* tend to be slimy (sometimes drying out), while those of *Chroogomphus* are made of fine fibrils (cortina-like), but are often rather obscure and only noticed when very young, or when forming a felty ring zone. Both genera appear to be parasitic on the mycelia of various boletes (not mycorrhizal, as was once thought). Each species appears to have a preference for a bolete host, and this can be a helpful identification feature. The table below reflects what we have observed on the Redwood Coast.

PARASITE	HOST	HABITAT
Gomphidius subroseus	*Suillus lakei*	Douglas-fir
Gomphidius smithii	*Suillus lakei*	Douglas-fir
Gomphidius glutinosus	*Suillus ponderosus*	Douglas-fir
Gomphidius oregonensis	*Suillus caerulescens*	Douglas-fir
Chroogomphus pseudovinicolor	*Suillus pseudobrevipes*	Pine
Chroogomphus vinicolor subsp. californicus	*Suillus pungens*	Monterey Pine and others
Chroogomphus ochraceus	*Suillus fuscotomentosus*	Three-needle Pines
Chroogomphus tomentosus	*Aureoboletus mirabilis*	Western Hemlock

Hygrophoropsis have whitish to light orange or bright orange, wavy, forking gills and central stipes. They have velvety caps and usually grow on wood chips or well-rotted logs.

Phylloporus fruitbodies have velvety-plush olive to tan-brown cap surfaces and brilliant yellow gills that usually stain slowly blue. The one *Tapinella* included in this section has very distinctive fruitbodies, with pale gills and dark brown, densely fuzzy stipes. *Paxillus* have strongly inrolled margins when young and forking gills that scrape off easily and stain brown when bruised.

Tapinella atrotomentosa
(Batsch) Šutara
VELVET-FOOTED TAP

CAP: 5–15 (20) cm across, convex with an inrolled margin at first, becoming uplifted, wavy and often lobed in age. Light brown, brown to rusty brown. Surface dry, velvety to matted-tomentose, often cracking in age. **GILLS:** Decurrent, close, often forking, narrow and scraping off easily. Creamy buff to light yellowish brown. **STIPE:** 2–10 cm long, 1.5–3.5 cm thick, central to off-center, rarely lateral, stout. Creamy beige to light brown at first, soon dark brown. Surface dry, distinctly velvety to matted-tomentose. **FLESH:** Thick, firm, somewhat fibrous. Pale yellowish to light brown, occasionally with vinaceous stains. **ODOR:** Indistinct. **TASTE:** Mild to bitter. **KOH:** Olive-yellow on all parts. **SPORE DEPOSIT:** Pale brown to yellowish brown. **MICROSCOPY:** Spores 4.5–6 x 3–4.5 μm, broadly elliptical, smooth, yellowish.

ECOLOGY: Solitary, scattered, or in clusters on rotting conifer stumps, logs, or buried roots. Typically fruiting in summer and early fall, with scattered fruitings later in fall. Widespread, but not especially common. Part of that may be because it fruits when few people are looking for mushrooms.

EDIBILITY: Not recommended, often bitter tasting. Commonly used by fiber artist for dyes.

COMMENTS: The large size, velvety brown cap, distinctly fuzzy brown stipe, and growth on wood help identify this species. *Paxillus* species generally grow on the ground, stain brown, and lack the fuzzy stipes. *Panus conchatus* has violet colors when young, but fades to beige, and has tough, leathery flesh.

Paxillus cuprinus
P. Jargeat, H. Gryta, J. P. Chaumeton & Vizzini
COPPERY PAX

CAP: 3–10 cm across, convex with a dramatically inrolled margin when young, expanding to plane, often with a depressed center, margin remaining distinctly inrolled. Dull brown, olive-brown, or yellow brown often mottled, becoming coppery with ocher stains in age. Surface densely tomentose when very young, soon matted-felty and often breaking up in a ribbed pattern near margin; usually becoming smooth and shiny in age. **GILLS:** Broadly attached to slightly decurrent, forking multiple times, occasionally forming irregular "pores" near top of stipe. Whitish cream when young, becoming dingy beige or mustard colored, staining reddish brown when handled. **STIPE:** 3–7 cm long, 1.5–2.5 cm thick, evenly cylindrical, stout. Dingy cream to beige or with olive-yellow tones, often light pinkish near apex, soon stained brown. Smooth to fibrillose, covered with a pale bloom. **FLESH:** Thick, firm, solid, marbled whitish and tan. Quickly dark reddish brown on gills, cap, and flesh when bruised or cut. **ODOR:** Indistinct. **TASTE:** Indistinct. **SPORE DEPOSIT:** Dingy ocher to reddish brown. **MICROSCOPY:** Spores 7–9 x 5–6 μm, almond shaped, smooth.

ECOLOGY: In small groups, clumps, or large arcs and rings on ground and in lawns surrounding planted Silver Birch (*Betula pendula*). Fruiting from late summer through fall.

EDIBILITY: Toxic, but sometimes slow to manifest (some people eat it for years before experiencing negative effects; others react badly right away). Contains a compound that hypersensitizes the immune system, leading it to attack and destroy the body's red blood cells.

COMMENTS: The distinctly inrolled margin, squat stature, coppery colors, and pale, forked gills that stain dramatically reddish brown separate it from most other species. However, there are at least two other species in our area, both undescribed. One is a poorly known species that grows primarily with pine and introduced cedar (*Cedrus* spp.) trees; the other is a larger species with a pale, whitish beige to tan cap that grows primarily with live oak in the winter.

MISAPPLIED NAME: *Paxillus involutus* (Batsch) Fr.

Gomphidius smithii

Singer

SMITH'S SLIME SPIKE

CAP: 1.5–7 cm across, rounded with an inrolled margin when young, becoming domed to broadly convex, expanding to flat and then uplifted and often slightly wavy in age. Pinkish, pinkish buff, pinkish lilac with a paler margin at first, becoming more dingy, to ochraceous in age. Surface smooth, very viscid and gooey when wet, sticky-slimy when dry. **GILLS:** Decurrent, slightly thick edged, widely spaced. White or creamy when young, soon becoming light gray or dingy beige-gray, and then smoky gray in age. **STIPE:** 5–8 cm long, 0.5–2 cm thick, cylindrical or tapered downward. White or off-white, base of stipe lacking, or just faintly washed with yellow, soon developing dark stains from base up in age. Smooth to fibrous, but covered in an extensive layer of sticky slime. **PARTIAL VEIL:** A translucent layer of slime covering gills when young, soon rupturing into fibrils forming a ring around top of stipe that becomes blackish as it collects dark spores. **FLESH:** Thin, fibrous, firm to soft, thick whitish throughout, sometimes with a hint of yellow in stipe base. **ODOR:** Indistinct. **TASTE:** Indistinct. **SPORE DEPOSIT:** Sooty gray to black. **MICROSCOPY:** Spores 14–18.5 x 4.5–6 μm, elongate, spindle shaped, smooth.

ECOLOGY: Solitary or in scattered groups in duff of conifer forest under Douglas-fir with *Suillus lakei* mycelium nearby. Uncommon to locally common on North Coast, south to Santa Cruz Mountains, but rare south of Sonoma County. Fruiting in fall and early winter.

EDIBILITY: Nontoxic, but gooey, bland, and known to concentrate heavy metals.

COMMENTS: The pinkish to pinkish buff cap (occasionally with lilac tones) and stipe that either lacks a yellow base or only has a blush of yellow help distinguish it from the brighter rosy pink *G. subroseus* and the smoky purple *G. glutinosus*. Older specimens with a stained stipe could be mistaken for *G. oregonensis*, which is typically much larger, often grows in clusters, and has black splotches on the cap and a yellow stipe base; it is much more common in most areas.

Gomphidius glutinosus

(Schaeff.) Fr.

PURPLE SLIME SPIKE

CAP: 4–7 cm across, rounded with an inrolled margin when young, expanding to bell shaped or broadly convex, then uplifted and wavy in age. Purple-gray to smoky gray when very young, becoming purplish brown to lilac-brown, or occasionally more bluish gray, to grayish buff in age. Surface smooth, extremely gooey-viscid when wet, and with a thick, moist, sticky layer even when dry. **GILLS:** Decurrent, slightly thick edged, widely spaced, often clean and neat looking. Waxy white, soon light gray, slowly becoming smoky gray. **STIPE:** 4–9 cm long, 1–3 cm thick, cylindrical or tapered toward the base. White with a brilliant chrome yellow stipe base, yellow extending farther upward in age, then developing a few dingy ocher-brown splotches. Surface smooth to scaly-fibrous but extensively covered in a thick layer of slime except near apex. **PARTIAL VEIL:** A thin layer of slime concealing gills when young, usually clear or milky translucent, soon rupturing and forming a ring of fine cobwebby fibers covered in dark spores around top of stipe. **FLESH:** Fibrous, firm to soft, white throughout except for brilliant yellow near base of stipe. **ODOR:** Indistinct. **TASTE:** Indistinct. **SPORE DEPOSIT:** Dark gray to black. **MICROSCOPY:** Spores 15–20 x 4–6 μm, elongate, smooth.

ECOLOGY: Solitary or scattered in troops in duff and humus, fruiting from midfall into early winter under Douglas-fir where mycelium of *Suillus ponderosus* is present (*Gomphidius* are thought to be parasitic on *Suillus*). Rare or absent south of Mendocino County, common on the Far North Coast.

EDIBILITY: Not recommended. Although nontoxic, it is extremely gooey and known to concentrate heavy metals.

COMMENTS: The slimy purplish cap, decurrent, widely spaced, pale to smoky gills, and viscid white stipe with a brilliant yellow base distinguish this species from all others in our area. Separating paler specimens from *G. oregonensis* can be tough, but that species often develops blackish blotches in age and is more likely to grow in clusters. Pinkish lilac variants of *G. glutinosus* can resemble *G. smithii*, which lacks the yellow stipe base, or *G. subroseus*, which is more deeply and evenly pinkish to rose colored and is often less brilliantly yellow at the stipe base.

Gomphidius oregonensis

Peck

BLACKENING SLIME SPIKE

CAP: (2) 4–15 cm across, rounded with an inrolled margin when young, expanding to broadly convex with an strongly incurved margin, then becoming flat or uplifted and often slightly wavy in age. Color extremely variable, ranging from dingy beige-white to tan with olive streaks when young, then becoming pinkish beige to tan, reddish tan, ocher-brown, usually developing sooty olive or blackish blotches and streaks, sometimes almost entirely black in age. Surface smooth, gooey-slimy when wet, often drying out and becoming shiny and metallic looking. **GILLS:** Decurrent, somewhat thick edged and widely spaced. Whitish to cream when young, becoming dingy gray and often blackening entirely in age. **STIPE:** 6–15 cm long, 1–4 cm thick, cylindrical or variably tapered or slightly enlarged at base, often curved or sinuous. White to dingy grayish beige at first and bright yellow over much of stipe base; developing sooty olive to black splotches. Smooth to fibrillose, covered in thin viscid layer or a sticky sheath of slime. **PARTIAL VEIL:** Thin membrane of translucent slime covering gills when young, breaking and then forming a ring of fibrils embedding in slime around apex of stipe that is soon blackish as it becomes covered in spores. **FLESH:** Thick, firm to soft, fibrous, solid, whitish throughout except for extensive yellow area near base of stipe. **ODOR:** Indistinct. **TASTE:** Indistinct. **SPORE DEPOSIT:** Sooty gray to black. **MICROSCOPY:** Spores 10.5–14 x 4.5–8 μm, elongate, spindle shaped, smooth.

ECOLOGY: Often in clumps with fused bases, or occasionally solitary in small troops in duff under Douglas-fir where *Suillus caerulescens* mycelium is present (the two sometimes fruit out of the same spot in the ground at the same time). Typically fruiting in fall and early winter, with scattered fruitings into spring.

EDIBILITY: Nontoxic, but rather gooey and often downright unattractive, and also known to concentrate heavy metals.

COMMENTS: Also known as The Hideous Gomphidius, this rather variable species can be recognized by its medium to large fruitbodies, decurrent white gills that turn gray and then black, bright yellow stipe base, and tendency to develop extensive black splotches in age. The color of the cap can overlap with other species *Gomphidius*, but none of the others are as likely to grow in fused clusters or develop black stains in age. *Gomphidius* (and their close relatives in *Chroogomphus*) are thought to be parasitic on the mycelium of *Suillus* (except for *C. tomentosus*, which appears to be associated with *Aureoboletus mirabilis*). Each species appears to have a preferred bolete partner, and this specificity can provide a helpful clue for identification.

Gomphidius subroseus

Kauffman

ROSY SLIME SPIKE

CAP: 1.5–7 cm across, rounded with an inrolled margin when young, becoming domed to broadly convex, expanding to flat and then uplifted and often slightly wavy in age. Dark rosy pink, reddish pink, or pastel pink, often quite evenly colored overall, sometimes more extensively creamy or grayish pink, paler toward margin and fading in age. Surface smooth, very viscid and gooey when whet, sticky-slimy when dry. **GILLS:** Decurrent, slightly thick edged, widely spaced. White or creamy when young, soon becoming light gray or dingy beige-gray, and then smoky gray in age. **STIPE:** 5–8 cm long, 0.7–1.5 (2.5) cm thick, cylindrical or tapered downward. White or off-white, base of stipe washed with yellow, usually fairly muted but sometimes brighter, becoming more yellowish upward in age. Smooth to fibrous, but covered in an extensive layer of sticky slime. **PARTIAL VEIL:** A translucent layer of slime covering gills when young, soon rupturing into fibrils forming a ring around top of stipe that becomes blackish as it collects dark spores. **FLESH:** Fibrous, firm to soft, thick white throughout except for yellow in stipe base. **ODOR:** Indistinct. **TASTE:** Indistinct. **SPORE DEPOSIT:** Sooty gray to black. **MICROSCOPY:** Spores 15–20 x 4.5–7 μm, elongate, spindle shaped, smooth.

ECOLOGY: Solitary or in scattered groups in duff of conifer forest under Douglas-fir with *Suillus lakei* mycelium nearby. *Gomphidius* are thought to be host-specific parasites on *Suillus* (see comments under *G. oregonensis*, p. 413). Common throughout most of our range (not yet recorded south of Monterey Bay). Fruiting from early fall through winter, occasionally in spring as well.

EDIBILITY: Nontoxic, but gooey, bland, and known to concentrate heavy metals.

COMMENTS: This beautiful species can be identified by its rosy cap, pallid to grayish, decurrent gills, slimy-gooey cap and stipe, and yellow flesh near the base of the fruitbody. *G. smithii* is very similar, but has a dingier or paler pink to beige-pink cap and mostly lacks yellow colors on the stipe base; it is uncommon in the northern part of our range. *G. pseudoflavipes* was described from under pine and true fir in the mountains of California, but it is a very poorly known species and may occur on the coast as well. It has an ocher-brown cap and bright yellow lower stipe, but the enormous spores (18–35 x 6–8 μm) are the most distinctive feature.

Chroogomphus vinicolor subsp. californicus

(Singer) Singer

PINE SPIKE

CAP: 3–8 cm across, convex with an incurved margin when young, becoming broadly convex, sometimes with a broad, pointed umbo, often uplifted in age. Mahogany brown, reddish to reddish purple, sometimes becoming bright magenta-purple in wet weather. Surface moist to viscid in wet weather. Otherwise dry and often metallic. Smooth or very tightly appressed-fibrillose. **GILLS:** Subdecurrent to decurrent, widely spaced, slightly thick edged. Pallid orange when young, becoming dingy reddish, then mottled dark olive grayish or dull blackish. **STIPE:** 5–11 cm long, 0.7–2 cm thick, usually fairly straight or slightly curved, cylindrical but usually tapered at base. Dull orangey yellow to orange at first, upper stipe with zone of dark fibrils (turning dark olive-gray with spore drop), sometimes with reddish chevrons of fibrils, extreme base brighter orange or yellow, sometimes with a tuft of light orangey rhizomorphs. Bruising orangey brown on stipe where handled. Surface fibrillose to fairly smooth. **FLESH:** Fibrous, fairly firm or spongy soft, distinctly orange throughout. **ODOR:** Indistinct. **TASTE:** Indistinct. **KOH:** Brilliantly magenta-purple, strongest on flesh. **SPORE DEPOSIT:** Dingy olive-gray to olive-black. **MICROSCOPY:** Spores 18–22 x 6–7.5 μm, long-elliptical, spindle shaped. Cystidia abundant on faces of gills, thick walled, cylindrical with a round tip.

ECOLOGY: Solitary, in arcs or in large groups, under pine, usually near *Suillus pungens* mycelium, and often fruiting at same time. Especially abundant near planted Monterey Pines. Common in appropriate habitat throughout our area, fruiting from fall into winter.

EDIBILITY: Edible but rather soft textured and bland. The bright magenta is attractive when the mushrooms are cooked, but David Arora notes that the cooked mushrooms can turn one's urine red!

COMMENTS: This mushroom is recognizable by its dark mahogany cap, decurrent gills, orange flesh, and growth under Monterey Pine with *Suillus pungens* in the area. The round-capped fruitbodies with downward-tapered stipes (shaped like a railroad stake) are characteristic of the genus. Separating this species from other *Chroogomphus* can be subtle. *C. ochraceus* is usually paler capped (with grayish green and orange tones dominating) and grows with *S. fuscotomentosus*. *C. pseudovinicolor* produces much larger fruitbodies with more consistently red-chevroned stipe ornamentation and grows with *S. pseudobrevipes* under Ponderosa Pine. An undescribed species with a slender stature and very dark cap grows with *S. umbonatus* under Shore Pine on the Far North Coast.

Chroogomphus ochraceus

(Kauffman) O. K. Mill.

OLIVE PINE SPIKE

CAP: 2.5–7 cm across, hemispheric to broadly convex with an incurved margin, becoming nearly plane, often with an acute to rounded umbo at all ages. Dingy ocher-olive to grayish tan (often with a lilac cast), usually mottled with a combination of all these colors. Edge of young caps often with a band of ocher fibrillose partial veil, especially in dry weather. Surface smooth, dry to moist or viscid in wet weather. Appearing metallic in dry conditions. **GILLS:** Decurrent, fairly widely spaced, sometimes forked near attachment to stipe, edges smooth. Dull ocher when young, becoming dull grayish ocher and then dark olive-gray. **STIPE:** 5–11 cm long, 1–2 cm thick, cylindrical or tapered to base, straight or slightly curved, sometimes slightly irregular or rooting into duff. Yellowish ocher to orange, sometimes marbled or streaked looking. Surface dry, upper portion sometimes with a slight ring zone of darker fibrils (soon disappearing). **FLESH:** Fibrous, distinctly ocher. **ODOR:** Indistinct. **TASTE:** Mild. **KOH:** Deep magenta. **SPORE DEPOSIT:** Dark olive-gray. **MICROSCOPY:** Spores 15–20 x 4–7 μm, long-ellipsoid, smooth. Cystidia long, abundant, thin walled.

ECOLOGY: Solitary or scattered in large groups near mycelium (and often fruitbodies) of *Suillus fuscotomentosus* under Monterey Pine and other three-needle pines. Fruiting from fall into winter. Common in appropriate habitat.

EDIBILITY: Edible.

COMMENTS: *C. ochraceus* is similar to other *Chroogomphus*, but has a tendency to show grayish olive and pale ocher cap colors and has a more slender stature and often more acutely pointed umbo. The association with *S. fuscotomentosus* is helpful as well, although the fruitbodies are not always present at the same time. The coloration of this species is duller than *C. vinicolor* and *C. pseudovinicolor*, usually lacking the strong red or mahogany tones of those taxa. The more northerly distributed *C. tomentosus* is entirely dull orange yellowish and has a more floccose-scaly cap surface, and is associated with *Aureoboletus mirabilis* under Western Hemlock.

Chroogomphus pseudovinicolor
O. K. Miller
GIANT PINE SPIKE

CAP: 5–15 cm across, buttons hemispherical, round-convex with an incurved margin, then broadly convex to plane, often uplifted and wavy in age. Mottled dingy reddish orange to tan or brownish red, sometimes with wine red or purplish blotches, tan or paler to ocher-orange toward margin. Surface usually dry, innately fibrillose, moist to nearly viscid in wet weather. **GILLS:** Broadly attached to decurrent, well spaced, often forked and branched near attachment to stipe, edges smooth. Dingy pallid orange when young, deeper orange to tan-brown or liver brownish in age, then dark olive-gray when spores are mature. **STIPE:** 8–20 cm long, 2.5–6 (10) cm thick, cylindrical, robust and somewhat tapered toward base (sometimes strongly so when young and then nearly triangular in cross section). Pale yellowish orange covered in fibrillose red or wine red chevrons overall, upper portion often with a ring zone with wispy remnants of partial veil (these becoming dark olive-gray when coated with spores). Surface dry, fibrillose. **FLESH:** Fibrillose and often rather firm, yellowish orange. **ODOR:** Indistinct. **TASTE:** Mild. **KOH:** Deep magenta-purple. **SPORE DEPOSIT:** Deep olive-gray to smoky black. **MICROSCOPY:** Spores 17–20 x 5–7 μm, long-ellipsoid. Cystidia large (70+ μm), cylindrical or shaped like bowling pins, sometimes with crystal incrustations over narrower upper portion.

ECOLOGY: Solitary or in small groups with *Suillus pseudobrevipes* under Ponderosa, Bishop, and Shore Pines. Fruiting from fall through early winter. Quite uncommon in our area, but locally abundant in a few spots in Santa Cruz Mountains and a handful of other places on the Redwood Coast.

EDIBILITY: Edible but bland. See comments for *C. vinicolor* (p. 415).

COMMENTS: The dark, reddish orange cap, large size, decurrent, widely spaced gills, and orange stipe extensively covered with reddish chevrons help distinguish this species. It produces the largest fruitbodies of any in its genus. *C. vinicolor* is similar, but is generally smaller, grows with *Suillus pungens*, and usually has a bare orange stipe.

Chroogomphus tomentosus
(Murrill) O. K. Mill.
WOOLLY PINE SPIKE

CAP: 2–10 cm across, peg-like when young with a rounded cap, becoming convex to plane, often with a broad umbo. Margin inrolled at first, often ruffled in age. Beige-orange to buff-orange, slightly darker toward center, sometimes developing vinaceous or purple stains in age. Surface moist to dry, appressed-tomentose to matted fibrous or finely scaly in dry conditions. **GILLS:** Subdecurrent to decurrent, thick, widely spaced. Buff-orange, orangish tan to brownish in age. **STIPE:** 5–15 cm long, 0.5–1.5 cm thick, more or less equal, but tapers to a point at base. Beige-orange to buff-orange. Surface moist to dry, finely tomentose to appressed-fibrillose. **PARTIAL VEIL:** Silky fibrillose when young, disappearing in age. **FLESH:** Thin, fibrous, pale beige-orange. **ODOR:** Indistinct, although may get an odor of rubbing alcohol in age. **TASTE:** Mild. **KOH:** Purple on all parts. **SPORE DEPOSIT:** Olive-black. **MICROSCOPY:** Spores 15–25 x 6–9 μm, long-ellipsoid, smooth.

ECOLOGY: Scattered or in small clusters on ground, often next to rotting logs. It appears to be associated with *Boletus mirabilis* and restricted to range of Western Hemlock. Very common on the Far North Coast, known from as far south as Sonoma County. Fruiting from first rains in fall into early winter.

EDIBILITY: Edible, but it cooks up to a slimy purple mess.

COMMENTS: The beige-orange colors, widely spaced decurrent gills, fibrillose veil when young, and black spores are important features to help identify this mushroom. Although *Cantharellus formosus* can have the same color, but the lack of a veil and ridgelike or veiny gills help differentiate it. Other *Chroogomphus* species are darker colored and grow under pines with *Suillus*.

Hygrophoropsis aurantiaca group
FALSE CHANTERELLE

CAP: 2–10 cm across, convex with an inrolled margin when young, becoming plane with a depressed center, to broadly funnel shaped in age. Typically bright to pale orange to creamy orange (*H. aurantiaca*–like) or orange-brown to brown with a paler margin, at times with an olivaceous tone when young (*H. rufa*–like). Occasionally creamy beige, with or without a darker center. Surface dry, velvety when young, more matted-tomentose to smooth in age. **GILLS:** Decurrent, crowded, forking repeatedly, often quite wavy and ruffled. Bright to dull orange, creamy orange, or buff to beige in some variants. **STIPE:** 2–7 (10) cm long, 0.3–1 (2) cm thick, more or less equal, occasionally tapered toward an often slightly rooting base. Orange, creamy orange, brown to beige; typically colored similar to cap, but usually with more orange tones. Surface dry, nearly smooth, base with whitish fuzzy mycelium and white rhizomorphs. **FLESH:** Thin to moderately thick, soft, stipe more fibrous. Whitish to pale orange. **ODOR:** Indistinct. **TASTE:** Indistinct. **SPORE DEPOSIT:** White. **MICROSCOPY:** Spores 5–8 x 3–5.5 μm, ellipsoid, smooth, dextrinoid.

ECOLOGY: Solitary or scattered on well-decayed logs, stumps, or humus in forested settings, or more often scattered in troops or clustered on wood chips, especially in urban areas. Very common and widespread, fruiting throughout the mushroom season.

EDIBILITY: Best avoided, as it causes adverse reactions in some people.

COMMENTS: The most telltale features of this group are the decurrent, often wrinkled gills that fork repeatedly and the growth on woody debris or thick humus. Based on preliminary data from Nhu Nguyen, it appears that we have at least five species in this complex in California, which helps explain the variability encountered with our False Chanterelles. The name *H. rufa* might be better suited for the species with orange-brown to brown caps often encountered in wood chips; other names or varieties have been applied, as well, such as *H. aurantiaca* var. *pallida* for the whitish gilled variant. Until this complex is sorted out, it is probably best to refer to it as the *H. aurantiaca* group. They only superficially resemble our real chanterelles, which have thicker, white stringy flesh and blunt, gill-like ridges.

Phylloporus arenicola

A. H. Sm. & Trappe

WESTERN GILLED BOLETE

CAP: 1–3.5 cm across, convex to nearly plane, slightly uplifted and wavy in age. Dark olive-green to lighter forest green with tan areas, older specimens more reddish brown, or tan-brown in drier weather. Surface dry, finely tomentose and suedelike, becoming appressed and smooth in age. **GILLS:** Variably attached, notched to slightly decurrent, more rarely nearly free. Thick, widely spaced, often forked near attachment to stipe, interveined or anastomose to sometimes truly poroid in areas. Bright yellow, duller in age, occasionally staining blue when damaged. **STIPE:** 3–9 cm long, 1–2.5 cm thick, cylindrical with a tapered base, often curved overall, base often with a tuft of fine yellow rhizomorphs. Bright to dull yellow, often streaked with tan or brown, and brownish punctate at apex. Surface dry, often with fine dots and streaks of brownish cells, to smooth. **FLESH:** Thin, firm to soft, yellow, occasionally bluing. **ODOR:** Indistinct. **TASTE:** Mild. **SPORE DEPOSIT:** Dull olive-brown. **MICROSCOPY:** Spores 9–12 x 4–5 μm, elongate-ellipsoid to somewhat spindle shaped, smooth.

ECOLOGY: Solitary or scattered in small groups of 3 or 4 fruitbodies, rarely more prolific. Occurring throughout our range in mixed woods, perhaps favoring habitats with Douglas-fir. Fruiting from fall into spring.

EDIBILITY: Edible.

COMMENTS: The greenish to olive-brown caps and strongly contrasting bright yellow gills that often bruise slowly blue distinguish this species. The gill morphology of the fertile surface obscures the fact that this species is a member of the bolete family; however, the colors, velvety cap texture, and yellowish rhizomorphs are reminiscent of *Xerocomus subtomentosus*, a close relative. The name *Phylloporus rhodoxanthus* has been applied to blue-staining specimens; it is our belief that we have a single, somewhat variable species in California.

MISAPPLIED NAME: *Phylloporus rhodoxanthus* (Schwein.) Bres.

16 • Boletes

The term bolete is used generally for those mushrooms with a stipe, a cap, a tube layer with pores on the underside of the cap, and soft or fleshy texture. Until fairly recently, only a few genera were included in the family Boletaceae, and most of the familiar boletes were in the genus *Boletus*. However, the family has undergone dramatic revision in recent years, and many species are now placed in smaller, separate genera (and more on the way). There are a few species with partially hypogeous and often semisequestrate or contorted growth forms (that is, *Gastroboletus*), but for the typical boletoid species, it's useful to learn the following major subgroups.

Boletus—"Core" boletes, including the King Bolete and its equally delicious edible relatives (this group is commonly called the Porcini or the Royalty). These species produce robust fruitbodies with small, densely packed pores that are white or light creamy yellowish when young, and have distinctly reticulate, club-shaped stipes.

B. smithii, B. coccyginus, **and** *B. mirabilis*—These are included here in "*Boletus*" (with quotation marks) to indicate that we know they aren't closely related to the Porcini (nor are they closely related to one another), but have yet to be reassigned to other genera.

Butyriboletus—"Butter Boletes" are similar to those in the Porcini group in that they have large, dense fruitbodies and reticulate stipes, but differ in having yellow flesh and yellow pores that bruise blue.

Caloboletus—"Bitter Boletes" are sometimes confused with Butter Boletes due to their large-statured fruitbodies and blue-bruising reactions (especially the flesh). They all have a sour-bitter flavor (nibble a small piece of the cap, then spit).

Aureoboletus—A small group of boletes with bright, citrine yellow pores, and often with tapered stipe bases; their caps and/or flesh taste lemony, which helps separate them from other bright yellow-pored boletes. *A. mirabilis* is the largest; has a cap that is covered with short, stiff hairs; and has a tall, club-shaped stipe.

Xerocomellus—Boletes with medium brown to dark gray to pinkish red or dark purplish black, often cracking caps, yellow pores that often stain blue, and red to reddish yellow stipes.

Xerocomus—Two species of boletes in our area with dry, velvety, brownish caps and rather coarse yellowish pores that sometimes slowly stain blue.

Although similar to *Xerocomellus*, the *Xerocomus* never have extensively red-colored stipes.

Chalciporus—Small to tiny boletes with coppery or pinkish tan pores, yellow mycelium at the base of the stipe, and a peppery taste. Usually found near *Amanita muscaria*.

Porphyrellus—Medium to large boletes with rich brown coloration overall, including dark chocolate brown pores.

Pulveroboletus and *Buchwaldoboletus*—Uncommon boletes with medium to large, brilliant yellow caps. The pores are bright yellow to reddish and stain blue. The former genus is mycorrhizal; the latter genus contains wood-decaying saprobic species (unusual among boletes in our area).

Rubroboletus and *Suillellus*—Boletes with bright red to dark red pore mouths that stain blue when bruised. *Rubroboletus* are the only dangerously toxic boletes in our area.

Tylopilus and *Gyroporus*—Mushrooms in these genera produce differently colored spore prints from all others in our area: pinkish tan in *Tylopilus*, straw yellow in *Gyroporus*. *Gyroporus* are the only boletes in our area with hollow, often fragile stipes.

Leccinum—Complicated genus, but relatively few occur in our area. The stipes are covered in upright, brownish to dark grayish or blackish tufts. The caps are usually orangey to reddish, and the pores range from grayish to pale.

Suillus—Very distinct group, commonly known as the jacks. They tend to be quite stout, and most have either a partial veil or sticky "glandular dots" on the stipe. The pore mouths tend to be coarse and irregular and vaguely radially arranged. All are mycorrhizal with trees in the pine family and are particularly common with Douglas-fir and pine in our area. Many are parasitized by *Chroogomphus* or *Gomphidius*.

Boletus edulis var. grandedulis

D. Arora & Simonini

CALIFORNIA KING BOLETE

CAP: (6) 10–20 (50!) cm across, hemispheric when young, expanding to domed or bun shaped, flattening out in age, sometimes uplifted, margin can be lobed or wavy at any time, extreme margin with a narrow flap of sterile tissue. Color variable within a well-defined range: creamy beige when young and buried in duff, soon becoming ocher-tan to warm brown, leather brown or rich reddish brown. Often mottled with some combination of these colors, usually darkest at center and paler toward margin. Surface viscid or more often greasy to dry, smooth or wrinkled. **TUBES:** Narrowly to broadly sunken around stipe, moderately thin at first, becoming quite long. **PORES:** Round, very small, whitish at first, soon creamy to straw yellow, becoming dingy brownish or olive-brown and then distinctly reddish brown in age. **STIPE:** 6–20 (25+) cm long, 3–6.5 cm thick, usually tall and stout with a club-shaped base, but can be quite curved (especially in thick duff). Whitish to cream, developing tan tones in age, upper part often dull warm brown. Surface covered in a fine, but well-developed pale reticulum, especially pronounced at apex. **FLESH:** Thick, firm, fleshy in cap, fibrous in stipe. White to cream, often marbled with light tan in stipe, slightly reddish under cap skin. Bruising slowly pale vinaceous pink. Occasionally light blue on cap flesh if pores are scraped away. **ODOR:** Indistinct to earthy, pleasant, can become unpleasant in age. **TASTE:** Mild to sweet (like carrot) or nutty. **SPORE DEPOSIT:** Olive-brown. **MICROSCOPY:** Spores 12–15.5 (17) x 4–6 μm, spindle shaped to long-ellipsoid, smooth.

ECOLOGY: Solitary, in pairs, or often scattered in arcs and troops. Abundant fruitings are produced along habitat edges, especially near trails, roads, or meadows. Often buried at first in thick duff, forming "mushumps" before breaking duff. Occurs in various coniferous forests, especially under Monterey and Bishop Pines, also Shore Pine on the Far North Coast, into the Pacific Northwest. Also occasionally associates with live oak and Madrone, more rarely with Valley Oak. Fruitings most abundant soon after first soaking rains in fall, with sporadic flushes into early winter and in spring. In some years, a significant second fruiting in midspring.

EDIBILITY: Edible and wonderful. This is one of the most sought-after mushrooms in California due to its rich flavor, large size, and firm texture. Thin slices of young fruitbodies are often eaten raw or marinated. Tubes of older specimens can be soft and mushy but are full of flavor; many people dry them separately for use as a rub or as a base for soup stock or gravy.

COMMENTS: The impressive size and stately posture of this species help distinguish them from many others. The pale, unstaining pores that develop reddish brown tones in age, warm ocher-tan to reddish brown cap, and reticulate stem help identify it from its close relatives. *B. barrowsii* has a dry, whitish cap and grows with oaks (at least in our area). *B. regineus* usually has a much darker cap with a prominent whitish bloom, yellow to yellowish green pores that don't turn reddish brown in age, and often a shorter stipe, and prefers hardwoods.

Boletus edulis var. *edulis*

Bull.

KING BOLETE

CAP: 5-20 (35) cm across, rounded to bun shaped at first, expanding to broadly convex, plane, or wavy in age. Color typically creamy beige or tan to light brown, more rarely dark brown (lacking the warm ocher to red-brown tones of var. *grandedulis*). Occasionally pale; white to whitish beige. **TUBES:** Sunken around stipe, fairly thin at first, becoming quite long in age. **PORES:** Stuffed with whitish tissue at first, opening as they mature, round and rather small. White when young, becoming creamy to yellow and then yellowish olive or discolored brownish in age (but not developing reddish tones). **STIPE:** (6) 8–17 (20+) cm long, 2–5 cm thick at apex, club shaped at first, but often becoming quite slender and elongate. Whitish to cream, developing tan tones in age. Surface covered in a fine to coarse reticulum, especially pronounced at the apex. **ODOR:** Indistinct. **TASTE:** Mild to sweet. **SPORE DEPOSIT:** Olive-brown. **MICROSCOPY:** Spores 15–17 x 6–7 μm, broadly ellipsoid with tapered ends, or spindle-shaped to long-ellipsoid, smooth.

ECOLOGY: Solitary or scattered in troops under Sitka Spruce (and possibly other northern conifers), almost always with *Clitopilus prunulus* in immediate vicinity. Like *B. edulis* var. *grandedulis*, it has a preference for forest edges, or disturbed areas (meadow edges, road or trail sides, or forest openings). Very common on the Far North Coast (continuing north into the Pacific Northwest), but known from as far south as Mendocino County. Fruiting in a large flush in early to midfall, with scattered fruitings into late fall or early winter.

EDIBILITY: Edible and delicious. Highly prized for their rich, nutty flavor, meaty texture and frequent abundance.

***

COMMENTS: This is the common Porcini of the Far North Coast; it is associated with fewer host trees than *B. edulis* var. *grandedulis*, and thus more northern in distribution (corresponding with the range of Sitka Spruce. On the northern coastal dunes, both varieties will grow alongside each other—*grandedulis* with Shore Pine, and *edulis* with spruce. Although *grandedulis* associates with a wider range of trees, (both hardwoods and conifers), it seems to be absent from spruce forests. The caps of *grandedulis* are more richly colored at maturity (warmer reddish, ocher, to orange-brown), and the pores become reddish to bronze-red in age.

Boletus fibrillosus

Thiers

FIB KING

CAP: 5–15 (20) cm across, rounded to domed when young, becoming broadly convex. Often dark brown when young, becoming brown to honey brown, with a darker center in age. Yellow to pale golden brown cap possible, but rare on the coast, more often occurring in the mountains. Surface dry, finely velvety at first, becoming matted-tomentose, and often breaking up into appressed-fibrillose tufts spreading out from disc, occasionally becoming cracked when dry. **TUBES:** Broadly attached to sunken around stipe, 0.5–2.5 cm long. **PORES:** Stuffed when young, small, round to slightly irregular. Creamy to pale yellowish at first, becoming yellow to yellowish olive in age. Generally not staining, but can discolor slightly greenish in age. **STIPE:** 7–15 (20+) cm long, 1.5–3 cm thick at apex, equal, or club shaped, becoming more enlarged or bulbous lower, but often with a rooting or pointed base. Often some shade of brown, with a paler band at apex, and a whitish base. Covered with coarse to fine reticulation; off-white at first, aging brown. **FLESH:** Thick, firm when young, becoming soft, stipe sometimes becoming hollow in age. White to beige in cap, white in stipe, not staining. **ODOR:** Indistinct. **TASTE:** Mild to sweet. **KOH:** Darkening, reddish brown on cap, no reaction elsewhere. **SPORE DEPOSIT:** Dark olive-brown. **MICROSCOPY:** Spores 13–16 x 5.5–6.5 μm, elongate to subelliptical, smooth.

ECOLOGY: Solitary or scattered on ground, in duff or moss under conifers, especially Grand Fir and Douglas-fir. Locally common north of San Francisco. Fruiting in fall, but continuing well into winter or spring in sporadic flushes.

EDIBILITY: Edible and very good, with a sweet, nutty flavor. Older specimens can get quite soft fleshed and are better suited to being dried and used as a powdered seasoning.

***

COMMENTS: This is one of the more distinctive members of the true *Boletus* (Porcini). The dark brown to honey brown matted-tomentose to fibrillose cap, a feature that is easier to see when the mushroom is dry, is unique on the West Coast. It also has creamy to pale yellowish olive pores and a brown stipe that is finely to coarsely reticulated with a paler, often rooting base. *B. regineus* also has brown cap, but has a powdery bloom when young, becoming smooth in age. *B. edulis* tends to be paler in color and has a smooth, thinly viscid cap.

Boletus regineus

D. Arora & Simonini

QUEEN BOLETE

CAP: 6–20 cm across, rounded to bun shaped at first, soon broadly convex to slightly flattening in age. Color variable, affected by weather, age, and degree to which pale bloom remains. Young caps covered in a very fine, powdery white bloom and often appear extensively or entirely white. Bloom remains in patches or as irregular pale mottling even on many dark older caps, making them appear paler. Without bloom, they are dark brown to mahogany brown, reddish brown, generally paler ocher-brown to tan-brown in age. Slightly viscid to dry, smooth to wrinkled and bumpy. **TUBES:** Separated from stipe by a narrow "gutter," moderately long. **PORES:** Small, round, stuffed with pith at first, soon opening. White when young, soon cool pale yellowish, then dingy yellow-olive, or slightly browner in age. Bruising slowly light brown on pores when scratched. **STIPE:** 6–15 cm long, 3–9 cm thick, cylindrical with a distinctly enlarged club-shaped base, sometimes more bulbous and squat. Whitish to creamy to beige-tan, especially in age, often stained olive-brown with spore deposit. Surface dry, with a whitish reticulum at least at apex and often over entire stipe. **FLESH:** Thick, firm, fleshy. Whitish throughout except for red-brown under cap skin. **ODOR:** Indistinct, pleasant, or strong and earthy in age. **TASTE:** Mild to slightly sweet (like cooked carrots). **SPORE DEPOSIT:** Olive-brown. **MICROSCOPY:** Spores 11.5–17 x 4–5.5 μm, long-elongate to somewhat spindle shaped, smooth.

ECOLOGY: Solitary or in small groups under hardwoods, especially with older live oaks, and in mixed Tanoak-Madrone forests. Also under pines in the Coast Ranges and Sierra Foothills. Uncommon overall, but can be locally abundant in early fall through midwinter in some years.

EDIBILITY: Edible and excellent. Opinions vary widely as to whether *B. edulis* and *B. barrowsii* are better.

COMMENTS: Young fruitbodies with a heavy white bloom might be confused with *B. barrowsii*, but will show a dark cap when the bloom is rubbed off. Most fruitbodies of *B. regineus* will show a darker cap than *B. edulis*, but older fruitbodies with pale caps and no trace of pale bloom can be hard to distinguish from those of *B. edulis*. The stipe of *B. regineus* is often fairly short compared to the width of the cap; combined with the bulbous base, this tends to produce a "squat" look seen in many Queen Boletes. Contrastingly, many King Boletes look taller and more statuesque with a "pin-headed" appearance when young. Ecology can also be helpful; on the coast, *B. regineus* is a hardwood associate and is much less likely to occur in mostly pure coastal stands of Bishop Pine or Monterey Pine. *B. fibrillosus* has a dark honey brown cap with a matted tomentose surface, and a brownish stipe that is often slightly rooting. Other boletes can be distinguished by their blue staining reactions, lack of reticulum on the stipe, or differently colored caps or stipes.

Boletus barrowsii

Thiers & A. H. Sm.

WHITE KING BOLETE

CAP: 5–15 cm across, bun shaped to convex at first, becoming broadly convex or unevenly flat and sometimes wavy in age. Dusty grayish white at first, then pale beige-white to beige-tan or pale tan-brown. Surface dry, smooth to slightly velvety with a bloom of whitish powder when young. **TUBES:** Rather thick, 1–2.5 cm long, separating easily from cap. **PORES:** Small and tightly packed, stuffed at first. White when young, becoming pale yellowish, yellowish tan or orangish tan, to dark brown to olive brown in age. Bruising reactions generally absent, very occasionally staining weak blue on pores or near junction of pores and cap. **STIPE:** 7.5–15 (20) cm long, 2.5–5 (9) cm thick, swollen to club shaped or bulbous toward base. Pale whitish beige or grayish white at first, becoming pale whitish tan, often developing peach tan or orangish tan tones. Surface reticulated with a raised fishnet pattern; usually more pronounced toward apex, but sometimes covering entire stipe. **FLESH:** Thick, firm when young, soft in age. Whitish; not staining. **ODOR:** Indistinct or pleasant. **TASTE:** Mild to sweet, or like carrots. **KOH:** No reaction. **SPORE DEPOSIT:** Olive-brown. **MICROSCOPY:** Spores 13–15 x 4–5 µm, long-elliptical to somewhat spindle shaped, smooth.

ECOLOGY: Often in clusters, small groups of individuals, or troops. Especially fond of warmer oak groves in the southern part of our range, but also occurring with Madrone and pine.

Absent from the North Coast. Fruiting in early to midfall, disappearing in winter, in some years fruiting during warm, moist spells in spring.

EDIBILITY: Edible and excellent! Very firm, with a lovely, sweet odor and nutty taste. As for most boletes, drying concentrates flavor and emphasizes rich earthy tones (at the expense of the delicate, sweet odor and flavors present when fresh).

COMMENTS: This bolete is unique in our area due to the pale coloration and lack of bruising reactions. Additionally, the mild taste and strong reticulation on the stipe help resolve any uncertainty in identification. *Caloboletus marshii* has a pale cap with a whitish bloom and fruits early in the season with oak, but has a yellow stipe without any fishnet pattern, turns blue rapidly when bruised, and tastes very bitter. *B. regineus* has a whitish bloom on the cap when young, but soon shows a dark mahogany brown to brownish orange cap, and its tubes show more greenish olive tones in age. Pale-capped *B. edulis* also look similar, but they generally have a taller stature and grow with spruce on the Far North Coast. *B. edulis* var. *grandedulis* has a more reddish brown to orange-brown cap that is viscid when wet, and a pore surface that becomes orange-brown to reddish brown in age.

Caloboletus conifericola

Vizzini

DARK BITTER BOLETE

CAP: 7–25 (30) cm across, rounded when young, becoming broadly convex to bun shaped in age. Margin inrolled at first, then downcurved well into maturely, to rounded or slightly uplifted in age. Dark olivaceous gray to dark gray at first, fading slightly as it ages, grayish brown dingy brown. Surface dry, finely velvety-tomentose, giving it a suede-leather feel, often becoming areolate cracked, with minute tomentose tufts in age, or with deep cracks when dry. **TUBES:** Broadly attached or with a slight notch, 0.3–3 cm long. **PORES:** Very small, angular to round. Dull "cool" yellow when young, soon yellow to pale olive-yellow, to dingy yellow-olive in age. Quickly staining blue when bruised, eventually bluing areas fading to dingy orangish brown. **STIPE:** 5–15 cm long, 2–5 cm thick at apex, often club shaped, with a thicker base when young, to more or less equal in age. Yellow overall at first, developing dingy grayish colors from base up as it matures, and sometimes a reddish base in age, staining blue when handled. Surface dry, covered with fine reticulation, at least at apex. **FLESH:** Very thick and firm when young, soft in age. Yellow to whitish, quickly staining blue when cut, slowly fading to grayish. Stipe often with reddish in base and around larva tunnels, at least when old. **ODOR:** Indistinct. **TASTE:** Extremely bitter. **KOH:** Brown on cap, orange on flesh, tubes, and upper stipe, brown on lower stipe. **SPORE DEPOSIT:** Olive-brown. **MICROSCOPY:** Spores (12) 14–17 x 4–6 μm, subcylindric to subellipsoid, smooth, ochraceous in KOH.

ECOLOGY: Solitary or scattered on ground, in moss or duff under conifers, especially Grand Fir and Western Hemlock. Uncommon, restricted to the Far North Coast, also common in the mountains. Often fruiting early in fall, soon after first rains.

EDIBILITY: Unknown, extremely bitter tasting.

COMMENTS: The dark olivaceous-gray cap when young, which becomes areolate cracked in age; yellow pores and flesh that blue quickly when damaged; yellow, finely reticulated stipe; and extremely bitter taste are the key features needed to identify this large, stately bolete. *C. rubripes* has a nonreticulate stipe with a reddish base and paler colors when young. Based on preliminary genetic analysis, it is possible that *C. frustosus* (better known as *Boletus calopus* var. *frustosus*) is the same species; since it is an older name, it would take precedence over *C. conifericola*.

SYNONYM: *Boletus coniferarum* E. A. Dick & Snell.

Caloboletus rubripes

(Thiers) Vizzini

RED-FOOTED BITTER BOLETE

CAP: 5–18 (28) cm across, rounded when young, becoming broadly convex to bun shaped in age, margin inrolled at first, then downcurved well into maturely, to rounded or slightly uplifted in age. Margin often with a narrow band of sterile tissue, which becomes rosy pink in age. Buff, or olivaceous buff when young, buff-brown to brown in age, bruising brown when rubbed. Surface dry, finely velvety-tomentose, giving it a suede-leather feel, often becoming areolate cracked, with minute tomentose tufts in age. **TUBES:** Broadly attached to the stipe, 0.3–2 cm long. **PORES:** Very small, angular to nearly round. Dull to bright yellow or golden yellow when young, to dingy yellow-olive in age. Quickly staining blue when bruised, eventually bluing areas fading to orangish brown. **STIPE:** 5–15 (20) cm long, 2–4.5 cm thick at apex, often club shaped, with a thicker base when young, to more or less equal in age. Yellow at apex, with a rosy red to red-blushed base when young, then redder from base up as it matures, becoming dingy red with a blackish red base in age. Staining blue when handled. **FLESH:** Very thick and firm when young, soft in age. Yellow to whitish at stipe apex, quickly staining blue when cut, slowly fading grayish. Stipe often with reddish in base and around larva tunnels. **ODOR:** Indistinct. **TASTE:** Slightly bitter, or sour. **KOH:** Brown on cap, orange on flesh, tubes, and upper stipe, brown on lower stipe. **SPORE DEPOSIT:** Olive-brown. **MICROSCOPY:** Spores 12.5–17.5 x 4–5 µm, subcylindric to subellipsoid, smooth, ochraceous in KOH.

ECOLOGY: Solitary or scattered on ground, in moss or duff under conifers, especially Grand Fir and Western Hemlock. Locally common from Sonoma County north, also common in the mountains. Fruiting in a large flush in fall, soon after the first rain, with scattered fruitings into winter.

EDIBILITY: Unknown. Although the taste is generally mildly bitter, it is not recommended one try this bolete, as there are reports of minor gastrointestinal distress from other *Caloboletus*.

COMMENTS: The buff cap that feels like suede leather (when dry), yellow pores, blue-staining flesh, nonreticulate stipe with a reddish base, and slightly bitter taste are distinctive. *C. conifericola* is very similar, but can easily be told apart when young by the dark olivaceous gray cap, finely reticulated stipe that does not have a red base, and extremely bitter taste. When *C. conifericola* is old, the cap color becomes lighter (versus becoming darker on *C. rubripes*), and the stipe might develop a dingy red color at the base. The rosy sterile margin on the mature caps of *C. rubripes* can help separate the two species at this stage. Other look-alikes are *C. marshii*, which has a beige to grayish buff cap and grows under oaks. *Boletus smithii* has a rosy red to pinkish velvety cap that fades to gray, yellow pores that become reddish blushed, a stipe with a red apex and yellow base, and a mild taste. *Xerocomellus* are smaller and have a mild or lemony taste.

SYNONYM: *Boletus rubripes* Thiers.

Caloboletus marshii

D. Arora, C. F. Schwarz & J. L. Frank

BEN'S BITTER BOLETE

CAP: 6–15 (20) cm across, bun shaped at first, becoming broadly convex. Margin often with a sterile flap of whitish tissue. Dusty grayish white when young, becoming beige, tan to pale brownish tan. Surface smooth to slightly velvety with a whitish bloom when young, dry, or slightly viscid when wet. **TUBES:** Very thin at first, soon elongating; up to 3 cm long. **PORES:** Small, round. Pale yellow to brighter yellow, quickly bruising blue when fresh, becoming dingy olive-gray in age. **STIPE:** 3–10 cm long, 2–7 cm thick, usually swollen and bulbous at base, reticulum absent at all stages. Pale whitish yellow to yellow, sometimes developing reddish dots and irregular splotches, especially where broken or grazed by invertebrates. **FLESH:** Thick and fairly firm when young, dingy whitish gray, bruising blue fairly rapidly when fresh (weakly in age). Stipe often with a sharply contrasting dull reddish brown portion below midpoint and yellow "rind" over exterior at base. **ODOR:** Distinctive, sour. **TASTE:** Very bitter when fresh, milder in old fruitbodies. **SPORE DEPOSIT:** Olive-brown. **MICROSCOPY:** Spore 11–14 x 4.5–6 μm, somewhat spindle shaped to long-ellipsoid, smooth.

ECOLOGY: Solitary, scattered or in small clusters on ground or in duff under live oak. Locally common within the range of Coast Live Oak, also rarely occurring with other oaks. Fruiting in late summer to fall, favoring dry weather before first fall rains, rarely persisting past November.

EDIBILITY: Nontoxic; at least a few people have eaten this species after having mistaken it for a Brown Butter Bolete. The strong bitter taste and sour odor discourage most people from attempting to eat it (but not the species' namesake, Ben Marsh, who repeatedly tried to make it palatable).

COMMENTS: Ben's Bitter Bolete is easily recognized by the pale cap, blue bruising reactions, growth with live oak, and bitter taste. *C. conifericola* and *C. rubripes* have more northerly distributions and are associated with conifers, and the latter shows more extensive red around the stipe base. *Boletus barrowsii* is also pale and bulky, but has a reticulate stipe, a sweet nutty taste, and white flesh that does not blue. *Butryiboletus persolidus* has a brown cap, a yellow reticulated stipe, and a mild taste.

Butryiboletus persolidus

D. Arora & J. L. Frank

BROWN BUTTER BOLETE

CAP: 7–20 (30) cm across, bun shaped to broadly convex, eventually flatter, sometimes partly uplifted in age. Margin often with a narrow, wavy, sterile band of tissue. Leather brown to dull tan-brown, cinnamon brown to yellowish brown. Surface dry to slightly tacky, often with a powdery bloom when young. **TUBES:** Very short and densely packed when young, 0.5–2 cm long when mature. **PORES:** Small, round. Pale yellow at first, becoming bright yellow to olive-yellow. Staining blue when damaged. Reaction sometimes weak in younger specimens, pronounced in older specimens. **STIPE:** 5–18 (25) cm long, 3–7 cm thick at apex, strongly swollen or bulbous toward base, stipe sometimes wider than cap when very young. Yellow, sometimes with pink or reddish splotches, or sometimes with a well-developed reddish band around base of stipe, often bluing when handled. Surface covered with a very fine reticulation over upper portion, obscurely reticulated or smooth below. **FLESH:** Thick, very firm and dense. Distinctly yellow, especially in stipe flesh, sometimes staining bluish when exposed. Often dull reddish brown around larva tunnels. **ODOR:** Indistinct. **TASTE:** Mild. **SPORE DEPOSIT:** Olive-brown. **MICROSCOPY:** Spores 12–15 x 4–5 μm, somewhat spindle shaped, smooth.

ECOLOGY: Solitary or in small groups of scattered individuals, fused clusters, or in troops under hardwoods. Especially favors oak and Tanoak, also found with Madrone and manzanita. Quite frequent in some years, especially those with warm early rain, rather scarce in other years. Occurring in the southern half of our range, absent from much of the North Coast. Fruiting from early to midfall, not usually persisting into winter.

EDIBILITY: Edible and very popular due to their firm texture. Some people have adverse reactions; make sure to cook it well.

COMMENTS: The brownish cap, yellow pores, finely reticulate yellow stipe, bluing reactions, and mild taste help distinguish this bolete. *B. querciregius* is very similar, but has a beautiful rosy pink cap when young; faded specimens are usually duller, but often show a distinct pink or yellowish pink hue. *Caloboletus* have pale yellow flesh that quickly blues and a bitter taste.

SYNONYM: *Boletus appendiculatus* (sensu Thiers).

Butyriboletus querciregius

D. Arora & J. L. Frank

PINK-CAPPED OAK BUTTER BOLETE

CAP: 5–18 (25) cm across, bun shaped to broadly convex, sometimes slightly wavy or lobed at margin. Pinkish red when young, often mixed with pinkish beige, tan, and yellow, pink tones fading in age. Surface finely velvety when young, soon becoming smooth, dry to slightly viscid. **TUBES:** Very short and densely packed when young, 0.5–2 cm long when mature. **PORES:** Small, round. Pale yellow at first, becoming bright yellow to olive-yellow. Weakly staining bluish gray when young, quickly bluing in age. **STIPE:** 5–12 (20) cm long, 2–6 cm thick, thinner near attachment to cap, often strongly swollen or bulbous toward base. Mostly lemon yellow to butter yellow, occasionally with a reddish blush. Surface dry, covered with a fine reticulation, strongest over upper part of stipe. Fresh specimens quickly bruising blue when handled. **FLESH:** Thick, very firm when young, distinctly yellow; especially in the stipe flesh, erratically staining bluish when cut. **ODOR:** Indistinct. **TASTE:** Mild. **SPORE DEPOSIT:** Olive-brown. **MICROSCOPY:** Spores 10–16 x 3.5–5 µm, spindle shaped, smooth.

ECOLOGY: Solitary or scattered in groups or arcs under live oak and rarely deciduous oaks. Widespread, occurring from the southern part of our range into the Pacific Northwest, rare north of Sonoma County, absent on the North Coast. Quite abundant some years (especially those with warm early rain), but can be rare to absent in others. Fruiting in early to midfall, smaller numbers found into spring.

EDIBILITY: Edible and very good, especially favored for the dense texture, but lacking the rich flavor of the Porcini (*Boletus* spp.).

COMMENTS: The cap of this species tends to be pinkish red to pinkish tan, whereas the cap of the very similar *B. autumniregius* tends to be more rosy or purplish pink. Other features that distinguish the latter include its association with conifers (especially Douglas-fir, also possibly Madrone and Tanoak) rather than live oak, and its longer and narrower spores. Older specimens of *B. querciregius* and *B. autumniregius* could be mistaken for *B. persolidus*, which lacks the pink and red tones. *B. abieticola* grows under firs on the Far North Coast and has a pale pinkish beige cap, often with small scales. *B. primiregius* has a dark red to vinaceous red cap and fruits in the spring in the mountains. Another red-capped, blue-bruising bolete under oak is *Xerocomellus dryophilus*, but that species has a much more velvety cap that often cracks extensively, lacks reticulation on the stipe, and is more slender. *Suillellus amygdalinus* has a deeper reddish to brownish red cap, orange or red pores, and a darker, much more intense blue bruising reaction.

MISAPPLIED NAME: *Boletus regius* (a name for a European species) had long been applied to three different red-capped butter boletes in California.

Butyriboletus abieticola

(Thiers) D. Arora & J. L. Frank

MOUNTAIN BUTTER BOLETE

CAP: 7–18 cm across, rounded at first, becoming broadly convex, sometimes slightly wavy or lobed in age. Dingy grayish beige to pinkish beige at first, becoming grayish pink, often with rosy splotches on the margin. Surface dry, covered with slightly darker, scattered fibrillose scales or patches. **TUBES:** Short and densely packed when young, 0.5–2 cm long when mature. **PORES:** Very small, round. Pale yellow to greenish yellow, staining deep blue when bruised. **STIPE:** 8–15 (20) cm long, 2–6 cm thick at apex, enlarged or bulbous toward base. Whitish yellow to pale lemon yellow, occasionally with a reddish blush, staining bluish when handled. Surface dry, covered with a fine reticulation, **FLESH:** Very thick and firm. Pale yellow, occasionally yellowish white in stipe, and reddish in stipe base, not staining, or erratically bluing when cut. **ODOR:** Indistinct. **TASTE:** Mild. **KOH:** Brownish on cap, blackish on tubes, orange on flesh. **SPORE DEPOSIT:** Olive-brown. **MICROSCOPY:** Spores 14–17.5 x 4.5–5.5 μm, elliptical to spindle shaped, smooth.

ECOLOGY: Solitary or scattered, rarely more than a few fruitbodies in any one area, in mixed coniferous forest, with Grand Fir. Locally common on the Far North Coast, common in Red Fir (*Abies magnifica*) forest in the mountains. Fruiting in the early to mid fall.

EDIBILITY: Edible and very good, possibly the tastiest butter bolete.

COMMENTS: The large dense fruitbodies, patchy scales on a grayish pink cap, blue-staining yellow pores, yellow, finely reticulated stipe, and growth under conifers are distinctive. It is often called the Mountain Butter Bolete since it is much more common in high-elevation fir forest. *B. autumniregius* has a darker rosy to purplish pink cap. *B. querciregius* can have a pale pinkish cap, but only occurs with oaks.

SYNONYM: *Boletus abieticola* Thiers.

"Boletus" coccyginus

Thiers

CAP: 2–6 (8) cm across, rounded-convex to broadly convex when young, to nearly plane and becoming irregular or wavy in age. Rosy red, red, pinkish red to pinkish, often cracking and showing yellow flesh in age. Surface dry to moist, pubescent when young, becoming smooth, often cracking in age. **TUBES:** Sunken around stipe and rather short at margin. **PORES:** Small and irregular at first, expanding in age. Dull yellow when young, to yellow or greenish yellow, becoming slight olive-yellow in age, not staining blue on younger specimens, occasionally with slight bluish green stains in age. **STIPE:** 1.5–7 cm long, 0.5–2 (3) cm thick, equal, peglike; tapering toward base to irregular. Pinkish red to pale reddish brown over a yellow base color. Surface dry, often streaked with longitudinal striations to appressed-fibrillose, finely punctate at apex, to smooth. **FLESH:** Firm, pale yellow, or yellowish brown in stipe base, unchanging when damaged. **ODOR:** Indistinct. **TASTE:** Mild. **KOH:** Dingy olive-green flash, quickly becoming golden orange on cap, dingy orange-brown on tubes, yellowish on stipe, no reaction on flesh. **SPORE DEPOSIT:** Olive-brown. **MICROSCOPY:** Spores 11–17.5 x 5–7 μm, cylindrical, ovoid to elliptical, variable in shape and size, smooth, moderately thick walled, ocher in KOH.

ECOLOGY: Solitary, scattered, or in small clusters on ground in mixed forests of Douglas-fir, Tanoak, and Coast Live Oak, fruiting in fall. Found from Santa Cruz County north into Oregon and east to Sierra Foothills, but very rare.

EDIBILITY: Unknown, probably edible. If found, it should be saved for the herbarium, not the table!

COMMENTS: This rare bolete can easily be told by the small size, rosy red to pink cap, yellow pores, and flesh that doesn't stain blue. It certainly doesn't belong in *Boletus*, but has yet to be formally placed in another genus. *Xerocomellus dryophilus* has a cracking burgundy-red to pinkish red cap that fades to reddish brown and has blue-staining pores and flesh; it is strictly a live oak associate.

"Boletus" smithii

Thiers

SMITH'S BOLETE

CAP: 5–15 (25) cm across, convex at first, becoming broadly convex to nearly plane, and often wavy and slightly uplifted in age. Dark to pale rosy pink when young, fading in a patchy manner, to yellow and then gray, but often with a combination of pink, yellow, and gray at the same time. Older caps can be extensively gray to olivaceous gray. Surface dry, finely velvety, or suedelike. **TUBES:** Often deeply sunken near stipe, 0.5–2 cm long. **PORES:** Small, round to irregular. Yellow at first, occasionally with a reddish blush to dingy yellow to reddish in age. Erratically and often slowly staining blue when bruised, in some specimens it can be quick, while others don't stain. **STIPE:** 5–18 cm long, 1–3.5 cm thick, equal or enlarged lower, often tapering at base. Rosy pink over upper portion, yellow to yellowish white at base. Surface dry, finely punctate with pinkish dots when young, mostly smooth in age. **FLESH:** Thick to thin, firm when young, soft in age. Yellow, but retaining thin pinkish zone under skin of cap, even in age. Not staining or erratically bluing. **ODOR:** Indistinct. **TASTE:** Mild. **SPORE DEPOSIT:** Olive-brown. **MICROSCOPY:** Spores 14.5–19 x 4–6 μm, spindle shaped to cylindrical, smooth.

ECOLOGY: Solitary, scattered, in small clusters on ground, in moss or duff under conifers, especially hemlock and spruce. Very common on the Far North Coast, occurring south to Mendocino County. Fruiting with a large flush in early fall, with scattered fruiting into winter.

EDIBILITY: Unknown. A closely related eastern counterpart is mildly toxic.

COMMENTS: This common North Coast bolete can be recognized by the rosy pink cap that fades irregularly to gray, with an intermediate yellow phase; yellow pores that become reddish blushed; and rosy pink stipe apex. It belongs in a small group of boletes that will soon be given their own genus; they are certainly distinct from the core *Boletus*. When exhibiting the rosy color, Smith's Bolete is probably most similar to *Butyriboletus autumniregius*, which has a rosy red, smooth (not velvety) cap and a yellow, finely reticulated stipe. Faded specimens could be mistaken for *Caloboletus rubripes*, which has a gray to grayish brown, soft suedelike cap, pale yellow flesh that quickly stains blue, a bitter taste, and red that is restricted to the base of the stipe when young, spreading upward in age. *Xerocomellus* are generally smaller, and most have cracking caps.

Aureoboletus mirabilis

(Murrill) Halling

ADMIRABLE BOLETE

CAP: 4–15 (22) cm across, domed at first, soon becoming broadly convex, margin with a narrow band of overhanging tissue, occasionally slightly uplifted in age. Dark reddish brown, maroon-brown to brown, often with a paler margin, and pale pinkish brown to whitish spots, especially in age. Surface moist to dry, distinctly roughened-tomentose to coarsely hairy, with stiff short hairs. **TUBES:** Attached or sunken around stipe, 1–4 cm long. **PORES:** Small to moderately large, round to irregular. Pale yellow to greenish yellow at first, becoming dingy olivaceous yellow in age. **STIPE:** 7–20 (25) cm long, 1–3 cm thick at apex, up to 7 cm thick at base, club shaped with a large, swollen base. Dark brown to reddish brown, spotted and streaked with paler pinkish buff to beige color. Surface with sparse reticulation over apex, smooth below, base often viscid. **FLESH:** Thin (cap is mostly tubes), soft in cap; firm, fibrous in stipe. Whitish, buff to pale yellow, often with vinaceous stains in age. **ODOR:** Indistinct. **TASTE:** Lemony. **SPORE DEPOSIT:** Olive-brown. **MICROSCOPY:** Spores 18–22 x 7–9 μm, spindle shaped to elongate-elliptical, smooth.

ECOLOGY: Solitary or scattered on well-decayed, moss-covered logs and stumps, occasionally on ground under Western Hemlock. Often found near *Chroogomphus tomentosus*, which is likely a parasite of this species. Known from Sonoma County northward, uncommon south of central Mendocino County. Fruiting from early fall into early winter.

EDIBILITY: Edible and good. Older caps can get quite mushy and waterlogged, but the stipes typically stay solid.

COMMENTS: One of our most distinctive boletes: the roughened-tomentose reddish brown cap, often with paler spots; long, club-shaped stipe; and growth on rotten stumps and logs readily distinguish this species. Although it has been classified as a *Boletus, Boletellus,* and *Xerocomus* in recent memory, genetic data suggests that it is closely related to *Aureoboletus*.

SYNONYM: *Boletus mirabilis* (Murrill) Murrill.

Aureoboletus citriniporus

(Halling) Klofac

CITRINE-PORED BOLETE

CAP: 3.5–11 cm across, round-convex to broadly convex to nearly flat in age. Dark brown or reddish brown when young, becoming lighter in age, often with olive tones. Thinly viscid when young, soon becoming dry, nearly smooth at first, showing radially arranged appressed fibrils or patches in age. **TUBES:** Thin, broadly attached to stipe or with a narrow notch. **PORES:** Small, round, or slightly irregular, intensely lemon yellow, remaining bright, even in age. **STIPE:** 4.5–8 cm long, 2–3.5 cm thick, apex narrowed near attachment to cap, base usually swollen to club shaped, sometimes slightly pointed. Yellowish to beige, occasionally with ocher-tan splotches. Sometimes appearing slightly reticulate at apex due to an elongation of pores. **FLESH:** Firm when young, mushy in age. Whitish with a narrow pinkish band above tube layer and under cap skin. **ODOR:** Indistinct. **TASTE:** Cap surface sour, acidic. **SPORE DEPOSIT:** Olive-brown. **MICROSCOPY:** Spores 10.5–13.5 x 3.5–4.5 µm, elliptical to spindle shaped, smooth.

ECOLOGY: Solitary or scattered, occasionally in small clusters on ground or in duff under live oak. Uncommon, known from Santa Cruz County north through the greater San Francisco Bay Area. Fruiting in fall and early winter.

EDIBILITY: Edible and pleasantly lemony, but rather uncommon and not generally sought after for food.

COMMENTS: This beautiful bolete is bound to make an impression when it is first turned over to reveal the blazing yellow pores. The dark reddish brown to dark brown cap, aforementioned pores, and lack of blue staining readily distinguish this species. *A. flaviporus* also has bright yellow pores, but has a more slimy or sticky cap that is paler pinkish tan to orange-brown, while its stipe is usually brighter whitish yellow and often has a rooting base with white rhizomorphs. Also similar is *Xerocomus subtomentosus*, which has a drier, more velvety cap surface, slightly duller yellow pores that slowly bruise blue, and a more slender stipe with a tuft of yellow rhizomorphs. The rare *Boletus amyloideus* has a moist to slightly viscid reddish brown cap, bright greenish yellow pores, a tan stipe, pink flesh, and a mild taste. It probably belongs in *Aureoboletus*, but we have yet to find it to confirm this.

Aureoboletus flaviporus

(Earle) Klofac

VISCID BOLETE

CAP: 3.5–8 cm across, rounded-convex to broadly convex or nearly flat in age. Medium tan-brown, pinkish tan to orange-brown. Surface smooth, quite viscid when moist, remaining sticky even in age. In dry weather it may not appear glutinous, but duff and debris should be tightly "glued" to cap. **TUBES:** Thin, sunken around stipe to decurrent. **PORES:** Fairly small and round at first, expanding in age and becoming slightly angular. Bright lemon yellow when young, becoming somewhat duller yellow with dingy grayish or ocher stains in age. **STIPE:** 5–9 cm long, 1.5–3.5 cm thick, apex usually narrowed, and base usually tapered, giving impression of a "swollen belly." Base often slightly rooting and with a conspicuous tuft of white rhizomorphs. Whitish or pale yellowish white at first, becoming pinkish tan or developing orangish brown discolorations. Surface often covered with an elongated or irregular bright yellow reticulum, especially at apex, from decurrent pores, to smooth. **FLESH:** Firm when young, becoming soft and mushy. Whitish to pinkish brown, often darker in stipe base. **ODOR:** Indistinct. **TASTE:** Acidic-lemony. **SPORE DEPOSIT:** Olive-brown. **MICROSCOPY:** Spores 12–15 x 5–6 µm, elongate to spindle shaped, smooth.

ECOLOGY: Solitary or scattered in troops under live oak, possibly with Tanoak and Madrone. Common from Mendocino County into southern California. Fruiting in fall and early winter.

EDIBILITY: Edible, but the slightly acidic taste, mushy flesh, and viscid cap aren't widely appealing.

COMMENTS: The viscid, orange-brown or pinkish tan cap and bright yellow pores are the distinctive features of this interesting bolete. *A. citriniporus* has a thinly viscid to dry, dark brown cap and a darker stipe that usually does not have whitish rhizomorphs. *Xerocomus subtomentosus* has a dry, darker brown, slightly velvety cap and bright yellow rhizomorphs.

Xerocomus subtomentosus group
BORING BROWN BOLETE

CAP: 4–15 cm across, convex at first, soon broadly convex to plane, occasionally uplifted, lobed or wavy in age. Dark brown to olive-brown or reddish brown but often with a pale sheen, becoming tan-brown with dark blotches, to paler overall in age. Surface dry, velvety to suedelike, often with cracks near edge of cap, becoming matted or smoother with age and sometimes nearly metallic in dry weather. **TUBES:** Sunken around stipe, or occasionally broadly attached, usually with a small zone of elongated pores running slightly down stipe. **PORES:** Round, irregular to angular, small to rather large, even on same fruitbody. Bright yellow to straw yellow, or becoming ocher-brown in age. Slowly bruising blue when damaged. **STIPE:** 7–12 cm long, 1–3 cm thick, cylindrical and often curving, with a tapered base attached to a tuft of yellow rhizomorphs. Surface white to yellowish and often extensively brown streaked in age. Surface smooth or with weak to prominent elongated brownish reticulum. **FLESH:** Whitish to pale yellowish, occasionally bluing slightly. **ODOR:** Indistinct. **TASTE:** Mild. **SPORE DEPOSIT:** Dull brown. **MICROSCOPY:** Spores 11.5–16 x 3.5–5 µm, somewhat spindle shaped to broadly cylindrical, smooth.

ECOLOGY: Solitary or scattered in troops on ground or in duff, most often with live oak and Tanoak, also occurring with conifers in the north. Common throughout our range. Fruiting in fall through midwinter.

EDIBILITY: Edible, but bland.

COMMENTS: The velvety brown cap, slender stipe, irregular and rather large pore mouths that slowly bruise blue, and yellow basal rhizomorphs distinguish this species. It could be confused for an *Aureoboletus*, but the cap is dry and velvety, the taste is mild, and the pores are duller yellow. *X. spadiceus* (sensu CA) is very similar, but has whitish mycelium at the stipe base and a sterile band of tissue on the cap margin, and usually grows under northern conifers. Murrill described *Ceriomyces oregonensis*, which is probably the appropriate western North American name for one member of the *X. subtomentosus* group, but more work is needed.

Xerocomellus "diffractus"
N. Siegel, C. F. Schwarz, & J. L. Frank, nom. prov.
CRACKED-CAP BOLETE

CAP: 3–10 cm across, rounded, bun shaped to broadly convex at first, becoming flat, occasionally slightly wavy in age. Olive-brown, leather brown to tan, becoming paler in age. Surface dry, very finely velvety, becoming cracked, at first showing whitish to pale yellow flesh in these cracks, becoming pinkish in age. **TUBES:** Sunken around stipe, 0.5–1.5 cm long. **PORES:** Very small and round at first, enlarging and becoming slightly angular. Pale yellow to dingy greenish yellow in age. Slowly staining bluish gray to blue. **STIPE:** 4–10 cm long, 0.8–2 cm thick, equal or club shaped with an enlarged base. Yellow to golden, extensively covered with fine red punctations, more concentrated toward base, often quite sparse at apex, becoming more red blushed in age. **FLESH:** Firm, fleshy. Whitish to pale yellow in cap, yellow in stipe, erratically and slowly bluing. **ODOR:** Indistinct. **TASTE:** Mild. **SPORE DEPOSIT:** Dull olive-brown. **MICROSCOPY:** Spores 12–17 x 4–6.5 µm, elongate to ellipsoid.

ECOLOGY: Solitary or scattered in troops under both conifers and hardwoods. Very common, occurring throughout our range. Fruiting in fall and early winter.

EDIBILITY: Edible.

COMMENTS: The Cracked-capped Boletes are a difficult group to fully grasp without a lot of field experience with all of California's species (even then, they can be very tough). There are three species that have been lumped under the incorrectly applied name *X. chrysenteron*, with *X. diffractus* being the most common and widespread of the bunch. *X. amylosporus* typically has a darker cap with more scattered and irregular cracks, the pores bruise dark inky blue, the stipe develops dingy brownish tones, and the spores are more reddish brown in color. *X. salicicola* is similar but much rarer, and can be told apart by its more-often cracked cap and growth with willow (*Salix* spp.). *X. mendocinensis* has quickly blue-staining pores and more coarsely punctate stipe that is often evenly red or with a distinct red belt near the apex.

MISAPPLIED NAME: *Xerocomellus (Boletus) chrysenteron* (Bull.) Fr.

"Xerocomellus" mendocinensis

(Thiers) N. Siegel, J. L. Frank & C. F. Schwarz comb. prov.

CAP: 3–10 cm across, domed or bun shaped when young, expanding to broadly convex, eventually nearly flat, occasionally wavy. Usually dark olive-brown when young (to almost blackish in one uncommon variant), soon olive-tan to medium brown, sometimes lighter gray in age. Surface dry, finely velvety, cracking near the margin, often only cracking on outer quarter of cap, but occasionally extensively cracked across entire cap. Flesh in cracks dull tan-brown or light olive to pallid creamy (rarely pinkish). **TUBES:** Sunken around stipe, 0.5–1.5 cm long. **PORES:** Small and irregularly round at first, soon moderately large and distinctly angular. Bright yellow to dull yellow, soon straw colored, often with ocher or reddish brown mouths in age, sometimes distinctly orange-red from start. Quickly bruising dark teal to navy blue where bruised. **STIPE:** 4–10 cm long, 1–2.5 cm thick, cylindrical, usually straight or slightly curved near base (sometimes squat and irregular). Yellow when very young, but soon extensively covered with a dense layer of reddish dots and often evenly washed red almost to top. Often with a strong, contrasting band or girdle of red pigment just below yellow area at extreme apex. **FLESH:** Yellow, marbled with creamy white in some areas, reddish to tan near base, slowly to quickly bruising blue. **ODOR:** Indistinct. **TASTE:** Mild. **KOH:** Blackish on cap, orange on flesh. **SPORE DEPOSIT:** Olive-brown. **MICROSCOPY:** Spores 12–15 x 4.5–6 μm, elongate, smooth, many with a strongly truncate (squared-off) end.

ECOLOGY: Solitary, scattered in troops or in small clumps, most common under live oak in the southern part of our range, but found in mixed forest of Douglas-fir and Tanoak northward. Quite common, fruiting in fall and early winter.

EDIBILITY: Edible. Texture is a bit soft, with an unremarkable flavor.

COMMENTS: Reliable identification of the cracked-cap boletes in our area is difficult and is more reliably done after experience with all of the species involved and the range of variation shown by each one. In comparison to the other *Xerocomellus* in our area, the relatively dark, moderately cracked cap, quick blue staining, and extensively red, punctate stipe are good clues. *X. diffractus* has a less extensively red stipe, an often paler cap, and stains blue more slowly. *X. "atropurpureus"* can appear similar, but has a dark blackish purple to wine red, wrinkled to smooth cap that doesn't become cracked and pores that rarely stain blue. *X. rainisiae* has a thick velvety-tomentose cap skin that only cracks in age, has a golden yellow stipe with a reddish blush at the base, and stains intensely dark blue-green when handled (especially on the stipe).

MISAPPLIED NAME: *Xerocomellus (Boletus) truncatus* is an eastern North American species.

Xerocomellus zelleri

(Murrill) Klofac

ZELLER'S BOLETE

CAP: 2–6 (9) cm across, convex to plane, occasionally uplifted and wavy in age. Dark vinaceous black, occasionally dark olivaceous black to brownish black, rarely deep reddish black; at all times with a paler, whitish beige to yellowish tan band around the margin. Surface smooth or finely roughened, wrinkled or pitted, finely, densely velvety when young, becoming more matted in age. **TUBES:** Sunken around stipe to slightly decurrent, short to moderately long. **PORES:** Small, round to slightly angular or irregular. Pale creamy yellow to dingy yellow, or pale yellowish olive when young, becoming dingy yellow to dingy yellow olive, occasionally developing reddish blushes in age. Not bruising blue; or occasionally in older, waterlogged specimens. **STIPE:** 2–7 cm long, 0.5–2 cm thick, cylindrical, or with a tapered base. Yellowish base color, covered extensive with fine rosy red punctations when young, becoming evenly rosy red to dark red in age. **FLESH:** Thin, firm, light creamy yellow, not bruising blue, or slightly so in older, specimens. **ODOR:** Indistinct. **TASTE:** Mild to lemony. **SPORE DEPOSIT:** Dull olive-brown. **MICROSCOPY:** Spores 12–15 x 4–5.5 μm, elongate to ellipsoid, smooth.

ECOLOGY: Solitary or scattered in moss, or on or around moss covered stumps and logs. Found extensively in mature and old growth coniferous forest. Rare in California; despite extensive collecting over a six year period, we only saw it three times; fruiting in the fall, from central Mendocino County northward. Not common anywhere, but more frequent in the Pacific Northwest.

EDIBILITY: Edible and good; it has a pleasant lemony taste.

COMMENTS: The dark cap with a velvety bloom, and a pale band around the margin, pallid yellow pores that rarely bruise blue (only in age), red stipe with crowded punctations when young, bleeding to a solid red color in age, and the relatively small size are helpful in distinguishing it from all other *Xerocomellus* except *X. "atropurpureus"*. *X. "atropurpureus"* has a variable cap color (usually with more red or vinaceous tones), is generally stockier in stature, and is much more common and widespread in California.

Xerocomellus "atropurpureus"

J. L. Frank, N. Siegel & C. F. Schwarz nom. prov.

CAP: 4–10 (12) cm across, rounded, bun shaped to broadly convex at first, becoming flat occasionally slightly wavy in age. Dark blackish purple, blackish, deep reddish purple to dark wine red. More rarely with extensive olivaceous-black tones. Surface dry to moist, usually bare or with a faint pale bloom, but not truly velvety, often extensively bumpy-wrinkled, occasionally smoother. **TUBES:** Sunken around stipe, 0.5–1.5 cm long. **PORES:** Moderately small, round to slightly angular. Pale dull yellow to dingy greenish yellow in age; occasionally reddish blushed, or rarely extensively red. Rarely bruising blue, and only in older and/or waterlogged specimens. **STIPE:** 4–9 cm long, 0.8–3 cm thick, equal or club shaped with an enlarged base. Surface densely covered with fine red punctations with a yellowish background, soon extensively red, often with slightly darker streaks. **FLESH:** Firm, fleshy, marbled yellow, unchanging or bluing slightly in older specimens. **ODOR:** Indistinct. **TASTE:** Mild or lemony. **SPORE DEPOSIT:** Dull olive-brown. **MICROSCOPY:** Spores 12–15 x 4–5.5 μm, elongate to ellipsoid, smooth.

ECOLOGY: Solitary or scattered in troops, on well-decayed, moss-covered logs and stumps or duff. Growing with a wide range of trees, both conifers and hardwoods. Very common and widespread. Mostly fruiting in fall and winter, with scattered spring fruitings on the North Coast.

EDIBILITY: Edible and good.

COMMENTS: The dark, wrinkly-bumpy or nearly smooth cap without extensive cracking, yellow pores, lack of blue staining, and evenly red stipe help distinguish this species from all but the much rarer *X. zelleri*. Reliable morphological criteria to separate these two species are provisional: *X. "atropurpureus"* more often has a reddish purple to wine-purple cap (usually lacking a distinct pale margin), whereas *X. zelleri* caps tend to be darker; blackish purple to olivaceous black with a distinctly paler margin. In addition, the young caps of *X. zelleri* are often slightly velvety (as opposed to glabrous on *X. "atropurpureus"*). For Californians, it is unfortunate that the common and widespread species that we knew as *zelleri*, is not conspecific with the true *X. zelleri*; which is more common in the Pacific Northwest.

MISAPPLIED NAME: *Xerocomellus (Boletus) zelleri.*

Xerocomellus dryophilus

(Thiers) N. Siegel, C. F. Schwarz & J. L. Frank

RED CRACKED-CAP BOLETE

CAP: 3–11 cm across, rounded, bun shaped to broadly convex at first, to nearly plane in age. Ruby red to vinaceous red or brick red when fresh, becoming dull tan-brown with pinkish or reddish tinges in dry weather. Dark brown to olive-brown forms can occur. Surface dry, velvety-tomentose, and usually extensively cracked into tomentose plaques or scales. Flesh in cracks pallid yellowish white at first, often becoming pinkish tinged. **TUBES:** Sunken around stipe. **PORES:** Moderately large, angular or irregularly shaped. Pale yellow to dingy yellow or lemon yellow, bruising blue readily. **STIPE:** 3–10 cm long, 1–2 cm thick, cylindrical straight or slightly curved (sometimes slightly sinuous), enlarged or club shaped at base, often tapered and sometimes slightly rooting at extreme base. Lemon yellow or dull yellow, downward becoming orange-red and then red to wine red near base. Reddish pigment sometimes appearing very finely punctate over surface, basal tomentum yellowish. **FLESH:** Firm, fleshy, yellow, bruising blue rather quickly when cut. **ODOR:** Indistinct. **TASTE:** Mild. **SPORE DEPOSIT:** Dull olive-brown. **MICROSCOPY:** Spores 11.5–16 x 5–6.5 µm, elongate to ellipsoid.

ECOLOGY: Solitary or scattered, often in drier forests under live oak. Very common in southern California, occasionally in southern part of our range, scarce north of Santa Cruz County. Fruiting from late fall into spring.

EDIBILITY: Edible, commonly consumed in some areas where better boletes are scarce. Prime specimens can be hard to find since mature specimens are fairly soft, and even young ones are commonly parasitized by *Hypomyces*.

COMMENTS: Identifiable by its growth with live oak, velvety red to pinkish-brown cap that cracks (and sometimes fades) as it ages, blue-staining pores and flesh, and typically bicolored (yellow and red) stipe that is often swollen and rounded lower, with a tapered base. *X. mendocinensis* has a much darker cap and more evenly red stipe, while *X. diffractus* has an olive-gray cap, and a slow bluing reaction on the pores.

SYNONYM: *Boletus dryophilus* Thiers.

Hypomyces microspermus

Rogerson & Samuels

BOLETE EATER

FRUITBODY: First appearing as a white mold on boletes, soon covering whole fruitbody, occasionally blistering. Bolete is generally still firm in this stage. Soon starts turning bright yellow to golden colored, with the host often collapsing and becoming quite decayed. Final (sexual) stage is rarely observed in the wild; a reddish brown, pimply crust. **ODOR:** Indistinct to putrid. **TASTE:** Not sampled. **SPORE DEPOSIT:** White in first stage, golden yellow in second stage. **MICROSCOPY:** Spores in white stage, 5–17 x 3–7 µm, oblong to broadly elliptic. In yellow stage, 8.5–15 µm, round, distinctly spiny, very thick walled. In sexual stage, 8–15 x 2.5–4.5 µm, spindle to lance shaped.

ECOLOGY: Very common on Xerocomoid boletes, especially *Xerocomellus dryophilus* and *Xerocomus subtomentosus*. Occurring throughout our range, possible whenever boletes are fruiting, which is usually fall into early winter.

EDIBILITY: Unknown, likely poisonous.

COMMENTS: This common bolete-parasitic mold goes through two asexual-sporulating stages leading up to the sexual stage, which is rarely observed (the host is often buried in duff by this point). *H. chrysospermus* also occurs in California, but is much less common and only reliably distinguished by its much larger spores measuring (10) 12–25 µm. *Dicranophora fulva* forms tiny, translucent white stalks holding up bright yellow, globular heads that completely engulf *Suillus* (and occasionally *Gomphidius*) fruitbodies. Unlike *Hypomyces*, it is a zygomycete fungus.

Pulveroboletus ravenelii

(Berk. & M. A. Curtis) Murrill

SULFUR-VEILED BOLETE

CAP: 4.5–15 cm across, rounded to tightly convex when young, expanding to broadly convex or nearly flat, sometimes waved or irregular. Covered with neon-bright chartreuse or yellow veil tissue at first. As cap expands, this veil thins out, showing underlying yellow to brownish yellow cap colors, often with red and green blotches, sometimes becoming entirely reddish in age. Surface dry, fibrillose to powdery when covered by veil fibrils, then smooth to innately fibrillose or with flattened scales, sometimes with brighter yellow remnants of veil adhering to surface. **TUBES:** Depressed around stipe, moderately thin and often irregular in thickness. **PORES:** Relatively small, round to slightly irregular. At first completely covered by curtain of cobwebby or nearly membranous veil fibrils that attach to pores themselves (veil doesn't simply hang over pores from edge of the cap to stipe). Pale yellowish beige when young, becoming brighter yellowish, and then dingy orange, finally turning dark reddish brown. Bruising blue, sometimes rather slowly, occasionally dingy red in age. **STIPE:** 7–13 cm long, 1–3 (5) cm thick, cylindrical or slightly irregular. Generally completely covered in bright chartreuse-yellow veil fibers well into age but often bare over apex, where it shows a paler dull beige-yellow color. Surface dry, roughened from persistent veil remnants. **FLESH:** Firm, fleshy in cap when young, soft, mushy in age. Stipe very firm, tough, sometimes corklike. Whitish to pale yellowish beige in cap, yellower downward, to yellow-orange near base of stipe. Bruising irregularly blue in cap when cut. **ODOR:** Indistinct. **TASTE:** Mild to slightly sweet. **KOH:** No reaction. **SPORE DEPOSIT:** Olive-brown. **MICROSCOPY:** Spores 8–11 x 4–6.5 μm, spindle shaped to ovoid, smooth.

ECOLOGY: Usually solitary but occasionally in small groups or with 2 or 3 fruitbodies in a cluster. It seems to have a preference for woods with Tanoak, but Douglas-fir is often also in areas where it is found. Uncommon to rare overall, but can be quite frequent in some areas during some years. Fruiting from fall into early winter.

EDIBILITY: Reportedly edible, but its rarity and bizarre combination of textures (cobwebby veil, mushy cap, and hard, woody stipe tissue) suggest that it's better appreciated for its strangeness and beauty rather than for its culinary characteristics.

COMMENTS: It seems impossible to mistake this astonishing species for any other on the Redwood Coast. The bright, almost neon-yellow veil tissue sets it apart from all other fungi in our area. Perhaps the only species with which it might be confused is *Buchwaldoboletus sphaerocephalus*; although similarly sulfur yellow overall with bluing pores that become dark reddish to brown in age, it lacks a partial veil tissue, is usually more squat, and grows in wood chips or on pine stumps and logs.

Buchwaldoboletus sphaerocephalus

(Barla) Watling & T. H. Li

CAP: 5–15 cm across, nearly round when young, becoming broadly convex, rarely flattening, margin incurved to inrolled at first, sometimes lobed and wavy in age, or retaining an inrolled margin with a narrow sterile edge extending beyond pores. Brilliant yellow as first, becoming golden with orangey, reddish, and brown blotches in age. Surface dry, suedelike at first, with broad appressed scales in age, and often cracking. **TUBES:** Narrowly joined to stipe, at first very shallow, expanding in age. **PORES:** Very small at first, becoming large and irregular or slightly angular. Pale yellow to bright yellow (rarely reddish) at first, mouths soon reddish, eventually entirely dark, dingy reddish brown, transitional areas of pore maturity showing orange tones. **STIPE:** 3–12 cm long, 3–8 cm thick, central to slightly or strongly off-centered, equal with a tapered to base. With prominent clumps of golden yellow rhizomorphic mycelium bound to base and clinging to substrate. Bright yellow to orangey yellow at first but soon mottled and dingy with reddish brown areas, becoming dark liver reddish brown overall in age. Bruising blue when damaged. **FLESH:** Thick, firm, fleshy, yellow to whitish yellow, bruising blue. **ODOR:** Musty. **TASTE:** Mild. **KOH:** Reddish on cap. **SPORE DEPOSIT:** Ocher-brown to cinnamon. **MICROSCOPY:** Spores 6–7.5 (9) x 3–4 µm, oblong to elliptical, smooth.

ECOLOGY: Solitary or in clusters, sometimes in impressive troops on piles of conifer wood chips, or on decaying logs or root balls of fallen or standing dead pines (especially Monterey Pine). Apparently saprotrophic. Other members of the genus have been noted to be closely tied to the occurrence of *Phaeolus schweinitzii*, although the exact relationship is unclear. Occurring throughout our range, but rather uncommon. Fruiting in summer and early fall.

EDIBILITY: Edible, but uncommon and extensive experience is lacking.

COMMENTS: The bright yellow cap and growth on wood chips or pine wood distinguish this species. A very similar species, *Boletus orovillus*, was described from Gray Pine (*Pinus sabiniana*), an inland tree in California, and has smaller spores and red pores when young. Species concepts in *Buchwaldoboletus* are not well defined, and the relationship between these two names as well as *B. hemichrysus* and *B. sulphureus* should be investigated.

Rubroboletus eastwoodiae

(Murrill) D. Arora, C. F. Schwarz & J. L. Frank

CALIFORNIA SATAN'S BOLETE

CAP: 4–30 cm across, usually quite round when young (although very small specimens may be shaped like flying saucers), expanding to convex to nearly plane. At first pallid whitish gray, becoming dusty pink or rose pink, then fading in age. Surface dry, textured like fine suede, often deeply cracked in dry weather, showing dingy pale yellow flesh. **TUBES:** Very short at first and densely packed, elongating up to 2 cm long. **PORES:** Very small, round, somewhat variably colored; at first usually an intense deep red, sometimes brighter blood red, becoming paler as pores expand, and then showing orange and yellow tones, especially near margin. In age dark dingy brownish with obscure red tones. Quickly staining dark blue when bruised. **STIPE:** 5–15 cm long, 5–15 cm thick, massively bulbous, very often wider than cap even when young, noticeably bulbous or swollen and club shaped with a narrow apex, even in age. Bubblegum pink to darker rose pink when young, yellowish to orange toward apex with maturity, usually with a distinct but fine layer of reddish pink dots or sometimes a very fine reticulation over upper stipe. Base of stipe often with a tuft of pale yellow rhizomorphs, especially visible when pulled from soft loam or sandy soil. **FLESH:** Thick and rather firm when young, becoming soft in age. Yellow to dingy pallid whitish gray frequently marbled in lower stipe, quickly staining bright electric blue when cut. **ODOR:** Indistinct to slightly sour. **TASTE:** Mild. **SPORE DEPOSIT:** Olive-brown. **MICROSCOPY:** 11–14.5 x 4–6 μm, elongate-elliptical to somewhat spindle shaped, smooth.

ECOLOGY: Solitary, in tight clusters or in large troops on ground under live oak. Common from Mendocino County south, often in relatively dry Coast Live Oak groves; much rarer to the north, under both live oak and decidous oaks. Occasionally fruiting in late summer, but most frequent in early fall.

EDIBILITY: Very poisonous! Causes severe diarrhea, vomiting, and abdominal cramps, sometimes requiring hospitalization.

COMMENTS: This unique species can be easily identified by the pink stipe, pallid whitish gray to rosy pink cap, blood red pores that quickly bruise dark blue, and ludicrous proportions, with the stipe often significantly more massive than the cap. It could be confused with *Butyriboletus querciregius*, which also shows pink colors and blue bruising, but that species has yellow pores and a yellow stipe. *Rubroboletus pulcherrimus* has a distinctly reticulated club-shaped stipe and grows with conifers on the North Coast.

SYNONYM: *Boletus eastwoodiae* (Murrill) Sacc. & Trotter.

MISAPPLIED NAME: *Boletus satanas*, a European species.

Rubroboletus pulcherrimus

(Thiers & Halling) D. Arora, N. Siegel & J. L. Frank

CAP: 8–20 (25) cm across, rounded when young, becoming broadly convex, margin with a narrow sterile band, incurved when young, rounded to slightly irregular in age. Color variable, often deep rosy red, burgundy to reddish brown when young, but can be olive to gray with a rosy pink blush on margin. Becoming more grayish olive overall in age. Surface dry, finely velvety to smooth, occasionally with flattened warts or patches, becoming finely cracked in age. **TUBES:** Sunken around stipe, 0.5–1.5 cm long. **PORES:** Very small, round to slightly irregular. Deep red at first, occasionally paler toward margin, becoming orange-red to reddish brown in age. Staining blue-black intensely and immediately when bruised. **STIPE:** 7–15 (20) cm long, 3–7 cm thick at apex, 5–12 thick at base, club shaped with an enlarged or bulbous base, occasionally equal or nearly so. Red, pink to orangish red, covered with deep red reticulation. Staining deep bluish to bluish black when handled, and discoloring from base up in age. **FLESH:** Thick, firm. Pale yellow, staining blue and fading to grayish yellow, often reddish in lower stipe and around larva tunnels. **ODOR:** Indistinct. **TASTE:** Mild. **KOH:** Brown on cap, tubes, and stipe, orangish brown on flesh. **SPORE DEPOSIT:** Dark olive-brown. **MICROSCOPY:** Spores 13–16 x 5.5–6.5 µm, elongate to subelliptical, smooth.

ECOLOGY: Solitary or scattered in small patches on ground under conifers, especially Grand Fir and Western Hemlock. Uncommon, occurring along coast from Sonoma County north. Fruiting in fall, soon after first rain, or late summer in areas with heavy fog drip.

EDIBILITY: Extremely poisonous, causing severe gastric distress and one death to date.

COMMENTS: This beautiful but highly toxic bolete is an uncommon resident of the North Coast coniferous forest. It can be recognized by the large size, rosy pink to red cap that fades in age, red pores, large, red reticulated stipe, and intense blue staining on all parts of the fruitbody. *Rubroboletus eastwoodiae* has a pinkish cap that fades to gray and a stout stipe with a large swollen base and narrow apex, and grows with live oak. *R. haematinus* is a similar conifer-associated species, but differs by having a pinkish cap that quickly goes brown as it ages and paler, orangish red pores when young. It is common in the summer and early fall in the mountains under fir. Also compare with *Suillellus amygdalinus*.

SYNONYM: *Boletus pulcherrimus* Thiers & Halling.

Suillellus amygdalinus

(Thiers) Vizzini, Simonini & Gelardi

LIVER BOLETE

CAP: 4–15 cm across, convex, broadly convex to plane, margin often wavy in age. Dingy pink to reddish or brownish red when young, developing olive-brown areas often becoming mustard colored, especially in dry weather. Surface dry, suedelike to smooth, often with cracked areas, bruising navy blue where damaged. **TUBES:** Sunken around stipe, rather short. **PORES:** Round to irregular, openings very small. Variably colored; at first yellowish orange to dark red, sometimes becoming dingy brick red, but often retaining yellow-orange color near margin. Bruising immediately deep blue. **STIPE:** 5–12 cm long, 3–6.5 cm thick, cylindrical to club shaped or tapered toward base. Pale yellowish to orangey with reddish dots or streaks to evenly dark red. **FLESH:** Thick, firm, yellow, with red in stipe base; immediately staining inky blue when cut. **ODOR:** Indistinct. **TASTE:** Mild. **KOH:** Orange on cap, flesh, and tubes. **SPORE DEPOSIT:** Olive-brown. **MICROSCOPY:** Spores 11–16 x 5–8 µm, elliptical to somewhat spindle shaped, smooth, thick walled.

ECOLOGY: Solitary or scattered in troops, sometimes in small fused clusters under hardwoods, especially live oak, particularly in drier and warmer microclimates from Mendocino County south. Absent from the Far North Coast, occasionally occurring in inland in the north. Fairly common most years from early fall to early winter, occasionally also fruiting in spring in smaller numbers.

EDIBILITY: Possibly toxic, although some people eat it with no issues. Probably best to avoid it, as other red-pored boletes in California are toxic.

COMMENTS: This highly variable bolete can be recognized by the brick red cap, red to orange pores, cylindrical or tapered stipe lacking reticulation, and intensely blue-staining flesh and pores. An undescribed *Suillellus* differs by having a brown to reddish brown cap, taller stature, and association with conifers. It's uncommon, occurring on the North Coast. Another similar undescribed species, this one belonging to the genus *Neoboletus*, differs by having a deep red, velvety cap, often with an olivaceous tone when young, and growth with spruce; it is rare in California, known only from a few sites on the Far North Coast. Both of these undescribed species were incorrectly lumped under the European name *Boletus erythnopos* in past guides. Fruitbodies of *Rubroboletus eastwoodiae* have bulbous stipes often wider than the caps and pinker coloration, and stain less intensely. They appear to be restricted to oak. *R. pulcherrimus* can look similar, but have reticulate stipes and grow with conifers. The pink-capped *Butyriboletus* have brighter caps, reticulate yellow stipes, and yellow pores that bruise less intensely blue.

SYNONYM: *Boletus amygdalinus* Thiers.

Gastroboletus turbinatus (sensu CA)

CAP: 3–8 cm across, irregular domed to convex at first, generally with a sunken center, with an uplifted and highly irregular margin in age. Reddish orange, reddish brown to yellowish brown when young, becoming dingy brown to dark brown, but often with a yellowish margin in age. Surface dry, velvety at first, matted-tomentose to appressed-fibrillose in age, bruising navy blue when damaged. **TUBES:** Distorted, often anastomosing, and pulling away from stipe, quite long. **PORES:** Highly irregular, fused, and facing many directions, mouths often sealed with tissue when young. Yellowish reddish to orange-red, quickly staining dark blue to bluish black, becoming quite dingy in age. **STIPE:** 1–3.5 (7) cm long, 0.7–2 cm thick, central to off-centered, equal or tapering toward base, often short. Yellowish to orange with a reddish base, darkening from base up in age. Quickly staining deep bluish black when handled. **FLESH:** Thin, firm, yellow, with red in stipe base and around larva tunnels, quickly bluing when cut. **ODOR:** Indistinct. **TASTE:** Mild. **KOH:** Orangish on cap, flesh, and tubes. **SPORE DEPOSIT:** None. **MICROSCOPY:** Spores 13.5–17 x 5–8.5 µm, elliptical to somewhat spindle shaped, smooth.

ECOLOGY: Solitary or scattered, often at least partially buried in duff under northern conifers, especially Grand Fir. Uncommon, known from southern Mendocino County northward. Fruiting from late summer into midfall.

EDIBILITY: Unknown.

COMMENTS: The often deformed fruitbodies, highly irregular tubes and pores, quick blue staining, and partially buried growth habit distinguish this species. The name *Gastroboletus turbinatus* is a grossly misapplied name in western North America—there are a number of undescribed species in this group (pertaining to multiple different genera), none of which are the "true" *G. turbinatus*. *Gastroboletus* is a highly artificial grouping; semisequestrate fruitbody morphologies have evolved within many different genera of boletes. In California, most of these species occur in the dry, high-elevation forests of the Sierra and Cascade ranges. There are gastroid Porcini (*Boletus subalpinus*), a gastroid *Xerocomellus* (*Gastroboletus xerocomoides*), a close relative of *B. smithii* (*Gastroboletus vividus*), an undescribed gastroid *Caloboletus*, a number gastroid *Suillus*, among others.

Chalciporus piperatus

(Bull.) Bataille

PEPPERY BOLETE

CAP: 2–9 cm across, rounded-conical to bun shaped when young, becoming broadly convex, to nearly plane and wavy around edge in age. Margin with a narrow band of overhanging tissue. Color variable, reddish brown, ocher-brown, brick red, light brown to pinkish tan with darker streaks, paler toward margin, and often pinkish ocher to beige in age. Surface viscid to moist, innately fibrillose to smooth. **TUBES:** Sunken around stipe, relatively thin. **PORES:** Rather large and angular. Coppery pinkish tan to pinkish or ocher-brown. Not staining, or just slightly reddish brown. **STIPE:** 4–10 cm long, 0.6–1.5 cm thick, cylindrical with an enlarged base. Pinkish beige to reddish tan or ocher-brown, with brilliant yellow mycelium at base. **FLESH:** Thin, soft. Pinkish buff to light brown, to pale ocher-brown in stipe. **ODOR:** Indistinct. **TASTE:** Acrid, peppery. **KOH:** Dark reddish brown on cap. **SPORE DEPOSIT:** Olive-brown. **MICROSCOPY:** Spores 8.5–12 x 3–4 µm, somewhat spindle shaped to elliptical, smooth, thick walled.

ECOLOGY: In scattered troops, sometimes in small clusters on ground, in duff, or in moss. Most often in coniferous forest, almost always near *Amanita muscaria* (and sometimes *A. gemmata* and *A. pantherina*). Common throughout our range. Fruiting from midfall through winter.

EDIBILITY: Best to avoid, possibly toxic (at least when raw).

COMMENTS: The fruitbodies of *Chalciporus* species are the smallest of any of the boletes in our area; in combination with the coppery pinkish tan to ocher pore mouths, peppery taste, and bright yellow mycelium at the stipe base, identification of the genus is straightforward. Those *Chalciporus* with blue-staining pores are usually called *C. piperatoides*, but a number of different species likely go by each of these names. All these *Chalciporus* usually fruit in close proximity to *Amanita* (those in the section *Amanita*, especially *A. muscaria*) and probably parasitize their mycelia.

Tylopilus humilis

Thiers

HUMBLE BOLETE

CAP: 5–8 (12) cm across, rounded-convex when young to broadly convex in age, margin inrolled and often crimped at first, lobed and generally not fully expanding in age. Reddish brown with an amethyst purple margin when young (occasionally more extensively amethyst purple), becoming pinkish brown to dark yellowish brown in age. Surface dry to moist, becoming finely suedelike to smooth. **TUBES:** Broadly attached or sunken around stipe, short, 0.5–1 cm long. **PORES:** Quite small, round to irregular. Pale whitish cream washed with pink when young, becoming dull straw yellow or brownish yellow in age. Bruising pinkish brown, especially distinct when young, sometimes slowly. **STIPE:** 3.5–5.5 cm long, 1.5–4.5 cm thick, short and sometimes quite irregular. Base often tapered and slightly rooting, often with a conspicuous tuft of white rhizomorphs. Whitish at first, becoming darker to tan in age, bruising dingy olive-brownish. Surface smooth, not reticulate. **FLESH:** Firm, pallid, slowly becoming pinkish when cut, sometimes with purplish stains immediately below cap surface, and occasionally showing bluing in the stipe in age. **ODOR:** Indistinct. **TASTE:** Sweetish, mild. **KOH:** Unknown. **SPORE DEPOSIT:** Olive-brown. **MICROSCOPY:** Spores 8–12 x 3–4 µm, cylindrical to spindle shaped, smooth, thin walled.

ECOLOGY: Solitary or scattered on ground in sandy soil. Rare, only known from a few locations, with pines or with manzanita, Madrone, and scrub oak, from Santa Cruz County north to Mendocino County. Fruiting in fall into early winter.

EDIBILITY: Likely edible, but too rare to be considered for the table.

COMMENTS: This modest bolete is quite unusual, especially for the West, which has very few *Tylopilus* species. The small size, often dumpy stature, mild taste, and pale pores that bruise pinkish brown are distinctive. *T. indecisus* is generally much larger and often has a paler cap and a reticulated stipe. It could also be mistaken for a small button of one of the King Boletes, but those species have well-developed, reticulate stipes and pores that do not bruise distinctly pinkish brown.

Tylopilus indecisus (sensu CA)

CAP: 8–25 cm across, bun shaped at first, often lobed, margin usually incurved, becoming broadly convex to flat, sometimes weakly uplifted or wavy in age. Usually blushed with amethyst purple, with whitish areas toward margin when very young, becoming dingy tan-brown to pinkish tan with beige areas and extensive brown streaking and a pale grayish to lilac-gray margin, eventually darker brown overall. Surface finely velvety to nearly smooth, slightly viscid or dry when young, soon viscid to moist. **TUBES:** Fairly thick, narrowly attached to stipe or with a narrow gutter around stipe apex. **PORES:** Small and nearly round at first, becoming elongated and coarser at maturity. Dingy whitish beige at first, soon pinkish tan, becoming pinkish brown in age. Quickly staining brown when scratched or rubbed. **STIPE:** 7–15 cm long, 2–5 cm thick, straight or curved, bulbous toward base but tapered at extreme base. Pale creamy beige to tan, developing brown blotches, sometimes stained dark blue-gray near base. Surface with extensive reticulum over most of length, at first pale but soon contrasting darker brown. **FLESH:** Thick, firm, fleshy, solid. Marbled with white, beige, and tan, sometimes browner with a lilac tone in cap. **ODOR:** Indistinct. **TASTE:** Mild to slightly sweet. **SPORE DEPOSIT:** Pinkish tan to reddish brown. **MICROSCOPY:** Spores 10.5–15 x 3–5 µm, cylindrical to elliptical, smooth.

ECOLOGY: Solitary or scattered in small troops, generally in sandy soil with live oak. Rare, known from the San Francisco Bay Area south to Santa Cruz County. Fruiting in late fall and early winter.

EDIBILITY: Edible and good, but rather rare.

COMMENTS: The size and stature, colors, and reticulate stipe are reminiscent of the King Boletes; however, the pore surface that soon becomes pinkish tan and quickly stains brown sets it apart, and the cap texture is also different (slightly velvety rather than smooth). A related species associated with Black Oak in California is *T. ammirattii;* it does not have a reticulate stipe and stains purplish where bruised. Also compare with the smaller, often misshapen *T. humilis,* which lacks stipe reticulation.

Porphyrellus porphyrosporus
(Fr. & Hök) E.-J. Gilbert
DARK BOLETE

CAP: 5–15 (18) cm across, rounded to bun shaped at first, becoming broadly convex to plane, occasionally weakly uplifted or wavy in age. Dark grayish black to dark vinaceous brown at first, fading to grayish brown, or dark gray. Often getting parasitized by *Hypomyces*, which starts as a whitish bloom over cap, before covering whole fruitbody with a white mold. Surface dry, finely velvety to suede-like. **TUBES:** Broadly attached to deeply grooved around stipe, short when young, up to 1.5 cm long in age. **PORES:** Small and angular at first, becoming moderately large and irregular in age. Color variable, typically deep reddish brown to reddish black at first, fading to grayish brown, reddish gray to gray. Staining dingy greenish blue to bluish black, then maroon-brown when bruised, but can skip blue phase and go straight to maroon-brown. **STIPE:** 6–15 cm long, 1–3 cm thick, often club shaped with a swollen lower portion, but tapering at extreme base, to more or less equal in age. Dark grayish black, deep reddish brown, grayish brown to gray with a whitish base, often staining reddish when scraped. Surface dry, not reticulated to extensive reticulum over upper portion. **FLESH:** Thick, firm, fleshy. Whitish, often staining bluish in cap, then reddish brown, typically unchanging, or reddening in stipe. **ODOR:** Indistinct. **TASTE:** Mild. **SPORE DEPOSIT:** Salmon buff to reddish brown. **MICROSCOPY:** Spores 12–18 x 6–7.5 μm, cylindrical to narrowly elliptical, smooth.

ECOLOGY: Solitary or scattered in troops on ground, moss, or duff, occasionally on decaying logs and stumps in coniferous forest, especially with Sitka Spruce, but also with pine on North Coast. Very common on the Far North Coast, rare south of Humboldt County. Fruiting in early fall into winter in south.

EDIBILITY: Reportedly edible, but local experience is lacking.

COMMENTS: The dark colors alone set this bolete apart. Other distinguishing features are the dry velvety cap, contrasting white flesh, and bluish to reddish staining reaction. There is much variation in the degree of staining reactions (or lack thereof), as well as degree of stipe reticulation. More work is needed to see if the western North American taxon is the same as the European *P. porphyrosporus*. If turns out to be a distinct, *P. olivaceobrunneus* would be the correct name for this species. *P. atrofuscus*, described from Point Reyes in Marin County, also appears to be a separate species. It has a distinctly olive-toned cap and a strongly reticulated stipe, and grows with pine.

Leccinum manzanitae

Thiers

MANZANITA BOLETE

CAP: 7–20 cm across, rounded when young, becoming broadly convex to bun shaped and then nearly plane and sometimes wavy. Margin with a thin flap of tissue overhanging and often folded under edge of pore layer. Bright red to orangey red to brick red or dark brownish red, often with a streaked appearance. Surface viscid to sticky or dry, fibrillose with flattened scaly patches, nearly smooth in patches, sometimes cracked near margin. **TUBES:** Broadly attached to stipe, rather long. **PORES:** Round, fairly small, whitish cream (in some collections dusky gray-brown at first but becoming paler as cap expands), then darkening to olive-brown in age. **STIPE:** 8–20 cm long, 1.5–4 cm thick, cylindrical with a club-shaped base or swollen and bulbous at midpoint. Whitish to pale cream or tan, but most of surface ornamented with pale to dark reddish brown or dark gray to nearly black scabers, sometimes forming a raised reticulum. Base with blotches of navy blue to blue-green stains, sometimes extensive. **FLESH:** Thick, firm, stipe tough, fleshy-fibrous. Whitish, slowly staining pinkish especially in flesh of cap and upper stipe when cut. **ODOR:** Indistinct. **TASTE:** Mild. **SPORE DEPOSIT:** Dull brown. **MICROSCOPY:** Spores 13–19 x 3–5.5 μm, spindle shaped to cylindrical, smooth, moderately thick walled.

ECOLOGY: Solitary, in groups or clumps under manzanita and Madrone. Very common throughout our range, fruiting in fall, with smaller numbers through winter, occasional in spring.

EDIBILITY: Edible with caution. Some people have adverse reactions. If you do choose to eat it, always cook it well.

COMMENTS: This species is by far the most common of its genus in our area, occurring south to San Luis Obispo County. *L. largentii* is practically identical, but supposedly associated only with Bearberry Manzanita (originally mistakenly described as associated with Toyon). Other similar taxa described from our area include *L. armeniacum*, which has a pinkish apricot cap, a white stipe with whitish scabers (never darkening), and slow, patchy fuscous staining to the flesh. Although *L. armeniacum* was originally described as occurring with Madrone, we have found it in Douglas-fir–redwood forest without any hardwoods present. *L. constans* has a pinkish buff cap with darker fibrils, a white stipe with whitish to pale brown scabers that darken in age, and whitish flesh that does not stain. There also appears to be a few undescribed *Leccinum* in our area, including a brown-capped species with dense black scabers and vinaceous pink-staining flesh, as well as a beautiful, all-white species.

Leccinum cyaneobasileucum

Lannoy & Estadès

BLUING BIRCH BOLETE

CAP: 4–12 cm across, hemispheric at first, soon bun shaped, to broadly convex, sometimes uplifted with a wavy margin in age. Dark gray to grayish brown or tan, sometimes paler beige, usually with a narrow white band around margin. Surface moist to dry, suedelike to finely tomentose, often bumpy and rough at first, becoming nearly smooth. **TUBES:** Deeply depressed at stipe, relatively shallow at first, becoming long. **PORES:** round, tiny at first, staying small in age. Creamy white to beige when young, soon pale buff, dingy in age. **STIPE:** 5–15 cm long, 1–3 cm thick, cylindrical with a swollen, club-shaped base, usually tall and straight. White to creamy white, but covered in small plaques and scabers; these are very pale and dense at first but soon become light gray to dusky gray and then nearly black and more spread out as stipe elongates. Base often with bright teal to blue-green blotches, and slowly bluing when scraped. **FLESH:** Whitish marbled with beige, solid, unchanging. **ODOR:** Indistinct. **TASTE:** Mild. **KOH:** Orange on cap, yellow in stipe flesh. **SPORE DEPOSIT:** Dusky brown. **MICROSCOPY:** Spores 11–21 x 3.5–7 μm, spindle shaped, smooth.

ECOLOGY: Solitary or scattered in troops, sometimes in large numbers on urban lawns under planted birches. Frequently found with *Russula versicolor*, *Paxillus cuprinus*, *Lactarius pubescens* var. *betulae*, and *Cortinarius hemitrichus*. Common in late summer and early fall.

EDIBILITY: Edible, and with a good flavor, but often quite slimy in texture.

COMMENTS: The habitat, grayish cap, pale stipe with grayish scabers, and blue-green stains at the base distinguish this species. Another introduced species, *L. scabrum*, has a dark brown cap when young and darker, denser scabers, and lacks the blue-green stains on the stipe; it also occurs under urban birches.

Gyroporus castaneus (sensu CA)

CHESTNUT BOLETE

CAP: 4–10 cm across, bun shaped to broadly convex, to nearly plane, slightly lobed or with uplifted areas in age. Warm brown to orange-brown, sometimes with rather bright orange areas, fading to brownish tan or ocher-tan, often with a narrow but strongly contrasting whitish beige margin. Surface dry, suede-like, to smoother in age. **TUBES:** Broadly to narrowly joined to stipe. **PORES:** Nearly round, quite small, white to creamy or pale straw colored, developing yellowish ocher and brown tones in age. **STIPE:** 5–10 cm long, 1.5–3 cm thick, equal, or often squat with a swollen, club-shaped base. Orange to ocher-brown, strongly contrasting with pores. Surface finely pruinose to smooth, often with multiple cracks around circumference in dry weather. **FLESH:** White throughout, mealy and brittle, more like a *Russula* than any other bolete in our area. Stipe stuffed with white pith when young, often hollow at maturity. **ODOR:** Pleasant, nutty. **TASTE:** Mild to sweet and nutty. **KOH:** Orange on cap. **SPORE DEPOSIT:** Creamy yellow to straw colored. **MICROSCOPY:** Spores 10–14 x 5–6 μm, elliptical, smooth, colorless to pale yellow in KOH.

ECOLOGY: Solitary, in groups or sometimes in arcs under Coast Live Oak. We saw it once under pine on the Far North Coast. Quite rare through most of our area, but patches seem to fruit every year and can be prolific in favorable weather, especially in warm and wet spells in fall and spring.

EDIBILITY: Edible and good, but rare. Apparently not as good as the East Coast species going by this name, according to Phil Carpenter.

COMMENTS: This is probably our most distinctive bolete, identified by the small size, orange-brown cap and stipe with contrasting creamy white pores, hollow stipe, and brittle, mealy texture. Genetic data show that the eastern North American taxon going by this name is different from the California collections.

Suillus brevipes

(Peck) Kuntze

SHORT-STALKED SLIPPERY JACK

CAP: 3–10 cm across, round to convex at first, becoming broadly convex to plane, margin downcurved when young, often with a thin, sterile band of tissue. Color variable, generally dark brown, but can be reddish brown, vinaceous brown to grayish brown when young, fading to cinnamon brown or even yellowish brown in age. Surface smooth, viscid, covered with glutinous slime when wet, shiny if dry. **TUBES:** Broadly attached to slightly decurrent, relatively shallow, 0.4–1 cm long. **PORES:** Small, round to slightly irregular. Creamy when young, soon yellow, becoming golden and developing an olivaceous tint as spores mature. **STIPE:** 2–6 cm long, 1–3 cm thick, short, stout, equal or swollen towards base. White, developing yellowish colorations in age. Surface dry, smooth to slightly pruinose, lacking glandular dots when young, occasionally scattered in age. **FLESH:** Firm when young, soft in age. White, but often developing yellow color in stipe in age. **ODOR:** Indistinct. **TASTE:** Mild. **KOH:** Dark gray on cap, grayish brown on tubes, lilac on flesh. **SPORE DEPOSIT:** Cinnamon brown. **MICROSCOPY:** Spores 7–9 (10) x 2.5–3.5 µm, elliptic to oblong, smooth, inamyloid. Pleurocystidia in bunches, cylindrical to club shaped, cheilocystidia cylindrical to broadly club shaped.

ECOLOGY: Solitary, scattered, or in small clusters under pines. Very common under Shore Pine on the Far North Coast, and Bishop Pine north of San Francisco. Also commonly found in urban areas under planted pines. Often fruits in large flushes in fall, but scattered specimens can pop up anytime, especially during times of fog drip.

EDIBILITY: Edible, one of the better Slippery Jacks. The slimy cap "skin" peels off easily, making it a much more desirable meal. Older specimens get really soft fleshed and cook up rather slimy, so stick with young buttons, or dry and powder older specimens.

COMMENTS: This is one of the ubiquitous Slippery Jacks in the coastal pine forest north of the Bay Area. It can be recognized by the variable-colored, viscid cap (often with some shade of dark brown to grayish brown) that fades in age; yellow pores; and short, white stipe. *S. quiescens* is very similar and is often tough to distinguish from *S. brevipes*, especially in age. However, the caps are light brown when young, and often with bluish green blotches, darkening in age; whereas *S. brevipes* usually fades in age. The stipe of *S. quiescens* is generally pale yellowish and develops fine glandular dots in age. It is locally common and widespread under pines, especially Bishop Pine. *S. pseudobrevipes* is very similar, but often has a paler cap and has a thin, evanescent partial veil. *S. pungens* has a whitish cap that becomes olive and finally orange-brown, and has dark glandular dots covering the stipe.

Suillus pungens
Thiers & A. H. Sm.

PUNGENT SLIPPERY JACK

CAP: 4–15 cm across, round to convex at first, becoming broadly convex to plane, margin incurved when young. Color extremely variable, milky to dingy white to pale olive at first, soon becoming olive, changing in a patchy manner to brownish to orange-brown to bright orange, finally orangish, cinnamon brown or even reddish brown. Surface smooth, viscid, covered with glutinous slime when wet, shiny if dry. **TUBES:** Broadly attached to slightly decurrent, relatively shallow, 0.2–1 cm long. **PORES:** Small, round to slightly irregular. Creamy white and covered with pinkish to pinkish buff droplets when young, becoming yellowish to dingy yellow in age. **STIPE:** 3–8 cm long, 1–3 cm thick, equal or club shaped, enlarged toward base. Whitish with pinkish glandular dots when young, developing yellow or dingy coloration or stains, dots darkening in age. Surface greasy to sticky, covered with glandular dots. **FLESH:** Firm and white when young, soft and becoming yellowish in age. **ODOR:** Pungent, citruslike or lemony, or fruity with a slightly disagreeable undertone. **TASTE:** Generally unpleasant, pungent, slightly acidic. **KOH:** Dark gray to blackish on cap, reddish on tubes, vinaceous on flesh. **SPORE DEPOSIT:** Brown. **MICROSCOPY:** Spores 9.5–10 x 2.8–3.5 μm, elliptic to subcylindric, smooth.

ECOLOGY: Solitary, scattered, clustered, or in large troops under Monterey Pine. It often grows alongside the Pine Spike (*Chroogomphus vinicolor* subsp. *californicus*). Very common throughout our range, though generally only with planted pines in northern part. Like many Slippery Jacks, it can fruit throughout the year, especially in irrigated areas, but is most abundant in fall or early winter.

EDIBILITY: Edible, but some people are put off by the pungent taste or slimy texture of older specimens. The slimy cap "skin" should be peeled before the mushroom is eaten. A very small percentage of people have adverse reactions to the glutinous Slippery Jacks, so they should be tried in moderation the first time.

COMMENTS: This is one of the most variably colored mushrooms: the cap can be milky white, olive, brown, orange, reddish brown, or yellow, or a combination of all these colors at once! Besides the variable cap color, it can be recognized by the creamy young pores with pinkish droplets, glandular dotted stipe, growth with Monterey Pine, and pungent odor and taste. Older *S. quiescens* are similar, but lack the pungent taste and generally only have fine glandular dots in age. *S. brevipes* has a dark brown cap at first and a short white stipe that lacks dots. *S. anomalus* has a dark brown to brownish orange cap and deeply decurrent pores. It is only known from a few sites in Siskiyou County and interior Humboldt County under Ponderosa Pine. *S. punctatipes* has a brown cap with darker streaks and slightly decurrent pores, and grows with hemlock and/or fir. It's rare in California, occurring in higher elevation forest in the coast ranges.

Suillus pseudobrevipes

H. Sm. & Thiers

CAP: 4–12 (15) cm across, round to convex at first, becoming broadly convex to plane, margin incurved and covered with partial veil tissue when young, expanding in age. Color variable, often dingy yellowish to honey brown at first, darkening to reddish brown to brown in age. Surface smooth, but often with appressed-fibrillose streaks, viscid, covered with glutinous slime when wet. **TUBES:** Sunken around stipe to decurrent, shallow, 0.5–1 cm long. **PORES:** Round to irregular, tiny at first, enlarging slightly in age. Yellow at first, becoming golden, then dingy dull ocher-yellow in age. **STIPE:** 2–8 cm long, 1–3 cm thick, equal, enlarged or tapered toward base. Whitish when young, becoming yellowish and often developing dingy yellowish brown discolorations or stains in age. **PARTIAL VEIL:** Thin, whitish, though sometimes tinged purplish, floccose to thinly membranous, leaving hanging remnants on cap margin and a simple, thin ring on stipe, which can disappear altogether in age. Surface greasy to sticky, lacking glandular dots. **FLESH:** Firm when young, soft in age. Whitish to pale yellow, discoloring pinkish around larva tunnels, and often orangish brown in stipe base in age. **ODOR:** Indistinct. **TASTE:** Mild to sour; lemony. **KOH:** Brownish on cap. **SPORE DEPOSIT:** Cinnamon buff to pale brown. **MICROSCOPY:** Spores 7–9 x 2.5–3 μm, oblong, smooth, walls slightly thickened.

ECOLOGY: Solitary, scattered, in small clusters or troops under pines. Very common under Ponderosa Pine in the Santa Cruz Sandhills and under Bishop and Shore Pines on the North Coast. Often fruiting soon after first rains in fall into spring, sporadically fruiting in areas of fog drip through summer.

EDIBILITY: Edible, but soft fleshed and slimy, and often sour tasting.

COMMENTS: The viscid yellowish to honey brown cap that darkens in age, yellow pores, stipe lacking glandular dots, and thin floccose veil that leaves remnants on the cap margin and a thin ring on the stipe are distinctive. *S. glandulosipes* is very similar in many regards, but has dark glandular dots all over the stipe. *S. luteus* has a thick membranous veil when young, collapsing to a thick, viscid, purplish ring. *S. pseudobrevipes* should be compared with *S. brevipes,* which is generally darker when young, has a short white stipe lacking the glandular dots, and does not have a veil. Also compare with the Douglas-fir–associated *S. caerulescens* and *S. ponderosus.*

Suillus tomentosus

(Kauffman) Singer

BLUING SLIPPERY JACK

CAP: 4–12 (15) cm across, round to convex at first, becoming broadly convex to plane, margin incurved and often fuzzy when young, becoming irregular and occasionally slightly uplifted in age. Base color yellowish to golden at first, partially to completely covered with grayish yellow to grayish brown fibrils or scales, base color darkening, golden orange, while scales become yellowish, red to reddish brown. Surface covered with fibrils or scales, but they wipe off easily, and cap can be completely smooth, especially in wet weather, viscid, but not glutinous. **TUBES:** Broadly attached to slightly decurrent at first, becoming sunken in age, 0.5–1.5 cm long. **PORES:** Tiny at first, enlarging in age, angular. Cinnamon brown to ocher at first, becoming ocher-yellow to yellow, to dingy olive-yellow in age. Slowly staining blue to bluish gray when bruised. **STIPE:** 2–12 cm long, 1–3 cm thick, equal, enlarged lower, but often tapered at base. Yellowish to ocher-orange, covered with whitish to pale gray glandular dots when young, dots darkening, grayish brown to reddish in age. Surface resinous, sticky, covered with glandular dots. **FLESH:** Firm when young, soft in age, whitish, yellow to ocher-orange, slowly staining bluish, especially in cap. Stipe flesh very soft, sometimes becoming hollow in age. **ODOR:** Indistinct. **TASTE:** Mild to sour; lemony. **KOH:** Vinaceous brown on flesh. **SPORE DEPOSIT:** Olive-brown to cinnamon brown. **MICROSCOPY:** Spores 7–10 (12) x 3–4 (5) μm, spindle shaped to elongate-ovoid, smooth, inamyloid, pale yellowish in KOH.

ECOLOGY: Solitary or scattered on ground or in duff under a host of different pines. Generally occurs with Bishop and Shore Pines on the coast, very common under Ponderosa Pine in the mountains. Very common throughout our range, often fruiting soon after first rains in fall into winter.

EDIBILITY: Edible, but really soft fleshed and sometimes sour tasting.

COMMENTS: This common *Suillus* can be recognized by the yellow to golden cap often completely covered in grayish yellow fibrils or scales when young, but the scales often scuff off and become darker, yellowish to reddish in age. The ocher to yellow pores, resinous to sticky stipe covered with glandular dots, and, most importuntly, blue-staining flesh make this Slippery Jack unmistakable. The staining can be faint at times, but is best observed by pinching a bit of flesh out of the center of the cap. Generally developing slowly, it can be anywhere from pale bluish gray to deep sky blue. Darker specimens could be mistaken for *S. fuscotomentosus*, but that has a dark brown cap when young and does not stain blue (or does so only slightly in the stipe base). *S. caerulescens* has a whitish partial veil, orangish brown-staining pores, and blue-staining flesh in the stipe base. It is common under Douglas-fir. *Xerocomus subtomentosus* has a dry, velvety cap, large angular blue-staining pores, and a dry (not resinous) stipe.

Suillus fuscotomentosus

(Kauffman) Singer

POOR MAN'S SLIPPERY JACK

CAP: 4–12 (15) cm across, round to convex at first, becoming broadly convex to plane, margin downcurved and often fuzzy at first, occasionally becoming irregular and slightly uplifted in age. Yellowish brown to dark brown, but often completely covered with brown, grayish brown to reddish brown fibrils and scales, base color becoming ocher-brown, cinnamon to yellow-brown, scales darkening slightly. Surface evenly covered with fibrils or scales, spreading out and becoming more appressed as cap expands, slightly viscid to dry. **TUBES:** Broadly attached to sunken around the stipe, 0.5–1.5 cm long. **PORES:** Angular, small at first, enlarging in age. Cinnamon gray and often with resinous milky droplets when very young, becoming orangish buff to ocher-orange, then more golden, and finally dingy olive-yellow to olive-orange in age. **STIPE:** 4–12 cm long, 1–3 cm thick, equal, or enlarged toward base. Yellowish, cinnamon to ocher, covered with grayish to pinkish buff glandular dots when young, dots darkening, grayish brown, vinaceous brown to almost black in age. Base occasionally with vinaceous red stains and sometimes staining blue. Surface resinous, sticky, covered with glandular dots. **FLESH:** Firm when young, very soft in age, yellow to ocher-orange, not staining. **ODOR:** Indistinct. **TASTE:** Mild to unpleasant. **KOH:** Dark gray to black on cap, pinkish at first, to pinkish gray on flesh. **SPORE DEPOSIT:** Olive-brown to cinnamon brown. **MICROSCOPY:** Spores 9–12 x 3–4 µm, spindle shaped to subcylindric, smooth, inamyloid, pale yellowish in KOH.

ECOLOGY: Solitary, scattered, in clusters or large troops on ground under three-needle pines (Knobcone, Monterey, and Ponderosa). Very common in the southern part of our range, becoming less abundant on the coast to the north, but is common inland. Often fruiting from after first fall rains, into winter.

EDIBILITY: Edible, but one of the worst-tasting Slippery Jacks (that's saying something).

COMMENTS: The dark cap covered with fibrils or scales; ocher to yellow pores; resinous to sticky stipe covered with glandular dots; growth with three-needle pines; and lack of a partial veil or blue staining make this *Suillus* distinctive. *S. tomentosus* is similar, but has a yellow to orangish cap with grayish yellow to reddish scales, and slowly stains blue, especially in the cap flesh. It grows with both two- and three-needle pines. *S. lakei* has a brown to reddish brown, scruffy, scaly cap, but has a partial veil and grows with Douglas-fir.

Suillus lakei

(Murrill) A. H. Sm. & Thiers

WESTERN PAINTED SUILLUS

CAP: 4–15 cm across, convex when young, becoming broadly convex to nearly plane. Margin inrolled and enclosed with the partial veil when young, often with thin hanging veil remnants as it expands, to bald in age. Color variable, reddish brown, orange-brown, dark brown to brick red with a brown to yellowish brown base color. Surface dry to thinly viscid, fibrillose to floccose-scaly when young, becoming more appressed-fibrillose, to occasionally nearly smooth with small scales in age. **TUBES:** Broadly attached to decurrent, shallow, 0.2–1 cm long. **PORES:** Small, occasionally radially arranged, angular. Yellow at first, becoming ochraceous orange, to ochraceous brown to dingy brown in age. Staining pinkish brown to brown when bruised. **STIPE:** 3–8 (12) cm long, 1–3 (4) cm thick, equal or with a swollen base. Yellow above the veil. Yellowish with reddish brown streaks or discoloration below veil at first, discoloring to dingy brown in age. Base occasionally staining bluish when young if scraped. Surface dry, often matted-fibrillose below veil, smooth above veil. **PARTIAL VEIL:** Thin, fibrillose to felty-tomentose, covering cap margin and forming a sheathing band around stipe when young, leaving ragged hanging remnants on margin and a bandlike ring on upper stipe, which can disappear in age. **FLESH:** Thick, firm. Yellowish to pinkish beige when young, slowly staining pinkish to reddish, darker, dingy yellowish in age, stipe more pallid, sometimes staining bluish green in stipe base when

young. **ODOR:** Indistinct. **TASTE:** Mild to slightly acidic. **KOH:** Unknown. **SPORE DEPOSIT:** Cinnamon brown. **MICROSCOPY:** Spores 7–11 x 3–4 (5) µm, subelliptic to subcylindric, smooth, inamyloid, colorless to pale greenish yellow in KOH.

ECOLOGY: Solitary, scattered, in small clusters or troops under Douglas-fir. Very common throughout our range. Fruiting from first rains in fall through winter. *Gomphidius subroseus* and *G. smithii* are often found growing along with it.

EDIBILITY: Edible, but not desirable.

COMMENTS: This fibrillose- to floccose-scaly–capped *Suillus* is distinctive when young, but if it loses the scales, as it sometimes does in age, it can be confused with a few other species. Other distinguishing features are the small angular yellow pores that stain pinkish brown when damaged, white veil, and growth with Douglas-fir. Older specimens are most likely to be mistaken for *S. caerulescens*, which has an ochraceous buff to yellowish brown cap, with scattered appressed fibrils, yellow pores that stain brownish, a dry whitish veil, and pronounced blue staining in the stipe base. *S. ponderosus* is also similar, but has a viscid orangish to reddish brown cap and viscid yellowish partial veil. Old *S. tomentosus* often have reddish scales on a yellowish to reddish brown cap, but lack a veil, and the flesh stains blue, especially in the cap. Although not known from California, the common Pacific Northwest species *S. cavipes* has a similar scaly cap, but has large angular, decurrent pores and a hollow stipe, and grows with larch (*Larix* spp.).

Suillus caerulescens

A. H. Sm. & Thiers

FAT JACK

CAP: 4–15 (20) cm across, convex when young, becoming broadly convex to nearly plane. Margin inrolled and enclosed with partial veil when young, often with thin hanging veil remnants as it expands, to bald in age. Ochraceous buff, cinnamon buff to yellowish brown, with a paler or yellow margin when young, generally more tawny brown to orangish brown in age. Surface viscid to dry, appressed-fibrillose with matted fibrils to somewhat streaked, occasionally with small scales. **TUBES:** Broadly attached to decurrent, 0.3–1 cm long. **PORES:** Moderately large, angular, occasionally radially arranged. Pale yellow at first, becoming golden yellow to dingy ochraceous orange in age. Staining cinnamon brown to orangish brown when bruised. **STIPE:** 3–10 (12) cm long, 1–3 (4.5) cm thick, equal, peg shaped (tapering toward base) or with a swollen lower portion but tapering below to base. Yellowish at first, quickly developing orangish brown to brown stains and discoloring to dingy yellowish brown in age. Base slowly staining bluish when scraped. Surface dry to slightly tacky, often matted-fibrillose below veil, dry and smooth above veil. **PARTIAL VEIL:** Fibrillose, felty-tomentose to thinly membranous, covering cap margin and forming a sheathing band around stipe, leaving hanging remnants on margin and a band-like ring on upper stipe. White to dingy whitish, discoloring brownish in age, typically not gelatinous or only slightly so. **FLESH:** Thick, firm when young, soft in age. Whitish to creamy at first, becoming pale yellow, discoloring reddish

brown around larva tunnels, slowly staining bluish in stipe base. **ODOR:** Indistinct. **TASTE:** Mild to slightly lemony. **KOH:** Unknown. **SPORE DEPOSIT:** Cinnamon brown. **MICROSCOPY:** Spores 7.5–11 x 3.5–5 μm, elliptic to nearly oblong, smooth, inamyloid, colorless to ochraceous in KOH.

ECOLOGY: Solitary, scattered, in clusters or troops under Douglas-fir. Very common through much of our range, but more scattered on the Far North Coast. Fruiting from first rains in fall through winter. *Gomphidius oregonensis* often grows along with it.

EDIBILITY: Edible. When young, one of the better *Suillus*.

COMMENTS: The medium to large size; ochraceous buff to yellowish brown, slightly viscid to dry cap; yellow pores that stain brownish; dry whitish veil when young; whitish to pale yellow flesh with bluish staining in the stipe base; and growth with Douglas-fir are important characters to look for when identifying this common *Suillus*. *S. ponderosus* is nearly indistinguishable from it in age, but can be told apart by the warmer, more orangish brown to reddish brown cap colors and, most importantly, the viscid yellow veil when young. It could also be mistaken for old or rain-beaten *S. lakei*, but that usually has some appressed scales remaining on the cap and has darker, dingy yellow flesh that slowly stains pinkish to reddish brown

SYNONYM: *Suillus imitatus* var. *imitatus* A. H. Sm. & Thiers.

Suillus ponderosus

A. H. Sm. & Thiers

SLIMY FAT JACK

CAP: 6–18 (25) cm across, convex when young, becoming broadly convex to nearly plane. Margin inrolled and enclosed with the partial veil when young, often with hanging veil remnants as it expands, generally bald in age. Cinnamon brown to orange-brown or occasionally yellow to yellow-brown when young, especially around margin. Becoming paler, cinnamon to ochraceous brown and often with dingy brownish stains in age. Young caps sometimes with sea green stains or discoloration. Surface smooth, glutinous to viscid, appearing streaked or occasionally with small scales in age. **TUBES:** Broadly attached to decurrent, 0.3–1.5 cm long. **PORES:** Moderately large, often radially arranged, angular. Pale yellow at first, becoming yellow to dingy ocher-yellow. Staining cinnamon brown when bruised. **STIPE:** 5–10 (15) cm long, 2–4 (6) cm thick, often widest in middle, and tapering toward base, to more or less equal. Yellow at first, quickly developing orangish brown stains and discoloring dingy brown. Base often staining bluish green when scraped. Surface viscid from veil, smooth and dry to slightly viscid above and below veil. **PARTIAL VEIL:** Membranous with a yellow to yellow-orange glutinous slime covering underside when young, becoming a cinnamon buff to reddish cinnamon, gelatinous, bandlike ring or dissolving in age. **FLESH:** Thick, firm when young, soft in age. Yellow, occasionally discoloring reddish brown around larva tunnels, slowly staining greenish to greenish blue in stipe base. **ODOR:** Indistinct. **TASTE:** Mild to acidic, sharp, lemony. **KOH:** Blackish on cap, vinaceous on flesh. **SPORE DEPOSIT:** Brown. **MICROSCOPY:** Spores 6–10 (12) x 3.8–5 µm, elliptic, spindle shaped or nearly oblong, smooth, colorless to yellowish in KOH.

ECOLOGY: Solitary, scattered, in clusters or troops under Douglas-fir. Very common from Sonoma County north, occasional south of Sonoma County. Fruiting from first rains in fall into early winter.

EDIBILITY: Edible. Good when young, mushy and slimy in age.

COMMENTS: The large size; cinnamon brown to reddish brown cap; yellow pores; viscid yellow veil when young; yellow flesh with greenish staining in the stipe base; and growth with Douglas-fir help set this *Suillus* apart from all but the very similar *S. caerulescens*. The latter differs by generally having a paler, yellowish brown cap when young, a white, flocculent to slightly membranous veil that is dry or only slightly viscid, and paler flesh. Older specimens are nearly indistinguishable from each other. *Gomphidius glutinosus* is often found growing next to *S. ponderosus*, especially in the northern part of our range, whereas *G. oregonensis* seems to occur with *S. caerulescens*. However, there may be some crossover.

SYNONYM: *Suillus imitatus* var. *viridescens* A. H. Sm. & Trappe.

Suillus luteus

(L.) Roussel

SLIPPERY JACK

CAP: 4–12 (15) cm across, round to convex at first, becoming broadly convex to plane. Color variable, dark brown, reddish brown to vinaceous brown at first, remaining these colors, or fading slightly, to yellow-brown, and often developing dark streaks and/or olivaceous tones in age. Surface smooth, viscid, covered with glutinous slime when wet, shiny if dry. **TUBES:** Sunken around stipe to broadly attached, shallow, 0.2–1 cm long. **PORES:** Round to irregular, tiny at first, enlarging slightly in age. Yellow when young, becoming golden, then dingy olive-yellow to orangish brown in age. **STIPE:** 3–8 cm long, 1–2.5 cm thick, equal or club shaped, enlarged toward base. Whitish to yellowish with darker glandular dots when young, often purplish around veil, developing dingy coloration or stains in age. **PARTIAL VEIL:** Thick and membranous, white with a purplish gelatinous underside, forming a flaring skirt at first, soon collapsing and dissolving in age. Surface greasy to sticky, covered with glandular dots, veil remnants slimy. **FLESH:** Firm when young, becoming soft. Creamy at first, becoming yellowish in age. **ODOR:** Indistinct. **TASTE:** Mild. **KOH:** Unknown. **SPORE DEPOSIT:** Brown to cinnamon. **MICROSCOPY:** Spores 7–9 x 2.5–3 µm, oblong, smooth.

ECOLOGY: Solitary, scattered, clustered, or occasionally in large troops under pine. It has been introduced to California; native to northern Europe. In California generally restricted to urban areas, especially with planted Scots Pine (*Pinus sylvestris*), but can occur with native pines. Fruiting in fall or early winter.

EDIBILITY: Edible. It has a pleasant taste, but is rather soft fleshed and cooks to a slimy texture. As for most Slippery Jacks, peeling the glutinous slime off the cap, or drying and powdering the mushrooms and using them as a seasoning (when you have forgotten how slimy they were) is the best way to consume them.

COMMENTS: The viscid brown cap, yellow pores, and thick membranous skirtlike veil when young are distinctive. The veil often has a purplish gelatinous underside, but occasionally can be all white. *S. pseudobrevipes* is similar, but is often paler. It also lacks the glandular dots on the stipe and has a thin veil that often disappears altogether in age. A similar montane species, *S. brunnescens*, is associated with Sugar Pine (*Pinus lambertiana*). Although uncommon, *S. brunnescens* is scattered through the Coast Ranges. It generally has a pale cap when young, darkening in age, and has a purplish partial veil that clings to the cap margin and doesn't form a ring on the stipe. *S. glandulosipes* has a glutinous, variable colored, cinnamon pink, yellowish to reddish brown cap at first, darkening in age to more vinaceous brown; a dotted white stipe; and a thick veil that clings to the cap margin and does not form a ring on the stipe.

Suillus umbonatus

E. A. Dick & Snell

SLIM JACK

CAP: 2–6 (10) cm across, conical to conical-convex at first, becoming broadly convex to plane with a low broad umbo, margin even, often with glutinous veil remnants when young, occasionally becoming wavy and slightly uplifted in age. Yellowish buff, grayish buff, to warm tan at first, covered with pale orangish brown glutinous slime, becoming olive tinged and slime discoloring reddish to orangish brown, often in streaks or stained over disc. Surface smooth, glutinous to viscid, appearing streaked in age. **TUBES:** Decurrent to broadly attached, 0.3–1 cm long. **PORES:** Moderately large, radially arranged, angular. Creamy yellow at first, becoming yellow to golden, to dingy golden in age. Often staining pinkish when bruised. **STIPE:** 3–8 cm long, 0.5–1 cm thick, equal, tapering downward or sometimes enlarged toward base. Whitish at first, becoming yellowish and developing dingy orangish brown stains, especially where handled and often with obscurer glandular dots. Surface viscid from veil, smooth and dry above, slightly tomentose to smooth below veil. Base often with white rhizomorphs. **PARTIAL VEIL:** Gelatinous membrane when young, forming a collarlike ring on stipe soon after it breaks free of cap margin, and dissolving into a viscid, bandlike ring in age. Pale translucent brown to orangish brown when young, becoming dingy cinnamon brown as it collects spore drop. **FLESH:** Thin, firm when young, very soft in age. Pale creamy yellow to dingy yellow, slowly staining light cinnamon brown. **ODOR:** Indistinct. **TASTE:** Often sour, lemony, or slightly unpleasant to mild. **KOH:** Reddish brown on cap. **SPORE DEPOSIT:** Olive-brown to cinnamon brown. **MICROSCOPY:** Spores 7–11 x 3.5–4.5 µm, narrowly elliptic to nearly oblong, smooth, yellowish in KOH.

ECOLOGY: Solitary or scattered, occasionally in small clusters on ground or in duff under Shore Pine. Found from coastal Sonoma County north. Fruiting soon after first rains in fall into winter. A small, dark (undescribed) *Chroogomphus* species often found growing with it.

EDIBILITY: Nontoxic, but so are Banana Slugs.

COMMENTS: The small size, often umbonate cap covered with translucent orangish brown glutinous slime, gelatinous partial veil, and restriction to Shore Pine habitat on the North Coast are distinctive. Coastal collections tend to be smaller and more slender than montane collections with Lodgepole Pine (*Pinus contorta* subsp. *murrayana*). There is still some question if the North American species is the same as the European *S. flavidus*. For the time being, we have chosen to continue using the North American name.

17 • Polypores and Allies

This section includes a wide array of species grouped here because they share a similarly structured fertile surface: a pored layer of tubes on the underside of the cap (although a few similarly shaped species with a smooth or wrinkled fertile surface are also included here).

Some species here have fruitbodies with distinct stipes and can easily be confused with the boletes. However, these "stalked polypores" tend to have a tough fibrous or leathery texture (rather than soft or fleshy). Of these, *Albatrellus* and its allies, *Boletopsis*, and *Coltricia*, are mycorrhizal.

The rest of the stalked polypores appear to be nonmycorrhizal; a few grow on wood, some grow in soil, and at least one emerges from a sclerotium.

Phaeolus, *Abortiporus*, *Onnia*, and *Bondarzewia* have "rosettelike" or amorphous fruitbodies that frequently develop overlapping tiers or lobes in age, and which often appear terrestrial (due to fruiting from buried wood or roots). Both *Phaeolus* and *Onnia* can also form shelves or conks.

Some polypores form hoof-shaped to fan-shaped, woody fruitbodies called conks. Although the fruitbodies of some are annual, those of the most common and conspicuous species are perennial and add new layers of tissue every year, giving many a layered appearance. Such fruitbodies are produced by *Pseudoinonotus*, *Bridgeoporus*, *Ganoderma*, *Rhodofomes*, *Fomitopsis*, *Laricifomes*, *Phellinus*, and *Porodaedalea*.

Other polypores with shelflike fruitbodies are more watery and fleshy, and often fruit in large overlapping tiers or groups on wood. These include *Amylocystis*, *Fistulina*, and *Laetiporus*, all of which form short-lived or annual fruitbodies. The smaller, thinner-fleshed shelving polypores are probably the most familiar and commonly encountered group. These include *Trametes*, *Trichaptum*, *Bjerkandera*, and *Postia*. A few oddball polypores with different fruitbody morphologies (crustlike, sheet forming, flabby, or small and spiky-fuzzy) are here as well. See *Fuscoporia*, *Gloeoporus*, *Phlebia*, and *Postia*.

Cryptoporus is more confusing: the rounded, polished ocher to dingy white fruitbodies have a pore layer hidden within the tough outer surface.

We've included a miscellaneous assortment of species in this section because of their superficial resemblance to polypores; all have smooth or weakly wrinkled fertile surfaces rather than pores. *Stereum* have thin, leathery, parchmentlike fruitbodies forming overlapping shelves on wood. *Cotylidia* and *Thelephora* have rosettelike, vase- or fan-shaped, terrestrial fruitbodies.

Albatrellopsis flettii

(Morse ex Pouzar) Audet

BLUE POLYPORE

CAP: 5–15 (25) cm across, variably shaped: round and convex with a downcurved margin at first, becoming uplifted with a lobed or wavy margin; when growing in clumps often fused and lobed, expanding into wavy-edged rosettes in age. Baby to sky blue or royal blue when young, often with darker "water spots," becoming mottled blue and pallid grayish or greenish blue, often developing orangish spots; occasionally extensively dingy gray to ocher-gray in age, but usually retaining some blue tones where cap is protected from elements. Surface dry, smooth to velvety or finely scaly, cracking in age. **PORES:** Slightly to deeply decurrent, small, round to slightly irregular, becoming ragged in age, tubes relatively short. Bright white at first, becoming creamy white, occasionally with orangish stains. **STIPE:** 5–15 cm long, 1–4 cm thick, central or off-center, equal to irregular. White when young, developing orangish stains, and often dingy grayish blue in age. **FLESH:** Thick, very firm and rubbery, but somewhat brittle in cap. White to cream. **ODOR:** Indistinct to slightly farinaceous. **TASTE:** Slightly farinaceous. **KOH:** Slight yellowing on cap. **SPORE DEPOSIT:** White. **MICROSCOPY:** Spores 4–5 x 3–3.5 μm, ellipsoid to ovoid, smooth, inamyloid.

ECOLOGY: Solitary or scattered, often in fused clusters on ground under conifers, most commonly with Western Hemlock, occasionally also in mixed Douglas-fir and Tanoak woods. Locally common north of San Francisco Bay Area, rare in Santa Cruz Mountains. First fruiting in early fall, but fruitbodies persist well into winter.

EDIBILITY: Edible, but not popular.

COMMENTS: This is one of the most beautiful and unusually colored mushrooms in our region. When young, it can easily be recognized by its large size, blue cap, and white pores. The blue color may be restricted to the creases or folds in the cap of older fruitbodies. *Neoalbatrellus subcaeruleoporus* also has a bluish cap, but is much smaller and has bluish to bluish gray pores.

SYNONYM: *Albatrellus flettii* Morse ex Pouzar.

Neoalbatrellus subcaeruleoporus

Audet & B. S. Luther

LITTLE BLUE POLYPORE

Cap: 1.5–5 cm across, convex to plane, often waxy in age. Light sky blue, baby blue to grayish blue, more rarely darker sky blue, fading and developing orangish stains in age. Surface dry, finely tomentose to smooth. **PORES:** Slightly decurrent, small, round, tubes very short. Sky blue to grayish blue. Stipe: 1–4 (6) cm long, 0.3–1 cm thick, central to off-center, equal or tapered downward. Sky blue to grayish blue, sometimes developing orangish stains in age. Surface dry, smooth. Flesh: Thin, brittle at first, somewhat corky in age. Whitish to pale bluish gray. Odor: Indistinct. Taste: Indistinct. KOH: No reaction. Spore deposit: White. Microscopy: Spores 4–5 x 3.5–4 μm, broadly ellipsoid to nearly round, smooth, inamyloid.

ECOLOGY: Solitary, scattered, or in small clusters on ground or moss under conifers, especially Western Hemlock. It has a preference for mossy banks on old road cuts or disturbed trailsides. Rare from Sonoma County northward. Fruiting in fall or early winter.

EDIBILITY: Reportedly edible, but rare and small. The counterpart species in eastern North America has a bitter-metallic taste.

COMMENTS: The small size, sky blue to grayish blue colors, small pores, and often off-center stipe are distinctive. *Albatrellopsis flettii* is another blue-capped species, but is much larger and has white pores.

MISAPPLIED NAME: *Neoalbatrellus* (=*Albatrellus*) *caeruleoporus* is an eastern species; our western species is a distinct species and was recently described.

Albatrellus avellaneus

Pouzar

CAP: 3–12 (18) cm across, convex with a downcurved margin at first, often irregular and lobed, becoming plane, uplifted and wavy in age. Whitish, creamy, or with vinaceous buff fibrils and scales at first, generally becoming pale buff to ochraceous, or occasionally darker vinaceous buff in age. Surface dry, appressed-fibrillose to finely scaly, cracking in age. **PORES:** Decurrent, small, round to slightly angular, at times becoming ragged in age. White at first, becoming creamy to pale yellowish, staining yellow when bruised. **STIPE:** 2–8 cm long, 1–4 cm thick, central to off-center, equal to club shaped. White to creamy when young, developing yellowish to ocher stains. Surface dry, smooth. **FLESH:** Thick and firm, but brittle in cap. White to cream, staining yellow. **ODOR:** Indistinct. **TASTE:** Mild. **KOH:** Yellow on all parts, brightest on flesh. **SPORE DEPOSIT:** White. **MICROSCOPY:** Spores 4.5–6 x 4–4.5 μm, nearly round to ovoid, smooth, inamyloid.

ECOLOGY: Often in clusters, occasionally solitary, usually in large troops and patches on mossy duff under Sitka Spruce. Very common on Far North Coast, fruiting from early fall into winter.

EDIBILITY: Edible and good. It has a firm, meaty texture and sweet flavor. It also turns a beautiful golden color when cooked.

COMMENTS: This polypore is one of the most prolific mushrooms of the coastal spruce forest, often growing in huge patches of hundreds of fruitbodies. It is easily recognized by its pallid colors with yellowish stains, white pores, and terrestrial growth. Some collections have smaller spores, within the size range of *A. ovinus*, suggesting that there may be cryptic species involved. *A. subrubescens* is very similar, but is usually slightly paler and often has a sweet odor; it is most reliably distinguished by its distinctly amyloid spores.

Scutiger pes-caprae

(Pers.) Bondartsev & Singer

CAP: 5–15 (20) cm across, convex with a downcurved and often lobed margin, becoming plane to uplifted with a wavy margin in age. Deep dark brown to reddish brown, occasionally with paler, orange-brown or ocher-gray patches. Surface dry, often soft and plush; covered with scales or clumped fibrils. **PORES:** Broadly attached to deeply decurrent, angular and in some places slightly mazelike, becoming ragged in age, tubes relatively shallow. White to creamy at first, becoming dingy in age and becoming olive greenish where bruised. **STIPE:** 2.5–8 cm long, 1–4 cm thick, lateral or off-center (rarely central), more or less equal to irregularly shaped. Whitish to yellowish when young, darkening to reddish brown in age. Surface dry, finely velvety, scaly or smooth, apex often with weak reticulum formed by rudimentary pores. **FLESH:** Thick, very firm, tough. White to cream, occasionally bruising pinkish. **ODOR:** Indistinct to strongly disagreeable, occasionally like rancid urine in age. **TASTE:** Mild to nutty when young, unpleasant in age. **SPORE DEPOSIT:** White. **MICROSCOPY:** Spores 8–11 x 6–8 μm, ellipsoid to teardrop shaped, smooth, inamyloid, colorless in KOH.

ECOLOGY: Solitary, in small clusters or scattered on ground in mixed hardwood-conifer forest, especially under Tanoak and Douglas-fir. Widespread, but generally uncommon. First fruiting in fall, but fruitbodies persist well into winter.

EDIBILITY: Edible when young, but flavor becomes poor to disgusting in age.

COMMENTS: The fibrous-scaly, reddish brown cap, white to creamy pores, off-center to lateral stipe, and growth on the ground are distinctive. *Scutiger ellisii* (=*Albatrellus ellisii*) is often paler, more coarsely scaly, and quicker to develop greenish stains over the whole fruitbody, not just the pores. It is not known from the coast, but is common in the Sierra Nevada and Cascade Range. *Jahnoporus hirtus* has a finely velvety brown to grayish brown cap, white pores, and a brownish gray stipe. It often grows on or around decaying stumps, and has a bitter-metallic or iodine taste with a lingering bitter aftertaste.

SYNONYM: *Albatrellus pes-caprae* (Pers.) Pouzar.

Jahnoporus hirtus

(Cooke) Nuss

BITTER IODINE POLYPORE

CAP: 4–12 (18) cm across, kidney shaped, spatula shaped, or irregularly round, convex to plane, margin wavy, lobed, often uplifted in age. Dark brown, warm brown, dull grayish brown, or tan. Surface dry, covered with stiff to velvety hairs. **PORES:** Broadly attached to deeply decurrent, angular, becoming ragged in age, tubes relatively short. Bright white at first, becoming creamy in age. **STIPE:** 2–12 cm long, 1–4 cm thick, often lateral or off-center, rarely central, equal to irregular. Whitish to gray, becoming brown. Surface dry, velvety. **FLESH:** Thick, very firm, tough. White to cream. **ODOR:** Mild to fragrant. **TASTE:** Extremely bitter, bitter metallic or iodine-like, persistent. **SPORE DEPOSIT:** White. **MICROSCOPY:** Spores 12.5–17 x 4.5–5.5 μm, spindle shaped to cylindrical, smooth, inamyloid.

ECOLOGY: Solitary or scattered, occasionally in small clusters. Often appearing terrestrial but emerging from buried roots around rotting stumps and dead trees, sometimes directly from stumps or logs. Widespread and common throughout our range. Fruiting from fall into spring.

EDIBILITY: Inedible.

COMMENTS: The gray or brown cap, tough stipe, and extremely bitter taste set this species apart. *Scutiger pes-caprae* is similar, but has a fibrous-scaly brown cap and creamy pores and a mild taste, and often stains slightly greenish. *Bondarzewia mesenterica* is often larger and thinner fleshed, usually lacks a distinct stipe, has a mild acrid or only slightly bitter taste, and has round, amyloid ornamented spores.

Boletopsis grisea

(Peck) Bondartsev & Singer

CAP: 5–15 (25) cm across, convex or lobed with a inrolled margin at first, expanding to plane or uplifted with an irregular, wavy margin in age. Off-white to pale gray at first, developing dark gray streaks and splotches as it ages, occasionally completely dark gray in age. Surface dry, smooth to finely fibrillose, occasionally cracking into small scales when dry. **PORES:** Slightly to deeply decurrent, small, round to slightly irregular, tubes relatively shallow. Bright white at first, often discoloring grayish to vinaceous gray in age. **STIPE:** 3–10 cm long, 1–5 cm thick, equal, tapered toward, or with, a rounded base. Off-white to pale gray, developing dark gray streaks. Surface dry, smooth. **FLESH:** Thick, firm, becoming brittle in cap. White to gray, slowly staining vinaceous gray when cut, occasionally with darker gray or vinaceous spots or stains in age, or with bluish green stains in stipe when waterlogged. **ODOR:** Indistinct to slightly farinaceous. **TASTE:** Mild to strongly bitter. **KOH:** Forest green on cap, grayish green on flesh, pores, and stipe. **SPORE DEPOSIT:** Light brown. **MICROSCOPY:** Spores 5–7 x 4–5 µm, irregular, angular (overall ellipsoid to globose in shape), with irregular warts and projections.

ECOLOGY: Solitary or scattered, often buried in needle duff, fruiting in fall and winter. Uncommon under Shore and Bishop Pines, patchily distributed from Mendocino County northward.

EDIBILITY: Edible but often bitter tasting. Some collections are mild enough to eat without special preparation, but this genus is not widely consumed in North America.

COMMENTS: This mushroom resembles a bolete or polypore, but is related to *Hydnellum*, *Sarcodon*, and *Thelephora*. It can be easily identified to genus by its stature, firm but often brittle white flesh, shallow white tubes, and growth on the ground. *B. grisea* appears to be limited to pine forest and has whitish to pale gray colors when young, developing darker streaks and stains in age. *B. leucomelaena* has a black cap, white to light gray pores, and a gray stipe (often with an orangish base), and grows with spruce. In addition to the two species mentioned above, we also have an undescribed species that has a gray cap when young, but darkens to vinaceous gray or almost black. It is rare, found in mixed forests of Madrone, Tanoak, and Douglas-fir. There is also a more lilac-gray species that grows with true firs, generally at higher elevations. Some species of *Albatrellus* resemble *Boletopsis*, but have white spores, generally more irregular growth forms, and/or a more scaly cap.

MISAPPLIED NAME: *Boletopsis subsquamosa* (L.) Kotl. & Pouzar.

Polyporus tuberaster group
TUBEROUS POLYPORE

CAP: 5–12 cm across, broadly convex, sometimes with a tightly inrolled margin at first, expanding to plane and often depressed at the center, sometimes slightly wavy, lobed, or occasionally forming weak rosettes. Surface moist to dry, margin often fringed with tufts of hairs. Whitish brown to grayish, surface velvety or with fine, dark, upright, and pointed tufts. **PORES:** More or less circular, quite small (sometimes angular-elongate or reticulate looking over lowest attachment to stipe), tubes shallow, forming a decurrent layer. White at first, becoming creamy or beige when dried out. **STIPE:** 5–10 cm long, 1–4 cm thick, central and cylindrical; sometimes forked or branched. Creamy whitish to beige, developing ocher or grayish areas upward from base. Surface fibrous, roughened. **SCLEROTIUM:** A mass of hard, rubbery tissue ranging from the size of a golf ball to nearly the size of a soccer ball, buried in ground and often overlooked. Nearly round with a dark grayish, wrinkled surface and a marbled white and ocher-yellow interior. **FLESH:** Thick, solid, firm, rubbery or leathery. Whitish. **ODOR:** Indistinct. **TASTE:** Indistinct. **SPORE DEPOSIT:** White. **MICROSCOPY:** Spores 10–16 x 4–7 μm, long-ellipsoid and tapered toward apiculus, smooth, thin walled. Clamps abundant in generative hyphae. Skeletal hyphae thick walled.

ECOLOGY: Generally solitary or in small clusters, fruiting from an underground (and usually overlooked) sclerotium. Most frequently encountered in mixed conifer and oak forests from San Francisco Bay Area south. Fruiting in early fall.

EDIBILITY: Inedible.

COMMENTS: The large, scaly caps and fairly long and/or thick stipes help distinguish this from other polypores in our area. A second, wood-dwelling form is usually lumped together under the name *P. tuberaster* by most identifiers in California, but it is almost certainly a different species (and it may be that neither is the "true" *tuberaster*). The wood-dwelling form has a creamy beige to ocher or orange-brown cap with large, brownish, and flattened fibrillose scales, sometimes with a shingled appearance. The stipe is off-center to lateral and often quite short; it emerges directly from wood (usually medium-size logs and thicker branches of hardwoods).

Polyporus badius (sensu CA)
BLACK-LEGGED POLYPORE

CAP: 2–10 (15) cm across, convex to plane with a sunken center and a downcurved margin at first, becoming wavy or uplifted in age. Dark brown, reddish brown to orange-brown, fading slightly in age. Surface smooth, dry. **PORES:** Very small, round, tubes very short. Creamy white to pale buff, staining slightly darker when fresh. **STIPE:** 1–5 cm long, 0.3–1.5 cm thick, central to off-center. Dark brown to blackish at base and pallid above at first, soon completely blackish. Surface dry, smooth. **FLESH:** Very thin, tough, leathery when fresh. **ODOR:** Indistinct. **TASTE:** Indistinct. **KOH:** No reaction. **SPORE DEPOSIT:** White. **MICROSCOPY:** Spores 7.5–9 x 3–5 μm, cylindrical, smooth, inamyloid.

ECOLOGY: Solitary or scattered on decaying hardwood branches or logs, rarely on conifers. Common, occurring throughout our range. Fruiting in fall, but fruitbodies persist for a long time.

EDIBILITY: Inedible.

COMMENTS: The smooth brown to orange-brown cap, dark stipe, and growth on wood are helpful identification features. *P. leptocephalus* is a similar, common species in our area that differs by its smaller size, paler, caramel-colored cap, and black color only on the lower half of the stipe. *P. tuberaster* is larger, has a scaly cap, and grows from a sclerotium.

Coltricia perennis group

CAP: 2–8 (10) cm across, circular to shallowly funnel shaped or plane with a depressed center, becoming ruffled, to ragged in age. Dark brown to reddish brown when wet, more orange-brown, ocher to cinnamon when dry, with darker and paler concentric zones and a pale margin. Surface dry, velvety, often wrinkled or roughened. **PORES:** Often slightly decurrent, small, very shallow, round to angular, creamy gray at first, soon cinnamon to brown, staining brown when bruised. Obscure near margin or present only as a weakly pitted surface. **STIPE:** 1.5–3.5 cm long, 0.2–1 cm thick, usually central, sometimes slightly off-center. Orange-brown, cinnamon to dark brown. **FLESH:** Thin, tough, leathery. **ODOR:** Indistinct. **TASTE:** Unpleasant. **KOH:** Black. **SPORE DEPOSIT:** Yellowish brown. **MICROSCOPY:** Spores 6–9 x 3.5–5 µm, ellipsoid to cylindrical, smooth, pale yellowish brown.

ECOLOGY: Solitary, scattered, or occasionally in fused clusters on ground or in moss under conifers, especially pine. It has a preference for recently disturbed or burned areas. Moderately common throughout our range. Fruiting in fall and winter, but fruitbodies persist for many months.

EDIBILITY: Unknown.

COMMENTS: The dark, velvety, zonate cap, thin, leathery flesh, central stipe, and terrestrial growth help distinguish this polypore; all other small central-stalked polypores grow on wood. *Coltricia* are ectomycorrhizal, unlike most other small polypores, which are wood decayers. We have more than one species going by the name *C. perennis*, but no modern work has been done on the group in western North America, and the name has probably been misapplied. The name *C. cinnamomea* also has probably been misapplied on the basis of slight differences in cap texture, color, and zonation. *Hydnellum subzonatum* can be indistinguishable from above, but the underside is coated in spines, not pores. The same goes for the somewhat similar *Phellodon tomentosus*.

Cotylidia diaphana

(Schwein.) Lentz

CAP: 1–3 cm across, funnel to vase shaped, often with a ruffled margin. Whitish to creamy, occasionally with a pinkish cast, in age pale grayish white, often with pale brownish zones. Surface dry to moist, fibrous. **UNDERSIDE:** Smooth to wrinkled, whitish cream. **STIPE:** 0.5–2.5 cm long, 0.2–0.4 cm thick, central to slightly off-center, irregular. Whitish cream. **FLESH:** Very thin, tough, leathery. **ODOR:** Indistinct. **TASTE:** Indistinct. **KOH:** No reaction. **SPORE DEPOSIT:** Whitish. **MICROSCOPY:** Spores 4–6 x 2–3.5 μm, ellipsoid, smooth.

ECOLOGY: Solitary, scattered in small troops, or occasionally in fused clusters, often on bare soil in disturbed areas under Coast Redwood. Uncommon, fruiting during the wet season, perhaps most common from late winter into spring. We have only seen it from the San Francisco Bay Area south, but it probably occurs on the North Coast as well.

EDIBILITY: Unknown.

COMMENTS: This oddball can be identified by its ruffled, funnel-shaped cap, smooth underside, pale colors, and terrestrial growth. *Stereopsis humphreyi* has a wavy white cap, a smooth underside, and a more distinct, slender stipe; it occurs in coastal spruce and hemlock forests in the Pacific Northwest but has yet to be recorded from California. *Thelephora* are quite similar but have much darker fruitbodies.

Thelephora terrestris group
FIBER FAN

CAP: 2–5 cm across, occasionally in larger fusing clusters, circular to fan shaped, nearly flat with a depressed center, to broadly funnel or vase shaped; occasionally with stacked, overlapping fans. Brown, reddish brown to tan, often with concentric bands, paler color toward a whitish margin when young, becoming darker overall in age. Surface dry, tomentose to roughly tomentose. **UNDERSIDE:** Smooth to wrinkled or finely veined. Grayish to grayish brown, paler, whitish toward margin at first, darkening in age. **STIPE:** Central to lateral, occasionally absent, often short and somewhat irregular. Brown. **FLESH:** Thin, tough, leathery. **ODOR:** Earthy. **TASTE:** Indistinct. **SPORE DEPOSIT:** Brown. **MICROSCOPY:** Spores 8–12 x 6–9 μm, angular-elliptical, spiny.

ECOLOGY: Solitary, scattered, or occasionally in fused clusters, often on disturbed ground or sandy soil under conifers. Occasionally surrounding small tree seedlings. Common and widespread. Fruiting in fall and winter.

EDIBILITY: Unknown.

COMMENTS: This *Thelephora* can be identified by its zoned, fan- to funnel-shaped cap, smooth or wrinkled underside, and large brown angular-elliptical spores. There are a number of similar and poorly understood *Thelephora* in California; no modern taxonomic work has been done on them. *T. caryophyllea* is usually smaller and has multiple stacked, overlapping, and often fringed caps arising from a central stipe. *T. multipartita* is very similar to *T. caryophyllea*, but is paler, and has a tomentose stipe.

Abortiporus biennis

(Bull.) Singer

FRUITBODY: 4–20 cm across, highly variable in shape: often a simple circular or semicircular blob, often with patches of pores on top; some specimens differentiated into a mass of fused rosettes or shelving clusters. Whitish to pinkish at first, soon staining dingy reddish, often red-brown overall with a paler margin in age. Surface moist to dry, often with red droplets when young, matted-tomentose, often irregular and knobby. Occasionally pores will be over the cap. **PORES:** Often quite irregular, angular to mazelike. Whitish to pale pinkish, staining reddish brown. **STIPE:** Usually absent, sometimes a short irregular, rooting mass. **FLESH:** Fibrous and watery, becoming tough. Whitish, staining pinkish to reddish. **ODOR:** Mild to sweet, sometimes rancid in age. **TASTE:** Mild. **SPORE DEPOSIT:** Whitish to pale yellow. **MICROSCOPY:** Spores 4–6.5 x 3.5–5 µm, broadly ellipsoid to ovoid, smooth, inamyloid. Asexual spores often present, 5–8.5 µm, globose.

ECOLOGY: Solitary or in scattered clusters on buried woody debris. Often in lawns, gardens, or disturbed areas around dying trees (growing from dead roots). Common in urban areas throughout our range, less common in forested habitat. Fruitbodies annual, mostly fruiting in fall and winter, also common in summer in irrigated areas.

EDIBILITY: Unknown.

COMMENTS: This very irregularly shaped polypore can resemble anything from a blob with red droplets to a shelving rosette. The most distinctive features are the pinkish red staining and angular to mazelike pores. It often grows around and engulfs plant stems and grass. Young specimens with red droplets might be mistaken for *Hydnellum peckii*, which has spines on the underside. Older fruitbodies have been mistaken for *Bondarzewia mesenterica,* a much larger polypore that does not become pinkish and has ornamented, amyloid spores. *Climacocystis borealis* has a creamy white hairy cap and angular pores; it is uncommon in California, occurring on the Far North Coast at the base of conifers, especially Western Hemlock.

Bondarzewia mesenterica

(Schaeff.) Kreisel

FRUITBODY: 5–35 (50) cm across, lumpy and irregular at first, expanding into a single or more often compound uplifted rosette of fan-shaped lobes. **UPPER SURFACE:** Brown to grayish brown at center, warmer and often paler outward; yellowish brown to ocher-brown, usually with a beige to whitish margin, often with light and dark zonations. Surface moist to dry, finely velvety with scattered felty patches or small hairy scales, smoother in age. **PORES:** Small, round at first, soon irregular, becoming rather jagged in age. White to cream, occasionally with pale yellowish buff stains in age. **STIPE:** Indistinct or central to lateral, often tapered downward and rooting. Brown when exposed, dingy whitish if buried. **FLESH:** Thick, tough, fibrous, whitish to creamy. **ODOR:** Pleasant when young, rancid in age. **TASTE:** Mild, acrid to slightly bitter. **SPORE DEPOSIT:** White. **MICROSCOPY:** Spores 6–8 x 5–7 µm, globose to subglobose, ornamented with strongly amyloid ridges and warts.

ECOLOGY: At base of dead or dying conifer trees or stumps. Generally solitary, occasionally a few fruitbodies growing around larger trees. Rather uncommon, preferring mature forests on the Far North Coast and in the mountains. Fruiting in fall, persisting well into winter.

EDIBILITY: Edible, but rather tough and often bitter.

COMMENTS: This uncommon large polypore is actually closely related to *Lactarius*! From the fruitbody morphology, one might have a hard time believing this, but the amyloid-ornamented spores look identical to many *Russula* and *Lactarius*. *Jahnoporus hirtus* has a finely velvety gray to brown cap and a more well-defined stipe, a bitter taste, and rather long, spindle-shaped to cylindrical spores. *Laetiporus* are bright golden yellow to orange when young.

Bridgeoporus nobilissimus

(W. B. Cooke) T. J. Volk, Burds. & Ammirati

NOBLE POLYPORE

FRUITBODY: 30–100 (150!) cm across, usually shelflike or hoof shaped; occasionally toplike with a central stipe when growing on top of cut stumps. **UPPER SURFACE:** Off-white, creamy to beige on growing margin, often with green algae, moss, and other debris accumulating on top. Surface coarsely hairy or fuzzy. **PORES:** Very small, round. White to creamy beige, occasionally buff when old. **STIPE:** Generally absent, except when fruiting on top of stumps. **FLESH:** Very thick, tough, fibrous, whitish. **ODOR:** Indistinct. **TASTE:** Not sampled. **KOH:** No reaction. **SPORE DEPOSIT:** White. **MICROSCOPY:** Spores 5.5–7 x 4–5 μm, subglobose to ovoid, smooth, inamyloid.

ECOLOGY: Growing at base of standing snags or large stumps of old-growth true firs. Apparently limited to trees with a trunk diameter of at least 1 m. Extremely rare in our area, known from a single location in northern Humboldt County (on Grand Fir).

EDIBILITY: Inedible.

COMMENTS: The large size, deeply fuzzy cap texture, and occurrence on old-growth firs distinguish this species. Although only one site is known in California, we include the Noble Polypore in the hope of leading to the identification of additional sites. Even after extensive surveying, only about a hundred fruitbodies of this species have ever been found, most of which are in the Cascade Range in northern Oregon. Sadly, many were growing on stumps of logged old-growth trees. This is one of the only fungi in North America to receive special protection, being a state-listed endangered species in Oregon, and it is considered a sensitive species by the U.S. Forest Service and Bureau of Land Management.

Pseudoinonotus dryadeus

(Pers.) T. Wagner & M. Fisch.

GIANT OAK POLYPORE

FRUITBODY: 20–70 cm across, semicircular to fan shaped, compressed and shelflike (2–6 cm thick at rounded margin) or fairly deep (up to 30 cm thick) near attachment to wood. **UPPER SURFACE:** Whitish to creamy or pale beige, mottled with areas of tan-brown to gray or darker brown (especially older parts near attachment to wood). Margin usually palest and covered in many shining droplets, smears, and puddles of ocher-brown to amber liquid, or pitted when dry. Rest of upper surface is dry, chalky to crumbly or hard and smooth or broken into plaques. Old fruitbodies are crumbly-woody and entirely dark reddish brown to blackish brown. **PORES:** Circular, very small, white, not strongly darkening when scratched; vertically arranged in a single layer, brown inside. Tube layer fairly thick (2–5 cm). **FLESH:** Thick, corky when fresh, to hard and woody in age. Mottled brownish flecked with white. **ODOR:** Indistinct to rancid, unpleasant. **TASTE:** Mild to sour. **SPORE DEPOSIT:** Whitish, copious, often covering entire area around older fruitbodies. **MICROSCOPY:** Spores 6–8 x 5–7 μm, irregularly ellipsoid to globose, smooth. Flesh with dark, thick-walled, hooked setae, hyphae without clamp connections.

ECOLOGY: Fruitbodies emerging in early fall and persisting for one to two years, at which point they crumble off the tree. Uncommon on oaks in most of our area; also occurring on conifers on the North Coast.

EDIBILITY: Inedible.

COMMENTS: These massive polypores are very impressive when fresh, rivaled in size only by the largest *Ganoderma* and nudged out of first by the much rarer *Bridgeoporus nobilissimus*. Unlike *Ganoderma*, this species coats its surroundings in a whitish spore powder (not rusty brown), and it has a mostly pale upper surface. *Bridgeoporus* has a deeply fuzzy surface and is restricted to true firs in the far northern part of our range and is extremely rare in California. Both of those genera produce longer-lasting (perennial) fruitbodies. *Inonotus arizonicus*, known from the San Francisco Bay Area southward, is much smaller and thinner, has yellow spores, and grows on California Sycamore (*Platanus racemosa*).

SYNONYMS: *Inonotus dryadeus* (Pers.) Murrill.

Phaeolus schweinitzii

(Fr.) Pat.

DYERS POLYPORE

FRUITBODY: 6–30 (45) cm across, starting as an irregular cushion-shaped blob, soon expanding into a circular disc and then forming a multishelved rosette or fused into a nearly circular, centrally depressed mass. Occasionally occurring as a shelflike bracket higher on trunks of trees. **UPPER SURFACE:** Extremely variable; golden-yellow with a whitish or pinkish margin at first, soon warm brown to dark brown with a paler margin and concentric bands, dark golden brown to reddish brown when mature, occasionally developing darker spots and stains, becoming evenly reddish brown to blackish brown in age. Surface dry, velvety to hairy at first, matted to bumpy in age. **PORES:** Small, angular to highly irregular, often weakly mazelike; with bands and zones of thicker walls. Pale olive-gray to dingy greenish yellow at first, becoming olive-brown, quickly staining dark brown to blackish brown when scratched or bruised. **STIPE:** Stout, irregular and slightly rooting, occasionally lateral or absent. **FLESH:** Soft and spongy at first, becoming tough, fibrous, or corky. **ODOR:** Indistinct. **TASTE:** Unpleasant, rancid or sour. **KOH:** Black. **SPORE DEPOSIT:** White. **MICROSCOPY:** Spores 6–9 x 2.5–5 μm, ellipsoid to ovoid, smooth, inamyloid.

ECOLOGY: Solitary or scattered around base of dead and dying conifer trees and stumps, occasionally on trunks or large downed logs. Very common throughout our range. The most abundant fresh fruitings occur in early fall (or late summer on the Far North Coast), with scattered fruitings into early winter. Rosettes are very persistent; their brittle, crumbly, blackened remains can be found any time of year.

EDIBILITY: Unknown, possibly toxic.

COMMENTS: Although highly variable in appearance, this polypore can be recognized by its cushion- to rosettelike shape, greenish yellow to olivaceous, irregularly shaped pores that stain dark brown, and growth on or around conifers. Highly prized by fiber artists, it yields bright golden yellow dye when paired with an alum mordant, and rich olive greens with an iron mordant. *Onnia triquetra* has a more evenly colored ocher-brown to caramel cap and pores that don't stain dark brown when bruised; it appears to be absent from Douglas-fir and spruce forests, instead growing with pines.

Mushrooms for Fiber Dye

A number of mushrooms can be used to produce safe, brilliant, colorfast dyes on protein fibers (wool, mohair, silk) and are highly prized by fiber artists for these qualities. With many of these mushrooms, it can be as simple as simmering the fiber with the fruitbodies in a pot of water, while other species require mordants (mineral salts) and modulation of the acidity (pH) of the dye bath. Water has a more or less neutral pH around 7. Vinegar will acidify the bath (lower pH), while ammonia can be used to make it more basic (higher pH). Mordants can increase colorfastness, as well as change or brighten the final colors. Alum (*aluminum sulfate*) is generally used to brighten colors, changing yellows to bright gold or turning pastel pinks into bright reds. The other commonly used mordant is iron (*ferrous sulfate*), often used to darken colors or give more earthy tones. For more information regarding mushroom dyes see Alissa Allen's website www.mycopigments.com.

Some of the Redwood Coast's common dye mushrooms:

Phaeolus schweinitzii (Dyer's Polypore)—Our most common dye mushroom and the easiest to use, this is a great species for beginners to try. The photo above shows a range of colors that can be yielded depending on the mordant used and the amount of fungal tissue. Without a mordant you can get pastel yellow to gold (brightened greatly by alum). Olive, army green, and brownish colors are easily achieved by using iron as a mordant. See facing page.

Gymnopilus ventricosus (Jumbo Gym)—This very common and abundant mushroom produces beautiful butter yellow dye. It is best used with an alum mordant and a pH around 4. See p. 132.

Cortinarius smithii (Western Red Dye)—This is considered the holy grail of California dye mushrooms, giving exceptionally bright red, pink, orange, or purple dyes. It is best used with a neutral pH (around 7) and an alum mordant for the red colors, or with a pH of 4 for bright orange colors. Other *Cortinarius* in the subgenus *Dermocybe* are also excellent dye mushrooms. See p. 172.

Hypomyces lactifluorum (Lobster Mushroom)—This yields colors ranging from peachy-orange to neon pink, depending on the pH (orange tones from more acidic baths, pink from more basic solutions). Mature or over-the-hill specimens give the best color. It is best used with alum or without any mordant. See p. 247.

Omphalotus olivascens (Western Jack O'Lantern)—This mushroom can produce violet to smoky purple colors with alum or without any mordant, but these colors can be tricky to achieve (sometimes you'll get underwhelming smoky gray to beige tones). With iron, it gives a very consistent dark forest green. The colors it yields are very similar to *Tapinella atrotomentosa* (p. 411). See p. 394.

Boletus edulis var. grandedulis (California King Bolete)—The mature tubes of all the Porcini produce bright golden dyes (a good way to use fruitbodies that are past their prime for eating). In addition, most of California's boletes will produce yellow to orange dyes. It is best used with an alum mordant and a pH of 4. See p. 422.

Hydnellum and *Sarcodon* spp.—Most of these tooth fungi produce blue-green dyes but are trickier to work with. All require mordants and a sustained pH of 9 for optimal colors. The closely related *Boletopsis grisea* (p. 462) also produces blue-green dyes. See p. 492–497.

Pisolithus arhizus (Dyer's Puffball)—This fungus can be very messy to use: it stains everything, and the spore mass is very powdery and hydrophobic (they float on water). Avoid inhaling spores by handling mature specimens gingerly and working in well ventilated areas or while wearing a mask. Younger (non-powdery) specimens are much easier to work with. They yield rich brown, reddish-brown and orange-brown colors, and they are a great choice for silk dyes. See p. 531.

Onnia triquetra

(Pers.) Imazeki

FALSE DYERS POLYPORE

FRUITBODY: 4–15 cm across, fan shaped or a semicircular bracket, occasionally top shaped with a central stipe. **UPPER SURFACE:** Ocher-brown when wet, paler caramel when dry, darkening in age. Surface dry, velvety-hairy, often with stiff clumps of hairs. **PORES:** Small, round to angular, tubes short. Whitish, buff to brown, often flashing different colors when viewed at different angles. **STIPE:** Absent or short and lateral when growing out of tree trunks, central to off-center when on buried roots. **FLESH:** Concolorous with cap, thick, tough, fibrous. **ODOR:** Indistinct. **TASTE:** Lemony. **KOH:** Blackish. **SPORE DEPOSIT:** Whitish. **MICROSCOPY:** Spores 5–6.5 x 3–4 µm, ellipsoid to ovoid, smooth, inamyloid. Setae 50–80 µm long, scattered, abundant, claw shaped with hooked ends.

ECOLOGY: Generally solitary, occasionally forming overlapping shelves on the trunk near the bases of pine trees, or appearing terrestrial when on buried roots. Fairly common in pine forest throughout our range. Fruit bodies annual, fruiting in the early fall and persisting through the wet season.

EDIBILITY: Unknown.

COMMENTS: This polypore can be recognized by its velvety-hairy, ocher-brown to caramel-colored cap, tough flesh, and growth on or near pines. Centrally stalked, top-shaped fruitbodies are often mistaken for *O. tomentosa*, which is generally thinner fleshed, grows in troops, and lacks the hooked setae; it appears to be restricted to spruce forest. Another species that commonly causes confusion is *Phaeolus schweinitzii*, which differs by its more irregular, slightly mazelike, pale greenish olive to grayish pores that quickly stain brown when bruised. Long known as *Inonotus* (=*Onnia*) *circinatus*, the type collection was found to be the same as *O. tomentosa*. *O. leporina* appears to be a misapplied name in our area.

Amylocystis lapponica

(Romell) Bondartsev & Singer

FRUITBODY: 5–15 cm across, thick, shelflike with a rounded and often lobed margin. **UPPER SURFACE:** Reddish brown to brown with a paler, pinkish white to pale ocher margin, darkening overall in age. Surface moist to dry, spiky-hairy to matted-tomentose or nearly smooth and slightly viscid when weathered by rain. **PORES:** Small, round to slightly irregular. Whitish to pale pinkish buff at first, darkening to buff, quickly staining ocher-brown to brown when handled. **STIPE:** Absent. **FLESH:** Thick, soft and watery, marbled buff, darker near cap surface. **ODOR:** Often strong, chemical or sweet, or at times more sour-earthy. **TASTE:** Mild to slightly bitter. **SPORE DEPOSIT:** Whitish. **MICROSCOPY:** Spores 8–11 x 2.5–3.5 µm, cylindrical, smooth, inamyloid. Cystidia abundant, large, strongly amyloid, tips incrusted.

ECOLOGY: Solitary, scattered, or in shelving clusters on logs, stumps, and standing snags of Sitka Spruce. Locally common on the Far North Coast. Fruiting in fall but persisting well into winter.

EDIBILITY: Unknown.

COMMENTS: The fleshy, reddish brown fruitbodies with a hairy surface, brown staining, and habitat on spruce are distinctive. *Ischnoderma benzoinum* has a more evenly brown, finely plush-velvety cap, tougher flesh, and a strong sweet anise odor. It also has smaller, narrowly cylindrical spores and lacks amyloid cystidia. *Fistulina hepatica* rarely grows on conifers, never has a hairy cap, and has streaked yellowish flesh that soon turns bloody red.

Fistulina hepatica

(Schaeff.) With.

BEEFSTEAK FUNGUS

FRUITBODY: 4–20 (40) cm across, kidney shaped to fan shaped or nearly circular, sometimes with a lateral stalk. **UPPER SURFACE:** Pinkish beige at first, soon blood red, staining reddish brown to vinaceous when bruised or in age. Pebbly, covered with bright golden yellow pimples when young, but in age often extremely wrinkled (like the skin of a Shar-Pei). Layer under cap skin becomes very gelatinous and gooey in age, and often drips off cap margin as reddish slime. **PORES:** Small, round, consisting of unfused, individual strawlike tubes of varying lengths. Whitish to light buff or pale pinkish at first, staining pinkish brown where scratched. **STIPE:** Absent to prominent, lateral when present, 2–10 cm long, 1–5 cm thick, stout, often somewhat rooting. **FLESH:** Thick, meaty, often bleeding juice when cut. Marbled yellowish and white when young (staining blood red after being cut), entirely red or reddish brown in age, marbled with many pale veins. **ODOR:** Indistinct. **TASTE:** Sour, lemony. **SPORE DEPOSIT:** Pinkish brown to pale rusty brown. **MICROSCOPY:** Spores 3.4–4.5 x 2.5–3 μm, ovoid to teardrop shaped, smooth.

ECOLOGY: Solitary or scattered at base of trees or stumps, almost always on Chinquapin or Pacific Wax Myrtle (*Myrica californica*), but we have seen it once on Western Hemlock! Generally uncommon, but can be locally abundant. Fruiting from late summer through early winter; fruitbodies are fairly persistent. Darvin DeShazer alerted us to an overlooked habitat for this species: swampy wax myrtle thickets at sea level on the Far North Coast.

EDIBILITY: Edible, and one of the few wild fungi frequently eaten raw. The lemony sour taste is not a good fit for most wild mushroom recipes, so this fungus should be treated specially.

COMMENTS: This is one of our more distinctive mushrooms; it resembles a bleeding slab of meat emerging from the base of trees. Compare with *Amylocystis lapponica*, a common spruce decayer.

Laetiporus gilbertsonii

Burds.

SULFUR SHELF

FRUITBODY: 5–20 (30) cm across, bloblike or cushion shaped at first, soon becoming fan shaped, often with a lobed and wavy margin, very often occurring in clusters of overlapping shelves. **UPPER SURFACE:** Sulfur yellow overall at first, becoming more orange to peachy orange, with paler concentric bands, light pinkish orange and then whitish when dried out. Surface dry, spongy-soft and slightly velvety when young, soon bald, bumpy. **PORES:** Very tiny, round, tube layer thin. Bright sulfur yellow to pale lemon yellow, fading to whitish when old. **FLESH:** Moderately thin to thick, soft and often juicy at first, becoming firm, meaty to corky, finally crumbly or chalky when old. Pale yellow to whitish. **ODOR:** Indistinct when young, unpleasant in age. **TASTE:** Mild to sour. **KOH:** No reaction. **SPORE DEPOSIT:** White. **MICROSCOPY:** Spores 5–6.5 x 3.5–4.5 μm, broadly oval, smooth, inamyloid.

ECOLOGY: In scattered, overlapping clusters or as solitary brackets emerging from wounds on living or dead trees, stumps, and large logs, especially those of eucalyptus and oak. Fresh fruitings start in summer and can continue into early winter. However, fruitbodies are quite persistent, and the crumbly remains can be found any time of year. Common in southern half of our range, rare on the conifer-dominated Far North Coast, where it is mostly replaced by *L. conifericola*.

EDIBILITY: Edible and very good. However, some people experience gastrointestinal distress after eating it. Be sure to cook it thoroughly, as that will decrease the chance of negative reactions. Only young fruitbodies (brightly colored, soft, spongy, and pliable) should be eaten.

COMMENTS: Also called Chicken of the Woods, this is one of two common species of *Laetiporus* in California. *L. gilbertsonii* is easily recognized by its bright colors and shelving growth on hardwoods. *L. conifericola* is generally slightly larger, has a richer orange cap, and is restricted to conifers. Like *L. gilbertsonii*, it can cause digestive distress (see edibility above).

MISAPPLIED NAME: Although both *L. gilbertsonii* and *L. conifericola* have long been lumped under the name *L. sulphureus*, that name refers to an eastern North American–European species.

Laetiporus conifericola

Burds. & Banik

CONIFER SULFUR SHELF

FRUITBODY: 5–20 (30) cm across, blob-like when first emerging, soon becoming fan shaped and lobed with a slightly to very wavy margin, often in large overlapping rows and clusters of shelves. **UPPER SURFACE:** Sulfur yellow, becoming more bright orange to carrot orange, margin bright lemon yellow when moist. Fading substantially when dry; older fruitbodies can be nearly white. Surface dry to moist, spongy-soft and slightly velvety when young, soon smooth or wrinkly. **PORES:** Tiny, round. Tube layer thin. Sulfur yellow to lemon yellow, fading in age. **FLESH:** Usually fairly thick, soft and juicy at first, becoming firm, meaty to corky, finally crumbly or chalky in age. **ODOR:** Indistinct when young, sour in age. **TASTE:** Mild to sour. **KOH:** No reaction. **SPORE DEPOSIT:** White. **MICROSCOPY:** Spores 6.5–8 x 4–5 μm, broadly ellipsoid, smooth, inamyloid.

ECOLOGY: In scattered, overlapping clusters or in clusters and swarms on dead conifers (including Coast Redwood!). Often on lower trunks of dead standing trees, or on fallen logs in dense rows and shelves. Fresh fruitings start in late summer and can continue into early winter. Crumbly whitish remains of older fruitbodies can be found into spring. Known from Santa Cruz County northwards, but most common on the Far North Coast and in the mountains.

EDIBILITY: Edible and very good. However, some people experience gastrointestinal distress after eating *Laetiporus*; see comments under *L. gilbertsonii* for more information.

COMMENTS: The brilliant orange, wavy, shelf-like fruitbodies often form stunning clusters. *L. gilbertsonii* is very similar, but grows on hardwoods, is often paler (less intensely orange), and has smaller spores.

MISAPPLIED NAME: Formerly called *Laetiporus sulphereus*, an eastern North American–European species that grows on hardwoods.

Fuscoporia gilva

(Schw.) T. Wagner & M. Fischer

FRUITBODY: 1–8 cm across, cushion-shaped blob at first, becoming shelflike to irregularly hoof shaped. **UPPER SURFACE:** Bright golden yellow to mustard yellow at first, bright colors soon restricted to margin, inner areas ocher-brown to deep reddish brown; eventually losing bright colors entirely. Surface dry, velvety at first, roughened and bumpy in age. **PORES:** Very small, round. Golden brown at first, ocher-brown to grayish brown in age. **STIPE:** Absent. **FLESH:** Thin to thick, tough, woody, ocher-brown to deep rusty brown. **ODOR:** Indistinct. **TASTE:** Indistinct. **KOH:** Red on yellow parts, otherwise black. **SPORE DEPOSIT:** Whitish to pale yellowish. **MICROSCOPY:** Spores 4–5 x 3–4 μm, ellipsoid to ovoid, smooth, inamyloid, colorless to yellowish.

ECOLOGY: In scattered masses or overlapping clusters, or occasionally solitary, on hardwood logs and stumps. Very common in the southern part our range, uncommon on the conifer-dominated Far North Coast. Fruitbodies annual but can persist for months.

EDIBILITY: Inedible.

COMMENTS: Through most of the year, this drab brown polypore is overlooked. However, it is eye-catching in fall, when the bright golden yellow fresh growth appears. *F. ferruginosa* is a common pinkish brown to rusty crustlike polypore found on hardwood branches (especially oak). *F. ferrea* is similar to the latter species, but has very slightly longer and narrower spores (5–7.5 x 2–2.5 μm, versus 5–7 x 3–3.5 μm for *F. ferruginosa*). *Inonotus andersonii* is thicker than the previous two species, forms thick, lumpy crusts, often with slightly projecting shelves, and has amazingly bright yellow spores that often coat the ground beneath it. *Pycnoporellus fulgens* has orange to rust-colored fan-shaped brackets, with large pale orangish, irregular pores, and a red KOH reaction.

SYNONYM: *Phellinus gilvus* (Schwein.) Pat.

Porodaedalea pini group
CONIFER MAZE POLYPORE

FRUITBODY: 2–20 (35) cm across, hoof shaped or irregularly shelflike, occasionally more fan shaped. **UPPER SURFACE:** Often with multiple grooves or tiers. Brown to reddish brown with a mustard to golden brown growth margin, becoming dark brown to nearly black. Surface dry, hairy, roughened. **PORES:** Generally round to angular at first, more highly irregular, often mazelike in age. Golden brown, grayish brown to rusty brown. **STIPE:** Absent. **FLESH:** Thin to thick, very tough, woody, reddish brown to rusty brown. **ODOR:** Indistinct. **TASTE:** Indistinct. **KOH:** Black. **SPORE DEPOSIT:** Pale yellowish to light brown. **MICROSCOPY:** Spores 4.5–7 x 3.5–5 μm, ovoid, smooth, inamyloid, colorless to yellowish.

ECOLOGY: Solitary or scattered on trunks of conifers, especially in "armpits" of broken branches. Very common throughout our range on a wide range of host conifers. Fruitbodies perennial, persisting for years.

EDIBILITY: Inedible.

COMMENTS: This common polypore can be recognized by its hoof-shaped or irregularly wavy fruitbodies with dingy olive gold, mazelike pores. *P. chrysoloma* is thinner fleshed and generally more crustlike, with a slightly shelving cap. Also very similar is *P. cancriformans*, which grows in tight, lobed clusters of overlapping caps. *Gloeophyllum sepiarium* is smaller, with elongated, gill-like pores. Fruitbodies of various *Phellinus*, *Inonotus*, and *Fuscoporia* (and others) may be similarly colored, but have round to slightly angular pores.

Phellinus arctostaphyli
(Long) Niemelä
MANZANITA CONK

FRUITBODY: 3–10 cm across, hoof shaped to semicircular, often with concentric ridges and longitudinal cracks. **UPPER SURFACE:** Gray to blackish with a paler growing margin. Surface dry, cracked, roughened, and often multitiered. **PORES:** Quite small, round; additional layers added each growing season. Grayish to light brown or coppery. **FLESH:** Tough, woody. Reddish brown or whitish near attachment to substrate. **ODOR:** Indistinct. **TASTE:** Indistinct. **KOH:** Blackish. **SPORE DEPOSIT:** Dark rust brown. **MICROSCOPY:** Spores 5–6 x 3.5–4.5 μm, ovoid, flattened on one side, smooth, inamyloid. Setae infrequent, awl shaped.

ECOLOGY: Solitary on trunks or branches of living or dead manzanita. Also reported on Chamise (*Adenostoma fasciculatum*). Locally common, but easily overlooked. Fruitbodies perennial, persisting for years.

EDIBILITY: Inedible.

COMMENTS: This small, hoof-shaped conk can be recognized by its cracking cap, pale pores, and growth on manzanita or Chamise. *Phellinus* are a tough group to identify to species, and learning to identify the host wood goes a long way toward being able to identify the fungus. *P. everhartii* is a larger hoof- to fan-shaped conk that grows on oak; *P. pomaceus* is a small conk that grows on *Prunus* trees (plum, cherry, peach, and others); *P. igniarius* grows on variety of hardwoods (especially willow and cottonwood) and has a dark grayish black hoof- to bracket-shaped cap and reddish brown flesh with whitish specks. *P. tremulae* is a slow-growing angular to hoof-shaped conk that often fruits out of wounds on Quaking Aspen (*Populus tremuloides*) in the mountains; it may grow on cottonwood in our area.

Laricifomes officinalis
(Vill.) Kotl. & Pouzar
AGARIKON

FRUITBODY: 5–30 (45) cm across, up to 120 cm long, cushion to hoof shaped at first, becoming multitiered, older specimens becoming cylindrical. One or two additional pore layers added each year. Whitish on fresh growth, grayish brown to ocher, beige, tan, or light brown on older parts, sometimes greenish with algal growth. Surface dry, roughened, often cracking. **PORES:** Very small, round to irregular. Fresh and living layers are white; older, inactive pores are pale beige-tan. **FLESH:** Thick, soft when fresh, soon corky, chalky or crumbly when dry or old. **ODOR:** Indistinct to farinaceous. **TASTE:** Very bitter. **SPORE DEPOSIT:** Whitish to creamy yellow. **MICROSCOPY:** Spores 6–9 x 3–4 μm, cylindrical, ellipsoid to ovoid, smooth, inamyloid.

ECOLOGY: Solitary or with a few scattered over trunks of conifers, usually fairly high up. It has a preference for old-growth Douglas-fir, but smaller conks can be found on younger trees and other conifers. Uncommon on West Coast, known from Santa Cruz County to Pacific Northwest. Fruitbodies perennial, very slow growing and persisting for years or decades.

EDIBILITY: Inedible, but prized for its medicinal properties. If these interest you, please stick to cultivated forms, and don't pick the conks in the wild, as a single conk can live for more than a hundred years.

COMMENTS: The large, pale, multitiered fruitbodies are almost unmistakable. Younger, hoof-shaped specimens are sometimes mistaken for *Fomitopsis pinicola*, but can be told apart by the paler colors, crumbly-chalky flesh, and extremely bitter taste. *Ganoderma brownii* can also have multitiered fruitbodies, but has woody, brown flesh and brown-staining pores. Also known as the Quinine Conk.

"Rhodofomes" cajanderi
(P. Karst.) comb. prov.
ROSY POLYPORE

FRUITBODY: 3–10 (15) cm across, hoof shaped at first, soon expanding into a fan-shaped bracket. **UPPER SURFACE:** Very young fruitbodies emerge as a blob of pinkish white fertile surface, often adorned with reddish droplets when wet. Mature specimens pinkish brown to brown or blackish brown in age, margin usually remaining pale pinkish. Surface usually dry, finely tomentose when young, lumpy and roughened to nearly smooth. **PORES:** Quite small, round, tube layer thin. Pale pink to rosy pink, becoming pinkish brown in age. **FLESH:** Moderately thin, fleshy when young, soon tough, woody. Pinkish brown. **ODOR:** Indistinct. **TASTE:** Indistinct. **KOH:** No reaction. **SPORE DEPOSIT:** White. **MICROSCOPY:** Spores 5–7 x 1.5–2 μm, cylindrical to sausage shaped (curved), smooth, inamyloid.

ECOLOGY: Solitary, scattered, or in overlapping clusters on dead conifers (rarely hardwoods). Widespread and fairly common. Fresh fruitings occur in fall or early winter, but some fruitbodies persist through summer.

EDIBILITY: Inedible.

COMMENTS: The pinkish pores and tough, woody flesh are distinctive. *Rhodofomes roseus* is generally larger and more distinctly hoof shaped and has paler pinkish pores and slightly wider cylindrical spores. *Leptoporus mollis* is whitish, but soon becomes sherbet to coral pink and has very soft flesh; it is uncommon to rare on dead conifers in our area.

Ganoderma brownii
(Murrill) Gilb.
WESTERN ARTIST'S CONK

FRUITBODY: 6–40 (120) cm across, 3–30 cm thick, projecting 5–25 (40) cm outward from substrate. Hoof shaped with many concentric zones or "stair steps," or broader and fan shaped. **UPPER SURFACE:** Dark brown to almost blackish (although it is often covered in rusty brown spores). Growing margin paler, whitish. Surface roughened, often multitiered. **PORES:** Very small, round, additional layers added on each growing season. Whitish to pale yellowish white at first, becoming dingy white to brownish when dormant, staining brown when scratched or bruised if fresh. **FLESH:** Thin to thick, tough, woody. Dark brown to purplish brown. **ODOR:** Not distinct. **TASTE:** Mild. **SPORE DEPOSIT:** Rusty brown. **MICROSCOPY:** Spores 11–12 x 7–8 μm, broadly ellipsoid with a truncate end and an apical germ pore, thick, double walled, inamyloid, brown.

ECOLOGY: Solitary or scattered on trunks stumps and logs of hardwoods, especially common on California Bay Laurel. Fruitbodies perennial, living for many years.

EDIBILITY: Not edible due to its woody texture, but some people collect it to make (purportedly) medicinal tinctures and teas. We discourage collecting it or any other perennial conk, since single fruitbodies can easily live for upward of fifty years!

COMMENTS: *G. brownii* can be recognized by its multitiered, hoof-shaped to fan-shaped fruitbodies, pale pores that immediately stain brown when bruised, and rusty brown spores (which often cover the cap and surrounding substrate). It is apparently restricted to hardwoods and is by far most abundant on California Bay Laurel. *G. applanatum* generally has flatter, fan-shaped fruitbodies, is not restricted to hardwoods, and is common on conifers on the Far North Coast; its spores are much smaller, 7–8 (9) x 4.5–6 μm.

Ganoderma oregonense

Murrill

WESTERN VARNISHED CONK

FRUITBODY: 10–60 (80) cm across, 10–30 cm thick, projecting 10–40 cm outward from substrate. Broadly fan shaped, often lobed or wavy, usually with concentric zones or "stair steps." **UPPER SURFACE:** Bright brick red to dark reddish brown or brownish burgundy (although often partially covered in dull brown spore powder). Growing margin, orange, yellow, or white. Surface smooth, dull when growing, soon varnished and shiny. **PORES:** Very small, round. White at first, becoming very dingy brick pink to brownish in age, staining brown when scratched or bruised when fresh. Pore surface often developing sunken pits or with areas of brown, sawdustlike frass (insect poop). **FLESH:** Thick, quite spongy when fresh, soon tough, but softer than many large polypores. Whitish, buff to honey brown. Skin a thin shellaclike crust. **ODOR:** Not distinct. **TASTE:** Mild to medicinal, somewhat astringent. **KOH:** No reaction. **SPORE DEPOSIT:** Dark rust brown. **MICROSCOPY:** Spores 13–17 x 8–10 µm, ellipsoid with a truncate end, thick, double walled, roughened, inamyloid, brown.

ECOLOGY: Solitary or sometimes in small groups or overlapping clusters on conifers (especially hemlock). Uncommon on the Redwood Coast, more common in the Pacific Northwest. Fruitbodies are annual, first emerging in late summer, mature by late fall, but beginning to break and become rotten by late winter, although recognizable fruitbodies often persist much longer.

EDIBILITY: Not edible due to the woody texture, but some people collect the fresh fruitbodies to make medicinal tinctures or teas. Please use restraint when collecting this species, as a single large fruitbody can produce over a *billion* spores a day!

COMMENTS: An attractive member of the northern polypore flora, this species is easily to recognize by virtue of the large size, shiny surface, and color. *G. tsugae*, may be a synonym, but Robert Gilbertson and Leif Ryvarden distinguished it by its slightly larger pore mouths, slightly smaller spores, and smaller fruitbodies; it may be primarily an East Coast species on hemlocks. Another varnished, red-capped species, *G. polychromum*, has dark and pale zones in the flesh. It is uncommon on oaks and occasionally other hardwoods in our area; it often has an antlerlike stipe with dark zones in the flesh.

Fomitopsis pinicola group
RED-BELTED CONK

FRUITBODY: 5–30 (75) cm across, cushion to hoof shaped at first, becoming more fan shaped in age. **UPPER SURFACE:** Typically whitish to creamy or pale pinkish at margin, then with a reddish to orangish zone inward, and dark brown to blackish near attachment to substrate. Some fruitbodies lack reddish band entirely. Surface dry, roughened. **PORES:** Very tiny, round. Whitish to creamy, slightly dingier in age, sometimes staining pale yellowish buff when bruised. **FLESH:** Thick, very tough corky to woody. Whitish to pale yellow. **ODOR:** Indistinct or fragrant. **TASTE:** Lemony sour to slightly bitter. **KOH:** Reddish to red-brown on flesh. **SPORE DEPOSIT:** Whitish to creamy yellow. **MICROSCOPY:** Spores 6–9 x 3.5–4.5 μm, cylindrical to ovoid, smooth, inamyloid.

ECOLOGY: Solitary or scattered on dead conifer trees and logs. Very common, occurring throughout our range. Fruitbodies perennial.

EDIBILITY: Inedible.

COMMENTS: This species can be recognized by its large size, often three-toned cap (black, red-orange, and pale), and creamy pores that don't stain brown when scratched. However, not all fruitbodies exhibit the reddish band, and this can lead to confusion with *F. ochracea*, which differs primarily by lacking the reddish color on the cap. Perhaps the easiest way to differentiate them is to take a lighter or a match to the cap: the surface of *F. pinicola* melts, whereas *F. ochracea* chars. Another species easily confused for members of this group is *Heterobasidion occidentale*, which generally has an irregularly roughened cap and often grows flattened against logs and trunks or forms a slightly projecting shelf. Other distinguishing features are its lack of a red KOH reaction on the flesh and its round to oval, minutely spiny amyloid spores. *H. irregulare* is nearly identical to *H. occidentale*, but has a preference for pine, whereas *H. occidentale* is more common on northern conifers (spruce, fir, hemlock). *Ganoderma oregonense* has soft flesh at first and an extensively dark red, varnished cap when mature. *G. applanatum* and *G. brownii* have brown flesh and brown staining pores (when fresh) and lack the reddish color on the cap; all *Ganoderma* have brown spores. *Phellinus* species usually have rusty brown flesh and a black KOH reaction. Recent genetic work has shown that we have *F. ochracea* and two other species going by the name *F. pinicola* in North America; neither is the same as the European species. However, no new names have been proposed, and it is not entirely clear how to recognize the differences without genetics data. Two common fungi often grow on old fruitbodies of *Fomitopsis* that have fallen off the tree: *Hypomyces aurantius*, which forms an orange, pimply crust, and *Hypocrea pulvinata*, which forms cushion-shaped ocher-yellow warts.

Cryptoporus volvatus

(Peck) Shear

VEILED POLYPORE

FRUITBODY: 1–6 (8) cm across, round to hoof shaped, pores covered with a thick "veil" forming a hollow pocket on underside. Mature specimens usually have a small round hole on bottom. Butterscotch to light ocher-brown or reddish brown, paler in age, whitish when dry. Surface dry, smooth, shiny, sometimes with a faint netted pattern or cracked into thin plaques. **PORES:** Tiny, round, extremely densely packed, tube layer thick. Whitish to pale buff or occasionally pinkish. **FLESH:** Moderately thick, rubbery at first, becoming tough, whitish to cream, sometimes pinkish. **ODOR:** Indistinct or rubbery or fragrant. **TASTE:** Slightly bitter. **MICROSCOPY:** Spores 12–16.5 x 4–5 µm, cylindrical, smooth.

ECOLOGY: Scattered or in large troops, rarely solitary, on recently dead conifers, especially pine. Fruitbodies usually situated in bark furrows up and down trunk. Common wherever there are dead pines, perhaps less frequent in the northern part of our range. This species is extremely common on other conifers (especially fir) in the mountains, especially after forest fires on burnt trees. Fruitbodies annual, first appearing in late summer (before fall rains) and persisting into the next year.

EDIBILITY: Inedible.

COMMENTS: This rounded to hoof-shaped polypore with pores encased inside a thick, rubbery "veil" is unique. On mature specimens, a small hole is almost always present on the underside; the reason for this is that *Cryptoporus* has a special relationship with bark beetles: As the fruitbody matures, the beetles eat a small hole through the veil and proceed to tunnel through the pores, becoming covered with spores in the process. After they exit the polypore, they burrow into trees to lay their eggs, infecting them with the *Cryptoporus* spores. The following year, *Cryptoporus* fruit out of these very same holes. This method of spore dispersal seems to have worked very well for this fungus, as it is one of the most commonly observed wood rotters in the western North American mountains.

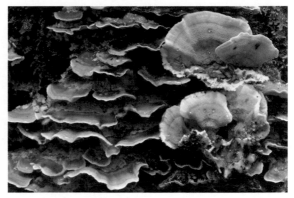

Trichaptum abietinum

(Dicks.) Ryvarden

PURPLE TOOTHED POLYPORE

FRUITBODY: 1–4 cm across (fused fruitbodies can form crusts or fused caps up to 10 cm across), protruding 0.5–3 cm. Bracket to fan-shaped shelves, occasionally more crustlike on undersides of fallen branches and logs, margin becoming ruffled or lobed in age. **UPPER SURFACE:** Beige to buff with a violet to royal purple growing margin at first, whitish when dry, often with a greenish layer of algae in age. Surface dry, coarsely hairy to matted-tomentose, occasionally with concentric zones. **PORES:** Round to angular and slightly mazelike at first, soon becoming toothlike to ragged in age. Violet at first, becoming buff to grayish, often with a violet margin (when still actively growing), dingy buff overall in age. **FLESH:** Thin, tough, leathery. **ODOR:** Indistinct. **TASTE:** Indistinct. **SPORE DEPOSIT:** Whitish. **MICROSCOPY:** Spores 6–7.5 x 2.5–3 μm, cylindrical to sausage shaped, smooth, inamyloid.

ECOLOGY: Often fruiting in masses of single, fused, or overlapping clusters on dead conifers (including redwood!). Very common throughout California, especially on pines and Douglas-fir. Fruitbodies mostly annual, emerging in early fall but persisting throughout year.

EDIBILITY: Unknown.

COMMENTS: When young, this polypore is easily recognized by the violet pores and growth on conifer wood. If it has lost the violet colors, the ragged pores and pale cap are useful clues. *T. fuscoviolaceum* is quite similar, but is less common; it has a crustlike to slightly reflexed fruitbody and often grows on pine. *T. biforme* grows on hardwoods; although it has been reported from California, we have never seen it here. *Chondrostereum purpureum* can look similar in color and shape, but has a smooth underside, grows on hardwoods (especially alder), and has a farinaceous odor.

Bjerkandera adusta

(Willd.) P. Karst.

SMOKY POLYPORE

CAP: 2–5 cm across, semicircular or fan shaped, occasionally crustlike with scattered projecting shelves. Ocher, beige, tan, or brown with a whitish to creamy growing margin, sometimes weakly zonate, developing dark stains in age. Surface moist to dry, obscurely matted-tomentose to smooth. **PORES:** Tiny, round, tube layer very thin. Pale whitish gray to dark smoky gray (paler toward margin), blackening slowly where damaged. **FLESH:** Thin to moderately thick, tough, leathery, white to grayish. **ODOR:** Mealy-mushroomy. **TASTE:** Mild to sour. **SPORE DEPOSIT:** White, scant. **MICROSCOPY:** Spores 5–6 x 2.5–3.5 μm, cylindrical, smooth, inamyloid.

ECOLOGY: Often in masses of overlapping shelves on dead hardwoods, especially alder. Locally common, especially north of San Francisco Bay Area. Fruitbodies annual, first fruiting in early fall but persisting into next year.

EDIBILITY: Inedible.

COMMENTS: The pale surface, smoky gray pores, and gray to blackish staining distinguish this species. *B. fumosa* is larger and thicker fleshed, and has buff to light gray pores. *Trametes* species can look very similar from above, but have white to creamy pores.

Trametes versicolor

(L.) Pilàt

TURKEY TAIL

FRUITBODY: 2–8 (10) cm across, semicircular, bracket to fan-shaped shelves, occasionally rosette clusters. **UPPER SURFACE:** Color quite variable, but usually a combination of dark and pale concentric zones with a white margin. Color of inner areas ranges from dark brown to grayish brown with pale grayish to buff zones, or reddish brown to orange-brown with tan to buff zones, or in some forms dark blue with paler gray zones. Fading when old or dry, often greenish due to algal growth on previous year's caps. Surface dry, distinctly tomentose on paler bands and smooth on darker concentric zones. **PORES:** Tiny, round, tube layer very thin. White to cream, pale buff or occasionally ocher in age. **STIPE:** Absent. **FLESH:** Thin to moderately thick, tough, leathery, white to creamy. **ODOR:** Indistinct. **TASTE:** Indistinct. **SPORE DEPOSIT:** White. **MICROSCOPY:** Spores 5–6 x 1.5–2.5 μm, cylindrical to sausage shaped, smooth, inamyloid.

ECOLOGY: In rows or masses of shelving or overlapping clusters, or occasionally as rosette clusters. Very common throughout our range, especially in live oak and Tanoak forest, but can be found on a tremendous range of hosts; we have even seen it on Monterey Cypress. Fruitbodies annual, first fruiting in early fall but persisting into next year.

EDIBILITY: Inedible due to the tough texture, but often dried and powdered to make teas and tinctures. Long considered a medicinal fungus in Eastern medicine, it has been and continues to be investigated for its immune-boosting and anticancer properties.

COMMENTS: This common polypore can be recognized by its gray to brown colors with paler concentric zones, tiny white pores, and growth on hardwoods. *T. betulina* is very similar from the top, but has gills instead of pores. *T. hirsuta* is usually thicker fleshed and more evenly pale grayish to whitish beige colored. *T. suaveolens* is very similar, but has a smooth to finely tomentose cap and sweet anise odor. *Trichaptum abietinum* has violet colors when fresh and larger pores that become ragged in age; it only grows on conifers. *Stereum hirsutum* has a variably colored, orangey buff cap and a smooth or wrinkled, orangey underside. *Bjerkandera adusta* has a grayish buff to light brown cap and smoky gray pores.

Trametes betulina

(L.) Pilát

GILLED POLYPORE

FRUITBODY: 2–12 (14) cm across, semicircular or fan shaped. Color of surface variable, usually a combination of brown, tan, grayish white, and beige, with many concentric zones. Brighter orange or yellow-capped forms are also occasionally found. Often fading in age and becoming greenish with algae. Surface dry, distinctly tomentose, becoming matted in age. **UNDERSIDE:** Made up of elongated, gill-like pores. Primarily regular and parallel, but in areas also forking, branched, or mazelike. At times more porelike with thick walls, especially when young or near attachment to substrate. White to creamy, occasionally pale buff in age. **FLESH:** Thin to moderately thick, tough, leathery, whitish. **ODOR:** Indistinct. **TASTE:** Indistinct. **SPORE DEPOSIT:** White. **MICROSCOPY:** Spores 5–6 x 2–3 μm, cylindrical to sausage shaped, smooth, inamyloid.

ECOLOGY: Often in rows or masses of shelving or overlapping clusters, or occasionally as rosette clusters on dead hardwoods. Very common throughout our range, especially in live oak and Tanoak forest. Fruitbodies annual, first fruiting in early fall but persisting into next year.

EDIBILITY: Inedible.

COMMENTS: This species looks similar to pale forms of the Turkey Tail (*T. versicolor*) from above, but is completely different as soon as it is flipped over to reveal the "gills" on the underside. The California version of this species is larger, thicker fleshed, and has a more irregular gill arrangement than its eastern North American counterpart, leading to confusion with *Daedalea quercina*. Most California collections we have seen labeled *D. quercina* are in fact *T. betulina*. The tomentose cap and tough, leathery texture for *T. betulina* set it apart from the bald cap and more woody flesh of *D. quercina*. *Gloeophyllum sepiarium* is similar in shape, but has orange to brown colors overall and grows on conifers.

SYNONYM: *Lenzites betulina* (L.) Fr.

Stereum hirsutum

(Willd.) Gray

FALSE TURKEY TAIL

CAP: 1–4 cm across (fused specimens often much larger), shelves protruding 0.5–3 cm. Bracket to fan shaped, occasionally crustlike on underside of logs but forming small shelves near sides. Occasionally funnel shaped when fruiting on top of logs. Margin usually ruffled and lobed in age. Orange to orange-brown to beige, with darker and paler concentric zones at first, becoming darker, with reddish brown tones and a yellowish orange margin; fading greatly in age. Surface dry, distinctly silky-hairy to tomentose, occasionally with concentric bald zones. **UNDERSIDE:** Smooth to slightly roughened, dull to bright orange to creamy orange, at times more grayish orange. **FLESH:** Thin, tough, cartilaginous or leathery. **ODOR:** Indistinct. **TASTE:** Indistinct. **KOH:** No reaction. **SPORE DEPOSIT:** White. **MICROSCOPY:** Spores 5.5–7 x 2.5–3.5 μm, cylindrical, smooth, weakly amyloid.

ECOLOGY: Often in masses of shelving or overlapping clusters on dead hardwoods. Very common throughout our range, especially on live oak and Tanoak. Fruitbodies mostly annual, occasionally perennial, first fruiting in early fall.

EDIBILITY: Nontoxic.

COMMENTS: One of our most common fungi, this species can be recognized by its orangey underside, hairy, zonate cap, smooth or wrinkly underside (lacking pores), and growth on hardwoods. *S. ochraceoflavum* has a dull orangish buff to tan underside, is often smaller, and grows on small hardwood branches and twigs. *S. sanguinolentum* is more crustlike, has a grayish brown underside, and quickly stains red when bruised; it is restricted to conifers. *S. gausapatum* is dingy orange to reddish brown and stains reddish when bruised; it grows on oaks. *Veluticeps abietina* has a dark reddish brown to blackish brown cap and a grayish lilac to purplish brown smooth or warty underside. It grows on conifers (especially true fir) on the Far North Coast, but is far more common in the mountains. Turkey Tails (*Trametes versicolor*) are frequently confused with *S. hirsutum*, but have small, whitish to creamy pores on the undersides. *Tremella aurantia* and *T. foliacea* commonly parasitize *S. hirsutum* and will often be found fruiting alongside them.

Postia caesia

(Schrad.) P. Karst.

BLUE CHEESE POLYPORE

FRUITBODY: 1–5 (7) cm across. A rather thick, fan-shaped to semicircular bracket. **UPPER SURFACE:** Whitish to pale bluish or bluish gray at first, becoming more grayish beige with a paler margin. Clumps of hairs usually bluish to navy blue. Surface hairy, moist to dry. **PORES:** Small, round to irregular, and angular or toothlike; tubes very short. White to grayish, occasionally staining pale blue. **FLESH:** Moderately thick to thin, rather soft, spongy, quite watery when wet. **ODOR:** Fragrant. **TASTE:** Sweet, but often astringent. **SPORE DEPOSIT:** Whitish to pale grayish blue. **MICROSCOPY:** Spores 5.5–7.5 x 1–2 μm, cylindrical to sausage shaped, smooth, inamyloid to weakly amyloid.

ECOLOGY: Solitary or scattered on decaying logs and branches of both hardwoods and conifers. Rather uncommon in the southern part of our range, quite common in the north, especially on Sitka Spruce. Fruiting throughout wet season.

EDIBILITY: Unknown.

COMMENTS: The soft, spongy texture and bluish colors are distinctive. *Tyromyces chioneus* is whitish to grayish with no blue coloration, and often has a slightly firmer, drier texture. *P. fragilis* is white to light brown and quickly stains yellowish to ocher, darkening to reddish brown where bruised.

Postia fragilis

(Fr.) Julich

RED-STAINING CHEESE POLYPORE

FRUITBODY: 1–7 cm across, a fan-shaped or semicircular bracket. **UPPER SURFACE:** Whitish to pale buff when young, to pale reddish brown in age. All parts quickly staining yellowish to ocher, then darkening to reddish brown when bruised. Surface moist to dry, velvety to finely tomentose. **PORES:** Small, round to irregular, tubes very short. White to pale buff, staining like cap. **FLESH:** Moderately thick to thin, rather soft, spongy, quite watery when wet, whitish. **ODOR:** Fragrant. **TASTE:** Mild. **SPORE DEPOSIT:** Whitish to pale grayish. **MICROSCOPY:** Spores 3–5 x 1–2 μm, cylindrical to sausage shaped, smooth, inamyloid, colorless.

ECOLOGY: Solitary or scattered on decaying logs and branches of conifers (rarely on hardwoods). Very common in the north, especially on pine and Sitka Spruce, occasional in the southern part of our range. Fruiting throughout wet season.

EDIBILITY: Unknown.

COMMENTS: The soft spongy texture, whitish to buff colors, and quick yellow to reddish brown staining are distinctive. *Tyromyces chioneus* is a similarly boring whitish polypore that does not stain. Another similar polypore is *Skeletocutis nivea*, which is especially common on the Far North Coast on small conifer branches. It has a bright white to reddish brown cap, bright white pores, and narrower sausage-shaped spores, 3–5 x 0.5–1.0 μm.

Postia ptychogaster

(F. Ludw.) Vesterh.

POWDERY POM POM POLYPORE

FRUITBODY: 0.7–5 cm across, in two distinct forms: a pale creamy buff, irregularly rounded, plush, fuzzy blob that soon becomes a light brown, powdery mass; or expanding and becoming irregularly bracket shaped, with a central to off-center stipe, and small, round to irregular pores. **STIPE:** Small, indistinct, often lateral or asymmetrical. **FLESH:** Rather thin, very soft. Whitish at first, either becoming powdery and light brown or more firm. **ODOR:** Indistinct. **TASTE:** Mild. **SPORE DEPOSIT:** Whitish from basidiospores, light brown asexual spores. **MICROSCOPY:** Basidiospores 4.5–5.5 x 2–3 μm, ellipsoid to cylindrical, thin walled, smooth, inamyloid, colorless. Asexual spores 5.5–10 x 3.5–7 μm, ellipsoid to oblong, often with truncate ends, thick walled, yellowish (brownish in mass).

ECOLOGY: Solitary or scattered on conifer wood, especially old, moss-covered Sitka Spruce branches. Common on the Far North Coast in fall, rarer southward, absent south of Mendocino County.

EDIBILITY: Unknown.

COMMENTS: Truly an oddball, this fungus either forms asexual spores (entirely becoming a powdery mass) or grows up into a more "normal" polypore. Young *Phlebia tremellosa* can also look like pale, fuzzy blobs when young, but soon expand into shelflike fruitbodies with a thin, rubbery texture and pinkish, wavy pores. *Climacocystis borealis* is larger and thicker and more shelflike, and has highly irregular pores.

Gloeoporus dichrous

(Fr.) Bres.

ELASTIC POLYPORE

FRUITBODY: Shelflike to broadly fan shaped, often in overlapping clusters, usually emerging from a crustlike layer of pores spreading over substrate. **UPPER SURFACE:** 1–5 (10) cm across, white to pale grayish white or pale buff. Moist to dry, finely tomentose at first, smooth or slightly bumpy in age. **PORES:** Round, small. Pinkish, lilac, vinaceous pink to vinaceous buff, margin broad, white, and without pores. **FLESH:** Very thin, pore layer distinctly gelatinous, stretching like a rubber band. **ODOR:** Indistinct. **TASTE:** Indistinct. **SPORE DEPOSIT:** Whitish. **MICROSCOPY:** Spores 3.5–5.5 x 0.7–1.5 μm, sausage shaped to cylindrical, smooth, inamyloid.

ECOLOGY: On well-rotted stumps and logs of hardwoods, especially live oak. Common in the southern part of our range but rarely collected. Fruiting from fall into spring.

EDIBILITY: Inedible.

COMMENTS: The combination of crustlike growth sporadically differentiated into perpendicular shelves, white cap, and pinkish buff elastic pore layer is distinctive. *G. taxicola* is a spreading crust with more ocher to orange-brown pores; it grows on conifers. *Phlebia tremellosa* is similarly colored, but has a veined underside and is more flabby than elastic.

Phlebia tremellosa

(Schrad.) Nakasone & Burds.

FRUITBODY: Forming a spreading crust with shelving projections, at times occurring in dense, overlapping clusters on sides of logs and stumps. Shelves 1–5 cm across, projecting 0.5–2.5 cm crustlike portion can be much larger. **UPPER SURFACE:** White, pinkish white to pale pinkish orange. Surface moist to dry, hairy and often ragged around edge. **FERTILE SURFACE:** Veined and wrinkled, often radially arranged, forming pseudopores. Dingy ocher, lilac-grayish, or dull pink to pinkish orange, occasionally blood red in age. **FLESH:** Thin, watery, slightly rubbery, soft and pliable. **ODOR:** Indistinct. **TASTE:** Indistinct. **SPORE DEPOSIT:** Whitish. **MICROSCOPY:** Spores 3.5–4.5 x 1–2 μm, cylindrical to bean shaped, smooth, inamyloid.

ECOLOGY: Often on barkless logs and stumps, occurring on both hardwoods and conifers. Widespread and common. Fruiting from fall into spring.

EDIBILITY: Inedible.

COMMENTS: Whether this species forms a spreading crust or a shelflike fruitbody, the telltale features are the pale, hairy cap and margin, pinkish veined underside, and thin, rubbery flesh. Crustlike fruitbodies should be compared with *P. radiata* (and others mentioned in the comments for that species). *Gloeoporus* can be quite similarly colored, but have more distinctly poroid undersides (with round mouths) and a drier, less flabby texture. *Byssomerulius corium* is a white, spreading crust with a much smoother, paler underside. It is common on small-diameter hardwood branches, especially in the northern part of our area.

18 • Crust Fungi

This is an artificial group encompassing a diverse assemblage of species from many different lineages of fungi (some quite closely related to the more typical mushrooms in this book). They are united by their simple, crustlike, thin, spreading fruitbodies with reduced or inconspicuous fertile surface structures. Some have pores, some have teeth, some have wrinkles or pimples, and some are entirely smooth. We include one ascomycete here (*Hypomyces rosellus*), but most of the crust fungi you'll encounter in our area are basidiomycetes. They are very ecologically important—some are plant parasites, most are ubiquitous wood rotters (practically any moist log you flip over in a forest will have one or more crust fungi growing beneath), and some are even ectomycorrhizal!

Their identification is nearly uniformly impossible without a microscope. Pay attention to the color, texture, and structure of the fruitbody, and subtle characteristics of the substrate (such as host identity, degree of decay, orientation). Even after microscopy, identification can be very difficult. Take note, however, that what these fungi lack in macroscopic charisma they make up for in interesting microstructures! Crystals, bizarre protrusions, encrusted pigmentation, aberrant clamp morphologies, and odd cystidia abound, making this a perfect group for the patient connoisseur with, shall we say, unique tastes. Taxonomic references for our area are lacking, but general references for the study of these "corticioid" fungi can be found (although they are usually expensive). Tom Bruns at the University of California, Berkeley, is leading a bit of a renaissance of interest in these amazingly ubiquitous and underappreciated fungi. Perhaps someday we will know as much about the crusts as we do about the larger mushrooms in our area.

Phlebia radiata

Fr.

SPREADING PHLEBIA

FRUITBODY: Spreading, knobby, pimpled crust with irregular projections, furrows, and ridges, size ranging from that of a coin to huge rows or sheets more than a meter across. Orange-buff to beige, pinkish orange or darker to reddish brown on older portion. **FLESH:** Thin, rubbery when wet. **ODOR:** Indistinct. **TASTE:** Indistinct. **SPORE DEPOSIT:** White. **MICROSCOPY:** Spores 4–5.5 x 1.5–2 µm, cylindrical, smooth, inamyloid.

ECOLOGY: On barkless logs or in cracks and furrows of bark and trunks of Tanoak and live oaks. Widespread and common. Fruiting from fall into spring.

EDIBILITY: Inedible.

COMMENTS: This common crust fungus can be found in most of our hardwood forests throughout the wet season. Although there a few similar *Phlebia*, this species can be recognized by the spreading growth habit and the irregular projections, giving it a bumpy appearance. *P. acerina* and *P. rufa* also form spreading crusts, but the hymenium is like a cross between *P. tremellosa* and *P. radiata*—bumpy with many veins and ridges, at times forming pseudopores. *P. acerina* has a yellowish brown fertile surface with no KOH reaction, while *P. rufa* has a reddish brown fertile surface that darkens with KOH. Both occur on hardwoods.

Ceriporia spissa

(Schwein.) Rajchenb.

ORANGE PORED CRUST

FRUITBODY: 4–12 cm across. Highly reduced; a soft, resupinate fertile surface. Finely but extensively convoluted-wrinkled at first, soon becoming poroid, eventually with a very shallow layer of tubes with tiny round mouths. Brilliant orange, becoming dingy reddish in age, margins with a narrow band of whitish or light apricot orange, finely cottony mycelium. **PORES:** Quite small, rounded. **FLESH:** Thin, orangish. **ODOR:** Indistinct. **TASTE:** Indistinct. **MICROSCOPY:** Spores 4–6 x 1.5–2 µm, cylindrical but slightly curved or bent, smooth. Clamps absent.

ECOLOGY: Usually forming single small to large patches on dead hardwood branches and logs; very fond of well-decayed California Bay Laurel branches. Fairly common south of Mendocino County, less frequent northward. Fruiting from early fall through spring.

EDIBILITY: Unknown.

COMMENTS: The orange poroid crust is very distinctive. *C. tarda* is similar, but often has poorly defined pores and is pinkish or light lilac-pink in color; it is uncommon in our area. There are many other crust fungi with pores, but none are as brightly colored as this species.

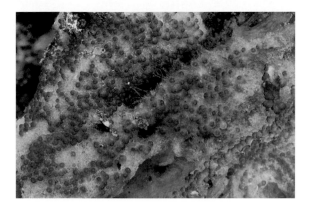

Hypomyces rosellus

(Alb. & Schwein.) Tul. & C. Tul.

PINK PIMPLE CRUST

FRUITBODY: 2–6 cm wide, very thin. Layer of pinkish to ruby or magenta-colored mycelium with white, cottony edges. Mature specimens studded with round, ruby red to rosy pink pimples (perithecia), in which sexual spores are produced. **FLESH:** Very thin, pinkish. **ODOR:** Indistinct. **TASTE:** Not sampled. **MICROSCOPY:** Sexual spores 25–30 x 4–5 μm, elongated and pointed at each end, knobby or constricted in outline, warted and often septate.

ECOLOGY: Forming small patches on surfaces of rotting fruitbodies of a wide range of fungi. In our area, we have seen it on *Amanita*, *Inocybe*, and especially on undersides of old, decayed polypores. Widespread but rarely encountered in most of our area (it is locally common in the Pacific Northwest and elsewhere). Fruiting any time during wet season.

EDIBILITY: Unknown.

COMMENTS: This is probably the most attractive *Hypomyces* in our area, but it tends to grow on more extensively decayed host fruitbodies than the others. See *H. microspermus* (p. 437) and *H. lactifluorum* (p. 247) for others.

Tomentella spp.

FRUITBODIES: 5–10 cm wide. Most species in this genus appear the same macroscopically: extremely reduced, thin, fuzzy-felty or flocculent sheets of fragile, cobwebby-membranous tissue, often appearing slightly bumpy. Color ranges from tan to dull pinkish brown to brownish gray or lilac-brown, often paler toward edge. **FLESH:** Very thin, soft. **ODOR:** Indistinct. **TASTE:** Indistinct. **MICROSCOPY:** Spores generally ellipsoid but irregular in outline with warts or small spines, often with large, "snoutlike" apiculi.

ECOLOGY: Found on undersides of logs, branches, under mats of pine duff or debris. Very common and widespread. Fruiting from fall into spring.

EDIBILITY: Unknown.

COMMENTS: Although rarely seen (consciously, anyway), members of this genus are very common and important ectomycorrhizal partners, ubiquitous in temperate forests around the world. They are practically impossible to identify without microscopic features and a serious dedication to the taxonomy of crust fungi.

Coniophora olivacea
OLIVE CRUST
(Fr.) P. Karst.

FRUITBODY: A thin, extensively wrinkled, corrugated, or bumpy-warted membrane forming spreading patches on wet, rotting wood. Center dingy mustard olive to brownish olive, margins whitish. Edges usually an appressed band of white, branching mycelial fibrils. **FLESH:** Very thin, soft to slightly rubbery. **ODOR:** Indistinct. **TASTE:** Indistinct. **MICROSCOPY:** Spores 14–17 x 10–12 μm, almond shaped to ellipsoid, smooth.

ECOLOGY: Forming small to large patches on dark, wet wood (often barkless logs and branches) of both hardwoods and conifers. Fruiting in wet periods almost any time of year. Very common throughout our area (and the world!), but usually overlooked.

EDIBILITY: Unknown.

COMMENTS: This crust fungus is recognizable by its mustard brown to olive-brown fruitbodies with white edges, bumpy texture, and growth on wood. However, other *Coniophora* have been distinguished as separate species based on subtle differences; *C. arida* is a common species with smaller spores (11–12 x 6–7 μm). Genetic evidence indicates that there are likely many other cryptic species.

"Purple Fuzz"

FRUITBODY: Cushiony or pillowy-looking mat of fuzzy or felted hyphae; more upright and spiky when young, becoming matted and felted in age or where touched. Royal purple to dusky lavender to bluish lilac, edges and tips usually paler to whitish. Frequently adorned with red droplets when fresh and sheltered. **ODOR:** Indistinct. **TASTE:** Not sampled. **MICROSCOPY:** Spores around 7 x 4 μm, but rather variable in size, ovoid to ellipsoid or barrel shaped with truncate ends. Clamps present, sometimes more than one at a single septum; hyphae thick walled, irregular, branching.

ECOLOGY: Forming small to very large patches (often with smaller outlying colonies) on dead hardwood logs and trunks. Mostly on fallen wood under bark layers that are beginning to separate, also on standing snags and decorticated (barkless) wood. Very common on live oak wood in fall and early winter; fresh fruitings occasionally found in spring.

EDIBILITY: Unknown.

COMMENTS: Once you learn to recognize the appropriate habitat, you'll find that this distinctive and striking species is very common. Despite our familiarity with this species, we have yet to find a definitive identification for it.

19 • Tooth Fungi

A number of different lineages of fungi have developed fruitbodies with a tooth-like fertile surface. Some have tiny, highly reduced fruitbodies composed only of the pointed fertile tissue, while others are more complex and stalked.

Auriscalpium are distinct by virtue of the right-angled stipes attached laterally to the cap and their growth exclusively on conifer cones! The wood-decaying *Mucronella* produce swarms of tiny white or yellow fruitbodies. *Hericium* are also wood decayers, but form large, fleshy white fruitbodies that look like pom-poms or spectacular branched cascades of icicles. Identification of species in these genera is relatively straightforward. For gelatinous toothed fungi, see *Pseudohydnum* (in the Jelly Fungi section).

The core diversity of the tooth fungi in our area lies in the mycorrhizal genera. *Hydnum* and *Sistotrema* are relatives of the chanterelles and produce pale fruitbodies with a relatively soft, fleshy texture. *Hydnellum* and *Sarcodon* have brown spore deposits and thick, tough-textured fruitbodies. *Phellodon* fruitbodies are also tough, but are usually smaller and thinner fleshed, and produce white spore deposits. They can be highly variable in appearance, differing greatly when young versus old, or wet versus dry. Important features to note when attempting to identify members of these genera include color of cap, stipe, and flesh (make a cross section); texture of cap; structure of flesh (a single layer or a duplex?); color of teeth; odor; taste (some are very acrid, so consider yourself warned); and associated trees.

Auriscalpium vulgare

Gray

CONIFER-CONE TOOTH

CAP: 1–3 cm across, convex to plane, round or kidney shaped, usually with a distinct "hump" or ridge at stipe attachment, almost always attached to stipe asymmetrically. Dark reddish brown, medium brown to ocher-brown with a much paler margin when young. Surface dry, finely hairy to densely covered in spiky fuzz. **SPINES:** Slender, conical. Pale tan, whitish gray, or brownish, usually paler than stipe. **STIPE:** 3–6.5 cm long, 0.2–0.4 cm thick, off-center to lateral, rarely central. Usually cylindrical and straight, but also often very irregular, sometimes with multiple upright branches. Quite variably colored, from blackish when moist to dark reddish brown, medium brown, ocher-brown, or pallid beige. Surface dry, entirely covered in short, dense fuzz. **FLESH:** Thin, very tough and persistent. **ODOR:** Indistinct. **TASTE:** Indistinct. **SPORE DEPOSIT:** Whitish. **MICROSCOPY:** Spores 4.5–5.5 x 3.5–5.5 µm, globose, weakly roughened with amyloid ornamentation.

ECOLOGY: Solitary or in small scattered groups on conifer cones. In our area, most commonly found on Douglas-fir cones, occasionally on pinecones, also rarely on redwood cones. We have seen it on well-decayed conifer wood, but this is a rare occurrence. Fresh fruiting usually occurs in early fall (a week or so after first steady rains) and continues throughout year. The tough fruitbodies may persist for a month or longer.

EDIBILITY: Likely nontoxic.

COMMENTS: Although difficult to find because of the small size and ability to blend into its surroundings, this species is easy to recognize by the toothed underside, off-center stipe, fuzzy overall texture, and growth on cones. For those having trouble finding it, the following method is fairly foolproof: Two weeks after the first 5 cm of fall rain, find a habitat with many Douglas-fir trees, preferably an older forest with many fallen cones. Slowly crawl around on your hands and knees under one of the larger trees, keeping a mental image of the dark, upright fruitbodies with caps projecting at right angles. *Strobilurus trullisatus* is a much more visible species that also grows on Douglas-fir cones and can be used as an indicator of appropriate habitat in which to find *A. vulgare*. Cones can be transported home and kept moist—your toothy pet will remain happy for weeks.

Phellodon tomentosus

(L.) Banker

OWL EYES

CAP: 2–6 cm across, flat to funnel shaped and often fused in irregular clusters with multiple fans or lobes, or occasionally a simple convex to plane cap with a sunken center. White to buff at first, soon white around margin, light to dark brown at center, often with concentric zones, dark brown to blackish in age. Paler white to creamy with a tan center if dry. Surface dry to moist, finely tomentose at first, soon bare, wrinkled. **SPINES:** 0.1–0.3 cm long, often shorter toward margin. White at first, becoming whitish buff to grayish white, staining orangish buff when bruised. **STIPE:** 2–5 (7) cm long, 0.1–0.6 cm thick, central to off-center, equal or tapered downward, often with a rooting base; arising from a buff mycelial mat. Light to dark brown, darkening slightly when handled. **FLESH:** Tough, thin, fibrous, engulfing organic matter. Whitish to brown in cap, brown to blackish in stipe. **ODOR:** Sweet, fragrant, like curry, fenugreek, or burnt sugar. **TASTE:** Mild to slightly bitter. **KOH:** Dingy gray to black. **SPORE DEPOSIT:** White. **MICROSCOPY:** Spores 3–4 x 3–4 μm, round or nearly so, finely spiny.

ECOLOGY: Often scattered in fused clusters or occasionally solitary on ground under conifers, especially pines and spruce, also with Chinquapin and Tanoak. Fruiting from fall into spring. Widespread and fairly common, but easily overlooked.

EDIBILITY: Unknown.

COMMENTS: The brownish caps with a white margin, whitish spines, thin stipe, and often sweet odor are distinctive. The common *P. melaleucus* is very similar, but the cap has a grayish to black disc and a white margin. The spines are white to pale grayish, and the stipe is grayish to black. It will often grow alongside *P. tomentosus*, especially in the northern part of our range. *Hydnellum* generally have thicker flesh and have brown spores.

Phellodon atratus

K. A. Harrison

CAP: 2–8 cm across, flat with a sunken center or top shaped at first, becoming more irregular or in fused clusters with multiple fans or lobes. Bluish to bluish black or black on disc; paler, bluish gray, to purplish gray on margin when wet. When dry bluish gray to silvery black with a paler margin. Surface dry to moist, finely velvety to tomentose at first, soon bare, wrinkled. **SPINES:** 0.1–0.3 cm long, often shorter toward margin. Bluish gray to bluish lilac at first, becoming grayish in age or when dry. **STIPE:** 2–6 cm long, 0.5–2 cm thick, central to off-center, equal, tapered downward, often with a rooting base, arising from a grayish mycelial mat. Bluish black, black, or dark gray. Surface dry, soft, matted-tomentose. **FLESH:** Tough, fibrous, thin to moderately thick, engulfing organic matter. Dark bluish black to black, with paler grayish flesh near cap surface (especially when dry). Stipe a duplex, with a tough black core and a softer bluish black to gray outer part. **ODOR:** Mild to earthy-fragrant. **TASTE:** Indistinct. **KOH:** Bluish black. **SPORE DEPOSIT:** White. **MICROSCOPY:** Spores 4.5–5 x 5 μm, round or nearly so, finely spiny.

ECOLOGY: Scattered, in groups or fused clusters, or occasionally solitary on ground under conifers, especially Sitka Spruce. Fruiting from early fall through winter. Very common on the Far North Coast, rare south of Mendocino County, has been recorded south to Santa Cruz County.

EDIBILITY: Unknown.

COMMENTS: The dark bluish gray color, dark flesh, and white spores help distinguish this species from other dark tooth fungi. *P. melaleucus* is similar, but is generally smaller and thin fleshed, and has with white spines when young and a cap without bluish tones. Older, faded specimens of *P. atratus* might be mistaken for it, but the thick flesh and duplex stipe help separate it. *Hydnellum cyanopodium* shares the same habitat, but is larger and has a roughened cap, bluish violet flesh, and brown spores. Other species of dark *Hydnellum* are generally larger, and all have brown spores.

Hydnellum cyanopodium

K. A. Harrison

BLEEDING BLUE TOOTH

CAP: 4–10 (15) cm across, convex at first, becoming plane with a sunken center and then often irregular in age. Color variable, generally silvery blue to blue-gray when young, becoming bluish black to purplish black at center but with a pale growing margin, occasionally developing brownish colors in age. Surface matted-tomentose and often pitted, becoming roughened, covered with coarse tufts or hairs, margin beaded with reddish orange droplets of juice when young. **SPINES:** 0.1–0.3 cm long, shorter toward margin. Blue to grayish blue when wet, silvery blue when dry, becoming more grayish or grayish brown in age. **STIPE:** 2–6 cm long, 1–3 cm thick, central to lateral, often tapering downward, occasionally rooting. Bluish to bluish black, occasionally bluish white over the lower portions. Surface dry to moist, velvety when young. **FLESH:** Tough, fibrous, engulfing organic matter, thick to moderately thin. Bluish to bluish black. **ODOR:** Earthy to sweet, currylike. **TASTE:** Earthy. **KOH:** Dark forest green to greenish black. **SPORE DEPOSIT:** Brown. **MICROSCOPY:** Spores 4–5 x 3.5–4.5 μm, angular to cross shaped, with 4–6 stout projections.

ECOLOGY: Solitary, scattered, or rarely clustered on ground under Sitka Spruce. Locally common but limited to wet spruce forest on the Far North Coast. Often first fruiting from fog drip in late summer, becoming more abundant after fall rains, and persisting through winter.

EDIBILITY: Inedible.

COMMENTS: The roughened-tomentose cap is extremely variable in color, depending on age and moisture content. It usually exhibits as silvery blue to blue-gray when young, becoming darker in age; that along with the bluish spines and flesh makes it unmistakable. This is one of a handful of *Hydnellum* that bleeds; it secretes red to reddish orange droplets that create pockmarks on the young caps or on the growing margins. *H. caeruleum* has a felty-tomentose cap that is sky blue when young, fading to orangish brown on the disc with a bluish margin, and then becoming orangish brown overall. The spines are light blue, whitish to grayish at first, becoming orangish brown in age. Besides the paler colors, *H. caeruleum* can be told apart by the orange flesh in the stipe that transitions into zonate blue, and the orange and white cap flesh. *H. regium* has a black cap (although it can be bluish black when young), dark grayish spines, and orangish flesh in the lower stipe. *Phellodon atratus* is generally smaller and thinner fleshed, has a smooth to finely tomentose, bluish black cap with a silvery blue growing margin, bluish gray spines, blackish flesh, and white spores. *Sarcodon fuscoindicus* has deep purplish brown colors and has a fleshy texture.

Hydnellum suaveolens

(Scop.) P. Karst.

SWEETGRASS HYDNELLUM

CAP: 4–15 (25) cm across, irregular to rounded at first, expanding to plane with a sunken center, and often developing additional fans or lobes, occasionally growing in fused clumps of multiple fruitbodies. White at first and staining brownish when handled, becoming tan to light brown on disc, with a whitish margin, often with concentric zones, then entirely grayish brown to brown in age. Surface finely velvety at first, soon bare, wrinkled, and occasionally developing irregular projections. **SPINES:** 0.1–0.5 cm long, often shorter toward margin. Whitish to pale gray at first, often with paler tips, becoming darker gray, to grayish brown in age. Staining grayish brown when bruised. **STIPE:** 1–5 (7) cm long, 1–3 cm thick, central to off-center, equal or swollen lower, occasionally rooting. Violet-blue to sky blue, darkening to navy blue when handled. Surface velvety, dry. **FLESH:** Tough, fibrous, engulfing organic matter, thick to moderately thin in age. Zoned white and blue in cap, deep navy blue in lower stipe. **ODOR:** Strong and fragrant, like sweetgrass (coumarin-like), or aniselike. May be faint at times, but can become overpowering, especially in an enclosed space. **TASTE:** Mild to sweet. **KOH:** Blue-green. **SPORE DEPOSIT:** Brown. **MICROSCOPY:** Spores 4.5–6 x 3–4 µm, subglobose to ellipsoid, covered with irregular projections or warts.

ECOLOGY: Solitary, scattered, or in fused clustered on ground under conifers, especially Sitka Spruce. Generally uncommon in California, restricted to the Far North Coast. Fruiting from early fall well into winter, but will sometimes first appear in summer in areas with heavy fog.

EDIBILITY: Unknown.

COMMENTS: The whitish to light brown cap, whitish to gray spines, blue stipe, zonate blue and white flesh, and sweet odor make this *Hydnellum* unmistakable. *H. caeruleum* has a bluish cap that fades as it ages and orange flesh in the stipe. *Phellodon tomentosus* has a similar-colored cap and a sweet, currylike odor, but is thin fleshed and lacks the blue stipe and flesh.

Hydnellum peckii

Banker

STRAWBERRIES AND CREAM

CAP: 2–15 cm across, rounded at first, becoming convex to plane to irregular, soon developing a sunken center. Pinkish to whitish when young, often covered in red droplets (or reddish brown pockmarks if dry). Becoming dark brown to reddish brown at center with a paler pinkish growing margin, brown overall in age. Surface soft, soon becoming roughen, covered with coarse tufts. **SPINES:** 0.1–0.6 cm long, shorter toward margin. Pinkish at first, becoming pinkish brown to brown in age, often with paler tips. Occasionally oozing red droplets when young. **STIPE:** 2–8 cm long, 1–3 cm thick, central to off-center, irregular, often tapering downward, occasionally rooting. Dingy dark brown, dark reddish brown, or almost black. Surface dry to moist, roughened, matted-tomentose. **FLESH:** Tough, fibrous, engulfing organic matter, thick to moderately thin. Pinkish to brownish, often faintly zoned in cap, darker brown to blackish in stipe. Slowly staining reddish brown. **ODOR:** Indistinct; earthy to sweet, or rancid. **TASTE:** Sickeningly acrid, sour, and farinaceous. **KOH:** Greenish black. **SPORE DEPOSIT:** Brown. **MICROSCOPY:** Spores 4.5–6 x 3.5–4.5 µm, globose to elliptical, almost appearing angular, covered with irregular projections.

ECOLOGY: Solitary, scattered, or occasionally clustered on ground, usually under conifers, especially pines and spruce. Like many of our toothed fungi, it also grows with Chinquapin. Widespread throughout our range, but generally uncommon to locally common. Fruiting in fall and winter, but will sometimes first appear in summer in areas with heavy fog.

EDIBILITY: Poisonous.

COMMENTS: When young and covered with red droplets, this mushroom is one of our most striking, bizarre, and beautiful fungi; in age, it becomes just another brown *Hydnellum*. Besides the red droplets (a feature shown by a number of *Hydnellum*), it can be recognized by the white to pinkish colors when young or at the growing margin in age, pinkish brown flesh, and horribly acrid taste. Although the spiciness varies between collections, one small bite will usually coat your mouth with a burning, astringent sensation. *H. scrobiculatum* (sensu CA) has a brown cap with a tan margin and lacks an acrid taste, like all other brown *Hydnellum* in our area.

Hydnellum subsuccosum

K. A. Harrison

CAP: 3–8 cm across, convex with a depressed center, becoming irregular, uplifted or funnel shaped, and occasionally developing additional fanlike caps, or occurring in clusters of multiple caps. Deep reddish, reddish brown to brown with darker concentric zones and a pinkish buff growing margin. Often with droplets or pools of reddish juice when wet, or brown with a tan margin and distinct zones if dry. Staining reddish brown when handled. Surface dry to moist, soft, finely velvety when young, often grooved or occasionally pitted, to somewhat smooth in age. **SPINES:** Deeply decurrent, 0.1–0.3 (0.5) cm long. Reddish to pinkish brown with grayish tips at first, becoming reddish brown to brown. Staining dark reddish brown when bruised. Occasionally with small reddish droplets when young. **STIPE:** 1–4 cm long, 1–2 cm thick, central to off-center. Reddish brown to brown. Surface dry, soft, velvety, but often mostly covered in spines. **FLESH:** Firm, thick to moderately thin, engulfing organic matter, somewhat soft when very young, becoming tough, fibrous, with a very tough core in older specimens. Dark reddish brown to brown, often with concentric lines in cap, dark reddish brown to almost black in stipe. **ODOR:** Indistinct to slightly farinaceous. **TASTE:** Mild to slightly farinaceous. **KOH:** Producing a quick violet flash, then turning forest green to greenish black. **SPORE DEPOSIT:** Brown. **MICROSCOPY:** Spores 4–5.5–7 x 3.5-4.5 μm, oblong to ellipsoid, covered with blunt, irregular projections.

ECOLOGY: Scattered in small to large troops, clustered, or occasionally solitary on ground under conifers, especially Sitka Spruce. Fairly common on the Far North Coast. Fruiting from fall into winter, but will sometimes first appear in summer in areas with heavy fog.

EDIBILITY: Unknown.

COMMENTS: The concentric-zoned reddish brown caps, reddish to pinkish brown spines with grayish tips (especially around the growing margin), red droplets, and mild taste are distinctive. The brown *Hydnellum* are a confusing swarm of species badly in need of modern taxonomic work. *H. scrobiculatum* (sensu CA) is a hardwood-associated species with a brown cap covered in bumps and projections. It is widespread in California, mostly occurring with oaks, and is almost certainly distinct from the true *H. scrobiculatum* of Europe. *H. concrescens* (=*H. zonatum*) is a thin-fleshed, conifer associate that is very similar to *H. subsuccosum*, but has brown, not reddish, spines when young, and is generally paler in color. *H. subzonatum* is a small, brown to gray-brown species that is common under Shore Pine; the small size, very thin flesh, and grayish to grayish brown spines help separate it. *H. peckii* is another brown-capped species that is often covered with red droplets when young; it has a strongly acrid taste.

Hydnellum aurantiacum

(Batsch) P. Karst.

ORANGE TOOTH

CAP: 3–15 (20) cm across, rounded to top shaped at first, expanding to convex or plane, or irregular and uplifted in age. Often with many knobby projections or developing additional fanlike caps. Whitish when young, soon orange with a whitish growing margin; becoming darker reddish orange to reddish brown. Staining orangish brown when young. Surface dry to moist, soft, plush, finely velvety, roughened, often very irregular with many knobs and projections. **SPINES:** 0.1–0.5 cm long, shorter toward margin. White at first, becoming grayish brown to brownish with white tips, to dingy brown with whitish to pale grayish tips in age. **STIPE:** 2–5 cm long, 1–3 cm thick, central to off-center, irregular, equal, tapering downward or bulbous toward base. Orange at first, becoming orangish to reddish brown. Surface dry, soft, velvety. **FLESH:** Tough, fibrous, engulfing organic matter, thick to moderately thin. Orange in cap, often with darker zones, reddish orange to reddish brown in stipe, slowly darkening when cut. **ODOR:** Indistinct to slightly fragrant. **TASTE:** Earthy to bitter-farinaceous. **KOH:** Blackish on cap, greenish black on flesh. **SPORE DEPOSIT:** Brown. **MICROSCOPY:** Spores 6–7 x 5–6 μm, subglobose to globose, covered with blunt, irregular projections.

ECOLOGY: Solitary or scattered in troops, rarely clustered, on ground under conifers, especially spruce. Widespread, but generally uncommon; grows north of the San Francisco Bay Area (but see comments). Fruiting in fall and winter, but will sometimes first appear in summer in areas with heavy fog drip.

EDIBILITY: Unknown.

COMMENTS: The orange cap and stipe, white spines, and dark orange reddish flesh are distinctive. However, the orange *Hydnellum* in California are a frustrating complex; our species concepts would benefit greatly from a modern genetic analysis. The description above refers to the species restricted to northern coniferous forest. However, with minor changes, it would fit a second common, widespread (and most likely undescribed) species on the California coast. It usually has a darker, dingy cap even when young and has orange spines that become orange-brown in age. It has been identified as *H. auratile* and *H. earlianum*, but both names are based on specimens from well outside our area, and neither concept matches our species well. *H. complectipes* is another member of this complex that generally forms caps with large fused lumps or multiple fanlike lobes and more brown to reddish brown colors with a paler margin; it has orange to grayish orange young spines and orangish brown to brown flesh. *H. geogenium* is a small species with a yellowish olive cap, yellow spines, and neon yellow mycelium. It is very rare in California, known only from a few locations on the Far North Coast.

Sarcodon stereosarcinon

Wehm.

CAP: 3–20 (25) cm across, at first simple, often 1–3 fused, convex to plane caps with a ruffled edge, developing additional lobes and fans and becoming an irregular fused cluster of wavy, overlapping caps. Warm brown, buff to orangish brown, with a paler margin when young, often with concentric zones, becoming darker brown on disc, and often with dingy brown spots or stains. Surface dry, smooth to finely pubescent, occasionally cracking in age. **SPINES:** 0.1–0.5 (0.8) cm long, shorter toward margin. Buff to orangish buff with paler tips at first, becoming buff-brown to light brown. Staining brown when handled. **STIPE:** 2–6 cm long, 1–3 cm thick, central to off-center, irregular, or tapering downward and occasionally with a rooting base. Buff to light brown, staining brown when handled. Surface dry, smooth to roughened from undeveloped spines. **FLESH:** Thick to moderately thin, firm, brittle in cap, stipe tougher, but still somewhat brittle. Off-white, buff to light brown, slowly staining brownish. **ODOR:** Strongly farinaceous. **TASTE:** Farinaceous. **KOH:** Slight olive reaction on all parts. **SPORE DEPOSIT:** Brown. **MICROSCOPY:** Spores 4.5–5 x 3.5–4.5 µm, subglobose to nearly oblong, with blunt, irregular projections.

ECOLOGY: Solitary, scattered, or in fused clusters on ground under Sitka Spruce. Common from Humboldt County north. Generally fruiting in early fall, persisting into winter.

EDIBILITY: Unknown.

COMMENTS: The warm brown colors, smooth, zonate caps in ruffled clusters, brittle flesh, and brown spores help identify this species. *S. scabrosus* has a pinkish brown cap with small scales, a bitter taste, and a bluish stipe base. *Bankera violascens* has a smooth, whitish to grayish beige cap, thicker, fragile flesh, white spores, and a sweet odor. Although it has been recorded in coastal Oregon, it has not been recorded from similar habitats in California.

Sarcodon fuscoindicus

(K. A. Harrison) Maas Geest.

VIOLET HEDGEHOG

CAP: 3–15 cm across, convex to plane, often with a sunken center, to irregular or uplifted in age. Dark grayish purple and sometimes with a powdery bloom when young, becoming royal purple, bluish purple, purplish black to brownish black in age. Paler, purplish brown or reddish brown around margin if dry. Surface dry to moist, roughened to appressed-fibrillose at first, often becoming scaly in age or when dry. **SPINES:** 0.1–0.5 (0.8) cm long, shorter toward margin. Grayish purple at first, becoming purplish brown to brownish in age. **STIPE:** 3–6 (10) cm long, 1–3 cm thick, central to off-center, tapering downward to a pointed base. Purplish gray to dingy purplish black, staining dark purple when handled. Surface dry, smooth to roughened from undeveloped spines. **FLESH:** Thick, firm, brittle in cap, fibrous in stipe. Deep purple, purple-gray, to purple-black, stipe generally slightly paler. All parts slowly staining dark purple. **ODOR:** Indistinct to slightly farinaceous or fragrant. **TASTE:** Mild to unpleasant, slightly acrid and farinaceous. **KOH:** No reaction. **SPORE DEPOSIT:** Brown. **MICROSCOPY:** Spores 5–6.5 x 4.5–5 µm, broadly ellipsoid to subglobose, with blunt, irregular projections.

ECOLOGY: Solitary or scattered on ground under both conifers and hardwoods. Widespread, locally common under Chinquapin, Tanoak, and Douglas-fir in southern half of our range. Typically in coniferous forest on the Far North Coast. Fruiting in fall through winter.

EDIBILITY: Edible, but rather poor tasting.

COMMENTS: The dark purplish cap, purplish gray to purplish brown spines, and deep purple to purplish black flesh make this mushroom unmistakable. This species, like many tooth fungi, produces an excellent wool-dyeing pigment, yielding blue colors in an alkaline dye bath. *S. scabrosus* has a finely scaly, pinkish brown cap, beige spines and flesh, and a dark blue-green stipe base. A species similar to *S. laevigatus* (but probably undescribed) has a dark vinaceous gray to blackish cap, gray spines, and pale gray flesh; it is common under spruce on the Far North Coast.

Sarcodon squamosus

(Schaeff.) Quél.

SCALY TOOTH

CAP: 4–20 (30+) cm across, convex to plane, often with a slightly depressed center, margin downcurved when young, becoming wavy or occasionally uplifted in age. Color variable, often dark vinaceous brown to purplish black, with a paler, buff-brown to grayish brown color between scales when wet; vinaceous brown to light brown with buff between scales when dry. Surface dry to slightly moist, covered with tufted or shingled scales, often with large, upright scales on disc, especially in age. **SPINES:** 0.3–1 (1.5) cm long, shorter toward stipe and margin. Beige to grayish at first, soon grayish brown with paler tips, to all brown in age. **STIPE:** 3–6 (9) cm long, 1–4 cm thick, central to off-centered, equal or tapering downward. Beige to pale dingy brown, and often darker brown at base, staining brown when handled. Surface dry, smooth. **FLESH:** Thick, firm, brittle in cap, tougher, more fibrous in stipe. Off-white to grayish beige, or darker; watery gray to grayish brown if wet. Often darker grayish brown around larva tunnels, especially in stipe, but lacks a bluish base. **ODOR:** Indistinct. **TASTE:** Mild, or occasionally slightly bitter-farinaceous and/or astringent. **KOH:** No reaction. **SPORE DEPOSIT:** Brown. **MICROSCOPY:** Spores 7–8 x 5–6.5 μm, subglobose or ellipsoid with blunt, irregular projections.

ECOLOGY: Solitary or in scattered troops, often forming arcs and rows on ground under conifers, especially pine on the North Coast. Also occurs with Chinquapin. Common and widespread. Generally fruiting from early fall into winter.

EDIBILITY: Edible, but not recommended; often bitter and slightly astringent. The similar Hawk's Wing (*S. imbricatus*) is commonly eaten in the Rocky Mountains, but Californian collections are often bitter.

COMMENTS: The vinaceous to brown scaly cap, grayish to brown spines, beige flesh that lacks bluish coloration at the stipe base, and large size help identify this *Sarcodon*. *S. imbricatus* is very similar, but often has a paler-colored cap with slightly larger scales. The cap also tends to be sunken or depressed at the disc. It also prefers (and is possibly restricted to) the Sitka Spruce forest in California. Another difference between these two species is that *S. squamosus* dyes wool bluish green, whereas *S. imbricatus* only gives a grayish beige color. *S. rimosus* is a large species with a smooth purplish gray blushed cap when young, which becomes brown and scaly in age. It has short, whitish to lilac-gray spines that become grayish brown in age, a strong farinaceous odor, a slightly bitter, farinaceous taste, and an olive-black KOH reaction on the cap. *S. scabrosus* is common under Shore Pine and often grows alongside *S. squamosus*. It has a finely scaly or cracked, pinkish brown cap when young, becoming scalier and brown in age; a bluish stipe base; and a strongly bitter taste.

Hydnum repandum

L.

HEDGEHOG

CAP: 4–12 (25) cm across, rounded, broadly convex to plane, often irregularly uplifted, wavy, or with smaller, fused caps in age. Pale peachy orange, creamy orange, creamy to beige, slowly staining orange when damaged. Surface dry, soft, smooth to bumpy, often with pits or cracks in age. **SPINES:** 0.2–1 cm long, densely covering underside, fragile, flaking off easily. White to pale peachy cream. **STIPE:** 2–10 (15) cm long, 1.5–4 (6) cm thick, central to strongly off-center, shape quite variable: short and stubby or rather elongate, base club shaped or tapered, often rooting. Creamy white to peachy white, slowly staining ocher where damaged or when handled. Surface dry, bumpy to smooth. **FLESH:** Thick, firm, often brittle in cap, rubbery (often squeaking when cut with a knife), creamy white to pale pinkish white, slowly staining orangish. **ODOR:** Indistinct to fruity. **TASTE:** Mild to sweet at first, becoming slightly acrid. **SPORE DEPOSIT:** White. **MICROSCOPY:** Spores 8–9 x 6–7 μm, globose to broadly ellipsoid, smooth, inamyloid.

ECOLOGY: Solitary or scattered in troops, clumps, arcs, or large fairy rings in coniferous woodlands, occasionally with Tanoak. Most abundant under coastal pines, especially under Bishop Pine north of San Francisco and in Sitka Spruce forest on the Far North Coast. Typically fruiting from early fall in north through winter or into spring in wet years. Very slow

growing; a single fruitbody can last for months, becoming larger and more irregular as season goes on. Occasionally, mushrooms will oversummer in the fog belt on the North Coast, giving rise to some huge, "mutant" fruitbodies.

EDIBILITY: Edible and treasured by many. The texture is firm and pleasant, and the flavor is similar to Chanterelles. The acrid taste disappears when cooked.

COMMENTS: The creamy cap, thick, white, ocher-staining flesh, and fragile spines on the underside make the Hedgehog Mushroom easy to identify. The fruitbodies of this slow-growing species can become huge, but these are less often seen because the mushrooms are often picked when still small. The similar *Hydnum umbilicatum* rarely poses an identification problem, since it is much smaller and more slender, has a "belly button" in the center of the mature cap, and tends to grow in large troops.

SYNONYM: *Dentinum repandum* (L.) Gray.

Hydnum umbilicatum (sensu CA)
BELLYBUTTON HEDGEHOG

CAP: 2–5 (8) cm across, convex to plane, becoming depressed at center, often with a deep "belly button." Margin downcurved at first, becoming wavy in age. Creamy beige, creamy orange to peachy orange, occasionally with concentric zones. Surface dry to slightly moist, soft, very finely tomentose to smooth. **SPINES:** 0.2–0.5 cm long, densely covering underside, decurrent or free of stipe, fragile, flaking off easily. Pale creamy white, peachy white to pale ocher. **STIPE:** 3–10 cm long, 0.4–1 (1.5) cm thick, central to slightly off-center, equal or enlarged slightly toward base, often wavy or bumpy. Creamy white to pale peachy orange, slowly staining orange to orange-brown where damaged or handled. Surface dry, often bumpy to smooth. **FLESH:** Thin, firm but brittle, breaking easily. Creamy white, slowly staining orangish. **ODOR:** Indistinct to fruity. **TASTE:** Mild to sweet. **SPORE DEPOSIT:** White. **MICROSCOPY:** (5.5) 6–9 x 6–8.5 μm, globose to subglobose, smooth, inamyloid.

ECOLOGY: Solitary or more often scattered in large troops on ground in duff, or on or near well-decayed logs and stumps in mixed woodlands. Generally in pine or Tanoak woods with a huckleberry understory. Very common north of the San Francisco Bay, rare south to the Santa Cruz Mountains. Typically fruiting in winter or early spring, with scattered fall fruitings northwards.

EDIBILITY: Edible and very good. There is much debate surrounding its culinary merits compared to its larger, fleshier cousin, *H. repandum*.

COMMENTS: This small, but often very abundant *Hydnum* is a staple winter edible of the North Coast, often fruiting alongside the Yellowfoot (*Craterellus tubaeformis*). The creamy orange cap with a "belly button," slender stipe, and spiny underside easily distinguish this species. *H. repandum* is larger (often much larger) and lacks the umbilicate center. The more richly ocher-colored variant growing with Western Hemlock on the Far North Coast looks much more similar to eastern North American specimens, suggesting that we may have two species going by this name in our area (and it's possible that neither is the true *H. umbilicatum*).

SYNONYM: *Dentinum umbilicatum* (Peck) Pouzar.

Sistotrema confluens

Pers.

CAP: 2–6 cm across, convex, plane to uplifted, often becoming more irregular in age, and occasionally with multiple caps fused together. Creamy white, beige to pale orangish beige, margin usually paler. Surface dry, finely tomentose to smooth. **UNDERSIDE:** With decurrent, short irregular spines, or sometimes smooth or with irregular veined areas or even slightly poroid. Creamy white. **STIPE:** 2–6 cm long, 0.2–0.7 cm thick, central to off-center, equal, irregular or enlarged toward base. Creamy white, discoloring orangish when handled. Surface dry, finely tomentose to smooth. **FLESH:** Thin, fragile, whitish, staining orangish. **ODOR:** Indistinct. **TASTE:** Mild. **SPORE DEPOSIT:** Whitish. **MICROSCOPY:** Spores 3.5–6 x 2–3.5 μm, elliptical to ovoid, smooth. Basidia mostly 6- to 8-spored, some 4-spored.

ECOLOGY: Solitary, in troops or fused clusters in duff and moss under Shore Pine on the Far North Coast, fruiting in winter. Rare in our area, only known from the Humboldt County dunes.

EDIBILITY: Unknown.

COMMENTS: This unusual fungus is closely related to *Hydnum* and *Cantharellus*. What's fascinating about this species is that the underside can occasionally resemble both, sometimes with irregular veins and fine spines on the same fruitbody! It can be recognized by the irregular, creamy white to orangish beige caps, orangish staining, and toothy-veiny underside. *Phellodon tomentosus* has a tough, leathery texture and a darker tan to brown cap with a white margin. *Hydnum umbilicatum* is more regular, with a distinctly spiny underside.

Hericium coralloides

(Scop.) Pers.

COMB TOOTH

FRUITBODY: 5–20 (30+) cm across, forming sparse clumps with multiple branches forking from a common base, covered with moderately short, hanging spines. Bright white to creamy white at first, occasionally developing dingy tones in age. **SPINES:** 0.2–1 (2) cm long, pointed, borne along branches, occasionally in tufts or small clusters. **FLESH:** Thin, firm, stringy and somewhat rubbery, white, branches hollow in age. **ODOR:** Fragrant-fungal. **TASTE:** Mild to sweet. **SPORE DEPOSIT:** White. **MICROSCOPY:** Spores 3.5–5.5 x 3–4.5 μm, globose to broadly ellipsoid, smooth to very finely roughened, amyloid.

ECOLOGY: Solitary or scattered on hardwoods, typically on fallen and decayed logs, also occurring on recently dead standing trees and logs. Common throughout our range on Tanoak and live oak. Fruiting from fall into spring.

EDIBILITY: Edible and excellent, but thinner fleshed than other *Hericium*. Small fragments make a nice visual and textural addition to miso or other clear soups.

COMMENTS: Comb Tooth is easily recognized by the rangy, multibranched fruitbodies with hanging spines and growth on hardwoods. *H. abietis* is very similar, but often forms larger fruitbodies with larger tufts and clumps of spines, and grows on conifers. It's rare on the coast, common in higher elevation fir and hemlock forest.

SYNONYM: *Hericium ramosum* (Bull.) Letellier.

Hericium erinaceus

(Bull.) Pers.

LION'S MANE

FRUITBODY: (5) 10–40 cm across, large clump of long, hanging, unbranched spines arising from a rounded mass of flesh. Rather compact at first, with dense spines, elongating on age. Clumps rather simple, or slightly lobed, occasionally stacked. White or blushed pinkish at first, becoming creamy white, discoloring ocher-buff to brownish in age. **SPINES:** 1–4 (7) cm long, pointed. **FLESH:** Thick, firm, stringy, somewhat rubbery, white. **ODOR:** Fragrant. **TASTE:** Mild to sweet. **SPORE DEPOSIT:** White. **MICROSCOPY:** Spores 5–6.5 x 4.5–5.5 μm, globose to broadly ellipsoid, smooth to very finely roughened, amyloid.

ECOLOGY: Usually solitary or occasionally clustered, fruiting from wounds on living oaks, standing dead trees, or fallen logs. Locally common in drier live oak woodlands, rare on the North Coast on Tanoak. Fruiting from late summer into winter, most abundant mid- to late fall.

EDIBILITY: Edible and excellent. The texture is reminiscent of crab meat, with a sweet flavor.

COMMENTS: The often large, rounded clumps of unbranched spines and growth on hardwoods set this species apart. The only species it is likely to be mistaken for is *H. abietis*, which differs by its shorter clumps of spines on multibranched fruitbodies and growth on conifers. It is rare on the Far North Coast and much more common in the mountains. There are unconfirmed sightings of *H. americanum* in California, another species that forms fruitbodies with many small, stacked, compact clumps of spines; it is rarely branched. All *Hericium* are great edibles. *Ramaria* and other coral fungi form clumps of upright branches, and most grow from the ground.

Mucronella fusiformis

(Kauffman) K. A. Harrison

PENDULOUS ICICLE FUNGUS

FRUITBODY: 0.5–1.5 (2) cm long, 0.05–0.3 (0.4) cm thick, pendulous, spindle shaped (with a tiny narrow base) but then wider and evenly tapering to the tip. Watery white to off-white, usually weakly but distinctly yellowish in age. **STIPE:** Distinctly narrowed, tiny, 0.05–0.3 cm long, less than 0.05 cm thick. Often pale yellow even when young (a bit darker than main part of fruitbody) and covered in very small pale hairs (hand lens!). **ODOR:** Indistinct. **TASTE:** Indistinct. **FLESH:** Watery, often dissolving when squeezed, very reminiscent of a citrus fruit vesicle. **KOH:** Pale yellow to yellow-orange. **SPORE DEPOSIT:** Usually not obtainable. **MICROSCOPY:** Spores 5.5–7.5 x 5–6 μm, ovoid to nearly globose, smooth.

ECOLOGY: Solitary or in loose groups on cut ends and undersides of large, well-rotted conifer logs, often Douglas-fir or redwood. Probably common, but very easy to overlook.

EDIBILITY: Unknown. Small and watery.

COMMENTS: The small size, pendulous shape with a distinct narrowed base, and translucent white color are distinctive. Look for it on the cut ends and undersides of very wet, rotting, dark reddish to black Douglas-fir logs. It usually grows in close proximity to other species of *Mucronella*, often fruiting simultaneously. *M. alba* (=*M. bresadolae*, sensu PNW) is slightly smaller than *M. fusiformis*, but is larger that *M. calva*; it lacks the distinct narrowed "stipe" of *M. fusiformis* and has larger spores, 6–7.5 (9) x 4.5–7 μm. *M. calva* is a smaller white species that lacks a narrowed stipe and grows in dense clusters. See *Mucronella* "Yellow" (p. 502) for small, clustered, lemon-colored fruitbodies. There are also a few coral and club fungi that can grow on mossy wood, but all are larger and upright.

MISAPPLIED NAME: *Mucronella bresadolae* (Quel.) Corner.

Mucronella calva

(Alb. & Schwein.) Fr.

CLUSTERED ICICLE FUNGUS

FRUITBODY: 0.2–0.5 (0.7) cm long, hanging icicle-shaped spines, in a clustered growth. Bright white to watery white, discoloring to slightly creamy in age. Often surrounded with fine whitish hairs at base. **ODOR:** Indistinct. **TASTE:** Indistinct. **FLESH:** Extremely thin, watery, whitish. **SPORE DEPOSIT:** Not obtainable. **MICROSCOPY:** Spores 4–6 x 2.5–3.5 µm, oblong to ellipsoid, smooth.

ECOLOGY: In clusters, sometimes covering a fairly large area on sides or undersides of large, well-rotted conifer logs, often Douglas-fir. Occurs throughout our range, fruiting in winter. Probably more common than we know, as it is very easy to overlook.

EDIBILITY: Unknown.

COMMENTS: The downward-facing white spines, small size, lack of a stipe, and clustered growth on well-rotted wood are distinctive. It is very often found on logs that simultaneously host fruitings of other species of *Mucronella*. M. "Yellow" has the same shape and fruits in clusters, but is bright yellow when young (it can fade in age to pale yellow). The fruitbodies of *M. fusiformis* are slightly larger, with a small but distinct narrowed base where they are attached to the wood substrate, and tend to be solitary or small tufts, not clusters. No modern work has been published on *Mucronella*, so it is possible that we have something other than the European *M. calva* in California.

Mucronella "Yellow"

YELLOW ICICLE FUNGUS

FRUITBODY: 0.2–0.5 (0.8) cm long, forming tiny cluster of hanging spines, tapered to a point. Bright lemon yellow to translucent yellow, a bit paler in age. **ODOR:** Indistinct. **TASTE:** Indistinct. **FLESH:** Vanishingly thin, watery, yellow. **SPORE DEPOSIT:** Usually not obtainable. **MICROSCOPY:** Spores 4–6.5 x 4-5.5 µm, subglobose to ovoid, smooth.

ECOLOGY: In clusters or elongate masses on sides, undersides, or cut ends of fairly large, well-rotted conifer logs, especially Douglas-fir and redwood. Not common, or at least not commonly found. Occurring throughout our range, a bit more common north of Sonoma County. When you find it, look for *M. calva* and *M. fusiformis*; the three often fruit together at the same time, on the same logs, but the bright color makes this species the easiest to find.

EDIBILITY: Unknown. Very small.

COMMENTS: If you can find it, you can identify it: a tiny, yellow cluster of hanging spines on rotten wood. It is more likely to be confused for a slime mold than anything else (some slime molds in our area are yellow and grow as small clustered cylinders on wood, but are upright, shiny, and round tipped). Although commonly referred to as *M. flava* and *M. pulchra*, neither of those names are a good fit for ours. *M. flava* was described from Iowa, is paler yellow, and has elongate-ellipsoid spores, 4–5 x 2.5–3 µm. *M. pulchra* was described from Pakistan and has nearly round to oval spores slightly larger, 5.5–7 x 4.7–6 µm, than those of *Mucronella* "Yellow". *M. calva* is similar in size and structure, but is white. *M. bresadolae* is whitish to pale watery yellow and a bit larger, has a narrowed stipe base, and does not form dense clumps.

20 • Coral and Club Fungi

Coral mushrooms comprise a number of unrelated lineages that have evolved upright, highly branched fruitbodies. Unlike the tooth fungi, the tips of these "coralloid" mushrooms point upward, rather than hanging downward. *Phaeoclavulina*, *Lentaria*, and *Artomyces* are duff and wood decayers; *Clavulina* and many *Ramaria* are mycorrhizal. Although a few *Ramaria* have slender, thin-fleshed fruitbodies, most have a thicker, fleshier texture than other genera of coral mushrooms. *Sparassis* have large fruitbodies that can resemble corals, but with flattened, wavy lobes, and are closely related to polypores.

Identifying coral mushrooms can be a truly difficult endeavor. Pay attention to the subtleties of stature and branching pattern, coloration (which in many species fades quickly and dramatically), and habitat. Microscopic examination is required for positive identification in many cases: the size and ornamentation of the spores, as well as the presence or absence of clamps at the base of the basidia, are among the most important features.

The Club Mushrooms are a sundry assortment of many different lineages that produce upright, cylindrical, oar-shaped, or spatula-shaped fruitbodies with little branching. They range from small to fairly large, dull to brightly colored, and fragile to fibrous, fleshy, leathery, pimpled, or rubbery-gelatinous.

Basidiomycete clubs—*Clavulinopsis* fruitbodies are yellow and mostly cylindrical or weakly branched. *Alloclavaria* fruitbodies are purple to lilac-gray, clustered, and cylindrical. *Clavaria* have simple, usually small fruitbodies ranging from whitish to tan or dull yellow or pinkish. *Macrotyphula* have extremely thin, cylindrical tan fruitbodies that grow in large troops.

Ascomycete clubs—*Onygena* fruitbodies have powdery heads and grow out of owl pellets and on bones. *Hypocrea* are pale with tiny freckles or pimples covering the heads. *Cordyceps* and *Tolypocladium* produce fruitbodies with pimpled heads; the former parasitizes insect larvae, the latter, truffles. *Heyderia*, *Microglossum*, and *Trichoglossum* produce smooth, paddle- or tongue-shaped fruitbodies with smooth, or wrinkled fertile heads and often grow in troops. *Cudonia* and *Leotia* both produce somewhat rubbery-gelatinous fruitbodies. *Xylaria* fruitbodies are thin, wiry, and mostly black (but with a powdery white coating of asexual spores near their tips).

Sparassis radicata

Weir

CAULIFLOWER MUSHROOM

FRUITBODY: 10–50 cm across, a rounded cushion of densely to loosely packed leafy, ruffled lobes. Creamy white to creamy pink, becoming pale creamy yellow, often more yellowish ocher in age. **UNDERSIDE:** Smooth, concolorus. **STIPE:** 5–15 cm long, 3–5 cm across, not distinctly set apart, tapering downward, often rooting. **FLESH:** Thin, crisp to slightly rubbery, creamy. **ODOR:** Fragrant when young, slightly disagreeable in age. **TASTE:** Sweet. **SPORE DEPOSIT:** Whitish to creamy yellow. **MICROSCOPY:** Spores 5–7 x 3–5 μm, ellipsoid, smooth, inamyloid.

ECOLOGY: Generally solitary, occasionally a few scattered clusters on ground around conifer trees and stumps. Most often with pine in southern part of our range, Douglas-fir or spruce in the north. Locally common, more infrequent north of the San Francisco Bay Area. Fruiting from midfall into winter, often in same place year after year.

EDIBILITY: Edible and excellent. The crunchy texture and distinctive flavor in combination with the frequently large size make this an ideal mushroom for the table. The only downside is that it is often dirty inside and can take awhile to clean. A few people seem to have allergic reactions, so cook it thoroughly and sample cautiously.

COMMENTS: This distinctive and highly prized edible mushroom can be easily identified by its resemblance to a giant head of cauliflower. Although it has a smooth fertile surface, it is related to the polypores. *Ramaria* species can be similarly colored, but are composed of clumps of upright branches, not ruffled lobes.

MISAPPLIED NAME: *Sparassis crispa* (Wulfen) Fr. is a European species with less ruffled lobes.

Ramaria sandaracina var. *chondrobasis*

Marr & D. E. Stuntz

FRUITBODY: 5–13 cm across, 7–15 cm tall, a rather crowded cluster of slender upright branches, forking near tips. Tips bright orange and remaining so in age, branches orange to dull orange, often with a yellow band on lower portion. **STIPE:** 1–3 cm long, 0.5–2 cm thick, rather short, tapering downward Whitish to pale orange, or with a yellow band at top. **FLESH:** Thin, somewhat gelatinous, branches slightly rubbery. Whitish in base, orange in branches and tips. **ODOR:** Indistinct. **TASTE:** Mild. **CHEMICAL REACTIONS:** Flesh inamyloid. **SPORE DEPOSIT:** Grayish yellow to grayish orange. **MICROSCOPY:** Spores 7–10 x 3.5–5 µm, broadly cylindrical to ellipsoid, ornamented with numerous warts. Basidia clamped.

ECOLOGY: Solitary, scattered, or in trooping masses. Very common on the Far North Coast under Sitka Spruce, occasionally elsewhere, especially when Western Hemlock is present. Fruiting in fall and early winter.

EDIBILITY: Unknown.

COMMENTS: One of many orange *Ramaria* in California, *R. sandaracina* var. *chondrobasis* can be identified by the medium to large size, bright orange tips with duller branches and moderately thin, slightly gelatinous flesh. *R. sandaracina* var. *sandaracina* is smaller and has a slender, upright stature, and var. *cuosma* has yellow tips and a sweet odor. All have shorter spores than other orange *Ramaria*. Another bright orange species is *R. gelatiniaurantia,* which has fleshier branches, a chunky white stipe, and gelatinous flesh; lacks clamps; and has broadly cylindrical spores (8–11 x 3.5–5 µm). It is common in both coniferous and hardwood forests south to Santa Cruz County. *R. flavigelatinosa* is very similar, but has orange-yellow to yellowish branches, yellow tips, a small stipe, and gelatinous flesh. Microscopic differences are slight; the spores average 9.6 µm long for *R. flavigelatinosa* versus 9.3 µm for *R. gelatiniaurantia. R. gelatinosa* var. *oregonensis* is generally dull orange to ocher-orange with yellowish orange tips, becoming ocher-buff in age. Its most distinctive feature is the clear gelatinous pockets in the stipe flesh; microscopically, it has clamped basidia and coarsely warted spores (7–10 x 4.5–6 µm). *R. aurantiisiccescens* differs from all the species mentioned above by having fleshy-fibrous (not gelatinous) flesh. The small size, orange tips, and bright yellow band on the upper stipe also help distinguish it. *R. largentii* is a large, bright orange to peachy orange species, with a massive whitish stipe and pinkish orange flesh in the branches.

Ramaria stuntzii

Marr

STUNTZ'S RED CORAL

FRUITBODY: 4–15 cm across, 6–17 cm tall, compact at first, expanding with multiple branches arising from chunky base, forming a spreading crown. Bright neon-pink, pinkish red to orange-red when young, fading to pinkish orange in age. Tips concolorous or slightly darker than branches in age. **STIPE:** 2–8 cm long, 1.5–5 cm thick, often chunky, bulbous or rooting. White on lower portion, pinkish or with an orangish band near branches. **FLESH:** Pinkish in tips, dingy pinkish in branches, white in base. **ODOR:** Indistinct. **TASTE:** Mild to slightly bitter. **CHEMICAL REACTIONS:** Strongly violet-gray on stipe flesh in Melzer's. **SPORE DEPOSIT:** Orange-yellow. **MICROSCOPY:** Spores 7–10 x 3–5 μm, broadly cylindrical, finely ornamented with lobed warts. Basidia lacking clamps.

ECOLOGY: Solitary or scattered in troops on ground or in duff under conifers (especially Western Hemlock) and possibly Tanoak. Very common on the Far North Coast, known from as far south as Sonoma County. Fruiting in fall.

EDIBILITY: Edible.

COMMENTS: The medium to large size, brightly colored fruitbody with a chunky white stipe, and amyloid reaction on the flesh help distinguish it from *R. araiospora*. Faded specimens could be mistaken for *R. cyaneigranosa*, which is light red to pinkish salmon from the start, lacks the amyloid flesh reaction, and has larger, warted spores (8–15 x 4–6 μm).

Ramaria araiospora var. rubella

Marr & D. E. Stuntz

NEON PINK CORAL

FRUITBODY: 2–7 cm across, 6–12 cm tall, compact, with multiple branches arising from a slender common base, branches forking repeatedly and becoming narrower toward tips. Branches bright crimson red to neon pink when young, fading slightly in age, becoming reddish pink to pink. Tips often forked, bright pink to crimson red, fading slightly, but remaining pink in age. **STIPE:** 1–4 cm long, 0.5–1.5 cm thick, often rooting, occasionally slightly bulbous. White to pinkish. **FLESH:** Pink to red in branches, white in base. **ODOR:** Indistinct. **TASTE:** Slightly peppery at first, becoming radishlike after a few seconds. **CHEMICAL REACTIONS:** No reaction from KOH or Melzer's. **SPORE DEPOSIT:** Yellowish. **MICROSCOPY:** Spores 8–14 x 3–5 μm, broadly cylindrical, finely ornamented. Basidia lacking clamps.

ECOLOGY: Solitary or scattered in small groups on ground in coastal conifer forest, especially in the spruce-hemlock zone on the Far North Coast. Known as far south as Sonoma County, but rare south of central Mendocino County. Fruiting throughout fall in north, into winter in south.

EDIBILITY: Edible.

COMMENTS: This small, brightly colored coral fungus can be told apart from other red *Ramaria* by the bright pink to red branches and tips that fade slightly with age but remain distinctly pinkish red. *R. araiospora* var. *araiospora* is very similar, but the tips go orange to yellow in age. *R. stuntzii* can be differentiated by its neon-pink colors (becoming bright red to orange-red), chunkier base, flesh that turns violet-gray in Melzer's, and smaller spores (3–5 x 7–10 μm). *R. cyaneigranosa* is generally more pinkish salmon and has slightly larger warted spores (8–15 x 4–6 μm).

Ramaria "Tanoak Red"

FRUITBODY: 5–(20) 25 cm across, 7–22 (27) cm tall, forming large clusters with a thick, chunky stipe. Branches relatively short, elongating somewhat in age. Tips wine red, pinkish red to vinaceous pink at first, becoming more dingy wine red to dingy orange-red. Branches whitish at first, occasionally with a wash of pink "bleeding" down from tips, developing yellowish ocher tones as spores mature. **STIPE:** 3–15 cm long, 2–9 cm thick, often cylindrical with a rounded or slightly bulbous base. White, not staining. **FLESH:** Thick, firm, fleshy. Pinkish in tips and upper branches, white in stipe. **ODOR:** Indistinct. **TASTE:** Bitter. **CHEMICAL REACTIONS:** Flesh weakly amyloid. **SPORE DEPOSIT:** Ocher. **MICROSCOPY:** Spores 10–12 x 4–5.5 µm, cylindrical, ornamented with longitudinal ridges.

ECOLOGY: Solitary, scattered, or in large troops on ground in Tanoak forest. Locally can be very common, especially in Mendocino County; surprisingly absent in similar habitat elsewhere. Occurring as far south as Santa Cruz Mountains. Fruiting from midfall into winter.

EDIBILITY: Edible, but often too bitter to eat.

COMMENTS: This large coral can be recognized by the wine red to vinaceous pink tips on relatively short, pale branches, large fleshy stipe, bitter taste, and growth with Tanoak. Unlike most related *Ramaria*, it retains the color into age. Probably most similar is *R. rubripermanens*, which has pinkish to reddish tips and creamy branches when young, but the pink color soon fades, and it becomes pale yellowish ocher overall. It also has a mild to nutty taste and grows with conifers. *R. rubrievanescens* has fleeting pinkish tips and develops brownish stains on the stipe. *R. botrytis* is generally smaller and has a more upright stature and paler pinkish magenta to pinkish blushed tips on creamy branches. *R. botrytoides* has bright rosy pink to pinkish tips and creamy branches, but lacks the large fleshy stipe of all the other species mentioned here.

Ramaria botrytis

(Pers.) Ricken

PINK-TIPPED CORAL

FRUITBODY: 4–12 cm across, 5–12 (15) cm tall, forming rather sparse crown with a chunky stipe. Branches relatively short, elongating in age. Tips generally vinaceous pink to wine red at first (although they can be light pink), soon fading. Branches whitish to creamy white, often with a pinkish blush when young, slowly bruising pale yellowish brown when damaged. **STIPE:** 2–7 cm long, 1.5–3.5 cm thick, often cylindrical with a rounded or slightly rooting base. Whitish, slowly bruising yellowish to pale brownish. **FLESH:** Thick, firm, fleshy, white. **ODOR:** Indistinct. **TASTE:** Mild. **CHEMICAL REACTIONS:** Flesh slowly and weakly amyloid. **SPORE DEPOSIT:** Ocher to pale orange. **MICROSCOPY:** Spores 11–17 x 4–6 µm, cylindrical, ornamented with spiraled to longitudinal ridges. Basidia clamped.

ECOLOGY: Solitary or scattered on ground, in moss, or in duff under conifers. Locally common on the Far North Coast, occasionally south to Sonoma County. Reported records from farther south all seem to be *R.* "Tanoak Red". Fruiting from midfall into early winter.

EDIBILITY: Edible and good, but very easily confused with bitter or bland species.

COMMENTS: The evanescent vinaceous pink color of the tips, moderately chunky stipe that bruises pale yellowish brown, and mild taste help distinguish it. Microscopically, it has longer, more ridged spores than other similar species. *R. botrytoides* is very similar in color, but lacks the chunky stipe, has much shorter spores, and lacks clamp connections. *Ramaria* "Tanoak Red" has darker tips, persistent wine red colors, and a bitter taste. Two very similar species that primarily differ microscopically are *R. rubripermanens*, with slightly darker colors and an unstaining stipe, and *R. rubrievanescens*, with paler colors and a brown-staining stipe. All of these species are edible, but *R.* "Tanoak Red" is usually too bitter to eat.

Ramaria formosa

(Pers.) Quel.

YELLOW-TIPPED CORAL

FRUITBODY: 5–14 cm across, 5–15 (20) cm tall, forming compact clusters with short, chunky and often tapered base. Branches short at first, rather upright, elongating in age. Tips yellow to pale yellow-ocher, transitioning into a peachy pink to salmon branches, to a whitish base. Becoming more uniform salmon buff in age and slowly staining brown when handled. **STIPE:** 2–5 cm long, 1.5–5 cm thick, stout, wide, tapering to a pointed base. Whitish to pinkish on upper part. Staining brownish when handled or in age. **FLESH:** Thick, firm, fleshy. White in stipe, pinkish in branches, yellowish to pinkish in tips. **ODOR:** Indistinct. **TASTE:** Mild. **CHEMICAL REACTIONS:** Flesh inamyloid. **SPORE DEPOSIT:** Golden yellow. **MICROSCOPY:** Spores 10.5–15.5 x 5.5–6.5 μm, ellipsoid to ovoid, ornamented with scattered warts. Basidia clamped.

ECOLOGY: Solitary, scattered, or in troops on ground, in moss, or in duff under conifers and Tanoak. Probably our most common and widespread *Ramaria* on the coast, occurring throughout our range. Fruiting from early fall in north well into winter in south.

EDIBILITY: Unknown; reported to have caused minor gastrointestinal problems. But *R. formosa* has been a catchall name for a lot of different species, so the exact identity of the culprit remains unclear.

COMMENTS: This common coral can be recognized by the peachy pink branches with yellowish tips, pale brownish staining, and large base. *R. leptoformosa* forms tall upright clusters with peachy orange branches and orange tips, and has clamp connections and narrow spores (8–13 x 3–5 μm). *R. conjunctipes* var. *tsugensis* has small to medium upright clusters, often with slender fused stipes; yellow to greenish yellow tips; and salmon- to peach-colored branches; it lacks clamps and has short spores (6–10 x 4–6 μm). *R. maculatipes* has the same colors as *R. formosa*, but has wine red stains on the stipe. *R. rubricarnata* is a spring-fruiting species with very similar colors, but usually has a yellow band of color on the upper stipe; it appears to be limited to higher-elevation montane forest.

Ramaria rubiginosa

Marr & D. E. Stuntz

FRUITBODY: 4–12 (15) cm across, 7–12 (20) cm tall, forming medium to large clusters and an indistinct to moderately chunky stipe. Branches compact at first, elongating in age. Tips pale yellow, branches pale creamy yellow to pale yellow at first, soon pale ocher-yellow as spores mature. **STIPE:** 1–7 cm long, 0.5–4 cm thick, slightly rooted to short and indistinct; often with a few branches fused together. Whitish to pale yellow, with distinct wine red stains, especially in age. **FLESH:** Moderately thick, firm, fleshy-fibrous. White, occasionally with wine red stains in stipe. **ODOR:** Indistinct. **TASTE:** Mild. **CHEMICAL REACTIONS:** Flesh inamyloid. **SPORE DEPOSIT:** Grayish ocher-yellow. **MICROSCOPY:** Spores 7–11 x 3.5–6 μm, cylindrical, ornamented with fine lobed warts. Basidia without clamps.

ECOLOGY: Solitary, scattered, or in troops on ground under Tanoak or conifers. Locally common, known from as far south as Sonoma County. Fruiting from midfall into winter.

EDIBILITY: Edible and good.

COMMENTS: The pale yellow colors, rather compact branches, and, most importantly, wine red stains on the stipe or lower branches help set this species apart. There are a number of other wine red–staining species of *Ramaria*: *R. rubribrunnescens* has salmon pink branches, yellowish tips, rosy red to wine red stains, and longer spores (10–14 x 3.5–5 μm). *R. maculatipes* has pale salmon orange branches and yellow to yellow-orange tips, deep red to wine red stains, and clamped basidia and spores measuring 9–11 x 4–5 μm. *R. vinosimaculans* has creamy to ivory colors when young, often with pale ocher tips, pronounced wine red stains over the lower portion in age, clamps on the basidia, and long ellipsoid spores (11–14 x 4–5 μm). Nonstaining look-alikes include *R. cystidiophora*, which is generally smaller, and has upright branches, a "cool" yellow color, and often a sweet or lemony scent; and *R. rasilispora*, a very common vernal-fruiting montane species that occasionally shows up on the coast in the fall. It has pale yellow to yellow-ocher colors, a chunky base, amyloid flesh, clamped basidia, and smooth spores measuring 8–11.5 x 3.5–4.5 μm.

Ramaria violaceibrunnea

(Marr & D. E. Stuntz) R. H. Petersen

VIOLET-BANDED CORAL

FRUITBODY: 1.5–8 cm across, 5–12 cm tall, generally with a slender upright crown and a distinct stipe. Branches violet overall when young, soon smoky beige to grayish brown with a violet band on upper stipe and lower branches. Developing ocher-yellow color (from spores) and becoming overall grayish tan when old. **STIPE:** 1–4 cm long, 0.5–2.5 cm thick, elongated, club shaped to slightly bulbous. Whitish at base, violet over upper portion. **FLESH:** Fibrous, white. **ODOR:** Indistinct. **TASTE:** Indistinct. **CHEMICAL REACTIONS:** KOH reddish on violet parts, flesh inamyloid. **SPORE DEPOSIT:** Light ocher-yellow. **MICROSCOPY:** Spores 9–13 x 4–5.5 μm, ellipsoid to ovoid, roughened with many small isolated warts. Basidia with clamps.

ECOLOGY: Solitary or scattered in troops on ground in mixed forest. With Tanoak through much of our range, but becoming more common in coniferous forest in the north. Occurring as far south as Santa Cruz County, rare south of the San Francisco Bay Area. Fruiting in fall and early winter.

EDIBILITY: Unknown.

COMMENTS: The small to medium, often slender, sparse crown and violet-gray colors make this common coral distinctive. Older specimens that have lost all their violet should be compared with *R. acrisiccescens*, a larger, bitter-acrid-tasting species. A large grape-juice purple species, *R. purpurissima*, retains its colors in age; it is rare, only known from the interior Coast Range in California. Another rare purple coral in California is an undescribed species in the *Clavaria zollingeri* group. It has deep violet colors when young, fading to lilac in age, and very fragile tips and branches; microscopically, it has small spores and four-spored basidia.

MISAPPLIED NAME: *Ramaria fennica.*

Ramaria acrisiccescens

Marr & D. E. Stuntz

BLAH CORAL

FRUITBODY: (2) 5–18 cm across, 5–25 cm tall, generally with tall slender upright branches and a short, slender stipe in clusters. Branches whitish to pale creamy beige at first, becoming more brownish in age. Tips concolorous or with a pinkish buff cast. **STIPE:** 1–7 cm long, 0.5–2.5 cm thick, slender and rooting, or short and indistinct. Whitish, staining brownish in age. **FLESH:** Fibrous, whitish to brownish. **ODOR:** Indistinct to musty. **TASTE:** Acrid-bitter. **CHEMICAL REACTIONS:** Flesh inamyloid. **SPORE DEPOSIT:** Grayish yellow. **MICROSCOPY:** Spores 8–14 x 4–6 μm, broadly cylindrical to ellipsoid, ornamented with distinct warts. Basidia without clamps.

ECOLOGY: Solitary or in scattered troops on ground under conifers on the North Coast. Common from central Mendocino County northward, not known from south of San Francisco Bay. Fruiting from fall into winter.

EDIBILITY: Unknown.

COMMENTS: This coral can be identified by the often large clumps of long, upright branches, lack of a chunky stipe, pallid colors, and acrid-bitter taste. Microscopically, the lack of clamps help differentiate it from *R. velocimutans*, a similar whitish to creamy beige species that stains brownish in age, and also has slightly more slender branches and narrower spores.

Ramaria apiculata var. *apiculata*

(Fr.) Donk

GREEN-TIPPED CORAL

FRUITBODY: 2–8 (12) cm across, 4–9 (15) cm tall, compact at first, branches slender, erect, tips often forked, pointed. Tips generally pale green to yellow-green (especially in cold weather), at time, more lime green transitioning into a buff, pale grayish brown to reddish brown branches. Occasionally losing green color from tips in age. **STIPE:** 0.5–1.5 cm long, 0.3–1 cm thick, stout, often with white mycelium at base. **FLESH:** Tough, leathery-fibrous, pallid. **ODOR:** Indistinct. **TASTE:** Slightly acrid and occasionally bitter. **CHEMICAL REACTIONS:** KOH brown on branches, yellowish on mycelial rhizomorphs. **SPORE DEPOSIT:** Buff. **MICROSCOPY:** Spores 7–11 x 3.5–5 μm, cylindrical to ellipsoid, ornamented with scattered fine warts.

ECOLOGY: Solitary or in scattered troops on well-decayed, and often buried wood or thick duff under conifers. Widespread, but uncommon outside spruce forest on the Far North Coast where it can be abundant. Fruiting from early fall into early winter.

EDIBILITY: Unknown.

COMMENTS: The thin, tough branches with greenish tips and growth on buried wood are helpful identifying features. In some collections, the green becomes more pronounced after collecting, while in others, it fades. *R. apiculata* var. *brunnea* is identical except that it lacks the greenish tips. *R. stricta* is another upright branched coral on wood; it differs by having a pinkish buff to yellowish buff colors. It is common throughout our range, fruiting from early fall into midwinter. *Phaeoclavulina myceliosa* and *P. abietina* are more ocher colored or beige; the latter stains bluish green on the stipe and lower branches in age. *Ramaria conjunctipes* var. *tsugensis* can also have greenish tips, but is overall fleshier and has pinkish salmon branches with a whitish base.

Ramaria rubella f. *rubella*

(Schaeff.) R. H. Petersen

FRUITBODY: 0.5–7 cm across, 3–7 (10) cm tall, forming rather sparse clusters of repeatedly forked branches with a short, slender stipe. Grayish pink to pale grayish red at first, becoming pinkish buff to warm grayish buff in age. **STIPE:** 0.5–2 cm long, 0.3–0.7 cm thick, stout, with whitish mycelial rhizomorphs at base. **FLESH:** Tough, leathery-fibrous, pallid. **ODOR:** Indistinct or slightly sweet. **TASTE:** Acrid. **CHEMICAL REACTIONS:** KOH bright pink on mycelial rhizomorphs. **SPORE DEPOSIT:** Golden yellow. **MICROSCOPY:** Spores 5.5–8 x 4–5.5 μm, ellipsoid, ornamented with numerous fine warts.

ECOLOGY: Solitary, scattered, or in troops on well-decayed wood or duff under rotting branches. Very common in Sitka Spruce forest on the Far North Coast, not known from south of Humboldt County. Fruiting from early fall into winter.

EDIBILITY: Unknown.

COMMENTS: This *Ramaria* is easily recognized by the pinkish gray colors, sparse crowns, tough flesh, growth on decayed wood, and pink KOH reaction of the rhizomorphs. *R. rubella* f. *blanda* lacks the KOH reaction on the rhizomorphs and has slightly shorter spores (7.1 μm on average versus 7.5 μm for f. *rubella*). *R. stricta* has more erect, creamy white, pinkish buff to yellowish buff, straight branches. *Lentaria* species generally have whitish pointed tips, beige to light brown lower branches, and smooth spores.

"Phaeoclavulina" myceliosa

(Peck) comb. prov.

FRUITBODY: 1.5–5 cm across, 2.5–6 cm tall, forming small upright to spreading clusters from a common base, branches often forking near pointed tips. Branches yellowish ocher to ocher-buff, occasionally olivaceous ocher, tips paler, yellowish. **STIPE:** Short, with numerous branches. Base with conspicuous white rhizomorphs. **FLESH:** Thin, tough, leathery. **ODOR:** Indistinct. **TASTE:** Bitter. **SPORE DEPOSIT:** Ocher. **MICROSCOPY:** Spores 4.5–6 x 2.5–3.5 μm, ellipsoid to tear shaped, finely spiny.

ECOLOGY: Often in large arcs, rings, or troops in duff and humus. Especially abundant under Monterey Cypress and redwood. Very common throughout our range, from early fall into spring.

EDIBILITY: Unknown.

COMMENTS: The tough texture, ocher-buff color, white rhizomorphs, and growth in duff are distinctive; it also tends to form conspicuous white mycelial mats. *P. abietina* is quite similar and occurs in the same habitats, but can easily be distinguished by the slow blue-green staining when damaged, especially on the stipe and lower branches. *Ramaria apiculata* has more erect, upright branches with greenish tips, while *R. stricta* has straighter, denser, creamy buff branches; both grow on decaying wood.

SYNONYM: *Ramaria myceliosa* (Peck) Corner.

Clavulina coralloides group
CRESTED CORAL

FRUITBODY: 2–5 cm across, 2–7 (10) cm tall, with many upright branches forking irregularly from a common base. Tips finely crested with many short spinelike branches when young, disappearing and becoming more blunt in age. Creamy white when young, graying slightly in age. Some forms have creamy tips and grayish branches. Commonly infected with *Helminthosphaeria clavariarum*, which distorts the fruitbody and turns it bluish gray to dark gray (see comments). Surface smooth, dry to moist. **STIPE:** Relatively short, or slightly rooted. **FLESH:** Thin, brittle to moderately tough, fibrous. **ODOR:** Indistinct. **TASTE:** Indistinct. **SPORE DEPOSIT:** Whitish. **MICROSCOPY:** Spores 7–11 x 6.5–10 µm, globose to subglobose, smooth, moderately thick walled.

ECOLOGY: In scattered clusters in duff or moss under conifers. Known from Santa Cruz County northward, but rare south of Mendocino County. One of the most abundant mushrooms in late fall and winter on the Far North Coast in Sitka Spruce forest. Fruiting from late fall into spring.

EDIBILITY: Edible.

COMMENTS: The Crested Coral is better known as *Clavulina cristata*; *C. coralloides* is an older name for the European species. That said, a number of species in this group occur in California, probably all unnamed. There are three distinct groups of *Clavulina*: *C. coralloides*, as described above; the *C. rugosa* group, with more simply structured fruitbodies with an upright stipe with a cluster of short, wrinkly branches near the tip; and the *C. cinerea* group, with grayish branches and blunt tips. To make matters more difficult, they are often parasitized by *Helminthosphaeria clavariarum,* a fungus that distorts the fruitbody and turns it bluish gray to dark gray in color (with embedded blackish perithecia), leading to great confusion distinguishing *C. cinerea* from *C. coralloides.* *Ramariopsis kunzei* forms clusters of white to pinkish, more regular upright branches arising from multiple bases; the tips are generally forked and pointed, but not nearly as fine as *C. coralloides. Lentaria* generally have simple pointed tips and grow on wood. Also compare with *Ramaria rubella*.

Ramariopsis kunzei group
WHITE CORAL

FRUITBODY: 2–9 cm across, 3–10 cm tall, branches upright to spreading, forking a few to many times. Tips rounded to blunt. White, creamy white, or with pinkish color on lower branches. **STIPE:** Generally indistinct, or short, slender, and rooting, often in fused clusters. **FLESH:** Thin, fragile. **ODOR:** Indistinct. **TASTE:** Indistinct. **SPORE DEPOSIT:** White. **MICROSCOPY:** Spores 3–5.5 x 2–4.5 µm, broadly ellipsoid to nearly globose, finely spiny.

ECOLOGY: Solitary, in fused clusters, or in scattered troops in duff under redwood and other conifers. More rarely in duff under hardwoods. Common, occurring throughout our range. Fruiting from fall into early winter on the Far North Coast, in winter and spring in south.

EDIBILITY: Edible.

COMMENTS: The whitish colors, rounded to blunt tips, and clustered growth in duff help distinguish this species. *Clavulina coralloides* has finer pointed tips (when young at least) and often forms more compact clumps with thinner branches. *Tremellodendropsis tuberosa* has tough flesh and sparse branches. Pale *Ramaria* species have thicker branches and fleshier stipes. We know that the California species is distinct from the European one, and it will be given a name soon.

Lentaria pinicola group
WHITE-TIPPED CORAL

FRUITBODY: 2–8 cm across, 3–9 cm tall, forming a sparse crown of forking upright branches from a common base. Tips pointed, bright white, transitioning into tan to buff-brown branches. **STIPE:** 0.5–2.5 cm long, 0.2–0.5 cm thick, often tough, light brown. Base with downy white mycelium and thin white rhizomorphs. **FLESH:** Tough, leathery-fibrous, pallid. **ODOR:** Indistinct. **TASTE:** Slightly bitter. **SPORE DEPOSIT:** Whitish. **MICROSCOPY:** Spores 7.5–9.5 x 4.5–6 µm, cylindrical to ellipsoid, smooth, thin walled.

ECOLOGY: Solitary or scattered in small troops on thick duff or woody debris, generally under redwood or other conifers. Uncommon, occurring throughout our range. Fruiting in winter.

EDIBILITY: Unknown.

COMMENTS: The tough, thin branches, bright white pointed tips, growth on woody debris, and smooth spores help distinguish the genus *Lentaria*. It appears we have a handful of undescribed species in California (including the one pictured above, which is genetically distinct from *L. pinicola*). However, until the species gets sorted out, it is probably best to refer to them as a group. *Ramaria* and *Phaeoclavulina*, which have a similar stature and texture, lack the white growing tips and have warted and/or spiny spores.

Artomyces piperatus

(Kauffman) Jülich

PEPPERY CROWN CORAL

FRUITBODY: 0.5–3 cm across, 2–8 cm tall, a sparse cluster of erect, upright branches forking in whorled rosettes. Ends often crowned with 2–4 pointed tips. Creamy white tips with brownish lower branches and base. Surface smooth, dry to moist. **STIPE:** Relatively short, tapering downward. **FLESH:** Thin, tough, fibrous. **ODOR:** Indistinct. **TASTE:** Acrid, peppery. **KOH:** No reaction. **SPORE DEPOSIT:** White. **MICROSCOPY:** Spores 4–5 x 3–3.5 µm, subglobose to broadly ovoid, roughened, amyloid.

ECOLOGY: In scattered clusters, often on sides of rotting coniferous logs or stumps. Uncommon, known from the San Francisco Bay Area north. Fruiting in winter.

EDIBILITY: Unknown.

COMMENTS: The sparse clusters of creamy-colored upright branches, forking in whorled rosettes (pyxidate) with crown-like tips; growth on rotting coniferous wood; and acrid taste are distinctive. *Ramaria stricta* has denser, thinner branches, a slightly darker color, and a bitter taste. *R. rubella* has pinkish branches with a blunt to pointed tip and an acrid taste, and grows on or near spruce branches.

Tremellodendropsis tuberosa

(Grev.) D. A. Crawford

FLAT-TIPPED CORAL

FRUITBODY: 0.5–4 cm across, 2–5 (7) cm tall, with sparse, erect, flattened, somewhat wavy and irregular branches arising from a common base. Tips whitish, lower portion whitish, pale buff to grayish. **STIPE:** 0.5–3 cm long, 0.1–0.3 cm thick, colored like lower branches. **FLESH:** Thin, tough, leathery. **ODOR:** Indistinct. **TASTE:** Indistinct. **SPORE DEPOSIT:** White. **MICROSCOPY:** Spores 12–24 x 4–9 µm, elongate-ellipsoid to spindle shaped, smooth, contents with oil droplets.

ECOLOGY: Solitary, clustered, or scattered in small troops, generally on bare soil or in disturbed areas. Common and widely distributed. Fruiting from late fall into spring.

EDIBILITY: Unknown.

COMMENTS: This small, tough corallike fungus is actually related to jelly fungi. It is easily identified by the sparse, often flattened branches joined to a common base, tough texture, and growth on the ground. Most of the similar coral fungi that grow on the ground are larger and/or have more fragile flesh (such as *Ramariopsis kunzei*).

Thelephora palmata

(Scop.) Fr.

STINKING EARTH FAN

FRUITBODY: 4–10 cm across, 4–8 cm tall, composed of many clustered branches fused into a common "trunk" or base. Branches usually flattened and fan shaped, occasionally many adjacent branches fused into sheets; tips squared to peglike or often more pointed in age; all surfaces slightly wrinkled. Branches mostly brown, dingy lilac-gray to brownish gray to tan, tips usually paler, tan to whitish beige. **UNDERSIDE:** Fertile surface smooth to wrinkled, often with tiny warts. Grayish to grayish brown, paler, whitish toward margin at first, darkening in age. **STIPE:** Usually central, sometimes narrow and distinct, in other cases indistinct and broad, made of loosely fused branches. Color like fertile surface, or darker reddish brown to nearly black. **FLESH:** Thin, quite tough, leathery. **ODOR:** Like rotting cabbage, garlic, or old broccoli, usually strong. **TASTE:** Unpleasant. **KOH:** Dark blue-green (appearing black). **SPORE DEPOSIT:** Brown. **MICROSCOPY:** Spores 8–10 x 6.5–9 µm, elliptical, slightly irregular in outline with many small, spiny warts.

ECOLOGY: Solitary or in clumps and arcs in mixed hardwood and conifer forest. Fruiting in late fall and winter. Widespread throughout our range but not particularly common. Likely very often overlooked, as it blends in with surrounding dark leaf litter.

EDIBILITY: Nontoxic, but the rank odor isn't very appealing.

COMMENTS: This species is recognizable by the multibranched structure with fan- or hand-shaped tips ("palmate"), dark colors, and unpleasant rotting cabbage odor. The fruitbody stature varies a great deal: sometimes loosely and finely branched, in other cases dense and blocky and forming tight masses with poorly defined branches. There may be more than one local species matching the above description, and none may be conspecific with the true European *T. palmata*. *T. anthocephala* is similar, but has more flattened branches and lacks an odor. Many other *Thelephora* occur in California, but most are smaller, more frilly edged, flattened, and sheetlike, and/or more rosette shaped.

Calocera viscosa

(Pers.) Fr.

YELLOW TUNING FORK

FRUITBODY: 1–4 cm across, 1–7 (10) cm tall, often in clusters of simple forking or antler-shaped clubs, occasionally with more elaborate branched or forked tips. Bright golden yellow to yellow-orange. Surface dull to shiny, smooth, often slightly viscid. **STIPE:** Short and inconspicuous to long, rooting, and sinuous. Concolorous with upper portion, occasionally, whitish at extreme base. **FLESH:** Thin, gelatinous to rubbery, somewhat tough. Translucent orange to yellow. **ODOR:** Indistinct. **TASTE:** Indistinct. **KOH:** No reaction. **SPORE DEPOSIT:** Pale yellowish. **MICROSCOPY:** Spores 8–12 (14) x 3.5–4.5 μm, cylindrical to slightly curved, smooth, colorless, becoming 1-septate at maturity.

ECOLOGY: Solitary, scattered, or in clusters on rotten conifer wood (sometimes buried). Common north of the San Francisco Bay Area, rare to occasional in the Santa Cruz Mountains. Fruiting from late fall into spring.

EDIBILITY: Unknown.

COMMENTS: The bright golden yellow color, forking or branched tips, rubbery texture, and growth on rotting conifer wood are helpful in distinguishing this jelly fungus. It is most likely to be confused with *Ramariopsis crocea*, which is often brighter orange colored, has a more branched crown, fibrous-brittle flesh, and tiny round spores. It also grows on soil, or in duff (especially under redwood). Orange and yellow *Ramaria* are fleshier and often brittle (not rubbery). *Calocera cornea* is much smaller, and has simple cylindrical clubs. *C. furcata* is also much smaller, but has multiple, often forking branches.

Calocera cornea

(Batsch) Fr.

FRUITBODY: 0.1–0.3 cm across, 0.3–1.5 (2) cm high, cylindrical to elongate clubs, sometimes forking at base, with blunt or pointed tips. Bright or dull orange to yellow. Surface dull to shiny, smooth. **STIPE:** Short, slightly tapered. Often slightly darker and a different tone from cap. **FLESH:** Very thin, gelatinous to rubbery. Translucent orange to yellow. **ODOR:** Indistinct. **TASTE:** Indistinct. **KOH:** No reaction. **SPORE DEPOSIT:** Pale yellowish. **MICROSCOPY:** Spores 7–10 x 2.5–4 μm, cylindrical to slightly curved, smooth, colorless, becoming 1-septate at maturity.

ECOLOGY: Occasionally scattered, but often in large troops on dead logs, trees, or branches. Prefers hardwoods and is very common on Live Oak but can be found on conifers as well. Generally starts fruiting soon after first rains in fall and persists into spring.

EDIBILITY: Unknown.

COMMENTS: This small but distinctive jelly fungus can be readily identified by the small cylindrical clubs, orange to yellow colors, gelatinous to rubbery texture, and growth on branches and logs. It is very common but easily overlooked and usually only gets noticed after rains or prolonged wet periods when the fruitbodies are plump with water. *C. furcata* is very similar, but has multiple, often forking branches. *C. viscosa* is a bright orange to yellow species common north of the San Francisco Bay Area. It differs by the larger size (1–7 cm high) and has slender clubs that become forked and develop a staghorn top. Also compare it with *Mucronella flava*, which has clusters of bright yellow, tiny downward-facing spines, but is very fragile and is not rubbery in texture.

Clavulinopsis corniculata

(Schaeff.) Corner

FRUITBODY: 0.2–1 cm across (crown can be up to 7 cm across), 2–10 cm tall. Highly variable in appearance; sometimes a simple nonbranching cylindrical club, or with a short forking or antlered tip, occasionally with a multiple branched spreading crown on a cylindrical stipe. Dull ocher-orange, yellow-orange to orangish beige. **STIPE:** Generally nondescript in club form; cylindrical on branched form. Concolorous when young, soon developing white fuzz from base. **FLESH:** Thin, somewhat tough, concolorous. **ODOR:** Farinaceous. **TASTE:** Bitter to farinaceous. **SPORE DEPOSIT:** Whitish. **MICROSCOPY:** Spores 4.5–7.5 µm, globose with a long apiculus, smooth, moderately thick walled, inamyloid.

ECOLOGY: Solitary or in scattered troops or small clusters in duff in mixed forest, especially under redwood. Common and widespread, occurring throughout our range. Fruiting from late fall through winter.

EDIBILITY: Edible.

COMMENTS: Typical California specimens of this species are rather simple cylindrical clubs with just a few branches and forked tips. However, in the Pacific Northwest, more densely branched clusters with spreading crowns are more common. The dull ocher-orange color helps distinguish the club-shaped forms from the bright yellow-orange *C. laeticolor*. *Ramariopsis crocea* can be distinguished from the branching form by its thinner branches, slightly rubbery flesh, and brilliant yellow-orange color. Two similar small, simple clubs are *C. gracillima*, which has dull ocher-yellow to peachy yellow colors, and *Clavaria flavipes*, which is dull yellow-tan with a brighter yellow lower stipe; both species are rare in our range.

Clavulinopsis laeticolor

(Berk. & M. A. Curtis) R. H. Petersen

YELLOW CLUB CORAL

FRUITBODY: 0.2–0.8 cm across, 2–10 cm tall, cylindrical to club shaped with a rounded to blunt, rarely pointed tip, occasionally longitudinally compressed or flattened in age. Bright yellow-orange, gold, orange to yellow. **STIPE:** Short; opaque orange with a pale base. **FLESH:** Thin, fragile, concolorous. **ODOR:** Indistinct. **TASTE:** Mild to slightly bitter. **SPORE DEPOSIT:** White. **MICROSCOPY:** Spores 5.5–9 x 3–5 µm, ellipsoid, oblong, globose to subtriangular, smooth, inamyloid.

ECOLOGY: Solitary or scattered, occasionally in small clusters in duff, moss, or soil in both hardwood and coniferous forest. Very common, especially under redwood, occurring throughout our range. Fruiting from fall through winter.

EDIBILITY: Edible.

COMMENTS: This common, brightly colored club can easily be identified by the yellow to orange color and simple club shape. Although it usually grows singly or scattered in small troops, it does occasionally fruit in clusters. The name *Clavulinopsis fusiformis* is often misapplied to the clustered forms in California. *Clavulinopsis gracillima* has dull ocher-yellow to peachy yellow colors. *Clavaria flavipes* is a small club-shaped coral with dull tan-beige colors and a small, yellower stipe. *Clavulinopsis corniculata* has dull yellowish orange to orangish beige colors, and often, but not always, forking tips.

Alloclavaria purpurea

(Müll.) Dentinger & D. J. McLaughlin

PURPLE CLUB CORAL

FRUITBODY: 0.2–0.7 cm across, 3–10 (13) cm tall, cylindrical with rounded to pointed tips, occasionally longitudinally compressed in age. Purple, lilac to purplish gray when young, often losing purple tones in age, becoming lilac-gray, gray, or grayish buff in age. **STIPE:** Indistinct. **FLESH:** Thin, fragile, concolorous. **ODOR:** Mild to slightly unpleasant. **TASTE:** Mild. **SPORE DEPOSIT:** White. **MICROSCOPY:** Spores (5.5) 7–11 (15.5) 3–5 µm, ellipsoid to oblong, smooth, inamyloid.

ECOLOGY: In scattered clusters or in large troops of many clusters, often on bare sandy soil, or in moss-covered sand. Locally common on forested dunes of the extreme Far North Coast. Fruiting from late fall into winter. It is probably mycorrhizal, but this has yet to be confirmed.

EDIBILITY: Edible.

COMMENTS: The simple, cylindrical shape, clustered growth, and purple to lilac-gray colors are distinctive. It is very common in the coastal dunes in Oregon, and high elevation spruce forests in the Cascades and Rocky Mountains. Although it can be common on the forested dunes north of Crescent City, it is not known to occur in similar habitat in Humboldt County. *Clavaria fumosa* has a similar shape and grows in clusters, but is smoky beige and usually grows under redwoods.

Clavaria rosea

Fr.

ROSY CLUB CORAL

FRUITBODY: 0.1–0.5 cm across, 1–5 cm tall, cylindrical to club shaped, enlarged toward top, with a rounded to flattened tip. Bright rosy pink to light pink, much paler in age. **STIPE:** Short, often darker and shinier than upper portion. **FLESH:** Thin, very fragile, concolorous. **ODOR:** Indistinct. **TASTE:** Mild. **SPORE DEPOSIT:** White. **MICROSCOPY:** Spores 5–8 x 2.5–3.5 µm, ellipsoid, smooth, inamyloid.

ECOLOGY: Solitary, scattered, or in small clusters in duff, moss, or grass, often in lightly disturbed areas, such as old roadsides or trail edges. Widespread, occurring throughout our range, but quite rare. Fruiting from late fall into early spring.

EDIBILITY: Tiny, rare, nontoxic.

COMMENTS: This strikingly beautiful species can be easily identified by the small size, rosy pink color, and solitary or weakly clumped cylindrical fruitbodies. *Alloclavaria purpurea* is quite a bit larger, purple to lilac-gray, and almost always clustered in growth.

Clavaria falcata

Pers.

FAIRY CLUB

FRUITBODY: 10.1–0.5 cm across, 1–5 cm tall, cylindrical to club shaped, enlarged toward top, with a rounded to flattened tip. White in color. **STIPE:** Short, distinct from fertile surface. Translucent watery white in color. **FLESH:** Thin, very fragile, concolorous. **ODOR:** Indistinct. **TASTE:** Mild. **SPORE DEPOSIT:** White. **MICROSCOPY:** Spores 7–10 x 5–9 µm, broadly ellipsoid, ovoid to subglobose, smooth, inamyloid.

ECOLOGY: Solitary, scattered, or in small tufts (rarely clustered) in disturbed soil along trails or in duff or moss. Widespread and fairly common throughout our range, but easily overlooked. Fruiting from late fall into winter.

EDIBILITY: Unknown.

COMMENTS: This species can be identified by the small white fruitbodies, habit of fruiting singly, and distinct translucent stipe. *C. fragilis* is larger and often clustered, and lacks the distinct stipe. *Macrotyphula juncea* forms extremely thin, tall, cylindrical fruitbodies that are yellowish buff in color. *Clavicorona taxophila* produces a tiny white, club-shaped fruitbody, with distinctly flattened to cup-shaped or slightly wavy-flattened tops.

SYNONYM: *Clavaria acuta* Pers.

Clavaria fragilis group
FAIRY FINGERS

FRUITBODY: 0.2–0.7 cm across, 3–10 (15) cm tall, cylindrical with rounded to pointed tips, occasionally splitting. White to creamy white, tips occasionally yellowing in age. Surface smooth, dry to moist. **STIPE:** Indistinct. **FLESH:** Thin, very fragile, concolorous. **ODOR:** Indistinct. **TASTE:** Mild. **SPORE DEPOSIT:** White. **MICROSCOPY:** Spores 5–7 x 3–4 µm, ellipsoid to teardrop shaped, smooth, inamyloid.

ECOLOGY: Scattered in small to large clusters, often in troops or arcs in duff or soil in a wide variety of forest types, especially dense redwoods. Very common throughout our range. Fruiting from early fall into winter.

EDIBILITY: Edible.

COMMENTS: The clusters of simple, white unbranching clubs with no distinct stipe are distinctive. Although usually identified as *C. fragilis* or *C. vermicularis*, our species is distinct and will be given a new name soon. It is also called the Worm Coral or White Club Coral. *C. fumosa* is quite similar, but has a smoky beige color; it is much less common. *C. acuta* is quite a bit smaller, rarely grows in such tight clusters, and has a narrower, distinct, translucent stipe.

Clavariadelphus occidentalis

Methven

WESTERN CLUB CORAL

FRUITBODY: 1–3 cm across, 4–15 cm tall, cylindrical, club shaped, sometimes branched, often flattened or fan shaped, broadest in middle or just below tip. Whitish to pale yellow at first, soon buff, yellowish, or ochraceous buff, becoming tan to grayish orange, paler at base, fading overall in dry weather. Surface dry, smooth, often wrinkled and vertically ridged. **FLESH:** Firm, stringy when young, soft, spongy in age. White, slowly staining light to dark brown. **ODOR:** Indistinct. **TASTE:** Mild to bitter. **KOH:** Golden orange. **SPORE DEPOSIT:** Whitish. **MICROSCOPY:** Spores 9.5–13 x 5.5–7 μm, broadly ovate to almond shaped, thin walled, smooth, inamyloid.

ECOLOGY: Solitary, scattered, or in small troops on ground or in thick duff in a wide variety of forest types. In the greater Bay Area generally fruits under live oak, but is common in mixed conifer or Tanoak-conifer forest on the North Coast. Fruiting in fall in north, winter into spring in south.

EDIBILITY: Edible, but often bland or bitter and tough.

COMMENTS: This distinctive mushroom is not common every year, but in some years can be locally abundant. The large, wrinkled, cylindrical to club-shaped, dull-colored fruitbodies help set them apart from other club fungi. The only similar species are other species of *Clavariadelphus*. The most similar is *C. subfastigiatus,* which differs by having pinkish buff to light cinnamon brick colors, a forest green KOH reaction, and smaller spores (8–10.5 μm x 5–6 μm). *C. caespitosus* often fruits in clusters of narrow, club-shaped, grayish red to dull red fruitbodies. *C. pallidoincarnatus* and *C. truncatus* have medium to large club-shaped fruitbodies with distinctly flattened or slightly rounded tops. They also have a sugary-sweet taste and are most common in the northern part of our range. *C. ligula* and *C. sachalinensis* are much smaller and grow in large troops under conifers; they are distinguished from each other only by spore size, greater than 18 μm long for *C. sachalinensis* and less than 12 μm long for *C. ligula.*

Clavariadelphus pallidoincarnatus

Methven

PALE CANDY CORAL

FRUITBODY: 1–2.5 cm across at the top, 7–15 cm tall, club shaped and often winkled, with a broad, flattened to rounded tip. Grayish orange to buff or ocher-beige with a bright orange to ocher-orange top, fading to whitish buff in age. Surface moist to dry, often wrinkled and vertically ridged, slowly staining brownish when rubbed. **FLESH:** Firm to soft. Whitish, slowly staining brownish when exposed. **ODOR:** Indistinct. **TASTE:** Sweet. **KOH:** Yellow. **SPORE DEPOSIT:** Whitish. **MICROSCOPY:** Spores 9–11.5 x 5.5–7 μm, broadly ovate to almond shaped, thin walled, smooth, inamyloid.

ECOLOGY: Solitary or scattered in troops on ground or in thick duff in northern coniferous forest, especially under Sitka Spruce on the Far North Coast. Fruiting in fall. Locally common, but with a very small range in California.

EDIBILITY: Edible; not as sweet tasting as *C. truncatus.*

COMMENTS: This species can be recognized by the baseball-bat shape, flattened to irregularly rounded top that is often contrastingly brighter colored, and growth in northern coastal coniferous forest. *C. truncatus* is larger and brighter orange; it occurs throughout our range under both conifers and hardwoods. It is often called the Candy Club because of its sweet, sugary taste. Also compare with *C. occidentalis.*

Macrotyphula juncea

(Alb. & Schwein.) Berthier

FAIRY HAIR

FRUITBODY: 0.05–0.3 cm across, 2–12 cm tall, tall, very thin cylindrical club with pointed to blunt tips. Yellowish buff to tan. Surface smooth, dry. **STIPE:** Short, often tapering downward, base with white rhizomorphs. Slightly darker in color than the fertile part. **FLESH:** Very thin, tough, fibrous, generally hollow, concolorous. **ODOR:** Mild to rancid. **TASTE:** Slightly sour to acrid. **SPORE DEPOSIT:** White. **MICROSCOPY:** Spores 7–10 x 3.5–4 µm, ellipsoid to almond shaped, smooth, inamyloid.

ECOLOGY: Often scattered in large troops in duff in a wide variety of forest types. We have seen large carpets under live oak and redwood soon after the first saturating rains in fall. Common throughout our range. Fruiting from early fall into winter.

EDIBILITY: Unknown.

COMMENTS: This small hairlike coral can be easily identified by the simple yellowish buff club with a slightly darker stipe, white rhizomorphs, and scattered growth in duff. *Typhula phacorrhiza* is very similar, but grows from a yellow-orange sclerotium on leaves and twigs of Bigleaf Maple and alder. Another species in that habitat is *T. erythropus*, which has a small, white club-shaped fruitbody with a reddish brown stipe. It also grows from a tiny sclerotium, often embedded into the leaf. Other *Typhula* species are smaller and grow from sclerotia. *Clavaria acuta* is smaller, more club shaped, and white in color, with a well-defined translucent stipe. The rare *M. fistulosa* is larger and club shaped (up to 30 cm tall, 0.2–1 cm thick) and generally ocher-yellow to ocher-brown.

Heyderia abietis

(Fr.) Link

TINY NEEDLE CLUB

FRUITBODY: 0.5–3.5 cm tall. Head 0.1–0.7 cm across, egg shaped to rounded-cylindrical, rarely round, sometimes flattened. Light tan to ocher-buff. **STIPE:** Thin, wiry, distinct from head. Brownish ocher. **FLESH:** Very thin, somewhat tough, rubbery to fibrous. **ODOR:** Indistinct. **TASTE:** Indistinct. **SPORE DEPOSIT:** Generally not obtainable. **MICROSCOPY:** Spores 10–14 x 2–2.5 µm, spindle shaped, often pointed at ends, smooth. Asci 8-spored.

ECOLOGY: Scattered in small to large troops and swarms on duff under conifers on the Far North Coast. Fruiting in the fall. Very common in north, known from as far south as Mendocino County.

EDIBILITY: Unknown. Dried specimens might be good as toothpicks.

COMMENTS: The tiny ocher-buff heads atop narrow, brownish stipes and growth on conifer duff are distinctive. They are most often overlooked, but carpets of fruitbodies can be can be quite apparent if one is actively looking for them. Often each individual club will be attached to a single needle.

SYNONYM: *Mitrula abietis* Fr.

Onygena corvina

Alb. & Schwein.

OWL-PELLET POWDERHEAD

FRUITBODY: 0.5–2.5 cm tall. Head 0.1–0.4 cm across, rounded. Pale buff to light brown. Surface dry, becoming powdery as spores mature. **STIPE:** 0.5–2 cm long, 0.1–0.2 cm thick, stout and equal at first, becoming irregular, curvy and often with an enlarge base in age. White to cream colored. **FLESH:** Very thin, soft to powdery in cap, fibrous in stipe. **ODOR:** Indistinct, but the substrate is usually stinky. **TASTE:** Not sampled. **SPORE DEPOSIT:** Buff to light brown. **MICROSCOPY:** Spores 6–8 x 2.5–3 µm, cylindrical to ellipsoid, smooth, with 1 or 2 oil droplets. Asci 8–spored, spherical, 10–16 x 10–12 µm.

ECOLOGY: Often in clusters on owl pellets, bird bones, and feathers. Also reported from animal hair. Rare, only known from a few collections in California, from Mendocino County northward. Fruiting in fall and winter

EDIBILITY: Unknown.

COMMENTS: This oddball has a very specialized habitat, as it almost always grows from owl pellets or on bird bones. Besides the habitat, it can be recognized by the rounded, buff to light brown head that becomes powdery and the white stipe. So far it is only known from four Californian collections (three found by Alissa Allen), growing on owl pellets. *O. equina* is very similar, but grows on rotting horns or hooves of grazing animals such as cattle, sheep, and deer. *Penicillium vulpinum* is similar in shape, but has a powdery bluish gray, irregularly rounded head, and a white to reddish brown stipe, and grows on dung. *Asterophora lycoperdoides* is much larger and grows on rotting mushrooms.

Cordyceps militaris

(L.) Fr.

TROOPING CORDYCEPS

FRUITBODY: 0.5-7 cm tall. Head 0.2–1 cm across, cylindrical, club shaped, or spindle shaped. Pale orange with bright orange, projecting perithecia. Surface dry, bumpy. **STIPE:** 0.5–4 cm long, 0.2–0.7 cm thick, equal or rooting. Whitish to pale orange. Surface dry. **FLESH:** Thin, whitish to pale orange. **ODOR:** Indistinct. **TASTE:** Indistinct. **MICROSCOPY:** Spores 300–500 x 1–1.5 µm, threadlike, but breaking up into barrel-shaped pieces 2–6 x 1–1.5 µm, smooth, thin walled. Asci 8-spored.

ECOLOGY: Solitary or in loose clusters, growing from insect cocoons buried in well-rotted wood or soil. Extremely rare in California, known only from a handful of collections. Fruiting in fall.

EDIBILITY: Much too rare in California to be considered. Commonly cultivated for its purported medicinal properties.

COMMENTS: The small bumpy, orange, club-shaped fruitbodies and growth on insect cocoons are distinctive. Although a common species in eastern North America, it is extremely rare in California. We know of only few collections from the state, on the Far North Coast and Sierra Foothills.

Hypocrea leucopus

(P. Karst.) H. L. Chamb.

FRUITBODY: 2–5 cm tall, 0.2–1 (1.5) cm across, club shaped, with a swollen head. Creamy white to pale ocher-buff. Surface dry, finely bumpy. **STIPE:** Narrower than cap at apex, often swollen toward base. White to pale cream. Surface dry, finely scruffy to smooth. **FLESH:** Thin, watery when wet, fibrous when dry. Whitish to cream. **ODOR:** Indistinct. **TASTE:** Indistinct. **MICROSCOPY:** Spores with two parts; upper part 2–4 x 2–3.5 μm, globose to subglobose, lower part 2–5 x 1.5–3.5 μm, wedge shaped to ellipsoid, finely spiny. Asci 8-spored.

ECOLOGY: Solitary or scattered in needle duff or moss in coniferous forest on the North Coast. Rare overall, but can be locally common in older spruce forests. Fruiting in late winter and spring.

EDIBILITY: Unknown.

COMMENTS: The club-shaped fruitbody with a creamy white head and bumpy surface, white stipe, and growth in duff help distinguish this species. *H. alutacea* is slightly larger, has a darker yellowish brown head, and grows on wood. *Cordyceps militaris* has a distinctly pimpled orange head and grows on insect cocoons.

SYNONYM: *Podostroma leucopus* P. Karst.

MISAPPLIED NAME: *Hypocrea alutacea* (Pers.) Ces. & De Not.

Tolypocladium capitatum
ROUND-HEADED TRUFFLE-CLUB

(Holmsk.) Quandt, Kepler & Spatafora

FRUITBODY: 3–12 cm tall overall. Head 0.8–2 cm across, egg shaped or rounded. Golden brown, dark brown to olive-brown. Surface dry, covered with tiny bumps, often crimped at stipe attachment. **STIPE:** 2.5–11 cm long, 0.4–1 cm thick, more or less equal. Golden yellow to yellowish olive, with darker scales, occasionally dingy olive in age. Surface dry, covered with chevronlike scales. **FLESH:** Firm, but somewhat fragile, fibrous in stipe. Whitish in color. **ODOR:** Indistinct. **TASTE:** Indistinct. **SPORE DEPOSIT:** Buff to light brown. **MICROSCOPY:** Spores threadlike but breaking up into cylindrical segments 8–25 (30) x 2.5–3 μm. Asci 350–550 x 10–12 μm, cylindrical.

ECOLOGY: Solitary or in small clusters, arising from an underground *Elaphomyces* truffle. Generally one or two clubs are attached to each truffle, but occasionally larger clusters are found. Uncommon, occurring north of San Francisco in mixed coniferous forests. Usually fruiting in fall and winter.

EDIBILITY: Unknown.

COMMENTS: The golden to dark brown head covered in tiny pimples, yellowish stipe, and growth from *Elaphomyces* fruitbodies make for a distinctive combination. The presence of the host truffle is very easy to miss, since the stipe base is often broken off when collected. A much rarer species in California, *T. ophioglossoides* (=*Elaphocordyceps ophioglossoides*), produces much smaller fruitbodies with black, cylindrical, or irregularly club-shaped heads, olive to blackish stipes, and vivid yellow mycelium and rhizomorphs attached to *Elaphocordyceps* truffle hosts.

SYNONYMS: *Elaphocordyceps capitata* (Holmsk.) G. H. Sung, J. M. Sung & Spatafora, *Cordyceps capitata* (Holmsk.) Link.

Xylaria hypoxylon

(L.) Grev.

CARBON ANTLERS

FRUITBODY: 2–8 (10) cm tall, 0.1–0.7 cm across, cylindrical at first, soon forking and branching at tips or becoming antler-like. Black on lower part, powdery, whitish on upper portion (rarely entirely) at first, becoming more grayish on tips, to all black when old. **FLESH:** Thin, tough, wiry. Whitish to pale gray at first, black in age. **ODOR:** Indistinct. **TASTE:** Not sampled. **MICROSCOPY:** Spores 11–14 x 5–6 µm, bean shaped, smooth, brownish black. Asexual spores ellipsoid, smooth, colorless.

ECOLOGY: In masses or loose clusters, occasionally solitary on stumps, logs, and other woody debris of hardwoods. Very common throughout our area. Fruitbodies are persistent and can be found throughout year.

EDIBILITY: Unknown.

COMMENTS: Shape alone is generally enough to recognized this tough, wiry, and often antler-shaped fungus. The often powdery tips and growth on wood are also helpful. The powdery white tips are asexual spores; sexual spores are produced from the pimplelike perithecia embedded in the surface. This fungus is also called Candlesnuff Fungus. Thicker, less branched, and/or more pimpled *Xylaria* fruitbodies are often encountered, and it is very likely that we are misapplying this name to some of the variants in our area, which could represent a handful of species.

Microglossum viride

(Pers.) Gillet

GREEN EARTH TONGUE

HEAD: 1–3 cm tall, 0.2–0.8 cm across, sometimes rounded when young, but usually elongate and grooved, sometimes branched or lobed, often paddle shaped at maturity. Shiny, grass green to teal-green or turquoise-green. Surface dry to moist, pruinose when young, becoming shiny in age. **STIPE:** 2–6 cm long, 0.2–0.4 cm thick, equal, or slightly enlarged toward head. Pale green or creamy with an aqua-green wash, usually evenly covered in irregular bands of dark aqua-green bracelets, chevrons, or dots. **FLESH:** Pallid greenish cream, rubbery. **TASTE:** Indistinct. **ODOR:** Indistinct. **KOH:** Unknown. **MICROSCOPY:** Spores 12–16 (22) x 5–6 µm, elongate-ellipsoid to cylindrical, smooth, with 2–4 oil droplets. Asci 8-spored.

ECOLOGY: Solitary, scattered, or in small clusters, often on exposed soil or in moss, usually around Tanoak. Fruiting in winter and early spring. Widespread in our area, but apparently only locally common, as it may be frequently overlooked.

EDIBILITY: Unknown.

COMMENTS: No other species really approaches the Green Earth Tongue in appearance. The green to teal colors and chevrons or dots on the stipe are distinctive. *M. olivaceum* has dingy olive-green to olive-beige colors and lacks the distinct chevrons and dots on the stipe. All other species of Earth Tongues (*Trichoglossum* and *Geoglossum*) in our area are dark brownish black to black. The name *M. viride* is applied to collections more or less matching this description from around the world, and it is likely that such broad usage will not be supported after closer scrutiny. The degree of ornamentation on the stipe and exact shade of green appear to vary substantially from region to region.

Trichoglossum hirsutum

(Pers.) Boud.

VELVETY-BLACK EARTH TONGUE

HEAD: 1–3 cm tall, 0.2–1.5 cm across, sometimes torpedo shaped when young, usually flattened or compressed, becoming more elongate and spatula shaped in age. Black to dark brownish black. Surface dry, finely fringed with projecting setae. **STIPE:** 2–6 cm long, 0.2–0.5 cm thick, cylindrical, straight. Black to dark brownish black. Surface dry, finely velvety. **FLESH:** Thin, tough, brownish black. **ODOR:** Indistinct. **TASTE:** Indistinct. **MICROSCOPY:** Spores 80–160 x 5–7 μm, very long and narrow, often with a swollen middle and tapering ends, smooth, 15-septate when mature. Asci 8-spored.

ECOLOGY: Solitary, scattered, or in small troops on ground, in moss or duff in mixed forest throughout our range. Very common. Fruiting from late fall into spring, peaking in midwinter.

EDIBILITY: Unknown.

COMMENTS: Of the many different species of black earth tongues in California, *T. hirsutum* appears to be our most common. *Trichoglossum* can be separated from *Geoglossum* macroscopically by the velvety stipes and finely hairy heads (due to the presence of projecting needle-shaped cells called setae). Beyond that, microscopic examination is needed to confirm species identification. *T. hirsutum* is distinct by having long, 15-septate spores in 8-spored asci. *T. velutipes* has 4-spored asci and up to 13-septate spores, with most being 9-septate. *Geoglossum* are a bit trickier to identify; generally they are more irregular in shape with a less defined head. The most common is *G. umbratile*, a dark blackish brown to black species that becomes slightly irregularly shaped in age. It has straight to curved, 7-septate spores, measuring 30–90 x 4.5–6.5 μm. *G. glabrum* is very similar, but has paraphyses that are more closely septate and generally longer than the asci. *G. fallax* is a dark brown to blackish brown species with spores that are generally colorless, measuring 45–110 x 5–7 μm, and are generally 7- to 12-septate when mature. We probably have additional species, but to our knowledge, nobody has intensively studied California's earth tongues. *Glutinoglossum* (=*Geoglossum*) *glutinosum* is distinctly viscid when wet and dark brown to black in color; it is fairly common on trail banks in mossy soil. *Hypomyces papulasporae* grows on the heads of *Trichoglossum* and *Geoglossum* fruitbodies, covering them with a white mold that develops pimplelike perithecia when mature; it is a rare species (we've only seen it once, in Mendocino County).

Cudonia circinans

(Pers.) Fr.

CAP: 0.5–1.5 (2) cm across, irregular-rounded to convex when wet, wrinkled when dry, margin inrolled at all stages. Pinkish buff, orangish buff to warm beige. Surface smooth, generally dull but can be shiny when wet. **UNDERSIDE:** Dull, smooth to wrinkled, concolorous with stipe. **STIPE:** 1.5–6 cm long, 0.3–1 cm thick, equal or enlarged toward base. Orangish buff to warm beige when young, becoming darker, smoky buff to vinaceous beige. Surface dry to moist, finely roughened or smooth. Solid when young, often hollow in age. **FLESH:** Thin, often rubbery to slightly gelatinous when wet, leathery if dry. Concolorous with cap and stipe. **ODOR:** Indistinct. **TASTE:** Indistinct. **MICROSCOPY:** Spores 28–46 x 2 μm, needle shaped, colorless. Asci 8-spored, club shaped.

ECOLOGY: Often in small or large clusters, rarely solitary, in moss, thick duff, and well-decayed woody debris, or on spruce cones. Locally common in Sitka Spruce forest of the Far North Coast, less frequent south to Mendocino County. Generally fruiting from late winter into spring, and even persists into summer during foggy years. Sometimes occurs in fall.

EDIBILITY: Avoid; there are reports of it being poisonous when raw.

COMMENTS: The irregular-rounded cap that becomes plump when wet, smooth or slightly veined underside, rubbery stipe, buff to beige colors, and growth under spruce (in California) help identify it. *Leotia lubrica* is very similar in stature, but is generally slightly larger, with more yellowish to yellow-green tones. *Pachycudonia monticola* is a similar montane species with a pinkish buff to grayish brown cap, a grayish brown stipe, and shorter spores (20–24 x 2 μm); it fruits in the spring or early summer. *Heyderia abietis* is a vaguely similar species with orangey brown fruitbodies the size and shape of matchsticks; it grows on conifer needles.

Leotia lubrica

(Scop.) Pers.

JELLY BABIES

CAP: 1–2.5 (4) cm across, irregular-rounded, convex, wrinkled or lobed, margin inrolled at all stages. Yellowish to greenish yellow or ocher-buff at first, occasionally becoming dark green in age (see comments). Surface smooth to wrinkled, viscid when wet, dull when dry. **STIPE:** 2–7 (10) cm long, 0.3–1 cm thick, equal or enlarged toward base. Golden, ocher-yellow to greenish yellow, base often paler. Surface viscid when wet, finely roughened over upper portion, to smooth. **FLESH:** Rubbery, with a gelatinous or watery core. Translucent yellowish. **ODOR:** Indistinct. **TASTE:** Indistinct. **MICROSCOPY:** Spores 16–25 x 4–6 μm, spindle shaped, smooth, colorless, 4–6 (8)-septate. Asci 8-spored.

ECOLOGY: Solitary, scattered, or in clusters on ground, in duff, in moss, or on well-decayed wood. Generally in mixed Tanoak forest, but occurs in coniferous forest as well. Very common throughout our range. Fruiting in late fall in north, winter into spring in the mid and southern part of our range.

EDIBILITY: Edible.

COMMENTS: The yellowish stipe, yellowish to dingy olive-gray head, plump cap, and rubbery texture are distinctive. One often comes across large patches of this mushroom while hunting Black Trumpets in the winter in Tanoak forests. The name *L. viscosa* has been applied to green-capped forms, but it is our belief that the Californian *Leotia* all represent a single species, regardless of cap color. It is not uncommon to find a green-capped fruitbody mixed into a cluster of ocher-buff and yellowish ones. *Cudonia circinans* has a very similar stature and rubbery texture, but has a pinkish buff cap and a darker stipe, and lacks the gelatinous core to the flesh.

21 • Puffballs, Earthballs, Earthstars, Stinkhorns, and Bird's Nests

These mushrooms are all basidiomycetes that don't forcibly discharge spores from their basidia. The interior spore masses of puffballs, earthballs, and earthstars become powdery at maturity and are then slowly dispersed by the energy of wind and rain droplets. The stinkhorns have gooey spore slime and are dispersed by insects. Bird's nest fungi disperse their spore packets, or "eggs," with the help of rain droplets (and often grazing animals). *Battarrea* have tall, upright fruitbodies similar to those of some stinkhorns, but their spore mass is powdery (not gooey), and their spores are wind dispersed.

This group comprises many different evolutionary lineages; even *Astraeus* and *Geastrum*, despite their very similar fruitbody morphologies, are not closely related! The former is related to boletes, while the latter is related to the coral fungi, *Ramaria*, and stinkhorns! *Astraeus*, *Pisolithus*, and *Scleroderma* are the only mycorrhizal genera in this group, and all three are more closely related to boletes than to the other species in this section. The puffballs (*Bovista*, *Calvatia*, and *Lycoperdon*) are closely related to the gilled fingi, *Agaricus* and *Lepiota*.

Although many species can be identified by details of fruitbody structure, coloration, and substrate, reliable identification sometimes requires microscopic inspection. The size and structure of the spores are particularly important, as are the structures of the internal hyphae (especially for the puffballs). Identification of stinkhorns is fairly straightforward.

Bovista pila

Berk. & M. A. Curtis

TUMBLING PUFFBALL

FRUITBODY: 2.5–9 cm across, round to cushion shaped, base pinched and often with a single, white, cordlike rhizomorph. White to pinkish lilac or grayish at first, outer skin smooth to slightly matted-tomentose, darkening to tan to brown, and peeling off in age, exposing a shiny, metallic brown skin that often tears to expose the spore mass in age. **INTERIOR:** Firm and whitish at first; very light and soft with a dark brown spore mass when mature. Sterile base absent. **ODOR:** Indistinct. **TASTE:** Indistinct. **MICROSCOPY:** Spores 3.5–4.5 µm, globose, with stalklike attachment, thick walled, smooth to finely roughened.

ECOLOGY: Solitary or scattered in grassy areas, in duff (especially under cypress), or around horse paddocks. Widespread and locally common. Fruiting throughout the wet season, more common in late spring.

EDIBILITY: Edible when young and white inside.

COMMENTS: The medium size, rounded shape, lack of a sterile base, and shiny brown fruitbodies that are very lightweight and blow around at maturity are distinctive. When young, the single rhizomorphic cord is also helpful in distinguishing it from similar *Bovista*. Another common grassland species, *B. plumbea*, is smaller (rarely more than 3 cm across) and a bit less metallic, and lacks the white cord at the base (but can have a tuft of mycelium). *B. aestivalis* is very similar to *B. plumbea*, but has a more floccose-felty outer surface when young, yellower colors in age, and spores with short stalklike attachments.

Calvatia fragilis

(Vittad.) Morgan

PURPLE-SPORED PUFFBALL

FRUITBODY: 4–9 cm across, 4–6 cm tall, round, cushion to top shaped, base pinched and often pleated to a short, rudimentary stipe. Whitish buff at first, soon brownish to straw colored, developing grayish violet and then dark purple tones. Outer skin felty to smooth, breaking up into patchy scales (like cracked mud), scuffing off in patches or sheets, exposing a purplish skin and purple, powdery spores. **INTERIOR:** Firm and whitish at first, soon soft and powdery, becoming violet to lilac, then deep purple to purplish brown as spores mature. Sterile base small, or lacking. **ODOR:** Indistinct. **TASTE:** Indistinct. **MICROSCOPY:** Spores 5.5–7.5 µm across, globose, thick walled, finely spiny.

ECOLOGY: Solitary or scattered, occasionally in rings on ground in grassy areas, more rarely in woodlands. Fruiting in winter and spring. Widespread but uncommon, absent from the North Coast.

EDIBILITY: Edible when young and pure white inside.

COMMENTS: This puffball can easily be identified by the thin skin that cracks into patches and the purplish spore mass. *Scleroderma* can also have purplish spores, but have much thicker skin, a very firm texture, and often well-developed rhizomorphs. *Bovista* species have smoother skin when mature and dark brown spores.

SYNONYM: *Calvatia cyathiformis* f. *fragilis* (Vittad.) A. H. Sm.

Calvatia pachyderma

(Peck) Morgan

ELEPHANT-SKIN PUFFBALL

FRUITBODY: 4–12 (15) cm across, 5–12 cm tall, round to cushion shaped, base slightly pinched and pointed downward, often with short mycelial threads. Whitish at first, discoloring in a patchy manner, becoming grayish to clay colored or brownish. Surface smooth, scaly or more often like cracked mud, breaking apart and exposing the spores, leaving a shallow cup-shaped base in age. **INTERIOR:** Outer rind thick and tough, interior very firm and white at first, soon soft, yellowish, golden brown to dark olive-brown as spores mature. **ODOR:** Indistinct or like Button Mushrooms when young. **TASTE:** Indistinct. **MICROSCOPY:** Spores 4–5.5 x 3.5–5 µm, globose to ellipsoid, smooth, thin walled.

ECOLOGY: Solitary, scattered, or in loose clusters, often in arcs or impressive rings in grassy areas. Common in the greater Bay Area. Fruiting in spring, smaller numbers in early fall.

EDIBILITY: Edible and very good. Only eat specimens with a firm, pure white interior.

COMMENTS: The medium to large size, thick skin, lack of a distinct sterile base, and golden brown to olive-brown spore mass are helpful identifying features for this puffball. *Mycenastrum corium* has a felty outer layer with a thicker, tough inner skin, a mass of mycelial threads at the base, and a dark brown to reddish brown spore mass; microscopically, it has larger spiny spores (8–12 µm across) and a barbwire-like capillitium (the hyphae in the spore mass). Another large white puffball found on the coast is *Lycoperdon utriforme* (=*Calvatia bovista*), which is thinner skinned and more pear shaped due to its large sterile base. *Calvatia booniana* is often larger and covered with flattened warts or scales, and has very finely roughened spores. It grows primarily in the mountains and high desert, but is occasionally found on the coast.

Lycoperdon pratense

Pers.

FRUITBODY: 2–4 cm across, 2–5 cm tall, broadly pear shaped, narrowing and often pinched toward base. White at first, becoming dingy yellowish to yellowish brown, then brown when mature. Surface covered in small spines and granules. Skin at top of fruitbody thin and soon tearing and breaking down, leaving a vase-shaped structure with a sterile base. **INTERIOR:** Firm and whitish at first, soon soft, becoming olive-brown, then dark brown when the spores mature. **ODOR:** Indistinct. **TASTE:** Indistinct. **MICROSCOPY:** Spores 3.5–4.5 x 4–5 µm, averaging 4 µm, subglobose to globose, smooth to finely roughened or warted.

ECOLOGY: Solitary, scattered, or in loose clusters, arcs, or rings in grassy areas. Uncommon, but widespread. Fruiting in fall and spring.

EDIBILITY: Edible when white inside, but small and rarely numerous.

COMMENTS: This species can be recognized by the small size, fine covering of spines and granules, distinct sterile base, and growth in grass. The spores of *Vascellum (Lycoperdon) lloydianum* average a bit larger (4.5 µm across); otherwise it is nearly indistinguishable. *Bovista aestivalis* looks very similar, but has a felty-powdery coating and lacks the sterile base. The rare *L. curtisii* is distinctly spiny, with the tips of the spines curved and joined, creating a starlike pattern.

SYNONYM: *Vascellum pratense* (Pers.) Kreisel.

Lycoperdon perlatum

Pers.

GEM-STUDDED PUFFBALL

FRUITBODY: 1–5 cm across, 1–7 (12) cm tall, pear shaped to oblong, with an oval or round top and an irregularly cylindrical stipe ranging from slender and elongate to occasionally bulbous. Surface dry, covered with pointed, pyramidal warts or spines and granular particles that leave oval scars or depressions when they fall off. Skin thin, cracking and tearing at apex. Creamy white, pinkish gray to pinkish buff or light brown, with slightly darker spines, overall golden brown to brown when old. **INTERIOR:** Firm and white at first, soon soft, becoming mushy, yellowish olive, then olive-brown to brown and powdery when spores mature. With a pronounced sterile base. **ODOR:** Indistinct. **TASTE:** Mild. **KOH:** Forest green on mature spore mass. **MICROSCOPY:** Spores 2.5–5 µm across, globose, thick walled, spiny.

ECOLOGY: Solitary, scattered, or in small to large clusters. It has a preference for disturbed ground such as roadsides and trail edges, also commonly occurs in duff and humus in forests. Widespread and common. Fruiting in fall and winter.

EDIBILITY: Edible. Eat only specimens that are pure white inside.

COMMENTS: The pear-shaped fruitbody and small, pyramidal spines help distinguish this species. When the larger spines fall off, they leave depressed oval scars, which helps distinguish this species from *L. pyriforme*; that species is typically smoother, grows in dense clusters on wood, and has white rhizomorphs at the base. *L. molle* is usually more grayish to grayish brown, and has short, granular spines that often are fused together at the tips. *L. utriforme* is considerably larger and has a granular surface. Darker specimens that share the pear shape of this species should be compared with *L. umbrinum*.

Lycoperdon pyriforme
Schaeff.
PEAR-SHAPED PUFFBALL

FRUITBODY: 1–3.5 cm across, 2–6 cm tall, pear shaped to oblong, often with round top, and a pinched, irregular pleated pseudostipe with a tuft of white rhizomes. Surface dry, granular to powdery at first, soon smooth or nearly so. Skin thin, cracking and tearing a apex when mature. Whitish, creamy beige, light tan to clay brown, often with a darker brown, obscure "nipple" at apex. Overall golden brown to brown when old. **INTERIOR:** Firm and white at first, soon soft and mushy, yellowish to yellowish olive, finally powdery olive-brown to brown as spores mature. Sterile base pronounced. **ODOR:** Indistinct. **TASTE:** Mild (when young). **KOH:** Forest green on mature spores. **MICROSCOPY:** Spores 2.5–3.5 x 2.5–4 µm, globose to subglobose, thick walled, roughened to smooth.

ECOLOGY: Often in large clusters, occasionally scattered or in small clusters, rarely solitary, on rotting wood, logs, or stumps (sometimes buried) of both hardwoods and conifers; occasionally on wood chips, rarely from soil. Widespread but uncommon south of the San Francisco Bay Area. Fruiting from fall through winter.

EDIBILITY: Edible and one of the better small puffballs. Eat only those that are completely white inside, as specimens with spores starting to mature (yellowish to olive tones) could make you sick.

COMMENTS: The pear-shaped fruitbody, whitish beige to warm brown colors, granular to smooth exterior, conspicuous white rhizomes, and clustered growth on wood are distinctive. Although a number of puffballs are pear shaped, *L. pyriforme* is the only one that routinely grows from wood and has white rhizomorphs. *L. molle* is grayish to grayish brown, and has very fine dark spines that often are fused together at the tips; it is locally common in oak duff. *L. perlatum* is covered with pyramidal spines or warts when young, but these often scuff off and leave round scars in age.

Lycoperdon umbrinum
Pers.
SHADOW PUFFBALL

FRUITBODY: 1–4 cm across, 3–5 cm tall, pear shaped overall, with an oval or round top, underside often pinched into a short pseudostipe. Surface dry, covered with very short spines that flake off in age, leaving small round to oval scars. Skin thin, cracking and tearing at apex when mature. Dark brown to blackish when young and covered in spines, patchily or extensively yellow-brown or metallic gold as spines wear off in age. **INTERIOR:** Firm and white at first, soon soft, becoming yellowish to yellowish olive and mushy, finally brown and powdery when spores mature. Sterile base pronounced. **ODOR:** Indistinct. **TASTE:** Mild. **KOH:** Forest green on mature spores. **MICROSCOPY:** Spores 3.5–5.5 µm across, globose, thick walled, roughened.

ECOLOGY: Solitary, scattered, or occasionally clustered on ground, duff, or moss in forest habitats, especially frequent under redwood. Widespread throughout our range. Fruiting from fall in the north to late winter in the south. Fruitbodies can persist into spring or summer.

EDIBILITY: Reportedly edible.

COMMENTS: The blackish brown to golden colors, short spines that leave only a slight scar when they fall off, and powdery brown mature spore mass are distinctive. *L. nigrescens* shares the same shape and colors, but has slightly longer, thicker spines that leave distinct oval craters when they fall off. It is far less common, occurring primarily in conifer forest on the Far North Coast, but is known south to the San Francisco Bay Area. *L. molle* is grayish to grayish brown and has very fine, dark spines that often are fused at the tips. It also differs from *L. umbrinum* by its smoother surface and paler, less golden colors. *L. pyriforme* grows on rotting wood, and has granular to smooth skin and conspicuous white rhizomes at the base.

Pisolithus arhizus group
DYER'S PUFFBALL

FRUITBODY: 4–12 (20) cm across, 6–30 cm tall, rounded-cylindrical or top shaped, occasionally cushion shaped; often with a distinct rooting stipe. Disintegrating and becoming quite irregular, turning into a powdery mass from top down. Surface beige, tan, grayish brown, orange-brown, or dull brown. Young fruitbodies occasionally metallic and bright golden. **INTERIOR:** Hard at first, slowly transitioning to crumbly-powdery in age. Made up of multicolored (gold, yellow, whitish, vinaceous, black, red, reddish brown, brown or tan) peridioles (small, oblong, discrete spore packets 0.2–0.7 cm long) embedded in a black matrix when young. Each one disintegrates into a reddish brown, rusty brown, or dull brown powdery mass of spores; this maturation progresses downward through each layer of peridioles. **ODOR:** Mild to unpleasant. **TASTE:** Not sampled. **MICROSCOPY:** Spores (6) 7–11 (12) µm (excluding spines, which are around 1 µm long), globose, thick walled, distinctly knobby-spiny.

ECOLOGY: Solitary, scattered, or in small clusters in nutrient-poor soil, disturbed areas, and roadsides, often around young planted trees. Very common with oaks and pines, but associating with a large variety of trees and shrubs in urban areas. Common and widespread, but more abundant in dryer environments. Fruiting in late summer, often before first rains, with fruitbodies persisting into late fall or early winter.

EDIBILITY: Unknown. The powdery spore mass can cause fits of sneezing.

COMMENTS: This puffball is easily recognized by the unattractive appearance (it is also called Dead Man's Foot and Dog Turd Fungus); multicolored peridioles (visible in cross sections of young fruitbodies) maturing into a powdery brown mass; and growth in nutrient-poor soil and disturbed areas. It is highly prized by fiber artists because it yields lightfast dyes on wool and silk, with colors ranging from golden to deep dark brown. We might have a number of different species going by this name in California; the common one with oaks is undescribed. It appears that a couple of species were introduced from Australia and grow with eucalyptus. One is probably *P. croceorrhizus*, with a paler skin, yellowish peridioles, light brown to yellow-brown spores, and golden rhizomorphs on the stipe; it was recently described from Queensland.

Scleroderma areolatum

Ehrenb.

COMMON CRACKING EARTHBALL

FRUITBODY: 4–9 cm across, 4–6 cm tall, round or cushion to top shaped, often with a distinct stipe and a large tuft of rhizomorphs and mycelium. Beige, light ocher-brown to light brown. Surface dry, cracking into small, flat scales that give it the appearance of dried mud. Tearing open at center when mature. **INTERIOR:** Very firm and white at first, staining pinkish red when young, soon maturing to dark, marbled purplish brown or purplish black, then purplish gray and powdery as spores mature. Stipe pallid whitish at first, becoming orange-brown in age. Base with numerous white rhizomorphs. **ODOR:** Mild to unpleasant. **TASTE:** Indistinct. **MICROSCOPY:** Spores 8–16 μm (excluding the spines, which are 1–2 μm long), globose, thick walled, distinctly spiny.

ECOLOGY: Solitary, scattered, or in small clusters on the ground, often in disturbed or urban areas. Grows with a number of introduced ectomycorrhizal hosts, especially Strawberry Tree (*Arbutus unedo*). Locally common. Fruiting in midsummer, then again from late fall through winter.

EDIBILITY: Toxic, causing severe gastrointestinal distress.

COMMENTS: All *Scleroderma* have a thick, tough skin, a very hard texture when young, a purplish mature gleba, and often extensive rhizomorphs at the base. All species are toxic and especially dangerous (sometimes even fatal) for dogs, which seem to like grazing on them. This species belongs to a very confusing group that includes *S. verrucosum*, which has straw yellow skin with darker, wartlike patches and a slightly more pear-shaped stature. *S. cepa* can be recognized by its relatively smooth, thick skin when young (cracked in age) and smaller (8–13 μm) globose spores. *S. hypogaeum* also has a smooth, thick skin, but generally grows partially buried in the soil under conifers and has large (15–25 μm), reticulated-spiny spores.

Scleroderma polyrhizum

(J. F. Gmel.) Pers.

EARTHSTAR EARTHBALL

FRUITBODY: 8–15 cm across, 5–10 cm tall, rounded, cushion to top shaped; lacking a distinct stipe, sometimes with mycelial rhizomorphs at base. Very tough and corky at first. Skin very thick, smooth or with small, scaly portions and often with adhering soil. Cracking and splitting into 4–9 rays and exposing spore mass, which is soon worn away, leaving only a leathery, star-shaped disc. Dingy whitish at first, then yellowish brown to grayish brown at maturity. **INTERIOR:** Very firm, whitish when young, maturing to a dark purplish brown or brownish black powdery mass (often retaining its shape for a bit after rays open). **ODOR:** Mild to unpleasant. **TASTE:** Not sampled. **MICROSCOPY:** Spores 6–9.5 (12) μm (excluding the spines, which are less than 1 μm long), globose, thick walled, finely spiny.

ECOLOGY: Solitary, scattered, or occasionally in small clusters, almost always in poor, sandy, or disturbed soil under conifers, especially pine. It seems to prefer edges of roads, trails, and pine forest. Widespread but only locally common; fairly frequently encountered in Santa Cruz Sandhills with Ponderosa Pine and on the dunes of the North Coast with Shore Pine. Fruiting from late summer into winter, but opened rays often persist for more than a year and are frequently covered in mold.

EDIBILITY: Toxic, causing severe gastrointestinal distress.

COMMENTS: The large size, firm texture, and thick tough skin that opens like a star are distinctive. *S. cepa* can occasionally open up into a starlike shape in age, but is much smaller, often grows under introduced trees, and has larger spores with longer spines. The rays of *Astraeus* species open to reveal a "puffball" (these species open and close repeatedly in response to wet-dry weather cycles). Also compare with *S. aerolatum* and its kin.

SYNONYM: *Scleroderma geaster* Fr.

Astraeus hygrometricus group
BAROMETER EARTHSTAR

FRUITBODY: 6–10 (12) cm across, rounded to cushion shaped at first, outer skin soon splitting and opening up into starlike rays, pushing up rounded puffball held inside, often holding it 5–9 cm high. Rays hygroscopic, opening in wet weather and closing when dry. Color quite variable. Exterior beige to golden brown or reddish brown when young, becoming smoother and dark brown to grayish brown. **INTERIOR:** Becomes upper surface after opening. Dull golden brown, reddish brown, or dark, dingy purplish gray. Surface of rays often with a golden or pale reddish brown, superficial upper layer that becomes cracked like dried mud (or with a reptile-skin appearance). Colors of all parts darker and duller in age. Thin skin of puffball portion typically gray to brown, but can be more brightly colored when it first opens. Spore mass whitish when young, soon reddish brown, then darker brown, becoming powdery; thin skin soon tearing at top, often overall disintegrating in age (but tough supporting arms will persist). **ODOR:** Indistinct. **TASTE:** Not sampled. **MICROSCOPY:** Spores 10–14 μm, globose, thick walled, finely spiny. Size highly variable between different variants; common medium-size variant with live oak has spores in this range.

ECOLOGY: Solitary, scattered, occasionally in small clusters under a wide variety of different trees. Most commonly encountered around edges of live oak forests and scrub. Very common from Sonoma County southward, uncommon on the Far North Coast. Typically fruiting in winter and spring, but rarely encountered until maturity. Expanded rays that have lost their spore mass can last for years.

EDIBILITY: Unknown, likely toxic.

COMMENTS: The thick skin, hygroscopic rays, puffball-like interior, and often large size are helpful in distinguishing *Astraeus*. It is unclear how many species we have in California; there appears to be at least three. What we do know is that the European species, *A. hygrometricus*, is not known from North America, and while there was some work done on North American species (recently describing two new species, *A. smithii* and *A. morganii*), it did not include western North American material. *A. pteridis*, a large species described from the Oregon coast, is the species people are most likely to run across on the Far North Coast; it has fruitbodies up to 15 cm across (when open). Dry, "clenched" *Astraeus* can be placed in a bowl with a wet paper towel or small amount of water, and the arms will slowly open. If your friends each bring their own, you can place bets on whose fruitbody will continue expanding after the most wet-dry cycles (it can take a few weeks to definitively determine the victor).

Geastrum fornicatum

(Huds.) Hook.

ARCHED EARTHSTAR

FRUITBODY: 2–4.5 cm across, an oblong egg with a soil-crusted exterior at first; soon splitting open and elevating a small puffball on 4–7 arched rays attached to edge of the cup from which they emerged. Puffball has a small mouth at apex and is perched atop a small platform. Opened fruitbodies 3–6 cm across, up to 8 cm tall. Rays dark brown, surface of puffball paler brown to light orange-brown, fading to beige. **INTERIOR:** Firm and white at first (but rarely observed in this stage), soon soft and powdery with mature chocolate brown spores. **ODOR:** Indistinct. **TASTE:** Indistinct. **MICROSCOPY:** Spores 3.5–4.5 μm globose, warted.

ECOLOGY: Solitary, scattered, or occasionally in small clusters in rich duff, especially under Monterey Cypress and *Ceanothus* spp. Locally common from the San Francisco Bay Area south, not currently known from the North Coast, but likely to show up under cypress. Fruiting from late summer into winter, but fruitbodies persist for some time.

EDIBILITY: Unknown.

COMMENTS: This earthstar is easily recognized by the three-tiered fruitbody: a basal cup, a "star" of upward-thrusted rays, and a puffball perched on top. *G. quadrifidum* is paler overall and lacks the basal cup, and has a small, distinct stipe that elevates the puffball.

Geastrum saccatum

Fr.

BOWL EARTHSTAR

FRUITBODY: 3–7 cm across, fig shaped and orange-brown to grayish brown when young, then cracking and opening up into a starlike disc with 4–9 grayish rays that often curve downward and curl inward. Puffball-like interior nestled in bowl-like center (without a stipe); pale brown with a darker, pointed mouth often with a pale ring around it. **INTERIOR:** Firm and white at first (but rarely observed in this stage), soon soft and powdery as spore mass matures, becoming chocolate brown. **ODOR:** Indistinct. **TASTE:** Indistinct. **MICROSCOPY:** Spores 3.5–5 μm, globose, warted.

ECOLOGY: Solitary, scattered, or occasionally in small clusters in duff or on ground under conifers, especially redwood and cypress. Widespread and common. Fruiting from fall into late winter in south, but fruitbodies persist throughout the year.

EDIBILITY: Unknown.

COMMENTS: This earthstar can be recognized by the puffball nestled in a bowl-like depression atop the curled rays, pointed mouth on the puffball, and nonhygroscopic rays (only opening once, unlike *Astraeus*). There are a number of very similar species: *G. fimbriatum* doesn't have the pale line surrounding the dark mouth of its puffball, and has smaller spores. *G. lageniforme* has longitudinal cracks on the underside of the rays, but is otherwise very similar. *G. rufescens* has darker brick pink rays and a distinctly felty underside when young. *G. coronatum*, which often grows in cypress duff, has flat rays that do not form a bowl-like sac around the puffball. *G. floriforme* is smaller, has hygroscopic rays (expanding and contracting in wet and dry conditions), and often grows in sandy areas or nutrient-poor soil. *G. triplex* has a distinct collar between the rays and the puffball. *Astraeus pteridis* has thicker skin, is often darker, and has hygroscopic rays and a simple tear at the top to release the spores. *Radiigera fuscogleba* could be mistaken for unopened specimens of *Geastrum*, but is more cushion shaped and has a small, knobbed, cylindrical columella inside.

Crucibulum crucibuliforme
(Scop.) V. S. White
WHITE-EGG BIRD'S NEST

NEST: 0.5–1 cm across, 0.3–1.2 cm tall, rounded at first, soon cylindrical with a cushion-shaped, coarsely hairy to smooth lid colored yellowish to cinnamon brown. Cup shaped after lid comes off. **OUTER SURFACE:** Yellowish, ocher to cinnamon brown, often with paler velvety hairs on lower portion. **INNER SURFACE:** Smooth, shiny, whitish, pale gray to beige. **EGGS:** Numerous, small, 0.1–0.2 cm across, flattened circular discs often embedded in a gel-like substance when young but soon dry. Attached to a very thin, white, elastic cord. **MICROSCOPY:** Spores 4.5–10 x 3.5–6 µm, averaging 8.8 x 4.4 µm, ellipsoid, smooth. Found inside mature eggs.

ECOLOGY: Solitary, scattered, or in large troops on small woody debris such as branches and wood chips or herbaceous stems, rarely on rich compost. Very common, occurring throughout our range. Fruiting in fall and winter, but persisting through the year.

EDIBILITY: Unknown.

COMMENTS: This bird's nest fungus can be recognized by the yellowish nest covered in coarse to velvety hairs when young and the whitish eggs attached to a thin cord. *Nidula* species are generally quite hairy on the exterior, but have grayish eggs that are not attached to a cord. *Cyathus* species have darker eggs.

SYNONYM: *Crucibulum laeve* (Huds.) Kambly.

Cyathus olla
(Batsch) Pers.
GRAY-EGG BIRD'S NEST

NEST: 0.8–1.5 cm across, 0.7–2 cm tall, rounded at first, soon cylindrical with a flaring top and a tapering, pointed base. Often broader and more bowl shaped after the lid comes off, with a flaring, wavy margin (occasionally splitting). Lid thin, finely silky-tomentose to smooth, whitish, pale gray to light brown. **OUTER SURFACE:** Gray to brown, finely tomentose at first, becoming more coarsely hairy in age. **INNER SURFACE:** Smooth, shiny, gray to grayish brown. **EGGS:** 1–10 (generally fewer than 5), 0.3–0.5 cm across, light to dark gray, plump or flattened discs, often embedded in a gel when young, soon dry, attached by a thin cord. **MICROSCOPY:** Spores 10–14 x 6–8 µm, broadly ellipsoid to oval, smooth. Found inside mature eggs.

ECOLOGY: Scattered or in large troops, occasionally solitary, on small woody debris, herbaceous stems, or rich soil. Especially common on wood chips and lignin-rich compost in garden beds. Found throughout our range, but more common in urban areas. Fruiting in fall and winter, but nests can persist for a long time.

EDIBILITY: Unknown.

COMMENTS: The large size (relative to other bird's nest fungi) and gray color of the eggs and the flaring rim of the "nest" are distinctive. *C. stercoreus* has smaller, black eggs and often grows in compost or dung-enriched soil, rarely on wood chips. The cups of *C. striatus* have a distinctly lined or grooved interior. *Crucibulum crucibuliforme* has whitish eggs, and *Nidula* have smaller whitish to gray eggs.

Cyathus stercoreus

(Schwein.) De Toni

DUNG-LOVING BIRD'S NEST

NEST: 0.5–1 cm across, 0.5–1.5 cm tall, rounded to cylindrical at first, cup to bowl shaped after lid comes off, often with a wavy margin in age. Lid thin, roughened-scaly, orangish brown to grayish. **OUTER SURFACE:** Covered with coarse orangish brown to grayish brown hairs or scales at first, becoming dark grayish brown to nearly black and smooth or nearly so in age. **INNER SURFACE:** Smooth to ribbed, shiny, light gray when young, dark gray to blackish in age. **EGGS:** Numerous flattened to irregular discs 0.1–0.25 cm across, attached to a very thin cord. Gray when young, soon black. **MICROSCOPY:** Spores 20–40 x 18–30 μm, highly variable in size, subglobose, smooth. Found inside mature eggs.

ECOLOGY: Scattered or in troops, occasionally solitary, in compost, manure, garden beds, or flower pots, rarely in wood chips. Occurring throughout our range, but seems restricted to areas with some degree of human disturbance. Fruiting in fall and winter, but nests can persist longer.

EDIBILITY: Unknown.

COMMENTS: The small, gray to black eggs, dark colors of the mature nest, and growth in compost or dung-enriched soil are distinctive. *C. olla* has larger gray eggs and a flaring margin on the nest. *C. striatus* has a brown, coarsely hairy exterior with a pale lid, a grooved interior with paler lines, and grayish brown eggs. Although uncommon in California (so far known from wood chips on the North Coast), it is likely to spread. *Nidula* have many small, whitish to gray eggs and hairy-felty exteriors.

Nidula candida

(Peck) V. S. White

NEST: 0.3–1 cm across, 0.5–2 cm tall, cylindrical with a rounded top at first, vase shaped after lid comes off, often with a narrow, flaring rim. Lid thick, coarsely hairy to scaly. Beige to light brown. **OUTER SURFACE:** Whitish, beige, or gray at first, then dingy brown or paler to whitish when old. Velvety-tomentose at first, covered with whitish hairs, becoming more coarsely hairy in age. **INNER SURFACE:** Smooth, shiny, paler toward margin, whitish to buff, lower parts light to dark brown or orangish brown. **EGGS:** Numerous, small flattened to plump circular discs, 0.1–0.3 cm across. Embedded in a gel when young, soon dry. Grayish to light brown. **MICROSCOPY:** Spores 8–10 x 4–6 μm, ellipsoid, smooth. Found inside mature eggs.

ECOLOGY: Scattered or in troops on small woody debris or herbaceous stems. Especially common on alder branches in wet areas, but occurring on other hardwood and conifer twigs and branches. Common in the northern part of our range, uncommon south of the San Francisco Bay Area. Fruiting in fall and winter, but nests can persist throughout year.

EDIBILITY: Unknown.

COMMENTS: The whitish to gray, fuzzy exterior, thick lid, and small, grayish to light brown eggs that are not attached to thin cords (but often embedded in a gel) are helpful features distinguishing this species from almost all other bird's nest fungi. *N. niveotomentosa* differs by its smaller size, more cylindrical shape, and bright white, fuzzy exterior. *Crucibulum crucibuliforme* has whitish eggs attached to thin cords and a coarsely hairy to smooth yellowish to yellowish brown exterior. *Nidularia farcta* is a rounded or irregularly shaped lump, with a tan skin that starts to dissolve as it matures, exposing a mass of reddish brown eggs. It is rare in California, occurring on the Far North Coast.

Battarrea phalloides group
SCALY-STALKED PUFFBALL

CAP: 3–6 cm across, domed to rounded, eventually broadly convex. At first appearing rather bright white and somewhat shiny, due to unbroken "eggshell" of universal veil, soon losing this thin, papery membrane to reveal a soft, extremely powdery rust brown to reddish brown spore mass. **STIPE:** 6–25 (50) cm long, 1–2.5 cm thick, cylindrical, usually very straight and tall, often thickest near midpoint. Light brown to creamy brown at first, soon reddish brown to orange-brown. Surface dry, variably smooth to lacerated-scaly and shaggy. Base embedded in a large, soft whitish volva, this often remaining in soil when fruitbodies are collected. **FLESH:** Thin, dry, and papery or slightly woody. **ODOR:** Indistinct. **TASTE:** Not sampled. **SPORE DEPOSIT:** Bright orange-brown. **MICROSCOPY:** Spores 5–7 μm, globose, with conspicuous warts and short, blunt spines. Specialized cells called elaters are interspersed in spore mass. These hyphae have bars of golden brown pigment on them and look a bit like ladders.

ECOLOGY: Solitary or in groups in disturbed areas, poor, sandy soil or thick duff under a variety of trees, especially Monterey Cypress and Brazilian Pepper Tree (*Schinus terebinthifolius*), but also pines and Ngaio (*Myoporum laetum*). Fresh fruitings typically appear in spring, summer, and fall, but fruitbodies are very resistant to decay and can be found any time of year.

EDIBILITY: Too tough to chew, and the copious powdery spore mass would likely cause fits of sneezing.

COMMENTS: Unique and truly bizarre, this fungus is a joy to find for those who have tired of more mundane fungi. Its papery-woody texture, odd anatomy, and orange-brown colors distinguish it readily. Recent genetic studies in Europe have shown the existence of several species going by the name *B. phalloides*; California material needs to be critically reexamined. Other stalked puffballs (*Tulostoma* species) in our area are much smaller and do not have an exposed orange-brown powdery spore mass at maturity. *Agaricus deserticola* is whitish and fleshier, and has a much darker spore mass.

Phallus hadriani

Vent.

PURPLE-EGG STINKHORN

CAP: 3–4 cm across, 3–6 cm tall, cylindrical when emerging, then egg shaped to somewhat conical, not expanding much, tip with a flattened and often slightly fringed and dimpled disc. Composed of an olive to greenish brown spore mass spread over a whitish to creamy tripelike or honeycombed membrane. Spore mass at first dense, but soon slimy and gooey (but still fairly cohesive, not dripping). **STIPE:** 7–15 cm long, 2–4 cm thick, cylindrical, soft, spongy, hollow. Whitish to cream, becoming dingy yellowish in age. Surface smooth at first but soon showing many small porous, irregularly angular cells (most visible when fully expanded). Edge of spore-bearing head not fused to stipe. **FLESH:** Quite firm and rubbery, but soon spongy; stipe with a gelatinous core when young, becoming hollow. Spore-bearing tissue of head thin and membranous. Whitish to creamy or yellowish, with lilac-magenta to purplish tones around edges in cross sections of "eggs." **UNIVERSAL VEIL:** Thin purple, lilac-pink, or magenta (occasionally more extensively white) membrane that surrounds entire fruitbody when young; in expanded fruitbodies remaining as a thin, membranous volva filled with clear to dingy gel around base of stipe. Extreme base with many whitish to purple elastic rhizomorphs that often penetrate surrounding substrate. **EGGS:** Cylindrical or oval, tapered at one end. Very firm and rubbery, filled with clear gel, interior showing a miniature, highly compressed version of adult stinkhorn (with olive spore mass tightly packed around edges). **ODOR:** Strong and distinct when young, like bean sprouts, funky buttermilk, or strongly spermatic, becoming fouler and more pervasive, like rotting cabbage in age. **TASTE:** Mild, but odor difficult to ignore. **SPORE DEPOSIT:** Not obtainable. **MICROSCOPY:** Spores 3–4.5 x 1.5–2 µm, ellipsoid to somewhat cylindrical, smooth.

ECOLOGY: Solitary or scattered in clumps or troops in gardens or wood chip beds, or along trails in parks. Can fruit in practically any urban setting at almost any time of year, but prefers rich or sandy soil and warm, humid weather. Uncommon in our range; common in southern California.

EDIBILITY: The young eggs are sometimes peeled, rinsed, and eaten. The gel can be an unpleasant textural hurdle (to say nothing of the smell). The foul odor is restricted to the spore mass in mature specimens; closely related species are washed and dried, and often available for sale in Chinese markets.

COMMENTS: The phallic shape, honeycombed head, purple-skinned egg from which the stipe emerges, and raunchy odor identify this species beyond much doubt. The slimy spore mass is dispersed by the insects that visit the fruitbody after being attracted by the smell. *P. impudicus* is very similar, but has a white-skinned egg without pink or purple tones. *P. ravenellii* is also quite similar (with a purple egg), but has a smooth cap membrane upon which the spore mass sits. To our knowledge, neither of these has yet been recorded from California.

Clathrus ruber

P. Micheli ex Pers.

RED BASKET STINKHORN

FRUITBODY: 5–15 (20) cm across, "eggs" round but often quite lumpy, often with a pattern of hexagonal depressions over surface, giving it the appearance of a small, pale soccer ball; surface smooth, dry, white, beige, or dingy tan. Soon rupturing to reveal a ball of wrinkly-spongy tissue with deep craters and pits; this cage of tissue rapidly expanding into a dramatic "basket." Color of tissue ranging from dull orange to pinkish beige, or more often coral red to pinkish red, paler on outer surface. Inner part of expanding cage covered with dingy gray to olive-tan or dark brown blobs and smears of gooey spore mass (gleba). **UNIVERSAL VEIL:** Remaining attached, surrounding base of cage as a whitish membrane, often filled with gel. **FLESH:** Cage crispy-spongy and very light; denser and more rubbery-gelatinous in egg stage. **ODOR:** Spermatic or radishlike in egg stage, slightly foul when very young, soon like rotting meat (like a dead rat). Wretched odor is produced by mature gleba. **TASTE:** Radishlike when young (in egg stage). **SPORE DEPOSIT:** Dark olive-brown. **MICROSCOPY:** Spores 4–6 x 1.5–2.5 μm, long-ellipsoid to rounded-cylindrical, smooth.

ECOLOGY: Usually fruiting in dense clumps and troops (almost never solitary) in wood chips and gardens during warm, wet weather. As is typical for a noninvasive introduced species, it is common and abundant at a few locations, but is almost entirely absent over the rest of our range. The vast majority of records are from counties surrounding the San Francisco Bay, with only a few from Sonoma and Santa Cruz Counties. However, it seems to have the potential to turn up in wood chips almost anywhere in the state. Native to the Mediterranean region of Europe.

EDIBILITY: Edible, or at least not harmful. When in the egg stage, it lacks the rotting meat stench, but Noah Siegel swears that when he ate it, the gleba matured in his stomach, and he ended up belching stinkhorn odors for hours. Bon appétit!

COMMENTS: No other mushroom in our area really looks like this species. *C. crispus* is similar, but has fewer, more evenly spaced pits, each with a circular "halo," overall giving it a rather different look. It has yet to be recorded from California (but is known from the Gulf Coast and may turn up here eventually, especially given the warming climate). *C. archeri* has a similar color scheme, smell, and habitat, but a very different shape, with 4–5 "tentacles."

Clathrus archeri

(Berk.) Dring

OCTOPUS STINKHORN

FRUITBODY: 5–11 cm across, composed of 4 or 5 "arms" at first held together in a tight, upright arrangement but very soon separating at tips and curving backward toward soil, becoming very open and somewhat uplifted. Mature fruitbodies resemble tentacles of a squid or octopus. Arms coral or reddish pink at first but can become paler to nearly whitish in age. Inner (when young) or upper (in age) ends of arms coated with a gooey olive-brown spore mass that becomes nearly black when dry. **FLESH:** Rubbery in egg stage, arms with a spongy, honeycombed texture and a bumpy, corrugated, or ribbed surface. **UNIVERSAL VEIL:** Thin, whitish outer membrane, a gelatinous interior, and most often a tuft of rubbery white rhizomorphs at base. **ODOR:** Mature gleba extremely foul, like rotting meat or fecal matter. **TASTE:** Mild to radishlike when in egg stage, not sampled when mature. **SPORE DEPOSIT:** Olive-brown. **MICROSCOPY:** Spores 4–7.5 x 2–2.5 μm, rounded-cylindrical, smooth.

ECOLOGY: Usually fruiting in groups (sometimes solitary) in wood chips, under bamboo, or in other lignin-rich organic matter. Mature fruitbodies often surrounded by buried "eggs." Almost entirely restricted to Santa Cruz County in our area, with one record for San Francisco. Fruiting mostly from summer into midfall, sometimes also abundant in spring. Like most stinkhorns, it can fruit opportunistically whenever weather is warm and humid. First recorded in California in the 1970s, this species is native to Australia and New Zealand.

EDIBILITY: Nontoxic. Some people have eaten specimens in the egg stage, but it seems likely to induce vomiting when the spore mass is at all mature, due to the nauseating smell (and presumably taste).

COMMENTS: Also known as the Stinky Squid, this species is as wildly weird and definitely disgusting as it is easy to distinguish: the tentacled structure of the fruitbody, pink colors, and foul odor are a unique combination among mushrooms in our area. *Lysurus mokusin* is a somewhat similar pinkish stinkhorn, but is shaped like a tall column with short arms that remain fused in an angular structure near the tip; the dark spore mass is on the exterior (rather than the inner surface) of these arms. *L. mokusin* is quite rare in our range, with records from Santa Cruz and Santa Clara Counties; it is much more common in southern California. *Clathrus ruber* is similarly colored and foul smelling, but is shaped like a cage.

22 • Truffles

Although they belong to many different evolutionary lineages, the species in this section are united by their highly reduced fruitbodies that are formed below ground (hypogeous). This syndrome of fruitbody morphology has independently evolved many times, and in many cases appears to be an adaptation to habitats that commonly experience rapid inversions of humidity and temperature. By sheltering their fertile surfaces inside protective outer layers and fruiting belowground (where temperature and humidity remain more stable), these fungi are insulated against harsh environmental conditions that can cause species with more exposed fruitbodies to cease spore production. The tradeoff is that their spores can no longer disperse as easily in wind currents. Consequently, many produce strong aromatic compounds that are attractive to small mammals; squirrels and voles dig up many of these fungi and eat them, and then disperse the spores in their feces. Some of the more aromatic truffles in our area are considered good edibles, but none have the culinary esteem of the white and black *Tuber* truffles of Europe.

Identification (even to genus) is often guesswork at best until the fertile surface and spores are examined under a microscope. However, with some practice and careful attention to coloration, internal structure, odor, surface texture, bruising reactions, presence or absence of latex, and habitat, you can learn to recognize many of the common local species.

Quite a few species appear to be "transitional" forms on an evolutionary trajectory toward a hypogeous, trufflelike fruiting syndrome. Such fruitbodies often have vestigial but recognizable gills, and frequently have contorted, short stipes that barely elevate the fruitbody above the duff. These are called secotioid or sequestrate mushrooms. Species with such fruitbodies often have been given their own genera, but genetic work has indicated that they should be placed in their "sister genera" with more typical gilled or pored morphology. Some the many examples from our area include *Thaxterogaster* (=*Cortinarius*), *Macowanites* (=*Russula*), *Arcangeliella* (=*Lactarius*), and *Longula* (=*Agaricus*).

Truncocolumella citrina

Zeller

FRUITBODY: 0.5–3 cm across, rounded, ovoid to irregular-lobed, often with a pinched basal pad or short stipe. Yellowish, grayish yellow to yellow-olive. Surface dry, smooth, occasionally cracking in age. **INTERIOR:** A firm, rubbery mass of tight, spongelike gleba, with tiny open chambers. Pale grayish at first, becoming grayish olive, to dark grayish olive and slightly gelatinous in age. Columella present as a yellowish mass at base with radiating veins into gleba. **ODOR:** Indistinct. **TASTE:** Mild to earthy. **KOH:** No reaction. **MICROSCOPY:** Spores 6.5–10 x 3.5–4.5 μm, elliptical to ovoid, smooth, thin walled, pale yellowish.

ECOLOGY: Solitary, scattered, or in small clusters, often buried or partially buried in soil or duff under Douglas-fir. Very common throughout our range. Generally fruiting from early fall into winter.

EDIBILITY: Edible, but not all that flavorful.

COMMENTS: The firm rubbery texture, yellowish color, and grayish olive gleba with a yellowish columella are distinctive. Because it often ruptures the surface when mature and is yellow in color, it is one of the most commonly collected truffles by casual mushroom hunters. *Rhizopogon truncatus* has a bright yellow exterior and a olive interior that lacks a columella. Other *Rhizopogon* lack the evenly yellow bright colors and lack a columella.

Rhizopogon occidentalis

Zeller & C. W. Dodge

WESTERN POGIE

FRUITBODY: 1–8 cm across, round, ovoid to irregular-lobed. Creamy yellow with golden yellow rhizomorphs at first, staining or becoming brighter yellow, orange to rosy red in irregular patches. Surface dry, covered with binding, minutely hairy rhizomorphs, which stain yellow, then slowly reddish when damaged. **INTERIOR:** Firm, rubbery mass of a tight, spongelike gleba, with tiny open chambers. Whitish to pale gray at first, becoming olive when mature. **ODOR:** Indistinct. **TASTE:** Indistinct. **KOH:** Reddish on exterior. **MICROSCOPY:** Spores 5.5–7 x 2–3 μm, oblong, pale yellowish.

ECOLOGY: Solitary, scattered, or in troops, buried in duff or soil, occasionally erupting through surface in areas with thin duff. Very common, occurring with pine throughout our range. Fruiting in fall through winter.

EDIBILITY: Edible.

COMMENTS: This rubbery truffle can be recognized by its pale yellow colors at first, yellow staining of the coating rhizomorphs, and reddening in age, and growth with pine. *R. ochraceorubens* is nearly indistinguishable; it generally exhibits more red or orange staining, but there seems to be overlap between these two species. *Rhizopogon* are easy to identify to genus, but narrowing them down to species can be next to impossible. They all exhibit a rubbery texture and bounce when dropped on the floor, have a firm mass of tight, spongelike gleba, and often have a coating of appressed rhizomorphs on the exterior. Other truffles that bounce include *Truncocolumella*, which have a translucent columella, and *Gautieria*, which generally have a pinkish to ochraceous chambered gleba, with a whitish or translucent columella.

Rhizopogon parksii

A. H. Sm.

DARK POGIE

FRUITBODY: 1–4 cm across, round, ovoid to irregular-lobed. Dingy whitish to grayish at first, staining dirty pinkish violet, darkening blackish in patches, to all black in age. Surface dry, covered with a matted tomentum and minutely hairy rhizomorphs. **INTERIOR:** A firm, rubbery mass of a tight, spongelike gleba, with tiny open chambers. Whitish to pale gray at first, becoming olive, and often developing dark stains in age. **ODOR:** Indistinct. **TASTE:** Indistinct **KOH:** Purplish black on exterior. **MICROSCOPY:** Spores 5–7 x 2.5–3 µm, elliptical to oblong, smooth, thin walled, colorless to pale yellowish.

ECOLOGY: Solitary, scattered, or in troops, buried in duff or soil, occasionally rupturing surface in areas with thin duff. Common, occurring with Douglas-fir throughout our range. Fruiting in fall through winter.

EDIBILITY: Edible.

COMMENTS: This Pogie can be told by its rubbery texture, dingy colors that darken with age, and growth with Douglas-fir. However, it is one of many similar *Rhizopogon*, most of which seemingly need a genetic sequence to reliably identify them.

Alpova diplophloeus

(Zeller & C. W. Dodge) Trappe & A. H. Sm.

ALDER TRUFFLE

FRUITBODY: 0.5–5 cm across, 0.5–4 cm tall, ovoid to round. Yellowish brown to pinkish brown when immature, darkening to reddish brown to brown. Surface moist to dry, finely matted-tomentose to smooth. **INTERIOR:** Firm, consisting of gelatinous gel-filled chambers, surrounded by pale yellowish veins. Chambers yellow-olive at first, soon reddish to orangish brown. **ODOR:** Mild to fruity in age. **TASTE:** Mild. **KOH:** Reddish on exterior. **MICROSCOPY:** Spores 4.5–5.5 x 2–3 µm, elliptical to oblong, colorless to pinkish buff, amyloid when immature, inamyloid when mature.

ECOLOGY: Solitary, scattered, or most often in "nest" of many fruitbodies. Usually buried in soil or duff at first, then rupturing surface as they mature. Growing under alder, usually in swampy areas or riversides, common in northern California, rare (or rarely found) south of Mendocino County, known at least to Santa Cruz. Fruiting from late summer into winter.

EDIBILITY: Reported to be bland. It has a downright weird texture.

COMMENTS: Since it is strictly associated with its host tree, *A. diplophloeus* is usually found in riparian areas or other constant sources of water where alders are common. In combination with its habitat, the reddish brown exterior and gel-like, chambered interior with pale yellowish veins are distinctive. Genetically, this species is closely related to *Rhizopogon*, and both are related to the boletoid genus *Suillus*. *Rhizopogon* (=*Alpova*) *olivaceotinctus* has a firm, rubbery texture, a yellow-brown to dark brown gel-like interior, and a yellowish brown exterior that darkens to olive-brown to dingy dark brown. It grows with conifers, especially in younger pine forest. Other *Rhizopogon* have a firmer texture and a more homogenous interior. *Melanogaster* have black, gel-like, chambered interiors with whitish lines. *Hysterangium* have firm, olive interiors with translucent, veinlike columellas. *Tubers* have hard, marbled interiors, and often pungent odors.

Melanogaster tuberiformis group

FRUITBODY: 0.7–3.5 cm across, round ovoid to irregular, firm. Reddish brown when young, becoming dark brown to blackish. Surface dry, sparsely covered with blackish, thread-like rhizomorphs. **INTERIOR:** Firm, consisting of gelatinous gel-filled chambers, surrounded by whitish veins. Chambers dark grayish olive to black when young, to deep black in age. **ODOR:** Strong, like true *Tuber* truffles, consisting of a complex mix of earthy, nutty, garlicky, and burnt rubber notes. **TASTE:** Indistinct. **KOH:** No reaction. **MICROSCOPY:** Spores 10–14 x 6–10 μm, ovoid to pear shaped, smooth, thick walled, dark vinaceous brown.

ECOLOGY: Solitary, scattered, or aggregated in mammal caches, often in the zone between humus and mineral soil, occasionally rupturing the surface, especially in hard-packed soil. Common under a wide variety of trees, both hardwoods and conifers. Fruiting throughout wet season, but appearing to be more common in spring.

EDIBILITY: The strong odor easily flavors foods such as cheese or butter with a pleasing, complex taste.

COMMENTS: The reddish brown exterior and black, gelatinous, chambered interior with whitish veins make *Melanogaster* an easy-to-identify to genus. *Melanogaster* consist of a number of species that differ mostly based on spore morphology. In addition, two of the names we commonly use, *M. tuberiformis* and *M. ambiguus*, are European names and, based on genetic studies, are different species in western North America. All *Melanogaster* appear to have culinary value, with a complex "truffle" odor and a pleasant flavor. *Alpova* differ by having brownish interiors and growth with alder.

Hysterangium separabile
Zeller

FRUITBODY: 0.5–3 cm across, irregular rounded to lobed or wrinkled, often with a short, white rhizomorph thread at base. Whitish to pale pinkish, staining pinkish to ocher-brown when bruised. Surface dry, smooth, skin separating from gleba easily. **INTERIOR:** A very firm, rubbery mass of tight, spongelike gleba, with gel-filled chambers when immature, and a translucent columella with a basal pad, and radiating veins into gleba. Grayish olive with translucent gel when young, dark olive in age. **ODOR:** Faint when immature, becoming pungent in age. **TASTE:** Mild to radishlike. **KOH:** No reaction. **MICROSCOPY:** Spores 12–19 x 6–8 μm, spindle shaped, smooth, thin walled.

ECOLOGY: Solitary or scattered, buried in soil under oaks. Common under Coast Live Oak from Mendocino County south. Fruiting in fall and winter.

EDIBILITY: Edible. All *Hysterangium* are edible, but older specimens develop a rather pungent odor. They are also closely related to stinkhorns, which should give you pause if you are contemplating eating them.

COMMENTS: The very firm rubbery texture, whitish exterior that stains pinkish to ocher-brown, easily separable skin, and olive gleba with a translucent columella are helpful identification features. *H. crassirhachis* generally has a paler-colored gleba and lacks inflated cells in the skin. *H. coriaceum* is another common species, especially under Tanoak. It is smaller and often more round to ovoid (less irregularly lobed), has a whitish skin that stains pinkish ocher, and smaller spores (8–12 x 3–4 μm). Another similar species, *Trappea darkeri*, is smaller; has dark olive to dark forest green gleba with a translucent columella; and has skin that is not easily separable and is often surrounded by thin whitish rhizomorphs in the soil around it. It also has much smaller, elliptic to oblong spores (4–5 x 2–3 μm). The unpleasant odor of *T. darkeri* makes it less than desirable for consumption.

Elaphomyces muricatus group
DEER TRUFFLES

FRUITBODY: 2–5 cm across, ovoid to round, or irregular-lobed, very firm. Deep ocher-orange to orangish brown. Surface covered orangish mycelium that binds to soil; when peeled away, reveals a spiny, roughened surface. **INTERIOR:** Stuffed with whitish, pithlike material at first, becoming a dark brown, powdery mass when mature, surrounded by a thick, marbled skin. **ODOR:** Mild, to funky in age. **TASTE:** Indistinct. **MICROSCOPY:** Spores 20–30 x 20–30 μm, globose, warted to spiny-warted, dark brown to blackish in KOH.

ECOLOGY: Solitary, scattered, or in troops, buried often rather deep in soil. Occurring north of the San Francisco Bay Area in coniferous forest, especially with Grand Fir and Western Hemlock. Uncommonly collected, usually only found when a conspicuous *Tolypocladium* is growing on the truffles. Fruiting from late summer through winter.

EDIBILITY: Inedible.

COMMENTS: These firm to hard truffles with a roughened-spiny exterior, a thick marbled skin, and a whitish pithlike young interior (or a powdery mature one) are distinctive. The *Elaphomyces granulatus* group (see photo on p. 522, where *Tolypocladium capitatum* is growing on it), has a thinner skin that lacks the marbled interior and has a granular exterior. Both of these *Elaphomyces* belong to large species complexes that consist of many difficult to distinguish, mostly undescribed species. For ease of identification, the thin to thick, granular-skinned species without a marbled interior are all put into the *E. granulatus* group, and the thick-skinned species with a marbled interior into the *E. muricatus* group.

Hydnangium carneum
Wallr.

FRUITBODY: 0.5–3 cm across, irregular, lobed, with a depressed center or grove, to rounded. Underside often pinched, forming a basal mycelial pad. Pinkish to pinkish buff, occasionally with darker, rosy pink colorations, especially when exposed. Surface dry, felty-fibrous to smooth. **INTERIOR:** Firm, marbled, consisting of small convoluted chambers when young, becoming soft and crumbly. Pinkish to pinkish buff, occasionally darkening to pinkish red. **ODOR:** Indistinct. **TASTE:** Mild. **MICROSCOPY:** Spores 10–14 x 10–14 μm (not counting spines), globose, roughly ornamented with rounded to pointed spines 1–2 μm long, inamyloid.

ECOLOGY: Solitary or scattered in soil under eucalyptus, *Corymbia* spp., and tea trees (*Melaleuca* spp.). Usually completely buried when young, occasionally rupturing the surface when mature. Common, growing with most eucalyptus, but rarely collected. Generally fruiting in winter and spring.

EDIBILITY: Edible, but small and not flavorful at all . . .

COMMENTS: The pinkish to pinkish buff colors, with pinkish red stains; irregular chambered interior; large, round spiny spores; and growth with eucalyptus are distinctive. This species was introduced from Australia and is only known to occur under Australian trees such as eucalyptus. *Hydnangium* are closely related to *Laccaria* and belong in the same genus. However, the name *Hydnangium* is older, meaning that unless a proposal to conserve the name *Laccaria* is accepted, our gilled *Laccaria* will become *Hydnangium*.

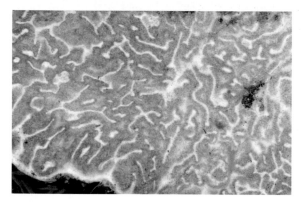

Tuber gibbosum group
OREGON WHITE TRUFFLES

FRUITBODY: 0.5–5.5 cm across, often highly irregular-lobed or furrowed, occasionally ovoid to round when small. Buff, light olive-brown to brownish, often with rusty to reddish stains or tones. Surface dry, smooth to minutely hairy in protected areas, often cracking in age. **INTERIOR:** Very firm, solid. Whitish to grayish brown when young; vinaceous brown to dark grayish brown when mature, marbled with whitish veins at all times. **ODOR:** Mild when immature, becoming pungent; a "truffly" mix of garlic spiciness, nuttiness, and cheese. **TASTE:** Mild to radishlike. **MICROSCOPY:** Asci (1) 2–4 (6)–spored. Spores highly variable in size, depending on number per asci; 21–60 x 16–38 µm (not counting spines), elliptical to broadly elliptical, ornamented with spiny reticulum, 3–4 (5) µm high, surrounding honeycomb-like pits, thick walled, yellowish.

ECOLOGY: Solitary or scattered, buried beneath duff or in soil under Douglas-fir. Prefers younger forest, especially plantations. Locally common, but tough to find without a trained dog. Fruiting from winter into spring.

EDIBILITY: Edible and choice. Although these truffles don't have the same acclaim as the European Truffles, the flavor can be on par. They developed a bad rap for "lacking flavor" from an influx of immature truffles on the market, collected by indiscriminate pickers raking the forest floor, causing considerable damage in the process. Ethical harvesting practices, by using trained dogs to sniff out only mature truffles, has helped alleviate these problems.

COMMENTS: The firm, buff to brownish, irregular-lobed fruitbody with a solid, grayish to vinaceous brown gleba marbled with whitish veins, and pungent truffle odor are distinctive features. *T. oregonense* is whitish at first, soon developing reddish brown to rusty orange-brown patches, to orangish brown overall in age. It generally fruits in the fall into winter. Four species are in the Oregon White Truffle complex and are separated by spore morphology (but more reliably separated by molecular analysis). *T. gibbosum* (two left photos) is described above. *T. castellanoi* differs by having slightly shorter spores that are broadly elliptical to globose (20–44 x 18–38 µm). Both *T. oregonense* (upper right photo) and *T. bellisporum* (lower right photo) have narrower spores (20–55 x 14–32 µm for the latter, 25–63 x 13–30 µm for the former). Because of variability in spore size and number of spores per asci, it can be time-consuming to measure enough spores to make a species determination. Don't fret too much, as they all have the same culinary properties. *T. californicum* has a very firm, irregularly lobed, pubescent fruitbody that is whitish to tan when young, becoming ocher-brown to brown in age; it also has a marbled, whitish to dark brown gleba with whitish veins, an herbal garlic-cheese odor, and globose spores. It is common in the late winter and spring under oaks and conifers. *T. candidum* has a reddish brown round fruitbody with a tan to orange-brown gleba, marbled with dark brown and white veins, and a mild to earthy odor. *Balsamia magnata* has an orangish pimpled exterior, a white marbled interior, and smooth elliptical spores.

23 • Jelly Fungi

As you can guess, the unifying feature of these fungi is the gelatinous texture of their fruitbodies. While all of the species included here are basidiomycetes, most have distinctive and oddly shaped basidia. Some resemble tuning forks, with two long, skinny sterigmata (projections on the basidia, bearing spores); others have vertical septations. Many jelly fungi form brainlike or leafy-lobed fruitbodies; others, such as *Pseudohydnum* and *Tremiscus*, form more elaborate stalked fruitbodies. *Calocera* form small club to coral-sampled fruitbodies and are included in that section. Most have smooth fertile surfaces, but *Pseudohydnum* have fine teeth on the underside of the caps, and *Tremiscus helvelloides* can have veins or ridges.

Some ascomycetes, such as *Bulgaria inquinans* and *Urnula padeniana*, have distinctly gelatinous flesh, but generally have more cup-shaped fruitbodies and are thus included in the Cup Fungi chapter.

Dacrymyces chrysospermus

Berk. & M. A. Curtis

ORANGE CONIFER JELLY

FRUITBODY: 1–6 cm across, 0.5–4 cm tall, often irregular, brainlike, lobed, rounded, or fan-shaped clumps or masses. Bright orange when young, sometimes discoloring reddish orange to orangish brown in age. **STIPE:** Generally indistinct or a short, tapered to somewhat rooting attachment to wood. Pale orange to whitish at point of attachment. **FLESH:** Thin to thick, gelatinous or jellylike. Translucent orange in color. **ODOR:** Indistinct. **TASTE:** Indistinct. **KOH:** No reaction. **SPORE DEPOSIT:** Orange to golden orange. **MICROSCOPY:** Spores 17–25 (28) x 6–8.5 μm, curved-cylindrical, thin walled, becoming 7-septate in maturity. Basidia distinctive, shaped like tuning forks.

ECOLOGY: Solitary, scattered, or in troops or masses on dead logs, trees, or branches of conifers, especially Sitka Spruce and Western Hemlock. Also occasionally occurring on hardwoods. Very common on the Far North Coast, less common across the rest of our range. Fruiting throughout the wet season.

EDIBILITY: Edible and harmless, but we have yet to find a way to make it appealing.

COMMENTS: This bright orange jelly fungus can be easily identified by the irregular brainlike, lobed, or fan-shaped masses, growth on conifers, and whitish base or attachment point. Most similar is *Tremella aurantia*, which grows on hardwoods (and often with *Stereum* growing next to it), is generally larger and more lobed or leafy in appearance, doesn't have a white point of attachment, and differs greatly microscopically. Other common species of *Dacrymyces* in our area include *D. stillatus*, which forms tiny round to top-shaped translucent yellow-orange jelly-drops on coniferous logs. *D. chrysocomus* forms small to medium top-shaped, conelike, or teardrop-shaped bright orange jelly-drops. It is common on the Far North Coast on hemlock and fir branches.

SYNONYM: It is called *Dacrymyces palmatus* in most field guides; however, *D. chrysospermus* is an older name and thus has precedence over *D. palmatus*.

Tremella aurantia

Schwein.

WITCH'S BUTTER

FRUITBODY: 1–8 cm across, 1–5 cm tall, often an irregular brainlike or lobed mass when young, becoming more lobed or leafy in age. Bright orange to yellow-orange when young, sometimes discoloring reddish orange when dry or becoming whitish in age. **STIPE:** Generally indistinct or a short, stout attachment. Concolorous with rest of the fruitbody. **FLESH:** Thin to thick, gelatinous or jellylike when wet, tough, rubbery if dry. Translucent orange in color. **ODOR:** Indistinct. **TASTE:** Indistinct. **KOH:** No reaction. **SPORE DEPOSIT:** Yellow. **MICROSCOPY:** Spores 7–10 x 6–8 μm, broadly elliptical to elliptical, smooth. Basidia 4-spored, ovate, club shaped to subglobose.

ECOLOGY: Solitary or scattered on dead logs, trees, or branches of hardwoods, especially Coast Live Oak and Tanoak. Almost always occurring with fruitbodies of *Stereum hirsutum* (p. 481), on which it is a parasite. Very common from Mendocino County south, uncommon to north. Often fruiting in late fall or winter but persisting well into spring.

EDIBILITY: Edible, but attempts to make a jelly candy with it have been less than appealing.

COMMENTS: This common jelly fungus can be identified by the orange to yellow-orange color, irregular brainlike, lobed, or leafy clumps, and growth on hardwoods. All *Tremella* are parasites on other fungi; this species grows on the hyphae of *Stereum hirsutum* and is commonly found on oak branches covered with *Stereum* fruitbodies. The similar-looking *T. mesenterica* is more common in the northern part of our range, generally on alder. It differs by having a simpler fruitbody of one to several leafy lobes (rarely a brainlike mass like *T. aurantia*). It usually is more yellow and is a parasite on *Peniophora* spp., which, for the most part are a nondescript crust fungus. *Dacrymyces chrysospermus* is another brainlike or lobed bright orange jelly fungus, but it usually grows on conifers and has a whitish base or attachment point. Other species of *Dacrymyces* are smaller, differ microscopically, and are not fungal parasites.

MISAPPLIED NAME: *Tremella mesenterica* Retz.

Tremella foliacea

Pers.

LEAFY JELLY

FRUITBODY: 2–15 (20+) cm across, 1–15 cm tall, an irregular mass of leafy lobes and folds. Dark brown to reddish brown when young, becoming paler, brown, orangish brown to beige, or occasionally violet-brown. **STIPE:** Absent; attached to wood with a small to broad attachment point. **FLESH:** Thin to thick, gelatinous or jellylike when wet, hard when dry. Translucent brown in color. **ODOR:** Indistinct. **TASTE:** Indistinct. **KOH:** No reaction. **SPORE DEPOSIT:** Whitish to pale yellowish. **MICROSCOPY:** Spores 7–11 x 6–9 µm, subglobose to elliptic, smooth.

ECOLOGY: Often solitary or scattered on dead logs, trees, or branches of hardwoods, especially Coast Live Oak. Like *T. aurantia*, it is a parasite of *Stereum hirsutum* (p. 481) and almost always occurs on logs with *Stereum* fruitbodies present. Common throughout our range on live oak in south, often on Black Oak and Tanoak in north. Often fruiting in late fall or winter but persisting well into spring.

EDIBILITY: Edible and often occurring in large clusters, but the gelatinized texture leaves something to be desired.

COMMENTS: The brown leafy-lobed, gelatinous-textured mass and growth on hardwoods are usually significant enough to identify this species. Some clumps can become quite large, reaching more than 20 cm across! It has the ability to dry out (and become rather hard), but then reconstitutes with the next rainy period. *Auricularia americana* has similar colors and texture, but differs by having a simpler, ear-shaped fruitbody, with a dull, finely tomentose top and a shiny underside. It is rare on the Californian coast (but common in the mountains), fruiting in fall on conifers.

Tremiscus helvelloides

(DC.) Donk

APRICOT JELLY

FRUITBODY: 1.5–8 (10) cm across, often spoon shaped to tongue shaped, with a lobed or wavy margin. Apricot, pinkish apricot to coral pink, underside usually smooth, but can be wrinkled or slightly veined. Generally dull and slightly paler than cap. Surface moist to dry, shiny when wet, dull when dry. **STIPE:** 2–10 cm long, 0.5–2 cm thick, lateral, often slightly grooved on one side. Concolorous with cap, but often paler, to whitish at base. **FLESH:** Thin to thick, gelatinous or jellylike when wet. Translucent pink to watery pink in color. **ODOR:** Indistinct. **TASTE:** Indistinct. **KOH:** No reaction. **SPORE DEPOSIT:** White. **MICROSCOPY:** Spores 9–12 x 5–6.5 µm, oblong to elliptic, smooth, colorless.

ECOLOGY: Solitary, scattered, or in small clusters on soil, duff, or well-decayed wood, often near disturbed areas such as trails or roadsides. Uncommon on the Far North Coast, rare south of Sonoma County. Fruiting in late fall in north, winter into spring in the central and southern part of range.

EDIBILITY: Nontoxic.

COMMENTS: The apricot to pinkish colors, tongue- or spoon-shaped cap with a lateral stipe, and gelatinized texture make this mushroom unmistakable. Although it usually grows solitary or scattered, we have observed some large clusters of fruitbodies. *Tremella foliacea* forms leafy clumps on hardwood branches, but is generally browner; it is unlikely to be mistaken for this species.

SYNONYM: There is some confusion about the correct name for this species. Most sources consider *Tremiscus* to be the correct genus; others use *Guepinia*. It has also been called *Phlogiotis helvelloides*.

Myxarium nucleatum
Wallr.
CRYSTAL BRAIN JELLY

FRUITBODY: 1–5 cm across, aggregated masses can be 10 cm across. Overall rounded, but with a bumpy or wrinkly, often lobed shape. Whitish to light gray, or pale tan to brownish. Usually quite translucent, showing multiple whitish to pale creamy internal nuclei (seedlike floating structures). **STIPE:** Absent. **FLESH:** Thin to relatively thick (for its size), gelatinous or somewhat jellylike, often watery and dissolving when rained on, or in age. **ODOR:** Indistinct. **TASTE:** Indistinct. **KOH:** No reaction. **SPORE DEPOSIT:** Pale, difficult to obtain. **MICROSCOPY:** Spores 9–12 x 4–5.5 μm, cylindrical and curved with rounded ends. Basidia divided into four sections, bases bulbous above a septum separating them from the stalk that bears each of them.

ECOLOGY: Scattered in groups or in large masses on rotting hardwood trunks and branches, especially those of Coast Live Oak. Common from Mendocino County south, occasional to the north. Fruiting almost anytime logs are wet enough from late fall well into spring.

EDIBILITY: Small, gooey, and flavorless.

COMMENTS: This inconspicuous but charming jelly fungus is distinguished handily by the pale nodules suspended in its gelatinous, translucent flesh. These are usually easily seen, but sometimes a close look is required to find them. Faded *Exidia glandulosa* can look very similar, and are often found in the same habitat; they can be told apart by their warted or bumpy exterior, and by their lack of pale internal nodules. *Tremella encephala* has a more firm, opaque, creamy white brainlike fruitbody with a tough white core; it is found on conifer wood associated with *Stereum sanguinolentum* in our area.

Exidia glandulosa group
BLACK JELLY ROLL

FRUITBODY: 1–10 (15+) cm across, 0.5–2.5 cm tall, forming brainlike, wrinkled, or lobed clumps or fused gelatinous masses when wet, drying to a thin, hard crust. Deep black to dark gray, or occasionally vinaceous black. Surface dull to shiny, covered with small punctate warts. **STIPE:** Absent. **FLESH:** Thin to thick, gelatinous or jellylike. Black to translucent gray. **ODOR:** Indistinct. **TASTE:** Indistinct. **KOH:** No reaction. **SPORE DEPOSIT:** Whitish. **MICROSCOPY:** Spores 10–16 x 4–5 μm, cylindrical to sausage shaped, smooth.

ECOLOGY: Often in troops or masses on dead logs, trees, or branches of hardwoods. Especially common on Coast Live Oak. Fruiting from fall to spring.

EDIBILITY: Nontoxic.

COMMENTS: This jelly fungus forms wrinkled or brainlike clumps or masses, has a finely warted surface (use a hand lens), and grows on hardwoods. These features, along with the dark colors, are usually enough to positivity identify it. *Myxarium nucleatum* (=*Exidia nucleata*) shares the same general shape and habitat, but is pale colored, translucent white to pale gray, becoming grayish beige to light orangish brown in age. It is uncommon, but widespread in California. The ascomycete jelly *Bulgaria inquinans* is often gumdrop to top shaped and has a roughened underside that is reddish brown when young, becoming darker in age.

NOMENCLATURE NOTE: *E. nigricans* is almost indistinguishably based on macro and micro morphology, but according to a recent study is widespread in North America and is often misidentified as *E. glandulosa*. It may turn out that is what we have in California.

Pseudohydnum gelatinosum group
JELLY HOGS

FRUITBODY: 1.5–8 (10) cm across, tongue shaped with a lateral stipe, more rarely fan shaped with a short or rudimentary stipe; margin even or wavy. Generally translucent white to pale grayish white, although some forms can be darker (see comments). Surface moist, dull, roughened or appearing finely hairy. Underside covered with short pointed spines, typically paler than cap. **STIPE:** 2–8 (11) cm long, 0.5–2.5 cm thick, lateral, often irregular, and slightly grooved on one side. Concolorous with cap, but often more watery at base. Surface finely roughened or bumpy. **FLESH:** Thin to thick, gelatinous or jellylike when wet, drying hard. Translucent white to watery gray in color. **ODOR:** Indistinct. **TASTE:** Indistinct. **KOH:** No reaction. **SPORE DEPOSIT:** Whitish. **MICROSCOPY:** Spores 5–7 μm, globose to subglobose, smooth, colorless. Basidia pear shaped to oblong, longitudinally septate.

ECOLOGY: Solitary, scattered, or in small clusters or troops on conifer branches or decayed logs. Common throughout our range, especially north of the San Francisco Bay Area. Fruiting in late fall or winter in north, winter into spring in south.

EDIBILITY: Edible.

COMMENTS: The tongue-shaped cap with a lateral stipe, spiny underside, and gelatinous texture are distinctive. The texture is reminiscent of a cat's tongue (hence one of the common names, Cat's Tongue). Although it is usually translucent white in color, darker forms occur, including a pleurotoid form that lacks a stipe or just has a short stipe. It generally has a darker gray to grayish brown cap and grows on coniferous logs or branches.

NOMENCLATURE NOTE: This is a species complex in western North America, with at least two species occurring in California, neither of which are the same as the "true" *P. gelatinosum* of Europe. Another common name is Jelly Tooth.

24 • Morels, False Morels, and Elfin Saddles

Thanks to the culinary reputation of morels and the unusual toxicity syndrome of some of the false morels, this is a well-known group. All are ascomycetes. The ecology of this group is varied. Many fruit in response to fire or disturbance; others occur in less disturbed forest settings. Some are saprobic; others are at least facultatively mycorrhizal. *Helvella* (the elfin saddles) are definitely mycorrhizal. Many *Gyromitra* (the false morels) and *Verpa* (the thimble morels) appear to be at least facultatively mycorrhizal.

Morchella (true morels) are easy to recognize by their honeycombed heads with distinct pits and cross walls, and hollow stipes. Fruitbodies of *Helvella* and *Gyromitra* usually have fertile heads with free or partially free edges (not fused to the stipes), and often have chambered or stuffed interiors. *Verpa* have simple, smooth stipes with distinct, easily detached brownish heads and stuffed or nearly hollow interiors.

Gyromitra are famous for containing molecular precursors of monomethylhydrazine (MMH)—when metabolized, these compounds can have severe negative effects on the central nervous system, leading to nausea and convulsions. Although this has led many to label the *Gyromitra* toxic (they should be avoided), keep in mind that morels also contain this compound, though in lower concentrations! As might be expected of a substance also used as rocket fuel, MMH is highly volatile and is driven off by high heat and drying, so always cook your morels thoroughly and do so in a well-ventilated area. Also keep in mind that some people appear to experience negative effects after eating morels, no matter how thoroughly they are cooked.

Important identifying features to note for these mushrooms are the structure of the cap (including subtle details of arrangement of pits and ridges, if present); degree and shape of attachment between cap and stipe; overall coloration (and any changes from young to old fruitbodies); details of interior structure (make a cross section); and substrate and surrounding trees.

Morchella importuna

M. Kuo, O'Donnell & T. J. Volk

INCONSIDERATE MOREL

CAP: 2–8 cm across at widest point, often tapering toward top, 3–15 cm tall, tall-conical, to more cylindrical in age. Vertical ridges markedly parallel, with deeply sunken pits and nearly perpendicular cross ridges, resulting in a ladder-like appearance. Brown to buff-brown, tan or grayish brown when young, becoming darker brown, ridges darkening in age, dark brown to blackish brown, pits lighter. **STIPE:** 3–10 cm long, 2–6 cm thick, often enlarged and pinched at base. Off-white, creamy to pale tan. Surface dry, smooth to slightly granular in age. **FLESH:** Thin, somewhat brittle to slightly rubbery, whitish to tan. Entirely hollow, with a single chamber extending through cap and stipe. **ODOR:** Indistinct to salty or slightly fishy smell, especially in age. **TASTE:** Indistinct. **SPORE DEPOSIT:** Ocher-buff, usually scant. **MICROSCOPY:** Spores 18–24 x 10–13 μm, elliptical, smooth.

ECOLOGY: Solitary, scattered, or in large troops in disturbed areas, often in wood chips. Also around recently poured concrete. Fruiting in spring, most often in urban areas.

EDIBILITY: Edible and very good; like all morels, it should be well cooked.

COMMENTS: As a group, morels can be identified by their conical, cylindrical to triangular, honeycomb-like cap with distinct pits, and their hollow interior with a single chamber through both cap and stipe. All morels are edible (for most people) when well cooked, undercooked or raw morels can make you sick. Some people appear to be allergic and experience minor to severe gastrointestinal distress after eating them. False morels (*Gyromitra* spp.) have convoluted, wrinkled, and/or lobed caps and are multichambered when cut in half; some species are potentially deadly poisonous, so learn to recognize them before picking true morels. *Morchella importuna* can be identified by the large size; inconsiderate habits (you never know when or where you might stumble upon this species); parallel vertical ridges with regular ladderlike cross ridges; and dark colors in age. The Woodchip Morel (*M. rufobrunnea*) grows in similar habitats (although never in sod), but has whitish gray to dirty blond ridges and often slowly stains orange-red when bruised or in age. Some of the black morels associated with forest fires can look similar, particularly *M. exuberans,* but that species only fruits in recently burned or clear-cut areas, and has less regularly arranged cross-ridges and often an olivaceous tone when young. There are at least two other black "burn morel" species in California, *M. sextelata* and *M. eximia,* but we have not seen them on the coast. Another postfire morel that has fruited on the coast is *M. tomentosa,* which is a distinctly tomentose and dark gray color when young. A smaller black morel, *M. brunnea,* grows under cottonwoods along rivers on the Far North Coast, as does the Western Half-Free Morel, *M. populiphila.* The Natural Black Morel, *M. snyderi,* has an often triangular cap and frequently a convoluted stipe base; it is a common montane species, usually growing under true firs (*Abies* spp.).

Morchella rufobrunnea

Guzmán & F. Tapia

WOODCHIP MOREL

CAP: 2–5 cm across at widest point, often tapering toward top, 3–10 (13) cm tall, conical to triangular-ovoid. Ridges vertical at first, becoming irregular, wrinkled, cross ridges highly irregular, with elongated pits at first, more angular in age. Ridges generally markedly paler than pits, whitish gray, beige to pale yellowish brown, pits dark brown to grayish brown. Becoming paler; yellowish brown overall in age, occasionally with reddish brown stains. All parts slowly staining orangish to reddish brown when fresh. **STIPE:** 2–9 cm long, 1–3 cm thick, often enlarged and pinched at base. Generally off-white to creamy brown, occasionally dark gray when young, fading in age. Surface dry, smooth to slightly granular in age. **FLESH:** Thin, somewhat brittle to slightly rubbery, whitish to tan. Entirely hollow, with a single chamber extending through cap and stipe. **ODOR:** Indistinct, becoming mealy in age. **TASTE:** Indistinct. **SPORE DEPOSIT:** Ocher-buff. **MICROSCOPY:** Spores 19–25.5 x 13–17 μm, ovoid, smooth.

ECOLOGY: Solitary, scattered, or in troops in wood chips, gardens, or disturbed soil. Can occur in large numbers in orchards, especially in the Central Valley. Most common in late winter and spring, but can fruit at any time during the year, especially in watered gardens.

EDIBILITY: Edible, but of inferior quality compared to other morels. It is also wise to consider carefully where the mushroom was growing, as some wood chips are treated with nasty chemicals and often border highly polluted urban areas.

COMMENTS: The pale ridges and often darker pits in young specimens, yellowish brown color in age, orangish red staining, and growth on wood chips or disturbed soil are distinctive. Originally described from Mexico, it has greatly expanded to the artificial wood chip habitat throughout California. Unlike other morels, which are restricted to spring fruitings, the Woodchip Morel commonly fruits in fall and winter. Another morel that inhabits wood chips is *Morchella importuna*, which is often larger and has distinct parallel vertical ridges with ladderlike cross ridges and has darker colors. The Blond Morel, *M. americana* (=*M. esculenta*, sensu NA) grows under cottonwood, ash, and fruit trees; it can easily be recognized by the yellowish to grayish tan colors and highly irregular pits (never strongly vertically arranged). *M. tridentina* is a yellowish-colored morel (but is genetically a black morel) that also blushes orange; it is a montane species found under various conifers.

Verpa bohemica

(Krombh.) J. Schröt.

WRINKLED THIMBLE-CAP

CAP: 1–7 cm across, 2–8 (10) cm tall, thimble shaped, cylindrical to bell shaped. Margin pinched against, but not attached to stipe when young, becoming flared or slightly uplifted in age. Surface dry, longitudinally ridged, deeply furrowed and wrinkled. Ocher-brown to brown. **UNDERSIDE:** Finely tomentose, tan to light brown. **STIPE:** 5–15 cm long, 1–3 cm thick, equal to club shaped. Whitish, buff to pale ocher. Surface dry, covered with fine whitish chevrons or granules. **FLESH:** Thin, very fragile. Stipe stuffed with whitish to pale pinkish orange pith, with hollow pockets, to completely hollow in age. **ODOR:** Indistinct. **TASTE:** Indistinct. **MICROSCOPY:** Spores 55–87 x 17–22 µm, cylindrical to elongate-elliptical, smooth, lacking oil droplets. Asci 2-spored.

ECOLOGY: Solitary or in scattered troops under cottonwoods. Uncommon in our range, mostly restricted to riparian zones on the Far North Coast. Fruiting in spring.

EDIBILITY: Edible.

COMMENTS: This species can be identified by the ridged, deeply furrowed and wrinkled thimble-shaped cap, attached only at the very top of a pith-filled stipe, and by the growth under cottonwoods. Most similar is an undescribed species of *Verpa* that grows with live oak; see *V. conica* (at right) for distinguishing features. *Morchella populiphila* also grows under cottonwoods, but differs by having a thimble-shaped to conical cap attached at about the midpoint and a single hollow chamber in both stipe and cap.

Verpa conica group

SMOOTH THIMBLE-CAP

CAP: 1–4 cm across, rounded, thimble shaped, bell shaped to broadly conical, margin often pinched to stipe when young, slightly flaring in age. Surface dry, often slightly wrinkled when young, to smooth, finely cracking in dry weather. Ocher-brown to brown, occasionally dark brown in age. **UNDERSIDE:** Finely tomentose, pale buff. **STIPE:** 3–12 cm long, 0.5–2 cm thick, equal to club shaped. Pale buff to pale ocher-buff. Surface dry, covered with fine chevrons or granules, often wrinkled or wavy. **FLESH:** Thin, very fragile. Stipe stuffed with whitish pith, with hollow pockets, to completely hollow in age. **ODOR:** Indistinct. **TASTE:** Indistinct. **MICROSCOPY:** Spores 29–34 x 17–20 µm, elliptical, smooth, lacking oil droplets. Asci 8-spored.

ECOLOGY: Solitary or scattered in duff under conifers. Uncommon on the coast, locally common in mountains. Fruiting in spring.

EDIBILITY: Edible.

COMMENTS: The slightly wrinkled to smooth thimble-shaped cap, pith-stuffed stipe, small granules or chevrons ornamenting the stipe surface, and growth with conifers are distinctive. More than one species goes under the name *V. conica* in California; one, *V. chicoensis*, was described from Chico in 1904 and promptly forgotten. Extensive collecting and modern taxonomic work are needed to fully understand these taxa. *V. bohemica* has a wrinkled and furrowed cap, grows with cottonwoods, and has much larger spores (two per ascus). There is a darker, distinctly wrinkly capped species under live oak that looks like a cross between *V. conica* and *V. bohemica* (and is often identified as the latter), but has smaller spores and 8-spored asci.

Helvella maculata

N. S. Weber

FLUTED BROWN ELFIN SADDLE

CAP: 1–6 cm across, irregular-lobed, folded, or occasionally saddle shaped. Margin often upturned or back folded, not attached to stipe. Grayish brown, brownish to gray, occasionally spotted. **UNDERSIDE:** Slightly roughened or scruffy to smooth in age. Whitish to pale gray. **STIPE:** (2) 4–10 cm long, 0.5–3 cm thick, equal to irregular, sometimes with a buttress base, ridged, furrowed or fluted. White or blushed with gray-brown to all gray-brown. **FLESH:** Thin, brittle. Stipe chambered, cartilaginous. **ODOR:** Indistinct. **TASTE:** Mild. **MICROSCOPY:** Spores 18–23.5 x 11–14 μm, broadly elliptical, smooth, with a large oil droplet.

ECOLOGY: Solitary or scattered in small troops on ground or in duff in mixed forest, generally with Douglas-fir. Fruiting from winter into spring. Known from throughout our range, but more common north of the San Francisco Bay Area.

EDIBILITY: Unknown.

COMMENTS: This species is distinguished from similar *Helvella* by the fluted stipe and irregular grayish brown cap that is not fused to the stipe. *H. vespertina* has a gray to black cap with a margin that attaches to the stipe and grows with conifers. *H. dryophila* is nearly identical to *H. vespertina* but grows with hardwoods. *H. compressa* is smaller and has a grayish saddle-shaped cap and a round or compressed (not fluted) white stipe.

Helvella compressa

(Snyder) N. S. Weber

COMPRESSED ELFIN SADDLE

CAP: 1–5 cm across, often distinctly saddle shaped with two pointed lobes, occasionally more irregularly lobed or folded cap. Gray to grayish brown. **UNDERSIDE:** Whitish to pale gray, finely pubescent. **STIPE:** 2–10 cm long, 0.3–1 (1.5) cm thick, equal or enlarged toward base, occasionally compressed or creased. White to cream, discoloring pale brown at base in age. Surface smooth to wavy, dry. **FLESH:** Thin, brittle, pale gray. Stipe hollow, white. **ODOR:** Indistinct. **TASTE:** Mild. **MICROSCOPY:** Spores 19.5–21 (24) x 12–14 μm, broadly elliptical, smooth, with a large oil droplet.

ECOLOGY: Solitary or scattered in small troops on soil or in duff or moss. Common under live oak in the southern part of our range, under Tanoak and possibly Douglas-fir in north. Fruiting in late winter and spring.

EDIBILITY: Unknown.

COMMENTS: The small, grayish saddle-shaped cap and white stipe are distinctive. The rare *H. albella* shares the same shape, but has a darker brown cap with a smooth underside. *H. elastica* is slightly smaller and has a pale grayish brown cap with more rounded lobes and a smooth underside. Also compare the cup-shaped *H. fibrosa* (and others), which can sometimes curl up and resemble *H. compressa*.

Helvella vespertina
N. H. Nguyen & Vellinga
WESTERN BLACK ELFIN SADDLE

CAP: 2.5–5 (12) cm across, irregular-lobed, folded, to brain-like. Margin typically fused to stipe; one or more lobes may be upturned. Typically dark gray, but can range from pale gray to black. **UNDERSIDE:** Ribbed to smooth, whitish to pale gray. **STIPE:** 4–15 (25) cm long, 1–5 (10) cm thick, irregularly grooved, ribbed and fluted, sometimes with a buttress base. Whitish, gray to blackish, occasionally with orangish stains in age. **FLESH:** Thin, brittle. Stipe chambered, cartilaginous. **ODOR:** Indistinct. **TASTE:** Mild. **MICROSCOPY:** Spores 15.5–21.5 x 10–12 (14) μm, broadly elliptical, smooth, with a large oil droplet.

ECOLOGY: Solitary or scattered in small troops on ground or in duff or moss under conifers. Fruiting in fall and winter. Very common throughout our range.

EDIBILITY: Edible, but cook well. Do not eat moldy specimens, as the common Helvella Mold, *Hypomyces cervinigenus*, is poisonous.

COMMENTS: Formerly called *Helvella lacunosa*, this elfin saddle in California was recently discovered to comprise at least four different species, none of which are the same as the European species. Two species, *H. dryophila* and *H. vespertina*, have been described and are common on the coast. *H. vespertina* fruits under conifers in fall and winter, whereas *H. dryophila* fruits under oaks in winter and spring. If you find them in mixed forest in the middle of the winter, you may be out of luck identifying them without a genetic sequence. *H. dryophila* typically has a darker black cap with a contrasting whitish stipe when young, and often develops orangish coloration on the stipe in age, but there is much overlap in appearance. The other two are montane species. All are commonly parasitized by whitish to pinkish *Hypomyces* molds. *Hypomyces cervinigenus* commonly occurs on *Helvella vespertina*; additional undescribed and difficult-to-distinguish species occur with other *Helvella* host species.

MISAPPLIED NAME: *Helvella lacunosa*.

Helvella dryophila
Vellinga & N. H. Nguyen
OAK-LOVING ELFIN SADDLE

CAP: 2–6 cm across, irregular-lobed and grooved, sometimes weakly saddle-shaped or brainlike; margin attached to, or free from stipe. Dark gray to jet black. **UNDERSIDE:** Ribbed to smooth, whitish to pale gray. **STIPE:** 2–12 cm long, 1–5 cm thick, irregularly grooved, ribbed and fluted, sometimes with a buttress base. Whitish, gray to blackish, occasionally with orangish stains in age. **FLESH:** Very thin, white, brittle to rubbery. Stipe chambered, cartilaginous. **ODOR:** Indistinct. **TASTE:** Mild. **MICROSCOPY:** Spores 15.5–19.5 x 10–12.5 μm, broadly elliptical, smooth, with a large oil droplet.

ECOLOGY: Solitary or in small troops in duff under oak. Common in appropriate habitat throughout our range, fruiting primarily in winter and spring.

EDIBILITY: Edible, but needs to be cooked well.

COMMENTS: Although the chambered, fluted, potholed stipe and dark brain-like or lobed cap clearly distinguish it as one of the black elfin saddles, it is very similar to *H. vespertina* and can be tough to tell apart. *H. dryophila* tends to be smaller (especially shorter), have a darker and more lobed or grooved cap, and to show more contrast between the cap and stipe. However, all of these characters are variable and are not reliable features on their own. Perhaps most helpful is habitat information: *H. dryophila* associates with oaks and fruits in the winter and spring, whereas *H. vespertina* is a conifer associate and fruits in the fall and winter. In areas where both kinds of hosts are present, it may be impossible to identify a collection without genetic data.

Gyromitra esculenta

(Pers.) Fr.

FALSE MOREL

CAP: 2–7 (10) cm across, irregular-lobed, convoluted, wrinkled, and folded. Margin hanging free or fusing to stipe. Reddish brown to dark vinaceous brown, occasionally paler red-brown, rusty red to yellow-brown. **UNDERSIDE:** Beige to tan, finely pubescent to smooth. **STIPE:** 2–10 cm long, 0.5–2.5 cm thick, irregular, often compressed, sometimes with a buttress base. Buff to vinaceous buff, often with a whitish base. Surface finely pubescent to smooth, dry. **FLESH:** Thin, brittle, pallid. Stipe stuffed with whitish pith, to nearly hollow in age. **ODOR:** Indistinct. **TASTE:** Mild. **MICROSCOPY:** Spores 17–28 x 10–14 μm, broadly elliptical to ovoid, smooth, with two oil droplets.

ECOLOGY: Solitary or scattered in small troops on the ground or in moss or duff under conifers. Uncommon on the coast, fruiting in late winter and spring. Common and locally abundant in the mountains.

EDIBILITY: Potentially deadly poisonous, especially if consumed raw. It contains gyromitrin, an unstable compound that hydrolyzes into monomethylhydrazine (a compound that has been used in rocket fuel). It is highly volatile and cooks off in the vapors, rendering the mushrooms "edible." Keep in mind that if the mushrooms are cooked in a closed pot, the toxins might not escape, and the vapors alone can cause violent illness. Even though some people consume the False Morel, we strongly recommend *not* eating this mushroom—it (or a related species) has caused a number of deaths in Europe.

COMMENTS: The reddish brown, convoluted cap, irregularly compressed stipe stuffed with pith when young, and growth in the late winter or spring are distinctive. *G. infula* is similar, but has a distinctly saddle-shaped cap and fruits in the fall and winter. *G. montana* is a large, ocher-brown montane species that fruits soon after snowmelt. Another montane species, *G. californica*, has a convoluted cap with a ribbed underside and free edge (not fused to the stipe). Similar *Helvella* species have fluted stipes. Morels (*Morchella* spp.) have honeycombed, conical to triangular caps and show a single hollow chamber from the stipe into the cap when cut in half.

Gyromitra infula

(Schaeff.) Quel.

SADDLE-SHAPED FALSE MOREL

CAP: 1–8 (12) cm across, often distinctly saddle shaped with two pointed lobes, at times with a more irregularly lobed cap. Margin pinched and attached to stipe. Ocher-brown to reddish brown, darkening to dark vinaceous brown in age. **UNDERSIDE:** Paler, finely pubescent. **STIPE:** 2–10 cm long, 1–3 cm thick, equal to irregular, often compressed, sometimes with a buttress base. Ochraceous to vinaceous brown and often with a whitish base at first, darkening in age. Surface finely powdery to bald, smooth, dry. **FLESH:** Thin, brittle to rubbery. Stipe hollow. **ODOR:** Indistinct. **TASTE:** Mild. **MICROSCOPY:** Spores 17–23 x 7–10 μm, narrowly elliptical, smooth, with two oil droplets.

ECOLOGY: Solitary or scattered in small troops on decaying logs, or woody debris, in duff or on ground. Common, especially in northern part of our range. Fruiting from fall into winter, rare in spring.

EDIBILITY: Poisonous. See the closely related *G. esculenta* (at left) for more details.

COMMENTS: The ocher-brown to reddish brown, saddle-shaped cap, irregular hollow stipe, and growth in the fall and winter are helpful in identification. *G. ambigua* is very similar, but is generally slightly darker colored with a more purple-brown stipe. It also has much larger spores (21–30 x 8–11 μm). *G. esculenta* is also similar and can have lobed caps, but is more convoluted and wrinkled and fruits in the late winter and spring. *Helvella compressa* is smaller and has a grayish saddle-shaped cap and a white stipe.

SYNONYM: *Helvella infula* Schaeff.

25 • Cup Fungi

This section contains those ascomycetes which have fruitbodies that are disclike, goblet shaped, or round. Ascomycetes with differently shaped fruitbodies can be found in other sections: Those with distinct stipes and honeycombed, brainlike or saddle-shaped heads are in the section on Morels, False Morels, and Elfin Saddles. Those with club-shaped fruitbodies (with relatively undifferentiated heads) are in the chapter on Club Fungi. A few hypogeous species are in the Truffles chapter.

When identifying any cup fungus, note its structure, coloration, and texture, as well as the substrate from which it is fruiting. For many of these species, identification with certainty requires examination of microscopic features—most importantly the shape, size, and ornamentation of the spores, the shape of the sterile cells (called paraphyses) between the asci, and whether or not the tips of the asci change color in Melzer's reagent.

The *Helvella* and *Plectania* in this section have dark blackish fruitbodies with thin central stipes (although this feature is easily overlooked in the *Plectania*).

Urnula and *Bulgaria* produce relatively large, fleshy fruitbodies with dark-colored fertile surfaces and gelatinous flesh.

Peziza and *Otidea* have beige to brown, bowl-shaped fruitbodies with fairly thin flesh and a fragile-brittle texture. *Peziza* tend to be most common on rotting logs in the forest and wood chips in urban habitats, while *Otidea* are mostly restricted to forest settings, usually in moss or humus.

Sowerbyella are bright yellow-orange, while *Sarcoscypha* are scarlet; both have bowl- to disc-shaped fertile surfaces with narrow central stipes. *Aleuria* have bright orange, shallow bowl-shaped fruitbodies that lack stipes.

A number of smaller, brightly colored discomycetes with small to tiny cup or disc-shaped fruitbodies also occur in our area: The bright red to orange, disc-shaped fruitbodies of *Scutellinia* have brown "eyelashes" around their edges and grow on wood. *Cheilymenia* produce round, orange fruitbodies in large swarms and are usually found on dung. *Bisporella* appear as swarms of tiny yellow discs on dead wood. *Chlorociboria* are small, beautiful aqua-green cups found on rotting wood.

Geopyxis have dull tan goblet-shaped fruitbodies and fruit from soil.

Annulohypoxylon form hard round fruitbodies on wood, and they are included here because they don't fit elsewhere.

Helvella acetabulum

(L.) Quel.

BROWN RIBBED ELFIN-CUP

FRUITBODY: 1.5–6 cm across, cup shaped to broadly bowl shaped, with ribs and veins running down the lower sides of the cup, down a ribbed stipe. **UPPER SURFACE:** Smooth, brown, grayish brown to buff. **UNDERSIDE:** Slightly paler, with whitish to pale buff ribs and veins, transitioning into a pale base. **STIPE:** 1–5 cm long, 0.5–3 cm thick, stout, equal or tapered downward, ribbed, creased and folded. White to pale buff. **FLESH:** Thin, brittle to rubbery, concolorus. **ODOR:** Indistinct. **TASTE:** Mild. **MICROSCOPY:** Spores 16–20 x 11–14 μm, broadly elliptical, smooth, with a single oil droplet.

ECOLOGY: Solitary or in small clusters scattered under live oak. Common from southwest Mendocino County south. Fruiting in late winter and spring.

EDIBILITY: Not recommended. There are mixed reports about this species. Some say it is poisonous (at least when raw or undercooked), while others say it is edible when well cooked. Experimentation not recommended.

COMMENTS: From the top, this species looks like just another cup fungus, but when flipped over, it has striking ribs that extend from the stipe midway up the cup. The only similar species have a darker color and ribs restricted to the stipe. *H. solitaria* has a slightly darker grayish brown to grayish buff cup and a whitish stipe. *H. leucomelaena* has a dark grayish brown to blackish cup and a whitish stipe.

Helvella fibrosa

(Wallr.) Korf

STALKED ELFIN-CUP

FRUITBODY: 1–4 cm across, cup shaped at first, becoming broadly cup shaped to almost disc shaped, with a short to long slender stipe. **UPPER SURFACE:** Smooth, dark gray or black when young, fading to grayish brown to pale gray. **UNDERSIDE:** Coarsely to finely pubescent, slightly paler than top. **STIPE:** 1–5 cm long, 0.2–0.5 cm thick, cylindrical. Black with a whitish base, becoming gray to all whitish. Surface dry, finely pubescent. **FLESH:** Very thin, fragile, light to dark gray. **ODOR:** Indistinct. **TASTE:** Mild. **MICROSCOPY:** Spores 16.5–20 x 10–12 μm, broadly elliptical, smooth, with a single oil droplet.

ECOLOGY: Solitary or in small troops on the ground or in moss or duff in mixed woods. Uncommon, fruiting in late winter and spring.

EDIBILITY: Unknown.

COMMENTS: The blackish to gray colors, slender stipe with a cup-shaped top, and pubescent underside are helpful in identifying members of this group. *H. macropus* is similar, but has larger, spindle-shaped, roughened spores (macroscopic differences are slight) and is generally slightly paler and more grayish buff in color. *H. solitaria* and *H. leucomelaena* are larger and have thicker, ribbed stipes.

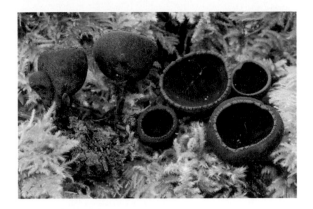

Plectania melastoma

(Sowerby) Fuckel

ORANGE-DUSTED CUP

FRUITBODY: 1–3 cm across, goblet shaped with an incurved margin at first, expanding to cup shaped. **UPPER SURFACE:** Smooth, dull dark brownish black to black when dry, shiny black when wet. **UNDERSIDE:** Finely tomentose, covered with rusty orange fibrils or granules, becoming mostly bald in age, but usually retaining orange granules around the rim. **STIPE:** Up to 1 cm long, central, consisting of fused black threads. **FLESH:** Thin to moderately thick, tough, with a gelatinous interior, at least when wet. **ODOR:** Indistinct. **TASTE:** Indistinct. **MICROSCOPY:** Spores 23–28 x 10–11 μm, spindle shaped, smooth, containing many oil droplets when young.

ECOLOGY: Solitary, scattered, or in small clusters on small buried woody debris, needles, or spruce cones. Especially common on the Far North Coast in mossy Sitka Spruce forest, rare elsewhere. Fruiting from midwinter into spring.

EDIBILITY: Unknown.

COMMENTS: When young, the black cup covered with orange granules is unmistakable. Older specimens lose most of the granules, but usually retain at least some orange flakes near the rim. *P. milleri* is a similar, common species that lacks the orange granules. See the comments under that species for additional similar black cup fungi.

Plectania milleri

Paden & Tylutki

FRUITBODY: 1–3 (4) cm across, cup shaped with a incurved margin at first, expanding to broadly cup shaped, occasionally disc shaped in age. Margin scalloped, with ragged toothlike tissue. **UPPER SURFACE:** Smooth, dull dark brownish black to black when dry. **UNDERSIDE:** Felty-tomentose, slightly roughened. **STIPE:** Short, indistinct. **FLESH:** Thin, somewhat brittle, with a slightly gelatinous core. **ODOR:** Indistinct. **TASTE:** Indistinct. **MICROSCOPY:** Spores 24.5–30 x 10.5–12.5 μm, elliptical, smooth.

ECOLOGY: Solitary, scattered, or in small clusters on the ground, in duff or moss, or in woody debris. Common throughout much of our range, but often overlooked. Fruiting from midwinter into spring.

EDIBILITY: Unknown.

COMMENTS: One of many small black cup fungi, *P. milleri* can be recognized by its scalloped margin, felty-tomentose underside, and long elliptical spores. *Pseudoplectania nigrella* has a slightly hairy margin (which can look scalloped), but has round spores. *Plectania melastoma* has orange granules covering the underside when young, usually restricted to the rim in age. *Pseudosarcosoma latahense* is distinctly rubbery (see comments under *Urnula padeniana*). *Plicaria endocarpoides* is a fragile, dark purplish brown to black cup that grows in burnt areas. *Donadinia nigrella* (=*Plectania nannfeldtii*) has a margarita-glass shape, with a cup perched atop a long, thin stipe; it is a common montane species that fruits near melting snowbanks in the spring.

Urnula padeniana

M. Carbone, Agnello, A. D. Parker & P. Alvarado

BLACK JELLY JUG

FRUITBODY: 5–12 cm across, generally cup shaped with a incurved margin at first, expanding to broadly cup shaped, occasionally top shaped when young. Often becoming disc shaped in age. **UPPER SURFACE:** Smooth, dull dark grayish black to black when dry, shiny black when wet. **UNDERSIDE:** Finely velvety, often wrinkled. Dark gray to black. **STIPE:** 2–8 cm long, 1–4 cm thick at apex, tapering downward, typically rather short and stout. Often ribbed or wrinkled. **FLESH:** Thick, rubbery, with a gelatinous core. Translucent gray when young, blackish in age. **ODOR:** Indistinct. **TASTE:** Indistinct. **MICROSCOPY:** Spores 23–30 x 11–13 μm, elliptical to sausage shaped, smooth, with one to three oil droplets.

ECOLOGY: Solitary, scattered, or in small clusters on buried woody debris or in duff. Uncommon, occurring from Santa Cruz County north. Fruiting from winter into spring.

EDIBILITY: Unknown.

COMMENTS: The black color and large cup- or disc-shaped fruitbodies with a rubbery texture and gelatinous flesh help distinguish this species. *Pseudosarcosoma latahense* is nearly identical, but is smaller and has a purplish black to black upper surface and a dark gray underside; microscopically, it has larger spores that lack interior oil droplets. The fruitbodies of *Bulgaria inquinans* have roughened, reddish brown undersides when young and grow in clusters on recently dead Tanoak and oak logs. *Pseudoplectania nigrella* has a hairy underside and has round spores. Also see *Plectania milleri* (p. 561) for other black cups.

Bulgaria inquinans

Pers.

BLACK JELLY DROPS

FRUITBODY: 1–4 (6) cm across, gumdrop to top shaped, often with a shallow cup at first, expanding to more convex, wavy or lobed. **UPPER SURFACE:** Smooth, black, occasionally dark olive-black or dark brown, darkening to black. **UNDERSIDE:** Roughened, dark brown to dark reddish brown at first, becoming black. **FLESH:** Gelatinous, rubbery. Somewhat translucent, dark brown to black. **ODOR:** Indistinct. **TASTE:** Indistinct. **KOH:** No reaction. **SPORE DEPOSIT:** Dark grayish black. **MICROSCOPY:** Spores 11–14 x 6–7 μm, kidney shaped, upper four spores in asci dark brown, lower four colorless, smooth, containing several small oil drops.

ECOLOGY: Often in rows, clumps, and scattered masses on Tanoak and live oak logs. Generally fruiting from bark furrows on recently fallen trees, at times carpeting the log. Fairly common throughout our range; fruiting almost any time from early fall into spring.

EDIBILITY: Inedible.

COMMENTS: This weird ascomycete can be recognized by its tough gummy-bear texture, smooth, dark olive to black fertile surface, reddish brown underside, and growth on hardwood logs, especially Tanoak. Unexpanded young specimens are more barrel shaped and mostly reddish brown with shallow reddish brown to black dimples of developing fertile surface. *Urnula padeniana* is larger and more distinctly cup shaped, and grows on well-rotted wood or duff. *Exidia glandulosa* forms brainlike or lobed masses and has colorless spores.

Peziza ammophila
Durieu & Lev.
SAND TULIPS

FRUITBODY: 2–8 cm across, at first a spherical to egg-shaped flask with a rooting stipe mostly buried in sand, with a small star-shaped opening above sand. Erupting farther out of sand and becoming more cup shaped with a large star-shaped opening, to broadly cup shaped in age. **UPPER SURFACE:** Smooth, dark brown with a tan to yellowish brown margin. **UNDERSIDE:** Pale ochraceous brown, but usually obscured by encrusted sand. **STIPE:** Varies from a fleshy rooting stipe up to 5 cm long to a small tuft of binding mycelium. **FLESH:** Thin, fragile at first, becoming somewhat tough. **ODOR:** Indistinct. **TASTE:** Indistinct. **MICROSCOPY:** Spores 14–18 x 8–10 μm, elliptical, smooth. Asci tips amyloid.

ECOLOGY: Solitary or scattered in coastal sand dunes around roots of dune grass (*Ammophila* spp.). Locally common within this habitat, mostly occurring on the Far North Coast, also reported from coastal dunes in Central California. First fruiting in late winter, more abundant in spring.

EDIBILITY: Unknown. Likely nontoxic, but it would be impossible to remove the sand.

COMMENTS: The coastal dune habitat, distinctive shape with a star-like mouth, dark brown interior, and sand-encrusted exterior are distinctive. Young specimens are often almost completely buried in the sand, with just the small opening projecting and visible. Some cup-shaped *Geopora* species also occur in sand, but have distinctly hairy exteriors. The Dutch call this fungus Zandtulpjes, meaning Sand Tulips.

SYNONYM: *Sarcosphaera ammophila* (Durieu & Lev.) Seaver.

Peziza arvernensis
Boud.
BROWN CUP

FRUITBODY: 3–9 (12) cm across, irregular cup shaped at first, soon broadly cup shaped, often flaring out in age. **UPPER SURFACE:** Smooth, light ocher-brown to light brown. **UNDERSIDE:** Finely tomentose to smooth, slightly paler than upper surface. **FLESH:** Thin, fragile. **ODOR:** Indistinct. **TASTE:** Indistinct. **MICROSCOPY:** Spores 14–20 x 8–10 μm, elliptical, ornamented with fine punctations, lacking oil droplets. Asci tips amyloid.

ECOLOGY: Often clustered or occasionally solitary on well-rotted woody debris or wood chips. Common across much of our range. Fruiting through wet season, more abundant in late winter and spring.

EDIBILITY: Unknown.

COMMENTS: This cup fungus can be recognized by its large size, pale ocher-brown color, and growth on woody debris. Microscopically, the ornamented spores help distinguish it from the smooth-spored *P. varia* group (=*P. repanda*), which are otherwise practically indistinguishable except for their slightly shorter spores (14–18 x 8–10.5 μm). *P. domiciliana* generally grows in wet basements, in bathrooms, or on wet carpets in houses. Microscopically, its finely warted spores with two small oil droplets measuring 12–15 x 6–9 μm, readily separate it from *P. varia*, which can also grow in domestic settings. *P. vesiculosa* has an incurved margin well into maturity and a distinctly roughened underside; it grows in dung-rich environments.

Peziza vesiculosa

Bull.

BLADDER CUP

FRUITBODY: 1–5 cm across, rounded to urn shaped with an incurved margin well into maturity, becoming more cup shaped when old. **UPPER SURFACE:** Smooth, tan, pale yellowish brown to light ocher-brown. **UNDERSIDE:** Roughened or scruffy, whitish to light ocher-brown. **FLESH:** Thin, fragile. **ODOR:** Indistinct. **TASTE:** Indistinct. **MICROSCOPY:** Spores 20–25 x 11–15 µm, elliptical, smooth. Asci tips amyloid.

ECOLOGY: Solitary, scattered, or in clusters on dung mixed with old straw or sawdust, or in compost piles mixed with dung, especially common around horse corrals. Also occurs in wood chips, especially in fertilized areas. Common in these habitats throughout our range. Mostly fruiting in spring, but can occur any time of year given sufficient moisture.

EDIBILITY: Unknown.

COMMENTS: The growth in dung-rich habitats, incurved margin into maturity, roughened underside, and relatively small size individual fruitbodies (sometimes forming large swarms) are distinctive. *P. arvernensis* and others often grow on woody debris or wood chips, are larger, and lack the distinctly incurved margin. *Tarzetta cupularis* has an inrolled, scalloped margin, but grows in duff or soil in forested habitats.

Peziza praetervisa

Bres.

VIOLET CUP

FRUITBODY: 1–4 (6) cm across, broadly cup shaped at first, becoming more irregular, to almost disc shaped in age. **UPPER SURFACE:** Smooth, light to dark purple, to purple-brown in age, occasionally pale violet when young. **UNDERSIDE:** Finely scruffy to smooth, lilac-gray to pale violet. **FLESH:** Thin, fragile. **ODOR:** Indistinct. **TASTE:** Indistinct. **MICROSCOPY:** Spores 11–15 x 6–8 µm, elliptical, finely warted, contents with two oil droplets. Asci tips amyloid.

ECOLOGY: Solitary, scattered, or in clusters in campfire pits or on burnt ground or burnt woody debris, occasionally on unburnt ground or in wood chips. Widespread but generally uncommon on coast. Fruiting any time of year, but most prolific the first spring after a forest fire.

EDIBILITY: Unknown.

COMMENTS: The handful of violet to purple *Peziza* species can be recognized by their pale violet, purple to purple-brown colors, relatively small size, and preference for burnt ground. *P. violacea* has been used as a catch-all name for the group, but microscopic features such as spore size and ornamentation help distinguish the different species. *P. praetervisa* (our most common species) has finely warted spores with two oil droplets. *P. sublilacina* has smooth, larger spores (13–17 x 7–10 µm) that lack oil droplets and prefers burnt ground. *P. gerardii* has much longer spores (25–35 x 9–12 µm), with low longitudinal ridges; it prefers disturbed soil (not necessarily burnt).

Otidea smithii

Kanouse

PURPLE-BROWN RABBIT'S EAR

FRUITBODY: 3–11 cm across, 5–11 cm tall, variable shaped, often a lopsided cup, shorter and slit or folded to one side (rabbit's ear shape). Margin folded in when young, flaring in age. **UPPER SURFACE:** Dark purplish brown to vinaceous brown; fading slightly purplish brown to slightly more orange-brown. Surface smooth to wrinkled, bare. **UNDERSIDE:** Concolorous, but slightly paler. **STIPE:** 1–2 cm long, 0.5–1 cm thick, stout, pinched. Pale whitish, dingy cream to light tan. **FLESH:** Thin, rubbery to somewhat fragile. **ODOR:** Indistinct. **TASTE:** Indistinct. **MICROSCOPY:** Spores (10) 12–15.5 x 6–8 μm, narrowly elliptical, smooth, with two oil droplets. Asci inamyloid.

ECOLOGY: Solitary, scattered, or in small clustered in humus or moss, occasionally in grassy areas with thick duff. Locally common on the Far North Coast, generally under conifers or Bigleaf Maple. Uncommon and more often found under live oak in southern part of our range. Fruiting in fall and winter.

EDIBILITY: Unknown.

COMMENTS: The relatively large, variably rabbit's-ear-shaped fruitbodies with dark purple-brown colors are distinctive. *O. alutacea* and *O. microspora* are generally smaller and more brown to tan in color; spore size is the best way to distinguish them: 9–11 μm long in *O. microspora*, 12–17 μm long in *O. alutacea*. *Peziza* are generally more broadly cup shaped and have amyloid asci.

Otidea tuomikoskii

Harmaja

TALL RABBIT'S EAR

FRUITBODY: 2–5 cm across, 5–15 cm tall, rabbit's ear shaped, narrowly folded when young, flaring in age. **UPPER SURFACE:** Smooth, brown at apex, otherwise light brown, fading to beige in age. **UNDERSIDE:** Concolorous, but slightly duller, slightly roughened when young. **STIPE:** 1–2 cm long, 0.3–1 cm thick, stout, pinched. Pale brown to beige. **FLESH:** Thin, rubbery to somewhat fragile. **ODOR:** Indistinct. **TASTE:** Indistinct. **MICROSCOPY:** Spores 8.5–11 x 5–6 μm, elliptical, smooth. Asci inamyloid.

ECOLOGY: Solitary, scattered, or in small clusters in humus or moss under conifers, especially Sitka Spruce. Locally common on the North Coast. Fruiting in fall and winter.

EDIBILITY: Unknown.

COMMENTS: The tall, elongated rabbit's-ear shape, light brown color (with a darker top when young), and slightly roughened underside are distinctive. *O. leporina* is generally shorter and has more yellow-brown colors. *O. onotica* (sensu CA) is very similar to *O. leporina*, but is often slightly more ocher and has a pinkish cast to the interior. Also compare with the more cup-shaped *Peziza*, which have amyloid asci.

Sowerbyella rhenana

(Fuckel) J. Moravec

STALKED ORANGE PEEL

FRUITBODY: 1–3 cm across, cup shaped to broadly cup shaped with a distinct stipe. **UPPER SURFACE:** Bright orange to yellow-orange, smooth. **UNDERSIDE:** Covered with downy whitish hairs at first, soon bare and pale orange to pale yellow-orange. **STIPE:** 0.5–4 cm long, 0.2–0.5 cm thick, equal, cylindrical, grooved, or wavy. Whitish to pale yellowish orange. **FLESH:** Thin, very fragile, orange in cup, paler in stipe. **ODOR:** Indistinct. **TASTE:** Indistinct. **MICROSCOPY:** Spores 18–26 x 9–12 µm, elliptical, ornamented with a coarse reticulum.

ECOLOGY: In tufts or clusters of multiple cups, often with fused stipes. Uncommon, found in mixed Tanoak and Douglas-fir forests from Santa Cruz Mountains northward. Fruiting in fall on the Far North Coast, through winter southward.

EDIBILITY: Unknown, likely nontoxic.

COMMENTS: The small orange cups with distinct stipes that grow in troops and clusters make this an easy fungus to identify. *Aleuria aurantia* is similar in color, but is generally larger, has shallow, irregularly shaped cups, and lacks a stipe.

SYNONYM: *Aleuria rhenana* Fuckel.

Aleuria aurantia

(Pers.) Fuckel

ORANGE PEEL

FRUITBODY: 1–10 cm across, cup shaped to broadly cup shaped, edges becoming wavy or irregular in age. **UPPER SURFACE:** Bright orange to yellow-orange, smooth. **UNDERSIDE:** Pale orange to whitish near base, finely powdery or downy. **STIPE:** Absent. **FLESH:** Thin, very fragile, orange. **ODOR:** Indistinct. **TASTE:** Indistinct. **MICROSCOPY:** Spores 17–24 x 9–11 µm, elliptical, containing two oil droplets, ornamented with a coarse reticulum and a projecting knob at each end.

ECOLOGY: Solitary, scattered, or clustered, often in troops and masses on bare or disturbed soil or thin duff. Especially common after logging on skid roads, on trails, or along roadsides. Occurring throughout our range. First fruiting in fall, often more abundant in late winter or early spring.

EDIBILITY: Edible and colorful, but not especially flavorful.

COMMENTS: The shallow, bright orange cup with a paler underside, lack of a distinct stipe, and growth in disturbed soil make this species practically unmistakable. *Caloscypha fulgens* has orange to yellowish cups, but stains greenish blue and fruits in spring; it is rare in our range, restricted to the Far North Coast (it is very common in the mountains). *Sowerbyella rhenana* has smaller, yellower cups with distinct stipes.

Sarcoscypha coccinea

(Jacq.) Lambotte

SCARLET CUP FUNGUS

FRUITBODY: 2–5 cm across, cup shaped with an incurved margin at first, becoming broadly cup shaped, edges wavy or irregular. **UPPER SURFACE:** Bright scarlet red, smooth. **UNDERSIDE:** Whitish, pinkish to pale red. Finely hairy. **STIPE:** Short, whitish, central, tapering, or frequently absent altogether. **FLESH:** Thin, fragile, red. **ODOR:** Indistinct. **TASTE:** Indistinct. **MICROSCOPY:** Spores 25–40 x (9) 12–15 µm, elliptical to cylindrical, containing multiple small oil droplets near ends.

ECOLOGY: Solitary, scattered, or in small clusters on woody debris and branches and twigs of hardwoods, especially live oak and California Bay Laurel. Uncommon from Santa Cruz to Mendocino Counties, rarer elsewhere. Fruiting from midwinter into spring.

EDIBILITY: Edible.

COMMENTS: This cup fungus brightens up the dark humus of the forest floor with its vivid scarlet red color. No other cup fungus in our area has such a bright red upper surface and pale underside. *Scutellinia* are much smaller and have dark "eyelashes" around their edges.

Scutellinia scutellata group

EYELASH CUP FUNGUS

FRUITBODY: 0.2–1.5 cm across, shallowly cup shaped to disc shaped, lower surface and margin distinctly hairy. **UPPER SURFACE:** Reddish orange, orange to yellow-orange, smooth. **UNDERSIDE:** Concolorous or slightly paler, covered with dark stiff, short hairs. Margin with longer, eyelashlike hairs. **FLESH:** Very thin and fragile. **ODOR:** Indistinct. **TASTE:** Indistinct. **MICROSCOPY:** Spores typically 15–20 x 12–16 µm (with lot of variation in size), elliptical, very finely roughened, but often appearing smooth.

ECOLOGY: Solitary, scattered, or in small troops on decaying logs, woody debris, soil, or wet mossy areas, such as stream banks. The group collectively is fairly common throughout our range. Fruiting from late winter into early summer.

EDIBILITY: Unknown.

COMMENTS: A large and confusing group of species, *Scutellinia* are recognizable by their orange to red disc-shaped fruitbodies with dark "eyelashes" around the margin and the growth on woody debris (occasionally soil). *S. scutellata* has been the catchall name for an untold number of Californian species. Identification to species is difficult (or impossible), since no real taxonomic work has been done on Californian specimens. *Cheilymenia* species are generally more orange to yellow in color, have sparser hairs, and grow on dung; see *C. granulata* (p. 568) for more details.

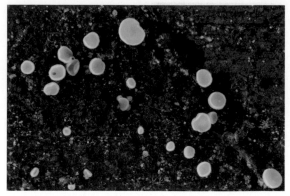

Cheilymenia granulata

(Bull.) J. Moravec

ORANGE DUNG DISC

FRUITBODY: 0.1–0.3 cm across, disc shaped to very shallowly cup shaped. **UPPER SURFACE:** Orange to orangish red, margin fringed. **UNDERSIDE:** Scruffy to finely granular, orange. **STIPE:** Absent. **FLESH:** Very thin, fragile, orange. **ODOR:** Indistinct. **TASTE:** Indistinct. **MICROSCOPY:** Spores 15–17 x 6.5–7.5 µm, elliptical, smooth, lacking oil droplets.

ECOLOGY: Scattered or in small masses on dung, most often that of cows or horses, but also on deer, elk, and bear droppings. Generally fruiting in late winter or spring, but can occur at any time if substrate is suitably moist.

EDIBILITY: Unknown.

COMMENTS: One of many small orange disc and cup fungi on dung, this species can be recognized by its orange color, scruffy underside, and lack of hairs around the margin. *C. theleboloides* is paler orangish yellow to yellow and has inconspicuous hairs at the margin. *C. stercorea* and *C. fimicola* are both orange, but have eyelash-like hairs on the margin (generally darker on the latter). If the substrate is overlooked, both could be mistaken for ground- or wood-dwelling *Scutellinia*. A number of *Lasiobolus* species have rather long, upward-pointing hairs and tiny fruitbodies (under 0.1 cm). *Ascobolus* species are a common group of cup- to disc-shaped dung dwellers that generally are greenish to yellowish when young, most maturing to brown. *Pseudombrophila* are small brown to yellow-brown cups. All of these (and many others) occur on dung, but require microscopic examination and a dedicated taxonomist to differentiate. Other similar orange disc and cup fungi include *Anthracobia macrocystis* and *A. melaloma*, which have orange to ocher-orange surfaces and brownish spots and tufts on the undersides. They grow on burnt ground following forest fires or in campfire pits. *Pyronema omphalodes* forms reddish orange masses of disc- to cushion-shaped fruitbodies, also in burnt areas. *Pithya cupressina* is a small orangish yellow disc with a tiny tapered stipe that grows on small twigs and needles of conifers (especially redwood); it has globose spores measuring 9–12 µm long. *Pithya vulgaris* is slightly larger and brighter orange, and has larger spores (12–14 µm).

Bisporella sulfurina

(Quél.) S. E. Carp.

LEMON DROPS

FRUITBODY: 0.1–0.3 cm across, shallow cup shaped to disc shaped, typically flattening out in age. **UPPER SURFACE:** Smooth, bright yellow to sulfur-yellow with a brighter or paler margin. **UNDERSIDE:** Powdery, very finely downy to smooth. Pale yellow. **STIPE:** Absent or very short. **FLESH:** Very thin and fragile. **ODOR:** Indistinct. **TASTE:** Indistinct. **MICROSCOPY:** Spores 8–11 x 1.5–2.5 µm, cylindrical, smooth, often with oil droplets.

ECOLOGY: Often in small masses of many fruitbodies or in scattered troops on decaying logs, branches, or woody debris, almost always with small, round, black fruitbodies of a Pyrenomycete present. Generally found on oak and Tanoak, but will occur on many different hardwoods. Common throughout our range. Fruiting through wet season.

EDIBILITY: Unknown.

COMMENTS: Lemon Drops are identified by their tiny, yellow, disc-shaped fruitbodies growing on wood near Pyrenomycetes (Flask Fungi). The latter are mostly found as tiny black bumps on logs; they are a highly diverse and easily overlooked group of fungi. The catchall name for any little yellow cup or disc matching the above description has been *B. citrina*, a very similar, widespread, and common species differentiated by its larger spores (9–14 x 3–5 µm) and growth on wood without a nearby Pyrenomycete. There are many other species of yellow disc fungi. Microscopic examination and technical literature are necessary to have even a hope of identification.

Geopyxis vulcanalis

(Peck) Sacc.

FAIRY FARTS

FRUITBODY: 0.3–1.5 (2) cm across, rounded cup shaped at first, soon urn shaped, to broadly cup shaped when old, margin scalloped to ragged. **UPPER SURFACE:** Smooth, orangish buff to pale orange, margin slightly paler. **UNDERSIDE:** With a powdery bloom to smooth, concolorous or slightly paler. **STIPE:** Short, central. **FLESH:** Thin, very fragile. **ODOR:** Distinctly of sulfur gas. **TASTE:** Unpleasant. **MICROSCOPY:** Spores 14–18 x 8–10 µm, elliptical with narrow ends.

ECOLOGY: Scattered, in small troops, or occasionally solitary on bare soil or in moss or duff. Common on the North Coast. Fruiting in the spring.

EDIBILITY: Unknown.

COMMENTS: This small cup can be recognized by its pale orangish buff colors, urnlike shape, and sulfurous odor (especially noticeable when crushed). *G. carbonaria* is darker, reddish tan to pale ocher-brown, and grows in burnt areas such as forest fire burns and campfire pits. *Tarzetta cupularis* is slightly larger and more goblet shaped, and has a roughened underside and a scalloped margin, which is cobwebby when young. *Humaria hemisphaerica* is similar, but has a brownish underside with distinct dark hairs and a smooth, silvery gray interior.

Chlorociboria aeruginascens

(Nyl.) Kanouse ex C. S. Ramamurthi, Korf & L. R. Batra

BLUE-GREEN WOOD CUP

FRUITBODY: 0.2–1 cm across, shallow cup shaped to irregularly disc shaped, margin often wavy. **UPPER SURFACE:** Smooth, bluish white when very young, soon turquoise to blue-green, often blotched and mottled with dark olive-blue. **UNDERSIDE:** Finely tomentose to bald, often wrinkled in age. Paler than surface. **STIPE:** Short, off-centered to central. **FLESH:** Thin, fragile. **ODOR:** Indistinct. **TASTE:** Indistinct. **MICROSCOPY:** Spores 5–7 (10) x (0.8) 1–2.5 µm, spindle shaped to elliptical with narrow ends.

ECOLOGY: Scattered or in small troops of single or clustered fruitbodies on decaying hardwoods; wood often extensively stained teal-blue or greenish. Uncommon throughout our range. Fruiting from winter into spring.

EDIBILITY: Unknown.

COMMENTS: The blue-green color, small size, and tendency for the mycelium of this fungus to stain its wood substrate blue-green are strong identifying features for the genus *Chlorociboria*. The frequently off-center stipe, wavy-edged fruitbody, and small spores help distinguish *C. aeruginascens* from the similar (and confusingly similarly named) *C. aeruginosa*. Fruitbodies of the latter species generally have a central stipe, more evenly cup-shaped fruitbodies, and larger spores (8–15 x 2–4 µm). The mycelium-stained wood substrate is encountered at least as often (perhaps more so) than the actual fruitbodies, suggesting that the fruiting periods for these species may be fairly short.

Annulohypoxylon thouarsianum

(Lev.) Y. M. Ju, J. D. Rogers & H. M. Hsieh

CARBON BALLS

FRUITBODY: 1–5 cm across, rounded to ovoid, occasionally lobed. Dark brownish black to grayish black when young, soon black. Surface dry, bumpy, covered with pimplelike perithecia. **INTERIOR:** Hard, shiny, charcoal-like. Dark gray to black, with vague zonations. **ODOR:** Indistinct. **TASTE:** Indistinct. **KOH:** Deep forest green (appearing black). **SPORE DEPOSIT:** Black. **MICROSCOPY:** Spores 14–24 x 4–5.5 µm, elliptical, with one side slightly flattened.

ECOLOGY: Solitary or in swarms on recently dead oak and Tanoak, especially those killed by Sudden Oak Death. Very common throughout our range. Fruiting in fall, fruitbodies persisting into the following years.

EDIBILITY: Inedible.

COMMENTS: This very common species can produce hundreds of fruitbodies on recently dead Tanoak and live oak trunks. If found when very young, the fruitbodies can be covered with an olive-gray powder made up of asexual spores. The life cycle of these fungi is quite fascinating: They invade young trees and spend most of their life as an endophyte (living within the plant cells), causing no apparent harm to their host. When the tree dies (from old age or Sudden Oak Death), these fungi fruit quickly and produce spores ready to disperse to a new host.

Sudden Oak Death (SOD)

Sudden Oak Death is a disease affecting Live Oaks and especially Tanoak throughout an area nearly exactly overlapping the Redwood Coast. It is caused by *Phytophthora ramorum*, a water mold (oomycete) organism, a pathogen that appears to have been introduced to California from Eurasia multiple times in the past few decades. It was first reported in 1995 and has now killed thousands of oaks and Tanoaks in California. Symptoms of SOD include oozing cankers and rapid leaf death. It is especially deadly to Tanoak, and the impacts of this epidemic on fungi associated with tanoaks is twofold: For wood-decomposers (such as *Stereum* and *Pluteus*) it is a boon, since it generates lots of dead logs and branches; however, the ectomycorrhizal associates of Tanoak are threatened by this loss of their hosts, and it may be that various species of *Cortinarius*, *Tricholoma*, *Ramaria*, and others have been extirpated from areas where Tanoaks have died out.

Acknowledgments

This book might never have come about (and if it had, it would've been a much lesser work) without the distilled knowledge, material and emotional support, advice (whether sage or incendiary), and companionship provided by a whole community of people.

First and foremost, our deepest gratitude to Else Vellinga, who was always there to answer any question we had, read over large portions of the manuscript, and helped make sure that we screwed up as little as possible. Profound thanks to Phil Carpenter, David Arora, Adam Ryszka, Joann and Eric Olson, Henry and Marje Young, Steve Trudell, Jonathan Frank, the late Bob Sellers, Joshua Birkebak, Penelope Gillette, Dr. David Largent, Rick Kerrigan, and Alfredo Justo. Your contributions to this book are immeasurable.

Special thanks to the Fungus Federation of Santa Cruz for funding a major sequencing project that informed this book in many ways, and for providing scholarships on a number of occasions that allowed us to explore California's mycoflora.

To Alissa Allen, Elizabeth Andrews, Chris Lay, and Caitlin Hannah: Thank you for your advice, companionship, and exceptional support even during the tough stretches.

To Dimitar Bojantchev, special thanks for your advice, expertise, and singular perspective, and for bringing California's *Cortinarius* into the modern era. Special thanks to Nathan Wilson, Jason Hollinger, and Ken-ichi Ueda for building an online community of amateur enthusiasts and helping to create new traditions of engaging with biodiversity.

To Erin Page Blanchard, Tom Bruns, Mark Benson, Darvin DeShazer, Reba and Milton Tam, Sava Krstic, Trent Pearce, Douglas Smith, Ron Pastorino, Nhu Nguyen, Adam Searcy, Thea Chesney, Pablo Alvarado, Mike Wood, Fred Stevens, Justin Pierce, Carl Atilano, Pat McKee, Andrew Smith, Eric Maklan, Daisy Austin, Danny Miller, Ian Gibson, Melina Kozanitas, Todd Osmundson, Daniel Nicholson, Ryane Snow, Wendy So, Brian Perry, Ron Lawrence, Jerry Cooper, Mark Carnessale and Melanie Onofrio, Dennis DesJardin, Tim Baroni, Brandon Matheny, Lee Summers, Barbara Banfield, and Tim Hyland, we say:

Thanks for including us in your projects, facilitating ours, reviewing our text, describing California's mushrooms, drawing our attention to new and interesting species (some right under our noses!), informing our understanding of status and distribution, encouraging us to see mushrooms in new ways, teaching us patiently, sharing your secret collecting grounds, and being generally life affirming.

To our editors Kate Bolen and Kelly Snowden, our designers Tatiana Pavlova and Angelina Cheney, and the rest of our team at Ten Speed Press; we give profound thanks for helping to make our vision a reality! You are outstanding!

And last but not least, to our families. Noah: Thanks to Kathy Morris and Larry Siegel for your support, guidance and love over the years (and not minding that the dining room table was always covered with mushrooms). Christian: Thanks to Mamita and Pop for your warmth, unwavering love, and unflinching (if bemused) acceptance of my bizarre hobby. To Uncle Mike and Bren, special thanks for your companionship, and for getting me interested in this world to begin with. I dedicate this book for my part to Aunt Barb, who gave me my first copy of *Mushrooms Demystified*. We miss you.

We hope to see you all in the woods!

About the Authors

NOAH SIEGEL's field mycology skills are extensive – he has spent two decades seeking, photographing, identifying, and furthering his knowledge about all aspects of macrofungi. He has hunted for mushrooms throughout the United States and Canada, as well as on multiple expeditions to New Zealand and Australia.

He is one of the premier mushroom photographers in the nation, having won numerous awards from the North American Mycological Association (NAMA) photography contest. His technique and attention to detail are unrivaled, arising from a philosophy of maximizing utility for identification purposes while maintaining a high degree of aesthetic appeal.

His photographs have appeared on the covers and have been featured in articles of multiple issues of *FUNGI Magazine* and *Mushroom the Journal*, the primary mushroom enthusiast magazines in the United States, numerous mushroom books, as well as NAMA and other club publications.

He is past president of the Monadnock Mushroomers Unlimited, (MMU) a mushroom club based out of Keene, NH, and is an active member of the Humboldt Bay Mycological Society and the Fungus Federation of Santa Cruz.

Noah travels and lectures extensively across America, following the mushrooms.

CHRISTIAN SCHWARZ is a naturalist interested in the diversity of living organisms in general, but the seemingly endless forms (whether grotesque, bizarre, or sublime) of fungi in particular stoke his curiosity. He spends most of his time teaching about natural history, collecting and photographing mushrooms, assembling an exhaustive mycoflora for Santa Cruz County (www.scmycoflora.org), and exploring wilderness around the world. He is particularly interested in the role of citizen scientists in the future of ecological and taxonomic research.

Future Directions

This book is almost entirely focused on alpha taxonomy: the discovery, naming, and identification of species. Mastering fungal taxonomy involves enough work to fill a few lifetimes. But there are many more broadly important questions and fields of study to which the application of taxonomic knowledge is vital. Below are some such fields.

Mycofloristics: Assembly of complete lists of fungal species in a given area requires strong taxonomic knowledge. A comprehensive mycoflora greatly facilitates the building of range maps and the gathering of phenological data (timing and patterns of fruiting). Together, these kinds of data allow us to monitor the health of ecosystems and their changing patterns of species composition over time.

Conservation: Well-documented mycoflora allow us to efficiently and credibly identify and prioritize regions for conservation. This is especially urgent, since practically no fungi are recognized for state or federal protection in the United States, even though many occur in at-risk habitats.

Climate change: Gathering baseline data on biogeography and phenology is extremely important in assessing the extent and character of climate change. Not only are fungal communities highly responsive to changing weather regimes, but the effect of changes in fungal community composition extends to impacts on mycorrhizal hosts. The potential for feedback effects is high, and much attention should be devoted to these types of ecological questions.

Biogeography: Appropriately splitting species that have long been considered conspecific with European and/or eastern North American species allows us to build more accurate models of biogeography and identify areas of high endemicity (like California!).

Systematics: Accurate family trees that reflect the paths of evolution are only possible when taxon sampling is thorough, which is greatly facilitated by extensive alpha-taxonomic and biogeographic work, and particularly by well-documented herbarium specimens.

Sequence database quality: The current state of databases for DNA sequences is chaotic and uneven. Only by the joined efforts of skilled amateur and professional taxonomists, sequencing technicians, and database architects will the quality of these databases increase. Copious photography, consistent inclusion of metadata, and data sharing will be among the most important methods of creating better databases.

Evolutionary and ecological synthesis: Well-resolved phylogenies are essential for inferring evolutionary outcomes of past ecological events and as context for future findings. What events in geologic history led to rapid diversification of fungi? What role did fungi play in global recovery from extinction events? One interesting current example is the exchange of mycorrhizal fungi between native and exotic/urban environments. For example, the mycorrhizal ecology of invasive *Amanita phalloides* and its hosts in the Americas should be closely monitored, as novel ecological interactions make their mark. Likewise, the range of native mycorrhizal fungi entering into symbiotic partnership with exotic hosts (especially *Eucalyptus* spp., *Arbutus unedo*, and *Acacia* spp.) is fertile ground for studies of ecology and evolution in action. These fungi and their tree partners are in the early days of their entwined evolutionary futures—the genomic and biochemical phenomena involved ought to be of interest to many!

In short, the world of fungi is ripe for many lines of curious questioning. As with any natural history investigation, learning to identify the actors involved is the essential first step. Basic taxonomy helps us do just that. Along with extensive time spent reading, discussing, and working in the lab or online, we especially encourage you to spend time in the woods, to become familiar with mushrooms by close and repeated scrutiny, and by creative engagement (using all your senses). In the words of natural history luminary Ken Norris:

"Always remember, the organism is the authority!"

Glossary

acidic: Sour tasting, like lemon juice or an unripe plum.

acrid: Burning-hot taste, but not the same sensation as that of capsaicin (hot peppers).

amatoxin: Small protein molecules produced by some mushrooms in the genera *Amanita, Conocybe, Galerina, Lepiota,* and others. Extremely toxic, inhibiting DNA replication. The primary compound responsible for mushroom-poisoning fatalities.

amyloid: Turning bluish to bluish black in Melzer's reagent, usually only visible microscopically.

anastomosis (adj. **anastomosing**): A pattern of veins joining, forking, and/or criss-crossing to form a complex webbed appearance; usually in reference to gills or veined fertile surfaces. Similar to *reticulate*, but more complex and irregular.

annulus: Membranous partial veil; can be thin or thick, hanging, flaring, or collarlike.

apical: Referring to a feature positioned at the top of another structure.

apiculus: Projecting appendage of a spore where it was attached to a sterigma of the basidium; also known as the hilar appendage. Very pronounced and snoutlike on some spores, absent in others.

appressed: Flattened against the surface; referring to fibrils or scales.

appressed-fibrillose: Fibrils, scales, or hairs that are pressed flat against the surface of a mushroom.

appressed-tomentose: Downy hairs or fibrils that are flattened, matted down, or felty.

areolate: Cracking pattern like that of dry mud.

ascus (pl. **asci**): Microscopic, cylindrical cell of an ascomycete, in which spores are produced (usually in groups of eight, more rarely in other multiples of four). For fertile cell of basidiomycetes, see *basidium*.

basal mycelium: Loose, usually pale (sometimes brightly colored), fuzzy coating of hyphae found at the stipe base of many mushrooms. See also *rhizomorphs*.

basidiomycete: A phylum of fungi that produce spores on basidia. Consisting of most of our fleshy fungi.

basidiospore: Dispersal unit of a basidiomycete fungus containing genetic material produced by sexual recombination.

basidium (pl. **basidia**): The microscopic, usually club-shaped cell on which spores are formed in basidiomycetes. For the fertile cell of ascomycetes, see *ascus*.

cap cuticle: Uppermost layer of tissue on the surface of the cap of a fruitbody. Often detectable by peeling the cap; varies from thin to thick, membranous to cartilaginous, and from gelatinous to dry, or viscid. The microscopic structure of this cuticle is the *pileipellis*, which can be an important identification feature.

capitate: Describing microscopic cells with a distinct "head," usually rounded.

caulocystidia: Sterile cells found on the stipe surface of some species, usually larger and with distinctive morphology. Important in identification of many genera, especially those with smaller fruitbodies. Presence of such cells often macroscopically visible as a fine powder or parallel vertical lines near the top of the stipe.

chlamydospore: Thick-walled, often darkly pigmented asexual spores with ability to resist heat and drought for long periods.

cheilocystidia: Sterile (non-spore-bearing) cells found at the edges of gills. The presence or absence of such cells, as well as their shape and characteristics, can be very important in identification.

clamp connection (clamp): Small structure spanning the septum between two hyphal cells, often described as looking like a kneecap. The presence or absence of such structures is often important in determining genus or species. Sometimes found at the base of basidia.

columella: Branched or columnar sterile structure found in the interior of some sequestrate and secotioid species; represents vestigial remnant of a stipe.

concolorous: Literally "colored the same," usually describing the color of a mushroom's stipe with reference to its cap.

conidial pegs/branches: Rare structure of stipe in which asexual spores (conidia) are produced on side branches emerging from the stipe; found only in *Dendrocollybia*.

cortina: Silky-fibrillose or cobwebby partial veil, especially prominent in the genus *Cortinarius*, but also found in many others.

cutis: Microscopic cap surface structure in which hyphae lie more or less flat. Opposite of *trichoderm*. See *pileipellis*.

cystidia: Microscopic structures; sterile cells that come in many shapes and sizes, and can be found on many different surfaces of a fruitbody—most notably the fertile tissue, cap, and stipe. Can be extremely useful in identification. See *caulocystidia, cheilocystidia, pleurocystidia*.

decurrent: Pattern of attachment between the fertile surface and stipe of a fruitbody in which the gills, veins, or tubes run down the top of the stipe (for example in *Chroogomphus*), or in extreme cases, runs down most of the length of the stipe (*Gomphus*).

disc: Central part of a mushroom cap.

dextrinoid: Reddish to red-brown color reaction of spores or tissues to Melzer's reagent. Usually only visible with a microscope.

ellipsoid/elliptical: Rounded, but not circular, with two major axes of symmetry.

farinaceous: Mealy odor or like fresh flour, sometimes described as vaguely cucumbery, but mustier. An important and frequently used descriptor for the odor of many mushrooms. There are frequent disagreements about what this odor really smells like.

fibril: Hairlike structure formed by single long hyphae or by bundles or chains of hyphae.

fibrillose: Covered in or composed of thin fibers.

floccose, flocculent: Having a surface texture of loosely clumped, deeply matted hairs or fluffy-powdery hyphae, the latter especailly found in many *Amanita* of Section *Lepidella*.

fruitbody: Sexual reproductive structure often called a mushroom. Fruitbodies come in many forms, from tiny black pimples on logs, to flattened crusts, to giant masses.

fusiform: Shaped like a spindle, with a swollen middle and tapered at each end.

germ pore: Small hole in the cell wall of a spore, through which the first hypha emerges upon germination. Not present in spores of all species.

gleba: Fertile tissue that appears as a continuous or weakly differentiated mass of spores (as in puffballs and stinkhorns).

globose: Globe shaped, round.

heterodiametric: Pertaining to the angular spores of the family Entolomataceae; length of spore distinctly greater than width.

hygrophanous: Tissues (usually the caps of some mushrooms) that change color as they dry. Generally darker and slightly translucent when wet, paler and more opaque when dry.

hymenium: Fertile surface; spore-producing tissue of a fungus.

hyphae: Cylindrical or threadlike cells that are the basic units of fungal anatomy. Many millions are woven together into fruitbodies.

hypogeous: Fruiting below surface of soil. Some species are only partially hypogeous (erupting through surface of the soil, also called *erumpent*).

inamyloid: Not reacting to Melzer's reagent, that is, neither dextrinoid nor amyloid.

isodiametric: Pertaining to the angular spores of the family Entolomataceae; length and width are equal or nearly so.

ixocutis: Microscopic cells of the cap surface horizontally arranged and embedded in a layer of slime.

KOH: Potassium hydroxide; a basic, high pH chemical used to produce distinctive and informative color changes on the surfaces of the fruitbodies of some mushrooms. Particularly useful for the genus *Cortinarius*. See p. 166.

latex: Milky or juicy secretion, mostly occurring in the genus *Lactarius*.

marginate: Referring to gills with constrastingly colored edges (margins); usually due to the presence of an abundant strip of cheilocystidia with internal pigments.

Melzer's reagent: Iodine-based chemical mixture used to determine amyloid, dextrinoid, or inamyloid reactions of hyphae and spore ornamentation.

micron: Abbreviated μm. One-thousandth of a millimeter (a millionth of a meter), used in reference to measurements of spores, cystidia, and other microscopic structures.

muscarine: Compound responsible for the toxicity of some *Clitocybe* and *Inocybe* species. Mimics activity of acetylcholine, a neurotransmitter.

muscimol: Compound responsible for the psychotropic effects of *Amanita muscaria* and its relatives. Generated by metabolism of ibotenic acid in the human body.

mycelium: Vegetative body of most fungi; a branching network of hyphae. See p. 3–4.

mycorrhizae: Symbiosis between a fungus and a plant, in which hyphae of mycelium and root tips become tightly entwined in order to accomplish efficient exchange of water and nutrients. See p. 3 (Ecology).

omphalinoid: Stature type of some mushrooms; refers to small fruitbodies with umbilicate caps, decurrent gills, thin stipes, and no veil tissue.

ovoid: Egg shaped; rounded but tapered to one end, with only one major axis of symmetry.

partial veil: Tissue protecting the gills or pores of young fruitbodies; ranges in structure and texture from membranous to slimy or cobwebby (see *annulus* and *cortina*).

peridium: Outer membrane of a puffball and trufflelike fruitbodies.

perithecia: Pimplelike structures on the fertile surfaces of some ascomycete fertile surfaces; each one contains a small flasklike cavity in which the spores are produced.

pileipellis: Layer of hyphae covering the cap of a fruitbody; microscopic structures that correlate with macroscopic texture of cap surface. There are many distinct kinds of pileipellis with their own names (see *cutis, trichoderm*). Details of structure can be extremely important in identification.

pileocystidia: Microscopic cells occurring on the cap, sometimes large enough to barely see with the naked eye.

pleurocystidia: Sterile cells found on the sides of gills that have a distinctive morphology, usually larger and clearly distinct from the basidiole "spacer cells." Presence or absence and morphology important as microscopic identification features. See the morphology diagram (p. 2) for terminology used to describe various common shapes.

poroid: Referring to a tissue composed of fused tubes with porelike openings. (sometimes small and round, sometimes large and angular or irregular).

pruinose: Covered in a fine powdery-looking "bloom" (usually pale), as if lightly dusted with flour.

pubescent: Covered in short, matted, and interwoven fibrils, resulting in a fuzzy appearance.

punctate: With a dotted pattern or appearance, usually due to clumps of contrastingly pale or pigmented cystidia.

resupinate: Crustlike fruitbody, flattened against the substrate.

reticulum (adj. **reticulate**): Raised lines or veins forming a fishnetlike pattern.

rhizomorphs: Stringlike bundles of hyphae at the base of the stipe, often slightly elastic or tough; easily visible. See also *basal mycelium*.

saprobe (adj. **saprobic**): Nutritional mode involving the decay of dead organic matter.

scaber: Upright clump of tissue, often pointed, and sometimes distinctly colored and contrasting with the background; found on the surface of the stipe of *Leccinum* boletes.

sclerotium (pl. **sclerotia**): Dense mass of hyphae forming a dormant "resting stage," usually a tuberlike structure from which mushrooms fruit during favorable weather. There is a tremendous range in size, shape, color, and texture of sclerotium. Some are tiny, pale, and fragile; others are dark, hard, and knobbed. Can occur singly or in clusters, either at the base of the stipe (as in *Collybia*), or sometimes buried in substrate and thus easily overlooked (*Dendrocollybia*, some *Polyporus*).

scrobiculate (pl. n. **scrobiculations**): Round, sunken markings on the surface of the stipe of a fruitbody; varying from small to large and from slightly depressed to deep potholes. Most commonly found on fruitbodies in *Lactarius*.

scurfy: Surface with many fine, small, often slightly uplifted scales or "scurfs"; usually in reference to the surface texture of the cap or stipe of a mushroom.

secotioid: Used to describe species with fruitbodies that are intermediate between fully exposed fertile surface (like most gilled and pored mushrooms) and sequestrate (like a puffball or hypogeous truffle).

septum (adj. **septate**): Cell wall separating two adjacent hyphae. A microscopic structure with a cross wall interrupting its length is said to be septate.

sinuate: Gill attachment characterized by a distinct notch or "s-curve" separating near the junction of gill and stipe.

spermatic: Common descriptor for the odor of many *Inocybe* (and some species in other genera); a mix of green corn, buttermilk, and bleach.

spore: Product of sexual recombination, the dispersal unit of most fungi. Similar to a seed of a plant, but with significant biological differences. See p. 4.

sterigma (pl. **sterigmata**): Long, thin projection from a basidium that supports the basidiospore; usually four per basidium (in some groups, each basidium has only one or two sterigmata).

stipe: Structure upon which the fertile surface of a fruitbody is perched or elevated.

striate: A pattern of radiating lines around the edge of the cap; these lines can be short or long, and usually correspond with the position of each gill on the underside of the cap. Or paler or darker, often twisting lines on a stipe. Thin-fleshed species where light shows through the cap between these lines are described as *translucent-striate*. See also *sulcate*.

subdecurrent: Gills that run down the top of the stipe for a very short distance; intermediate between broadly attached and truly decurrent.

subdistant: Arbitrary and imprecise descriptor of the spacing of a mushroom's gills; an arbitrary intermediate between gills that are closely packed and those that are very widely spaced.

subglobose: Rounded but not quite spherical, slightly asymmetrical.

subviscid: Slightly slick or slippery, but without a distinct slime layer. See *viscid*.

sulcate: Furrowed or grooved, especially with regard to cap margins (often less precisely used interchangeably with *striate*, which refers to the pattern of gills visible through the cap tissue as radial lines, but without grooves in between).

taxon (pl. **taxa**): Unit in a taxonomic hierarchy. Most often used in this book to refer to species-level units, but can refer to one or more genera or families.

tomentose: Surface densely covered in short hairs, appearing finely fuzzy-felty.

tomentum: Fuzzy layer of interwoven hyphae. See *basal mycelium*.

translucent-striate: Pattern of darker or lighter lines, often observed on the cap margin when a mushroom is wet.

trichoderm: Microscopic cap surface structure in which hyphae are arranged more or less vertically. Opposite of *cutis*. See *pileipellis*.

µm: Micron symbol. See *micron*.

umbilicate: Posessing a small, distinct depression, or "belly button," usually in reference to the center of the cap.

umbo (adj. **umbonate**): Bump, protrusion, or "nipple" at the center of a mushroom's cap. Can be small or large, round or pointed, distinct or low and inconspicuous.

universal veil: Membranous tissue completely surrounding the young fruitbodies of some mushroom species. Often leaving remnants at the stipe base (volva) and/or on the cap (warts or as a "skull patch").

velvety-tomentose: Surface with very dense, short, fine hairs, giving the appearance and feel of suede leather.

ventricose: Swollen, especially in the middle.

virgate: Cap surface with a subtle pale or silvery sheen, usually appearing radially arranged. Especially common in *Tricholoma*.

viscid: With a slimy coating. Ranges in character from thin and watery to thick and sticky (but varies with degree of hydration or dessication).

volva: Saclike remnant of universal veil surrounding the base of the stipe in some genera of mushrooms, such as *Amanita*, *Volvariella*, and *Volvopluteus*. Sometimes indistinct and hard to see.

zonate: Patterned with concentric rings of pigmentation; the caps of many *Lactarius* show such a pattern.

Resources for the Mycophile

Cyberliber
www.cybertruffle.org.uk/cyberliber
Thanks to Cyberliber, digital copies of many journals, old books, and other mycology print resources are available for free online. The site is a bit clunky, and the scanned resources are not searchable and often must be paged through one page-scan image at a time, but this is an invaluable resource for engaging primary literature. We frequently use the scanned versions of past issues of *Mycologia*.

Index Fungorum
www.indexfungorum.org
Like MycoBank, another nomenclature database that acts as a repository for species names and the various kinds of data associated with them. Great for researching synonymy and priority naming issues. Remarkably up-to-date.

Mushroom Observer
www.mushroomobserver.org
One of the better resources for amateur mycology. Users can post and discuss pictures of and observations about mushrooms from anywhere in the world, making for an ever-growing database of information. Very useful for finding out which species grow in your area and at what time of year. Better yet, you can add to the knowledge of mushrooms in your area by posting pictures of your finds!

MycoBank
www.mycobank.org
Repository of mushroom names, including their authors, dates and details of publication, and often much more (brief descriptions, images of microscopic details). This resource is still being built, but is very useful when researching the origins of a mushroom name. Will be indispensable if and when it links to outside resources such as DNA sequences at Genbank.

Mycology Collections Portal
www.mycoportal.org/portal
Although still incomplete, MyCoPortal is becoming an outstanding resource for North American fungi. By digitizing and georeferencing the data associated with new and especially historical herbarium specimens, the site will make it possible to assemble species lists for particular regions supported by vouchers. Such data is essential for building a reliable mycoflora for North America. Corresponding images and links to genetic data are continually added.

The Mycota of Santa Cruz County
www.scmycoflora.org
Although this resource is focused on Santa Cruz County, a great many of the species found on the Redwood Coast occur in the county, making this website a great resource for photographs, genetic data, discussion, detailed keys to genera, and links to other resources. Website is slowly growing toward accumulating a complete mycoflora of this small but diverse county.

MykoWeb
www.mykoweb.com
Lots of great mycological resources such as digital copies of out-of-print books and journal articles. Additionally, it has "The Fungi of California," which currently details more than 650 species: www.mykoweb.com/CAF/new.html.

California Mushroom Clubs
There is no better way to learn about mushrooms than by surrounding yourself with the knowledgeable people of California's many mushroom clubs. Most have monthly meetings throughout the mushroom season and offer local as well as long-distance forays to collect and identify fungi.

Humboldt Bay Mycological Society
hbmycologicalsociety.org/wp/

Sonoma County Mycological Association
somamushrooms.org

Mycological Society of San Francisco
mssf.org

Bay Area Mycological Society
bayareamushrooms.org

Fungus Federation of Santa Cruz
ffsc.us

Los Angeles Mycological Society
lamushrooms.org/index.html

San Diego Mycological Society
sdmyco.org

Bibliography

Field Guides and Taxonomic References

Arora, David. *Mushrooms Demystified.* 2nd ed. Berkeley, CA: Ten Speed Press, 1986.

Benjamin, D. R. *Mushrooms: Poisons and Panaceas.* New York: Freeman, 1995.

Bessette, Alan E., Arleen R. Bessette, William C. Roody, and Steven A. Trudell. *Tricholomas of North America: A Mushroom Field Guide.* Austin: University of Texas Press, 2013.

Bessette, Alan E., William C. Roody, and Arleen R. Bessette. *North American Boletes: A Color Guide to the Fleshy Pored Mushrooms.* Syracuse, NY: Syracuse University Press, 2000.

Beug, Michael, Alan E. Bessette, and Arleen Bessette. *Ascomycete Fungi of North America: A Mushroom Reference Guide.* Austin: University of Texas Press, 2014.

Bigelow, H. E. *North American Species of Clitocybe.* Part 1. Vaduz, West Germany: J. Cramer, 1982.

———. *North American Species of Clitocybe.* Part 2. Vaduz, West Germany: J. Cramer, 1985.

The Bird's Nest Fungi. Toronto and Buffalo: University of Toronto Press, 1975.

Breitenbach, J., and F. Kränzlin. *Fungi of Switzerland.* Vol. 1, *Ascomycetes.* Lucerne, Switzerland: Edition Mykologia, 1984.

———. *Fungi of Switzerland.* Vol. 3, *Boletes and Agarics First Part.* Lucerne, Switzerland: Edition Mykologia, 1991.

———. *Fungi of Switzerland.* Vol. 4, *Agarics Second Part.* Lucerne, Switzerland: Edition Mykologia, 1995.

Castellano, M., Efrén Cázares, Brian Fondrick, and Tina Dreisbach. *Handbook to Additional Fungal Species of Special Concern in the Northwest Forest Plan.* General Technical Report PNW-GTR-572. United States Department of Agriculture, 2003.

Castellano, M., Jane E. Smith, Thom O'Dell, Efrén Cázares, and Susan Nugent. *Handbook to Strategy 1 Fungal Species in the Northwest Forest Plan.* General Technical Report PNW-GTR-476. United States Department of Agriculture, 1999.

Desjardin, Dennis E. *The Agaricales (Gilled Fungi) of California 7. Tricholomataceae I. Marasmioid Fungi: The Genera Baeospora, Crinipellis, Marasmiellus, Marasmius, Micromphale, and Strobilurus.* Eureka, CA: Mad River Press, 1987.

Desjardin, D. E., M. G. Wood, and F. A. Stevens. *California Mushrooms: The Comprehensive Identification Guide.* Portland, OR: Timber Press, 2015.

Exeter, Ron, Lorelei Norvell, and Efrén Cázares. *Ramaria of the Pacific Northwestern United States.* USDI BLM/OR/WA /PT-06/050-1792, Salem, OR, 2006.

Franklin, Wilfred A. "An Alpha-Taxonomic Study of Hydnellum and Sarcodon for Northern California." Master's thesis, Humboldt State University, 1999.

Gilbertson, R. L., and L. Ryvarden. *North American Polypores.* Vol. 1. Port Jervis, NY: Lubrecht & Cramer Ltd., 1986.

———. *North American Polypores.* Vol. 2. Port Jervis, NY: Lubrecht & Cramer Ltd., 1987.

Gregory, Denise C. "The Genus Clitocybe in California." Master's thesis, San Francisco State University, 2007.

Halling, Roy E. *The Genus Collybia (Agaricales) in the Northeastern United States and Adjacent Canada.* Braunschweig, Germany: J. Cramer, 1983.

Harrison, Kenneth A. *The Stipitate Hydnums of Nova Scotia.* Canadian Department of Agriculture Research Branch, 1961.

Hausknecht, Anton. *Fungi Europaei Volume 11: A Monograph of the Genera Conocybe and Pholiotina in Europe.* Alassio, Italy: Edizioni Candusso, 2010.

Hesler, L. R. *North American Species of Gymnopilus.* New York: Hafner Publishing, 1969.

Hesler, L. R., and A. H. Smith. *North American Species of Crepidotus.* New York: Hafner Publishing, 1965.

———. *North American Species of Hygrophorus.* Knoxville: University of Tennessee Press, 1963.

———. *North American Species of Lactarius.* Ann Arbor: University of Michigan Press, 1979.

Jarvis, Stephanie Shea. "The Lycoperdaceae of California." Master's thesis, San Francisco State University, 2014.

Kerrigan, Richard W. *The Agaricales (Gilled Fungi) of California 6. Agaricaceae.* Eureka, CA: Mad River Press, 1986.

Knudsen, Henning, and Jan Vesterholt, eds. *Funga Nordica: Agaricoid, Boletoid and Cyphelloid Genera.* Copenhagen, Denmark: Nordsvamp, 2012.

Largent, D. L. *The Agaricales (Gilled Fungi) of California 5. Hygrophoraceae.* Eureka, CA: Mad River Press, 1994.

———. *Entolomatoid Fungi of the Western United States and Alaska.* Eureka, CA: Mad River Press, 1995.

Lloyd, C. G. *Mycological Writings of C. G. Lloyd* 4 Cincinnati, OH: printed by author (1916): 531.

———. *Mycological Writings of C. G. Lloyd* 5 Cincinnati, OH: printed by author, 1916.

———. *Mycological Writings of C. G. Lloyd* 6 Cincinnati, OH: printed by author, 1916.

Methven, Andrew S. *The Agaricales (Gilled Fungi) of California 10. Russulaceae II. Lactarius.* Eureka, CA: Mad River Press, 1997.

Moser, M. *Keys to Agarics and Boleti.* Translated by S. Plant. London: Roger Phillips, 1983.

Noordeloos, Machiel E. *Fungi Europaei.* Vol 13, *Strophariaceae s.l.* Alassio Italy: Edizioni Candusso, 2011.

Norvell, Lorelei L., and Ronald L. Exeter. *Phaeocollybia of Pacific Northwest North America.* Salem, OR: U.S. Department of Interior, Bureau of Land Management, 2008.

Perry, Brian A. "A Taxonomic Investigation of Mycena of California." Master's thesis, San Francisco State University, 2002.

Roody, William C. *Mushrooms of West Virginia and the Central Appalachians.* Lexington: University of Kentucky Press, 2003.

Shanks, Kris M. *The Agaricales (Gilled Fungi) of California.* Vol. 11, *Tricholomataceae II Tricholoma.* Eureka, CA: Mad River Press, 1997.

Singer, Rolf. *The Agaricales in Modern Taxonomy.* 4th ed. Koenigstein, Germany: Koeltz Scientific Books, 1986.

Smith, Alexander H. *North American Species of Mycena.* Ann Arbor: University of Michigan Press, 1947.

——. *The North American Species of Psathyrella.* Vol. 24, *Memoirs of The New York Botanical Garden.* New York: New York Botanical Garden, 1972.

Smith, Alexander H., and L. R. Hesler. *North American Species of Pholiota.* New York: Hafner Publishing, 1968.

Stamets, Paul. *Psilocybin Mushrooms of the World.* Berkeley, CA: Ten Speed Press, 1996.

Thiers, Harry D. *The Agaricales (Gilled Fungi) of California.* Vol. 9, *Russulaceae I. Russula.* Eureka, CA: Mad River Press, 1997.

——. *California Mushrooms: A Field Guide to the Boletes.* New York: Hafner Press, 1975.

Trappe, Matt, Frank Evans, and James Trappe. *Field Guide to North American Truffles: Hunting, Identifying, and Enjoying the World's Most Prized Fungi.* Berkeley, CA: Ten Speed Press, 2007.

Trudell, Steve, and Joe Ammirati. *Mushrooms of the Pacific Northwest.* Portland, OR: Timber Press, 2009.

Periodical Articles

Antonín, V., R. E. Halling, and M. E. Noordeloos. "Generic Concepts within the Groups of *Marasmius* and *Collybia* sensu lato." *Mycotaxon* 58 (1997): 359–68.

Arora, David, and Giampaolo Simonini. "California Porcini: Three New Taxa, Observations on Their Harvest, and the Tragedy of No Commons." *Economic Botany* 62, no. 3 (2008): 356–75.

Arora, David, and Jonathan L. Frank. "Clarifying the Butter Boletes: A New Genus, *Butyriboletus*, Is Established to Accommodate *Boletus* sect. *Appendiculati*, and Six New Species Are Described." *Mycologia* 106, no. 3 (2014): 464–80.

Audet, Serge. "Essai de Découpage Systématique du Genre *Scutiger* (Basidiomycota): *Albatrellopsis, Albatrellus, Polyporoletus, Scutiger* et Description de Six Nouveaux Genres." *Mycotaxon* 111 (2010): 431–64.

Banerjee, Partha, and Walter J. Sundberg. "The Genus *Pluteus* (Pluteaceae, Agaricales) in the Midwestern United States." *Mycotaxon* 53 (1995): 189–246.

——. "Reexamination of *Pluteus* Type Specimens: Types Housed at the New York Botanical Garden." *Mycotaxon* 49 (1993): 413–35.

——. "Three New Species and a New Variety of *Pluteus* from the United States." *Mycotaxon* 47 (1993): 389–94.

Baroni, Timothy J. "A revision of the Genus *Rhodocybe* Maire (Agaricales)." *Beih. Nova Hedwigia* 67 (1981): 1–194.

Baroni, Timothy J., Valerie Hofstetter, David L. Largent, and Rytas Vilgalys. "*Entocybe* Is Proposed as a New Genus in the Entolomataceae (Agaricomycetes, Basidiomycota) Based on Morphological and Molecular Evidence." *North American Fungi* 6, no. 12 (2011): 1–19.

Baroni, Timothy J., and David L. Largent. "The Genus *Rhodocybe*: New Combinations and a Revised Key to Section *Rhodophana* in North America." *Mycotaxon* 84, no. 1 (1989): 47–53.

Baumgartner, Kendra, and David M. Rizzo. "Distribution of *Armillaria* Species in California." *Mycologia* 93, no. 5 (2001): 821–30.

Beug, Michael W. "NAMA Toxicology Committee Report for 2010 North American Mushroom Poisonings." *McIlvainea* 20 (2010): 1–11.

Bojantchev, Dimitar. 2013. "*Cortinarius* of California: Eight New Species in subg. *Telamonia*." *Mycotaxon* 123 (2013): 375–402.

Bojantchev, Dimitar, and R. Michael Davis. "*Amanita augusta*, a New Species from California and the Pacific Northwest." *North American Fungi* 8, no. 5 (2013): 1–11.

——. "*Cortinarius callimorphus*, a New Species from Northern California." *Mycotaxon* 117 (2011): 1–8.

——. "*Cortinarius xanthodryophilus* sp. nov.—A Common *Phlegmacium* under Oaks in California." *Mycotaxon* 116 (2011): 317–28.

Bojantchev, Dimitar, Shaun R. Pennycook, and R. Michael Davis. "*Amanita vernicoccora* sp. nov.—The Vernal Fruiting 'Coccora' from California." *Mycotaxon* 117 (2011): 485–97.

Bonito, Gregory, James M. Trappe, Pat Rawlinson, and Rytas Vilgalys. "Improved Resolution of Major Clades within *Tuber* and Taxonomy of Species within the *Tuber gibbosum* Complex." *Mycologia* 102, no. 5 (2010): 1042–57.

Borovička, Jan, Alan Rockefeller, and Peter Werner. "*Psilocybe allenii*—A New Bluing Species from the Pacific Coast, USA." *Czech Mycology* 64, no. 2 (2012): 181–95.

Burlingham, Gertrude S. "The Lactariaeae of the Pacific Coast." *Mycologia* 5 (1913): 305–11.

——. "New or Noteworthy Species of *Russula* and *Lactaria*." *Mycologia* 28 (1936): 253–67.

Carbone, Matteo, Carlo Agnello, and Pablo Alvarado. "Phylogenetic Studies in the Family Sarcosomataceae (Ascomycota, *Pezizales*)." Ascomycete.org 5, no. 1 (2013): 1–12. http://www.ascomycete.org/Portals/0/Volumes/AscomyceteOrg%2005-01%201-12.pdf

Carbone, Matteo, Carlo Agnello, and Andrew Parker. "*Urnula padeniana* (*Pezizales*) sp. nov. and the Type Study of *Bulgaria mexicana*." Ascomycete.org 5(1) (2013): 13–24.

Co-David, D., D. Langeveld, and M. E. Noordeloos. "Molecular Phylogeny and Spore Evolution of *Entolomataceae*." *Persoonia* 23 (2009): 147–76.

Denison, William C. "The Genus *Cheilymenia* in North America." *Mycologia* 56 (1964): 718–37.

Dunham, Susie M., Thomas E. O'Dell, and Randy Molina. "Analysis of nrDNA Sequences and Microsatellite Allele Frequencies Reveals a Cryptic Chanterelle Species *Cantharellus cascadensis* sp. nov. from the American Pacific Northwest." *Mycological Research* 107, no. 10 (2003): 1163–77.

Garnica, Sigisfredo, Phillip Spahn, Bernhard Oertel, Joseph Ammirati, and Franz Oberwinkler. "Tracking the Evolutionary History of *Cortinarius* Species in the Section *Calochroi*, with Transoceanic Disjunct Distributions." *BMC Evolutionary Biology* 11 (2011): 213.

Ginns, J. "The Taxonomy and Distribution of Rare or Uncommon Species of *Albatrellus* in Western North America." *Canadian Journal of Botany* 75 (1997): 261–73.

———. "Typification of *Cordyceps canadensis* and *Cordyceps capitata*, and a New Species, *C. longisegmentis*." *Mycologia* 80, no. 2 (1988): 217–22.

Grob, C. S., A. L. Danforth, G. S. Chopra, M. Hagerty, C. R. McKay, A. L. Halberstadt, and G. R. Greer. "Pilot Study of Psilocybin Treatment for Anxiety in Patients with Advanced-Stage Cancer." *Arch Gen Psychiatry* 68, no. 1 (2011): 71–78.

Guzmán, Gastón, and Clark L. Ovrebo. "New Observations on Sclerodermataceous Fungi." *Mycologia* 92, no. 1 (2000): 174–79.

Hall, D., and D. E. Stuntz. "Pileate Hydnaceae of the Puget Sound Area. I. White-Spored Genera: *Auriscalpium*, *Hericium*, *Dentinum*, and *Phellodon*." *Mycologia* 63, no. 6 (1971): 1009–27.

———. "Pileate Hydnaceae of the Puget Sound Area. III. Brown-Spored Genus: *Hydnellum*." *Mycologia* 64 (1973): 560–90.

Hallen, Heather E., Roy Watling, and Gerard C. Adams. "Taxonomy and Toxicity of *Conocybe lactea* and Related Species." *Mycological Research* 107, no. 8 (2003): 969–79.

Halling, R. E. "*Collybia fuscopurpurea* in the Americas." *Mycological Research* 94, no. 5 (1990): 671–74.

Hansen, Karen, K. F. LoBuglio, and D. H. Pfister. "Evolutionary Relationships of the Cup-Fungus Genus *Peziza* and Pezizaceae Inferred from Multiple Nuclear Genes: RPB2, Beta-Tubulin and LSU rDNA." *Molecular Phylogenetics and Evolution* 36, no. 1 (2005): 1–23.

Harmaja, Harri. "*Amylolepiota*, *Clavicybe* and *Cystodermella*, New Genera of the Agaricales." *Karstenia* 42 (2002): 39–48.

Harrison, K. A. "New or Little Known North American Stipitate *Hydnums*." *Canadian Journal of Botany* 42 (1964): 1205–33.

Holec, Jan, and Miroslav Kolařík. "Notes on the Identity of *Hygrophoropsis rufa* (Basidiomycota, *Boletales*)." *Czech Mycology* 65, no. 1 (2013): 15–24.

Hughes, Karen W., David Mather, and Ronald H. Petersen. "A New Genus to Accommodate *Gymnopus acervatus* (Agaricales)." *Mycologia* 102, no. 6 (2010): 1463–78.

Johannesson, Hanna, Svengunnar Ryman, Hjördis Lundmark, and Eric Danell. "*Sarcodon imbricatus* and *S. squamosus*— Two Confused Species." *Mycological Research* 103 (1999): 1447–52.

Ju, Yu-Ming, and Jack D. Rogers. "A Revision of the Genus *Daldinia*." *Mycotaxon* 61 (1997): 243–93.

Justo, Alfredo, Ekaterina Malysheva, Tatiana Bulyonkova, Else C. Vellinga, Gerry Cobian, Nhu Nguyen, Andrew M. Minnis, and David S. Hibbett. "Molecular Phylogeny and Phylogeography of Holarctic Species of Pluteus Section Pluteus (Agaricales: Pluteaceae), with Description of Twelve New Species." *Phytotaxa* 180, no. 1 (2014): 1–85.

Kerrigan, Richard. "*Agaricus* section *Xanthodermatei*: A Phylogenetic Reconstruction with Commentary on Taxa." *Mycologia* 97, no. 6 (2005): 1292–1315.

Kropp, Bradley R., P. Brandon Matheny, and Leonard J. Hutchison. "*Inocybe* Section *Rimosae* in Utah: Phylogenetic Affinities and New Species." *Mycologia* 105, no. 3 (2013): 728–47.

Kuo, Michael, Damon R. Dewsbury, Kerry O'Donnell, M. Carol Carter, Stephen A. Rehner, John David Moore, Jean-Marc Moncalvo et al. "Taxonomic Revision of True Morels (*Morchella*) in Canada and the United States." *Mycologia* 104, no. 5 (2012): 1159–77.

Liimatainen, K., T. Niskanen, B. Dima, I. Kytövuori, J. F. Ammirati, and T. G. Frøslev. "The Largest Type Study of *Agaricales* Species to Date: Bringing Identification and Nomenclature of *Phlegmacium* (*Cortinarius*) into the DNA Era." *Persoonia* 33 (2014): 98–140.

Matheny, P. Brandon, Judd M. Curtis, Valérie Hofstetter, M. Catherine Aime, Jean-Marc Moncalvo, Zai-Wei Ge, Zhu-Liang Yang et al. "Major Clades of Agaricales: A Multilocus Phylogenetic Overview." *Mycologia* 98, no. 6 (2006): 982–95.

Matheny, P. Brandon, Else C. Vellinga, Neale L. Bougher, Oluna Ceska, Pierre-Arthur Moreau, Maria Alice Neves, and Joseph F. Ammirati. "Taxonomy of Displaced Species of *Tubaria*." *Mycologia* 99, no. 4 (2007): 569–85.

Methven, Andrew. "*Flammulina* RFLP Patterns Identify Species and Show Biogeographical Patterns within Species." *Mycologia* 92, no. 6 (2000): 1064–1070.

———. "Notes on *Clavariadelphus*. III. New and Noteworthy Species from North America." *Mycotaxon* 34, no. 1 (1989): 153–79.

Miller, Orson K. "A Revision of the Genus *Xeromphalina*." *Mycologia* 60 (1968): 156–88.

Minnis, Andrew M., and Walter Sundberg. "*Pluteus* Section *Celluloderma* in the U.S.A." *North American Fungi* 5, no. 1 (2010): 1–107.

Moncalvo, Jean-Marc, Rytas Vilgalys, Scott A. Redhead, James E. Johnson, Timothy Y. James, M. Catherine Aime, Valérie Hofstetter et al. "One Hundred and Seventeen Clades of Eugarics." *Molecular Phylogenetics and Evolution* 23 (2002): 357–400.

Morgado, L. N., M. E. Noordeloos, Y. Lamoureux, and J. Geml. "Multi-Gene Phylogenetic Analyses Reveal Species Limits, Phylogeographic Patterns, and Evolutionary Histories of Key Morphological Traits in *Entoloma* (*Agaricales, Basidiomycota*)." *Persoonia* 31 (2013): 159–78.

Moser, M. M., and J. F. Ammirati. "Studies on North American *Cortinarii* IV. New and Interesting *Cortinarius* Species (Subgenus *Phlegmacium*) from Oak Forests in Northern California." *Sydowia* 49, no. 1 (1997): 25–48.

———. "Studies on North American *Cortinarii* V. New and Interesting *Phlegmacia* from Wyoming and the Pacific Northwest." *Mycotaxon* 72 (1999): 289–321.

———. "Studies in North American *Cortinarii* VI. New and Interesting Taxa in Subgenus *Phlegmacium* from the Pacific States of North America." *Mycotaxon* 74 (2000): 1–36.

Mueller, G. M. "New North American Species of *Laccaria* (*Agaricales*)." *Mycotaxon* 20, no. 1 (1984): 101–16.

Murrill, W. A. "The Agaricaceae of the Pacific Coast—I." *Mycologia* 4 (1912): 205–17.

———. "The Agaricaceae of the Pacific Coast—II." *Mycologia* 4 (1912): 231–62.

———. "The Agaricaceae of the Pacific Coast—III." *Mycologia* 4 (1912): 294–308.

———. "The Agaricaceae of the Pacific Coast—IV." New Species of *Clitocybe* and *Melanoleuca*." *Mycologia* 5 (1913): 206–23.

Nakasone, Karen K., and Kenneth J. Sytsma. "Biosystematic Studies on *Phlebia acerina*, *P. rufa*, and *P. radiata* in North America." *Mycologia* 85, no. 6 (1993): 996–1016.

Nishida, Florence H. "Key to the Species of *Inocybe* in California." *Mycotaxon* 34, no. 1 (1989): 181–96.

———. "New Species of *Inocybe* from Southern California." *Mycotaxon* 33 (1988): 213–22.

Niskanen, Tuula, Kare Liimatainen, Ilkka Kytövuori, and Joseph F. Ammirati. "New *Cortinarius* Species from Conifer-Dominated Forests of North America and Europe." *Botany* 90 (2012): 743–54.

Niskanen, Tuula, Kare Liimatainen, Joseph F. Ammirati, and Karen Hughes. "*Cortinarius* Section *Sanguinei* in North America." *Mycologia* 105, no. 2 (2013): 344–56.

Nguyen, Nhu H., Fidel Landeros, Roberto Garibay-Orijel, Karen Hansen, and Else C. Vellinga. "The *Helvella lacunosa* Species Complex in Western North America: Cryptic Species, Misapplied Names and Parasites." *Mycologia* 105 (2013): 1275–86.

Nguyen, Nhu H., Jennifer Kerekes, Else C. Vellinga, and Thomas D. Bruns. "Synonymy of *Suillus imitatus*, the Imitator of Two Species within the *S. caerulescens/ponderosus* Complex." *Mycotaxon* 122 (2012): 389–98.

Norvell, L. L., S. A. Redhead, and J. F. Ammirati. "*Omphalina* Sensu Lato in North America 1: *Omphalina wynniae* and the Genus *Chrysomphalina* 2: *Omphalina* sensu Bigelow." *Mycotaxon* 50 (1994): 379–407

Nuytinck, Jorinde, and Annemieke Verbeken. "Morphology and Taxonomy of the European Species in *Lactarius* Sect. *Deliciosi* (Russulales)." *Mycotaxon* 92 (2005): 125–68.

Nuytinck, Jorinde, Steven L. Miller, and Annemieke Verbeken. "A Taxonomical Treatment of the North and Central American Species in *Lactarius* Sect. *Deliciosi*." *Mycotaxon* 96 (2006): 261–307.

Nuytinck, Jorinde, Annemieke Verbeken, and Steven L. Miller. "Worldwide Phylogeny of *Lactarius* Sect. *Deliciosi* Inferred from Its ITS and Glyceraldehyde-3-Phosphate Dehydrogenase Gene Sequences." *Mycologia* 99, no. 6 (2007): 820–32.

Nuytinck, Jorinde, and Joseph Ammirati. "A New Species of *Lactarius* Sect. *Deliciosi* (Russulales, Basidiomycota) from Western North America." *Botany* 92 (2004): 767–74.

Ortiz-Santana, Beatriz, and Ernst E. Both. "A Preliminary Survey of the Genus *Buchwaldoboletus*." *Bulletin of the Buffalo Society of Natural Sciences* 40 (2011): 1–14.

Osmundson, Todd W., Cathy L. Cripps, and Gregory M. Mueller. "Morphologic and Molecular Systematics of Rocky Mountain Alpine *Laccaria*." *Mycologia* 97, no. 5 (2005): 949–72.

Otrosina, William J., and Matteo Garbelotto. "*Heterobasidion occidentale* sp. nov. and *Heterobasidion irregulare* nom. nov.: A disposition of North American *Heterobasidion* Biological Species." *Fungal Biology* 114 (2010): 16–25.

Padamsee, Mahajabeen, P. Brandon Matheny, Bryn T. M. Dentinger, and David J. McLaughlin. "The Mushroom Family Psathyrellaceae: Evidence for Large-Scale Polyphyly of the Genus *Psathyrella*." *Molecular Phylogenetics and Evolution* 46 (2008): 415–29.

Peck, C. H. "Annual Report of the New York State Museum of Natural History" 29: 38. (1878) [1876].

———. "Annual Report of the State Botanist." New York State Museum, 1906.

———. "New Species of Fungi." *Bulletin of the Torrey Botanical Club* 25 (1898): 321–28.

———. "New Species of Fungi." *Bulletin of the Torrey Botanical Club* 27 (1900): 14–21.

———. "The New York Species of *Russula*." *New York State Museum Bulletin* 116 (1907): 67–98.

Phosri, Cherdchai, Maria P. Martín, and Roy Watling. "*Astraeus*: Hidden Dimensions." *IMA Fungus* 4, no. 2 (2013): 347–56.

Redhead, S. A. "*Arrhenia* and *Rimbachia*, Expanded Generic Concepts, and a Re-Evaluation of *Leptoglossum* with Emphasis on Muscicolous North American Taxa." *Canadian Journal of Botany* 62 (1984): 865–92.

Redhead, S. A., Glenn R. Walker, Joseph F. Ammirati, and Lorelei L. Norvell. "*Omphalina* sensu lato in North America 4: *Omphalina rosella*." *Mycologia* 87, no. 6 (1995): 880–85.

Robertson, Christie P., Leesa Wright, Sharmin Gamiet, Noelle Machnicki, Joe Ammirati, Joshua Birkebak, Colin Meyer, and Alissa Allen. "*Cortinarius rubellus* Cooke from British Columbia, Canada and Western Washington, USA." *Pacific Northwest Fungi*. 1, no. 6 (2006): 1–7.

Rogerson, Clark T. "A New Species of *Hypomyces* on *Helvella*." *Mycologia* 63 (1971): 416–21.

Rogerson, Clark T., and Gary J. Samuels. "Boleticolous Species of *Hypomyces*." *Mycologia* 81 (1989): 413–32.

Schwarz, Christian F. "*Entoloma medianox*, a New Name for a Common Species on the Pacific Coast of North America." *Index Fungorum* 220 (2015); 1. See also: http://scmycoflora.org/documents/Entoloma-medianox.pdf.

———. "*Pseudobaeospora deckeri* sp. nov.—A New Agaric from Central California." *Mycotaxon* 119 (2012): 459–65.

Seitzman B. H., A. Ouimette, R. L. Mixon, E. A. Hobbie, D. S. Hibbett "Conservation of Biotrophy in Hygrophoraceae Inferred from Combined Stable Isotope and Phylogenetic Analyses". *Mycologia* 103, no. 2 (2011): 280–90.

Sewell, R. A., J. H. Halpern, H. G. Pope Jr. "Response of Cluster Headache to Psilocybin and LSD." *Neurology* 66, no. 12 (2006): 1920–2.

Shaffer, R. L. "Notes on the Subsection *Crassotunicatinae* and Other Species of *Russula*." *Lloydia* 33 (1970): 49–96.

———. "The Subsection *Compactae* of *Russula*." *Brittonia* 14 (1962): 254–84.

Shanks, Kris M. "New Species of *Tricholoma* from California and Oregon." *Mycologia* 88, no. 3 (1996): 497–508.

Singer, Rolf. "Additional Notes on the Genus *Leucopaxillus*." *Mycologia* 39 (1947): 725–36.

———. "Type Studies on Agarics." *Lloydia* 5 (1942): 97–135.

———. "Type Studies on Basidiomycetes I." *Mycologia* 34 (1942): 60–93.

———. "Type Studies on Basidiomycetes II." *Mycologia* 35 (1943): 142–63.

———. "Type Studies on Basidiomycetes III." *Mycologia* 39 (1947): 171–89.

Singer, Rolf and A. H. Smith. "A Monograph on the Genus *Leucopaxillus* Boursier." *Papers of the Michigan Academy of Sciences and Letters* 28 (1943): 85–132.

Smith, Alexander H. "Notes on Agarics from the Western United States." *Bulletin of the Torrey Club* 64 (1937): 477–87.

———. "Studies in the Genus *Cortinarius* I." *Contributions from the University of Michigan Herbarium*, no. 2 (1939).

———. "Studies of North American Agarics—I." *Contributions from the University of Michigan Herbarium*, no. 5 (1941).

Smith, Alexander H., and Rolf Singer. "A Monograph on the Genus *Cystoderma*." *Papers of the Michigan Academy of Science, Arts and Letters* 29, no. 1 (1944): 71–124.

———. "Notes on the Genus *Cystoderma*." *Mycologia* 40 (1948): 454–60.

Tulloss, Rodham E., Janet E. Lindgren, David Arora, Benjamin E. Wolfe, and Cristina Rodríguez. "*Amanita pruittii*—A New, Apparently Saprotrophic Species from US Pacific Coastal States." *Amanitaceae* 1, no. 1 (2014): 1–9.

Vellinga, Else C. "*Lepiota clypeolaria*, I Presume?" *McIlvainea* 14, no. 2 (2000): 41–45.

———. "*Lepiota* in California: Species with a Hymeniform Pileus Covering." *Mycologia* 102, no. 3 (2010): 664–74.

———.. "Lepiotaceous Fungi in California, U.S.A.—1. *Leucoagaricus amanitoides* sp. nov." *Mycotaxon* 98 (2006): 197–204.

———. "Lepiotaceous Fungi in California, U.S.A.—3. Pink and Lilac Species in *Leucoagaricus* sect. *Piloselli*." *Mycotaxon* 98 (2006): 213–24.

———. "Lepiotaceous Fungi in California, U.S.A.—4. Type Studies of *Lepiota fumosifolia* and *L. petasiformis*." *Mycotaxon* 98 (2006): 225–32.

———. "Lepiotaceous Fungi in California, U.S.A.—5. *Lepiota oculata* and Its Look-Alikes." *Mycotaxon* 102 (2007): 267–80.

———. "Lepiotaceous Fungi in California, U.S.A.—6. *Lepiota castanescens*." *Mycotaxon* 103 (2008): 97–108.

———. "Lepiotaceous Fungi in California, U.S.A: *Leucoagaricus* Sect. *Piloselli*." *Mycotaxon* 112 (2010): 393–444.

———. "*Leucoagaricus decipiens* and *La. erythrophaeus*, a New Species Pair in Sect. *Piloselli*." *Mycologia* 102, no. 2 (2010): 447–54.

———. "Notes on *Lepiota* and *Leucoagaricus* Type Studies on *Lepiota magnispora*, *Lepiota barssi*, and *Agaricus americanus*." *Mycotaxon* 76 (2000): 429–38.

———. "Studies in *Lepiota* IV. *Lepiota cristata* and *L. castaneidisca*." *Mycotaxon* 80 (2001): 297–306.

———. "Type Studies in Agaricaceae. *Chlorophyllum rachodes* and Allies." *Mycotaxon* 85 (2003): 259–70.

———. "Wood Chip Fungi: *Agrocybe putaminum*." *Fungi* 1, no. 4 (2008): 5, 37–39.

Vilgalys, Rytas, Ajiri Smith, Bao Lin Sun, and Orson K. Miller Jr. "Intersterility Groups in the *Pleurotus ostreatus* Complex from the Continental United States and Adjacent Canada." *Canadian Journal of Botany* 71 (1993): 113–28.

Voitk, Andrus. "The Genus *Fomitopsis* in Newfoundland & Labrador." *Omphalina* 4, no.7 (2013): 14–17.

Volk, Thomas J., Harold H. Burdsall Jr., and Mark T. Banik. "*Armillaria nabsnona*, a New Species from Western North America." *Mycologia* 88, no. 3 (1996): 484–91.

Wagner, Tobias, and Michael Fischer. "Proceedings Toward a Natural Classification of the Worldwide Taxa *Phellinus* s.l. and *Inonotus* s.l., and Phylogenetic Relationships of Allied Genera." *Mycologia*. 94, no. 6 (2002): 998–1016.

Watling, Roy, and Jeremy Milne. "The Identity of European and North American *Boletopsis* spp. (Basidiomycota; Thelephorales, Boletopsidaceae)." *North American Fungi* 3, no. 7 (2008): 5–15.

Weber, Nancy Smith. "Notes on Western Species of *Helvella*." *Beih. Nova Hedwigia* 51 (1975): 25–38.

Index

GENUS AND SPECIES INDEX

Page numbers in boldface
indicate pages with
mushroom photographs.

Russula, continued
  rhodopus, 230
  sanguinea, **230**, 231
  semirubra, 236
  silvicola, 229
  simillima, 228
  smithii, 238
  stuntzii, **241**, 243
  turci, 232
  urens, 238
  versicolor, **240**, 447
  veternosa, 228
  xerampelina, 230, 231, 233, 234, 235, **236–37**

S

Sarcodon, 462, 469, 489
  fuscoindicus, 492, **496**
  imbricatus, 497
  laevigatus, 496
  rimosus, 497
  scabrosus, 496, 497
  squamosus, **497**
  stereosarcinon, **496**
Sarcomyxa serotina, 395, **405**
Sarcoscypha, 559
  coccinea, **567**
Sarcosphaera ammophila, 563
Schizophyllum commune, **406**
Scleroderma, 526, 527, 532
  areolatum, **532**
  cepa, 532
  geaster, 532
  hypogaeum, 532
  polyrhizum, **532**
  verrucosum, 532
Scutellinia, 559, 567
  scutellata, **567**
Scutiger
  ellisii, 461
  pes-caprae, **461**
Sequoia sempervirens, 5–6, 10
Serpula, 5
Simocybe, 125, 143, 409
  centunculus, **143**, 392
  haustellaria, 143
  sumptuosa, 143
Sistotrema, 489
  confluens, **500**
Skeletocutis nivea, 482
Sowerbyella, 559
  rhenana, **566**

Sparassis, 503
  crispa, 504
  radicata, **504**
Squamanita, 5, 56
  paradoxa, 56
Stereopsis humphreyi, 465
Stereum, 458, 548, 570
  gausapatum, 481
  hirsutum, 480, **481**, 548, 549
  ochraceoflavum, 481
  sanguinolentum, 481, 550
Strobilurus, 297
  albipilatus, 336
  diminutivus, 336
  occidentalis, 335, **336**
  trullisatus, 335, **336**, 490
Stropharia, 92, 105
  aeruginosa, 124
  albivelata, 104
  ambigua, 121, **122**, 123
  aurantiacum, 120
  caerulea, **124**
  coronilla, **123**
  dorsipora, 111
  hornemannii, 122, **123**
  kauffmannii, 108, **124**
  pseudocyanea, 124
  rugosoannulata, 123
Suillellus, 421, 442
  amygdalinus, 429, 441, **442**
Suillus, 3, 35, 410, 413, 421, 437, 443, 543
  anomalus, 449
  brevipes, **448**, 449, 450
  brunnescens, 456
  caerulescens, 3, 410, 413, 450, 451, 453, **454**, 455
  cavipes, 453
  flavidus, 457
  fuscotomentosus, 410, 415, 416, 451, **452**
  glandulosipes, 450, 456
  imitatus, 455
  lakei, 410, 412, 414, 452, **453**, 454
  luteus, 450, **456**
  ponderosus, 410, 412, 450, 453, 454, **455**
  pseudobrevipes, 410, 415, 417, 448, **450**, 456
  punctatipes, 449
  pungens, 410, 415, 417, 448, **449**

  quiescens, 448, 449
  tomentosus, **451**, 452, 453
  umbonatus, 415, **457**

T

Tapinella, 410
  atrotomentosa, 409, **411**, 469
  panuoides, 405, **409**
Tarzetta cupularis, 564, 569
Tephrocybe anthracophila, 347
Tetrapyrgos, 397
  subdendrophora, 323, **403**
Thaxterogaster, 541
Thelephora, 30, 458, 462, 465, 514
  anthocephala, 514
  caryophyllea, 465
  multipartita, 465
  palmata, **514**
  terrestris, **465**
Tolypocladium, 503
  capitatum, **522**, 545
  ophioglossoides, 522
Tomentella, **487**
Trametes, 458, 479
  betulina, 480, **481**
  hirsuta, 480
  suaveolens, 480
  versicolor, **480**, 481
Trappea darkeri, 544
Tremella, 548
  aurantia, 481, **548**
  encephala, 550
  foliacea, 481, **549**
  mesenterica, 548
Tremellodendropsis tuberosa, 512, **514**
Tremiscus, 547
  helvelloides, 547, **549**
Trichaptum, 458
  abietinum, **479**, 480
  biforme, 479
  fuscoviolaceum, 479
Trichoglossum, 346, 503, 523, 524
  hirsutum, **524**
  velutipes, 524
Tricholoma, 53, 107, 206, 296, 298, 362, 570
  aestuans, 368
  arenicola, 385
  arvernense, **368**, 370
  atrosquamosum, 373, 375, 376

The information in this book is accurate to the best of the authors' knowledge. However, neither the authors nor the publisher are responsible for mistakes in identification or idiosyncratic reactions to mushrooms. It is certainly not necessary to eat mushrooms in order to enjoy them. People who choose to eat mushrooms do so at their own risk.

Published in the United States by Ten Speed Press, an imprint of the Crown Publishing Group, a division of Penguin Random House LLC, New York.
www.crownpublishing.com
www.tenspeed.com

Ten Speed Press and the Ten Speed Press colophon are registered trademarks of Penguin Random House LLC.

Library of Congress Cataloging-in-Publication Data

Schwarz, Christian, 1988- author.
  Mushrooms of the redwood coast : a comprehensive field guide to the fungi of coastal northern California / Christian Schwarz and Noah Siegel. — First edition.
      pages cm
  Includes bibliographical references and index.
  1. Mushrooms—California, Northern—Identification. I. Siegel, Noah, 1982- author. II. Title.

  QK605.5.C2S38 2016
  579.609794--dc23
                          2015027853

Trade Paperback ISBN: 978-1-60774-817-5
eBook ISBN: 978-1-60774-818-2

Printed in China

Design by Tatiana Pavlova and Angelina Cheney

10 9 8 7 6 5 4 3

First Edition